Handbook of Experimental Pharmacology

Volume 164

Editor-in-Chief
K. Starke, Freiburg i. Br.

Editorial Board
G.V.R. Born, London
M. Eichelbaum, Stuttgart
D. Ganten, Berlin
F. Hofmann, München
B. Kobilka, Stanford, CA
W. Rosenthal, Berlin
G. Rubanyi, Richmond, CA

Springer

*Berlin
Heidelberg
New York
Hong Kong
London
Milan
Paris
Tokyo*

Tachykinins

Contributors
P.L.R. Andrews, S. Boyce, S.D. Brain, N.W. Bunnett,
B.J. Canning, J.M. Conlon, S.K. Costa, J. Donnerer,
X. Emonds-Alt, C.A. Gadd, R.G. Hill, T. Hökfelt, P. Holzer,
S.P. Hunt, G.F. Joos, J.K. Kerns, E. Kuteeva, F. Lembeck,
Å. Ljungdahl, C.A. Maggi, S.B. Mazzone, R. Patacchini,
A. Rawlingson, D. Roosterman, J.A. Rudd, W.L. Rumsey,
N.M.J. Rupniak, D. Stanic, M. Sukumaran

Editor
Peter Holzer

Springer

Professor
Peter Holzer
Department of Experimental
and Clinical Pharmacology
Medical University of Graz
Universitätsplatz 4
8010 Graz
Austria
e-mail: peter.holzer@meduni-graz.at

With 56 Figures and 28 Tables

ISSN 0171-2004

ISBN 3-540-20690-6 Springer-Verlag Berlin Heidelberg New York

Library of Congress Cataloging-in-Publication Data
Tachykinins / contributors, P.L.R. Andrews ... [et al.] ; editor, Peter Holzer. p. ; cm. – (Handbook of experimental pharmacology ; v. 164)
Includes bibliographical references and index.
ISBN 3-540-20690-6 (alk. paper)
1. Tachykinins–Physiological effect–Handbooks, manuals, etc. 2. Tachykinins–Agonists–Therapeutic use–Handbooks, manuals, etc. 3. Tachykinins–Antagonists–Therapeutic use–Handbooks, manuals, etc. I. Holzer, Peter, Mag. rer. nat. Dr. phil. II. Series. [DNLM: 1. Tachykinins. 2. Receptors, Tachykinin. WL 104 T117 2004]
OP905.H3 vol. 164   [QP552.T33]   615'.1s–dc22   [612'.015756]   2003069359

This work is subject to copyright. All rights are reserved, whether the whole or part of the material is concerned, specifically the rights of translation, reprinting, re-use of illustrations, recitation, broadcasting, reproduction on microfilm or in any other way, and storage in data banks. Duplication of this publication or parts thereof is permitted only under the provisions of the German Copyright Law of September 9, 1965, in its current version, and permission for use must always be obtained from Springer-Verlag. Violations are liable to Prosecution under the German Copyright Law.

Springer-Verlag is a part of Springer Science+Business Media
springeronline.com

© Springer-Verlag Berlin Heidelberg 2004
Printed in Germany

The use of general descriptive names, registered names, etc. in this publication does not imply, even in the absence of a specific statement, that such names are exempt from the relevant protective laws and regulations and free for general use.

Product liability: The publishers cannot guarantee the accuracy of any information about dosage and application contained in this book. In every individual case the user must check such information by consulting the relevant literature.

Editor: Dr. R. Lange
Desk Editor: S. Dathe

Cover design: design & production GmbH, Heidelberg
Typesetting: Stürtz AG, 97080 Würzburg

Printed on acid-free paper   27/3150 hs – 5 4 3 2 1 0

Preface

More than 70 years have elapsed since U.S. von Euler and J.H. Gaddum discovered an unidentified depressor substance in the brain and gut. The effects of the powdery extracts were marked as 'P' on the kymograph tracings, and the nondescript name of 'substance P' still carries the breath of this adventurous period. In the 1960s, substance P returned in another disguise, staging as a hypothalamic peptide that causes copious salivary secretion (see chapter by F. Lembeck and J. Donnerer). This time, though, the mysterious substance was tracked down by S.E. Leeman and her collaborators as an undecapeptide, after it had eluded its identification for some 40 years. Substance P turned out to be the mammalian counterpart of a family of peptides which had been extracted from amphibian and nonvertebrate species and which had been given the name 'tachykinins' by V. Erspamer. Soon novel members of this peptide family were discovered, and in mammals substance P was joined by neurokinin A and neurokinin B. The presence of tachykinins in frog skin as well as in venoms and toxins of microbes and arachnids raises the possibility that these peptides represent an old system of biological weapons that have been transformed to a particular messenger system in mammals. The first two tachykinin genes, now abbreviated *TAC1* and *TAC3* by the Human Genome Organisation (HUGO) Nomenclature Committee, were cloned in the late 1980s, and the pharmacological evidence for a heterogeneity of tachykinin receptors was corroborated by the identification of three tachykinin receptor genes, now termed *TACR1*, *TACR2* and *TACR3*.

Unfortunately, the quest for a unified nomenclature of tachykinins and their receptors was not appropriately met by the scientific community. Disaster struck at the Substance P Meeting in Montreal in 1986 where the convention voted to name the peptide family tachykinins and, logically, the receptors mediating their actions tachykinin TK_1, TK_2 and TK_3 receptors. Illogically, though, the peers of the meeting got entangled in heated discussions, with the result that the receptors kept their group name 'tachykinin receptors' but were abbreviated as NK_1, NK_2 and NK_3. Explain this to a newcomer in the field! Despite the confusion that followed (see chapter by R. Patacchini and C.A. Maggi on the tachykinin nomenclature), no revision of the troubled tachykinin nomenclature has since come about. It is for this reason and the sake of consistency that I have also enforced the 'Montreal nomenclature' throughout this book, although it is my hope that the wise decision of the HUGO Nomenclature Committee to abbreviate the tachykinin genes *TAC* and the receptor genes *TACR* is now going to solve the confusion and to cause the NK misacronym to be abandoned.

These drawbacks on the semantic side did not, fortunately, stand in the way of formidable triumphs in preclinical tachykinin pharmacology. By the beginning of the 1990s the first highly potent nonpeptide antagonists for the tachykinin NK_1 receptor became available, being followed by tens of compounds selective for either the NK_1, the NK_2 or the NK_3 receptor. With these tools it became possible to screen the physiological implications of tachykinins and to obtain important hints of their involvement in experimental models of disease. The tachykinins now certainly represent one of the most thoroughly investigated family of neuropeptides, justifying substance P to be celebrated as a pioneer amongst neuropeptides. Frustration, though, spread when clinical trials showed that tachykinin (mostly NK_1) receptor antagonists lack efficacy in the treatment of inflammation, hyperalgesia, migraine and asthma. While annoying the scientific community and pharmaceutical industry alike, this outcome has led to a more sophisticated consideration of the potential of tachykinin receptor antagonists in a number of diseases.

The ups and downs of the substance P saga are reflected in many chapters of this volume. Nonetheless, research on tachykinins continues to be in full swing, as there is good reason to predict that tachykinin receptor antagonists are beneficial in the treatment of nausea and emesis, affective disorders, chronic obstructive airway disease and irritable bowel syndrome. It thus seems to be a particularly good time to review this field of research, in which such a wealth of information has been accumulated over the past decades, and to discuss its implications in the development of new medicines. As this volume goes to press, the tachykinin field has regained enormous momentum from two major breakthroughs that have just been accomplished. In 2003 the first NK_1 receptor antagonist was successfully turned into a drug, after aprepitant was approved by both FDA and EMEA for the combination treatment of chemotherapy-induced emesis. The excitement about this much-awaited milestone is rightly reflected in a core chapter of this volume in which P.L.R. Andrews and J.A. Rudd outline the broad antiemetic potential of tachykinin NK_1 receptor antagonists. The other major advance that we are just witnessing relates to the discovery of novel mammalian tachykinins such as hemokinin-1, endokinin A and endokinin B, which are encoded by *TAC4*, and the chromosome 14 tachykinin-like peptide. As we are going to learn more about the role of these peptides in health and disease, even further tachykinin receptors may come to light.

It is with such considerations that I hand over this volume to the reader. Its chapters have been written by most eminent investigators in the field, and I am most grateful to all of them for giving so freely of their expertise and time. Special thanks go to Dr. Moira A. Vekony and to Susanne Dathe and Helmut Schwaninger of Springer-Verlag for their invaluable help in seeing this project through the press. While all efforts have been made to cover the major issues of contemporary research in experimental and clinical tachykinin pharmacology, the field is so extensive that a few aspects could not be dealt with to an equal extent. Although the chapters have been edited for consistency in nomenclature, there may be some redundancies in the historical and general introductions of the individual articles. These overlaps are not out of place, because a handbook is usually not read like a novel from the beginning to the end but consulted in quest of a certain topic. In addition to its encyclopedic character, I wish that the book also provides gratifying reading in less well-known areas of the field. After all, did you know that the pleasure of massaging may be related to the cutaneous release of substance P (see chapter by A. Rawlingson, S.K. Costa and S.D. Brain).

Graz, Austria, January 2004   Peter Holzer

List of Contributors

(Addresses stated at the beginning of respective chapters)

Andrews, P.L.R.   359

Boyce, S.   441
Brain, S.D.   459
Bunnett, N.W.   141

Canning, B.J.   245
Conlon, J.M.   25
Costa, S.K.   459

Donnerer, J.   1

Emonds-Alt, X.   219

Gadd, C.A.   297

Hill, R.G.   441
Hökfelt, T.   63
Holzer, P.   511
Hunt, S.P.   297

Joos, G.F.   491

Kerns, J.K.   273
Kuteeva, E.   63

Lembeck, F.   1
Ljungdahl, Å.   63

Maggi, C.A.   121, 173
Mazzone, S.B.   245

Patacchini, R.   121, 173

Rawlingson, A.   459
Roosterman, D.   141
Rudd, J.A.   359
Rumsey, W.L.   273
Rupniak, N.M.J.   341

Stanic, D.   63
Sukumaran, M.   297

List of Contents

History of a Pioneering Neuropeptide: Substance P 1
    F. Lembeck, J. Donnerer

The Tachykinin Peptide Family, with Particular Emphasis on Mammalian
Tachykinins and Tachykinin Receptor Agonists........................ 25
    J. M. Conlon

The Histochemistry of Tachykinin Systems in the Brain................ 63
    T. Hökfelt, E. Kuteeva, D. Stanic, Å. Ljungdahl

The Nomenclature of Tachykinin Receptors........................... 121
    R. Patacchini, C. A. Maggi

The Mechanism and Function of Agonist-Induced Trafficking
of Tachykinin Receptors 141
    D. Roosterman, N. W. Bunnett

Tachykinin NK_1 Receptor Antagonists 173
    R. Patacchini, C. A. Maggi

Tachykinin NK_2 Receptor Antagonists 219
    X. Emonds-Alt

Tachykinin NK_3 Receptor Antagonists 245
    S. B. Mazzone, B. J. Canning

Combined Tachykinin NK_1, NK_2, and NK_3 Receptor Antagonists........... 273
    W. L. Rumsey, J. K. Kerns

Pre-protachykinin and Tachykinin Receptor Knockout Mice 297
    C. A. Gadd, M. Sukumaran, S. P. Hunt

Therapeutic Potential of Tachykinin Receptor Antagonists
in Depression and Anxiety Disorders................................ 341
    N. M. J. Rupniak

The Role of Tachykinins and the Tachykinin NK_1 Receptor in Nausea
and Emesis 359
    P. L. R. Andrews, J. A. Rudd

Substance P (NK$_1$) Receptor Antagonists—Analgesics or Not? 441
   S. Boyce, R. G. Hill

Role of Tachykinins in Neurogenic Inflammation of the Skin
and Other External Surfaces 459
   A. Rawlingson, S. K. Costa, S. D. Brain

Role of Tachykinins in Obstructive Airway Disease. 491
   G. F. Joos

Role of Tachykinins in the Gastrointestinal Tract 511
   P. Holzer

Subject Index 559

History of a Pioneering Neuropeptide: Substance P

F. Lembeck · J. Donnerer

Institut für Experimentelle und Klinische Pharmakologie, Universitätsplatz 4,
8010 Graz, Austria
e-mail: fred.lembeck@uni-graz.at

1	Extraction, Biological Testing and Structural Elucidation............	1
2	Masanori Otsuka and the Action of Substance P in the Spinal Cord.......	8
3	Radioimmunoassay and Immunohistochemistry of Substance P.........	8
4	Nicolas Jancsó and Capsaicin............................	10
5	Peptide and Nonpeptide Transmitters in Primary Afferent Neurons......	12
6	Involvement of Substance P in Autonomic and Neuroendocrine Reflexes...	13
7	Tachykinin Genes, Precursors and Receptors...................	15
8	Colocalization of Substance P with Peptides and Amino Acids..........	15
9	Substance P and Neurokinin A in the Gut.....................	17
10	Tachykinin Receptor Antagonists: The Role of Substance P Revisited.....	18
References..		19

Keywords Substance P · Discovery · Pain transmission · Autonomic reflexes · Neurogenic inflammation

1
Extraction, Biological Testing and Structural Elucidation

The initial discovery of an 'unidentified depressor substance in certain tissue extracts' by von Euler and Gaddum (1931) has been described in detail by von Euler (1977):

The discovery of substance P (SP) was unexpected but not wholly fortuitous. As a postgraduate student in 1930 I was allowed to work in H. H. Dale's laboratory in Hampstead, London. At that time acetylcholine (ACh) was of primary interest in the laboratory, and I was given the task of trying to demonstrate a release of ACh from the intestine on vagus stimulation. This led to some experiments in which an intestinal extract was tested for biological activity on the isolated jejunum of the rabbit. Since the contraction observed was not inhibited by atropine and since histamine—another of the favourite substances of the laboratory—did not contract the rabbit's jejunum, it appeared that the effect observed was due to a new or at least unknown principle. Dale suggested that the work should be continued jointly with J. H. Gaddum, senior assistant in the laboratory. As a result the effect observed was systematically investigated, and after some time we were satisfied that the active agent was not identical with any of the compounds then known that had a stimulating action on the gut. A study of extracts from various organs also showed that this substance was present in appreciable quantities in the brain. We also observed that all extracts that contracted the gut also lowered the blood pressure, especially in the rabbit. In order to make quantitative estimations we used a purified standard preparation simply referred to as 'P' on the tracings and in the protocols. Some 30 years later, Gaddum wrote "We concentrated the active substance in the form of a stable dry powder known as preparation P (which probably contained about 15 units/mg). It is impossible to justify the widespread custom, started by Gaddum and Schild, of calling the active principle itself substance P, but it is probably too late to change that now." The active principle is now generally called substance P or SP, somewhat in analogy to PG for prostaglandins, which has the advantage of allowing convenient addition of qualificatory suffixes.

The substance could be distinguished from acetylcholine as it was not inhibited by atropine, and also from histamine because this amine did not contract the rabbit jejunum, which was used as the test organ. The substance was found only in extracts from gut and brain. All the extracts also lowered the blood pressure, especially in the atropinized rabbit.

Gaddum and Schild (1934) excluded adenosine derivatives in this extract, found its acid stability, its solubility in acetone and ethyl alcohol, and described it as a basic compound. They also called the stable dry precipitate, a powder, the first time 'substance P'. Von Euler (1936) removed large proteins from the tissue extracts by boiling at pH 4, followed by salting out the peptides by ammonium sulfate. This stable dry powder, which probably contained 15 units/mg substance P, was used in all the earlier experiments. The peptide nature of the active principle was indicated by inactivation with trypsin.

All the earlier papers measured the amount of substance P in biological units, although an International Standard has never been established. After the synthesis of substance P, an activity of about 200,000 biological units/mg was established, thus 1 unit is roughly equal to 5 ng substance P.

It is worth remembering what Dale (1933) said. In 1933, nearly at the end of his third Dohme lecture, Dale made a statement which is as topical now as it was then and may serve as a kind of warning. He said:

The discovery, in artificial extract from an organ or tissue, of a substance which on artificial injection produces a pharmacodynamic effect provides only a first item of presumptive evidence in support of a theory that the action of this substance plays a part in normal physiology. Much more evidence is required before we can attribute clearly defined functions to such a substance, as we can now do in the cases of histamine and acetylcholine. But even where this is already possible, we have still no evidence to justify the assumption that the substance comes naturally into action in the body in the free condition in which we isolate and identify it in the laboratory after various unnatural chemical procedures.

Not much attention was paid to substance P for the next 20 years. At that time a number of peptides such as the hormones insulin, vasopressin and oxytocin, and other agents like bradykinin or angiotensin were known. But no methods were yet available for their isolation and establishing their structure.

Lembeck's interest in substance P emerged from other findings. Hellauer, the youngest coworker of Otto Loewi in Graz, published two papers with Loewi in 1940, in which it was shown that the dorsal roots of the spinal cord, in contrast to the ventral roots, do not contain acetylcholine. Long ago, Stricker (1876) had made an unusual observation: when he stimulated the peripheral endings of the cut dorsal roots in a dog, he observed a peripheral vasodilatation, measured by a mercury thermometer between the toes. This was a contradiction to the Law of Bell and Magendie which says that the dorsal roots contain only afferent nerve fibers. Stricker's observation was fully confirmed by Bayliss (1901) and was thereafter mentioned as an unexplained curiosity by the name 'antidromic vasodilatation' in physiology textbooks.

In his Walter Dixon Memorial Lecture, Dale (1935) commented:

When we are dealing with two different endings of the same sensory neuron, the one peripheral and concerned with vasodilatation and the other at a central synapse, can we suppose that the discovery and identification of a chemical transmitter of axon-reflex vasodilatation would furnish a hint as to the nature of the transmission process at a central synapse? The possibility has at least some value as a stimulus to further experiment.

Hellauer and Umrath (1948) injected subcutaneously crude extracts of ventral and dorsal roots into the ears of rabbits in order to find the transmitter substance of sensory fibers. The differences, observed only visually—as no suitable equipment to measure vasodilatation was available in these post-war years—seemed to Lembeck, having been trained in bioassay in Gaddum's laboratory, not at all convincing. But the approach raised his interest. Lembeck's laboratory facilities were also fairly primitive. The isolated organ bath to suspend a guinea pig ileum was home made, the guinea pigs were home bred and the kymograph was a heirloom of Otto Loewi's time. Lembeck prepared substance P containing ammonium precipitate and found that it caused contractions in the presence of atropine and an antihistamine. Therefore he interpreted it as the effect of substance P. He went to the slaughterhouse to collect ventral and dorsal roots and made simple extracts. And indeed there was at least ten times more activity in the dorsal as compared to the ventral roots (Lembeck 1953a). The dorsal roots had the activity of 4 units substance P/mg wet tissue, thus equaling roughly 20 ng substance P. Lembeck had not been aware before that the guinea pig ileum, used for the first time as a substance P bioassay, reacted to such a small amount. A proteolytic destruction of the activity of the dorsal and ventral roots could also be shown, which was a strong argument for substance P. As a second bioassay he used the decrease in blood pressure of the atropinized rabbit, but there was no difference in the activity.

Holton and Holton (1952) found that the antidromic stimulation of the n. auricularis induced the release of ATP from the rabbit ear: the decrease in blood pressure caused by ventral and dorsal root extracts was about equal. But ATP contracted the gut only in amounts above 10 µg and could therefore not be involved in contraction of the guinea pig ileum induced by the released material. The old findings of Holton and Holton (1952) and Holton and Perry (1951) found confirmation in recent years, as ATP is a co-transmitter in many synapses (see Hökfelt et al. 1980).

When Lembeck showed his results to Gaddum, Gaddum critically asked "Could you exclude serotonin?" and Lembeck's reply was also a question "What is serotonin?" Gaddum thereafter informed Lembeck about the recent isolation of serotonin by Rapport, Green and Page (1948) and the previous work of Erspamer. Fisher of the Upjohn Company had just synthesized serotonin and sent Lembeck a 10-mg sample. Desensitizing the guinea pig ileum to serotonin by the large dose of 50 µg/ml did not influence the effects of the dorsal root extracts; therefore serotonin could be excluded.

In parallel with these investigations, Kopera and Lazarini (1953) studied the distribution of substance P in the bovine and feline brain. Simple extracts were prepared and tested on guinea pig ileum in the presence of atropine and an antihistamine. A high amount of substance P in the substantia nigra was noted. These results were confirmed and extended by investigations of Pernow (1953), Zetler and Schlosser (1955), and others (for a review see Pernow 1983).

All of these findings evoked a certain amount of interest in substance P. Changes in different functional stages of the brain were investigated by Zetler and Ohnesorge (1957), and investigations of the functions on the gut and on the central nervous system (CNS), even a search of antagonists under arbitrarily selected compounds (Stern and Huković 1961), led to a first Symposium on Substance P, held on 9 and 10 June 1961 in Sarajevo (Fig. 1 and Fig. 2). At the end of the symposium, all participants agreed that further progress would depend on the expected isolation of substance P. Pernow (1953) had already found that substance P could be absorbed on alumina and eluted with decreasing concentrations of aqueous methanol; in this way the activity was increased from a few biological units per milligram to about 3,000 units per milligram. Franz, Boissonas and Stürmer (1961) of Basel reported at the symposium that they had reached an activity of 30,000–35,000 units per milligram by chromatography and electrophoresis.

At a symposium of the New York Academy of Sciences under the title 'Structure and Function of Biologically Active Peptides: Bradykinin, Kallidin and Congeners' in 1963, 30 papers were devoted to bradykinin, for which the greatest progress including isolation and synthesis was achieved, and only eight to substance P (Whipple et al. 1963). The contributions from Basel on the isolation of substance P were the most remarkable ones, by Boissonas and Stürmer of

Fig. 1 Participants of the Symposium on Substance P, organized by P. Stern and held on 9–10 June 1961 in Sarajevo. The photograph was taken by U.S. von Euler (see Fig. 2). *Front row seated, from right*: K. Umrath, J.H. Gaddum, G. Zetler, V. Varagic, H. Caspers, E. Stürmer. *Second row seated*: B. Pernow (*third from right*), F. Lembeck (*second from left*). *Standing, from right*: S. Hukovic (*second*), W.A. Krivoy (*third*), K. Lissak (*fourth*), P. Stern (*fifth*), Marthe Vogt (*ninth, half-covered*)

Fig. 2 U.S. von Euler. The photograph was taken by F. Lembeck at the occasion of the Symposium on Substance P in Sarajevo, 9–10 June 1961

Sandoz, by Vogler and coworkers of Roche, and by Zuber of Ciba-Geigy—they all had almost reached the goal, but the essential hit came from somewhere else.

Erspamer had found that the salivary gland of a mollusc, *Eledone moschata*, contains besides serotonin the peptide eledoisin, and the skin of a South American frog, *Physalemus fuscomaculata*, contains the peptide physalaemin. Both peptides occur in large amounts and therefore their isolation and elaboration of their chemical structure was no problem. Lembeck once asked Erspamer about the motivation behind these experiments. His reply was short and precise: "It was just serendipity, nothing else". The strategy of his work was to test the activity of eledoisin, physalaemin and substance P on many bioassay preparations in vitro and in vivo. Ten of the amino acids in physalaemin were present in the not yet completed analysis of substance P as shown by the researchers in Basel. In a paper of Bertaccini et al. (1965) we find the following:

But the door still remained closed for several years. I think the next important step emerged from the Mediterranean Sea, like Aphrodite, but in the shape of a mollusk, *Eledone moschata*. Erspamer discovered eledoisin in its salivary glands and among so many other interesting peptides, shortly thereafter, physalaemin in the skin of a South American frog. The isolation and careful pharmacological analysis of these peptides led to a statement by Erspamer in 1965 (Bertaccini et al. 1965) which is worth quoting: It may be seen that as many as ten of the amino acids found in substance P are also present in eledoisin, and eight in physalaemin. The only amino acid lacking in the molecule of substance P is apparently the tremendously important methioninamide. The suspicion seems to be justified that the labile methioninamide residue has escaped the attention of research workers, who have isolated and studied substance P. We would suggest that this possibility be checked.

At this time, Lembeck compared eledoisin and physalaemin with substance P in terms of their effects on salivary secretion. This led to a profound speculation: could substance P, when released from the gut into the circulation, induce increased salivary secretion like secretin released from the gut induces increased fluid secretion from the pancreas? Lembeck and Starke measured salivary secretion after intravenous injection of the substance P preparation into chicken, rat and dog, the species used by Bertaccini et al. (1965) and observed a pronounced increase in salivary secretion (Lembeck and Starke 1968). In addition, they simultaneously recorded the blood pressure in the dog: substance P, at a dose that induced only a weak salivary secretion, evoked a profound decrease in blood pressure. From common experience it is known that the smell of a delicious meal can induce salivation, however, fortunately no fainting—so this romantic speculation had to be abandoned.

A few days later, on 4 May 1967, Lembeck made an important discovery in an unusual place: sitting in a convenient Intercity train, reading the *Frankfurter Allgemeine Zeitung*, he found a short note about the isolation of a 'salivary secretion stimulating factor' called 'sialogen' in peptide extracts from hypothalamus by Leeman and Hammerschlag (1967), presented at a Federation Meeting. Lembeck had the feeling that he was the only really fascinated reader of this message, written by an unknown reporter. After his return to Tübingen, he sent Leeman all the results on salivation by air mail. Samples were exchanged and there was soon enough evidence that sialogen and substance P were identical.

Susan Leeman's work was a spin-off product of a rather complicated endocrinological test. Glucocorticoids are synthesized and released, but not stored in the adrenal cortex. When their synthesis is increased, the high amount of ascorbic acid in the rat adrenal cortex decreases. This reduction in ascorbic acid, which can easily be measured, was therefore a parameter for the assay of glucocorticoid production such that the effect of adrenocorticotrophic hormone (ACTH)—released from the pituitary and stimulating the synthesis of glucocorticoids—could be quantified. The neurohormonal control of the ACTH secretion is based on the release of the corticotropin-releasing factor (CRF). Leeman's project was to purify this factor (Leeman et al. 1962). The presence of CRF in a crude hypothalamic extract was verified and the extract was subjected to gel exclusion chromatography on Sephadex G-75 as the next step in a purification procedure. Each fraction from the column had to be injected into an anesthetized rat to measure the biological activity. Material from certain fractions rapidly caused a copious secretion of saliva. The sharp cut-off of sialogogic activity in the elution profile was shown to be caused most probably by the presence of vasopressin, which produced a marked blanching in the test rats and effectively inhibited the sialogogic response. Michael Chang joined the project and they turned from CRF to the isolation of the newly discovered hypothalamic peptide. The initial large-scale tissue extraction (the extract from 200 bovine hypothalami was applied to one column!) outgrew the facilities of the laboratory. Further steps were ion exchange chromatography and paper electrophoresis (Chang and Leeman 1970). The amino acid sequence of substance P was published in 1971 (Chang et al. 1971). Tregear et al. (1971) were able to synthesize the peptide and to produce Tyr^8-substance P which enabled the labeling with ^{125}I and thus permitted the development of a radioimmunoassay (Powell et al. 1973). Studer et al. (1973) showed that the substance P in gut was identical with that in brain.

This story is strongly reminiscent of what Gaddum wrote in 1954, and which will, hopefully, remain true in the future:

The most interesting discoveries (in pharmacology) are the unexpected ones, and in spite of the fact that research is much more organized today than it was once, unexpected discoveries are still made.

The availability of synthetic substance P dramatically changed the scenery in this field of research, and the number of publications on substance P increased steeply between 1970 and 1980 (see Pernow 1983). Previous material, compiled by Lembeck and Zetler in reviews 1962 and 1971 needed to be confirmed and extended. New methods allowed unexpected guidelines to a much better insight into the distribution of substance P in tissues and conclusions about its function. In the following only the keys to these new methods are described, while the full gain of the new results is described in the other chapters.

2
Masanori Otsuka and the Action of Substance P in the Spinal Cord

At the International Congress of Pharmacology 1972 in San Francisco Lembeck was elected Secretary General of the IUPHAR. His involvement in administration had considerably diluted his attention to the scientific program. When Lembeck met Marthe Vogt she asked whether he knew Masanori Otsuka, what Lembeck denied. "But he has confirmed your results on substance P from 1953", she said, and brought Otsuka and Lembeck together.

Otsuka had found that dorsal roots contain 20 times more substance P than ventral roots, and he first observed—by the use of a bioassay—that the substance P content of the dorsal horn declined after section of the dorsal roots; this means that substance P is produced in the dorsal root ganglion und transported by the central fibers of afferent neurons into the dorsal horn (Otsuka et al. 1972). Lembeck returned from California with a feeling like that of the gold digging people. Another confirmation came from B. Pernow: While the substance P content of the rat ventral horn is 134±33 pmol/g substance P, that of the dorsal horn is 1070±160 pmol/g wet tissue (see Pernow 1983).

Konishi and Otsuka (1974a) investigated the effects of substance P and other peptides on the isolated spinal cord of the frog. Thereafter they used the spinal cord of newborn rats (Konishi and Otsuka 1974b). The dorsal root was placed in a suction electrode that was stimulated and the ventral root was placed in a recording suction electrode. Depolarization in the ventral root was measured after the addition of substance P and L-glutamate, and the activity of substance P was estimated to be about 200 times higher than that of glutamate which was regarded as the leading candidate for the excitatory transmitter in the spinal cord (Curtis and Johnston 1974). It was not before 1994 (Ueda et al. 1994) that the co-release of substance P and glutamate was shown. Yanagisawa et al. (1984) developed the isolated spinal cord tail preparation of the newborn rat, in which they could demonstrate capsaicin-induced nociceptive responses. (For the neurotransmitter functions of mammalian tachykinins, see also the review by Otsuka and Yoshioka 1993.)

3
Radioimmunoassay and Immunohistochemistry of Substance P

At Nobel Symposium 37 (1976, Stockholm; organized by U.S. von Euler and B. Pernow) under the title 'Substance P', (Fig. 3), the dominant role of the immunological methods based on the radioimmunoassay of Powell et al. (1973) was evident. Susan Leeman agreed to accept Rainer Gamse as coworker at the Harvard Medical School in order to learn the radioimmunoassay. When he returned to Graz after a year, he opened the way to many new projects. Thomas Hökfelt presented, based on 23 impressive slides, the distribution of substance P in the central and peripheral nervous system and made comments on the results which are worth reading even today.

Fig. 3 Participants of the Nobel Symposium 37, entitled 'Substance P', organized by U.S. von Euler and B. Pernow and held in Stockholm in June 1976. *Lowest step, at right*: E.G. Erdös. *Second step, from right*: U.S. von Euler (*second*), G.F. Erspamer (*third*), S.E. Leeman (*fourth*), K. Krnjevic (*fifth*). *Third step, from right*: B. Pernow (*second*), W.A. Krivoy (*third*), G. Zetler (*fourth*), P. Oehme (*second from left*), F. Lembeck (*first from left*). *Fourth step, from right*: S. Rosell (*second*), J.L. Henry (*second from left*), R. Gamse (*first from left*). *Fifth step, from right*: M. Otsuka (*third from left*), T. Hökfelt (*second from left*). *Last step, from right*: K. Folkers (*first*), A. Carlsson (*second*)

The word 'tachykinin' was a creation of Erspamer. On isolated smooth muscle preparations eledoisin, physalaemin and substance P caused faster contractions than bradykinin, and to designate this difference at a time when none of these peptides had been isolated he created these descriptive names which are still used today. Two peptides closely related to substance P, neurokinin A and neurokinin B, were discovered in mammals, and many further tachykinins were isolated from lower animals.

The immunohistochemical technique, including the immunofluorescence technique, the peroxidase technique, the very sensitive peroxidase-antiperoxidase technique and the 'double staining techniques' allowed the identification of more than 20 peptides in neurons of the brain, spinal cord and in the periphery. In several cases peptides occur together with a 'classical' transmitter in the same neuron (Hökfelt et al. 1980). Calcitonin gene-related peptide (CGRP) was an unexpected finding and remarkable, because its distribution and functions are closely related to those of substance P (for a review see Wimalawansa 1996). Somatostatin was also found concomitantly with substance P (Gamse et al. 1981). The concentration of these and many other peptides in the CNS is about 1,000 times lower than that of monoamines and 100,000 times lower than that of amino acids. But peptides may activate their receptors at a much lower concentration than classical transmitters. Classical transmitters are synthesized not only in the cell body, but also at the nerve endings, and a reuptake mechanism exists. Peptides are produced only in the cell body and are carried by the axonal transport to the nerve terminals where no reuptake mechanism seems to exist. Classical transmitters usually cause a rapid response of short duration, whereas the peptides are responsible for a long-lasting effect. It has to be emphasized that the new findings of Hökfelt and many others provided a completely new view of the afferent, autonomic and central nervous systems. The findings also include new aspects of the organization of neuronal pathways within the CNS. Finally the coexistence of peptides with classical transmitters expanded the original principle of Dale (i.e., one neuron–one transmitter) to a system which seems to be adequate for the great variety of neuronal functions, especially within the CNS (for a review see Hökfelt et al. 1980).

4
Nicolas Jancsó and Capsaicin

It was in 1967 when Lembeck attended the Annual Meeting of the Hungarian Pharmacological Society in Budapest. He extended the journey to Szeged where he met Nicolas Jancsó in his laboratory, which was full of all kind of experimental equipment and had hundreds of bottles on the shelves. Lembeck knew Jancsó's work on the extinction of pain by capsaicin (Jancsó et al. 1959). Lembeck invited Jancsó to Tübingen for a seminar, and Jancsó arrived a few months later with his wife and Thuranszky, the owner of a pre-war Mercedes car. Jancsó gave his seminar on a very hot summer day, at two o'clock in the afternoon. He carefully explained his simple, but conclusive experiments, and Lembeck was probably the only one who was convinced that he was right. After the availability of synthetic substance P there were so many projects to deal with, that only with some delay we started to unravel the functional implications of substance P-containing afferent fibers by the action of capsaicin. The initial stimulus for this work was the work of Nicolas Jancsó, Aurelia Jancsó-Gábor and János Szolcsányi (1967) who reported:

1. By antidromic electrical stimulation of the sensory nerves (saphenous or trigeminal) of rats, the following signs of an inflammatory response could be elicited: arteriolar vasodilatation, enhancement of vascular permeability, protein exudation, fixation of injected colloidal silver onto the walls of venules and, later, their storage in histiocytes.
2. The inflammatory response induced by electrical stimulation could not be altered by parenterally administered atropine, physostigmine, hexamethonium, phentolamine, dibenamine, propranolol, promethazine, chloropyramine, or methysergide.
3. After the degeneration of the sensory nerve, capsaicin, xylene, mustard oil and ω-chloroacetophenone did not evoke inflammation. Hence, these substances induce inflammation purely or dominantly through the involvement of sensory nerves.

4. Capsaicin desensitization inhibited the signs of inflammation induced both by antidromic stimulation of the sensory nerve and by orthodromic stimulation of pain sensitive nerve terminals with irritants.
5. The experiments suggest that a mediator substance, a neurohumor, is released by orthodromic or antidromic stimulation of pain sensitive nerve terminals and that this substance is responsible for the signs of inflammation produced by some substances.

The question was whether the proposed 'neurohumor' was indeed substance P. The paper by Gamse, Holzer and Lembeck (1980) supplied a clear answer: "Familiar with antidromic vasodilatation from the work of Hellauer and Umrath (1948), we checked whether antidromic vasodilatation, so far only shown by Bayliss (1901) in the dog, could be reproduced in the rat hind leg. By means of a hand-made drop-recorder, we found that the venous outflow from the hind leg was increased by a 1-min stimulation of the saphenous nerve for about 10 min, comparable to the result of Bayliss (1901) in the dog." In capsaicin-pretreated rats an almost complete inhibition was observed (Lembeck and Holzer 1979). The postocclusive vasodilatation was also inhibited after capsaicin pretreatment (Lembeck and Donnerer 1981). Janscó had demonstrated that the 'neurogenic inflammation' consisted of vasodilatation and plasma extravasation. Substance P also evoked a plasma extravasation by histamine release from nearby mast cells, whereas CGRP did not. These events seemed to be closely related to the 'nocifensor system', which Sir Thomas Lewis had found by studies on human skin (see Lembeck 1985). It consisted of a peripheral neurogenic vasodilatation and plasma extravasation (flare and wheal response), providing a defense mechanism which removes exogenous or endogenous toxic material by increase of blood flow and lymph drainage (see also Lembeck 1988). When selective substance P (NK_1) receptor antagonists became available, it was definitively proven that substance P is the mediator of neurogenic plasma extravasation (Lembeck et al. 1992).

Vasodilatation and plasma protein extravasation by substance P are, together with the contraction of various smooth muscles, non-neural effects. Could substance P, released in regions of the CNS, induce vasodilatation besides the stimulation of defined neurons? Lembeck and Starke (1963) tested purified substance P extracts, which were free of amines and ATP, containing different amounts of substance P. The extract of substantia nigra contained enough substance P to produce a pronounced increase in capillary permeability, whereas the other extracts were correspondingly less active. It was concluded that those parts of the brain which are rich in substance P might have a particularly high exchange rate between the capillaries and the surrounding tissue. Hypothalamic blood flow was measured in conscious rabbits by the ^{133}xenon washout technique. Intrahypothalamic injections of substance P at doses of 50 and 500 ng were performed; the results suggested that substance P may cause an increase in blood flow via the endogenous release of acetylcholine which in turn stimulates an intracerebral noradrenergic pathway (Klugmann et al. 1980). Neither publication found an echo and no further experiments of this kind were carried out.

5
Peptide and Nonpeptide Transmitters in Primary Afferent Neurons

Of the primary afferent neurons, Aβ fibers detect innocuous stimuli applied to skin, muscle and joints. Stimulation of Aβ fibers can reduce pain, such as when these fibers are activated by rubbing one's hand. Thinly myelinated Aδ fibers and the most slowly conducting C fibers transmit the rapid, acute, sharp pain, and the delayed, more diffuse, dull pain, respectively. Aδ nociceptors respond to intense mechanical stimuli and to intense heat. Most C fibers are polymodal, responding to noxious thermal, chemical and mechanical stimuli, whereas others are mechanically insensitive, but respond to noxious heat; however, most C fiber nociceptors respond to chemical stimuli such as capsaicin or acid. Tissue injury might sensitize so-called 'silent' receptors. The peptidergic group of unmyelinated C fibers contains tachykinins (substance P and neurokinin A) and expresses trkA (high affinity tyrosine kinase) receptors for nerve growth factor. The second population expresses P2X receptors, specific ATP-gated ion channels. Further peptides found in C fibers include CGRP, released concomitantly with substance P (for a review see Holzer 1992), somatostatin, vasoactive intestinal polypeptide (VIP) and galanin, with much less well defined functions. Besides peptides, glutamate is most probably the predominant excitatory neurotransmitter in all primary afferent neurons.

In contrast to vision, olfaction or taste, sensory nerve endings that detect painful stimuli ('nociceptors') are not localized to a particular anatomical structure, but are dispersed over the body, innervating skin, muscle, joints and internal organs. The transmission of pain has raised the greatest interest in recent years. The vanilloid receptor VR1 was cloned and functionally characterized; it is activated by noxious heat, capsaicin and acid (for a review see Szallasi and Blumberg 1999). The VRL-1 receptor, expressed in a subset of medium- to large-diameter myelinated neurons, is activated by extreme noxious heat. VR1 and VRL-1 belong to a larger family of transient receptor potential channels (for review see Julius and Basbaum 2001).

Primary afferent C fibers are stimulated at their peripheral endings by many exogenous and endogenous substances such as capsaicin, mustard oil, xylol and other compounds which easily penetrate the skin. Histamine, acetylcholine and serotonin, jointly occurring in the stings of nettle, belong to this group of compounds. When the sting tip breaks off, the mixture is released into the dermis. Following stimulation, substance P is released from the peripheral terminals of sensory neurons, as shown by the vasodilatation and plasma extravasation due to antidromic electrical nerve stimulation. Plasma extravasation may be augmented by the release of histamine from mast cells, specifically by basically charged peptides, like substance P, but not by CGRP. This recalls the 'nocifensor system' claimed by Sir Thomas Lewis. He showed the cutaneous 'wheel and flare response' to be of neurogenic origin, absent after chronic denervation, in clinical studies on the human skin. As antidromic vasodilation can only be evoked by nerve stimulation, it cannot be regarded as physiological. But under physiological conditions two actions occur: firstly transmitter release at the peripheral terminals, and secondly neuronal conduction to the central terminals of the dorsal root fibers within the CNS. At this site, the release of substance P takes place, and the amount of substance P is increased under inflammatory conditions in the periphery. This is therefore a nociceptive pathway. But the capsaicin-sensitive C fiber afferents serve not only the signaling of pain—they are much more the afferent part of the autonomic nervous system (which we erroneously regard as being only efferent).

Alarming messages recruiting one or the other defense reaction usually arise from higher brain centers, but the initiating information is signaled to the brain from the periphery. The signal might activate a single, unconscious response or recruit more than one of the possible defense reactions. The peripheral pathways which signal the alarming event to the CNS seem to be mainly afferent neurons which share certain common properties:

1. They are small diameter, unmyelinated C-fibers.
2. They synthesize substance P and other peptides, and transport them to the terminals where they are released by nervous impulses.
3. These neurons are capsaicin-sensitive, i.e., a small dose of capsaicin causes the release of the peptide, and systemic treatment with a very large dose of capsaicin causes these neurons to loose their function or to degenerate.
   The loss of function after systemic capsaicin pretreatment can therefore be regarded as evidence for the involvement of these afferent fibers in the response. It does not allow neurochemical definition of the type of neuron involved. Therefore it has been agreed to speak of 'capsaicin-sensitive neurons'. This limitation is essential because other C-fibers exist which are not capsaicin-sensitive: e.g., cold fibers (Petsche et al. 1983) and osmosensitive fibers in the hepatic portal vein (Stoppini et al. 1984).

6
Involvement of Substance P in Autonomic and Neuroendocrine Reflexes

Orthodromic conduction and transmitter release from capsaicin-sensitive neurons can evoke several reflex responses:

1. At the spinal level it induces the scratching syndrome in mice (Piercey et al. 1985).
2. Cardiovascular reflexes mediated by capsaicin-sensitive afferent fiber stimulation are set in action either by a stimulation or a withdrawal of the noradrenergic vasoconstrictor tone (Juan and Lembeck 1974; Donnerer et al. 1988).

3. Stimulation by heat induces a heat loss reaction in rats (Cormareche-Leyden et al. 1985; Donnerer and Lembeck 1983).
4. The micturition reflex in rats caused by distension of the urinary bladder involves capsaicin-sensitive afferents (Holzer-Petsche and Lembeck 1984; Maggi and Meli 1988) whose activation might depend on ATP.

Endocrine regulations influenced by capsaicin-sensitive afferents and possibly comediated by tachykinins have also been revealed:

1. The hepatic portal vein of rats contains osmoreceptors and glucoreceptors (Stoppini et al. 1984) which are sensitive to hypoglycemia and signal this information to hypothalamic centers via capsaicin-sensitive afferents, which in turn leads to release of adrenaline from the adrenal medulla to upregulate the blood glucose concentration (Amann and Lembeck 1986; Donnerer 1988).
2. The release of ACTH from the pituitary can be evoked either by emotional or various peripheral 'stressors'. Capsaicin-sensitive neurons seem to convey such peripheral signals to the CNS. Cold exposure of rats is known to cause the release of ACTH; in capsaicin-treated rats, this response is abolished. Other peripheral stimuli like intraperitoneal administration of formalin, which is painful, or intravenous injection of isoprenaline induce the release of ACTH; the effect of these treatments is abolished in capsaicin-treated rats. Emotional stress, induced by restraint conditions, which is not conveyed by capsaicin-sensitive afferents but by other central pathways, remains unchanged in capsaicin-treated rats (Amann and Lembeck 1986; Lembeck and Amann 1986; Donnerer and Lembeck 1988). Capsaicin-pretreated rats have normal levels of plasma ACTH, which means a normal function of the endocrine regulation of glucocorticoids. Only the stimulation of the immediate release of ACTH, emerging from afferent neuronal signals, is under the influence of capsaicin-sensitive afferents.
3. In the rat the fertilized egg cell develops to a blastocyte within 3 days—during transport into the rat uterine cavity . In the meantime endocrine influences transform the uterine mucosa to the decidua, an essential requirement for nidation. These endocrine adaptations via hypothalamus and pituitary are set into action by afferent messages originating from the stimulation of vagina and cervix during copulation . These messages run via capsaicin-sensitive neurons to the CNS (Traurig et al. 1984a, 1984b). The existence of this neural pathway had already been shown by Marthe Vogt in 1933.
4. Lactation in the rat is regulated by prolactin and oxytocin. The stimulation of oxytocin release is induced by ultrasonic vocalization of the pups, by pheromones and by suction on the nipples which are richly innervated by substance P-containing neurons. When normal pups were fed by capsaicin-treated dams, the amount of milk they drank and their gain in weight were

both reduced by 20% (Traurig et al. 1984c). This is another example of the involvement of capsaicin-sensitive afferents in an endocrine regulation.

These few examples of autonomic reflexes and endocrine regulations present evidence that the afferent part is mediated by capsaicin-sensitive neurons. It is known that the n. vagus or the n. splanchicus, described in textbooks as 'efferent', contain a high amount of afferent fibers. Immunohistochemistry showed that substance P fibers account for about 10% of all fibers in the cat vagus nerve (Gamse et al. 1979).

7
Tachykinin Genes, Precursors and Receptors

A new era in tachykinin research began in the 1980s through application of recombinant DNA technology. Nakanishi and his coworkers (1983–1986) opened the way to investigate the biosynthesis of substance P and first isolated its gene and precursor protein containing 384 amino acid residues (Nawa et al. 1983; Kawaguchi et al. 1986). Currently the list of mammalian tachykinins is extended by the discovery of hemokinins and endokinins. In addition, there is a large number of nonmammalian tachykinins including eledoisin in mollusca, scyllorinin I and II in fishes, physalaemin and others in amphibia. They all share the same carboxyl terminal with the mammalian tachykinins.

The *Xenopus* oocyte expression system combined with electrophysiogical measurements was the key to the molecular structure and function of the three tachykinin receptors. Intranuclear injection of oocytes with cDNA encoding the receptor, or injection of cRNA into the cytoplasm of an oocyte produces a functional, foreign receptor–channel complex in the membrane of the oocyte. After application of its specific ligand, an electrophysiological response can be recorded. It was shown that the preferred receptors for substance P, neurokinin A and neurokinin B are encoded by different mRNAs. The investigation of the relative activities on various test preparations showed that the functional site of the NK_1 receptor is activated predominantly by substance P, that of the NK_2 receptor by neurokinin A, and that of the NK_3 receptor by neurokinin B (see Henry et al. 1987). The membrane topology of all three receptor molecules is that of rhodopsin-type receptors, consisting of seven hydrophobic membrane-spanning domains with an extracellular amino terminus and a cytoplasmic carboxyl terminus (Nakanishi 1991).

8
Colocalization of Substance P with Peptides and Amino Acids

Like other neurotransmitters, substance P is stored mainly in the synaptosomal fraction of nerves and in microsomal fractions of brain homogenates. The binding sites of substance P in synaptic vesicles are extractable with ether and chloroform. These and other observations suggest that the specific substance P binding is phosphatidyl serine, which is of interest since this lipid is a constituent of synaptic membranes and vesicles.

Until the 1970s it had been a widespread belief that a neuron contains and releases only one transmitter according to Dale's concept of one neuron–one transmitter. Hökfelt et al. (1978) showed the co-existence of substance P and serotonin in the same neuron, a finding later confirmed by Cuello et al. (1982). Numerous examples of the coexistence of substance P and serotonin exist not only within the CNS, but also in the periphery. Thus, substance P, neurokinin A and serotonin coexist in carcinoid tumors (Bergström et al. 1995). When Lembeck (1953b) first found serotonin in a carcinoid tumor, he noted another gut-stimulating compound with a different Rf-value. By desensitization with large doses of substance P and serotonin, respectively, he found an indication for the presence of both compounds. But the methods available at that time were not sufficient for a clear proof and he mentioned this finding only in the discussion but not in the summary of the paper.

Numerous examples of the coexistence of peptides with other peptides, amines and amino acid neurotransmitters are now established. The functional significance of more than one transmitter is still not fully understood. If a neuron stores and releases a fast-acting transmitter (e.g., an amino acid) and a slow-acting transmitter (e.g., a peptide), then upon excitation it will produce a fast excitatory or inhibitory postsynaptic potential (EPSP or IPSP) and in addition slow EPSPs and IPSPs in postsynaptic cells (Otsuka and Yanagisawa 1987). For example, if γ-aminobutyric acid and substance P are co-released, a short-lasting inhibition and a prolonged excitation will be produced (Otsuka and Yoshioka 1993). In some circumstances, multiple transmitters released from a nerve terminal may diffuse across some distance and selectively act on cells equipped with the respective receptors, called 'chemically addressed transfer of information' (Iversen 1986).

An interesting issue is the co-release of substance P and glutamate. De Biasi and Rustioni (1988) first showed the coexistence of glutamate and substance P in dorsal root ganglion neurons and at their central terminals. Donnerer and Amann (1994) demonstrated that blockade of ionotropic glutamate receptors of the N-methyl-D-aspartate (NMDA) type inhibits afferent nerve-mediated autonomic reflexes. Ueda et al. (1994) found with a continuous on-line monitoring method that glutamate is released from capsaicin-sensitive primary afferent fibers in the rat. The inhibition—by the NMDA receptor antagonist MK-801—of the central transmission of reflexes mediated by capsaicin-sensitive afferents points to an essential role of glutamate in these reflex responses, whereas substance P antagonists had no effect (Juranek and Lembeck 1996). A decrease of the capsaicin-induced release of glutamate from afferent nerve terminals by clonidine (Ueda et al. 1995a) and by morphine (Ueda et al. 1995b) has also been demonstrated, confirming and explaining earlier findings of Donnerer et al. (1988) on the depressor effect. A facilitation of the glutamate release by co-released substance P was assumed (Juranek and Lembeck 1997).

The life cycle of substance P ends by internalization of the receptor-bound peptide and enzymatic breakdown. The degradation of substance P is difficult to define, as several organs (kidney, spleen, liver and intestine homogenates) have a high capacity to inactivate it. One has to differentiate between elimination of substance P released into the circulation, and substance P destroyed locally such as in the brain. By using the rat salivary response as a bioassay, Lembeck et al. (1978) showed that the highest degree of elimination of substance P occurs in the liver and hind limbs, followed by the kidney. More than 80% of substance P infused into the portal vein is degraded by the liver (Lembeck et al. 1978). Several cytosolic and membrane-bound peptidases capable of degrading substance P have been extracted from the brain. A partly purified neutral endopeptidase was first described by Benuck and Marks (1975). Degradation of substance P by a neutral metallopeptidase system of synaptosomal fractions of brain also exists (Berger et al. 1979). Neutral endopeptidase subsequently turned out to significantly inhibit the transmitter action of substance P.

9
Substance P and Neurokinin A in the Gut

Although substance P was discovered in 1931 in the gut and brain, by a fall in the blood pressure and by the contraction of the rabbit ileum, no-one asked what its physiological role might be in the latter. Acetylcholine and noradrenaline were accepted as gastrointestinal neurotransmitters, but about a dozen other compounds were known to contract intestinal smooth muscle, such as histamine, oxytocin, later on serotonin, bradykinin and other compounds. The first hint that substance P is involved in neuronal functions was the finding of Ehrenpreis and Pernow (1953) that the aganglionic, inactive part of the rectosigmoid in Hirschsprung's disease contained significantly less substance P than control pieces, whereas the proximal, hyperactive part showed normal levels of substance P. Substance P was found in all parts of the intestine of animal species and man. The guinea pig ileum was later used as a bioassay for substance P, as it is particularly sensitive to this peptide. The method of Holzer and Lembeck (1979) to record peristalsis in the isolated guinea pig ileum offered experimental possibilities to investigate the action of substance P on propulsive motility.

The essential breakthrough came from immunohistochemistry. Two fundamental types of neurons supply the digestive tract: an extensive population of neurons contained within the gut wall (intrinsic neurons) and an extrinsic innervation by cholinergic and noradrenergic autonomic fibers as well as by substance P-containing sensory neurons from dorsal root ganglia and the vagus nerve. The substance P-containing nerve cells in the myenteric ganglia issue a very dense network of varicose nerve fibers to other myenteric ganglia, as well as fibers to the longitudinal and circular muscle layers, to the submucous ganglia and to the mucosa. The number of substance P cell bodies in the myenteric plexus has been underestimated. Only 2.8%–3.5% of the myenteric neurons contained substance P, but in segments of the intestine in a culture medium with colchicine for 24 h, about 20% of the total population are substance P-containing neurons. The proportion of substance P neurons in the submucous plexus is about 11%. In the human intestine, the substance P neurons have a similar distribution as those in the guinea pig gut (Holzer and Holzer-Petsche 1997a).

The extrinsic sensory substance P neurons are fewer than the intrinsic neurons. The extrinsic neurons are capsaicin-sensitive whereas the intrinsic substance P neurons are not. They sense intestinal pain (colic) or they induce reflexes (vomiting) or diarrhea to expel toxic material (for a review see Holzer and Holzer-Petsche 1997b).

The enteric nervous system, already separated from the autonomic nervous system by Langley (1932), consists of a large number of neurons, in the range of the number of neurons within the spinal cord. Immunohistochemistry led to the detection of many other neuropeptides in the gut, besides those mentioned above, such as neuropeptide Y, somatostatin, cholecystokinin, VIP, and others (for review see Holzer and Holzer-Petsche 1997a, 1997b).

10
Tachykinin Receptor Antagonists: The Role of Substance P Revisited

Antagonists of substance P, acting specifically like atropine on muscarinic acetylcholine receptors, were an old dream of pharmacologists. The attempts to find such a compound on a simple preparation like the guinea pig ileum were not successful. A new approach was chosen by Snider et al. (1991) and led to the discovery of the first specific nonpeptide antagonist CP96345. It was found by a robot and Lembeck was keen to see this assembly. So he went to the Pfizer Company in Groton, CT, and saw how 96-hole plates with cells containing substance P receptors were taken from the deep-freeze and brought to room temperature. Radioactively-labeled substance P was injected into the samples, 96 different compounds were added and left for a suitable time in the wells. Thereafter, the fluid of the wells was removed by suction, the probes dried by microwave and transferred to a scanner.

The scanner measured the amount of labeled substance P, and the result was a print-out in gray shades on a piece of paper. Binding of substance P to the receptor was 'black', full inhibition of the binding by the antagonist was 'white'. The most cumbersome work was the selection of the many substances and the preparation of suitable dilutions. The man who had set up this robot left the company after the screening of some 20,000 compounds, being unaware that among them the antagonist CP96345 had already been detected. The antagonist was specifically active on the fall in blood pressure induced by injection of substance P and neurokinin A, but not on that by VIP or CGRP. It inhibited the plasma protein extravasation induced by mustard oil in rats when injected intravenously or given orally. On intestinal smooth muscles specific effects of the antagonist were shown at concentrations below 1 µM, whereas at higher concentrations nonspecific effects were seen also. No effect was observed on the peristaltic reflex. Cardiac effects indicated an effect of both enantiomers on calcium channels.

We probably made an error. The substance P antagonist completely inhibited the effects of substance P released by antidromic nerve stimulation or capsaicin from the peripheral terminals of primary afferent fibers. We knew that substance P is also released from the central terminals of these fibers in the substantia gelatinosa of the dorsal horn of the spinal cord, as shown in vitro on spinal cord slices of the rat (Gamse et al. 1979) and in vivo in the cat (Duggan et al. 1987). We assumed that several reflexes and endocrine regulations, which were blocked following capsaicin pretreatment, should also be blocked by the substance P antagonist. But we did not find an action of the substance P antagonist. At the same time the co-release of substance P and glutamate was shown (Ueda et al. 1994). All the centrally mediated effects evoked or blocked by capsaicin were shown to be inhibited by the NMDA receptor blocker MK-801, therefore being mediated by glutamate (Juranek and Lembeck 1997). The question remains what the physiological effect of substance P at these synapses might be. In the meantime many new and better antagonists blocking NK_1 receptors have been synthesized or isolated from plants, e.g., Compositae (Yamamoto et al. 2002).

As a summary it can be stated that decades of intense research in the field of substance P, tachykinins, and their receptors have pioneered significant advances in the field of physiology, pathophysiology and pharmacology, yet the great pharmacological opportunities are still to come.

References

Amann R, Lembeck F (1987) Stress-induced ACTH release in capsaicin-pretreated rats. Br J Pharmacol 90:727–731

Amann R, Lembeck F (1986) Capsaicin-sensitive afferent neurons from peripheral glucose receptors mediate the insulin-induced increase in adrenaline secretion. Naunyn-Schmiedeberg's Arch Pharmacol 334:71–76

Bayliss WM (1901) On the origin from the spinal cord of the vasodilator fibres of the hind limb, and on the nature of these fibres. J Physiol London 26:173–209

Benuck M, Marks N (1975) Enzymatic inactivation of substance P by a partially purified enzyme from rat brain. Biochem Biophys Res Commun 65:153–160

Berger H, Fencher K, Albrecht E, Niedrich H (1979) Substance P: in vitro inactivation by rat brain fractions and human plasma. Biochem Pharmacol 28:3173–3180

Bergström M, Theodorsson E, Norheim I, Öberg K (1995) Immunoreactive tachykinins in 24-h collections of urine from patients with carcinoid tumors: characterization and correlation with plasma concentrations. Scand J Clin Lab Invest 55:679–689

Bertaccini G, Cei JM, Erspamer V (1965) Br J Pharmacol 25:380–391

Chang MM, Leeman SE (1970) Isolation of a sialogogic peptide from bovine hypothalamus and its characterization as substance P. J Biol Chem 245:4784–4790

Chang MM, Leeman SE, Niall HD (1971) Amino-acid sequence of substance P. Nature New Biol 232:86–87

Cormareche-Leyden M, Shimada SG, Stitt JT (1985) Hypothalamic thermosensitivity in capsaicin-desensitized rats. J Physiol 363:227–236

Cuello AC, Priestley JV, Matthews MR (1982) Localization of substance P in neuronal pathways. In: Porter R, O'Connor M (eds) Substance P in the nervous system. Pitman, London, pp 55-83

Curtis DR, Johnston GA (1974) Amino acid transmitters in the mammalian central nervous system. Ergeb Physiol 69:97-188

De Biasi S, Rustioni A (1988) Glutamate and substance P coexist in primary afferent terminals in the superficial laminae of the spinal cord. Proc Natl Acad Sci USA 85:7820-7824

Donnerer J, Lembeck F (1983) Heat loss reaction to capsaicin through a peripheral site of action. Br J Pharmacol 79:719-723

Donnerer J, Lembeck F (1988) Neonatal capsaicin pretreatment of rats reduces ACTH secretion in response to peripheral neuronal stimuli but not to centrally acting stressors. Br J Pharmacol 94:647-652

Donnerer J (1988) Reflex activation of the adrenal medulla during hypoglycemia and circulatory dysregulations is regulated by capsaicin-sensitive afferents. Naunyn-Schmiedeberg's Arch Pharmacol 338:282-286

Donnerer J, Amann R (1994) Effect of NMDA receptor blockade on afferent nerve stimulation-induced autonomic reflexes. Naunyn-Schmiedeberg's Arch Pharmacol 349:R85

Donnerer J, Zhao Y, Lembeck F (1988) Effects of clonidine and yohimbine on a C-fibre-evoked blood pressure reflex in the rat. Br J Pharmacol 94:848-852

Duggan AW, Morton CR, Zhao ZQ, Hendry IA (1987) Noxious heating of the skin releases immunoreactive substance P in the substantia gelatinosa of the cat: a study with antibody microprobes. Brain Res 403:345-349

Ehrenpreis T, Pernow P (1953) On the occurrence of substance P in the rectosigmoid in Hirschsprung's disease. Acta Physiol Scand 27:380-388

Franz J, Boissonnas RE, Stürmer E (1961) Isolierung von Substanz P aus Pferdedarm und ihre biologische und chemische Abgrenzung gegenüber Bradykinin. Helv Chim Acta 44:881-883

Gaddum JH, Schild H (1934) Depressor substances in extracts of intestine. J Physiol Lond 83:1-14

Gamse R, Molnar A, Lembeck F (1979) Substance P release from spinal cord slices by capsaicin. Life Sci 25:629-636

Gamse R, Holzer P, Lembeck F (1980) Decrease of substance P in primary afferent neurons and impairment of neurogenic plasma extravasation by capsaicin. Br J Pharmacol 68:207-213

Gamse R, Leeman SE, Holzer P, Lembeck F (1981) Differential effects of capsaicin on the content of somatostatin, substance P, and neurotensin in the nervous system of the rat. Naunyn-Schmiedeberg's Arch Pharmacol 317:140-148

Gamse R, Lembeck F, Cuello AC (1979) Substance P in the vagus nerve. Immunochemical and immunohistochemical evidence for axoplasmic transport. Naunyn-Schmiedeberg's Arch Pharmacol 306:37-44

Hellauer HF, Umrath K (1948). Pflüger's Arch ges Physiol 249:619

Henry JL, Couture R, Cuello AC, Pelletier M, Quirion R, Regoli D (1987) Substance P and Neurokinins. Springer, New York

Hökfelt T, Johansson O, Ljungdahl A, Lundberg JM, Schultzberg M (1980) Peptidergic neurons. Nature 284:515-521

Hökfelt T, Ljungdahl A, Steinbusch H, Verhofstad G, Nilsson E, Brodin E, Pernow P, Goldstein M (1978) Immunohistochemical evidence of substance P-like immunoreactivity in some 5-hydroxytryptamine-containing neurons in the rat central nervous system. Neuroscience 3:517-538

Holton FA, Holton P (1952) The vasodilator activity of spinal roots. J Physiol Lond 118:310-327

Holton P, Perry WLM (1951) On the transmitter responsible for antidromic vasodilatation in the rabbit's ear. J Physiol Lond 114:240-251

Holzer P (1992) Peptidergic sensory neurons in the control of vascular functions: mechanisms and significance in the cutaneous and splanchnic vascular beds. Res Physiol Biochem Pharmacol 121:49–146
Holzer P, Holzer-Petsche U (1997a) Tachykinins in the gut. Part I. Expression, release and motor function. Pharmacol Ther 73:173–217
Holzer P, Holzer-Petsche U (1997b) Tachykinins in the gut. Part II. Roles in neural excitation, secretion and inflammation. Pharmacol Ther 73:219–263
Holzer P, Lembeck F (1979) Effect of neuropeptides on the efficiency of the peristaltic reflex. Naunyn-Schmiedeberg's Arch Pharmacol 307:257–264
Holzer-Petsche U, Lembeck F (1984) Systemic capsaicin pretreatment impairs the micturition reflex in the rat. Br J Pharmacol 83:935–941
Iversen LL (1986) Introduction. In: Iversen LL, Goodman E (eds) Fast and slow chemical signalling in the nervous system. University Press, Oxford, pp xi–xiii
Jancsó N, Jancsó-Gábor A (1959) Dauerausschaltung der chemischen Schmerzempfindlichkeit durch Capsaicin. Naunyn-Schmiedeberg's Arch exp Pathol Pharmacol 236:142–145
Jancsó N, Jancsó-Gábor A, Szolcsányi J (1967) Direct evidence for neurogenic inflammation and its prevention by denervation and by pretreatment with capsaicin. Br J Pharmacol Chemother 31:138–151
Juan H, Lembeck F (1974) Action of peptides and other algesic agents on paravascular pain receptors of the isolated perfused rabbit ear. Naunyn-Schmiedeberg's Arch Pharmacol 283:151–164
Julius D, Basbaum AI (2001) Molecular mechanisms of nociception. Nature 413:203–210
Juranek I, Lembeck F (1996) Evidence for the participation of glutamate in reflexes involving afferent, substance P-containing nerve fibres in the rat. Br J Pharmacol 117:71–78
Juranek I, Lembeck F (1997) Afferent C-fibres release substance P and glutamate. Can J Physiol Pharmacol 75:661–664
Kawaguchi Y, Hoshimaru M, Nawa H, Nakanishi S (1986) Sequence analysis of cloned cDNA for rat substance P precursor: existence of a third substance P precursor. Biochem Biophys Res Commun 139:1040–1046
Konishi S, Otsuka M (1974a) The effects of substance P and other peptides on spinal neurons of the frog. Brain Res 65:397–410
Konishi S, Otsuka M (1974b) Excitatory action of hypothalamic substance P on spinal motoneurones of newborn rats. Nature Lond 252:734–735
Kopera H, Lazarini W (1953) Zur Frage der zentralen Übertragung afferenter Impulse. IV. Die Verteilung der Substanz P im Zentralnervensystem. Arch Exp Pathol Pharmakol 219:214–222
Langley JN (1932) Antidromic action. J Physiol Lond 57:428–446
Leeman SE, Glenister DW, Yates FE (1961) Characterization of a calf hypothalamic extract with adrenocorticotropin-releasing properties: evidence for a central nervous system site for corticosteroid inhibition of adrenocorticotropin release. Endocrinology 70:249–262
Leeman SE, Hammerschlag R (1967) Stimulation of salivary secretion by a factor extratced from hypothalamic tissue. Endocrinology 81:803–810
Lembeck F (1953a) Zur Frage der zentralen Übertragung afferenter Impulse. III. Mitteilung. Das Vorkommen und die Bedeutung der Substanz P in den dorsalen Wurzeln des Rückenmarks. Arch Exp Pathol Pharmakol 219:197–213
Lembeck F (1953b) 5-Hydroxytryptamine in a carcinoid tumor. Nature Lond 172:910
Lembeck F (1954) Über den Nachweis von 5-Oxytryptamin (Enteramin, Serotonin) in Carcinoid. Naunyn-Schmiedeberg's Arch exp Pathol Pharmacol 221:50–66
Lembeck F (1985) Sir Thomas Lewis's nocifensor system, histamine and substance P-containing primary afferent nerves. In: Bousfield D (ed) Neurotransmitters in Action. Elsevier, Amsterdam, pp173–179

Lembeck F (1988) The 1988 Ulf von Euler lecture. Substance P: from extract to excitement. Acta Physiol Scand 133:435-454

Lembeck F, Donnerer J (1981) Postocclusive cutaneous vasodilatation mediated by substance P. Naunyn-Schmiedeberg's Arch Pharmacol 316:165-171

Lembeck F, Donnerer J, Tsuchiya M, Nagahisa A (1992) The non-peptide tachykinin antagonist, CP-96,345, is a potent inhibitor of neurogenic inflammation. Br J Pharmacol 105:527-530

Lembeck F, Gamse R (1982) Substance P in peripheral sensory processes. In: Poter R, O'Connor M (eds) Substance P in the Nervous System, Ciba Foundation Symposium 91. Pitman, London, pp 35-54

Lembeck F, Holzer P (1979) Substance P as a neurogenic mediator of antidromic vasodilation and plasma extravasation. Naunyn-Schmiedeberg's Arch Pharmacol 310:175-183

Lembeck F, Holzer P, Schweditsch M, Gamse R (1978) Elimination of substance P from the circulation of the rat and its inhibition by bacitracin. Naunyn-Schmiedeberg's Arch Pharmacol 305:9-16

Lembeck F, Starke K (1963) Substanz P content and effect of capillary permeability of extract of various parts of human brain. Nature (London) 199:1295-1296

Lembeck F, Starke K (1968) Substanz P und Speichelsekretion. Naunyn-Schmiedeberg's Arch exp Pathol Pharmacol 259:375-385

Lembeck F, Zetler G (1962) Sustance P: a polypeptide of possible physiological significance, especially within the nervous system. Int Rev Neurobiol 4:159-215

Lembeck F, Zetler G (1971) Substance P. In: Peters G (ed) International Encyclopedia of Pharmacological Therapy, section 72, vol 1. Pergamon Press, Oxford, pp 29-71

Maggi A, Meli A (1988) The sensory-efferent function of capsaicin-sensitive sensory neurons. Gen Pharmacol 19:1-43

Nakanishi S (1991) Mammalian tachykinin receptors. Ann Rev Neurosci 14:123-136

Nawa H, Hirose T, Takashima H, Inayama S, Nakanishi S (1983) Nucleotide sequences of cloned cDNAs for two types of bovine brain substance P precursor. Nature Lond 306:32-36

Otsuka M, Yanagisawa M (1987) Does substance P act as a pain transmitter. Trends Pharmacol Sci 8:506-510

Otsuka M, Konishi S, Takahashi T (1972) A further study of the motoneuron-depolarizing peptide extracted from dorsal roots of bovine spinal nerves. Proc Jpn Acad 48:747-752

Otsuka M, Yoshioka K (1993) Neurotransmitter functions of mammalian tachykinins. Physiol Rev 73:229-308

Pernow B (1953) Studies on substance P. Acta Physiol Scand 29:Suppl 105

Pernow B (1983) Substance P. Pharmacol Rev 35:85-141

Piercey MF, Moon MW, Blinn JR (1985) The role of substance P as a central nervous system neurotransmitter. In: Jordan CC, Oehme P (eds) Substance P, metabolism and biological actions. Taylor and Francis, London, pp 165-176

Powell D, Leeman S, Tregear GW, Niall HD, Potts JT (1973) Radioimmunoassay for substance P. Nat New Biol 241:252-254

Rapport MM, Green AA, Page IH (1948). J Biol Chem 174:735

Snider RM, Constantine JW, Lowe JA, Longo KP, Lebel WS, Woody HA, Drozda SE, Desai MC, Vinick FJ, Spencer RW, Hess HJ (1991) A potent nonpeptide antagonist of substance P (NK_1) receptor. Science 251:435-437

Stern P, Hukovic S (1961) Specific antagonists of substance P. In: Stern P (ed) Symposium on substance P. Scientific Soc Bosnia Herzegovina Proc vol 1, pp 83-88

Stricker S (1876) Untersuchungen über die Gefäßwurzeln des Ischiadicus. Sitzungsber Kaiserl Akad Wiss Wien 3:173-185

Stoppini L, Barju F, Mathison R, Baertschi I (1984) Spinal substance P transmits bradykinin but not osmotic stimuli from hepatic portal vein to hypothalamus in rat. Neuroscience 11:903–912

Studer RO, Trzeciak A, Lergier W (1973) Isolierung und Aminosäuresequenz von Substanz P aus Pferdedarm. Helv Chim Acta 56:860–866

Szallasi A, Blumberg PM (1999) Vanilloid (capsaicin) receptors and mechanisms. Pharmacol Rev 51:159–212

Traurig H, Saria A, Lembeck F (1984b) The effects of neonatal capsaicin pretreatment on growth and subsequent reproductive function in the rat. Naunyn-Schmiedeberg's Arch Pharmacol 327:254–259

Traurig H, Saria A, Lembeck F (1984a) Substance P in primary afferent neurons of the female rat reproductive system. Naunyn-Schmiedeberg's Arch Pharmacol 326:343–346

Traurig H, Papka RE, Lembeck F (1984c) Substance P immunoreactivity in the rat mammary nipple and the effects of capsaicin pretreatment on lactation. Naunyn-Schmiedeberg's Arch Pharmacol 328:1–8

Tregear GW, Niall HD, Potts JD, Leeman SE, Chang MM (1971) Synthesis of substance P. Nature New Biol 232:87–89

Ueda M, Kuraishi Y, Sugimoto K, Satoh M (1994) Evidence that glutamate is released from capsaicin-sensitive primary afferent fibres in rats: study with on-line continuous monitoring of glutamate. Neurosci Res 20:231–237

Ueda M, Sugimoto K, Oyama T, Kuraishi Y, Satoh M (1995b) Opioidergic inhibition of capsaicin-evoked release of glutamate from rat spinal dorsal horn slices. Neuropharmacology 34:303–308

Ueda M, Oyama T, Kuraishi Y, Akaike A, Satoh M (1995a) Alpha 2-adrenoceptor-mediated inhibition of capsaicin-evoked release of glutamate from rat spinal dorsal horn slices. Neurosci Lett 188:137–139

Vogt M (1933) Über den Mechanismus der Auslösung der Gravidität und Pseudogravidität. Naunyn-Schmiedeberg's Arch exp Pathol Pharmacol 170:72–82

Von Euler US, Gaddum JH (1931) An unidentified depressor substance in certain tissue extracts. J Physiol London 72:74–87

Von Euler US (1936) Preparation of substance P. Scand Arch Physiol 73:142–144

Von Euler US (1977) Historical notes. In: Von Euler US, Pernow P (eds) Substance P, Nobel Symposium 37. Raven press, New York, pp 1–3

Wimalawansa SJ (1996) Calcitonin gene-related peptide and its receptors: molecular genetics, physiology, pathophysiology, and therapeutic potentials. Endocrine Rev 17:533–585

Whipple HE, Silverzweig S, Erdös EG (1963) Structure and function of biologically active peptides. Annals NY Acad Sci 104:1–464

Yamamoto A, Nakamura K, Furukawa K, Konishi Y, Ogino T, Higashiura K, Yago H, Okamoto K, Otsuka M (2002) A new nonpeptide tachykinin NK1 receptor antagonist isolated from the plants of Compositae. Chem Pharm Bull Tokyo 50:47–52

Yanagisawa M, Murakoshi T, Tamai S, Otsuka M (1984) Tail-pinch method in vitro and the effects of some antinociceptive compounds. Eur J Pharmacol 106:231–239

Zetler G, Schlosser L (1955) Über die Verteilung von Substanz P und Cholinacetylase im Gehirn. Naunyn-Schmiedeberg's Arch exp Pathol Pharmakol 224:159–175

Zetler G, Ohnesorge G (1957) Die Substanz P-Konzentration im Gehirn bei verschiedenen Funktionszuständen des Zentralnervensystems. Naunyn-Schmiedeberg's Arch exp Pathol Pharmakol 231:199–210

The Tachykinin Peptide Family, with Particular Emphasis on Mammalian Tachykinins and Tachykinin Receptor Agonists

J. M. Conlon

Department of Biochemistry, Faculty of Medicine and Health Sciences,
United Arab Emirates University, 17666 Al-Ain, United Arab Emirates
e-mail: jmconlon@uaeu.ac.ae

1	**Biosynthesis of the Tachykinins**	26
1.1	The Preprotachykinin A (ppt-a) Gene	26
1.2	The Preprotachykinin B (ppt-b) Gene	28
1.3	The Preprotachykinin C (ppt-c) Gene	28
1.4	Evolution of the Preprotachykinin Genes	29
2	**Naturally Occurring and Synthetic Tachykinin Receptor Agonists**	30
2.1	Preprotachykinin A-Derived Peptides	30
2.1.1	Substance P	30
2.1.2	Neurokinin A	33
2.1.3	Neuropeptide K	34
2.1.4	Neuropeptide γ	35
2.2	Preprotachykinin B-Derived Peptides	36
2.2.1	Neurokinin B	36
2.3	Preprotachykinin C-Derived Peptides	37
2.3.1	Hemokinin 1	37
3	**Actions of the Tachykinins**	38
3.1	Preprotachykinin A-Derived Peptides	38
3.1.1	Substance P	38
3.1.2	Neurokinin A	42
3.1.3	Neuropeptide K	44
3.1.4	Neuropeptide γ	45
3.2	Preprotachykinin-B Derived Peptides	46
3.2.1	Neurokinin B	46
3.3	Preprotachykinin C-Derived Peptides	48
3.3.1	Hemokinin-1	48
	References	49

Abstract The tachykinins, substance P (SP), neurokinin A (NKA), neuropeptide γ (NPγ) and neuropeptide K (NPK) are encoded by the single copy preprotachykinin A (PPT-A) gene on chromosome 7. Four tachykinin precursors (α-, β-, γ- and δ-PPT-A) arise from this gene by an alternative RNA splicing mechanism. The preprotachykinin B gene on chromosome 12 encodes the precursor of neurokinin B (NKB). The preprotachykinin C gene on chromosome 17 encodes a protein that contains the sequence of hemokinin 1(HK-1) and two peptides with limited structural similarity to SP, termed endokinin C and endokinin D. The actions of the tachykinins are mediated through activation of the NK_1, NK_2 and NK_3 receptors. However, NH_2-terminal fragments of SP such as SP (1–7) exert behavioral and other effects when administered centrally that appear not to be mediated through direct interaction with these receptors. Analogs of SP such as [Sar^9]SP sulfone, and peptidase-resistant agonists such as GR73632, show greater potency and selectivity than SP for the NK_1 receptor. Septide {[$pGlu^6,Pro^9$]SP(6–11)}, although a poor competitor for radiolabeled SP in ligand binding assays, is a potent and selective agonist in many functional assays that involve activation of NK_1 receptors. Septide is believed to bind to a distinct 'septide sensitive' site on the NK_1 receptor. Analogs such as [$Lys^5,MeLeu^9,Nle^{10}$] NKA(4–10) and [βAla^8]NKA(4–10) show higher affinity and selectivity than NKA for the NK_2 receptor and [$MePhe^7$]NKB and senktide {succinyl-[Asp^6, $MePhe^8$]SP(6–11)} are more selective than NKB for the NK_3 receptor. NPK and NPγ are preferred agonists for the NK_2 receptor. The NH_2-terminal extension to the NKA sequence in these tachykinins has only minor effects on potency in vitro but the greater stability of these agonists in vivo results in a different spectrum of activities. The physiological role of HK-1 remains to be firmly established but the peptide has a pharmacological profile similar to that of SP. Endokinins C and D display only very weak tachykinin-like bioactivity. Orthologs of the mammalian tachykinins have been isolated from the tissues of nonmammalian vertebrates ranging from birds to lampreys and analysis of their amino acid sequences and properties gives insight into structure–activity relationships.

Keywords Tachykinin · Substance P · Neurokinin · Hemokinin · Neuropeptide γ · Neuropeptide K · Septide · Cardiovascular regulation

1
Biosynthesis of the Tachykinins

1.1
The Preprotachykinin A (ppt-a) Gene

The synthesis of the tachykinins, substance P (SP), neurokinin A (NKA), neuropeptide K (NPK) and neuropeptide γ (NPγ), in mammals is directed by a single copy gene (known as ppt-a or TAC1) encoding preprotachykinin A (Fig. 1). The gene is located on chromosome 7 (7q21–7q22) and comprises seven exons. Sequence analysis of cloned cDNAs from various human and animal tissues have identified mRNAs directing the synthesis of four biosynthetic precursors of SP (α-, β-, γ- and δ-preprotachykinin A) that arise from the preprotachykinin A gene by an alternative RNA splicing mechanism.

As shown in Fig. 2, the mRNA encoding β-preprotachykinin A is derived from transcription of all seven exons of the gene so that β-preprotachykinin A

Substance P	RPKPQQFFGLM.NH$_2$
Neurokinin A	HKTDSFVGLM.NH$_2$
Neuropeptide K	DADSSIEKQVALLKALYGHGQISHKRHKTDSFVGLM.NH$_2$
Neuropeptide γ	DAGHGQISHKRHKTDSFVGLM.NH$_2$
Neurokinin B	DMHDFFVGLM.NH$_2$

Fig. 1 A comparison of the primary structures of substance P and the neurokinins that have been identified in mammalian tissues. The peptides are aligned so as to emphasize the regions of sequence identity. The symbol ".*NH$_2$*" indicates that the peptides contain a C-terminally α-amidated amino acid residue

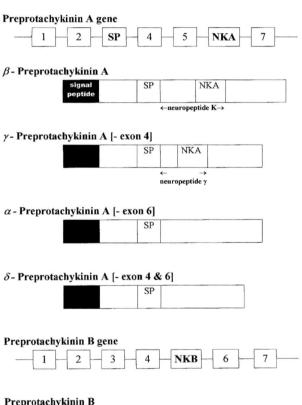

Fig. 2 A schematic representation of the organization of the preprotachykinin A and preprotachykinin B genes and the biosynthetic precursors of the mammalian tachykinins, substance P (*SP*), neurokinin A (*NKA*) and neurokinin B (*NKB*)

contains SP, NKA and its 36 amino acid residue NH_2 terminally extended form, NPK (Nawa et al. 1984; Harmar et al. 1986; Carter and Krause 1990). The mRNA encoding γ-preprotachykinin A lacks exon 4 so that γ-preprotachykinin A contains the sequence of SP, NKA and its 21 amino acid residue NH_2 terminally extended form, NPγ (Kawaguchi et al. 1986; Krause et al. 1987; Magert et al. 1993). The mRNA encoding α-preprotachykinin A lacks exon 6, which precisely specifies the NKA region (Nawa et al. 1984; Krause et al. 1987), and the mRNA encoding δ-preprotachykinin A lacks exons 4 and 6 (Harmar et al. 1990; Khan and Collins 1994) so that these biosynthetic precursors contain the sequence of SP only.

1.2
The Preprotachykinin B (ppt-b) Gene

The synthesis of NKB is directed by a single copy gene (ppt-b; TAC3) that is located on human chromosome 12 (12q13-q21). The organization of the preprotachykinin B gene is similar to that of the preprotachykinin A gene, comprising seven exons interrupted by six introns (Kotani et al. 1986; Page et al. 2000) (Fig. 2). However, preprotachykinin B contains only the single tachykinin sequence of NKB and overall sequence similarity between the ppt-a and the ppt-b genes is low. The primary structure of preprotachykinin B, deduced from the nucleotide sequences of cDNAs and/or genomic fragments, is known for the ox (Kotani et al. 1986), rat (Bonner et al. 1987), mouse (Kako et al. 1993) and human (Page et al. 2000).

1.3
The Preprotachykinin C (ppt-c) Gene

Analysis of a cDNA library derived from the murine pre-B cell line 70Z/3 led to the identification of a 1.2-k cDNA molecule encoding a 128-amino-acid protein that was termed preprotachykinin C (PPT-C) (Zhang et al. 2000). This cDNA shows no significant structural similarity with the other preprotachykinins except in the region encoding an 11-amino-acid peptide with limited structural similarity to SP that has been termed hemokinin 1 (HK-1) or tachykinin 4 (Fig. 3). Subsequently, the human and rat orthologs of the murine ppt-c gene were identified by searching the expressed sequence tag (EST) and genomic sequence databases (Kurtz et al. 2002). The human ppt-c gene (TAC4) is localized on chromosome 17. The open reading frame of the rat ppt-c gene encodes for a precursor of 170 amino acids on three exons whereas the corresponding human ppt-c gene encodes for a truncated precursor of only 68 amino acids on two exons. Rat HK 1 is identical to the mouse ortholog but sequence similarity with the human peptide is confined to the C-terminal region (Fig. 3). The PPT-C precursor does not contain a region with structural similarity to NKA.

More recently (Page et al. 2003) it has been shown that the human ppt-c (TAC4) gene transcript can generate four splice variants, termed α-, β-, γ- and

```
Mouse/rat PPT-C      ...LKLQELKRSRTRQFYGLMGKRV....   Hemokinin 1
Human PPT-C          ...LQLQEVKTGKASQFFGLMGKRV....   Hemokinin 1
                     ...QPRRKKAYQLEHTFQGLLGKRS....   Endokinin C
                     ...MGKRVGAYQLEHTFQGLLGKRS....   Endokinin D
Human PPT-A          ...LLQRIARRPKPQQFFGLMGKRD....
```

Fig. 3 Amino acid sequences in the region of mouse/rat preprotachykinin C (*PPT-C*) containing the hemokinin 1 sequence and human PPT-C containing the hemokinin 1 endokinin C and endokinin D sequences compared with the corresponding region of human/rat/mouse preprotachykinin A (*PPT-A*) containing the substance P sequence. The putative tachykinins are *underlined*

δ-TAC4, that are encoded on a combination of five different exons. The γ-precursor, which has lost exon 4, and the δ-precursor, which has lost exons 3 and 4, contain the sequence of HK-1 only. The α-precursor, derived from all five exons, contains the sequence of an additional tachykinin-like peptide, termed endokinin C, that is located C-terminally to the HK-1 sequence. The β-precursor, which has lost exon 3, contains the sequence of the N-terminally modified endokinin D. The predicted endokinins terminate in the sequence Phe-Gln-Gly-Leu-Leu.NH$_2$ (Fig. 3).

1.4
Evolution of the Preprotachykinin Genes

The presence of *Hox* gene clusters, a network of regulatory genes responsible for establishing cellular identity during embryogenesis, on human chromosomes 2, 7, 12, and 17 has led to the hypothesis that the entire genome duplicated twice early in vertebrate history (Garcia-Fernandez and Holland 1994; Larhammar et al. 2002). It has been proposed that the first putative polyploidization event occurred before the divergence of the Agnatha (hagfish and lampreys) but after the appearance of the cephalocordates (amphioxus), and the second in the gnathostome lineage shortly after the appearance of the Agnatha but before the appearance of the Elasmobranchii (sharks and rays) (Sidow 1996). In support of this proposal, Lundin (1993) has presented several examples of possible paralogous genes on the *Hox*-bearing human chromosomes and the localization of the genes encoding members of the neuropeptide Y family in different taxa may be explained in terms of multiple polyploidization events occurring during vertebrate evolution (Söderberg et al. 2000).

The similar organization of the preprotachykinin genes suggests that they are evolutionarily related (homologous) and the locations of the ppt-a (TAC1) gene on chromosome 7, the ppt-b (TAC3) gene on chromosome 12, and the ppt-c gene (TAC4) on chromosome 17 are consistent with the proposal that the genes may have arisen as a result of two rounds of whole genome duplication. The lack of a high degree of nucleotide sequence identity among the three paralogous genes is consistent with the idea that the divergence of the genes was ancient. The implication of this theory is either that a fourth preprotachykinin gene remains to be identified on human chromosome 2 or that this gene has been lost during the course of vertebrate evolution. It has been claimed that the rate of gene loss following polyploidization is such that if two whole genome duplications did occur in the time frame suggested (between 530 and 750 million years ago), not more than 4%–9% of the human proteome would show the effect of these events (Hughes 1999). Although this rate estimate may be overestimated, the alternative hypothesis that the three preprotachykinin genes duplicated independently at different times and underwent translocation events cannot be rejected.

2
Naturally Occurring and Synthetic Tachykinin Receptor Agonists

2.1
Preprotachykinin A-Derived Peptides

2.1.1
Substance P

The history of the discovery (von Euler and Gaddum 1931) and rediscovery (Chang and Leeman 1971) of SP has been often described. At the time of writing, a search of the Medline literature database using the term 'substance P' unearths a total of 17,064 articles demonstrating that SP is among the most intensively studied regulatory peptides. Structure–activity studies employing SP fragments (Regoli et al. 1984), alanine-substituted (Couture et al. 1979) and D-amino acid-substituted (Duplaa et al. 1991; Wang et al. 1993a) analogs have indicated that changes in the NH_2-terminal region of SP have only minor effects on the potency of the peptide in activating the NK_1 receptor. For example, the COOH-terminal hexapeptide [SP(6–11)] is at least equipotent with SP in increasing neuronal firing in rat supraspinal neurons (Jones and Olpe 1982) and SP (5–11) produces the same pattern of central cardiovascular and behavioral responses as SP and also retains the ability to desensitize the NK_1 receptor like SP (Tschope et al. 1995). The SP fragment [pGlu^5]SP(6–11) has been isolated from an extract of a midgut carcinoid tumor expressing the ppt-a gene (Conlon et al. 1985; Roth et al. 1985) and in vitro conversion of SP to this fragment by the dipeptidylaminopeptidase IV activity in human (Conlon and Sheehan 1983) and rat (Conlon and Goke 1984) plasma has been demonstrated.

Because of their potential as therapeutic agents for treatment of depression and anxiety disorders, irritable bowel syndrome, urinary incontinence and for use as anti-emetics, particularly in patients with chemotherapy-induced nausea and vomiting, most recent work, especially from pharmaceutical companies, has been directed towards the synthesis of novel NK_1 receptor antagonists. Nevertheless, considerable progress has been made in the design of potent and selective NK_1 receptor agonists. Modifications at position 9 appear to be important as replacement of Gly^9 by sarcosine (Drapeau et al. 1987b) improves selectivity towards NK$_1$ receptors. Selectivity is further refined by oxidation of the COOH-terminal methionine residue to its sulfone derivative so that [Sar^9, Met(O$_2$)^{11}]SP represents an analog that may be radiolabeled with ^{125}I-Bolton-Hunter reagent to produce a reagent with high selectivity for NK$_1$ receptors in binding studies (Tousignant et al. 1990; Lew et al. 1990). Fragments such as acetyl[Arg^6,Sar^9, Met(O$_2$)^{11}]SP(6–11) have also been used in functional studies as selective NK$_1$ receptor agonists (Nguyen-Le et al. 1996). Selectivity towards NK$_1$ receptors is also conferred by substitution of Gly^9 by L-proline (Lavielle et al. 1986b; Petitet et al. 1991) and septide {[pGlu^6,Pro^9]SP(6–11)} has found widespread utility as an agonist with greater selectivity than SP towards the NK$_1$ receptor (see Sect. 2.1.1). The septide-related ligand ALIE-124 {propionyl-[Met(O$_2$)^{11}]SP(7–11)} may be labeled with [^{3}H] and has been used to characterize 'septide-sensitive' binding sites in rat submandibular glands (section 2.1.1) (Sagan et al. 1997). [Gly^9-psi(CH2-CH2)-Leu^{10}]SP is a potent and selective NK$_1$ receptor agonist that has been prepared by substituting the peptide bond between the Gly^9 and Leu^{10} residues of SP with a carba (CH$_2$–CH$_2$) bond (Lavielle et al. 1993). Substitution of peptide bonds at the NH$_2$-terminal region of SP (residues 1–6) by psi[CH$_2$–CH$_2$] resulted in only minor changes in affinity for the NK$_1$ receptor except for the sevenfold reduction in the case of Pro^4–Gln^5 (Rivera Baeza and Unden 1991).

Peptidase-resistant NK$_1$ receptor-selective agonists, such as GR73632 {δ-aminovaleryl-[Pro^9,N-MeLeu^{10}]SP(7–11)}(Hagan et al. 1991) are particularly active both in vitro and in vivo. For example, this compound was 120-fold more potent than SP in potentiating the contractile response to field stimulation in the guinea pig vas deferens (Hall and Morton 1991) and 200-fold more potent than SP in inducing behavioral responses (scratching, biting and licking) when injected intrathecally into mice (Sakurada et al. 1999). Replacement of Gly^9 in SP by [(R)-β2-HAla] led to an agonist that was as potent as SP but more resistant to degradation by angiotensin-converting enzyme (Sagan et al. 2003).

The pharmacological properties of several analogs of SP modified at position 11 indicate that the NK$_1$ receptor has limited tolerance to substitution at this site. COOH-terminal alkyl esters of SP show moderately high affinity and greater selectivity for the NK$_1$ receptor than SP but SP free acid is >1000-fold less potent than SP (Watson et al. 1983) and substitution of Met^{11} by D-Met has a drastic effect of potency (Duplaa et al. 1991). Substitution of Met^{11} in [Orn^6]SP(6–11) by glutamate γ-t-butyl ester generates an agonist that is more potent than the parent hexapeptide in stimulating motility of the guinea pig ileum and rat colon (Karagiannis et al. 1991) but replacement of the –SCH$_3$ group in Met^{11} by charged groups markedly reduced activity (Poulos et al. 1987). The actions of NK$_1$ receptor agonists are mediated through the activation of several second messenger signaling systems which include stimulation of phosphoinositol turnover and elevation of intracellular Ca^{2+} concentrations, cAMP accumulation and arachidonic acid mobilization (Quartara and Maggi 1997). Interestingly, replacement of Met^{11} in SP (labeled at its NH$_2$ terminus by biotinyl sul-

Ox	RPKPQQFFGLM	Chang and Leeman 1971
Chicken	--R--------	Conlon et al. 1988b
Tortoise	--R----Y---	Wang et al. 1999b
Frog (R. ridibunda)	K-N-ER-Y---	O'Harte et al. 1991
Frog (R. catesbeiana)	K-S-DR-Y---	Kozawa et al. 1991
Frog (B. marinus)	K-R-D--Y---	Conlon et al. 1998
Amphiuma	DN-SVG--Y---	Waugh et al. 1995a
Goldfish	K-R-H--I---	Lin and Peter 1997
Cod	K-R----I---	Jensen and Conlon 1992
Trout	K-R-H------	Jensen and Conlon 1992
Sturgeon	K---H------	Wang et al. 1999a
Dogfish	K-R-G------	Waugh et al. 1993
Lamprey	RK-H-KE-V---	Waugh et al. 1994

Fig. 4 Naturally occurring orthologs of substance P from a range of vertebrate taxa; *dashes* denote residue identity

phone-5-aminopentanoic acid) by the photoreactive amino acid para-benzoylphenylalanine gave the analog Bapa^0[pBzl]Phe^{11})SP that was an agonist in the phospholipase C pathway but an antagonist in the adenylate cyclase pathway (Sachon et al. 2002). SP analogs containing the conformationally constrained methionines, (2S,3S)- and (2S,3R)-prolinomethionines have been used to characterize the signal transduction pathway associated with the 'septide sensitive' binding site on the NK$_1$ receptor (Sagan et al. 1999).

Naturally occurring orthologs of SP have been isolated from neuronal tissues of a wide range of nonmammalian vertebrates and a comparison of their primary structures provides insight into structure–activity relationships (Fig. 4). The amino acid sequence of SP has been rather poorly conserved during the evolution of vertebrates. However, evolutionary pressure has acted to conserve those residues at the COOH-terminal region of the peptide (Phe^7, Gly^9, Leu^{10}, and Met^{11}) that are known to be important in the activation of tachykinin receptors. With the exception of the peptide from the amphiuma, the Pro^4 residue is invariant and it has been shown that this amino acid is important in conferring selectivity towards NK$_1$ receptors (Cascieri et al. 1992). A peptide with the amino acid sequence Ala-Lys-Phe-Asp-Lys-Phe-Tyr-Gly-Leu-Met.NH$_2$, termed scyliorhinin-1, was isolated from the intestine of the spotted dogfish, *Scyliorhinus canicula* (Conlon et al. 1986). A related peptide Ser-Lys-Ser-His-Gln-Phe-Tyr-Gly-Leu-Met.NH$_2$ was isolated from an extract of the stomach of the primitive actinopterygian fish, the bowfin *Amia calva* (Waugh et al. 1995b). These tachykinins resemble SP only at their COOH-terminal region and their evolutionary relationship to the mammalian tachykinins is unclear.

2.1.2
Neurokinin A

NKA (originally termed neurokinin α, substance K, and neuromedin L) was first isolated from an extract of porcine spinal cord using the guinea pig ileum bioassay for detection (Kimura et al. 1983; Minamino et al. 1984). Deletion of the first three residues of NKA (His-Lys-Thr) produces a full agonist with only minor changes in potency compared with the native peptide at the NK_2 receptor in human (Warner et al. 2001), rabbit (Rovero et al. 1989) and rat (Fisher and Pennefather 1998) tissues so that most subsequent structure-activity studies have been carried out using NKA(4–10). The COOH-terminal fragments NKA(3–10) and NKA(4–10) occur naturally (Theodorsson-Norheim et al. 1987) and the in vitro conversion of NKA to NKA(3–10) by aminopeptidases present in the longitudinal muscle layer of the guinea pig small intestine has been demonstrated (Nau et al. 1986).

An alanine-scan of NKA(4–10) identified Asp^4, Phe^6, Val^7, Leu^9 and Met^{10} as critical residues in tachykinin receptor activation (Rovero et al. 1989). In both binding and functional studies, replacement of any of the residues of NKA(4–10), except for Ser^5, with alanine decreased the affinity of the peptide for the NK_2 receptor (Matuszek et al. 1999). Substitution of Gly^8 by [Ala], [D-Ala] or aminoisobutyrate (Aib) results in a loss of potency but replacement by the more conformationally flexible β-alanine residue gives an analog with increased potency and greater selectivity for the NK_2 receptor (Rovero et al. 1989; Maggi et al. 1990a). Similarly, N-methylation of the leucine residue at position 9 in NKA(4–10) produced an analog that was more potent in binding and functional studies in rabbit pulmonary artery (Burcher et al. 1993) and human colon circular muscle (Warner et al. 2002). As in the case of activation of the NK_1 receptor by SP (see Sect. 2.1.1), the COOH-terminal residue of NKA is important in conferring selectivity. Replacement of the C terminally α-amidated methionine in NKA(4–10) by norleucine amide improved selectivity for the NK_2 receptor in the rabbit pulmonary artery (Drapeau et al. 1987b) whereas replacement by the free acid form of the COOH-terminal methionine produced an analog that was inactive at the NK_2 receptor in the hamster trachea (Patacchini et al. 1993). The Ser^5 residue in NKA(4–10) appears to be relatively unimportant in NK_2 receptor activation although replacement of the Ser^5 residue by either Lys or Arg gives an agonist that is fivefold more potent in binding and functional studies using the rat gastric fundus (Comis and Burcher 1999). Combining these advantageous structural features in the analog [Lys^5,$MeLeu^9$,Nle^{10}]NKA(4–10) gave a compound with very high potency and selectivity for the NK_2 receptor (Chassaing et al. 1991) and the related compound [^{125}I][Lys^5,$Tyr(I_2)^7$,$MeLeu^9$, Nle^{10}]NKA(4–10) has been employed as a highly selective radioligand for binding to NK_2 receptors in the rat gastric fundus (Burcher et al. 1993). Potent peptidase-resistant analogs with selectivity for NK_2 receptors have been developed by introducing conformational restrictions at the COOH-terminal region of NKA, e.g., a lactam bridge in GR64349 {[Lys^3,Gly^8-R-γ-lactam-Leu^9]NKA(3–10)} (Hagan et al. 1991).

Pig	HKTDSFVGLM	Kimura et al. 1983
Frog (X. laevis)	TLTTGKD-----	Johansson et al. 2002
Frog (R. ridibunda)	--L---I---	Wang et al. 1992a
Frog (R. Catesbeiana)	-NPA--I---	Kozawa et al. 1991
Amphiuma	--*-A-I---	Waugh et al. 1995a
Trout	--IN------	Jensen and Conlon 1992
Lamprey	-F*-E-----	Waugh et al. 1995a

Fig. 5 Naturally occurring orthologs of neurokinin A from a range of vertebrate taxa; *dashes* denote residue identity and *asterisks* denote a residue deletion

As shown in Fig. 5, naturally occurring orthologs of NKA have been isolated from the tissues of a range of nonmammalian vertebrates. In contrast to SP, the primary structure of the peptide has been fully conserved among birds (chicken, Conlon et al. 1988) and reptiles (alligator, Wang et al. 1992b; python, Conlon et al. 1997; tortoise, Wang et al. 1999b). [Leu^3,Ile^7]NKA, isolated from the brain of the European green frog *Rana ridibunda* (Wang et al. 1992a) showed moderate affinity but a lack of selectivity towards NK$_1$, NK$_2$ and NK$_3$ receptors in rat tissues. Similarly, [Ile^3,Asn^4]NKA from trout, cod (Jensen and Conlon 1992) and goldfish (Lin and Peter 1997) showed relatively low affinity and specificity for the NK$_2$ receptor in rat fundus that was ascribed to the substitution Asp^4→Asn (Badgery-Parker et al. 1993).

2.1.3
Neuropeptide K

Neuropeptide K was first isolated from an extract of porcine brain using a chemical assay that detected the presence of a C-terminally α-amidated amino acid (Tatemoto et al. 1985). Subsequent studies identified the peptide in guinea pig sensory neurons (Hua et al. 1985), rat central nervous system (CNS) tissues (Valentino et al. 1986; Arai and Emson 1986), human cerebrospinal fluid (Toresson et al. 1994) and dental pulp (Casasco et al. 1990), and in relatively high concentrations in plasma and tumor tissue extracts of patients with carcinoid tumors (Theodorsson-Norheim et al. 1985; Conlon et al. 1987). NPK is converted into NKA and the NH$_2$-terminal fragment NPK(1–24) in neurons of the myenteric plexus-containing longitudinal muscle layer of the guinea pig small intestine (Deacon et al. 1987) and in the circulation (Martling et al. 1987b). NPK(1–24) was measured by specific radioimmunoassay in tumor tissue and plasma of a patient with a carcinoid tumor of the midgut (Conlon et al. 1988a). Orthologs of NPK have not yet been isolated from the tissues of a nonmammalian vertebrate.

Although systematic structure–activity studies of NPK have not been carried out, it appears that, in an in vitro system, the NH_2-terminal extension to the NKA sequence does not influence greatly interaction with the NK_2 receptor. Thus in the isolated hamster urinary bladder NPK was equipotent with NKA in binding studies using ^{125}I-NKA, stimulating phosphatidylinositol turnover, and contracting isolated smooth muscle (van Giersbergen et al. 1992). Similarly, NPK was equipotent with NKA in producing relaxations of precontracted cerebral arteries from a range of species (Jansen et al. 1991). However, as discussed in Sect. 3.1.3, the appreciable longer half-life of NPK in plasma and in the CNS compared with SP and NKA results in a distinct spectrum of biological activities in vivo.

2.1.4
Neuropeptide γ

Neuropeptide γ was first isolated from an extract of rabbit intestine using an antiserum directed against the COOH-terminal region of NKA in a radioimmunoassay to facilitate detection (Kage et al. 1988). In the rat, the peptide is present in all tissues in which the ppt-a gene is expressed although in lower concentrations than SP and NKA (Takeda et al. 1990). The NH_2-terminal fragment of NPγ [NPγ(1–9)] was detected by specific radioimmunoassay in extracts of rat brain and gastrointestinal tract but lacked myotropic activity on a range of smooth muscle targets (Wang et al. 1993b).

NPγ, along with NKA and NPK, is regarded as an endogenous ligand for the NK_2 receptor (Dam et al. 1990). In common with NPK (Sect. 2.1.3), the NH_2-terminal extension to the NKA sequence in NPγ has little effect upon the interaction with the NK_2 receptor in vitro. In the isolated hamster urinary bladder, NPγ was equipotent with NKA in competing with ^{125}I-NKA for binding sites in a crude membrane preparation, stimulating phosphatidylinositol turnover, and in contracting isolated smooth muscle (van Giersbergen et al. 1992). This conclusion was supported by a structure–activity study demonstrating that the N-acetylated fragments (3–21)NPγ, (5–21)NPγ, (7–21)NPγ, and (9–21)NPγ were equipotent with each other and with NKA for binding to NK_1 receptor sites in the rat submandibular gland and to NK_2 receptor sites in the rat gastric fundus. The affinities of the analogs for the NK_3 receptor site in rat brain were significantly less than that of NKA consistent with a preference for smaller ligands such as senktide (Badgery-Parker et al. 1993). In the isolated human bronchus, however, NPγ is between threefold (Qian et al. 1994) and tenfold (Burcher et al. 1991) more potent than NKA in inducing contraction.

As shown in Fig. 6, numerous naturally occurring orthologs of NPγ have been isolated from the tissues of a wide range of nonmammal vertebrates. Evolutionary pressure has acted to conserve only the functionally important COOH-terminal region of the peptide whereas the amino acid sequence of the NH_2-terminal extension is highly variable.

```
Rabbit      DAG***HGQISHKRHKTDSFVGLM    Kage et al. 1987
Tortoise    ---***Y-----------------    Wang et al. 1999b
Alligator   ---***Y-----------------    Wang et al. 1992b
Python      ---***-SPL--------------    Conlon et al. 1997
Goldfish    SPA***NA--TR----IN------    Conlon et al. 1991
Trout       SSA***NP--T-----IN------    Jensen et al. 1993
Sturgeon    SSA***NR--TG--Q-IN------    Wang et al. 1999a
Bowfin      SGAPQ*TVPLGR----GEM-----    Waugh et al. 1995b
Shark       ASGPTQAGIVGR--Q-GEM-----    Waugh et al. 1995a
```

Fig. 6 Naturally occurring orthologs of neuropeptide γ from a range of vertebrate taxa; *dashes* denote residue identity. Gaps denoted by *asterisks* have been introduced into some sequences in order to maximize sequence similarity

2.2 Preprotachykinin B-Derived Peptides

2.2.1 Neurokinin B

Neurokinin B (originally termed neurokinin β or neuromedin K) was first isolated from an extract of porcine spinal cord on the basis of its ability to contract the guinea pig ileum (Kangawa et al. 1983; Kimura et al. 1983). Subsequently, NKB was identified in CNS tissues of the rat (Tateishi et al. 1989) and guinea pig (Too et al. 1989) using antisera raised against NKB but claims that the ppt-b gene is (Tateishi et al. 1990; Kishimoto et al. 1991) and is not (Too et al. 1989) expressed in peripheral tissues are complicated by the lack of completely specific antisera, the difficulty of separating NKB and NPK by HPLC and the fact that the hydrophobic NKB is extracted poorly by aqueous solvents (Conlon 1991). Recent work has indicated that the ppt-b gene is expressed in a subpopulation of enteric nerves in the rat ileum that also produce ppt-a-derived tachykinins (Yunker et al. 1999) and in the rat uterus (Pinto et al. 2001). Implications of the synthesis of NKB by the human placenta are discussed in Sect. 3.2.1. Evidence for the expression of the ppt-b gene in human neoplastic tissues has been provided by the identification of NKB in extracts of a pheochromocytoma (Kage and Conlon 1989) and neuroblastomas (McGregor et al. 1990). Although data are sparse, the primary structure of NKB appears to have been strongly conserved during vertebrate evolution. For example, NKB with an amino acid sequence identical to that from mammals, was isolated from the brain of the frog *Rana ridibunda* (O'Harte et al. 1991).

The insolubility of NKB in physiological buffers has inhibited investigators from carrying out extensive structure–activity studies. Selectivity and high affinity for the NK_3 receptor in rat portal vein was achieved by replacement of Val^7 by N-methylated Phe in both NKB and NKB(4–10) (Drapeau et al. 1987a).

Eledoisin	pGlu-Pro-Ser-Lys-Asp-ala-Phe-Ile-Gly-Leu-Met.NH$_2$
Scyliorhinin II	Ser-Pro-Ser-Asn-Ser-Lys-Cys-Pro-Asp-Gly-Pro-Asp-Cys -Phe-Val-Gly-Leu-Met.NH$_2$
PG-KII	pGlu-Pro-Asn-Pro-Asp-Glu-Phe-Val-Gly-Leu-Met.NH$_2$
[MetPhe^7]NKB	Asp-Met-His-Asp-Phe-Phe-MePhe-Gly-Leu-Met.NH$_2$
Senktide	Suc-Asp-Phe-MePhe-Gly-Leu-Met.NH$_2$

Fig. 7 Naturally occurring and synthetic peptides with selectivity towards the NK$_3$ receptor. *Suc*, Succinyl; *pGlu*, pyroglutamyl; *MePhe*, N-methylated phenylalanine

A structurally related analog, senktide {succinyl-[Asp^6,MePhe^8]SP(6–11)} has found wide applicability in pharmacological studies that require an NK$_3$-selective agonist (Wormser et al. 1986) (Fig. 7). [^{3}H]senktide has been used as a selective radiolabel for identification of NK$_3$ receptors (Guard et al. 1990). NK$_3$ receptor selectivity is also conferred by the substitution of Val^7 by Pro in NKB (Lavielle et al. 1990) and by introduction of a cystine bridge in the analog [Cys^2,Cys^5]NKB (Lavielle et al. 1986; Ploux et al. 1987).

Among the naturally occurring peptides, eledoisin, isolated from the salivary gland of the Mediterranean octopus *Eledone moschata* and the first tachykinin to be characterized structurally (Erspamer and Anastasi 1962), shows some selectivity for the NK$_3$ receptor and [^{125}I]eledoisin proved to be a valuable reagent in early studies demonstrating the heterogeneity of tachykinin receptors (Beaujouan et al. 1984). Scyliorhinin II, first isolated from an extract of the intestine of the dogfish *S. canicula* (Conlon et al. 1986) also shows limited selectivity for NK$_3$ receptors (Buck and Krstenansky 1987) and [^{125}I]-Bolton-Hunter scyliorhinin II has been used as an NK$_3$-selective radioligand (Mussap and Burcher 1990). PG-KII, a tachykinin isolated from the skin of the Australian frog, *Pseudophryne guntheri* (Myobatrachidae) also shows selectivity for the NK$_3$ receptor in guinea pig ileum (Improta et al. 1996) (Fig. 7).

2.3
Preprotachykinin C-Derived Peptides

2.3.1
Hemokinin 1

The rat, mouse and human HK-1 peptides are flanked at their COOH termini by the same Gly-Lys-Arg processing/amidation site present in PPT-A but at their NH$_2$ termini by a single Lys putative processing site rather than the single Arg processing site at the NH$_2$ terminus of substance P (Fig. 3). Processing at single lysyl sites in vertebrate prepropeptides, although not unprecedented, is much less common than at single arginyl sites so the pathway of post-translational processing of PPT-C remains to be established by amino acid sequence analysis of the isolated hemokinins.

The strong expression of the ppt-c gene in hematopoietic cells led to its discovery (Zhang et al. 2000) but expression profiling of PPT-C by RT–PCR using gene-specific, intron-spanning primers has demonstrated moderate expression in human heart, skeletal muscle, skin and thyroid (Kurtz et al. 2002). Similarly, in the mouse, moderately strong signals were detected in the brain, spleen, stomach, skin and lactating breast. Of five EST clones for mouse PPT-C in the GenBank EST databases, two are from pooled brain regions and one from the pineal gland providing further evidence for expression of the ppt-c gene in tissues of the CNS (Kurtz et al. 2002).

In the human, the ppt-c (TAC4) gene is expressed predominantly in peripheral tissues (Page et al. 2003). Expression of the different TAC4 transcripts is tissue specific with the adrenal gland being the only tissue to express all four transcripts. The α-transcript has the most restricted distribution, found only in the adrenal gland, fetal liver and spleen. β-TAC4 is expressed in heart, liver, bone marrow, prostate, adrenals and testis. The γ- and δ-transcripts are expressed strongly in the placenta leading to the suggestion that HK-1 may be involved in regulating blood flow during pregnancy (Page et al. 2003).

3
Actions of the Tachykinins

3.1
Preprotachykinin A-Derived Peptides

3.1.1
Substance P

The physiological and pharmacological actions of the tachykinins have been reviewed in several comprehensive articles (Holzer and Holzer-Petsche 1997a, 1997b; Quartara and Maggi 1997, 1998; Lecci et al. 2000; Snijdelaar et al. 2000; DeVane 2001; Harrison and Geppetti 2001; Hökfelt et al. 2001; Severini et al. 2002). The roles of SP and other tachykinins in nociception and in the functions and diseases of the CNS, gastrointestinal tract and airways are described in other chapters in this volume. Consequently, this section of the article will highlight only some studies in which selective synthetic and naturally occurring NK_1 receptor selective agonists have been used.

[$Sar^9,Met(O_2)^{11}$]SP ([Sar^9]SP sulfone) has been used by several groups in functional studies that require an agonist with selectivity for the NK_1 receptor and the peptide is both more potent and effective than SP in several systems. In urethane anesthetized rats, intravenous injection of [Sar^9]SP sulfone evoked a significantly greater vasodepressor response than SP (697% vs.100%) and produced an initial increase in heart rate followed by long-lasting bradycardia (Couture et al. 1989). Administered intracerebroventricularly (i.c.v.) to the unanesthetized rat, [Sar^9]SP sulfone increased arterial blood pressure and heart rate (Cellier et al. 1999). [Sar^9]SP sulfone was 10–100 times more potent than SP in increasing plasma protein extravasation in rat skin after intradermal injection (Ahluwalia et al. 1995). In the anesthetized rabbit, [Sar^9]SP sulfone was unexpectedly more potent than either NPγ or the NK$_2$ receptor agonist [Lys^5, MeLeu^9,Nle^{10}]NKA(4–10) in producing bronchomotor effects in the absence of the peptidase inhibitor phosphoramidon but this agent markedly enhanced the responses to NPγ (Yuan et al. 1998).

With respect to behavioral effects, [Sar^9]SP sulfone was the most effective agent in promoting scratching and grooming when injected i.c.v. in mice (Ravard et al. 1994). Intracerebroventricular septide was more potent than [Sar^9]SP sulfone in producing locomotor hyperactivity in guinea pigs but less potent in inducing grooming behavior and wet-dog shakes (Piot et al. 1995). [Sar^9]SP sulfone, as well as the peptidase-resistant selective NK$_1$ receptor agonist GR73632, induced pronounced chromodacyorrhea in gerbils following intravenous infusion and repetitive hind paw tapping following i.c.v. injection (Bristow and Young 1994). These responses were not observed after administration of [β-Ala^8]NKA(4–10) (NK$_2$ selective agonist) or senktide (NK$_3$ selective agonist). Infusion of GR73632 into the ventral tegmental area (A10) or the nucleus accumbens of the rat activated the mesolimbic dopamine pathway leading to increased basal locomotor activity (Elliott et al. 1992). This effect appeared to be mediated by the NK$_1$ receptor as the NK$_2$ selective agonist GR64349 and senktide were inactive.

The lack of correlation of binding affinity and functional activity of the NK$_1$ receptor agonist septide and structurally related ligands in a range of systems has suggested the possibility of either a distinct 'septide-selective' receptor or two classes of 'septide-sensitive' and 'septide-insensitive' NK$_1$ receptors. For example, septide potently contracts the guinea pig ileum (Petitet et al. 1992) and stimulates inositol phophospholipid hydrolysis in rat urinary bladder (Torrens et al. 1995) but has low affinity for binding sites for ^{3}H-[Pro^9]SP on membranes from these tissues. The septide-induced contractions of the guinea pig ileum are also more sensitive to a range of nonpeptide NK$_1$ receptor antagonists than those produced by SP (Nguyen-Le et al. 1996) and only septide-induced responses are potentiated by tetrodotoxin (Burcher and Stamatakos 1994). Similarly, septide is 130-fold less potent than SP in displacing [^{3}H]SP from binding sites in the guinea pig lung but it is 14-fold more potent than SP as a bronchoconstrictor (Floch et al. 1994). In contrast, the NK$_1$ receptor mediating relaxation of the guinea pig trachea via nitric oxide release was classified as 'septideinsensitive' as the action of [Sar^9]SP sulfone was not mimicked by septide (Figini et al. 1996).

Studies with cloned recombinant NK$_1$ receptors expressed in COS cells do not support the idea that the actions of septide are mediated by a separate septide-selective receptor. Septide and SP elicited similar maximal increases in inositol monophosphate accumulation in cells transfected with the rat NK$_1$ receptor and septide was a weak competitor of [^{3}H]SP binding (Pradier et al. 1994). It was concluded that septide was a potent functional agonist at the NK$_1$ receptor but differing effects of the antagonist RP67580 on septide- and SP-induced responses suggested that the agonists were acting at different sites on the receptor. A later study using radiolabeled septide in a homologous binding assay demonstrated that septide, and also NKA, had a high affinity for the recombinant NK_1 receptor but was a poor competitor for SP (Hastrup and Schwartz 1996). A similar conclusion was drawn from a study using the human cloned NK_1 receptor expressed in CHO cells and [^{3}H]ALIE-124 as radioligand (Sagan et al. 1997). Two point mutations in the cloned human NK_1 receptor (E193L and V195R) suppress the functional activity of septide but not SP suggesting that this region of the receptor, located at the end of the second intracellular loop, may be part of a binding domain for septide that is distinct from the SP binding site (Wijkhuisen et al. 1999). Conversely, the mutation (G166C) in the NK_1 receptor increases the apparent affinity and binding capacity for septide in competition assays using radiolabeled SP (Ciucci et al. 1998). It has been suggested that the most abundant binding site on NK_1 receptor transfected cells (80%–85% of the total receptor population) binds ligands such as [Sar^9]SP sulfone and is positively coupled to adenylate cyclase whereas the less abundant binding site recognizes septide and NKA and is coupled to phospholipase C (Sagan et al. 1999).

As discussed in Sect. 2.1.1, the COOH-terminal domains of the tachykinins (Phe-Xaa-Gly-Leu-Met.NH$_2$) are responsible for binding to the NK_1, NK_2 and NK_3 receptors and NH$_2$-terminal fragments do not activate these receptors. In this light, the extensive literature relating to behavioral and other effects of centrally administered NH$_2$-terminal fragments of SP appears paradoxical. SP(1–7) fragment is produced from SP in relatively high concentrations in CNS tissues by the action of endopeptidase 24.11 and other peptidases (Sakurada et al. 1985; Michael-Titus et al. 2002). Intrathecal administration of this peptide reduced fighting in mice made aggressive by prolonged isolation (Hall and Stewart 1984), produced a transient decrease in reaction time in the rat (Cridland and Henry 1988) and mouse (Stewart et al. 1982) tail flick tests and displayed a dose-dependent antinociceptive effect in the in the mouse hot-plate assay (Mousseau et al. 1994). It has been suggested that proteolytic conversion of SP to its (1–7) fragment may be a prerequisite for development of its antinociceptive action (Cridland and Henry 1988; Kreeger et al. 1994). In rats, SP(1–7) lowered blood pressure and heart rate when microinjected into the nucleus of the solitary tract (NTS) (Hall et al. 1989a), enhanced rearing, sniffing and locomotor activity following injection into the substantia nigra (Hall and Stewart 1992) and produced anxiolytic-like effects when administered into the nucleus basalis area of the ventral pallidum (Nikolaus et al. 2000). Significantly, the cardiovascular effects produced by NTS administered SP, but not SP(1–7), were blocked by intra-NTS injection of phosphoramidon, an inhibitor of endopeptidase 24.11, implying that proteolytic cleavage of SP was necessary for activity (Hall et al. 1989b). In mice, pretreatment of mice with SP(1–7) potentiated the behavioral activity produced by intrathecal kainic acid (Larson and Sun 1992) and the peptide administered intraperitoneally attenuated the development of tolerance to the analgesic effect of morphine and inhibited the expression of withdrawal (Kreeger and Larson 1996). SP has been implicated in memory function that involves glutamate transmission mediated through the *N*-methyl-D-aspartate (NMDA) receptor (McRoberts et al. 2001). The importance of the NH_2-terminal region is indicated by the observation that i.c.v. injections of SP(1–7) in rats enhance memory and learning (Huston and Hasenohrl 1995) and upregulate expression of the NMDA receptor subunits in certain regions of the brain (Zhou et al. 2000). In contrast to its agonistic actions, SP (1–7) or SP (1–8), when co-administered intrathecally with SP, reduced the SP-induced behavioral responses of scratching, biting and licking (Sakurada et al. 1990) and SP(1–7) injected into the rat substantia nigra potently antagonized the contralateral rotation produced by intranigral SP (Herrera-Marschitz et al. 1990).

At this time, a specific receptor that recognizes and mediates the effects of NH_2-terminal fragments of SP has not been fully characterized. However, specific binding sites for SP(1–7) have been identified in rat brain membranes by using [^{3}H]SP(1–7) in a radioreceptor assay (Igwe et al. 1990). Binding was inhibited by [D-Ala^2,NMePhe^4,Gly-ol] enkephalin (DAMGO) but not by NK_1, NK_2 and NK_3 receptor agonists. Alternatively, the observation that SP(1–7) will induce internalization of the NK_1 receptor in rat striatum has suggested that NH_2-terminal SP fragments have affinity for a conformer of the receptor that is different from the one that recognizes SP (Michael-Titus et al. 1999). It seems probable that the actions of exogenously administered SP(1–7) may involve, at least in part, indirect mechanisms such as the inhibition of degradation of endogenous SP (Hall et al. 1989b) and regulation of expression of NK_1 receptors in the CNS (Velasquez et al. 2002).

The biological actions of several of the naturally occurring SP-related agonists listed in Fig. 4 have been studied in their species of origin. Bolus intra-arterial injection of python SP ([Arg^3,Tyr^8]SP) into the anesthetized python, *Python regius* produced concentration-dependent decreases in arterial blood pressure and systemic peripheral resistance concomitant with increases in cardiac output and stroke volume but with only minor effects on heart rate. The python cardiovascular system was extremely sensitive to the effects of the peptide with a dose as low as 0.01 nmol/kg producing significant effects (Wang et al. 2000). In contrast, injections of trout SP ([Lys^1,Arg^3,His^5]SP) into unanesthetized rainbow trout *Oncorhynchus mykiss* produced an increase in both systemic and coelic resistances leading to hypertension, bradycardia and a decrease in cardiac output (Kågström et al. 1996). A similar cardiovascular response to the bowfin SP-related peptide was seen in the unanesthetized bowfin *Amia calva*. Following bolus injections into the bulbus arteriosus, a dose-dependent rise in vascular resistance and arterial blood pressure and fall in cardiac output was seen but there was no change in heart rate (Waugh et al. 1995b). Intra-arterial injections of high doses (10–50 nmol/kg) of dogfish SP ([Lys^1,Arg^3,Gly^5]SP) into the unanesthetized dogfish *S. canicula* (Elasmobranchii) produced a slight pressor response but low doses were without effect on blood pressure or heart rate (Waugh et al. 1993). Studies in vitro with isolated trout intestinal smooth muscle and the vascularly perfused trout stomach have demonstrated that trout SP increases motility in a concentration-dependent manner (Jensen et al. 1993).

The pharmacological actions of the frog tachykinin ranakinin ([Lys^1,Asn^3, Glu^5,Arg^6,Tyr^8]SP) on the adrenal gland of the green frog *Rana ridibunda* have been studied in detail. The peptide stimulates corticosterone and aldosterone release from perfused adrenal gland slices by a mechanism that involved activation of the arachidonic acid cascade (Leboulenger et al. 1993). Subsequent studies demonstrated that ranakinin caused mobilization of calcium from intracellular stores in cultured frog adrenal chromaffin cells by activation of a phospholipase C via a pertussis toxin-sensitive G protein (Kodjo et al. 1995a, 1995b). It was suggested that the stimulatory effect of ranakinin on corticosteroid secretion is indirect being mediated via presynaptic activation of adrenochromaffin cells but involving a receptor whose ligand binding properties differ substantially from the mammalian NK$_1$ receptor subtype (Kodjo et al. 1996).

Radiolabeling of the toad tachykinin, bufokinin ([Lys^1,Arg^3,Asp^5,Tyr^8] SP) with [I^{125}]Bolton-Hunter reagent provided a reagent to characterize specific binding sites in the small intestine of the cane toad *Bufo marinus* (Liu et al. 1999). The rank order of binding affinity of a range of tachykinins indicated that the putative toad receptor had similar binding properties to the mammalian NK$_1$ receptor. The peptide stimulated inositol monophosphate formation suggesting that the tachykinin receptor in the toad gut is coupled to phosphoinositol hydrolysis. Bufokinin produced sustained concentration-dependent contractions of both circular and longitudinal smooth muscle from the toad small intestine. In a further study (Liu et al. 2000) it was shown that intravenous injections of bufokinin caused a dose-dependent fall in systemic blood pressure (maximum fall of 20 mmHg) with a 50% excitatory dose of 2.9 pmol. At higher doses, the effect was prolonged and blood pressure did not return to baseline within 60 min. There was no significant change in heart rate associated with hypotension.

3.1.2
Neurokinin A

The airways and urinary bladder represent important sites of action of NKA. Bronchial smooth muscle contraction is mediated through activation of NK$_2$ receptors (Dion et al. 1990) but the presence of NK$_1$ receptors in human lung means that SP is more potent in producing vasodilatation, increasing vascular permeability and in affecting secretory responses (Frossard and Advenier 1991). The observation that the rank order of potency of the tachykinins in stimulating the motility of the human urinary bladder is NKA>NKB>SP indicates the presence of NK$_2$ receptors (Dion et al. 1990) and NK$_2$ receptor agonists are the most potent in activating the micturition reflex following intravenous infusion (Lecci and Maggi 2001). The NK$_2$ receptor selective agonists [βAla^8]NKA(4–10) (Maggi et al. 1990a) and [Lys^5,MeLeu^9,Nle^{10}]NKA(4–10) (Floch et al. 1994) were the most potent tachykinins tested in producing bronchospasm in guinea pigs and contraction of the rat bladder but the former agonist was only very weakly active in producing hypotension or plasma extravasation in rats (Maggi et al.

1990a). In contrast, [βAla^8]NKA(4–10) was effective in lowering blood pressure in the guinea pig and the effect was attenuated by NK$_2$ receptor antagonists, indicative of a species difference in the distribution of receptors (Floch et al. 1996). The potency of the NK$_2$ selective agonist [Nle^{10}]NKA(4–10) was greater than SP and NKB in producing contraction of the isolated human urinary bladder (Dion et al. 1988) and [Lys^5,MeLeu^9,Nle^{10}]NKA(4–10) potently stimulated inositol monophosphate formation in the rat bladder (Torrens et al. 1995). Intravenous injection of NKA(4–10) into anesthetized rats produced only a very weak vasodepressor response but the peptide evoked tachycardia that was blocked by the β-adrenergic receptor antagonist, sotalol and by a combination of guanethidine and bilateral adrelectomy indicative of the involvement of catecholamines (Couture et al. 1989).

In the gastrointestinal tract, [βAla^8]NKA(4–10) produced contractions of both circularly-orientated (Maggi et al. 1989) and longitudinally orientated (Maggi et al. 1990b) smooth muscle strips from the human small intestine with the same potency as NKA. A cytoprotective role for gastric mucosa has been proposed for the NKA released from primary afferent terminals. Studies with capsaicin-treated rats have shown that formation of gastric mucosal lesions in response to ulcerogenic factors is significantly enhanced (Holzer and Sametz 1986). Consistent with this, subcutaneous pretreatment of rats with NKA(4–10) reduced the degree of gastric lesions induced by oral administration of ethanol (Evangelista et al. 1989). SP and [Me-Phe^7]NKB were ineffective. Similarly, peripheral or central administration of [Ala^5]NKA(4–10) inhibited gastric secretion and ulcer formation in pylorus-ligated rats (Improta et al. 1997).

The presence of NK$_2$ receptors in the brain of mammals is controversial but binding sites for [^{125}I]NKA with the pharmacological properties of NK$_2$ receptors were identified in the hippocampus, thalamus and septum of the rat (Saffroy et al. 2001). However, the effects of centrally administered NKA-related peptides are generally much weaker than those of NK$_1$ receptor agonists. For example, i.c.v. administration of [Lys^5,MeLeu^9,Nle^{10}]NKA(4–10) to guinea pigs did not produce a characteristic behavioral response in contrast to the locomotor hyperactivity produced by NK$_1$ receptor agonists and the wet-dog shakes produced by senktide (Piot et al. 1995).

Several studies have investigated the biological activities of naturally occurring agonists of the NK$_2$ receptor in the species of origin. Trout NKA ([Ile^3, Asn^4]NKA), given intra-arterially at a dose of 1 nmol/kg, was equally effective as trout SP in increasing both coelic and systemic vascular resistance in the rainbow trout and was approximately equipotent with trout SP in increasing the dorsal aortic vascular resistance in an in vitro perfusion system (Kågström et al. 1996). In contrast, trout NKA was 14 times less potent than trout SP in stimulating the motility of isolated trout intestinal smooth muscle and 28 times less potent in stimulating the motility of the vascularly perfused trout stomach (Jensen et al. 1993). In the amphibia, the NKA-related peptide from the clawed frog *Xenopus laevis* was equipotent with *Xenopus* SP (bufokinin) in contracting isolated strips of circular smooth muscle from *Xenopus* stomach but produced a significantly greater maximum response (Johansson et al. 2002). [Leu^3,Ile^7]NKA, isolated from the green frog *Rana ridibunda* stimulated corticosterone release from the frog's adrenal gland but with an effectiveness that was less than that of ranakinin (Leboulenger et al. 1993).

3.1.3
Neuropeptide K

Although it has been shown that the NH$_2$-terminal extension to NPK does not influence binding to the NK$_2$ receptor appreciably (Sect. 2.1.3), this structural feature is responsible for an increased stability of NPK relative to NKA and so modulates the biological actions of the peptide in vivo. Pharmacokinetic studies in the guinea pig involving intravenous infusions of tachykinins have shown that the half-life of NKA in plasma is less than 2 min whereas the clearance of NPK was biphasic with apparent half-lives of 0.9 and 6 min. The half-life of SP was too short to measure (Martling et al. 1987b). Consequently, under these conditions the hypotensive and bronchoconstrictor actions of NPK are more intense and more prolonged than those of NKA and SP. In contrast, NKA is more potent than NPK and SP on isolated human bronchi in an in vitro system (Martling et al. 1987a). In related studies, NPK, administered by intravenous infusion, is more potent than NKA in producing a decrease in pulmonary conductance in anesthetized mechanically ventilated guinea pigs (Shore et al. 1993) and stimulating salivary gland secretion in rats (Takeda and Krause 1989). Intravenous injection of NPK into urethane-anesthetized rats produced sustained hypotension, ascribed to a direct action on arterial blood vessels, an initial tachycardia, ascribed to nonreflex activation of the sympathoadrenal system, followed by long-lasting bradycardia possibly arising from a vagal reflex (Decarie and Couture 1992). The fact that NPK is the most potent circulating tachykinin acting on the cardiovascular system has suggested that release of this peptide by carcinoid tumors, particularly those of the midgut region that have metastasized to the liver, is responsible for the cutaneous flushing in at least some of the patients with carcinoid syndrome (Norheim et al. 1986; Conlon et al. 1987; Balks et al. 1989). Similarly, increased secretion of NPK into the intestinal lumen has been implicated (along with serotonin) in contributing to the enhanced intestinal motility and secretory diarrhea seen in many of these patients (Makridis et al. 1999).

The greater stability of NPK in the CNS compared with other tachykinins (Michael-Titus et al. 2002) is reflected by the fact that NPK, but not SP and NKA, is detectable in freshly taken cerebrospinal fluid from healthy human subjects (Toresson et al. 1994). Intracerebroventricular administration of NPK to unanesthetized rats induced tachycardia and concentration-dependent increases in arterial blood pressure (Prat et al. 1994). Concurrently, behavioral responses such as increased frequency of face-washing, head scratching, grooming and wet-dog shakes were observed. Both the cardiovascular and behavioral responses were blocked by prior administration of an NK$_1$ receptor antagonist (CP96345 or RP67580) but not by NK$_2$ or NK$_3$ receptor antagonists. Other behavioral effects of centrally administered NPK include suppression of copulatory behavior in male rats presented with receptive females (Kalra et al. 1991), disruption of maternal behavior in female rats primed by pregnancy, termination and estrogen injection (Sheehan and Numan 1997), and enhancement of memory retention in mice after foot-shock avoidance training (Flood et al. 1990). Intracerebroventricular injection of NPK potently inhibited drinking in water-deprived rats and rats stimulated to drink by i.c.v. administered angiotensin II or by subcutaneous injection of saline (Achapu et al. 1992). At high doses, NPK inhibited food intake (Sahu et al. 1988; Achapu et al. 1992) but this effect may be a nonspecific consequence of the intense grooming behavior evoked.

In general, effects of central administration of NPK on the neuroendocrine system are more pronounced than those seen with other preprotachykinin A-derived peptides. Intracerebroventricular injection of NPK evoked a delayed stimulation of growth hormone release in ovariectomized rats that was not reproduced by NKA or NPγ (Debeljiuk et al. 1995) and NPK, administered centrally or peripherally, was more potent than NKA and NPγ in stimulating adrenal corticosterone release in gonad-intact and castrated male rats (Kalra and Kalra 1993). Effects of central administration of NPK on leutenizing hormone (LH) release are complex. Intracerebroventricular injection of NPK in ovariectomized rats resulted in a significant suppression of circulating LH concentrations whereas NKA was inactive and NPK suppressed the LH surges induced by progesterone in ovariectomized estrogen-treated rats (Sahu and Kalra 1992). In contrast, i.c.v. injections of NPK in intact male rats induced a significant increase in circulating LH but a significant decrease in orchidectomized rats (Kalra et al. 1992). It was concluded that NPK (and to a lesser extent NPγ and NKA) have a suppressive effect on LH release at the level of the hypothalamus but the response is affected by the steroid environment (Debeljuk and Lasaga 1999).

3.1.4
Neuropeptide γ

The differential effects on in vitro and in vivo potency and effectiveness of the NH_2-terminal extension to the NKA sequence in NPK (Sect. 3.1.3) are also seen in NPγ although effects upon peptide stability are not as great. Intravenous injections of NPγ to anesthetized guinea pigs resulted in a fall in blood pressure that was mediated by interaction with NK_1 receptors (Yuan et al. 1994). The rank order of potency was [Sar^9,Met(O$_2$)^{11}]SP>NPγ>> [Lys^5,MeLeu^9,Nle^{10}]NKA(4–10) and the response was attenuated by the NK_1 receptor antagonist CP96345. In the same study, NPγ increased total lung resistance and decreased dynamic lung compliance exclusively via interaction with NK_2 receptors. Intracerebroventricular administration of NPγ to the unanesthetized rat evoked dose-dependent increases in mean arterial blood pressure and heart rate that were accompanied by similar behavioral responses seen following central injection of NPK (face washing, head scratching, grooming and wet-dog shake) (Hagio et al.

1991; Picard and Couture 1996). Both cardiovascular and behavioral responses were attenuated by the NK_2 receptor antagonist SR48968 but the NK_1 receptor antagonist RP67580 was without effect indicating that central actions of NPγ are mediated, at least in part, through interaction with NK_2 receptors. Using the same experimental protocol, intrathecal administration of NPγ produced dose-dependent increases in blood pressure and heart rate but the pressor response was converted to a vasodepressor response by pretreatment with phentolamine and the chronotropic response was attenuated by pretreatment with phentolamine, thereby indicating an involvement of catecholamines released from sympathetic fibers or the adrenal medulla (Poulat et al. 1996).

NPγ shares with NPK the ability to stimulate salivary secretion in the rat and it has been shown that, in assays employing [^{125}I]NKA as radiolabel, both agonists bind with high affinity to the 'septide-sensitive' binding site associated with the NK_1 receptor in the rat submaxillary gland (Sect. 3.1.1) (Beaujouan et al. 1999). Similarly, like NPK, pulse i.c.v. injection of NPγ in the rat potently inhibited drinking induced by angiotensin II, subcutaneous hypertonic NaCl and water deprivation but had no effect on food intake (Poliodori et al. 1995). The antidipsogenic effect was blocked by the NK_1 receptor antagonist WIN 62577. NPγ was more effective than [Nle^{10}]NKA(4–10) in producing contractions of isolated longitudinal muscle strips from rat duodenum and more effective in vivo in disrupting the migrating myoelectric complex and inducing irregular spiking in the rat small intestine (Rahman et al. 1994).

3.2
Preprotachykinin-B Derived Peptides

3.2.1
Neurokinin B

The selective agonist senktide has been used in several studies designed to define the role of NK_3 receptor activation in various physiological processes. Following intravenous injection into urethane anesthetized rats, senktide was significantly more potent than SP in evoking a depressor response whereas the responses produced by [MePhe^7]NKB and [βAsp^4,MePhe^7]NKB(4–10) were not different from SP (Couture et al. 1989). These NK_3 receptor-selective agonists produced a rapid and marked bradycardia that was blocked by hexamethonium, methylatropine and by bilateral vagotomy. In the venous mesenteric vasculature of the rat, [MePhe^7]NKB induced a dose-dependent pressor effect that was not reproduced by NK_1 and NK_2 receptor selective agonists (D'Orleans-Juste et al. 1991). Intracerebroventricular injection of senktide into unanesthetized chronically instrumented rats produced a marked increase in heart rate and frequently produced the 'wet-dog shake' response (Itoi et al. 1992).

Senktide and the naturally occurring NK_3 receptor selective agonist PG-KII were weak (compared with carbachol and caerulein) stimulants of amylase release from isolated pancreatic lobules of the guinea pig (Linari et al. 2002).

Similarly, i.c.v. injections of both senktide and PG-KII in rats induced dose-dependent inhibition of colonic propulsion measured as an increase in the expulsion time of a glass bead placed in the distal colon (Broccardo et al. 1999). SP, NKA and NKB had much weaker actions. Senktide was the most potent tachykinin amongst a range of NK_1, NK_2 and NK_3 receptor agonists in stimulating the motility of isolated longitudinally oriented smooth muscle from the guinea pig common bile duct. The mechanism was neurogenic involving the release of endogenous acetylcholine and tachykinins, the latter acting, in turn, on postjunctional tachykinin NK_1/NK_2 receptors (Patacchini et al. 1997). Given i.c.v. to rats, senktide potently reduced the acid response to histamine, but not to pentagastrin and bethanechol (Improta and Broccardo 1991).

Effects of NK_3 receptor agonists on the airways are complex. Senktide and [MePhe^7]NKB, administered as an aerosol to guinea pigs that had been pretreated with phosphoramidon and salbutamol, did not produce bronchoconstriction. However, airway hyper-responsiveness 24 h later was induced as displayed by an exaggerated response to the bronchoconstrictor effect of intravenous acetylcholine (Daoui et al. 2000). In the same animal model, these NK_3 receptor agonists were the most potent tachykinins tested in potentiating the histamine-induced increase of microvascular permeability in the airways (Daoui et al. 2001).

In behavioral studies with mice in the elevated plus-maze test, i.c.v. administration of senktide evoked responses indicative of an anxiolytic action (Ribeiro et al. 1999). Conversely, the NK_3 receptor antagonist [Trp^7,β-Ala^8]NKA(4–10) evoked an anxiogenic response. The involvement of NK_3 receptors in tachykinin-induced antinociception was suggested by the observation that intrathecal administration of [βAsp^4,MePhe^7]NKB(4–10) to rats increased tail-flick latency (Laneuville et al. 1988). It was suggested that the effect was mediated by the spinal release of an opioid. Analgesia towards a radiant heat stimulus was also evoked by i.c.v. injection of PG-KII but the effect was not blocked by naloxone (Improta and Broccardo 2000).

Interest in NKB has been stimulated by recent data demonstrating that the human placenta is an abundant source of NKB (Page et al. 2000). Preprotachykinin B mRNA expression is restricted to the outer syncytiotrophoblasts and enhanced release of placental NKB into the maternal circulation during the third trimester of pregnancy has been found in women suffering from pre-eclampsia (Page et al. 2001). As circulating tachykinins are generally regarded as hypotensive in mammals, implication of NKB as the causative agent in the life-threatening hypertension of pre-eclampsia appears, at first sight, to be surprising. However, intravenous infusions of relatively high concentrations of NKB into unrestrained female rats produced a transient rise in arterial blood pressure (Page et al. 2000) and, as previously remarked (Sect. 3.2.1), [MePhe^7] NKB produced a dose-dependent pressor response in the rat venous mesenteric vasculature (D'Orleans-Juste et al. 1991). Pre-eclampsia is also associated with tissue edema and intravenous infusions of NKB into mice produced edema (assessed as plasma extravasation) in skin, uterus, liver and particularly in the lung (Grant et al. 2002).

There have been few studies of the effects of NK_3 receptor agonists in non-mammalian vertebrates but it is noteworthy that scyliorhinin II (Sect. 2.2.1), a peptide isolated from the intestine of the spotted dogfish *S. canicula*, stimulates the secretory activity of the rectal gland in its species of origin (Anderson et al. 1995).

3.3
Preprotachykinin C-Derived Peptides

3.3.1
Hemokinin-1

Although HK-1 has yet to be isolated from tissue extracts and characterized structurally, a synthetic replicate of the putative 11 amino-acid mouse HK-1 demonstrated biological properties similar to those of SP, such as induction of plasma extravasation and mast cell degranulation but, in contrast to SP, HK-1 stimulated the proliferation of interleukin 7-expanded B cell precursors and promoted the survival of freshly isolated bone marrow B lineage cells and cultured lipopolysaccharide-stimulated pre-B cells. Taken together with the observation that the tachykinin receptor antagonist, N-acetyl-L-tryptophan-3,5-bis-trifluoromethyl benzyl ester promotes apoptosis in cultured pre-B cells, the data suggest that HK-1 may be an autocrine factor that is important for B-cell development (Zhang et al. 2000).

Both human and mouse HK-1 show pharmacological properties that are very similar to SP with respect to binding affinity to Chinese hamster ovary cells stably expressing the human NK_1, NK_2 and NK_3 tachykinin receptors (Kurtz et al. 2002). Both peptides bind with highest affinity to the NK_1 receptor [50% inhibitory concentration (IC_{50}) values: human HK-1, 1.8 nM, mouse HK-1, 0.13 nM, SP, 0.12 nM] are 60–200-fold less potent than NKA at the NK_2 receptor. Human HK-1 bound with approximately eightfold higher affinity than mouse HK-1 and twofold higher affinity than SP at the NK_3 receptor (IC_{50} values: human HK-1, 370 nM, mouse HK-1, 3200 nM, SP, 780 nM). Both human and mouse HK-1 peptides are full agonists at the three human NK receptor subtypes, as measured by calcium mobilization and are full agonists and equipotent with SP at the NK_1 receptor in terms of their abilities to generate inositol monophosphate.

The pharmacological profile of mouse HK-1 in functional in vitro and in vivo assays is similar, but not identical, to that of SP (Bellucci et al. 2002). In the rat urinary bladder, HK-1 was a full agonist but approximately threefold less potent than SP and in the rabbit pulmonary artery and in the guinea pig ileum HK-1 was a full agonist but approximately 500-fold less potent than NKA and NKB respectively. Intravenous administration of mouse HK-1 produced a dose-dependent hypotensive response in anesthetized guinea pigs and stimulated salivary secretion in anesthetized rats with potencies similar to that of SP. These effects were blocked by the NK_1 receptor antagonist SR140333. Similar results were obtained by Camarda et al. (2002) who reported that, in three tachykinin receptor systems (rabbit jugular vein for NK_1, rabbit pulmonary artery for NK_2, and rat portal vein for NK_3 receptors), mouse HK-1 behaved as a full agonist with similar potencies to SP. In the rabbit jugular vein preparation, SR140333 and the NK_1 receptor antagonist MEN11467 blocked the effects of HK-1 and SP with similarly high potencies.

Recent cardiovascular studies in unanesthetized rats using the decapeptide $GKASQFFGLM.NH_2$ [referred to as endokinin A/B but equivalent to human HK-1(2–11)] have shown that this peptide is equally effective as SP in producing dose-dependent falls in arterial blood pressure associated with tachycardia, mesenteric vasoconstriction and hindquarters vasodilatation (Page et al. 2003). In contrast, endokinin C and D had no hemodynamic effects at doses below 1 nmol/kg but at higher doses (10 and 100 nmol/kg) both peptides produced a hypotensive response. Studies in vitro have shown that neither endokinin C nor endokinin D, at concentrations up to 600 μM, produced changes in intracellular calcium ion concentration in U373 MG human glioblastoma cells that express high levels of the endogenous NK_1 receptor. Similarly, in a radioligand binding assay using transfected cells expressing the NK_1, NK_2 and NK_3 receptors, endokinins C and D functioned as very weak partial agonists in all three systems (Page et al. 2003). It remains to be established whether high-affinity receptors selective for these endokinins are present in human tissues.

References

Achapu M, Pompei P, Polidori C, de Caro G, Massi M (1992) Central effects of neuropeptide K on water and food intake in the rat. Brain Res Bull 28:299–303

Ahluwalia A, Giuliani S, Maggi CA (1995) Demonstration of a 'septide-sensitive' inflammatory response in rat skin. Br J Pharmacol 116:2170–2174

Anderson WG, Conlon JM, Hazon N (1995) Characterization of the endogenous intestinal peptide that stimulates the rectal gland of *Scyliorhinus canicula*. Am J Physiol 268:R1359–R1364

Arai H, Emson PC (1986) Regional distribution of neuropeptide K and other tachykinins (neurokinin A, neurokinin B and substance P) in rat central nervous system. Brain Res 399:240–249

Badgery-Parker T, Lovas S, Conlon JM, Burcher E (1993) Receptor binding profile of neuropeptide γ and its fragments: comparison with the non-mammalian peptides carassin and ranakinin at three mammalian tachykinin receptors. Peptides 14:771–776

Balks HJ, Conlon JM, Creutzfeldt W, Stockmann F (1989) Effect of a long-acting analogue (Octreotide) on circulating tachykinins and the pentagastrin-induced carcinoid flush. Eur J Clin Pharmacol 36:133–137

Beaujouan J, Torrens C, Viger A, Glowinski J (1984) A new type of tachykinin binding site in the rat brain characterized by specific binding of a labeled eledoisin derivative. Mol Pharmacol 26:248–254

Beaujouan JC, Saffroy M, Torrens Y, Sagan S, Glowinski J (1999) Pharmacological characterization of tachykinin septide-sensitive binding sites in the rat submaxillary gland. Peptides 20:1347–1352

Bellucci F, Carini F, Catalani C, Cucchi P, Lecci A, Meini S, Patacchini R, Quartara L, Ricci R, Tramontana M, Giuliani S, Maggi CA (2002) Pharmacological profile of the novel mammalian tachykinin, hemokinin 1. Br J Pharmacol 135:266–274

Bonner TI, Affolter H-U, Young AC, Young WS (1987) A cDNA encoding the precursor of the rat neuropeptide, neurokinin B. Mol Brain Res 2:243–249

Bristow LJ, Young L (1994) Chromodacryorrhea and repetitive hind paw tapping: models of peripheral and central tachykinin NK_1 receptor activation in gerbils. Eur J Pharmacol 253:245–252

Broccardo M, Improta G, Tabacco A (1999) Central tachykinin NK_3 receptors in the inhibitory action on the rat colonic propulsion of a new tachykinin, PG-KII. Eur J Pharmacol 376:67–71

Buck SH, Krstenansky JL (1987) The dogfish peptides scyliorhinin I and scyliorhinin II bind with differential selectivity to mammalian tachykinin receptors. Eur J Pharmacol 144:109–111

Burcher E, Stamatakos C (1994) Septide but not substance P stimulates inhibitory neurons in guinea-pig ileum. Eur J Pharmacol 258:R9–10

Burcher E, Alouan LA, Johnson PR, Black JL (1991) Neuropeptide γ, the most potent contractile tachykinin in human isolated bronchus, acts via a 'non-classical' NK_2 receptor. Neuropeptides 20:79–82

Burcher E, Badgery-Parker T, Zeng X-P, Lavielle S (1993) Characterisation of a novel, selective radioligand, [^{125}I]Lys^5,Tyr(I$_2$)^7,MetLeu^9,Nle^{10}]neurokinin A(4–10) for the tachykinin NK-2 receptor in rat fundus. Eur J Pharmacol 233:201–207

Carter MS, Krause JE (1990) Structure, expression, and some regulatory mechanisms of the rat preprotachykinin gene encoding substance P, neurokinin A, neuropeptide K, and neuropeptide γ. J Neurosci 10:2203–2214

Camarda V, Rizzi A, Calo G, Guerrini R, Salvadori S, Regoli D (2002) Pharmacological profile of hemokinin 1: a novel member of the tachykinin family. Life Sci 71:363–370

Casasco A, Calligaro A, Springall DR, Casasco M, Poggi P, Valentino Kl, Polak JM (1990) Neuropeptide K in human dental pulp. Arch Oral Biol 35:33–36

Cascieri MA, Huang RRC, Fong TM, Cheung AH, Sadowski S, Ber E, Strader CD (1992) Determination of the amino acid residues in substance P conferring selectivity and specificity for the rat neurokinin receptors. Mol Pharmacol 41:1096–1099

Cellier E, Barbot L, Iyengar S, Couture R (1999) Characterization of central and peripheral effects of septide with the use of five tachykinin NK_1 receptor antagonists in the rat. Br J Pharmacol 127:717–728

Chang MM, Leeman SE (1971) Isolation of a sialogogic peptide from bovine hypothalamic tissue and its characterization as substance P. J Biol Chem 245:4784–4790

Chassaing G, Lavielle S, Locuillet D, Robilliard P, Carruette A, Garret C, Beaujouan J-C, Saffroy M, Petitet F, Torrens Y, Glowinski J (1991) Selective agonists of NK-2 binding sites highly active on rat portal vein (NK-3 bioassay). Neuropeptides 19:91–95

Ciucci A, Palma C, Manzini S, Werge TM (1998) Point mutation increases a form of the NK_1 receptor with high affinity for neurokinin A and B and septide. Br J Pharmacol 125:393–401

Comis A, Burcher E (1999) Structure-activity studies at the rat tachykinin NK_2 receptor: effect of substitution at position 5 of neurokinin A. J Pept Res 53:337–342

Conlon JM (1991) Measurement of neurokinin B by radioimmunoassay. In: Conn PM (ed) Methods in Neurosciences, Vol 6. Academic Press, San Diego, pp 221–231

Conlon JM, Goke B (1984) Metabolism of substance P in human plasma and in the rat circulation. J Chromatogr 296:241–247

Conlon JM, Sheehan L (1983) Conversion of substance P to C-terminal fragments in human plasma. Regul Pept 7:335–345

Conlon JM, Schafer G, Schmidt WE, Lazarus LH, Becker HD, Creutzfeldt W (1985) Chemical and immunochemical characterization of substance P-like immunoreactivity and physalaemin-like immunoreactivity in a carcinoid tumour. Regul Pept 11:117–132

Conlon JM, Deacon CF, O'Toole L, Thim L (1986) Scyliorhinin I and II, two novel tachykinins from dogfish gut. FEBS Lett 200:111–116

Conlon JM, Deacon CF, Richter G, Stockmann F, Creutzfeldt W (1987) Circulating tachykinins (substance P, neurokinin A, neuropeptide K) and the carcinoid flush. Scand J Gastroenterol 22:97–105

Conlon JM, Deacon CF, Grimelius L, Cedermark B, Murphy RF, Thim L, Creutzfeldt W (1988a) Neuropeptide K-(1–24)-peptide: storage and release by carcinoid tumours. Peptides 9:859–866

Conlon JM, Katsoulis S, Schmidt WE, Thim L (1988b) [Arg^3]substance P and neurokinin A from chicken small intestine. Regul Pept 20:171–180

Conlon JM, O'Harte F, Peter RE, Kah O (1991) Carassin: a tachykinin related to neuropeptide gamma from the brain of the goldfish. J Neurochem 56:1432–1436

Conlon JM, Adrian TE, Secor SM (1997) Tachykinins (substance P, neurokinin A and neuropeptide γ) and neurotensin from the intestine of the Burmese python, *Python molurus*. Peptides 18:1505–1510

Conlon JM, Warner FJ, Burcher E (1998) Bufokinin: a substance P-related peptide from the gut of the toad, *Bufo marinus* with high binding affinity but low selectivity for mammalian tachykinin receptors. J Peptide Res 51:210–215

Couture R, Fournier A, Magnan J, St-Pierre S, Regoli D (1979) Structure-activity studies on substance P. Can J Physiol Pharmacol 57:1427–1436

Couture R, Laneuville O, Guimond C, Drapeau G, Regoli D (1989) Characterization of the peripheral action of neurokinins and neurokinin receptor selective agonists on the rat cardiovascular system. Naunyn Schmiedeberg's Arch Pharmacol 340:547–557

Cridland RA, Henry JL (1988) N- and C-terminal fragments of substance P: spinal effects in the rat tail flick test. Brain Res Bull 20:429–432

Dam TV, Takeda Y, Krause JE, Escher E, Quirion R (1990) γ-Preprotachykinin-(72–920-peptide amide: an endogenous preprotachykinin I gene-derived peptide that preferentially binds to neurokinin-2 receptors. Proc Natl Acad Sci USA 87:246–250

Daoui S, Naline E, Lagente V, Emonds-Alt X, Advenier C (2000) Neurokinin B- and specific tachykinin NK$_3$ receptor agonists-induced airway hyperresponsiveness in the guinea-pig. Br J Pharmacol 130:49–56

Daoui S, Ahnaou A, Naline E, Emonds-Alt X, Lagente V, Advenier C (2001) Tachykinin NK$_3$ receptor agonists induced microvascular leakage hypersensitivity in the guinea-pig airways. Eur J Pharmacol 433:199–207

Decarie A, Couture R (1992) Characterization of the peripheral action of neuropeptide K on the rat cardiovascular system. Eur J Pharmacol 213:125–131

Deacon CF, Agoston DV, Nau R, Conlon JM (1987) Conversion of neuropeptide K to neurokinin A and vesicular colocalization of neurokinin A and substance P in neurons of the guinea pig small intestine. J Neurochem 48:141–146

Debeljuk L, Rettori V, Bartke A, McCann S (1995) In vivo and in vitro effects of neuropeptide K and neuropeptide γ on the release of growth hormone. Neuroreport 6:2457–2460

Debeljuk L, Lasaga M (1999) Modulation of the hypothalmo-pituitary-gonadal axis and the pineal gland by neurokinin A, neuropeptide K and neuropeptide γ. Peptides 20:285–299

DeVane CL (2001) Substance P: a new era, a new role. Pharmacotherapy 21:1061–1069

Dion S, Corcos J, Carmel M, Drapeau G, Regoli D (1988) Substance P and neurokinins as stimulants of the human isolated urinary bladder. Neuropeptides 11:83–87

Dion S, Rouissi N, Nantel F, Drapeau G, Regoli D, Naline E, Advenier C (1990) Receptors for neurokinins in human bronchus and urinary bladder are of the NK-2 type. Eur J Pharmacol 178:215–219

D'Orleans-Juste P, Claing A, Telemaque S, Warner TD, Regoli D (1991) Neurokinins produce selective venoconstriction via NK-3 receptors in the rat mesenteric vascular bed. Eur J Pharmacol 12:329–334

Drapeau G, D'Orleans-Juste P, Dion S, Rhaleb NE, Regoli D (1987a) Selective agonists for neurokinin B receptors. Eur J Pharmacol 136:401–403

Drapeau G, D'Orleans-Juste P, Dion S, Rhaleb NE, Rouissi NE, Regoli D (1987b) Selective agonists for substance P and neurokinin receptors. Neuropeptides 10:43–54

Duplaa H, Chassaing G, Lavielle S, Beaujouan JC, Torrens Y, Saffroy M, Glowinski J, D'Orleans Juste P, Regoli D, Carruette A et al. (1991) Influence of the replacement of amino acid by its D-enantiomer in the sequence of substance P. 1. Binding and pharmacological data. Neuropeptides 19:251–257

Erspamer V, Anastasi A (1962) Structure and pharmacological actions of eledoisin, the active endecapeptide of the posterior salivary glands of *Eledone*. Experientia 18:58–59

Evangelista S, Lippe IT, Rovero P, Maggi CA, Meli A (1989) Tachykinins protect against ethanol-induced gastric lesions in rats. Peptides 10:79–81

Figini M, Emanueli C, Bertrand C, Javdan P, Geppetti P (1996) Evidence that tachykinins relax the guinea-pig trachea via nitric oxide release and by stimulation of a septide-insensitive NK_1 receptor. Br J Pharmacol 117:1270–1276

Fisher L, Pennefather JN (1998) Structure-activity atudies of analogues of neurokinin A mediating contraction of rat uterus. Neuropeptides 32:405–410

Flood JF, Baker ML, Hernandez EN, Morley JE (1990) Modulation of memory retention by neuropeptide K. Brain Res 520:284–290

Floch A, Fardin V, Cavero I (1994) Characterization of NK_1 and NK_2 tachykinin receptors in guinea-pig and rat bronchopulmonary and vascular systems. Br J Pharmacol 111:759–768

Floch A, Thiry C, Cavero I (1996) Pharmacological evidence that NK-2 tachykinin receptors mediate hypotension in the guinea pig but not in the rat. Fundam Clin Pharmacol 10:337–343

Frossard N, Advenier C (1991) Tachykinin receptors and the airways. Life Sci 49:1941–1953

Garcia-Fernandez J, Holland PWH (1994) Archetypal organization of the amphioxus *Hox* gene cluster. Nature 370:563–566

Grant AD, Akhtar R, Gerard NP, Brain SD (2002) Neurokinin B induces oedema formation in mouse lung via tachykinin receptor independent mechanisms. J Physiol 543:1007–1014

Guard S, Watson SP, Maggio JE, Too HP, Watling KJ (1990) Pharmacological analysis of [^{3}H]-senktide binding to NK_3 tachykinin receptors in guinea-pig ileum longitudinal muscle-myenteric plexus and cerebral cortex membranes. Br J Pharmacol 99:767–773

Hagan RM, Ireland SJ, Jordan CC, Beresford IJ, Deal MJ, Ward P (1991) Receptor-selective, peptidase-resistant agonists at neurokinin NK-1 and NK-2 receptors: new tools for investigating neurokinin function. Neuropeptides 19:127–135

Hagio T, Takano Y, Nagashima A, Nakayama Y, Tateishi K, Kamiya H (1991) The central pressor actions of a novel tachykinin peptide, γ-preprotachykinin-(72–92)-peptide amide Eur J Pharmacol 192:173–176

Hall JM, Morton IK (1991) Novel selective agonists and antagonists confirm neurokinin NK_1 receptors in guinea-pig vas deferens. Br J Pharmacol 102:511–517

Hall ME, Stewart JM (1984) Modulation of isolation-induced fighting by N- and C-terminal analogs of substance P: evidence for multiple recognition sites. Peptides 5:85–89

Hall ME, Stewart JM (1992) The substance P fragment SP(1–7) stimulates motor behavior and nigral dopamine release. Pharmacol Biochem Behav 41:75–78

Hall ME, Miley F, Stewart JM (1989a) Cardiovascular effects of substance P peptides in the nucleus of the solitary tract. Brain Res 497:280–290

Hall ME, Miley F, Stewart JM (1989b) The role of enzymatic processing in the biological actions of substance P. Peptides 10:895–901

Harmar AJ, Armstrong A, Pascall JC, Chapman K, Rosie R, Curtis A, Going J, Edwards CRW, Fink G (1986) cDNA sequence of human β-preprotachykinin, the common precursor to substance P and neurokinin A. FEBS Lett 208:67–72

Harmar AJ, Hyde V, Chapman K (1990) Identification and cDNA sequence of δ-preprotachykinin, a fourth splicing variant of the rat substance P precursor. FEBS Lett 275:22–24

Harrison S, Geppetti P (2001) Substance P. Int J Biochem Cell Biol 33:555–576

Hastrup H, Schwartz TW (1996) Septide and neurokinin A are high-affinity ligands on the NK-1 receptor: evidence from homologous versus heterologous binding analysis. FEBS Lett 399:264–266

Herrera-Marschitz M, Terenius L, Sakurada T, Reid MS, Ungerstedt U (1990) The substance P(1–7) fragment is a potent modulator of substance P actions in the brain. Brain Res 521:316–20

Hökfelt T, Pernow B, Wahren J (2001) Substance P: a pioneer amongst neuropeptides. J Intern Med 249:27–40

Holzer P, Holzer-Petsche U (1997a) Tachykinins in the gut. I. Expression, release and motor function. Pharm Ther 73:173–217

Holzer P, Holzer-Petsche U (1997b) Tachykinins in the gut. II. Roles in neural excitation, secretion and inflammation. Pharm Ther 73:219–263

Holzer P, Sametz W (1986) Gastric mucosal protection against ulcerogenic factors in the rat mediated by capsaicin-sensitive afferent neurons. Gastroenterology 91:975–981

Hua XY, Theodorsson-Norheim E, Brodin E, Lundberg JM, Hökfelt T (1985) Multiple tachykinins (neurokinin A, neuropeptide K and substance P) in capsaicin-sensitive sensory neurons in the guinea-pig. Regul Pept 13:1–19

Hughes AL (1999) Phylogenies of developmentally important proteins do not support the hypothesis of two rounds of genome duplication early in vertebrate history. J Mol Evol 48:565–576

Huston JP, Hasenohrl RU (1995) The role of neuropeptides in learning: focus on the neurokinin substance P. Behav Brain Res 23:117–27

Igwe OJ, Kim DC, Seybold VS, Larson AA (1990) Specific binding of substance P aminoterminal heptapeptide [SP(1–7)] to mouse brain and spinal cord membranes. J Neurosci 10:3653–3663

Improta G, Broccardo M (1991) Inhibitory role on gastric secretion of a central NK-3 tachykinin receptor agonist, senktide. Peptides 12:1433–1434

Improta G, Broccardo M (2000) Effects of supraspinal administration of PG-SPI and PG-KII, two amphibian tachykinin peptides, on nociception in the rat. Peptides 21:1611–1616

Improta G, Broccardo M, Severini C, Erspamer V (1996) In vitro and in vivo biological activities of PG-KII, a novel kassinin-like peptide from the skin of the Australian frog, *Pseudophryne guntheri*. Peptides 17:1003–1008

Improta G, Broccardo M, Tabacco A, Evangelista S (1997) Central and peripheral antiulcer and antisecretory effects of [Ala^5]NKA(4–10), a tachykinin receptor agonist, in rats. Neuropeptides 31:399–402

Itoi K, Tschope C, Jost N, Culman J, Lebrun C, Stauss B, Unger T (1992) Identification of the central tachykinin receptor subclass involved in substance P-induced cardiovascular and behavioral responses in conscious rats. Eur J Pharmacol 219:435–444

Jansen I, Alafaci C, McCulloch J, Uddman R, Edvinsson L (1991) Tachykinins (substance P, neurokinin A, neuropeptide K, and neurokinin B) in the cerebral circulation: vasomotor responses in vitro and in vivo. J Cereb Blood Flow Metab 11:567–575

Jensen J, Conlon JM (1992) Substance P-related and neurokinin A-related peptides from the brain of the cod and trout. Eur J Biochem 206:659–664

Jensen J, Olson KR, Conlon JM (1993) Primary structures and effects on gastrointestinal motility of tachykinins from the rainbow trout. Am J Physiol 265: R804–R810

Johansson A, Holmgren S, Conlon JM (2002) The primary structures and myotropic activities of two tachykinins isolated from the African clawed frog, *Xenopus laevis*. Regul Pept 108:113–121

Jones RS, Olpe HR (1982) A structure-activity profile of substance P and some of its fragments on supraspinal neurones in the rat. Neurosci Lett 33:67–71

Kage R, Conlon JM (1989) Neurokinin B in a human pheochromocytoma measured with a specific radioimmunoassay. Peptides 10:713–716.

Kage R, McGregor GP, Thim L, Conlon JM (1988). Neuropeptide γ: a peptide isolated from rabbit intestine that is derived from gamma-preprotachykinin. J Neurochem 50:1412–1417

Kågström J, Holmgren S, Olson KR, Conlon JM, Jensen J (1996) Vasoconstrictive effects of native tachykinins in the rainbow trout, *Oncorhynchus mykiss*. Peptides 17:39–45

Kako K, Munekata E, Hosaka M, Murakami K, Nakayama K (1993) Cloning and sequence analysis of mouse cDNAs encoding preprotachykinin A and B. Biomed Res 14:253–259

Kalra PS, Kalra SP (1993) Neuropeptide K stimulates corticosterone release in the rat. Brain Res 610:330–333

Kalra P, Sahu A, Bonavera J, Kalra S (1992) Diverse effects of tachykinins on luteinizing hormone release in male rats: mechanism of action. Endocrinology 131:1195–1201

Kalra SP, Dube MG, Kalra PS (1991) Neuropeptide K (NPK) suppresses copulatory behavior in male rats. Physiol Behav 49:1297–1300

Kangawa K, Minamino N, Fukuda A, Matsuo H (1983) Neuromedin K: a novel mammalian tachykinin identified in porcine spinal cord. Biochem Biophys Res Commun 114:533–540

Karagiannis K, Manolopoulou A, Stavropoulos G, Poulos C, Jordan CC, Hagan RM (1991) Synthesis of a potent agonist of substance P by modifying the methionyl and glutaminyl residues of the C-terminal hexapeptide of substance P. Structure-activity relationships. Int J Pept Protein Res 38:350–356

Kawaguchi Y, Hoshimaru M, Nawa H, Nakanishi S (1986) Sequence analysis of cloned cDNA for rat substance P precursor. Biochem Biophys Res Commun 139:1040–1046

Khan I, Collins SM (1994) Fourth isoform of preprotachykinin messenger RNA encoding for substance P in the rat intestine. Biochem Biophys Res Commun 202:796–802

Kimura S, Okada M, Sugita Y, Kanazawa I, Munekata E (1983) Novel neuropeptides, neurokinins α and β, isolated from porcine spinal cord. Proc Jpn Acad Ser B 59:101–104

Kishimoto S, Tateishi K, Kobayashi H, Kobuke K, Hagio T, Matsuoka Y, Kajiyama G, Miyoshi A (1991) Distribution of neurokinin A-like and neurokinin B-like immunoreactivity in human peripheral tissues. Regul Pept 36:165–171

Kodjo MK, Leboulenger F, Conlon JM, Vaudry H (1995a) Effect of ranakinin, a novel tachykinin, on cytosolic free calcium in frog adrenochromaffin cells. Endocrinology 136:4535–4542

Kodjo MK, Leboulenger F, Porcedda P, Lamacz M, Conlon JM, Pelletier G, Vaudry H (1995b) Evidence for the involvement of chromaffin cells in the stimulatory effect of tachykinins on corticosteroid secretion by the frog adrenal gland. Endocrinology 136:3253–3259

Kodjo MK, Leboulenger F, Morra M, Conlon JM, Vaudry H.(1996) Pharmacological profile of the tachykinin receptor involved in the stimulation of corticosteroid secretion in the frog *Rana ridibunda*. J Steroid Biochem Mol Biol 57:329–335

Kotani H, Hoshimaru M, Nawa H, Nakanishi S (1986) Structure and gene organization of the bovine neuromedin K precursor. Proc Natl Acad Sci USA 83:7074–7078

Kozawa H, Hino J, Minamino N, Kangawa K, Matsuo H (1991) Isolation of four novel tachykinins from frog (*Rana catesbeiana*) brain and intetine. Biochem Biophys Res Commun 177:588–595

Krause JE, Chirgwin JM, Carter MS, Xu ZS, Hershey AD (1987) Three rat preprotachykinin mRNAs encode the neuropeptides, substance P and neurokinin A. Proc Natl Acad Sci USA 84:881–885

Kreeger JS, Larson AA (1996) The substance P amino-terminal metabolite substance P(1–7), administered peripherally, prevents the development of acute morphine tolerance and attenuates the expression of withdrawal in mice. J Pharmacol Exp Ther 279:662–667 Kurtz M, Wang R, Clements M, Cascieri M, Austin C, Cunningham B, Chicchi G, Liu Q (2002) Identification, localization and receptor characterization of novel mammalian substance P-like peptides. Gene 296:205–212

Laneuville O, Dorais J, Couture R (1998) Characterization of the effects produced by neurokinins and three agonists selective for neurokinin receptor subtypes in a spinal nociceptive reflex of the rat. Life Sci 42:1295–1305

Larhammar D, Lundin LG, Hallböök F (2002) The human Hox-bearing chromosome regions did arise by block or chromosome (or even genome) duplications. Genome Res 12:1910–1920

Larson AA, Sun X (1992) Amino terminus of substance P potentiates kainic acid-induced activity in the mouse spinal cord. J Neurosci 12:4905–4910

Lavielle S, Chassaing G, Besseyre J, Marquet A, Bergstrom L, Beaujouan JC, Torrens Y, Glowinski J (1986a) A cyclic analogue selective for the NKB specific binding site on rat brain synaptosomes. Eur J Pharmacol 128:283–285

Lavielle S, Chassaing G, Julien S, Marquet A, Bergstrom L, Beaujouan JC, Torrens Y, Glowinski J (1986b) Specific recognition of SP or NKB receptors by analogs of SP substituted at positions 8 and 9. Eur J Pharmacol 125:461–462

Lavielle S, Chassaing G, Loeuillet D, Convert O, Torrens Y, Beaujouan JC, Saffroy M, Petitet F, Bergstrom L, Glowinski J (1990) Selective agonists of tachykinin binding sites. Fundam Clin Pharmacol 4:257–268

Lavielle S, Chassaing G, Brunissen A, Rodriguez M, Martinez J, Convert O, Carruette A, Garret C, Petitet F, Saffroy M (1993) Importance of the leucine side-chain to the spasmogenic activity and binding of substance P analogues. Int J Pept Protein Res 42:270–277

Leboulenger F, Vaglini L, Conlon JM, Homo-Delarche F, Wang Y, Kerdelhue B, Pelletier G, Vaudry H (1993) Immunohistochemical distribution, biochemical characterization, and biological action of tachykinins in the frog adrenal gland. Endocrinology 133:1999–2008

Lecci A, Maggi CA (2001) Tachykinins as modulators of the micturition reflex in the central and peripheral nervous system. Regul Pept 101:1–18

Lecci A, Giulani S, Tramontana M, Carini F, Maggi CA (2000) Peripheral actions of tachykinins. Neuropeptides 34:303–313

Lew R, Geraghty DP, Drapeau G, Regoli D, Burcher E (1990) Binding characteristics of [125I]Bolton-Hunter [Sar^9,Met(O$_2$)^{11}]substance P, a new selective radioligand for the NK$_1$ receptor. Eur J Pharmacol 184:97–108

Lin XW, Peter RE (1997) Goldfish gamma-preprotachykinin mRNA encodes the neuropeptides substance P, carassin, and neurokinin A. Peptides 18:817–824

Linari G, Broccardo M, Nucerito V, Improta G (2002) Selective tachykinin NK$_3$-receptor agonists stimulate in vitro exocrine pancreatic secretion in the guinea pig. Peptides 23:947–953

Liu L, Warner FJ, Conlon JM, Burcher E (1999) Pharmacological and biochemical investigation of receptors for the toad gut tachykinin peptide, bufokinin, in its species of origin. Naunyn Schmiedeberg's Arch Pharmacol 360:187–195

Liu L, Shang F, Comis A, Burcher E (2000) Bufokinin: actions and distribution in the toad cardiovascular system. Clin Exp Pharmacol Physiol 27:911–916

Lundin LG (1993) Evolution of the vertebrate genome as reflected in paralogous chromosomal regions in man and the house mouse. Genomics 16:1–19

Magert HJ, Heitland A, Rose M, Forssmann WG (1993) Nucleotide sequence of the rabbit gamma-preprotachykinin I cDNA. Biochem Biophys Res Commun 195:128–131

Maggi CA, Patacchini R, Santicioli P, Giuliani S, Turini D, Barbanti G, Giachetti A, Meli A (1989) Human isolated ileum: motor responses of the circular muscle to electrical field stimulation and exogenous neuropeptides. Naunyn Schmiedebergs Arch Pharmacol 341:256–261

Maggi CA, Giuliani S, Ballati L, Rovero P, Abelli L, Manzini S, Giachetti A, Meli A (1990a) In vivo pharmacology of [βAla8]neurokinin A-(4–10), a selective NK-2 tachykinin receptor agonist. Eur J Pharmacol 177:81–86

Maggi CA, Patacchini R, Santicioli P, Giuliani S, Turini D, Barbanti G, Beneforti P, Misuri D, Meli A (1990b) Human isolated small intestine: motor responses of the longitudinal muscle to field stimulation and exogenous neuropeptides. Naunyn Schmiedebergs Arch Pharmacol 339:415–423

Makridis C, Theodorsson E, Akerstrom G, Oberg K, Knutson L (1999) Increased intestinal non-substance P tachykinin concentrations in malignant midgut carcinoid disease. J Gastroenterol Hepatol 14:500–507

Martling C-R, Theodorsson-Norheim E, Lundberg JM (1987a) Occurrence and effects of multiple tachykinins; substance P, neurokinin A and neuropeptide K in human lower airways. Life Sci 40:1633–1643

Martling C-R, Theodorsson-Norheim E, Norheim I, Lundberg JM (1987b) Bronchoconstrictor and hypotensive effects in relation to pharmokinetics in the guinea-pig—evidence for extraneuronal cleavage of neuropeptide K to neurokinin A. Naunyn-Schmiedeberg's Arch Pharmacol 336:183–189

Matuszek MA, Comis A, Burcher E (1999) Binding and functional potency of neurokinin A analogues in the rat fundus: a structure–activity study. Pharmacology 58:227–235

McGregor GP, Gaedicke G, Voigt K (1990) Neurokinin-immunoreactivity in human neuroblastomas. Evidence for selective expression of the preprotachykinin (PPT) II gene. FEBS Lett 277:83–87

McRoberts JA, Coutinho SV, Marvizon JC, Grady EF, Tognetto M, Sengupta JN, Ennes HS, Chaban VV, Amadesi S, Creminon C, Lanthorn T, Geppetti P, Bunnett NW, Mayer EA (2001) Role of peripheral N-methyl-D-aspartate (NMDA) receptors in visceral nociception in rats. Gastroenterology 120:1737–1748

Michael-Titus AT, Blackburn D, Connolly Y, Priestley JV, Whelpton R (1999) N- and C-terminal substance P fragments: differential effects on striatal [^{3}H]substance P binding and NK_1 receptor internalization. Neuroreport 10:2209–2213

Michael-Titus AT, Fernandes K, Setty H, Whelpton R (2002) In vivo metabolism and clearance of substance P and co-expressed tachykinins in rat striatum. Neuroscience 110:277–286

Minamino N, Kangawa K, Fukuda A, Matsuo H (1984) Neuromedin L: a novel mammalian tachykinin identified in porcine spinal cord. Neuropeptides 4:157–166

Mousseau DD, Sun X, Larson AA (1994) An antinociceptive effect of capsaicin in the adult mouse mediated by the NH2-terminus of substance P. J Pharmacol Exp Ther 268:785–790

Mussap CJ, Burcher E (1990) [^{125}I]-Bolton-Hunter scyliorhinin II: a novel, selective radioligand for the tachykinin NK_3 receptor in rat brain. Peptides 11:827–836

Nau R, Schafer G, Deacon CF, Cole T, Conlon JM (1986) Proteolytic inactivation of substance P and neurokinin A in the longitudinal muscle layer of the guinea small intestine. J Neurochem 47:856–864

Nawa H, Kotani H, Nakanishi S (1984) Tissue-specific generation of two preprotachykinin mRNAs from one gene by alternative RNA splicing. Nature 312:729–734

Nguyen-Le XK, Nguyen QT, Gobeil F, Jukic D, Chretien L, Regoli D (1996) Neurokinin receptors in the guinea pig ileum. Pharmacology 52:35–45

Nikolaus S, Huston JP, Hasenohrl RU (2000) Anxiolytic-like effects in rats produced by ventral pallidal injection of both N- and C-terminal fragments of substance P. Neurosci Lett 283:37–40

Norheim I, Theodorsson-Norheim E, Brodin E, Oberg K (1986) Tachykinins in carcinoid tumors: their use as a tumor marker and possible role in the carcinoid flush. J Clin Endocrinol Metab 63:605–612

O'Harte F, Burcher E, Lovas S, Smith DD, Vaudry H, Conlon JM (1991) Ranakinin: a novel NK_1 tachykinin receptor agonist isolated with neurokinin B from the brain of the frog, *Rana ridibunda*. J Neurochem 57:2086–2091

Page NM, Bell, NJ, Gardiner SM, Manyonda IT, Brayley KJ, Strange PG, Lowry PJ (2003) Characterization of the endokinins: human tachykinins with cardiovascular activity. Proc Natl Acad Sci USA 100:6245–6250

Page NM, Woods RJ, Gardiner SM, Lomthaisong K, Gladwell RT, Butlin DJ, Manyonda IT, Lowry PJ (2000) Excessive placental secretion of neurokinin B during the third trimester causes pre-eclampsia. Nature 405:797–800

Page NM, Woods RJ, Lowry PJ (2001) A regulatory role for neurokinin B in placental physiology and pre-eclampsia. Regul Pept 98:97–104

Patacchini R, Quartara L, Rovero P, Goso C, Maggi CA (1993) Role of C-terminal amidation on the biological activity of neurokinin A derivatives with agonist and antagonist properties. J Pharmacol Exp Ther 264:17–21

Patacchini R, Bartho L, Maggi CA (1997) Characterization of receptors mediating contraction induced by tachykinins in the guinea-pig isolated common bile duct. Br J Pharmacol 122:1633–1638

Petitet F, Beaujouan J-C, Saffroy M, Torrens Y, Chassaing G, Lavielle S, Besseyre J, Garret C, Curruette A, Glowinski J (1991) Further demonstration that [Pro^9]-substance P is a potent and selective ligand of NK-1 tachykinin receptors. J Neurochem 56:879–889

Petitet F, Saffroy M, Torrens Y, Lavielle S, Chassaing G, Loeuillet D, Glowinski J, Beaujouan JC (1992) Possible existence of a new tachykinin receptor subtype in the guinea pig ileum. Peptides 13:383–838

Picard P, Couture R (1996) Intracerebroventricular responses to neuropeptide γ in the conscious rat: characterization of its receptor with selective antagonists. Br J Pharmacol 117:241–249

Pinto FM, Cintado CG, Devillier P, Candenas ML (2001) Expression of preprotachykinin-B, the gene that encodes neurokinin B, in the rat uterus. Eur J Pharmacol 425:R1–R2

Piot O, Betschart J, Grall I, Ravard S, Garret C, Blanchard JC (1995) Comparative behavioural profile of centrally administered tachykinin NK_1, NK_2 and NK_3 receptor agonists in the guinea-pig. Br J Pharmacol 116:2496–2502

Ploux O, Lavielle S, Chassaing G, Julien S, Marquet A, d'Orleans-Juste P, Dion S, Regoli D, Beaujouan JC, Bergstrom L et al. (1987) Interaction of tachykinins with their receptors studied with cyclic analogues of substance P and neurokinin B. Proc Natl Acad Sci USA 84:8095–8099

Polidori C, Staffinati G, Perfumi MC, de Caro G, Massi M (1995) Neuropeptide γ: a mammalian tachykinin endowed with potent antidipsogenic action in rats. Physiol Behav 58:595–602

Poulat P, de Champlain J, Couture R (1996) Cardiovascular responses to intrathecal neuropeptide γ in conscious rats: receptor characterization and mechanism of action. Br J Pharmacol 117:250–257

Poulos C, Stavropoulos G, Brown JR, Jordan CC (1987) Structure-activity studies on the C-terminal hexapeptide of substance P with modifications at the glutaminyl and methioninyl residues. J Med Chem 30:1512–1525

Pradier L, Menager J, Le Guern J, Bock MD, Heuillet E, Fardin V, Garret C, Doble A, Mayaux JF (1994) Septide: an agonist for the NK_1 receptor acting at a site distinct from substance P. Mol Pharmacol 45:287–293

Prat A, Picard P, Couture R (1994) Cardiovascular and behavioural effects of centrally administered neuropeptide K in the rat: receptor characterization. Br J Pharmacol 112:250–256

Qian Y, Advenier C, Naline E, Bellamy JF, Emonds-Alt X (1994) Effects of SR 48968 on the neuropeptide γ-induced contraction of the human isolated bronchus. Fundam Clin Pharmacol 8:71–75

Quartara L, Maggi CA (1997) The tachykinin NK$_1$ receptor. Part I: ligands and mechanisms of cellular activation. Neuropeptides 31:537–563

Quartara L, Maggi CA (1998) The tachykinin NK$_1$ receptor. Part II: Distribution and pathophysiological roles. Neuropeptides 32:1–49

Rahman M, Lordal M, al-Saffar A, Hellstrom PM (1994) Intestinal motility responses to neuropeptide γ in vitro and in vivo in the rat: comparison with neurokinin 1 and neurokinin 2 receptor agonists. Acta Physiol Scand 151:497–505

Ravard S, Betschart J, Fardin V, Flamand O, Blanchard JC (1994) Differential ability of tachykinin NK-1 and NK-2 agonists to produce scratching and grooming behaviours in mice. Brain Res 651:199–208

Regoli D, Mizrahi J, D'Orleans-Juste P, Escher E (1984) Receptors for substance P. II. Classification by agonist fragments and homologues. Eur J Pharmacol 97:171–177

Ribeiro SJ, Teixeira RM, Calixto JB, De Lima TC (1999) Tachykinin NK$_3$ receptor involvement in anxiety. Neuropeptides 33:181–188

Rivera Baeza C, Unden A (1991) Investigation of the importance of the N-terminal peptide bonds of substance P by synthetic pseudopeptide analogs. Neuropeptides 20:83–86

Roth KA, Makk G, Beck O, Faull K, Tatemoto K, Evans CJ, Barchas JD (1985) Isolation and characterization of substance P, substance P 5–11, and substance K from two metastatic ileal carcinoids. Regul Pept 12:185–199

Rovero P, Pestellinin V, Rhaleb N-E, Dion S, Rouissi N, Tousignant C, Telemaque S, Drapeau G, Regoli D (1989) Structure-activity studies of neurokinin A. Neuropeptides 13:263–270

Sachon E, Girault-Lagrange S, Chassaing G, Lavielle S, Sagan S (2002) Analogs of substance P modified at the C-terminus which are both agonist and antagonist of the NK-1 receptor depending on the second messenger pathway. J Pept Res 59:232–240

Saffroy M, Torrens Y, Glowinski J, Beaujouan JC (2001) Presence of NK$_2$ binding sites in the rat brain. J Neurochem 79:985–996

Sagan S, Beaujouan JC, Torrens Y, Saffroy M, Chassaing G, Glowinski J, Lavielle S (1997) High affinity binding of [3H]propionyl-[Met(O$_2$)^{11}]substance P(7–11), a tritiated septide-like peptide, in Chinese hamster ovary cells expressing human neurokinin-1 receptors and in rat submandibular glands. Mol Pharmacol 52:120–127

Sagan S, Karoyan P, Chassaing G, Lavielle S (1999) Further delineation of the two binding sites (R*(n)) associated with tachykinin neurokinin-1 receptors using [3-Prolinomethionine(11)]SP analogues. J Biol Chem 274:23770–23776

Sagan S, Milcent T, Ponsinet R, Convert O, Tasseau O, Chassaing G, Lavielle S, Lequin O (2003) Structural and biological effects of a beta2- or beta3-amino acid insertion in a peptide. Eur J Biochem 270:939–949

Sahu A, Kalra S (1992) Effects of tachykinins on luteinizing hormone release in female rats: potent inhibitory action of neuropeptide K. Endocrinology 130:1571–1577

Sahu A, Kalra PS, Dube MG, Kalra SP (1988) Neuropeptide K suppresses feeding in the rat. Regul Pept 23:135–143

Sakurada T, Le Greves P, Stewart J, Terenius L (1985) Measurement of substance P metabolites in rat CNS. J Neurochem 44:718–22

Sakurada T, Tan-No K, Yamada T, Sakurada S, Kisara K, Ohba M, Terenius L (1990) N-terminal substance P fragments inhibit the spinally induced, NK 1 receptor mediated behavioural responses in mice. Life Sci 47:109–113

Sakurada C, Watanabe C, Inoue M, Tan-No K, Ando R, Kisara K, Sakurada T (1999) Spinal actions of GR73632, a novel tachykinin receptor agonist. Peptides 20:301–304

Severini C, Improta G, Falconieri-Erspamer G, Salvadori S, Erspamer V (2002) The tachykinin peptide family. Pharmacol Rev 54:285–322

Sheehan TP, Numan M (1997) Microinjections of the tachykinin neuropeptide K into the ventromedial hypothalamus disrupts the hormonal onset of maternal behavior in female rats. J Neuroendocrinol 9:677–687

Shore SA, Sharpless C, Drazen JM (1993) Bronchoconstrictor activities of NP γ and NPK in anaesthetized guinea-pigs: effect on NEP inhibition. Pulm Pharmacol 6:143–147

Sidow A (1996) Gen(om)e duplications in the evolution of early vertebrates. Curr Opin Genet Dev 6:715–722

Snijdelaar DG, Dirksen R, Slappendel R, Crul BJ (2000) Substance P. Eur J Pain 4:121–135

Söderberg C, Wraith A, Rigvall M, Yan YL, Postlethwait JH, Brodin L, Larhammar D (2000) Zebrafish genes for neuropeptide Y and peptide YY reveal origin by chromosome duplication from an ancestral gene linked to the homeobox cluster. J Neurochem 75:908–918

Stewart JM, Hall ME, Harkins J, Frederickson RC, Terenius L, Hokfelt T, Krivoy WA (1982) A fragment of substance P with specific central activity: SP(1–7). Peptides 3:851–857

Takeda Y, Krause JE (1989) Neuropeptide K potently stimulates salivary gland secretion and potentiates substance P-induced salivation. Proc Natl Acad Sci USA 86:392–396

Takeda Y, Takeda J, Smart BM, Krause JE (1990) Regional distribution of neuropeptide γ and other tachykinin peptides derived from the substance P gene in the rat. Regul Pept 28:323–333

Tateishi K, Matsuoka Y, Hamaoka T (1989) Establishment of highly specific radioimmunoassays for neurokinin A and neurokinin B and determination of tissue distribution of these peptides in rat central nervous system. Regul Pept 24:245–257

Tateishi K, Kishimoto S, Kobayashi H, Kobuke K, Matsuoka Y (1990) Distribution and localization of neurokinin A-like immunoreactivity and neurokinin B-like immunoreactivity in rat peripheral tissue. Regul Pept 30:193–200

Tatemoto K, Lundberg JM, Jornvall H, Mutt V (1985) Neuropeptide K: isolation, structure and biological activities of a novel brain tachykinin. Biochim Biophys Res Commun 128:947–953

Theodorsson-Norheim E, Norheim I, Oberg K, Brodin E, Lundberg JM, Tatemoto K, Lindgren PG (1985) Neuropeptide K: a major tachykinin in plasma and tumor tissues from carcinoid patients. Biochem Biophys Res Commun 131:77–83

Theodorsson-Norheim E, Jornvall H, Andersson M, Norheim I, Oberg K, Jacobsson G (1987) Isolation and characterization of neurokinin A, neurokinin A(3–10) and neurokinin A(4–10) from a neutral water extract of a metastatic ileal carcinoid tumour. Eur J Biochem 166:693–697

Too H-P, Cordova JL, Maggio JE (1989) A novel radioimmunoassay for neuromedin K. I. Absence of neuromedin K-like immunoreactivity in guinea pig ileum and urinary bladder. II. Heterogeneity of tachykinins in guinea pig tissues. Regul Pept 26:93–105

Toresson G, Brodin E, de las Carreras C, Nordin C, Zachau AC, Bertilsson L (1994) Quantitation of N-terminally extended tachykinins in cerebrospinal fluid from healthy subjects. Regul Pept 24:185–191

Torrens Y, Beaujouan JC, Saffroy M, Glowinski J (1995) Involvement of septide-sensitive tachykinin receptors in inositol phospholipid hydrolysis in the rat urinary bladder. Peptides 16:587–594

Tousignant C, Guillemette G, Drapeau G, Telemaque S, Dion S, Regoli D (1990) ^{125}I-BH[Sar^9,Met(O$_2$)^{11}]-SP, a new selective ligand for the NK-1 receptor in the central nervous system. Brain Res 524:263–270

Tschope C, Jost N, Unger T, Culman J (1995) Central cardiovascular and behavioral effects of carboxy- and amino-terminal fragments of substance P in conscious rats. Brain Res 690:15–24

Valentino KL, Tatemoto K, Hunter J, Barchas JD (1986) Distribution of neuropeptide K-immunoreactivity in the rat central nervous system. Peptides 7:1043–1059

Van Giersbergen PL, Shatzer SA, Burcher E, Buck SH (1992) Comparison of the effects of neuropeptide K and neuropeptide γ with neurokinin A at NK$_2$ receptors in the hamster urinary bladder. Naunyn Schmiedeberg's Arch Pharmacol 345:51–56

Velazquez RA, McCarson KE, Cai Y, Kovacs KJ, Shi Q, Evensjo M, Larson AA (2002) Upregulation of neurokinin-1 receptor expression in rat spinal cord by an N-terminal metabolite of substance P. Eur J Neurosci 16:229–241

Von Euler US, Gaddum JH (1931) An unidentified depressor substance in certain tissue extracts. J Physiol 72:74–87

Wang Y, Badgery-Parker T, Lovas S, Chartrel N, Vaudry H, Burcher E, Conlon JM (1992a) Primary structure and receptor-binding properties of a neurokinin A-related peptide from frog gut. Biochem J 287:827–832

Wang Y, O'Harte F, Conlon JM (1992b) Structural characterization of tachykinins (neuropeptide γ, neurokinin A and substance P) from a reptile, *Alligator mississipiensis*. Gen Comp Endocrinol 88:277–286.

Wang JX, Dipasquale AJ, Bray AM, Maeji NJ, Spellmeyer DC, Geysen HM (1993a) Systematic study of substance P analogs. II. Rapid screening of 512 substance P stereoisomers for binding to NK_1 receptors. Int J Pept Protein Res 42:392–399

Wang Y, Bockman CS, Lovas S, Abel PW, Murphy RF, Conlon JM (1993b) Neuropeptide γ-(1–9)-peptide: a major product of the post-translational processing of γ-preprotachykinin in rat tissues. J Neurochem 61:1231–1235

Wang Y, Barton BA, Nielsen PF, Conlon JM (1999a) Tachykinins (substance P and neuropeptide γ) from the brains of the pallid sturgeon, *Scaphirhynchus albus* and the paddlefish, *Polyodon spathula* (Acipenseriformes). Gen Comp Endocrinol 116:21–30

Wang Y, Lance VA, Nielsen PF, Conlon JM (1999b) Neuroendocrine peptides (insulin, pancreatic polypeptide, NPY, galanin, somatostatin, substance P, neuropeptide γ) from the desert tortoise, *Gopherus agassizii*. Peptides 20:713–722

Wang T, Axelsson M, Jensen J, Conlon JM (2000) Cardiovascular actions of python bradykinin and substance P in the anesthetized python, *Python regius*. Am J Physiol 279:R531–R538

Warner FJ, Mack P, Comis A, Miller RC, Burcher E (2001) Structure-activity relationships of neurokinin A(4–10) at the human tachykinin NK_2 receptor: the role of natural residues and their chirality. Biochem Pharmacol 61:55–60

Warner FJ, Miller RC, Burcher E (2002) Structure–activity relationships of neurokininA(4–10) at the human tachykinin NK_2 receptor: the effect of amino acid substitutions on receptor affinity and function. Biochem Pharmacol 63:2181–2186

Watson SP, Sandberg BE, Hanley MR, Iversen LL (1983) Tissue selectivity of substance P alkyl esters: suggesting multiple receptors. Eur J Pharmacol 87:77–84

Waugh D, Wang Y, Hazon N, Balment RJ, Conlon JM (1993) Primary structures and biological activities of substance P-related peptides from the brain of the dogfish, *Scyliorhinus canicula*. Eur J Biochem 214:469–474

Waugh D, Sower S, Bjenning C, Conlon JM (1994) Novel tachykinins from the brain of the sea lamprey, *Petromyzon marinus*, and the skate, *Raja rhina*. Peptides 15:155–161

Waugh D, Bondareva V, Rusakov Y, Bjenning C, Nielsen PF, Conlon JM (1995a) Tachykinins with unusual structural features from a urodele, the amphiuma, an elasmobranch, the hammerhead shark and an agnathan, the river lamprey. Peptides 16:615–621

Waugh D, Groff KE, Platzack B, Youson JH, Olson KR, Conlon JM (1995b) Isolation, localization and cardiovascular activity of tachykinins from the stomach of the bowfin, *Amia calva*. Am J Physiol 269:R565–R571

Wijkhuisen A, Sagot MA, Frobert Y, Creminon C, Grassi J, Boquet D, Couraud JY (1999) Identification in the NK_1 tachykinin receptor of a domain involved in recognition of neurokinin A and septide but not of substance P. FEBS Lett 447:155–159

Wormser U, Laufer R, Hart Y, Chorev M, Gilon C, Selinger Z (1986) Highly selective agonists for substance P receptor subtypes. EMBO J 5:2805–2808

Yuan L, Burcher E, Nail BS (1994) Use of selective agonists and antagonists to characterize tachykinin receptors mediating airway responsiveness in anaesthetized guinea-pigs. Pulm Pharmacol 7:169–178

Yuan L, Burcher E, Nail BS (1998) Characterization of tachykinin receptors mediating bronchomotor and vasodepressor responses to neuropeptide gamma and substance P in the anaesthetized rabbit. Pulm Pharmacol Ther 11:31–39

Yunker AM, Krause JE, Roth KA (1999) Neurokinin B- and substance P-like immunoreactivity are co-localized in enteric nerves of rat ileum. Regul Pept 80:67–74

Zhang Y, Lu L, Furlonger C, Wu GE, Paige CJ (2000) Hemokinin is a hematopoietic-specific tachykinin that regulates B lymphopoiesis. Nat Immunol 1:392–397

Zhou Q, Le Greves P, Ragnar F, Nyberg F (2000) Intracerebroventricular injection of the N-terminal substance P fragment SP(1–7) regulates the expression of the N-methyl-D-aspartate receptor NR1, NR2A and NR2B subunit mRNAs in the rat brain. Neurosci Lett 291:109–112

The Histochemistry of Tachykinin Systems in the Brain

T. Hökfelt · E. Kuteeva · D. Stanic · Å. Ljungdahl

Department of Neuroscience, Karolinska Institutet, Stockholm, Sweden
e-mail: Tomas.Hokfelt@neuro.ki.se

1	**Introduction**	64
1.1	Historical Aspects	64
1.2	Members of the Tachykinin Family	65
1.3	Scope of the Present Chapter	66
2	**Aspects on Methodology**	67
2.1	Immunohistochemistry	67
2.2	In Situ Hybridization	68
2.3	Visualizing Cell Bodies and Tracing Pathways	68
2.4	Specificity	69
3	**Distribution of Tachykinins**	70
3.1	Rat Brain	70
3.1.1	Overview—SP/NKA and NKB Systems	70
3.1.2	Distribution of SP/NKA Cell Bodies and Nerve Terminals	75
3.1.3	Fiber Tracts	81
3.1.4	Identified SP/NKA and NKB Pathways	81
3.2	The Human Brain	83
3.2.1	Telencephalon (Without Basal Ganglia)	85
3.2.2	Basal Ganglia	87
3.2.3	Diencephalon	87
3.2.4	Lower Brain Stem	88
3.3	Monkey Brain	89
3.4	Other Species	90
3.5	Coexistence of SP and Classical Transmitters	91
4	**Distribution of Tachykinin Receptors**	93
4.1	Rat Brain	93
4.1.1	Overview	93
4.1.2	Distribution of NK_1 and NK_3 Receptors	96
4.1.3	Comparison Between Distribution of SP, NK_1 and NK_3 Receptors	97
4.1.4	Distribution of NK_2 Binding Sites	97
4.2	Other Species	99
5	**General Remarks**	101
5.1	The Histochemical Methods	101
5.2	Tachykinin Receptors	102
5.3	The Mismatch Issue	102
5.4	Tachykinin Circuitries	103
5.5	Clinical Aspects	103
References		104

Abstract After more than 70 years of research it seems an appropriate moment to summarize the progress in the substance P (SP)/tachykinin field, not at least because of the recent FDA approval of an SP (NK_1) antagonist for treatment of chemotherapy-induced emesis/nausea. In the present review a wide range of histochemical studies based especially on immunohistochemistry and in situ hybridization, but also ligand binding autoradiography have been compiled to give an overview of the distribution of the three cloned tachykinins, SP, neurokinin A (NKA) and neurokinin B (NKB), as well as their three receptors (as so far known), NK_1, NK_2 and NK_3. The main focus is on the rat brain, and to a lesser extent on the human brain, but several other species are also briefly mentioned. All tachykinins and their receptors are widely distributed in the brain, often with a strong similarity between species. The SP/NKA and NKB systems are often complementary, for example NKB is strongly expressed in cortical areas, where SP and NKA are comparatively less abundant. In addition the receptors show complementary distribution patterns, and there is often an overlap between ligands and receptors. However, cases of apparent mismatch can also be found. The wide distribution suggests that tachykinin systems can be involved in a large number of brain functions. It will be an important task to further correlate such functions with distribution patterns, in particular in the human brain and especially with regard to possible involvements of tachykinin systems in brain disease. Here further detailed histochemical analyses can provide important information to better understand such processes.

Keywords Immunohistochemistry · In situ hybridization · Neuropeptide · Receptor · Substance P

Our co-author Å. Ljungdahl passed away on 24 November 2003. We deeply regret the loss of a fine scientist, an outstanding clinician and a wonderful human being.

1
Introduction

1.1
Historical Aspects

Substance P (SP) was discovered and named by Ulf von Euler and John Gaddum more than 70 years ago (von Euler and Gaddum 1931). Forty years later Susan Leeman and her collaborators were able to identify SP as an undecapeptide (Chang et al. 1971). Moreover, her group synthesized SP (Tregear et al. 1971) and developed a radioimmunoassay (RIA) (Powell et al. 1973). This breakthrough initiated a new wave of SP research, and a check in PubMed in August 2003 revealed 17,307 entries under the keyword SP, reflecting the intense interest in this peptide over the last 30 years. In fact, SP has in many respects been a model with regard to research strategies in the neuropeptide field, including defining further members of the tachykinin (TK) family, identifying TK receptors and developing drugs that act on these receptors (see Hökfelt et al. 2001). Just recently a SP (NK$_1$) receptor antagonist, Aprepitant (Emend), has been approved by the Food and Drug Administration in the USA for treatment of chemotherapy-induced nausea and emesis. The same compound is also being investigated in the clinic as a possible candidate for treatment of major depression (Kramer et al. 1998).

After the original discovery, SP research advanced slowly; for example, von Euler published some 10 years later a purification procedure (von Euler 1942), Bengt Pernow continued this work in von Euler's laboratory and presented his doctoral thesis another decade later (Pernow 1953). Here further purification steps were presented, and the distribution of SP and its biological actions in brain and peripheral tissues of the dog were reported. In fact, Pernow's study represents the first mapping of SP, demonstrating, for example, high concentrations in substantia nigra and hypothalamus as well as in the dorsal horn of the spinal cord. This was elegantly complemented by Lembeck's analysis of the dorsal roots, providing the first evidence that SP is a sensory neurotransmitter (Lembeck 1953). In the 1960s this work was continued in particular by Zetler who also described further extraction procedures and carried out distribution studies in the human brain. His results agreed well with those of Pernow on the dog brain, showing, for example, the highest concentrations of SP in the substantia nigra and low levels in cortical areas (Zetler 1970).

1.2
Members of the Tachykinin Family

Two further members of the mammalian TK family were identified, neurokinin A (NKA; also called substance K or neuromedin L) and neurokinin B (NKB; also known as neuromedin K) (see Nakanishi 1987; Maggio 1988). This was of significance for immunohistochemical studies on the distribution of SP, since they all have three C-terminal amino acids with C-terminal amidation in common, which raises the possibility that antibodies against SP may cross react with the other two TKs. It was then realized that there are two genes encoding mammalian TKs, the pre-pro-TK (PPT) A (PPT-A; now called *TAC1*) gene and the PPT-B (*TAC3*) gene, whereby the former gives rise to SP and NKA (Nawa et al. 1983, 1984; Kawaguchi et al. 1986; Krause et al. 1987), and the second gene encodes NKB (Kotani et al. 1986; Bonner et al. 1987). Other members of the TK family include neuropeptide K and neuropeptide γ which are elongated forms of NKA (see Nakanishi 1987; Maggio 1988). Recently a third TK gene, PPT-C (*TAC4*) has been discovered (Zhang et al. 2000), but has so far been little explored in the central nervous system (CNS) and will not be dealt with here. For information on this new gene and nomenclature, as well as on the present status of the field, see the chapter on TK nomenclature by Patacchini and Maggi in this volume.

1.3
Scope of the Present Chapter

In the following we will describe in some detail the distribution of SP/NKA and NKB in rat and human brain as shown by histochemistry, and will briefly mention some other species. We will discuss methodological aspects. No complete maps of SP in the mouse brain have been reported, but we will use some of our own unpublished material from mouse brain illustrating the distribution of SP-like immunoreactivity (LI) in rodents. In fact, the rat and mouse appear to have many SP systems in common. With regard to TK receptors in the rat, we will briefly summarize recent findings based on immunohistochemistry and in situ hybridization and, again, show some unpublished examples. We note that TK receptors have been studied in several species, including man, most recently in the living human brain with positron emission tomography (PET) (Hargreaves 2002). Taken together, this chapter is an overview and compilation of papers published on the distribution of TKs and their receptors in the rat brain, material that is supplemented by also mentioning some other species, especially human, one aim being to help the reader to find original papers to enable more detailed reading. In spite of the long reference list many interesting studies, often dealing with subregions in the brain, had to be left out, for which we apologize.

We will not discuss here SP and TK receptors in spinal and sensory systems, including the primary afferent branches in the dorsal horn of the spinal cord, but refer to several reviews on this topic (see Dalsgaard 1988; Rustoni and Weinberg 1989; Cuello et al. 1993; Ribeiro-da-Silva and Hökfelt 2000; Mantyh 2002; Todd 2002), and to some more general reviews (see Pernow 1983; Otsuka and Yoshioka 1993). Information on the gastrointestinal tract can be found in a book by Furness and Costa (1987) and in reviews by Sternini (1991) and Holzer (this volume), and on the retina by Karten and Brecha (1983) and Kolb et al. (1995). Nor will we deal with studies focusing on regulation of SP and other TKs, for example in the striatum, where it is well known that SP, present in the large population of medium spiny γ-aminobutyric acid (GABA)-positive neurons projecting to the substantia nigra, is regulated by dopamine (see Graybiel 1990; Gerfen 1992). For developmental aspects of SP, that is ontogenetic brain patterns, we refer to Inagaki et al. (1982a, 1982b) and Sakanaka et al. (1982a) and to a review by Sakanaka (1992) who also describes the ontogeny of SP in several other species. Nässel (1999) has described TK-related peptides in invertebrates, and a broad review on TKs in the animal kingdom has been published by Severini et al. (2002). Finally, the vast majority of studies focuses on neuronal SP, but there is evidence for this peptide also being expressed in glial cells (see Ubink et al. 2003), another aspect not dealt with here.

2
Aspects on Methodology

2.1
Immunohistochemistry

The generation of antibodies to SP allowed analysis of the distribution of SP using RIA, originally developed by Berson, Yalow and collaborators (1956), as well as immunohistochemistry. The main methodology for mapping TK neurons is the indirect immunohistochemical method originally described by Coons and collaborators (see Coons 1958) and several subsequent modifications including the peroxidase (PAP) technique developed by Nakane and Pierce (1967), Avrameas (1969) and Sternberger et al. (1970) as well as recent amplification methods, for example the tyramide signal amplification (TSA) (Adams 1992). The generation of monoclonal SP antibodies by Cuello et al. (1979), that also allowed internal labeling (Cuello et al. 1982b), was an important methodological advancement. The higher sensitivity of the recent methods, especially when combined with incubation of free-floating sections with antibody, makes it highly desirable to 'remap' the TK (and other) systems. More recently, the cloning of the TK receptors made it possible to raise specific and sensitive antibodies for immunohistochemical analysis (see below).

Both RIA and immunohistochemistry for SP were explored early and studies using RIA were published by Brownstein et al. (1976) and Kanazawa and Jessell (1976). The punch technique developed by Palkovits, combined with a highly sensitive RIA, provided quantitative information on SP levels in discrete brain regions (Brownstein et al. 1976). In parallel, the indirect immunofluorescence technique described the cellular distribution of SP (Nilsson et al. 1974; Hökfelt et al. 1975). Using this approach a widespread distribution of SP-positive nerve terminals was found in the brain, which roughly paralleled the distribution monitored by RIA. However, it was clear that the sensitivity of the histochemical technique was inferior to the RIA, since the latter approach demonstrated presence of SP in cortical areas, where at that time no, or only single, fibers were encountered at the microscopic level. It must be borne in mind that for RIA, even in areas with a low innervation, many nerve endings (each perhaps with a low peptide content) are 'pooled' for monitoring, whereas with immunohistochemistry the content in a single nerve ending has to be visualized.

A further methodological approach is radioimmunocytochemistry, where incubation with the primary anti-SP antibody is followed by tritium- (Glazer et al. 1984) or iodine- (McLean et al. 1985a, 1985b) labeled antibodies, providing high sensitivity and excellent overviews. In the study by Glazer et al. (1984) it was reported that five times more dilute primary antibodies could be used in comparison to adjacent sections processed according to the PAP procedure.

Sutoo et al. (2002) have developed a quantitative 'brain mapping analyzer' which combined with a microphotometry system can monitor immunohistochemical intensities in very small areas; in fact, a rat brain section is divided into up to 300,000 micro areas at 20-μm intervals.

2.2
In Situ Hybridization

The cloning of the first two PPT genes (*TAC1*, *TAC3*; see below) made possible histochemical analysis of the transcripts using in situ hybridization. This technique offers a major advantage, allowing visualization of the cell bodies that produce TKs without colchicine treatment, and often shows more cell bodies than immunohistochemistry even in colchicine treated animals (Warden and Young 1988; Harlan et al. 1989; Lucas et al. 1992; Marksteiner et al. 1992). In a similar way, the cloning of the tachykinin receptors (see below) allowed mapping of the distribution of NK_1 (Maeno et al. 1993) and NK_3 (Shughrue et al. 1996) receptor transcripts in the rat with in situ hybridization. This technique can be used on human brains, in spite of post mortem delay, with high sensitivity, a good example being the study by Chawla et al. (1997) describing the localization of SP/NKA and NKB transcripts in the hypothalamus and basal forebrain. On the other hand, in situ hybridization has the disadvantage of showing only cell bodies, an intrinsic deficiency overcome with immunohistochemistry which reveals the distribution of peptides and receptor proteins in single nerve terminals and, sometimes, axons and often in dendritic processes. For autoradiographic ligand binding techniques, see comments below.

2.3
Visualizing Cell Bodies and Tracing Pathways

In the very early histochemical studies it was not possible to visualize the cell bodies producing SP, and therefore it was difficult to understand the origin of the widespread nerve terminal networks and thus which pathways utilized SP as a messenger molecule. This problem was in part solved by the use of the mitosis inhibitor colchicine, which arrests axonal transport, mostly without affecting synthesis (exceptions are present). Thus, SP (and other peptides), which are ribosomally produced and stored in so called large dense core vesicles, builds up and accumulates in the cell bodies and can be more easily visualized. Still, levels in axons are often low, and it is difficult to follow the projections of SP neurons, regardless of whether normal or colchicine treated animals are used. To overcome this problem lesions and (retrograde and anterograde) tracing techniques have been used (see Heimer and Zaborszky 1989). For example, retrograde tracer is injected into an area with SP nerve terminals, and after allowing retrograde transport to the cell bodies (from a few days to several weeks), the animals are treated with colchicine and then processed for immunohistochemical visualization of SP-positive cell bodies (see Hökfelt et al. 1983). If double-labeled cells can be observed in a certain area, that is containing both retrograde tracer and SP, it can be concluded that the SP cell group projects to the injection area. This, of course, provided that the tracer is neither taken up by axons nor is anterogradely or transsynaptically transported. The pathway can then be confirmed by lesioning the cell body area and establishing loss of terminals in the projection area. This careful approach was applied early in TK research by in particular Masaya Tohyama and collaborators in Japan (see below).

2.4
Specificity

Immunological techniques, particularly immunohistochemistry, which work with much higher antibody concentrations than RIA, are always prone to nonspecificity. However, it seems reasonable that most mapping studies carried out so far using antibodies to SP in fact show genuine SP systems, although the antibodies also may react with NKA, which mostly is coexpressed with SP. Routine controls encompass adsorption experiments, where the peptide is added to the antiserum at high concentrations (10^{-5} or 10^{-6} M), which should result in deletion of staining. Another possibility is to compare immunohistochemical distribution patterns with results obtained with in situ hybridization, that is a comparison between localization of cell bodies stained with antiserum for the peptide with those containing the transcript after in situ hybridization. It is sometimes possible to perform both techniques on the same section and thus demonstrate to what extent the two techniques show the same cell bodies. If this cannot be achieved, it is also valuable to show overlapping distribution patterns on separate sections. More recently transgenic animals, in the present case mice with the deleted PPT-A (*TAC1*) gene (Cao et al. 1998; Zimmer et al. 1998) or the NK_1 receptor gene (De Felipe et al. 1998), can be used to test specificity of antisera. This appears to represent the most reliable and the ultimate approach.

The peptide mapping studies are based on the assumption that the peptide is produced in the same cells that contain the mRNA. Thus, it is assumed that peptides after release are degraded and broken down by extracellular peptidases, in contrast to classical transmitters which via specific transporter molecules are taken up into nerve endings and storage vesicles (see Hökfelt et al. 2003). Such cell membrane transporters for a certain transmitter are not necessarily confined to those neurons which use this molecule as a transmitter, and may therefore not be 'specific' markers. However, there is some evidence that the C-terminal fragment SP (5–11) accumulates in glial cells as well as nerve terminals, suggesting presence of a high affinity uptake mechanism for this fragment (Inoue et al. 1984). Even if such an uptake mechanism has not been confirmed by other laboratories, this possibility should not be overlooked.

Since SP and other TKs in general are conserved to a large extent during evolution, it is mostly possible to use the same antiserum for several species with good results. An interesting cross-reactivity may have affected primary sensory neurons which contain high concentrations of SP and where staining with antiserum to ovine corticotropin releasing factor (CRF) seems to represent cross-reactivity with SP and not genuine CRF (Berkenbosch et al. 1986). It cannot be excluded that similar types of unexpected—and so far not explained—cross-reactivities occur in the literature.

3
Distribution of Tachykinins

3.1
Rat Brain

3.1.1
Overview—SP/NKA and NKB Systems

After the preliminary reports in the mid 1970s two major mapping studies were published on the distribution of the SP in the brain and spinal cord (Cuello and Kanazawa 1978; Ljungdahl et al. 1978), describing a widespread distribution along the rostrocaudal axis, and using colchicine treatment more than 30 cell groups could be identified (Ljungdahl et al. 1978), many of which can be seen in Fig. 1. Overall SP appeared mainly to be a subcortical peptide with few immunoreactive structures in neocortical and hippocampal areas. Other early immunohistochemical mapping papers dealing with major parts of the brain were carried out by Inagaki et al. (1982a) (forebrain and upper brain stem), Sakanaka et al. (1982b) (lower brain stem), and Panula et al. (1984) (hypothalamus). Shults et al. (1984) provided SP immunohistochemistry combined with a discrete regional SP RIA analysis and SP receptors using ^{125}I-Bolton–Hunter labeling, where immunohistochemistry and RIA correlated well. Neuropeptide K, the 36-amino acid elongated form of NKA, has been mapped in the rat brain (Valentino et al. 1986), providing similar distribution patterns as the SP/NKA studies. A more recent detailed semiquantitative description of SP/NKA peptides is found in Nakaya et al. (1994) who published primarily a detailed distribution of NK_1 receptor immunoreactivity (see below), but where the endogenous ligand (SP) for comparison was mapped on adjacent sections and where the results were included in a comprehensive table (no micrographs or maps shown).

After the cloning of the first two PPT genes (*TAC1*, *TAC3*; Nawa et al. 1983; Kawaguchi et al. 1986; Kotani et al. 1986; Bonner et al. 1987; Krause et al. 1987) the distribution of SP and NKA gene transcripts was mapped with in situ hybridization (Warden and Young 1988; Harlan et al. 1989). The NKB neurons have also been mapped in detail using this technique (Warden and Young 1988), and with immunohistochemistry (Merchenthaler et al. 1992) as well as with in situ hybridization combined with immunohistochemistry (Lucas et al. 1992; Marksteiner et al. 1992). Arai and Emson (1986) described the regional distribution of four TKs (SP, NKA, NKB and neuropeptide K) in the rat brain based on RIA.

The text in Sect. 3.1.2 is based on Ljungdahl et al. (1978) and Nakaya et al. (1994) with regard to fibers (immunohistochemistry) and on Harlan et al.

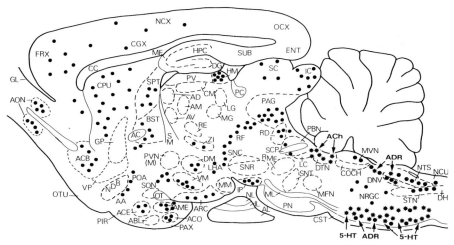

Fig. 1 Schematic illustration of the distribution of major SP-positive cell groups (*dots*) in a sagittal section of the rat brain. This map represents multiple parasagittal levels through the rat brain and was originally reconstructed by Khachaturian et al. (1985) using the rat brain atlas of Paxinos and Watson (1982), and has been expanded from Hökfelt et al. (1987) by addition of information on SP/NKA and SP/NKA mRNA-positive cells published by Harlan et al. (1989). Cell groups that have been shown to also contain a classical transmitter (indicated by *abbreviations* and *arrows*) are labeled by a *star*. *AA*, Anterior amygdala; *ABL*, basolateral nucleus of amygdala; *AC*, anterior commissure; *ACB*, nucleus accumbens; *ACE*, central nucleus, amygdala; *ACO*, cortical nucleus, amygdala; *AD*, anterodorsal nucleus, thalamus; *AL*, anterior lobe, pituitary; *AON*, anterior olfactory nucleus; *ARC*, arcuate nucleus; *AV*, anteroventral nucleus, thalamus; *BST*, bed nucleus of stria terminalis; *CC*, corpus callosum; *CGX*, cingulate cortex; *CM*, central-medial nucleus, thalamus; *COCH*, cochlear nucleus complex; *CPU*, caudate-putamen; *CST*, corticospinal tract; *DH*, dorsal horn, spinal cord; *DG*, dentate gyrus; *DM*, dorsomedial nucleus, hypothalamus; *DNV*, dorsal motor nucleus, vagus; *DTN*, dorsal tegmental nucleus; *ENT*, entorhinal cortex; *FRX*, frontal cortex; *GL*, glomerular layer, olfactory bulb; *GP*, globus pallidus; *HM*, medial habenular nucleus; *HPC*, hippocampus; *IC*, inferior colliculus; *IL*, intermediate lobe, pituitary; *IP*, interpeduncular nuclear complex; *LC*, locus coeruleus; *LG*, lateral geniculate nucleus; *LHA*, lateral hypothalamic area; *MF*, mossy fibers, hippocampus; *MFN*, motor facial nucleus; *MG*, medial geniculate nucleus; *ML*, medial lemniscus; *MM*, medial mammillary nucleus; *MVN*, medial vestibular nucleus; *NCU*, nucleus cuneatus; *NCX*, neocortex; *NDB*, nucleus of diagonal band; *NL*, neural lobe of pituitary; *NRGC*, nucleus reticularis gigantocellularis; *NTS*, nucleus tractus solitarii; *OCX*, occipital cortex; *OT*, optic tract; *OTU*, olfactory tubercle; *PAG*, periaqueductal gray; *PAX*, periamygdaloid cortex; *PBN*, parabrachial nucleus; *PC*, posterior commissure; *PIR*, piriform cortex; *PN*, pons; *POA*, preoptic area; *PV*, periventricular nucleus, thalamus; *PVN(M)*, paraventricular nucleus, pars parvocellularis; *RD*, nucleus raphe dorsalis; *RE*, nucleus reuniens, thalamus; *RF*, reticular formation; *RME*, nucleus raphe medianus; *SC*, superior colliculus; *SCP*, superior cerebellar peduncle; *SM*, stria medullaris thalami; *SNC*, substantia nigra, pars compacta; *SNR*, substantia nigra, pars reticulata; *SNT*, sensory nucleus of trigeminal tract; *SON*, nucleus supraopticus; *SPT*, septum; *SUB*, subiculum; *VM*, ventromedial nucleus, hypothalamus; *VP*, ventral pallidum; *ZI*, zona incerta

(1989) for SP/NKA cell bodies (mRNA, and thus summarizes early results). There are more detailed studies on TK systems in discrete brain regions which, however, have not been incorporated here. Figure 1 shows the distribution of SP/NKA cell bodies in a sagittal section of the rat brain. We will illustrate our description with unpublished immunofluorescence micrographs from mouse

Table 1 Comparison of distribution patterns/levels for SP/NKA and NKB (modified after Lucas et al. 1992)

Telencephalon	SP/NKA system				NKB system	
	SP/NKA mRNA (Harlan et al. 1989)	SP-LI cb/f (Ljungdahl et al. 1978)	SP pg/μg protein (Shults et al. 1984)	NKB mRNA (Lucas et al. 1992)	P2-LI cb/f (Lucas et al. 1992)	
Main olfactory bulb	0					
Periglomerular cells		0/+	0.72	+++	0/+	
External tufted cells		0/+		++	+/+	
Mitral cells		0/+		+	+/+	
Internal granule cells		0/+	0.62	+	0/+	
Anterior olfactory nucleus	+++	0/+	0.41	+	+/+	
Olfactory tubercle			1.15			
Layer I	+++	0/+ +		0	0/+++	
Layer II	+++	0		0	0/+	
Islands Calleja				0	0/+	
Non-island clusters	+++	++/+ to +++		+++	0/+++	
Nucleus accumbens	+++	++/+ to +++		+++	0/+	
Caudate-putamen	0	0/++	0.74–1.08	+	0/++	
Globus pallidus	++	0/++	1.60	++	+/+	
Frontal cortex				++	+/+	
Other cortex	0	++/+	0.12–0.70	+	0/+	
Tenia tecta	++	+++/+ to ++	0	+	0/+	
Diagonal band Broca	+	+/+ to +++	2.20	++	0/+	
Medial preoptic area	++	+++/++	3.85	0	0/+	
Lateral preoptic area	+	0/+ to ++	3.78	+	0/+++	
Lateral septum	++	0/+ to +++	3.18	+	0/++	
Medial septum			1.72			
Bed nucleus of stria terminalis			3.66	+++	++/+++	
Amygdala						
Central	+	+++	2.13	+++	+++/+++	
Medial	++	+++/+ to +++	2.72	+	0/+	

Table 1 (continued)

	SP/NKA system			NKB system	
	SP/NKA mRNA (Harlan et al. 1989)	SP-LI cb/f (Ljungdahl et al. 1978)	SP pg/µg protein (Shults et al. 1984)	NKB mRNA (Lucas et al. 1992)	P2-LI cb/f (Lucas et al. 1992)
Telencephalon					
Cortical	++	0/+	0.72	+	0/+
Basal	0	+/+		++	0/++
Hippocampus	+	0/+	<0.13	+	+/+
Diencephalon					
Anterior hypothalamic area	++	+++/++	2.82	++	0/+
Suprachiasmatic nucleus	0	+/+++	1.50	+	+/+++
Supraoptic nucleus	−	+/+	1.70	0	0/++
Perifornical area	++	+++/+	−	+	0/+++
Paraventricular nucleus	0	+/+	2.25	+	0/+
Medial forebrain bundle	++	+/++	−	+	0/++
Zona incerta (rostral)	+	+/+	−	+	0/+
Ventromedial nucleus	++	+++/+	1.31	++	0/+
Dorsomedial nucleus	+	+++/++	2.03	++	0/++
Arcuate nucleus (caudal)	++	0/+	1.70	++	++/+++
Dorsal premammillary nucleus	+	0/+	−	0	0/+
Ventral premammillary nucleus	0	+++/+++	−	+	0/0
Supramammillary nucleus	+	+++/+	−	0	0/+
Posterior hypothalamic nucleus	+	++/++	1.43	0	0/++
Medial habenular nucleus	+++	+++/+++	1.43	+++	+++/+++
Mesencephalon					
Periaqueductal gray	++	+++/+++	−	++	0/++
Deep mesencephalic nucleus	+	0/++	−	0	0/++
Superior colliculus			0.84		
Superficial gray	+	0/+++	−	0	0/0
Deep gray	+	++/++	−	+	0/+
Cuneiform nucleus	+	0/+	−	0	0/+

Table 1 (continued)

Telencephalon	SP/NKA system			NKB system	
	SP/NKA mRNA (Harlan et al. 1989)	SP-LI cb/f (Ljungdahl et al. 1978)	SP pg/μg protein (Shults et al. 1984)	NKB mRNA (Lucas et al. 1992)	P2-LI cb/f (Lucas et al. 1992)
Dorsal raphe	+	+/+ to ++	1.53	0	0/++
Edinger-Westphal nucleus	++	++/+	–	0	0/++
Median raphe	+	0/+	–	0	0/+
Inferior colliculus	++	0/++ to +++	0.81	+	0/+ to +++
Interpeduncular nucleus (most rostral)	+	+++/+++	2.32	0	0/+++
Rhombencephalon					
Dorsal tegmental nucleus	++	0/+	1.43	0	0/+
Dorsolateral tegmental nucleus	++	+++/+	–	0	0/+
Locus coeruleus	0	+/+ to ++	1.57	0	0/+
Parabrachial nucleus	+	0/++	2.39–3.04	0	0/+++
Parvocellular reticular nucleus	++	0/+	–	0	0/++
Vestibular nucleus					
Medial	++	0/+	–	0	0/+
Spinal	+	+++/+	–	0	0/0
Nucleus of the solitary tract	+++	+/++	4.30	0	0/+++
Gigantocellular reticular nucleus	++	0/+	1.12	+	0/+
Facial nucleus/around nucleus of the spinal trigeminal tract	++	0/++	1.51	0	0/+
Oral	+	0/+	–	0	0/+
Interpolar	++	++/+	–	0	0/+
Caudal	0	0/+	–	+	0/++
Raphe nuclei					
Magnus	+++	+++/+	1.11	0	0/+
Pallidus	+++	+++/+	–	0	0/+
Pontis	0	+/+	–	0	0/+
Obscurus	+++	+++/+	–	0	0/+
Commissural nucleus	+	+++/+++	–	0	0/+++

cb, Cell bodies; f, fibers.

brains (Figs. 2–4), as this offers some new information as such, but also because mouse and rat appear to be very similar with regard to SP systems. One reason that these mouse staining patterns appear stronger than those in Ljungdahl et al. (1978) is the use of the sensitive TSA method. The NKB system in rat will only be described in Table 1, where we have summarized results from Lucas et al. (1992), who compared the distribution of SP-LI (Ljungdahl et al. 1978) with that of SP/NKA mRNA (Harlan et al. 1989), NKB mRNA (Lucas et al. 1992) and NKB-LI with antiserum P2 directed against a 30-amino acid peptide of a non-TK portion of the NKB precursor peptide (Lucas et al. 1992). In addition, we have added a fifth column listing RIA values for SP from Shults et al. (1984) in corresponding regions (whenever available). To facilitate reading we have omitted the results on NKB mRNA by Warden and Young (1988) from Table 1 (these can be found in Lucas et al. 1992), which in general agree with those by Lucas et al. (1992).

3.1.2
Distribution of SP/NKA Cell Bodies and Nerve Terminals

Telencephalon. High numbers of SP cell bodies are observed in the anterior olfactory nucleus, olfactory tubercle, nucleus accumbens and the caudate putamen and moderate numbers in frontal cortex, the medial and cortical amygdaloid nuclei, the lateral septum, the bed nucleus of stria terminalis and the medial preoptic area; even the hippocampal formation has some cell bodies, especially in the mouse (Fig. 3B, F).

SP-immunoreactive (ir) fibers are in general low in the olfactory bulb, but the mitral cell layer in the accessory olfactory bulb shows a strong signal. A low density of fibers, or even only single fibers, are found in most cortical areas as well as in hippocampus, although moderate numbers are observed in the pre- and infralimbic as well as cingulate cortices (Fig. 2A, B). In the amygdaloid complex a very high fiber density is observed in the medial and a moderate to high density in the central nucleus, whereas other nuclei have low density networks (Fig. 2D). In the caudate nucleus and nucleus accumbens mostly medium densities are observed, with the strongest signal in the ventral aspects, continuing into the ventral pallidum and the olfactory tubercle (Fig. 2C, E). A dense SP fiber network is found in the globus pallidus with lower densities in its dorsal and lateral parts, and with the fibers appearing to follow dendritic processes. These fibers are continuous in the ventral direction running below the anterior commissure into the ventral pallidum. The septal complex in general contains moderately dense SP fiber networks also forming a continuous band into the ventral pallidum (Fig. 2C). In the lateral septal nucleus fibers are strongly fluorescent, often surrounding cell bodies in a basket-like manner (Fig. 2C). The bed nucleus of the stria terminalis has mostly a medium dense network, some parts even a high density (Fig. 2F). The substantia innominata is characterized by a high density fiber network in continuity with the one present in the ventral

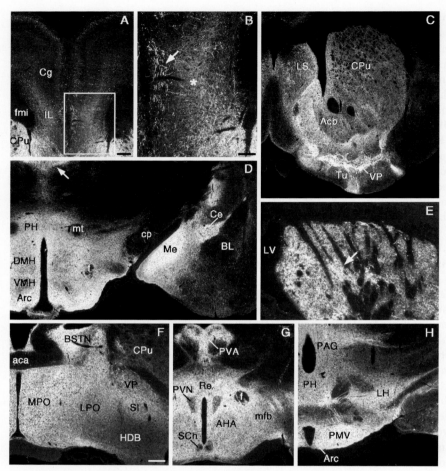

Fig. 2A–H Immunofluorescence micrographs of coronal sections of normal or colchicine treated, formalin–picric acid perfused mouse brain incubated with SP antiserum. (**B** shows *boxed area* in **A**); Bregma 1.42 mm according to Paxinos and Franklin (2001). Cingulate cortex (*Cg*) showing a weak network in lamina II extending into secondary motor cortex with a network of strongly fluorescent fibers in the ventral aspects, including fibers running in a dorsal–ventral direction in the superficial layers (*arrow*), as well as a dense plexus of fine, weakly fluorescent fibers (*asterisk*) in the deeper layers extending into the infralimbic cortex (*IL*); *fml*, fascicularis medialis longitudinalis. **C, E** Bregma 0.98 mm. Overview of the central forebrain demonstrating moderate to dense fiber networks in the lateral septum (*LS*), caudate putamen (*CPu*) and nucleus accumbens (*Acb*) with higher density in the shell extending into the ventral pallidum (*VP*) and the olfactory tubercle (*Tu*). Arrow in **E** points to patch with higher density of SP fibers; *LV*, lateral ventricle. **D** Bregma 1.82 mm. Mid-hypothalamus and thalamus and the amygdaloid complex. A high density of SP-positive fibers is seen in most hypothalamic nuclei/areas with a lower density in the ventromedial (*VMH*) and dorsomedial (*DMH*) nuclei extending into the posterior hypothalamic area (*PH*). In the thalamus moderate to high fiber densities are seen in the midline areas (*arrow*), whereas remaining thalamic regions are fairly devoid of SP fibers. In the amygdaloid complex a very high density is seen in the medial (*Me*) and central (*Ce*) nuclei, whereas only few fibers are seen in the basal lateral nucleus (*BL*); *f*, fornix; *mt*, fasciculus mammillo-thalamicus. **F** Bregma 0.02 mm. At the

pallidum. A dense network is seen in the entopeduncular nucleus. In most parts of the medial preoptic area moderate densities are seen (Fig. 2F).

Diencephalon. In the thalamus a very high number of SP-positive cell bodies is found in the medial habenula. In the hypothalamus moderate numbers are present in the anterior hypothalamic area, the perifornical area, the medial forebrain bundle area, the ventromedial nucleus and the arcuate nucleus.

With regard to nerve fibers the thalamus shows low to moderate densities in the midline areas (Fig. 2D, G), including the paraventricular thalamic nucleus and the rhomboid nucleus with lower levels in the nucleus reuniens. A moderate density is also seen in the lateral habenular nucleus and the posteromedianus nucleus, with only scattered fibers in most other thalamic nuclei. In the hypothalamus high or medium densities of SP-ir fibers are seen in many areas including the anterior hypothalamic nucleus, the dorsal part of the supraoptic nucleus and the periventricular nucleus, with lower densities in the paraventricular and suprachiasmatic nuclei (Fig. 2G, H). High or moderate densities are also observed in most other mid- and posterior hypothalamic nuclei including the arcuate nucleus, with the exception of the ventromedial nucleus and the median eminence which only have low to moderate numbers of fibers (Fig. 2D, F, H). The mammillary nuclei also have very few fibers, although the supramammillary area has a moderate density.

Mesencephalon. A moderate number of substance P-ir cell bodies is seen in the periaqueductal gray (Fig. 3B, D), the Edinger-Westphal nucleus and inferior colliculus, whereas low numbers of cells are found in many other areas including the deep mesencephalic nucleus, the superior colliculus, the cuneiform nucleus, the dorsal and median raphe, the parabigeminal nucleus and the interpeduncular nucleus.

A very high density of SP-ir fibers is observed in the zona reticulata of the substantia nigra partly extending into zona compacta (Fig. 3B, C, G). A similarly high density is also seen in the lateral aspects of the interpeduncular nucleus

◀──

preoptic hypothalamic level a high to very high density is seen in the various parts of the medial preoptic complex (*MPO*) and the bed nucleus of stria terminalis (*BSTN*) with lower densities in the lateral preoptic area (*LPO*) and very low densities in the horizontal limb of the diagonal band nucleus (*HDB*). Note high densities in VP and substantia innominata (*SI*); *aca*, anterior commissure. **G** At the anterior level (Bregma 0.58 mm) there are high densities of SP fibers in the anterior hypothalamic region area (*AHA*; including the medial forebrain bundle area, *mfb*), with lower densities in the suprachiasmatic nucleus (*SCh*) and paraventricular (*PVN*) nuclei. The thalamic highest densities are in the midline regions including nucleus reuniens (*Re*) as well as the thalamic paraventricular nucleus (*PVA*) and adjacent nuclei. **H** At the posterior hypothalamic level (Bregma 2.46 mm) a strong immunofluorescence is seen in the ventral premammillary nucleus (*PMV*) and arcuate nucleus (*Arc*) as well as in the lateral hypothalamus (LH). Moderate densities are seen in the posterior hypothalamic area (*PH*) and the periaqueductal central gray (*PAG*). Scale bars, 200 μm (**A=C=D=G**; **F=H**) and 100 μm (**B, E**)

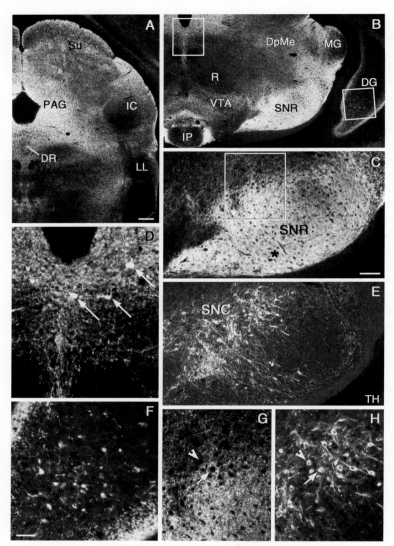

Fig. 3A–H Immunofluorescence micrographs of coronal sections of normal or colchicine treated, formalin–picrid acid perfused mouse brain incubated with SP antiserum. **A** Bregma 4.6 mm. At the posterior pontine level a strong immunofluorescence is seen in the periaqueductal gray (*PAG*) extending ventrolaterally into the pontine reticular formation. There is also a moderately dense fiber plexus in the superior colliculus (*Su*) with indications of layering, and a less dense fiber network in the raphe regions (*DR*, dorsal raphe nucleus). The inferior colliculus (*IC*) at this level and the lateral lemniscus (*LL*) have only a low number of fibers. Some SP-positive cell bodies can be seen in the dorsal raphe complex (*arrows*). **B–H** Bregma 3.64 mm. At the mesencephalic level a very high density of fibers is seen in the substantia nigra pars reticulata (*SNR*) extending into the zona compacta (*SNC*) and in the lateral interpeduncular nucleus (*IP*). This can also be seen at higher magnification (**C, G**). Note partial overlap between SP and dopamine dendrites/cell bodies as visualized with tyrosine hydroxylase (**E, H**) after double staining of the same section (**G, H**). **G** and **H** show *boxed area* in **C**; compare **C** and **G** with **E** and **H**. Note that

(Fig. 3B). The remaining parts of the zona compacta as well as the ventral tegmental area have a moderate fiber density (Fig. 3B, G), as can be seen also in the mouse when sections are double-stained for tyrosine hydroxylase (Fig. 3E, H), here a marker for dopamine neurons. The periaqueductal central gray has a moderate fiber density, extending from the posterior hypothalamic levels continuing into mesencephalon and pons (Fig. 3A). A moderately dense network is observed in the cuneiform nucleus extending ventrally along the dorsal aspects of the lateral lemniscus, and is also seen in some layers of the superior colliculus (Fig. 3A). A very high density is seen in the parabigeminal nucleus. Other mesencephalic areas have low fiber density or even single fibers.

Rhombencephalon. In the pons/medulla oblongata high numbers of SP cell bodies are observed in the solitary tract nucleus and in the raphe regions (raphe magnus, pallidus and obscurus) (Fig. 4D, F), with moderate numbers in the dorsal and dorsal lateral tegmental nuclei, the parvocellular reticular nucleus, the gigantocellular reticular nucleus, the lateral paragigantocellular nucleus, areas surrounding the facial nucleus, as well as parts of the interpolar spinal trigeminal nucleus. In several other regions low number of cell bodies are seen. Figure 4 (B, E) shows cell bodies in the mouse vestibular complex.

Moderate fiber densities are seen in the parabrachial complex, the periaqueductal gray, the locus coeruleus and parts of the pontine raphe regions (Fig. 4A). The caudal reticular pontine nucleus has a moderate fiber density extending outside the nucleus in dorsal and lateral directions. In other pontine areas the fiber density is low. In the medulla oblongata a very high density of SP-ir fibers is observed in the superficial layers of the caudal (Fig. 4D), but not rostral (Fig. 4C), spinal trigeminal nucleus. Moderate and even high densities are observed in the solitary tract nucleus, at some levels extending into the dorsal vagal nucleus as well as ventrolaterally into the parvocellular reticular nucleus, and with a high density in the lateral aspects of the lateral reticular nucleus (Fig. 4C, D). The remaining medullary reticular formation and cerebellum have low density fiber networks.

⬅ ───

dendrites and ventral cell bodies (*asterisk*) are surrounded by this high fiber density. In **G** and **H** some cell bodies (*arrowheads*) are not, others (*arrow*) are, surrounded by a high number of SP terminals. The high fiber density extends dorsolaterally into the pars lateralis as well as into the area adjacent to the mediate geniculate body (*MG*), whereas the adjacent area of the deep mesencephalic nucleus (*DpMe*) only has moderate density and the magnocellular red nucleus complex (*R*) only very low SP levels (*B*). The ventral tegmental area (*VTA*) has a moderate fiber density (*B*). There are some SP-positive cell bodies in the DR (*arrows* in **D**). Note distinct signal in the granular cell layer of the dentate gyrus (*DG*), where also numerous positive cell bodies can be seen in the polymorph layer after colchicine treatment (**F**). **D** and **F** show, respectively, *left* and *right* boxed area in **B**. *Scale bars,* 200 μm (**A, B**), 100 μm (**C–E**) and 50 μm (**F–H**)

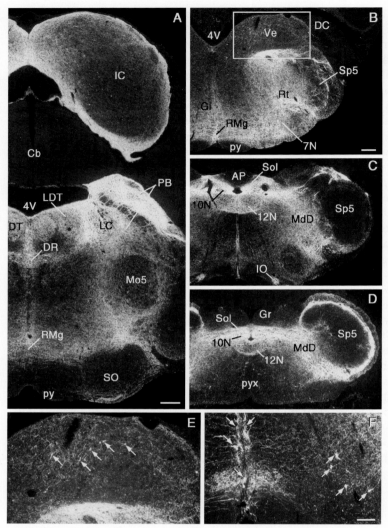

Fig. 4A–F Immunofluorescence micrographs of coronal sections of normal or colchicine treated, formalin–picric acid perfused mouse brain incubated with SP antiserum. **A** Bregma 5.34 mm. At the caudal pontine level a very dense fiber network is seen in the parabrachial nucleus (especially its dorsal parts; *PB*) and in the lateral parts of the periventricular central gray, including locus coeruleus (*LC*), with somewhat lower densities in the lateral dorsal (*LDT*) and dorsal tegmental (*DT*) nuclei. The peripheral aspects of the inferior collicle (*IC*) have a moderate to high density, as have the dorsal (*DR*) and magnocellular (*RMg*) raphe nuclei. The pontine reticular area, the motor trigeminal nucleus (*Mo5*) and the superior olive (*SO*) have only low number of fibers, and the cerebellum (*Cb*) at this level virtually lacks SP-positive fibers. *4V*, Fourth ventricle; *py*, pyramidal tract. **B, E** Bregma 6.00 mm. Dense SP-positive networks are seen in the reticular formation (*Rt*), moderate to dense networks in the RMg and in the gigantocellular reticular nucleus alpha (*Gi*) extending dorsally in the midline. The facial nucleus (*7N*) also has a dense innervation. The vestibular complex (*Ve*) has a sparse fiber network as well as some SP-positive cell bodies after colchicine treatment (**E**, *boxed area* in **B**). **C** Bregma 7.56 mm. A very high density of

3.1.3
Fiber Tracts

Weakly fluorescent, often densely packed SP-ir thin, nonvaricose fibers, apparently forming bundles and representing tracts, are observed in several brain regions (Ljungdahl et al. 1978). They include the following tracts: the striohypothalamic tract, stria terminalis, stria medullaris, zona incerta, capsula interna, fasciculus retroflexus, and the spinalis trigeminal tract. The exact origin and termination of these presumable tracts were at that time difficult to estimate, but using modern tracing techniques more exact information has been obtained, as discussed below.

3.1.4
Identified SP/NKA and NKB Pathways

The early work on pathways was summarized by Cuello et al. (1982a; see also Ribeiro-da-Silva and Hökfelt 2000).

Primarily Descending Projections. The first major SP pathway to be identified was the habenulo-interpeduncular pathway (Hong et al. 1976; Mroz et al. 1976; Emson et al. 1977; Hökfelt et al. 1977; Hamill and Jacobowitz 1984; Artymyshyn and Murray 1985; Groenewegen et al. 1986; Contestabile et al. 1987), in agreement with the many SP-ir cell bodies present in the medial habenula and the very dense innervation of the lateral interpeduncular nucleus. Medial habenular SP neurons also project to the medial part of the lateral habenular nucleus (Shinoda et al. 1984) and to the dorsal raphe nucleus (Neckers et al. 1979). SP fibers in the lateral part of the lateral habenular nucleus originate in the rostral entopeduncular nucleus and the adjacent area (Shinoda et al. 1984).

A powerful SP pathway from the striatum to the substantia nigra zona compacta was also rapidly identified and its relation/identity to the well established parallel GABA pathway was demonstrated (Brownstein et al. 1977; Gale et al. 1977; Hong et al. 1977; Kanazawa et al. 1977; Mroz et al. 1977; Jessell et al. 1978; McLean et al. 1985a, 1985b; Christensson-Nylander et al. 1986; Bolam and Smith 1990; Graybiel 1990; Gerfen 1992). Also in this case the projection area, substan-

---

fibers is seen in the solitary tract (*Sol*) and dorsal vagal motor (*10N*) nuclei extending laterally into the dorsal medullary reticular nucleus (*MdD*). The hypoglossal nucleus (*12N*) has a moderate to high density. Moderate fiber densities are seen in the raphe regions. Note high density in the lateral tip of the subnucleus A of the inferior olive (*IO*). At this level the spinal trigeminal nucleus (*Sp5*) has a limited innervation. **D** Bregma 7.92 mm. Also at this level, Sol and 10N have a very high density extending laterally into the MdD continuing into the superficial layer of the caudal part of the Sp5. **F** Bregma 7.56 mm. SP-positive cells (*arrows*) are seen in the nucleus raphe obscurus and pallidus and in the adjacent region; *pyx*, crossing of the pyramidal tract. *Scale bars*, 200 μm (**B–D**) and 100 μm (**E, F**)

tia nigra zona reticulata, had a very high fiber density, but SP-ir cell bodies in the striatum were very difficult to visualize (Ljungdahl et al. 1978), and the extent of the SP cell bodies in the striatum was perhaps only recognized when the in situ hybridization results came along. Kohno et al. (1984) described a topographic organization, whereby SP-positive neurons in the lateral, ventral part of the anterior caudate putamen project to the pars compacta and pars reticulata, whereas the posterior portion projects to the pars lateralis. Striatonigral SP neurons send collaterals to the striatum, presumably to NK_1 receptor containing neurons (Lee et al. 1997), to the entopeduncular nucleus (Paxinos et al. 1978; Kanazawa et al. 1980), to the cholinergic basal forebrain neurons (Henderson 1997), to the globus pallidus (see Reiner and Anderson 1990) and to the ventral pallidum and nucleus accumbens (Napier et al. 1995). The striatonigral neurons contain both SP and NKA (Lee et al. 1986). NKB neurons are widely distributed in the caudate-putamen, and almost 60% of these neurons have SP mRNA (Burgunder and Young 1989). The NKB neurons project to the globus pallidus but not to the substantia nigra (Burgunder and Young 1989), and to the substantia innominata (Furuta et al. 2000). A 'parallel' pathway is represented by SP neurons in the nucleus accumbens projecting to the ventral pallidum and ventral tegmental area (Lu et al. 1998).

SP neurons in the lateral bed nucleus of the stria terminalis and in the central amygdaloid nucleus project to the dorsal vagal complex (Gray and Magnuson 1987), and SP neurons in the lateral dorsal tegmentum project to the dorsal raphe nucleus (Sim and Joseph 1992). Cells in the midbrain periaqueductal gray project to the solitary tract nucleus (Li et al. 1992) and to the magnocellular reticular nucleus pars alpha (Zeng et al. 1991).

There are several supraspinal descending pathways. Thus, SP cell bodies in the midline periaqueductal central gray (Edinger-Westphal nucleus) project to the spinal cord (Skirboll et al. 1983). A major SP cell group in the medullary raphe nuclei and adjacent structures projects to the spinal cord as shown both with immunohistochemistry (Hökfelt et al. 1978) and RIA (Gilbert et al. 1982). This bulbospinal projection has been conclusively demonstrated with retrograde tracing studies (Bowker et al. 1981; Helke et al. 1982; Charlton and Helke 1987). It also contains 5-hydroxytryptamine (5-HT) and is present in many species (see below).

Primarily Ascending Projections. Fibers in the cortex, in particular cingulate and prefrontal cortex, were thought to originate caudal to the mesencephalon (Paxinos et al. 1978), and are now known to arise, at least in part, from the lateral dorsal tegmental nucleus (Sakanaka et al. 1983; Vincent et al. 1983). Tachykinins in the hippocampal formation may originate in the septum (Vincent and McGeer 1981), whereby cells in the lateral region of the medial septum project to a portion of CA2/3a fields (Peterson and Shurlow 1992). Cells in the supramammillary area also project to the hippocampus (Borhegyi and Leranth 1997). Sakanaka et al. (1981, 1982b) have demonstrated a SP pathway from the lateral dorsal tegmental nucleus and from an area between the anterior hypothalamic nucleus and the lateral hypothalamus to the lateral septum, but the dense innervation of the lateral septal area also derives from several other hypothalamic and tegmental nuclei as well as intrinsic septal neurons (Szeidemann et al. 1995). SP neurons project from the solitary tract nucleus to the parabrachial nucleus (Milner et al. 1984; Riche et al. 1990) and from the parabrachial area to the central amygdaloid nucleus (Yamano et al. 1988; Block et al. 1989). The nucleus accumbens receives bilateral projections from the solitary tract nucleus (Li et al. 1990a) and from the periaqueductal gray, with an ipsilateral dominance (Li et al. 1990b). The hypothalamus receives SP inputs from many regions, such as the ventromedial hypothalamic nucleus, from the ventrolateral subnucleus and ventral portion of the medial subnucleus of the periaqueductal gray (Li et al. 1991), and the paraventricular and supraoptic nuclei from cell bodies in the lower brain stem (Bittencourt et al. 1991). The dense SP fiber network in the preoptic area seems to originate in the ventral lateral part of the anterior hypothalamic nucleus (Takatsuki et al. 1983a, 1983b). Another short intrahypothalamic projection runs from the ventromedial nucleus to the medial preoptic area (Yamano et al. 1986). There is a long ascending SP pathway that projects from the spinal cord to the thalamus (Nahin 1987; Battaglia and Rustioni 1992).

Other Projections. The nucleus raphe magnus receives SP afferents from widespread areas including the paragigantocellular reticular nucleus, cuneiform nucleus, solitary tract nucleus, trigeminal subdivision of the lateral reticular nucleus, superior central raphe nucleus and the oral pontine nucleus (Beitz 1982). Takasuki et al. (1983a, 1983b) have described pathways from the ventral reticular formation to the lateral lemniscus and lateral parabrachial area. SP nerve terminals in the facial nucleus originate from the contralateral dorsal medial part of the medullary reticular formation (Senba and Tohyama 1985). Bennett-Clarke et al. (1989) have reported that neurons in the superior colliculus are the sole source of SP fibers in the parabigeminal nucleus. The SP input to the cerebellum has been described by Inagaki et al. (1982a).

3.2
The Human Brain

The distribution of SP has been extensively studied in the human brain. Most studies focus on particular regions, but there are RIA (Ghatei et al. 1984), immunohistochemical (Del Fiacco et al. 1984; Mai et al. 1986; Bouras et al. 1986a, 1986b; Nomura et al. 1987) and in situ hybridization (Figs. 5, 6) (Hurd et al. 1999) reports that cover major parts of the brain, showing marked regional distribution patterns, often similar to those seen in the rat. For example, particularly dense innervations are found in the substantia nigra, the superficial portions of the colliculus inferior, the central gray matter, the parabrachial area, the dorsal tegmental nucleus of Gudden, the reticular tegmental pontine nucleus, the solitary tract nucleus, the dorsal motor nucleus of the vagus nerve and the spinal trigeminal tract nucleus.

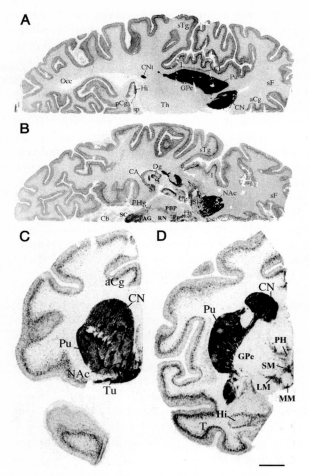

Fig. 5A–D Distribution of preprotachykinin-A (*PPT-A*) mRNA expression in whole human hemisphere, horizontal cryosections showing two dorsoventral levels (**A**, 71 mm; **B**, 86 mm) and of whole hemisphere, coronal sections of the cynomolgus monkey brain (**C, D**) at two different rostrocaudal levels, (**C**) showing the most rostral one. **A, B** Note very strong signal in caudate nucleus (*CN*) and putamen (*Pu*) as well as in nucleus accumbens (*Nac*), strong signals in hippocampus (*Hi*) and dentate gyrus (*Dg*), superior colliculus (*SC*), periaqueductal gray (*PAG*) and some hypothalamic regions (*Hy*). Cortical areas also have a distinct signal, especially in deeper layers. *aCg*, Anterior cingulate gyrus; *Occ*, occipital cortex; *Cnt*, tail of caudate nucleus; *pCg*, posterior cingulate gyrus; *sp*, splenum; *Th*, thalamus; *GPe*, external globus pallidus; *Cl*, claustrum; *sTg*, superior temporal gyrus; *sF*, superior frontal cortex; *I*, insula; *CA*, cornu Ammonis; *Cb*, cerebellum; *PHg*, parahippocampal gyrus; *PBP*, parabrachial pigmental nucleus; *RN*, red nucleus; *Pn*, paranigralis; *SI*, substantia innominata; *Ug*, uncal gyrus. **C, D** Similarly, in the monkey brain the strongest signal is seen in the CN and Pu as well as NAc and tuberculum olfactorium (*Tu*). Also, a distinct signal is encountered in several hypothalamic regions as well as in Hi and temporal cortex (*T*). *Acg*, anterior cingulate gyrus; *PH*, posterior hypothalamus; *LM*, lateral mammillary nucleus; *MM*, medial mammillary nucleus. *Scale bars*, 20 mm (**A, B**) and 5 mm (**C, D**). (From Hurd et al. 1999, with permission)

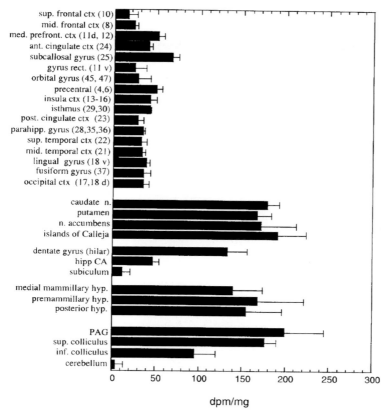

Fig. 6 PPT-A mRNA measured by in situ hybridization in the human brain. The *bar graphs* represent disintegrations per minute per milligram (dpm/mg values, subtracted from background white matter levels) measured within the total specified brain area in whole human hemisphere sections from four subjects. *Numbers in parenthesis* correspond to Brodmann cerebral cortex nomenclature; *d*, dorsal portion of the specified brain region; *v*, ventral portion of the specified brain region. For abbreviations, see Legend to Fig. 5. (From Hurd et al. 1999, with permission)

3.2.1
Telencephalon (Without Basal Ganglia)

In general SP seems to be more abundant in human cortical (Hökfelt et al. 1976; Sakamoto et al. 1985; Taquet et al. 1988; Jansen et al. 1991; Hurd et al. 1999) and hippocampal (Del Fiacco et al. 1987; Lotstra et al. 1988; Hurd et al. 1999) areas than observed in the rat, but the adult human cerebellum is low in SP as in other species (Del Fiacco et al. 1988). Biochemical analysis of cortex revealed a decrease in SP gradient from the frontal to the occipital pole (Ghatei et al. 1984). In the olfactory bulb SP fibers were seen in all layers, and SP-positive multipolar neurons were found in the anterior olfactory nucleus (Smith et al. 1993). Higher

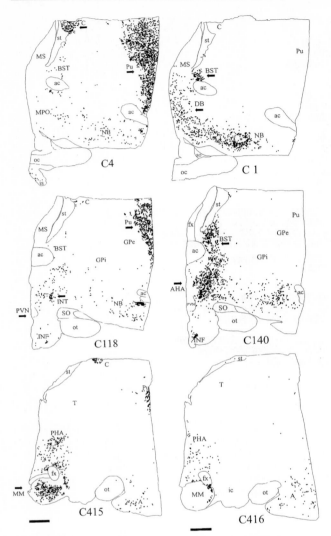

Fig. 7 Coronal maps of neurons containing SP mRNA (C4, C118, C415, left) and NKB mRNA (C1, C140, C416, right). The maps are plotted at successive anterior-to-posterior intervals with the section number at the bottom of each map. Each *symbol* represents one labeled neuron. Note the anteroposterior gradients with high number of SPmRNA-positive cells in the putamen (*Pu*), posterior hypothalamic area (*PHA*) and medial mammillary area (*MM*) and of NKB mRNA in the nucleus basalis/diagonal band nucleus (*NB/DB*), bed nucleus of the stria terminalis (*BST*) and anterior hypothalamic area (*AHA*). A, Amygdala; *ac*, anterior commissure; *AHA*, anterior hypothalamic area; *BST*, bed nucleus of the stria terminalis; *C*, caudate; *fx*, fornix; *GPe*, globus pallidus, external segment; *GPi*, globus pallidus, internal segment; *ic*, internal capsule; *INF*, infundibular nucleus; *INT*, intermediate nucleus; *is*, infundibular stalk; *lv*, lateral ventricle; *MPO*, medial preoptic area; *MS*, medial septal nuclei; *mt*, mammillothalamic tract; *oc*, optic chiasm; *ot*, optic tract; *ovlt*, organum vasculosum of the lamina terminalis; *PM*, premammillary nucleus; *POG*, paraolfactory gyrus; *PVN*, paraventricular nucleus; *SO*, supraoptic nucleus; *st*, stria terminalis; *T*, thalamus; *VMN*, ventromedial nucleus. *Scale bar*, 3 mm. (From Chawla et al. 1997, with permission)

numbers of SP-ir cell bodies were found in the deep layers of the subicular complex and entorhinal cortex, as well as in the stratum oriens and hilus of the dentate gyrus, with nerve terminals widely and unevenly distributed, being particularly abundant in the pyramidal CA2 and CA3 subfields and the dentate granule cell layer.

The basal forebrain has been analyzed with immunohistochemistry (Haber and Watson 1985) and in situ hybridization (Fig. 7) (Chawla et al. 1997). In the latter study many NKB, but few SP mRNA neurons were identified in the magnocellular basal forebrain. In the bed nucleus of the stria terminalis SP was mainly found in fibers in the lateral and medial subdivisions (Walter et al. 1991), with many NKB and few SP mRNA-positive cells (Chawla et al. 1997).

3.2.2
Basal Ganglia

Particular interest has been paid to the human basal ganglia which have been analyzed both with immunohistochemistry and in situ hybridization. Nuclease protection analysis for human PPT mRNA showed mainly β-PPT-A (β-*TAC1*; 80–85%), the remaining being γ-PPT-A (γ-*TAC1*) but no α-PPT-A (α-*TAC1*) (Bannon et al. 1992). The immunohistochemical analysis revealed that in the dorsal caudate nucleus SP followed the classical striosome/matrix organization, whereas this was less obvious in the ventromedial caudate nucleus/nucleus accumbens regions (Beach and McGeer 1984; Manley et al. 1994; Holt et al. 1997). PPT-A mRNA was present in clusters of cell bodies more prominent in the caudate nucleus than in putamen, mostly confined to the striosomes, and with a more even distribution in the nucleus accumbens (Fig. 5 A, B) (Chesselet and Affolter 1987; Hurd and Herkenham 1995; Hurd et al. 1999). In the substantia nigra dense SP-positive fibers were found surrounding the dopamine cell bodies and dendrites (Haber and Groenewegen 1989; Marksteiner et al. 1994; Reiner et al. 1999; Sutoo et al. 1999).

3.2.3
Diencephalon

This region has been studied by several groups. In the thalamus no immunoreactive cell bodies were seen, but fibers were found in most intralaminar nuclei, the posterior complex and tentatively in the nucleus submedius (Hirai and Jones 1989). In the hypothalamus RIA showed high SP concentrations in the pituitary stalk and the mediobasal hypothalamus (Helme and Thomas 1986). Immunohistochemistry showed cell bodies in the median eminence, lateral hypothalamus, ventromedial nucleus, periventricular, and perifornical areas, whereby SP terminals appose luetinizing hormone-releasing hormone neurons (Helme and Thomas 1986; Dudas and Merchenthaler 2002). Chawla et al. (1997) have shown extensive, mostly non-overlapping, complementary cell groups, expressing, respectively, SP mRNA and NKB mRNA, the former being highly concentrated in

Table 2 Distribution of cells containing substance P and neurokinin B mRNAs in the human hypothalamus and adjacent structures (modified after Chawla et al. 1997)

Region	Structure	SP	NKB
Basal forebrain	Medial septum	–	+
	Diagonal band Broca	+	+++
	Nucleus basalis of Meynert	+	+++
Chiasmatic	Supraoptic nucleus, dorsolateral	–	–
	Supraoptic nucleus, ventromedial	–	–
	Medial preoptic area	+	++
	Anterior hypothalamic area	+	++
	Intermediate nucleus	++	+
	Suprachiasmatic nucleus	–	–
	Paraventricular nucleus	–	+
	Lamina terminalis (LT)	–	–
	Organum vasculosum of LT	–	–
Tuberal	Dorsomedial nucleus	+	–
	Ventromedial nucleus	++	–
	Infundibular nucleus	++	++
	Premammillary nucleus	++	–
	Lateral tuberal nucleus	–	–
	Tuberomammillary nucleus	–	–
Mammillary	Medial mammillary nucleus	+++	+
	Supramammillary nucleus	+++	–
	Lateral mammillary nucleus	–	–
	Posterior hypothalamic area	++	+
Adjacent regions	Central nucleus of the amygdala	+	++
	Bed nucleus of stria terminalis	+	++
	Caudate nucleus	+++	–
	Putamen	+++	–
	Globus pallidus	–	–
	Paraolfactory gyrus	+	+

Density of cells in an area or nucleus: –, none; +, <5% of cells; ++, 5–30%; +++, >30%.
SP, substance P; NKB, neurokinin B.

the posterior hypothalamus, e.g., the mammillary complex and posterior hypothalamic area, and NKB mRNA mainly in the anterior hypothalamic area (Fig. 7, Table 2).

3.2.4
Lower Brain Stem

Several studies have focused on the lower brain stem. In the dorsal vagal complex and the solitary tract nucleus extensive and highly regional fiber plexuses were observed, and cell bodies were most frequent in the nucleus gelatinosus with lower numbers in the medial and intermediate subnuclei of the solitary tract nucleus (Fodor et al. 1994; Marksteiner et al. 1994; McRitchie and Törk 1994). The parabrachial nucleus was rich in SP fibers (Fodor et al. 1992), whereas only few SP fibers were found in the locus coeruleus (Pammer et al. 1990). In a series of three impressive papers the Adelaide (Australia) group analyzed the distribution of SP in the lower brain stem, including quantitative data and relations to monoamine systems (Halliday et al. 1988, 1990; Baker et al. 1991). In the lower brain stem the majority of SP immunoreactive neurons with varying morphology was found in four main regions: the lateral medulla, the dorsal medial medulla, the spinal trigeminal nucleus and the raphe nuclei (Halliday et al. 1988). These authors found 90,000 SP neurons throughout the adult human medulla oblongata with distribution patterns significantly different from those in the rat and cat. Many SP-positive neurons in the raphe region also contained 5-HT. In a subsequent study Halliday et al. (1990) counted almost 200,000 neurons in the tegmentum, concentrated in three main regions, the mesencephalic ventricular formation, the central gray and the pontine reticular formation. Two types of neurons were found, large neurons mainly in the caudal midbrain, and small ones located more rostrally, and many were presumably cholinergic. In the third study the dorsal raphe nuclei were analyzed and found to contain around 75,000 SP neurons, most of them in the rostral part and many coexpressing 5-HT (Baker et al. 1991).

3.3
Monkey Brain

Several species have been analyzed including rhesus, macaque, and squirrel monkeys. These studies deal with: (1) multiple brain areas (Fig. 5C, D) (Hurd et al. 1999); (2) olfactory bulb (Baker 1986; Sanides-Kohlrausch and Wahle 1991); (3) cortical and hippocampal areas using immunohistochemistry (Amaral and Campbell 1986; Hayashi and Oshima 1986; Hendry et al. 1988; Jones et al. 1988; Iritani et al. 1989; Carboni et al. 1990; Ong and Garey 1990; Beal et al. 1991; Nitsch and Leranth 1994; Jakab et al. 1996, 1997; Seress and Leranth 1996), in

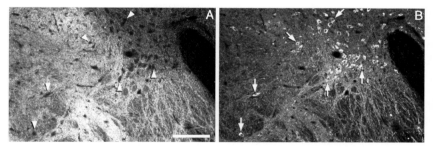

Fig. 8A, B Immunofluorescence micrographs of a coronal section of a normal, formalin–picric acid perfused monkey brain, double labeled with antisera against SP (**A**) and tyrosine hydroxylase (*TH*) (**B**). A strong network of fibers immunoreactive for SP is seen throughout the region of the substantia nigra (**A**) and an overlapping dense network of TH-ir dendrites is seen ventral to the TH-ir cells in the substantia nigra pars reticulata. *Arrows* in **B** indicate TH-ir cells in the substantia nigra zona compacta surrounded by more or less dense SP-ir fibers (corresponding *arrowheads* in **A**). Scale bars, 200 μm (for **A** and **B**)

situ hybridization (Fig. 5C, D) and RIA (Hayashi and Oshima 1986); (4) the caudate putamen/striatum and nucleus accumbens (Fig. 5C, D) (Haber et al. 1990; Martin et al. 1991; Bennett and Bolam 1994; Ikemoto et al. 1995; Hurd et al. 1999); (5) globus pallidus and substantia nigra (Fig. 8A, B) (Haber and Elde 1981; DiFiglia et al. 1982; Reiner et al. 1999); (6) thalamus (Molinari et al. 1987); (7) various hypothalamic regions (Hökfelt et al. 1978; Ronnekleiv et al. 1984; Mick et al. 1992; Nitsch and Leranth 1994); (8) the dorsal raphe nucleus (Charara and Parent 1998); and (9) the solitary tract nucleus (Maley et al. 1987).

3.4
Other Species

Gallager et al. (1992) analyzed the guinea pig brain and found that SP was in general distributed in a similar way as in the rat. The major difference was found in forebrain regions, where the guinea pig cortex contained many more SP-ir cells and fibers in layers II–VI and the hippocampus contained particularly strong staining in the CA1–3 pyramidal cell layers and in the dentate gyrus, with SP immunoreactive cells in the hilus region. Gall and Selawski (1984) reported that most SP fibers in the guinea pig hippocampus, including a dense supragranular plexus, originate in the supramammillary region (see also Drake et al. 1997). Other areas analyzed for SP in the guinea pig are the olfactory bulb (Baker 1986; Matsutani et al. 1989), hypothalamus (Nielsen and Blaustein 1990; Airaksinen et al. 1992; Ricciardi and Blaustein 1994; Dufourny et al. 1998, 1999), midbrain central gray (Turcotte and Blaustein 1997) and medulla oblongata (Chiba and Masuko 1989).

No complete mapping of TK systems has, to our knowledge, been performed on the cat brain, although many studies have dealt with discrete regions, e.g., olfactory systems (Wahle and Meyer 1986; Baker 1986; Wahle et al. 1990), visual cortex (Gu et al. 1994), the basal forebrain (Rieck et al. 1995), striatum (Graybiel et al. 1981; Beckstead and Kersey 1985; Bolam et al. 1988), substantia nigra (Inagaki and Parent 1984), interpeduncular nucleus (Kapadia and de Lanerolle 1984a, 1984b), diencephalon (Sugimoto et al. 1984; Burgos et al. 1988; Yanagihara and Niimi 1989; Battaglia and Rustioni 1992; Battaglia et al. 1992; Velasco et al. 1993), superior colliculus (Hutsler and Chalupa 1991), pontine tegmentum (Leger et al. 1983), solitary tract nucleus (Maley et al. 1983), periaqueductal gray (Moss and Basbaum 1983), lower brain stem (Triepel et al. 1985), parabrachial nucleus (Blomqvist and Mackerlova 1995), and ventral medulla oblongata (Ciriello et al. 1988; Holtman 1988; Pretel and Ruda 1988); including the finding of SP in pyramidal, presumably glutamatergic cortical neurons (Conti et al. 1992a, 1992b).

In the mouse, SP has been studied in the hippocampal formation and cortex (Diez et al. 2000), striatum (Ariano et al. 2002; Sun et al. 2002), suprachiasmatic nucleus/geniculate body (Silver et al. 1999; Abrahamson and Moore 2001; Piggins et al. 2001) and inferior olive (Gregg and Bishop 1997).

In the rabbit, SP has been analyzed in the olfactory bulb (Baker 1986), hypothalamus (Caballero-Bleda et al. 1993) and brain stem (Gingras et al. 1995; Kyrkanides et al. 2002).

In the hamster, SP is expressed in the olfactory bulb (Davis et al. 1982; Baker 1986; Kosaka et al. 1988), subcortical visual centers (Hartwich et al. 1994), hypothalamus, bed nucleus of stria terminalis and amygdala (Neal et al. 1989; Neal and Newman 1991; Debeljuk et al. 1992; Morin et al. 1992; Piggins et al. 2001), subcommissural organ (Nurnberger and Schoniger 2001) and medulla oblongata (Hancock 1984; Davis and Kream 1993).

In the ferret, SP neurons have been studied in the caudate nucleus (Izzo et al. 1987; Bolam et al. 1988) and lower medulla oblongata (Boissonade et al. 1993, 1996).

3.5
Coexistence of SP and Classical Transmitters

As most other neuropeptides, SP is almost always present in neurons that also contain a classical transmitter (see Hökfelt 1991), also often together with one or more other additional peptides. We will not deal with the latter situation that occurs very frequently, but only briefly mention cell bodies of some systems that contain both SP and one (or perhaps even two) classical transmitters in the mammalian brain. The earliest established coexistence for SP with a classical transmitter was in the rat medullary raphe nuclei (raphe magnus, pallidus and obscurus), and adjacent areas, where many neurons express both SP and serotonin (Chan-Palay et al. 1978; Hökfelt et al. 1978; Gilbert et al. 1982; Helke et al. 1982; Kachidian et al. 1991; Nevin et al. 1994). A quantitative evaluation along the rostrocaudal axis was presented by Johansson et al. (1981). These neurons also contain NKA (Nevin et al. 1994). The notion that these neurons also may be glutamatergic (Kaneko et al. 1990; Nicholas et al. 1992; Gras et al. 2002; Schäfer et al. 2002) is supported by functional studies (Johnson 1994; Li and Bayliss 1998) (see Hökfelt et al. 2000). This SP–5-HT coexistence can also be found in the corresponding medullary neuronal cell bodies in other species, including guinea pig (Chiba and Masuko 1989), hamster (Hancock 1984), cat (Marson and Loewy 1985; Ciriello et al. 1988; Arvidsson et al. 1994), monkey (Nicholas et al. 1992), and man (Halliday et al. 1988). In contrast, it has been thought that the rat dorsal raphe 5-HT neurons do not express SP, but some studies report that around 10% of these 5-HT neurons, especially in the ventrolateral part of the periaqueductal gray and dorsal raphe nucleus contain SP (Magoul et al. 1986; Ma and Bleasdale 2002). In the monkey (Charara and Parent 1998) and human (Baker et al. 1991; Sergeyev et al. 1999) coexistence of SP and 5-HT in the raphe complex is more common (Fig. 9A–C). In another rat bulbospinal system (Lorenz et al. 1985) and in an ascending system (Bittencourt et al. 1991) SP coexists with adrenaline. A major ascending rat cholinergic system originates in the peduncular pontine nucleus and coexpresses SP (Vincent et al. 1983, 1986). A few dopamine neurons in the rat ventral tegmen-

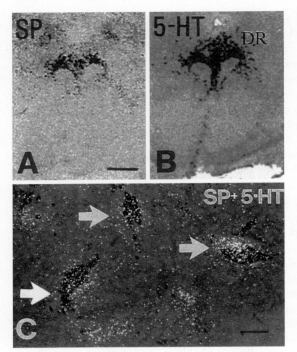

Fig. 9A–C Film autoradiographs of two adjacent sections of the human dorsal raphe showing transcripts for SP (**A**) and 5-HT transporter (**B**), as well as combined bright and darkfield micrographs (**C**) of a single section of the human dorsal raphe nucleus after double hybridization with probes complementary to SP mRNA labeled with digoxigenin (*dark precipitates*) or radioactively labeled 5-HT transporter (*bright grains*). Note overlap of the two transcripts seen on the film autoradiographs (compare **A** with **B**) as well as several double-labeled cells (*arrows* in **C**). *Scale bars*, 1 mm (**A, B**) and 25 μm (**C**). (From Sergeyev et al. 1999, with permission)

tal area contain SP (Seroogy et al. 1988). In the rat entopeduncular nucleus SP coexists with GABA (Murakami et al. 1989), and in the rat caudate nucleus virtually all projection neurons are GABAergic (Kita and Kitai 1988), and therefore most SP neurons in the striatum are GABAergic, projecting to the substantia nigra zona reticulata as well as to the globus pallidus, as shown for the cat (Beckstead and Kersey 1985; Penny et al. 1986a) and rat (Penny et al. 1986a; Gerfen and Young 1988; Reiner and Anderson 1990). Also in the rat neocortex GABA and SP coexist (Penny et al. 1986b). Finally, GABA and SP coexist in the hamster main olfactory bulb (Kosaka et al. 1988).

4
Distribution of Tachykinin Receptors

4.1
Rat Brain

4.1.1
Overview

Early binding studies revealed the existence of subtypes of TK receptors (Teichberg et al. 1981; Lee et al. 1982). Subsequently three TK receptors were cloned, and called NK_1, NK_2 and NK_3 receptors, the first one by Nakanishi and collaborators (Masu et al. 1987; Sasai and Nakanishi 1989; Yokota et al. 1989; Hershey and Krause 1990; Shigemoto et al. 1990). They are preferred receptors for the endogenous ligands SP, NKA and NKB, respectively. Mapping studies on the localization of the receptors were carried out with tritiated and iodinated ligands (Quirion et al. 1983; Rothman et al. 1984; Beaujouan et al. 1986; Buck et al. 1986; Danks et al. 1986; Mantyh et al. 1989). Over the years selective ligands for the three receptors have been developed, allowing demonstration of regional distribution patterns of the three receptors with ligand binding autoradiography, the latest comparative study on the rat being recently published by Saffroy et al. (2003) (Fig. 10A–D), which will be discussed briefly below.

The cloning also opened up the possibility to study the localization of the receptor synthesizing cell bodies, first with RNase protection analysis (Tsuchida et al. 1990). Subsequently the cellular distribution patterns of the NK_1 transcript in the whole brain were published by Maeno et al. (1993) (Fig. 11, left column) and of the NK_3 transcript by Shughrue et al. (1996) (Fig. 11, right column).

Of perhaps even more importance, it also became possible to produce antibodies against these receptors and to study their localization with immunohistochemistry. Thus complete mappings of the NK_1 receptor protein in the CNS have been published by Nakaya et al. (1994) and Ribeiro-da-Silva et al. (2000), and for the NK_3 receptor by Ding et al. (1996) and Mileusnic et al. (1999). Some of these papers contain extensive tables showing the localization of the receptors with semiquantitative evaluations as well as corresponding data for the respective endogenous ligands, that is SP/NKA and NKB. Both receptors are widely spread in the nervous system with distinct and discrete, often complementary distribution patterns (Figs. 12–14). Antibodies against the tachykinin receptors have also allowed interesting histochemical studies on internalization of both the NK_1 and NK_3 receptor (see below).

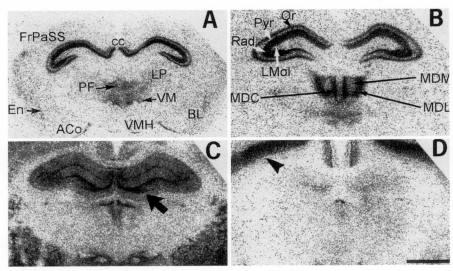

Fig. 10A–D Autoradiographic localization of NK_2-sensitive [^{125}I]-labeled NKA (A, B), NK_1 [^{125}I]-labeled Bolton–Hunter substance P (**C**) and NK_3 [^{125}I]-labeled Bolton–Hunter eledoisin (**D**) binding sites in coronal sections of rat brain (**A** shows whole brain, **B–D** hippocampus and thalamus) processed according to the autoradiographic ligand binding technique and exposed to tritium-sensitive hyperfilm (Leica). For details see Saffroy et al. (2003). Note that NK_2 receptor binding sites are seen mainly in the hippocampus and medial thalamic structures (**A, B**). Also NK_1 receptor binding sites are seen in the hippocampus (**C**), whereas NK_3 receptor binding is very low in these structures (**D**). Note strong NK_3 receptor binding in deep cortical layers (*arrowhead* in **D**). Note high NK_2 receptor binding in stratum oriens (*Or*) and radiatum (*Rad*), while NK_1 receptor binding is strongest in the polymorph and granular layers of the dentate gyrus (*arrow* in **C**). *Scale bar,* 5 mm (**A**) and 2 mm (**B–D**). *ACo,* Anterior cortical amygdaloid nucleus; *BL,* basolateral amygdaloid nucleus; *cc,* corpus callosum; *En,* endopiriform nucleus; *FrPaSS,* frontal parietal cortex, somatosensory, area 2; *LP,* lateral posterior thalamic nucleus; *PF,* parafascicular thalamic nucleus; *VM,* ventromedial thalamic nucleus; *VMH,* ventromedial hypothalamic nucleus; *LMol,* lamina molecularis; *MDC,* medial dorsal thalamic nucleus central part; *MDL,* mediodorsal thalamic nucleus, lateral part; *MDM,* mediodorsal thalamic nucleus, medial part; *Or,* striatum oriens; *Pyr,* pyramidal cell layer; *Rad,* striatum radiatum. (From Saffroy et al. 2003, with permission)

Fig. 11 Schematic representation of coronal sections, from rostral to caudal, showing the distribution of NK_1 receptor mRNA (*left;* from Maeno et al. 1993) and NK_3 receptor mRNA (*right;* from Shughrue et al. 1996) in the rat brain. In the diagrams on the *left, large dots* show positive cells with a strong signal, while *small dots* indicate cells with a moderate to weak signal. In the diagrams on the *right, small dots* indicate five labeled cells, *large dots* 30 or more labeled cells. Note differential distribution of the two transcripts, the NK_3 receptor mRNA having a particular strong expression in cortical areas including hippocampus (*HIF*), amygdaloid nuclei and cerebellum. Note also strong NK_3 receptor expression in the substantia nigra zona compacta (*SNC*). NK_1 receptor mRNA is present in the mitral cells, the caudate putamen/nucleus accumbens/olfactory tubercle, as well as in the dentate gyrus and dorsal tegmental nucleus. Both transcripts are present in the central gray at the posterior hypothalamic level, the inferior collicus, the parabrachial nucleus and the dorsal vagal complex. *10,* Dorsal motor nucleus of vagus; *12,* hypoglossal nucleus; *ac,* anterior commissure; *Acb,* accumbens nucleus; *Ahi (Ahia),* amygdalohippocampal area; *Amb,* ambiguus nucleus; *AOB,* accessory olfactory bulb; *AP,* area postrema; *Apir,* amygdalopiri-

The Histochemistry of Tachykinin Systems in the Brain

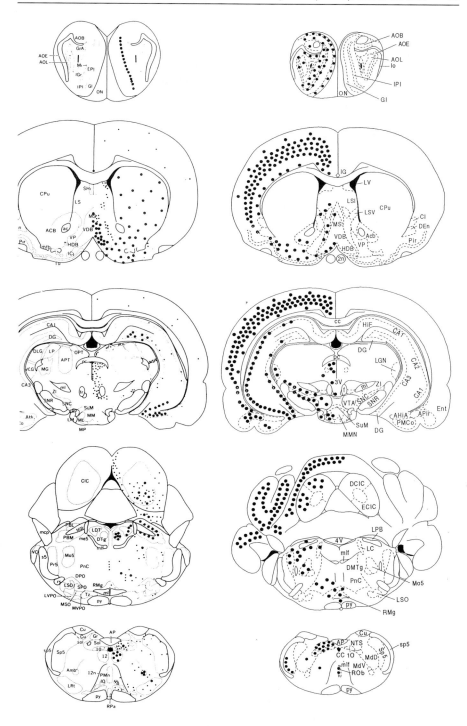

4.1.2
Distribution of NK$_1$ and NK$_3$ Receptors

In the telencephalon NK$_1$ receptor mRNA, as described by Maeno et al. (1993), is present in mitral cells in the olfactory bulb, in some cortical neurons and in the granule cell layer of the dentate gyrus. A lower number of cells is found in the hippocampal pyramidal cell layers. Many NK$_1$-positive cell bodies are present in the ventral forebrain, ventral and horizontal limbs of the diagonal band nucleus, olfactory tubercle and striatum. High numbers are present in the medial amygdaloid nucleus and the amygdalo-hippocampal area extending into the ventral CA1 region. In the diencephalon many cells are seen in the hypothalamic paraventricular nucleus extending into the zona incerta and to the midline posterior thalamic nuclei, continuing into the periventricular periaqueductal central gray. In the lower brain stem numerous cell bodies are seen in the superior and inferior colliculi as well as in the parabrachial nucleus, locus coeruleus and in the lateral and the lateral dorsal tegmental nuclei. At more caudal levels cells are seen in the medullary reticular formation and in the nucleus ambiguus and the hypoglossal nucleus.

These findings by Maeno et al. (1993) are in good agreement with the NK$_1$ receptor protein distribution showing a very strong expression in cell bodies and dendritic processes (Nakaya et al. 1994), as summarized in the following. Areas with a high density of NK$_1$-LI include the glomerular and external plexiform layer in the olfactory bulb, the accessory olfactory bulb, the bed nucleus of the anterior olfactory tract, the hilus of the dentate gyrus, the medial and dorsal part of the lateral septal nucleus, the nucleus of the diagonal band, the septofimbrio nucleus, parts of the bed nucleus of the stria terminalis, the medial preoptic area, the medial forebrain bundle area, the cortical amygdaloid nucleus and parts of the central nucleus as well as the amygdala hippocampal area, the medial and lateral habenular nuclei, the anterodorsal thalamic nucleus, the parafascicular nucleus, parts of the zona incerta, the posterior hypothalamic nucleus and lateral part of the medial mammillary nucleus, the later part of the supramammillary nucleus, many regions of the colliculi, the dorsal and ventral parts of the parabigeminal nucleus, the periaqueductal gray and the dorsal raphe nucleus, the lateral parabrachial nucleus, the subpeduncular tegmental nucleus, the sphenoid nucleus and locus coeruleus, the rostral ambiguus nucleus, the solitary tract nucleus, the hypoglossal nucleus and the medial accessory nucleus of the inferior olive and lamina I of the spinal trigeminal nucleus.

The NK$_3$ mRNA transcripts, as described by Shughrue et al. (1996) (Fig. 11F–J), are particularly abundant in cortical areas including the cerebellum (granular layer), the olfactory bulb and hippocampus, with very high numbers in the subiculum, the basal amygdaloid nuclei, and the ventral aspects of the forebrain including the diagonal band nucleus. High numbers of transcript-positive cell bodies are found in the supraoptic and magnocellular paraventricular nuclei, lateral and dorsal hypothalamic areas, perifornical nucleus, medial mammillary nucleus and zona incerta, substantia nigra pars compacta and ventral tegmental area, interpeduncular nucleus, superior colliculus, pontine nucleus, vagal dorsal motor nucleus, solitary tract nucleus and superficial laminae of the spinal trigeminal nucleus (Shughrue et al. 1996). The distribution patterns of NK_3-LI, not described here, are in good agreement (Ding et al. 1996) (see also below).

4.1.3
Comparison Between Distribution of SP, NK_1 and NK_3 Receptors

We have compared, on three adjacent, 50-μm-thick sections, the distribution of NK_1 receptor-LI, SP- and NK_3 receptor-LIs in a limited number of regions (see also above). In the dorsal dentate gyrus a strong NK_1 receptor-LI is seen in the polymorph-granular layers with many processes extending into the molecular layer (Fig. 12A). Here very few SP-positive nerve terminals are seen (Fig. 12B) with a small number of NK_3-positive cells on the border of the granular cell layer (Fig. 12C). In the lateral hippocampus long NK_1 processes traverse the stratum radiatum, but less so in stratum oriens (Fig. 12D); in the latter layer there is the highest density of SP-positive fibers (Fig. 12E), whereas NK_3-LI is strongest in radiatum (Fig. 12F). In the amygdaloid complex a strong NK_1-LI is seen in the medial and cortical nuclei extending into the medial basal nucleus (Figs. 13A, 14A) with a good overlap with SP-LI, especially medially (Figs. 13B, Fig. 14B), whereas NK_3-LI is present mainly deep in the basomedial amygdala and in the bed nucleus of stria terminalis (Fig. 13C). In the central nucleus SP-LI is strong (Fig. 13E), whereas NK_1 receptor-LI (Fig. 13D) and NK_3 receptor-LI (Fig. 13F) are very weak, apparently almost absent. In the hypothalamus, SP-LI is abundant in most places (Fig. 14B, E), except the ventromedial nucleus, NK_3 receptor-LI being present mainly in dorsal lateral and medial perifornical areas (Fig. 14C, F), with some overlap but mainly complementary to NK_1 receptor-LI (Fig. 14A, D).

4.1.4
Distribution of NK_2 Binding Sites

As described by Saffroy et al. (2003) the density of NK_2 binding sites in the rat brain is weak when compared to the NK_1 and NK_3 binding, and only a few structures have specific NK_2-sensitive [^{125}I]-NKA binding sites with a markedly differential distribution. Thus, high labeling is observed in strata oriens and radiata of the CA1 and CA2 fields, the deep layers of some limbic cortices and the lateral parts of the mediodorsal thalamic nucleus (Fig. 10A, B). This pattern is distinctly different from the distribution of NK_1 (Fig. 10C) and NK_3 (Fig. 10D) binding sites. Moderate NK_2 receptor labeling is observed in parts of the nucleus accumbens, the ventral dentate gyrus, the dorsal lateral septal nucleus and some cortical and thalamic regions. Many other areas, including the hypothalamus and the lower brain stem, are devoid of labeling, except the caudal interpeduncular nucleus.

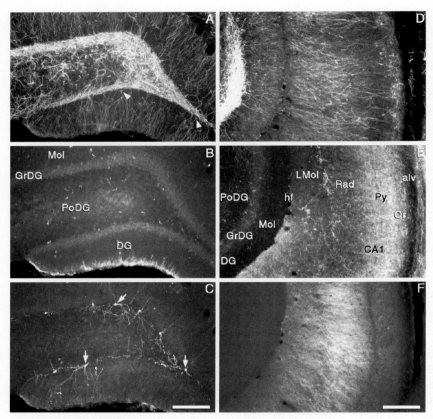

Fig. 12A–F Immunofluorescence micrographs of three adjacent, 50-μm-thick coronal sections of a normal, formalin–picric acid perfused rat brain after free-floating incubation with antiserum against NK_1 (**A**, **D**), SP (**B**, **E**) and NK_3 (**C**, **F**). **A–C** At the level of the rostral dorsal hippocampus (Bregma 3.14 mm; Paxinos and Watson 1986) a dense network of strongly fluorescent NK_1 processes and cells is seen within the polymorph layer of the dentate gyrus (*PoDG*; **A**), and sparsely distributed NK_3-ir cells (*arrows*) are located along the border of the granular layer (*GrDG*), with dendrites in the molecular layer (*Mol*; **C**). *Arrowheads* point to some NK_1-ir cells with dendrites in the Mol. In the GrDG SP-LI is weak. **D–F** In the caudal, ventral aspect of field CA1 of the hippocampus (*CA1*) (Bregma 6.04 mm), a rich network of NK_1-ir cells and processes is seen within the PoDG (**D**), which is very weakly SP-ir (**E**) and lacks NK_3 fibers (**F**). The GrDG and Mol contain NK_1-LI with dendrites in Mol, and a relatively weak immunoreactivity for SP but none for NK_3 receptors. Lateral to the hippocampal fissure (*hf*), the lacunosum moleculare layer of the hippocampus (*LMol*) is dense in NK_1 processes, has moderate SP-LI ventrally, and a low density of weakly fluorescent NK_3 processes. The stratum radiatum (*Rad*) contains both NK_1 cells and a high density of NK_1 and NK_3 processes, the latter possibly dendrites from pyramidal cells (*Py*). In Py and stratum oriens (*Or*) there is strong SP-LI, with a sparse NK_1- and only weak NK_3-LI. *Scale bars*, 200 μm (**A**=**B**=**C**; **D**=**E**=**F**)

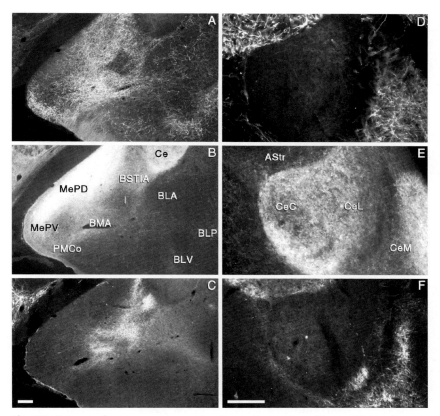

Fig. 13A–F Immunofluorescence micrographs of coronal rat sections (Bregma 3.14 mm) processed as described in Fig. 12 after incubation with antiserum against NK_1 (**A, D**), SP (**B, E**) and NK_3 (**C, F**). **A–C** In the amygdaloid complex, a very dense network of NK_1-ir cells and fibers is seen within the bed nucleus of the striata terminalis, intra-amygdaloid division (*BSTIA*) and the basomedial amygdaloid nucleus, anterior part (*BMA*; **A**), where SP-LI is moderately dense (**B**) and NK_3-LI strong (**C**). The medial amygdaloid nucleus, posterodorsal part (*MePD*) contains very strong SP-LI, along with a high density of NK_1 processes, but relatively few NK_3 processes. A network of strongly fluorescent NK_1 fibers is also located in the basolateral amygdaloid nucleus, posterior (*BLP*) and ventral (*BLV*) parts; however SP- and NK_3-LIs are weak. Within the basolateral amygdaloid nucleus, anterior part (*BLA*), there is a low density of strongly NK_1-ir fibers, a weak NK_3-LI, and apparently no SP-LI. In the intercalated nucleus of the amygdala (*I*), there is strong NK_3 receptor-LI, less immunoreactivity for NK_1 receptors and weak SP-LI. **D–F** In the central amygdaloid nucleus, capsular part (*CeC*) and in its lateral division (*CeL*), SP-LI is strong but NK_1- and NK_3-LIs are virtually absent. However, the medial division (*CeM*) has a dense network of NK_1 (**D**), NK_3 (**F**) and SP (**E**) fibers. *Astr*, amygdalostriatal transition area. *Scale bars*, 200 μm (**A=B=C; D=E=F**)

4.2
Other Species

NK_1 receptors have been analyzed with immunohistochemistry in monkey striatum (Parent et al. 1995; Jakab and Goldman-Rakic 1996; Jakab et al. 1996), and

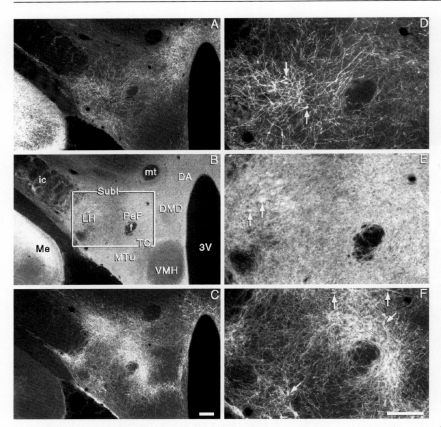

Fig. 14A–F Immunofluorescence micrographs of coronal rat sections (Bregma 3.14 mm) processed as described in Fig. 12 after incubation with antiserum against NK_1 (**A, D**), SP (**B, E**) and NK_3 (**C, F**). **D–F** Higher magnification of **A–C**, respectively, as indicated by *box* in **B**. **A–C** A strong network of NK_1-ir cells and processes is seen within the lateral (*LH*) and dorsal hypothalamic (*DA*) areas (**A**), which both have a dense network of SP-ir (**B**) and moderately dense NK_3-ir (**C**) processes. Intense NK_3-ir cells and fibers, with overlapping immunoreactivity for SP, were observed within the subincertal nucleus (*SubI*), periformical nucleus (*PeF*) and tuber cinereum area (*TC*), while NK_1-LI is weak. The dorsomedial hypothalamic nucleus, dorsal part (*DMD*) contains strong SP-LI, with sparse NK_1-LI in more dorsal regions, and relatively few NK_3 processes throughout. **D–F** There is strong SP-LI in the LH/PeF area together with a higher density of NK_3 than NK_1 cells and processes. NK_1 and NK_3 cells do, however, largely not overlap. *Arrows* in **D–F** point to cells labeled by antiserum for NK_1, SP and NK_3, respectively. The density of NK_1-, SP- or NK_3-ir processes is very low in the mammillothalamic tract (*mt*), fornix (*f*), ventromedial hypothalamic nucleus (*VMH*) and internal capsule (*ic*). *MTu*, medial tuberal nucleus. *3V*, 3rd ventricle. *Scale bars*, 200 μm (**A=B=C; D=E=F**)

a PET study with a ^{11}C-labeled NK_1 receptor antagonist showed the highest specific binding in the striatum (Bergström et al. 2000). NK_1 receptors have also been studied in the human brain with PET (Hargreaves 2002). The NK_1 receptor has been studied in the cat hypothalamus (Echevarria et al. 1997; Yao et al. 1999) and in the mouse diencephalon (Piggins et al. 2001; Liu et al. 2002). The

NK_3 receptor has been mapped in the human brain with immunohistochemistry (Mileusnic et al. 1999).

5
General Remarks

The studies reviewed in this chapter show widespread distribution patterns of all three major TKs (SP/NKA and NKB) and two of the three TK receptors (NK_1 and NK_3). Only autoradiographic ligand binding studies are available for the NK_2 receptor, and they suggest a more limited distribution of this receptor, notably being very rare in hypothalamus and the lower brain stem. However, for a final judgment in situ hybridization and immunohistochemical analyses have to be awaited. This broad presence of TK systems suggests involvement in virtually all type of brain functions. We have here focused on results based on immunohistochemistry and in situ hybridization, the latter being a superior technique to demonstrate cell bodies (but only the very cell soma), in which these molecules are produced. Immunohistochemistry provides information on the distribution of the peptide/receptor protein both in cell bodies (but with a lower sensitivity) and processes, including nerve terminals, axons and dendrites.

5.1
The Histochemical Methods

Over the last 30 years or so immunohistochemical procedures have been continuously improved, and it would be desirable to repeat the early and even later studies on TKs using, for example, the TSA method and free-floating sections. SP cell bodies and fibers have more recently been described in rat cortex, and as shown here this is the case also in mouse cortex, suggesting presence of several SP systems also in rodent neocortical areas. It is also clear that excellent antibodies have been produced against the NK_1 and NK_3 receptors (Shigemoto et al. 1993; Vigna et al. 1994; Ding et al. 1996; Griffond et al. 1997), and indeed very informative (and beautiful) staining patterns have been published with such antisera by these and other authors (see also Figs. 12–14 based on the generously donated antisera mentioned above). A correspondingly powerful antibody to the NK_2 receptor has to our knowledge not been generated, and the distribution patterns of this receptor based on immunohistochemistry are therefore still missing, as are NK_2 transcript mappings.

When comparing results from different studies, including semiquantitative estimations of histochemical distribution patterns, it is important to remember that the thickness of the sections may differ, which is especially important for immunohistochemistry. For example in many studies 10–14-µm-thick sections thawed onto the slides are incubated with antibody (see Figs. 2–5), whereas in others 40–50-µm-thick free-floating sections are processed (see Figs. 12–14). Naturally the thicker sections contain more immunoreactive structures and will show more impressive and quantitatively higher numbers of immunoreactive structures. Also, the free-floating approach is clearly more sensitive, allowing more efficient penetration of antibodies.

5.2
Tachykinin Receptors

Based on histochemical analyses the NK_1 and NK_3 receptors seem to represent mainly postsynaptic receptors, since cell bodies and dendritic processes are preferentially stained. It can, however, not be excluded that the NK_1 and NK_3 receptors are transported into the axon to act as presynaptic receptors. Thus, in studies in our own laboratory on the neuropeptide Y (NPY) Y1 receptor we originally assumed that this receptor is a postsynaptic receptor based on our staining patterns and in agreement with the original definition by Wahlestedt et al. (1986) of the Y1 receptor being postsynaptic and the Y2 receptor presynaptic. However, using an improved technique we and others have been able to demonstrate that this receptor is also transported into the axon and possibly acts as a presynaptic receptor (see Brumovsky et al. 2002 and references therein).

The availability of powerful NK_1 receptor antibodies has allowed most interesting in vivo studies on internalization processes (see Mantyh 2002), first in endothelial cells (Bowden et al. 1994), then in the striatum (Mantyh et al. 1995a) and the spinal cord (Mantyh et al. 1995b). In fact, this appears to be such a sensitive approach that it is a way to monitor endogenous SP release (Abbadie et al. 1997). It has also been used to internalize a toxin (saporin) conjugated to SP, in this way selectively ablating certain neuron populations (Mantyh et al. 1997; Gray et al. 2001; Truitt and Coolen 2002), providing important information on the functional role of these neurons.

5.3
The Mismatch Issue

A much discussed topic over the last decades has been the apparent mismatch between distribution of receptor and endogenous ligand (see Herkenham 1987) and a closely related process, so-called volume transmission (see Fuxe and Agnati 1991), that is the idea that messengers can diffuse over long distances to reach their receptor. Clearly, detailed studies, preferably at the ultrastructural level, on the relation between ligand and receptor should be able to clarify these questions, but it is of utmost importance that a very high sensitivity is achieved in the immunohistochemical analysis to avoid false negatives. Here, when comparing studies cited above, both NK_1 and NK_3 receptors mostly show very extensive distribution patterns, and they are largely complementary (see Figs. 12–14) with acceptable correlation between the distribution of the receptors and the respective preferred ligand. However, several examples of mismatch may exist. Here we take as an example most of the central amygdaloid nucleus which at the level shown in Fig. 13D–F has a very dense SP innervation but no distinct

NK_1- or NK_3-LIs (although weak, diffuse NK_1- and NK_3-LI can be seen under the microscope). NK_2 binding is also weak in the amygdaloid complex (Saffroy et al. 2003). A partially reversed situation seems to exist in the dorsal hippocampal (Fig. 12A–C), where there are strong NK_1-positive cell bodies with extensive ramifications, as well as some distinct NK_3-positive neurons, but apparently with only a very sparse SP innervation. However, NKB mRNA-positive cells are present (Shughrue et al. 1996). In this context it should be remembered that even if SP is the preferred ligand for the NK_1 receptor, SP can also act on the other receptors (Nakanishi 1991; Quartara and Maggi 1997; Holzer and Holzer-Petsche 2001). For example, the substantia nigra zona reticulata has one of the highest densities of SP terminals surrounding the dopamine neurons. However, these neurons do not express NK_1 receptors (Maeno et al. 1993) but most of them are NK_3-positive (Ding et al. 1996; Shughrue et al. 1996). Is it possible that the concentrations of released SP are high enough to activate the NK_3 receptors? Finally, electrophysiology will have the last word, since it seems likely that even with improved histochemical methods it may not be possible to visualize single receptor molecules which still may be of functional significance.

5.4
Tachykinin Circuitries

Much work remains to be done, in spite of the many published papers on TK circuitries (many more than cited in this chapter). An important issue will be to define in more detail the target neurons for TKs and circuitries involving SP/NKA and NKB and their receptors. This is now possible by combining double-labeling, e.g., showing the presence of NK_3 receptors in melanin-concentrating hormone (MCH) neurons in the lateral hypothalamus (Griffond et al. 1997), showing that NK_1 receptors are present in the cholinergic striatal and basal forebrain neurons (Gerfen 1991), and by combining double-labeling with tract tracing methodologies as done for NK_1 neurons in the dorsal horn of the spinal cord (Mantyh 2002; Todd 2002).

5.5
Clinical Aspects

SP has over the years continued to be, and still is, an attractive target for many research groups with almost 1,000 entries in PubMed for each of the last 5 years. This work has, in a way, culminated in the clinical approval of the NK_1 antagonist Aprepitant (Emend) for the treatment of chemotherapy-induced emesis, and this compound is also in Phase III trials for major depression[1]. However, NK_2 and NK_3 antagonists are also of interest for indications including asthma, anxiety (NK_2) and schizophrenia (NK_3) (see Hökfelt et al. 2003). It will be of in-

---

[1] In December 2003 we were informed that these clinical trials have been discontinued because of the drug's apparent lack of efficacy.

terest to follow these compounds. Clearly, it will be important to understand the mechanism(s) underlying these drug effects, and their site of action, which may shed light on the disease being treated. Therefore, further histochemical work, of course along with other methodologies, is needed, in particular analyses of the human brain under normal and pathological conditions. As mentioned, in situ hybridization offers good possibilities to study cellular transcript levels in various brain regions possibly involved in disease processes, but so far little has been done with regard to immunohistochemical receptor analysis in humans. Also, should the initial therapeutic effects of NK_1 antagonists in major depression (Kramer et al. 1998) be confirmed, it seems likely that there is an underlying, in any case secondary, derangement of a SP/ NK_1 system, e.g., increased peptide synthesis/release or increased NK_1 receptor levels or affinity. It has been suggested that the antidepressive effects of the NK_1 receptor antagonist is exerted in the amygdala (Kramer et al. 1998), but little is known about the tachykinin systems in this brain region of humans. For example, Hurd et al. (1999) reported 'PPT-A mRNA-positive cells scattered throughout the amygdala nuclei'. It should be possible to approach questions of this type in the future with histochemical methods.

Acknowledgements. This study was supported by the Swedish Research Council, The Marianne and Marcus Wallenberg Foundation, The Knut and Alice Wallenberg Foundation and an Unrestricted Bristol-Myer Squibb Neuroscience Grant. The NK_1 antiserum used in this review (Figs. 12A, D; 13A, D; 14A, D) was generously donated by Dr. R. Shigemoto, Kyoto University, Kyoto, Japan, the NK_3 antiserum (Figs. 12C, F; 13C, F; 14C, F) by Dr. P. Ciofi, Faculte de Medicine, Besancon, France, and the SP antiserum (Figs. 2A–H; 3A–H; 4A–F; 8A; 12B, E; 13B, E; 14B, E) by Dr. L. Terenius, Karolinska Institutet, Stockholm and Dr. Ingrid Nylander, Uppsala University, Uppsala, Sweden. We also thank Drs. Y. Hurd, Karolinska Institutet, Stockholm, Sweden (Figs. 5, 6), N. Rance, University of Arizona, College of Medicine, Tucson, AZ, USA (Fig. 7), J. Glowinski and J.-C. Beaujouan, College de France, Paris, France (Fig. 10), M. Toyhama, Osaka University, Osaka, Japan (Fig. 11A–E) and I. Merchenthaler and P.J. Shughrue, The Women's Health Laboratory Institute, Wyeth-Ayerst Research Radnor, PA, USA (Fig. 12F–J) for generously making their published material available for this chapter. Finally, we thank Dr. J. Gong, The Neuroscience Institute, The Fourth Military Medical University, Xian, P.R. China for supply of formalin-fixed monkey tissue (Fig. 8).

References

Abbadie C, Trafton J, Liu H et al. (1997) Inflammation increases the distribution of dorsal horn neurons that internalize the neurokinin-1 receptor in response to noxious and non-noxious stimulation. J Neurosci 17:8049–8060

Abrahamson EE, Moore RY (2001) Suprachiasmatic nucleus in the mouse: retinal innervation, intrinsic organization and efferent projections. Brain Res 916:172–191

Adams JC (1992) Biotin amplification of biotin and horseradish peroxidase signals in histochemical stains. J Histochem Cytochem 40:1457–1463

Airaksinen MS, Alanen S, Szabat E et al. (1992) Multiple neurotransmitters in the tuberomammillary nucleus: comparison of rat, mouse, and guinea pig. J Comp Neurol 323:103–116

Amaral DG, Campbell MJ (1986) Transmitter systems in the primate dentate gyrus. Hum Neurobiol 5:69–80
Arai H, Emson PC (1986) Regional distribution of neuropeptide K and other tachykinins (neurokinin A, neurokinin B and substance P) in rat central nervous system. Brain Res 399:240–249
Ariano MA, Aronin N, Difiglia M et al. (2002) Striatal neurochemical changes in transgenic models of Huntington's disease. J Neurosci Res 68:716–729
Artymyshyn R, Murray M (1985) Substance P in the interpeduncular nucleus of the rat: normal distribution and the effects of deafferentation. J Comp Neurol 231:78–90
Arvidsson U, Cullheim S, Ulfhake B et al. (1994) Quantitative and qualitative aspects on the distribution of 5-HT and its coexistence with substance P and TRH in cat ventral medullary neurons. J Chem Neuroanat 7:3–12
Avrameas S (1969) Coupling of enzymes to proteins with glutaraldehyde. Use of the conjugates for the detection of antigens and antibodies. Immunohistochemistry 6:43–47
Baker H (1986) Species differences in the distribution of substance P and tyrosine hydroxylase immunoreactivity in the olfactory bulb. J Comp Neurol 252:206–226
Baker KG, Halliday GM, Hornung JP et al. (1991) Distribution, morphology and number of monoamine-synthesizing and substance P-containing neurons in the human dorsal raphe nucleus. Neuroscience 42:757–775
Bannon MJ, Poosch MS, Haverstick DM et al. (1992) Preprotachykinin gene expression in the human basal ganglia: characterization of mRNAs and pre-mRNAs produced by alternate RNA splicing. Mol Brain Res 12:225–231
Battaglia G, Rustioni A (1992) Substance P innervation of the rat and cat thalamus. II. Cells of origin in the spinal cord. J Comp Neurol 315:473–486
Battaglia G, Spreafico R, Rustioni A (1992) Substance P innervation of the rat and cat thalamus. I. Distribution and relation to ascending spinal pathways. J Comp Neurol 315:457–472
Beach TG, McGeer EG (1984) The distribution of substance P in the primate basal ganglia: an immunohistochemical study of baboon and human brain. Neuroscience 13:29–52
Beal MF, Walker LC, Storey E et al. (1991) Neurotransmitters in neocortex of aged rhesus monkeys. Neurobiol Aging 12:407–412
Beaujouan JC, Torrens Y, Saffroy M et al. (1986) Quantitative autoradiographic analysis of the distribution of binding sites for [^{125}I]Bolton Hunter derivatives of eledoisin and substance P in the rat brain. Neuroscience 18:857–875
Beckstead RM, Kersey KS (1985) Immunohistochemical demonstration of differential substance P-, met-enkephalin-, and glutamic-acid-decarboxylase-containing cell body and axon distributions in the corpus striatum of the cat. J Comp Neurol 232:481–498
Beitz AJ (1982) The nuclei of origin of brain stem enkephalin and substance P projections to the rodent nucleus raphe magnus. Neuroscience 7:2753–2768
Bennett BD, Bolam JP (1994) Localisation of parvalbumin-immunoreactive structures in primate caudate-putamen. J Comp Neurol 347:340–356
Bennett-Clarke C, Mooney RD, Chiaia NL et al. (1989) A substance P projection from the superior colliculus to the parabigeminal nucleus in the rat and hamster. Brain Res 500:1–11
Bergström M, Fasth KJ, Kilpatrick G et al. (2000) Brain uptake and receptor binding of two [^{11}C]labelled selective high affinity NK$_1$-antagonists, GR203040 and GR205171— PET studies in rhesus monkey. Neuropharmacology 39:664–670
Berkenbosch F, Schipper J, Tilders FJ (1986) Corticotropin-releasing factor immunostaining in the rat spinal cord and medulla oblongata: an unexpected form of cross-reactivity with substance P. Brain Res 399:87–96
Berson SA, Yalow RS, Bauman A et al. (1956) Insulin-I^{131} metabolism in human subjects: Demonstration of insulin binding globulin in the circulation of insulin-treated subjects. J Clin Invest 35:170–190

Bittencourt JC, Benoit R, Sawchenko PE (1991) Distribution and origins of substance P-immunoreactive projections to the paraventricular and supraoptic nuclei: partial overlap with ascending catecholaminergic projections. J Chem Neuroanat 4:63–78

Block CH, Hoffman G, Kapp BS (1989) Peptide-containing pathways from the parabrachial complex to the central nucleus of the amygdala. Peptides 10:465–471

Blomqvist A, Mackerlova L (1995) Spinal projections to the parabrachial nucleus are substance P-immunoreactive. NeuroReport 6:605–608

Boissonade FM, Davison JS, Egizii R et al. (1996) The dorsal vagal complex of the ferret: anatomical and immunohistochemical studies. Neurogastroenterol Motil 8:255–272

Boissonade FM, Sharkey KA, Lucier GE (1993) Trigeminal nuclear complex of the ferret: anatomical and immunohistochemical studies. J Comp Neurol 329:291–312

Bolam JP, Izzo PN, Graybiel AM (1988) Cellular substrate of the histochemically defined striosome/matrix system of the caudate nucleus: a combined Golgi and immunocytochemical study in cat and ferret. Neuroscience 24:853–875

Bolam JP, Smith Y (1990) The GABA and substance P input to dopaminergic neurones in the substantia nigra of the rat. Brain Res 529:57–78

Bonner TI, Affolter HU, Young AC et al. (1987) A cDNA encoding the precursor of the rat neuropeptide, neurokinin B. Mol Brain Res 2:243–249

Borhegyi Z, Leranth C (1997) Distinct substance P- and calretinin-containing projections from the supramammillary area to the hippocampus in rats; a species difference between rats and monkeys. Exp Brain Res 115:369–374

Bouras C, Magistretti PJ, Morrison JH (1986a) An immunohistochemical study of six biologically active peptides in the human brain. Hum Neurobiol 5:213–226

Bouras C, Vallet PG, Dobrinov H et al. (1986b) Substance P neuronal cell bodies in the human brain: complete mapping by immunohistofluorescence. Neurosci Lett 69:31–36

Bowden JJ, Garland AM, Baluk P et al. (1994) Direct observation of substance P-induced internalization of neurokinin 1 (NK_1) receptors at sites of inflammation. Proc Natl Acad Sci USA 91:8964–8968

Bowker RM, Steinbusch HW, Coulter JD (1981) Serotonergic and peptidergic projections to the spinal cord demonstrated by a combined retrograde HRP histochemical and immunocytochemical staining method. Brain Res 211:412–417

Brownstein MJ, Mroz EA, Kizer JS et al. (1976) Regional distribution of substance P in the brain of the rat. Brain Res 116:299–305

Brownstein MJ, Mroz EA, Tappaz ML et al. (1977) On the origin of substance P and glutamic acid decarboxylase (GAD) in the substantia nigra. Brain Res 135:315–323

Brumovsky PR, Shi TJ, Matsuda H et al. (2002) NPY Y1 receptors are present in axonal processes of DRG neurons. Exp Neurol 174:1–10

Buck SH, Helke CJ, Burcher E et al. (1986) Pharmacologic characterization and autoradiographic distribution of binding sites for iodinated tachykinins in the rat central nervous system. Peptides 7:1109–1120

Burgos C, Aguirre JA, Alonso JR et al. (1988) Immunocytochemical study of substance P-like fibres and cell bodies in the cat diencephalon. J Hirnforsch 29:651–657

Burgunder JM, Young WS 3rd (1989) Distribution, projection and dopaminergic regulation of the neurokinin B mRNA-containing neurons of the rat caudate-putamen. Neuroscience 32:323–335

Caballero-Bleda M, Lagares C, Fernandez B et al. (1993) A chemoarchitectonically similar internal extension connects the rabbit intergeniculate leaflet to midline dorsal thalamaic nuclei. J Hirnforsch 34:35–42

Cao YQ, Mantyh PW, Carlson EJ et al. (1998) Primary afferent tachykinins are required to experience moderate to intense pain. Nature 392:390–394

Carboni AA, Lavelle WG, Barnes CL et al. (1990) Neurons of the lateral entorhinal cortex of the rhesus monkey: a Golgi, histochemical, and immunocytochemical characterization. J Comp Neurol 291:583–608

Chan-Palay V, Jonsson G, Palay SL (1978) Serotonin and substance P coexist in neurons of the rat's central nervous system. Proc Natl Acad Sci USA 75:1582–1586

Chang MM, Leeman SE, Niall HD (1971) Amino-acid sequence of substance P. Nature New Biol 232:86–87

Charara A, Parent A (1998) Chemoarchitecture of the primate dorsal raphe nucleus. J Chem Neuroanat 15:111–127

Charlton CG, Helke CJ (1987) Substance P-containing medullary projections to the intermediolateral cell column: identification with retrogradely transported rhodamine-labeled latex microspheres and immunohistochemistry. Brain Res 418:245–254

Chawla MK, Gutierrez GM, Young WS 3rd et al. (1997) Localization of neurons expressing substance P and neurokinin B gene transcripts in the human hypothalamus and basal forebrain. J Comp Neurol 384:429–442

Chesselet MF, Affolter HU (1987) Preprotachykinin messenger RNA detected by in situ hybridization in striatal neurons of the human brain. Brain Res 410:83–88

Chiba T, Masuko S (1989) Coexistence of varying combinations of neuropeptides with 5-hydroxytryptamine in neurons of the raphe pallidus et obscurus projecting to the spinal cord. Neurosci Res 7:13–23

Christensson-Nylander I, Herrera-Marschitz M, Staines W et al. (1986) Striato-nigral dynorphin and substance P pathways in the rat. I. Biochemical and immunohistochemical studies. Exp Brain Res 64:169–192

Ciriello J, Caverson MM, Calaresu FR et al. (1988) Neuropeptide and serotonin immunoreactive neurons in the cat ventrolateral medulla. Brain Res 440:53–66

Contestabile A, Villani L, Fasolo A et al. (1987) Topography of cholinergic and substance P pathways in the habenulo-interpeduncular system of the rat. An immunocytochemical and microchemical approach. Neuroscience 21:253–270

Conti F, De Biasi S, Fabri M et al. (1992a) Substance P-containing pyramidal neurons in the cat somatic sensory cortex. J Comp Neurol 322:136–148

Conti F, Fabri M, Minelli A (1992b) Numerous SP-positive pyramidal neurons in cat neocortex are glutamate-positive. Brain Res 599:140–143

Coons AH (1958) Fluorescent antibody methods. In Danielli JF (ed) General Cytochemical Methods. Academic Press: New York, pp 399–422

Cuello AC, Galfre G, Milstein C (1979) Detection of substance P in the central nervous system by a monoclonal antibody. Proc Natl Acad Sci USA 76:3532–3536

Cuello AC, Kanazawa I (1978) The distribution of substance P immunoreactive fibers in the rat central nervous system. J Comp Neurol 178:129–156

Cuello AC, Priestley JV, Matthews MR (1982a) Localization of substance P in neuronal pathways. Ciba Found Symp, pp 55–83

Cuello AC, Priestley JV, Milstein C (1982b) Immunocytochemistry with internally labeled monoclonal antibodies. Proc Natl Acad Sci USA 79:665–669

Cuello AC, Ribeiro-da-Silva A, Ma W et al. (1993) Organization of substance P primary sensory neurons: ultrastructural and physiological correlates. Regul Pept 46:155–164

Dalsgaard C-J (1988) The sensory system. In Björklund A, Hökfelt T, Owman C (eds) Handbook of Chemical Neuroanatomy, Vol 6: The Peripheral Nervous System. Elsevier: Amsterdam, pp 599–636

Danks JA, Rothman RB, Cascieri MA et al. (1986) A comparative autoradiographic study of the distributions of substance P and eledoisin binding sites in rat brain. Brain Res 385:273–281

Davis BJ, Burd GD, Macrides F (1982) Localization of methionine-enkephalin, substance P, and somatostatin immunoreactivities in the main olfactory bulb of the hamster. J Comp Neurol 204:377–383

Davis BJ, Kream RM (1993) Distribution of tachykinin- and opioid-expressing neurons in the hamster solitary nucleus: an immuno- and in situ hybridization histochemical study. Brain Res 616:6–16

De Felipe C, Herrero JF, O'Brien JA et al. (1998) Altered nociception, analgesia and aggression in mice lacking the receptor for substance P. Nature 392:394-397
Debeljuk L, Villanua MA, Bartke A (1992) Substance P variations in the hypothalamus of golden hamsters at different stages of the estrous cycle. Neurosci Lett 137:178-180
Del Fiacco M, Dessi ML, Levanti MC (1984) Topographical localization of substance P in the human post-mortem brainstem. An immunohistochemical study in the newborn and adult tissue. Neuroscience 12:591-611
Del Fiacco M, Levanti MC, Dessi ML et al. (1987) The human hippocampal formation and parahippocampal gyrus: localization of substance P-like immunoreactivity in newborn and adult post-mortem tissue. Neuroscience 21:141-150
Del Fiacco M, Perra MT, Quartu M et al. (1988) Evidence for the presence of substance P-like immunoreactivity in the human cerebellum. Brain Res 446:173-177
Diez M, Koistinaho J, Kahn K et al. (2000) Neuropeptides in hippocampus and cortex in transgenic mice overexpressing V717F beta-amyloid precursor protein–initial observations. Neuroscience 100:259-286
DiFiglia M, Aronin N, Leeman SE (1982) Immunoreactive substance P in the substantia nigra of the monkey: light and electron microscopic localization. Brain Res 233:381-388
Ding YQ, Shigemoto R, Takada M et al. (1996) Localization of the neuromedin K receptor (NK_3) in the central nervous system of the rat. J Comp Neurol 364:290-310
Drake CT, Chavkin C, Milner TA (1997) Kappa opioid receptor-like immunoreactivity is present in substance P-containing subcortical afferents in guinea pig dentate gyrus. Hippocampus 7:36-47
Dudas B, Merchenthaler I (2002) Close juxtapositions between LHRH immunoreactive neurons and substance P immunoreactive axons in the human diencephalon. J Clin Endocrinol Metab 87:2946-2953
Dufourny L, Warembourg M, Jolivet A (1998) Multiple peptides infrequently coexist in progesterone receptor-containing neurons in the ventrolateral hypothalamic nucleus of the guinea-pig: an immunocytochemical triple-label analysis of somatostatin, neurotensin and substance P. J Neuroendocrinol 10:165-173
Dufourny L, Warembourg M, Jolivet A (1999) Quantitative studies of progesterone receptor and nitric oxide synthase colocalization with somatostatin, or neurotensin, or substance P in neurons of the guinea pig ventrolateral hypothalamic nucleus: an immunocytochemical triple-label analysis. J Chem Neuroanat 17:33-43
Echevarria D, Matute C, Albus K (1997) Neurons in the rat occipital cortex co-expressing the substance P-receptor and GABA: a comparison between in vivo and organotypic cultures. Eur J Neurosci 9:1530-1535
Emson PC, Cuello AC, Paxinos G et al. (1977) The origin of substance P and acetylcholine projections to the ventral tegmental area and interpeduncular nucleus in the rat. Acta Physiol Scand Suppl 452:43-46
Fodor M, Görcs TJ, Palkovits M (1992) Immunohistochemical study on the distribution of neuropeptides within the pontine tegmentum–particularly the parabrachial nuclei and the locus coeruleus of the human brain. Neuroscience 46:891-908
Fodor M, Pammer C, Görcs T et al. (1994) Neuropeptides in the human dorsal vagal complex: an immunohistochemical study. J Chem Neuroanat 7:141-157
Furness JB, Costa M (1987) The Enteric Nervous System. Churchill Livingstone: New York
Furuta T, Mori T, Lee T et al. (2000) Third group of neostriatofugal neurons: neurokinin B-producing neurons that send axons predominantly to the substantia innominata. J Comp Neurol 426:279-296
Fuxe K, Agnati LF, eds (1991) Volume Transmission in the Brain. Raven Press: New York
Gale K, Hong JS, Guidotti A (1977) Presence of substance P and GABA in separate striatonigral neurons. Brain Res 136:371-375

Gall C, Selawski L (1984) Supramammillary afferents to guinea pig hippocampus contain substance P-like immunoreactivity. Neurosci Lett 51:171–176

Gallagher AW, Chahl LA, Lynch AM (1992) Distribution of substance P-like immunoreactivity in guinea pig central nervous system. Brain Res Bull 29:199–207

Gerfen CR (1991) Substance P (neurokinin-1) receptor mRNA is selectively expressed in cholinergic neurons in the striatum and basal forebrain. Brain Res 556:165–170

Gerfen CR (1992) The neostriatal mosaic: multiple levels of compartmental organization. Trends Neurosci 15:133–139

Gerfen CR, Young WS 3rd (1988) Distribution of striatonigral and striatopallidal peptidergic neurons in both patch and matrix compartments: an in situ hybridization histochemistry and fluorescent retrograde tracing study. Brain Res 460:161–167

Ghatei MA, Bloom SR, Langevin H et al. (1984) Regional distribution of bombesin and seven other regulatory peptides in the human brain. Brain Res 293:101–109

Gilbert RF, Emson PC, Hunt SP et al. (1982) The effects of monoamine neurotoxins on peptides in the rat spinal cord. Neuroscience 7:69–87

Gingras JL, Long WA, Segreti T et al. (1995) Pre- and postnatal effects of chronic maternal hypoxia on substance-P immunoreactivity in rabbit brainstem regions. Dev Neurosci 17:350–356

Glazer EJ, Ramachandran J, Basbaum AI (1984) Radioimmunocytochemistry using a tritiated goat anti-rabbit second antibody. J Histochem Cytochem 32:778–782

Gras C, Herzog E, Bellenchi GC et al. (2002) A third vesicular glutamate transporter expressed by cholinergic and serotoninergic neurons. J Neurosci 22:5442–5451

Gray PA, Janczewski WA, Mellen N et al. (2001) Normal breathing requires preBötzinger complex neurokinin-1 receptor-expressing neurons. Nature Neurosci 4:927–930

Gray TS, Magnuson DJ (1987) Neuropeptide neuronal efferents from the bed nucleus of the stria terminalis and central amygdaloid nucleus to the dorsal vagal complex in the rat. J Comp Neurol 262:365–374

Graybiel AM (1990) Neurotransmitters and neuromodulators in the basal ganglia. Trends Neurosci 13:244–254

Graybiel AM, Ragsdale CW Jr, Yoneoka ES et al. (1981) An immunohistochemical study of enkephalins and other neuropeptides in the striatum of the cat with evidence that the opiate peptides are arranged to form mosaic patterns in register with the striosomal compartments visible by acetylcholinesterase staining. Neuroscience 6:377–397

Gregg KV, Bishop GA (1997) Peptide localization in the mouse inferior olive. J Chem Neuroanat 12:211–220

Griffond B, Ciofi P, Bayer L et al. (1997) Immunocytochemical detection of the neurokinin B receptor (NK_3) on melanin-concentrating hormone (MCH) neurons in rat brain. J Chem Neuroanat 12:183–189

Groenewegen HJ, Ahlenius S, Haber SN et al. (1986) Cytoarchitecture, fiber connections, and some histochemical aspects of the interpeduncular nucleus in the rat. J Comp Neurol 249:65–102.

Gu Q, Liu Y, Cynader MS (1994) A study of tachykinin-immunoreactivity in the cat visual cortex. Brain Res 640:336–340

Haber S, Elde R (1981) Correlation between Met-enkephalin and substance P immunoreactivity in the primate globus pallidus. Neuroscience 6, 1291–1297.

Haber SN, Groenewegen HJ (1989) Interrelationship of the distribution of neuropeptides and tyrosine hydroxylase immunoreactivity in the human substantia nigra. J Comp Neurol 290:53–68

Haber SN, Watson SJ (1985) The comparative distribution of enkephalin, dynorphin and substance P in the human globus pallidus and basal forebrain. Neuroscience 14:1011–1024

Haber SN, Wolfe DP, Groenewegen HJ (1990) The relationship between ventral striatal efferent fibers and the distribution of peptide-positive woolly fibers in the forebrain of the rhesus monkey. Neuroscience 39:323–338

Halliday GM, Gai WP, Blessing WW et al. (1990) Substance P-containing neurons in the pontomesencephalic tegmentum of the human brain. Neuroscience 39:81–96

Halliday GM, Li YW, Joh TH et al. (1988) Distribution of substance P-like immunoreactive neurons in the human medulla oblongata: co-localization with monoamine-synthesizing neurons. Synapse 2:353–370

Hamill GS, Jacobowitz DM (1984) A study of afferent projections to the rat interpeduncular nucleus. Brain Res Bull 13:527–539

Hancock MB (1984) Substance P-immunoreactive processes on 5HT/SP-immunoreactive medullary cells. Brain Res Bull 13:559–563

Hargreaves R (2002) Imaging substance P receptors (NK_1) in the living human brain using positron emission tomography. J Clin Psychiatry 63 Suppl 11:18–24

Harlan RE, Garcia MM, Krause JE (1989) Cellular localization of substance P- and neurokinin A-encoding preprotachykinin mRNA in the female rat brain. J Comp Neurol 287:179–212

Hartwich M, Kalsbeek A, Pevet P et al. (1994) Effects of illumination and enucleation on substance-P-immunoreactive structures in subcortical visual centers of golden hamster and Wistar rat. Cell Tissue Res 277:351–361

Hayashi M, Oshima K (1986) Neuropeptides in cerebral cortex of macaque monkey (Macaca fuscata fuscata): regional distribution and ontogeny. Brain Res 364:360–368

Heimer L, Zaborszky L, eds (1989) Neuroanatomical Tract-Tracing Methods 2. Plenum Press: New York

Helke CJ, Neil JJ, Massari VJ et al. (1982) Substance P neurons project from the ventral medulla to the intermediolateral cell column and ventral horn in the rat. Brain Res 243:147–152

Helme RD, Thomas K (1986) Substance P in human hypothalamus. Clin Exp Neurol 22:97–101

Henderson Z (1997) The projection from the striatum to the nucleus basalis in the rat: an electron microscopic study. Neuroscience 78:943–955

Hendry SH, Jones EG, Burstein N (1988) Activity-dependent regulation of tachykinin-like immunoreactivity in neurons of monkey visual cortex. J Neurosci 8:1225–1238

Herkenham M (1987) Mismatches between neurotransmitter and receptor localizations in brain: observations and implications. Neuroscience 23:1–38

Hershey AD, Krause JE (1990) Molecular characterization of a functional cDNA encoding the rat substance P receptor. Science 247:958–962

Hirai T, Jones EG (1989) Distribution of tachykinin- and enkephalin-immunoreactive fibers in the human thalamus. Brain Res Rev 14:35–52

Holt DJ, Graybiel AM, Saper CB (1997) Neurochemical architecture of the human striatum. J Comp Neurol 384:1–25

Holtman JR Jr (1988) Immunohistochemical localization of serotonin- and substance P-containing fibers around respiratory muscle motoneurons in the nucleus ambiguus of the cat. Neuroscience 26:169–178

Holzer P, Holzer-Petsche U (2001) Tachykinin receptors in the gut: physiological and pathological implications. Curr Opin Pharmacol 1:583–590

Hong JS, Costa E, Yang HY (1976) Effects of habenular lesions on the substance P content of various brain regions. Brain Res 118:523–525

Hong JS, Yang HY, Racagni G et al. (1977) Projections of substance P containing neurons from neostriatum to substantia nigra. Brain Res 122:541–544

Hurd YL, Herkenham M (1995) The human neostriatum shows compartmentalization of neuropeptide gene expression in dorsal and ventral regions: an in situ hybridization histochemical analysis. Neuroscience 64:571–586

Hurd YL, Keller E, Sotonyi P et al. (1999) Preprotachykinin-A mRNA expression in the human and monkey brain: An in situ hybridization study. J Comp Neurol 411:56–72

Hutsler JJ, Chalupa LM (1991) Substance P immunoreactivity identifies a projection from the cat's superior colliculus to the principal tectorecipient zone of the lateral posterior nucleus. J Comp Neurol 312:379–390

Hökfelt T (1991) Neuropeptides in perspective: The last ten years. Neuron 7:867–879

Hökfelt T, Arvidsson U, Cullheim S et al. (2000) Multiple messengers in descending serotonin neurons: localization and functional implications. J Chem Neuroanat 18:75–86

Hökfelt T, Bartfai T, Bloom F (2003) Neuropeptides: opportunities for drug discovery. Lancet Neurol 2:463–472

Hökfelt T, Johansson O, Holets V et al. (1987) Distribution of neuropeptides with special reference to their coexistence with classic transmitters. In Meltzer HY (ed) Psychopharmacology: The Third Generation of Progress. Raven Press: New York, pp 401–416

Hökfelt T, Johansson O, Kellerth J-O et al. (1977): Immunohistochemical distribution of substance P. In Euler US von, Pernow B (eds) Substance P. Raven Press: New York, pp 117–145

Hökfelt T, Kellerth JO, Nilsson G et al. (1975) Substance P: localization in the central nervous system and in some primary sensory neurons. Science 190:889–890

Hökfelt T, Ljungdahl A, Steinbusch H et al. (1978) Immunohistochemical evidence of substance P-like immunoreactivity in some 5-hydroxytryptamine-containing neurons in the rat central nervous system. Neuroscience 3:517–538

Hökfelt T, Meyerson B, Nilsson G et al. (1976) Immunohistochemical evidence for substance P-containing nerve endings in the human cortex. Brain Res 104:181–186

Hökfelt T, Pernow B, Wahren J (2001) Substance P: a pioneer amongst neuropeptides. J Intern Med 249:27–40

Hökfelt T, Skagerberg G, Skirboll L et al. (1983) Combination of retrograde tracing and neurotransmitter histochemistry. In Björklund A, Hökfelt T (eds) Handbook of Chemical Neuroanatomy, Vol 1, Methods in Chemical Neuroanatomy. Elsevier: Amsterdam, pp 228–285

Ikemoto K, Satoh K, Maeda T et al. (1995) Neurochemical heterogeneity of the primate nucleus accumbens. Exp Brain Res 104:177–190

Inagaki S, Parent A (1984) Distribution of substance P and enkephalin-like immunoreactivity in the substantia nigra of rat, cat and monkey. Brain Res Bull 13:319–329

Inagaki S, Sakanaka M, Shiosaka S et al. (1982b) Experimental and immunohistochemical studies on the cerebellar substance P of the rat: localization, postnatal ontogeny and ways of entry to the cerebellum. Neuroscience 7:639–645

Inagaki S, Sakanaka M, Shiosaka S et al. (1982a) Ontogeny of substance P-containing neuron system of the rat: immunohistochemical analysis. I. Forebrain and upper brain stem. Neuroscience 7:251–277

Inoue A, Nakata Y, Yajima H et al. (1984) Active uptake system for substance P carboxy-terminal heptapeptide (5–11) into a fraction from rabbit enriched in glial cells. Jpn J Pharmacol 36:137–145

Iritani S, Fujii M, Satoh K (1989) The distribution of substance P in the cerebral cortex and hippocampal formation: an immunohistochemical study in the monkey and rat. Brain Res Bull 22:295–303

Izzo PN, Graybiel AM, Bolam JP (1987) Characterization of substance P- and [Met]enkephalin-immunoreactive neurons in the caudate nucleus of cat and ferret by a single section Golgi procedure. Neuroscience 20:577–587

Jakab RL, Goldman-Rakic P (1996) Presynaptic and postsynaptic subcellular localization of substance P receptor immunoreactivity in the neostriatum of the rat and rhesus monkey (Macaca mulatta). J Comp Neurol 369:125–136

Jakab RL, Goldman-Rakic P, Leranth C (1997) Dual role of substance P/GABA axons in cortical neurotransmission: synaptic triads on pyramidal cell spines and basket-like innervation of layer II-III calbindin interneurons in primate prefrontal cortex. Cereb Cortex 7:359–373

Jakab RL, Hazrati LN, Goldman-Rakic P (1996) Distribution and neurochemical character of substance P receptor (SPR)-immunoreactive striatal neurons of the macaque monkey: accumulation of SP fibers and SPR neurons and dendrites in "striocapsules" encircling striosomes. J Comp Neurol 369:137–149

Jansen KL, Faull RL, Dragunow M et al. (1991) Distribution of excitatory and inhibitory amino acid, sigma, monoamine, catecholamine, acetylcholine, opioid, neurotensin, substance P, adenosine and neuropeptide Y receptors in human motor and somatosensory cortex. Brain Res 566:225–238

Jessell TM, Emson PC, Paxinos G et al. (1978) Topographic projections of substance P and GABA pathways in the striato- and pallido-nigral system: a biochemical and immunohistochemical study. Brain Res 152:487–498

Johansson O, Hökfelt T, Pernow B et al. (1981) Immunohistochemical support for three putative transmitters in one neuron: coexistence of 5-hydroxytryptamine, substance P- and thyrotropin releasing hormone-like immunoreactivity in medullary neurons projecting to the spinal cord. Neuroscience 6:1857–1881

Johnson MD (1994) Synaptic glutamate release by postnatal rat serotonergic neurons in microculture. Neuron 12:433–442

Jones EG, DeFelipe J, Hendry SH et al. (1988) A study of tachykinin-immunoreactive neurons in monkey cerebral cortex. J Neurosci 8:1206–1224

Kachidian P, Poulat P, Marlier L et al. (1991) Immunohistochemical evidence for the coexistence of substance P, thyrotropin-releasing hormone, GABA, methionine-enkephalin, and leucin-enkephalin in the serotonergic neurons of the caudal raphe nuclei: a dual labeling in the rat. J Neurosci Res 30:521–530

Kanazawa I, Emson PC, Cuello AC (1977) Evidence for the existence of substance P-containing fibres in striato-nigral and pallido-nigral pathways in rat brain. Brain Res 119:447–453

Kanazawa I, Jessell T (1976) Post mortem changes and regional distribution of substance P in the rat and mouse nervous system. Brain Res 117:362–367

Kanazawa I, Mogaki S, Muramoto O et al. (1980) On the origin of substance P-containing fibres in the entopeduncular nucleus and the substantia nigra of the rat. Brain Res 184:481–485

Kaneko T, Akiyama H, Nagatsu I et al. (1990) Immunohistochemical demonstration of glutaminase in catecholaminergic and serotoninergic neurons of rat brain. Brain Res 507:151–154

Kapadia SE, de Lanerolle NC (1984a) Substance P neuronal organization in the median region of the interpeduncular nucleus of the cat: an electron microscopic analysis. Neuroscience 12:1229–1242

Kapadia SE, de Lanerolle NC (1984b) Populations of substance P, Met-enkephalin and serotonin immunoreactive neurons in the interpeduncular nucleus of cat: cytoarchitectonics. Brain Res 302:33–43

Karten HJ, Brecha N (1983) Localization of neuroactive substances in the vertebrate retina: evidence for lamination in the inner plexiform layer. Vision Res 23:1197–1205

Kawaguchi Y, Hoshimura M, Nawa H et al. (1986) Sequence analysis of cloned cDNA for rat substance P precursor: Existence of a third substance P precursor. Biochem Biophys Res Commun 139:1040–1046

Khachaturian H, Lewis ME, Schäfer MKH et al. (1985) Anatomy of CNS opioid systems. Trends Neurosci 8:111–119

Kita H, Kitai ST (1988) Glutamate decarboxylase immunoreactive neurons in rat neostriatum: their morphological types and populations. Brain Res 447:346–352

Kohno J, Shiosaka S, Shinoda K et al. (1984) Two distinct strio-nigral substance P pathways in the rat: an experimental immunohistochemical study. Brain Res 308:309–317

Kolb H, Fernandez E, Ammermuller J et al. (1995) Substance P: a neurotransmitter of amacrine and ganglion cells in the vertebrate retina. Histol Histopathol 10:947–968

Kosaka K, Hama K, Nagatsu I et al. (1988) Possible coexistence of amino acid (gamma-aminobutyric acid), amine (dopamine) and peptide (substance P); neurons containing immunoreactivities for glutamic acid decarboxylase, tyrosine hydroxylase and substance P in the hamster main olfactory bulb. Exp Brain Res 71:633–642

Kotani H, Hoshimura M, Nawa H et al. (1986) Structure and gene organization of bovine neuromedin K precursor. Proc Natl Acad Sci USA 83:7074–7078

Kramer MS, Cutler N, Feighner J et al. (1998) Distinct mechanism for antidepressant activity by blockade of central substance P receptors. Science 281:1640–1645

Krause JE, Chirgwin JM, Carter MS et al. (1987) Three rat preprotachykinin mRNAs encode the neuropeptides substance P and neurokinin A. Proc Natl Acad Sci USA 84:881–885

Kyrkanides S, Tallents RH, Macher DJ et al. (2002) Temporomandibular joint nociception: effects of capsaicin on substance P-like immunoreactivity in the rabbit brain stem. J Orofac Pain 16:229–236

Lee CM, Iversen LL, Hanley MR et al. (1982) The possible existence of multiple receptors for substance P. Naunyn-Schmiedeberg's Arch Pharmacol 318:281–287

Lee JM, McLean S, Maggio JE et al. (1986) The localization and characterization of substance P and substance K in striatonigral neurons. Brain Res 371:152–154

Lee T, Kaneko T, Shigemoto R et al. (1997) Collateral projections from striatonigral neurons to substance P receptor-expressing intrinsic neurons in the striatum of the rat. J Comp Neurol 388:250–264

Leger L, Charnay Y, Chayvialle JA et al. (1983) Localization of substance P- and enkephalin-like immunoreactivity in relation to catecholamine-containing cell bodies in the cat dorsolateral pontine tegmentum: an immunofluorescence study. Neuroscience 8:525–546

Lembeck F (1953) Zur Frage der zentralen Übertragung afferenter Impulse. III. Mitteilung. Das Vorkommen und die Bedeutung der Substanz P in den dorsalen Wurzeln des Rückenmarks. Naunyn-Schmiedeberg's Arch Pharmacol 219:197–213

Li YQ, Rao ZR, Shi JW (1990a) Substance P-like immunoreactive neurons in the nucleus tractus solitarii of the rat send their axons to the nucleus accumbens. Neurosci Lett 120:194–196

Li YQ, Rao ZR, Shi JW (1990b) Midbrain periaqueductal gray neurons with substance P- or enkephalin-like immunoreactivity send projection fibers to the nucleus accumbens in the rat. Neurosci Lett 119:269–271

Li YQ, Zeng SL, Dong YX et al. (1991) Serotonin-, substance P- and tyrosine hydroxylase-like immunoreactive neurons projecting from the periaqueductal gray to the ventromedial hypothalamic nucleus in the rat. Neurosci Lett 134:33–36

Li YQ, Zeng SL, Rao ZR et al. (1992) Serotonin-, substance P- and tyrosine hydroxylase-like immunoreactive neurons projecting from the midbrain periaqueductal gray to the nucleus tractus solitarii in the rat. Neurosci Lett 134:175–179

Li YW, Bayliss DA (1998) Presynaptic inhibition by 5-HT$_{1B}$ receptors of glutamatergic synaptic inputs onto serotonergic caudal raphe neurones in rat. J Physiol 510:121–134

Liu HL, Cao R, Jin L et al. (2002) Immunocytochemical localization of substance P receptor in hypothalamic oxytocin-containing neurons of C57 mice. Brain Res 948:175–179

Ljungdahl A, Hökfelt T, Nilsson G (1978) Distribution of substance P-like immunoreactivity in the central nervous system of the rat. I. Cell bodies and nerve terminals. Neuroscience 3:861–943

Lorenz RG, Saper CB, Wong DL et al. (1985) Co-localization of substance P- and phenylethanolamine-N-methyltransferase-like immunoreactivity in neurons of ventrolateral medulla that project to the spinal cord: potential role in control of vasomotor tone. Neurosci Lett 55:255–260

Lotstra F, Mailleux P, Vanderhaeghen JJ (1988) Substance P neurons in the human hippocampus: an immunohistochemical analysis in the infant and adult. J Chem Neuroanat 1:111–123

Lu XY, Ghasemzadeh MB, Kalivas PW (1998) Expression of D1 receptor, D2 receptor, substance P and enkephalin messenger RNAs in the neurons projecting from the nucleus accumbens. Neuroscience 82:767–780

Lucas LR, Hurley DL, Krause JE et al. (1992) Localization of the tachykinin neurokinin B precursor peptide in rat brain by immunocytochemistry and in situ hybridization. Neuroscience 51:317–345

Ma QP, Bleasdale C (2002) Modulation of brain stem monoamines and gamma-aminobutyric acid by NK_1 receptors in rats. NeuroReport 13:1809–1812

Maeno H, Kiyama H, Tohyama M (1993) Distribution of the substance P receptor (NK-1 receptor) in the central nervous system. Mol Brain Res 18:43–58

Maggio JE (1988) Tachykinins. Annu Rev Neurosci 11:13–28

Magoul R, Onteniente B, Oblin A et al. (1986) Inter- and intracellular relationship of substance P-containing neurons with serotonin and GABA in the dorsal raphe nucleus: combination of autoradiographic and immunocytochemical techniques. J Histochem Cytochem 34:735–742

Mai JK, Stephens PH, Hopf A et al. (1986) Substance P in the human brain. Neuroscience 17:709–739

Maley B, Mullett T, Elde R (1983) The nucleus tractus solitarii of the cat: a comparison of Golgi impregnated neurons with methionine-enkephalin- and substance P-immunoreactive neurons. J Comp Neurol 217:405–417

Maley BE, Newton BW, Howes KA et al. (1987) Immunohistochemical localization of substance P and enkephalin in the nucleus tractus solitarii of the rhesus monkey, Macaca mulatta. J Comp Neurol 260:483–490

Manley MS, Young SJ, Groves PM (1994) Compartmental organization of the peptide network in the human caudate nucleus. J Chem Neuroanat 7:191–201

Mantyh PW (2002) Neurobiology of substance P and the NK_1 receptor. J Clin Psychiatry 63 Suppl 11:6–10

Mantyh PW, Allen CJ, Ghilardi JR et al. (1995a) Rapid endocytosis of a G protein-coupled receptor: substance P evoked internalization of its receptor in the rat striatum in vivo. Proc Natl Acad Sci USA 92:2622–2626

Mantyh PW, DeMater E, Malhotra R et al. (1995b) Receptor endocytosis and dendrite reshaping in spinal neurons after somatosensory stimulation. Science 268:1629–1632

Mantyh PW, Gates T, Mantyh CR et al. (1989) Autoradiographic localization and characterization of tachykinin receptor binding sites in the rat brain and peripheral tissues. J Neurosci 9:258–279

Mantyh PW, Rogers SD, Honore P et al. (1997) Inhibition of hyperalgesia by ablation of lamina I spinal neurons expressing the substance P receptor. Science 278:275–279

Marksteiner J, Saria A, Hinterhuber H (1994) Distribution of secretoneurin-like immunoreactivity in comparison with that of substance P in the human brain stem. J Chem Neuroanat 7:253–270

Marksteiner J, Sperk G, Krause JE (1992) Distribution of neurons expressing neurokinin B in the rat brain: immunohistochemistry and in situ hybridization. J Comp Neurol 317:341–356

Marson L, Loewy AD (1985) Topographic organization of substance P and monoamine cells in the ventral medulla of the cat. J Auton Nerv Syst 14:271–285

Martin LJ, Hadfield MG, Dellovade TL et al. (1991) The striatal mosaic in primates: patterns of neuropeptide immunoreactivity differentiate the ventral striatum from the dorsal striatum. Neuroscience 43:397–417

Masu Y, Nakayama K, Tamaki H et al. (1987) cDNA cloning of bovine substance-K receptor through oocyte expression system. Nature 329:836–838

Matsutani S, Senba E, Tohyama M (1989) Distribution of neuropeptidelike immunoreactivities in the guinea pig olfactory bulb. J Comp Neurol 280:577–586

McLean S, Bannon MJ, Zamir N et al. (1985a) Comparison of the substance P- and dynorphin-containing projections to the substantia nigra: a radioimmunocytochemical and biochemical study. Brain Res 361;185–192

McLean S, Skirboll LR, Pert CB (1985b) Comparison of substance P and enkephalin distribution in rat brain: an overview using radioimmunocytochemistry. Neuroscience 14:837–852

McRitchie DA, Törk I (1994) Distribution of substance P-like immunoreactive neurons and terminals throughout the nucleus of the solitary tract in the human brainstem. J Comp Neurol 343:83–101

Merchenthaler I, Maderdrut JL, O'Harte F et al. (1992) Localization of neurokinin B in the central nervous system of the rat. Peptides 13:815–829

Mick G, Najimi M, Girard M et al. (1992) Evidence for a substance P containing subpopulation in the primate suprachiasmatic nucleus. Brain Res 573:311–317

Mileusnic D, Lee JM, Magnuson DJ et al. (1999) Neurokinin-3 receptor distribution in rat and human brain: an immunohistochemical study. Neuroscience 89:1269–1290

Milner TA, Joh TH, Miller RJ et al. (1984) Substance P, neurotensin, enkephalin, and catecholamine-synthesizing enzymes: light microscopic localizations compared with autoradiographic label in solitary efferents to the rat parabrachial region. J Comp Neurol 226:434–447

Molinari M, Hendry SH, Jones EG (1987) Distributions of certain neuropeptides in the primate thalamus. Brain Res 426:270–289

Morin LP, Blanchard J, Moore RY (1992) Intergeniculate leaflet and suprachiasmatic nucleus organization and connections in the golden hamster. Vis Neurosci 8:219–230

Moss MS, Basbaum AI (1983) The peptidergic organization of the cat periaqueductal gray. II. The distribution of immunoreactive substance P and vasoactive intestinal polypeptide. J Neurosci 3:1437–1449

Mroz E, Brownstein MJ, Leeman SE (1977) Distribution of immunoassayable substance P in the rat brain: Evidence for the existence of substance P-containing tracts. In von Euler US, Pernow B (eds) Substance P. Raven Press: New York, pp 147–154

Mroz ED, Brownstein MJ, Leeman SE (1976) Evidence for substance P in the habenulointerpeduncular tract. Brain Res 113:597–599

Murakami S, Kubota Y, Kito S et al. (1989) The coexistence of substance P- and glutamic acid decarboxylase-like immunoreactivity in entopeduncular neurons of the rat. Brain Res 485:403–406

Nahin RL (1987) Immunocytochemical identification of long ascending peptidergic neurons contributing to the spinoreticular tract in the rat. Neuroscience 23:859–869

Nakane PN, Pierce GB (1967) Enzyme-labelled antibodies for the light- and electron-microscopic localization of tissue antigens and antibodies. J Cell Biol 33:307–311

Nakanishi S (1987) Substance P precursor and kininogen: their structures, gene organizations, and regulation. Physiol Rev 67:1117–1142

Nakanishi S (1991) Mammalian tachykinin receptors. Annu Rev Neurosci 14:123–136

Nakaya Y, Kaneko T, Shigemoto R et al. (1994) Immunohistochemical localization of substance P receptor in the central nervous system of the adult rat. J Comp Neurol 347:249–274

Napier TC, Mitrovic I, Churchill L et al. (1995) Substance P in the ventral pallidum: projection from the ventral striatum, and electrophysiological and behavioral consequences of pallidal substance P. Neuroscience 69:59–70

Nawa H, Hirose T, Takashima H et al. (1983) Nucleotide sequences of cloned cDNAs for two types of bovine substance P precursor. Nature 306:32–36

Nawa H, Kotani H, Nakanishi S (1984) Tissue-specific generation of two preprotachykinin mRNAs from one gene by alternative RNA splicing. Nature 312:729–734

Neal CR Jr, Newman SW (1991) Prodynorphin- and substance P-containing neurons project to the medial preoptic area in the male Syrian hamster brain. Brain Res 546:119–131

Neal CR Jr, Swann JM, Newman SW (1989) The colocalization of substance P and prodynorphin immunoreactivity in neurons of the medial preoptic area, bed nucleus of the stria terminalis and medial nucleus of the amygdala of the Syrian hamster. Brain Res 496:1–13

Neckers LM, Schwartz JP, Wyatt RJ et al. (1979) Substance P afferents from the habenula innervate the dorsal raphe nucleus. Exp Brain Res 37:619–623

Nevin K, Zhuo H, Helke CJ (1994) Neurokinin A coexists with substance P and serotonin in ventral medullary spinally projecting neurons of the rat. Peptides 15, 1003–1011

Nicholas AP, Pieribone VA, Arvidsson U et al. (1992) Serotonin-, substance P- and glutamate/aspartate-like immunoreactivities in medullo-spinal pathways of rat and primate. Neuroscience 48:545–559

Nielsen KH, Blaustein JD (1990) Many progestin receptor-containing neurons in the guinea pig ventrolateral hypothalamus contain substance P: immunocytochemical evidence. Brain Res 517:175–181

Nilsson G, Hökfelt T, Pernow B (1974) Distribution of substance P-like immunoreactivity in the rat central nervous system as revealed by immunohistochemistry. Med Biol 52:424–427

Nitsch R, Leranth C (1994) Substance P-containing hypothalamic afferents to the monkey hippocampus: an immunocytochemical, tracing, and coexistence study. Exp Brain Res 101:231–240

Nomura H, Shiosaka S, Tohyama M (1987) Distribution of substance P-like immunoreactive structures in the brainstem of the adult human brain: an immunocytochemical study. Brain Res 404:365–370

Nurnberger F, Schoniger S (2001) Presence and functional significance of neuropeptide and neurotransmitter receptors in subcommissural organ cells. Microsc Res Tech 52:534–540

Nässel DR (1999) Tachykinin-related peptides in invertebrates: a review. Peptides 20:141–158

Ong WY, Garey LJ (1990) Pyramidal neurons are immunopositive for peptides, but not GABA, in the temporal cortex of the macaque monkey (Macaca fascicularis). Brain Res 533:24–41

Otsuka M, Yoshioka K (1993) Neurotransmitter functions of mammalian tachykinins. Physiol Rev 73:229–308

Pammer C, Görcs T, Palkovits M (1990) Peptidergic innervation of the locus coeruleus cells in the human brain. Brain Res 515:247–255

Panula P, Yang HY, Costa E (1984) Comparative distribution of bombesin/GRP- and substance-P-like immunoreactivities in rat hypothalamus. J Comp Neurol 224:606–617

Parent A, Cicchetti F, Beach TG (1995) Striatal neurones displaying substance P (NK_1) receptor immunoreactivity in human and non-human primates. NeuroReport 6:721–724

Paxinos G, Emson PC, Cuello AC (1978) The substance P projections to the frontal cortex and the substantia nigra. Neurosci Lett 7:127–131

Paxinos G, Franklin KBJ (2001) The Mouse Brain in Stereotaxic Coordinates. Academic Press: San Diego

Paxinos G, Watson C (1982) The Rat Brain in Stereotaxic Coordinates. Academic Press: New York

Paxinos G, Watson C (1986) The Rat Brain in Stereotaxic Coordinates, Second Edition. Academic Press: San Diego

Penny GR, Afsharpour S, Kitai ST (1986a) The glutamate decarboxylase-, leucine enkephalin-, methionine enkephalin- and substance P-immunoreactive neurons in the neostriatum of the rat and cat: evidence for partial population overlap. Neuroscience 17:1011-1045

Penny GR, Afsharpour S, Kitai ST (1986b) Substance P-immunoreactive neurons in the neocortex of the rat: a subset of the glutamic acid decarboxylase-immunoreactive neurons. Neurosci Lett 65:53-59

Pernow B (1953) Studies on substance P: purification, occurrence, and biological actions. Acta Physiol Scand 29 Suppl 105:1-90

Pernow B (1983) Substance P. Pharmacol Rev 35:84-141

Peterson GM, Shurlow CL (1992) Morphological evidence for a substance P projection from medial septum to hippocampus. Peptides 13:509-517

Piggins HD, Samuels RE, Coogan AN et al. (2001) Distribution of substance P and neurokinin-1 receptor immunoreactivity in the suprachiasmatic nuclei and intergeniculate leaflet of hamster, mouse, and rat. J Comp Neurol 438:50-65

Powell D, Leeman SE, Tregear GW et al. (1973) Radioimmunoassay for substance P. Nature New Biol 241:252-254

Pretel S, Ruda MA (1988) Immunocytochemical analysis of noradrenaline, substance P and enkephalin axonal contacts on serotonin neurons in the caudal raphe nuclei of the cat. Neurosci Lett 89:19-24

Quartara L, Maggi CA (1997) The tachykinin NK_1 receptor. Part I: ligands and mechanisms of cellular activation. Neuropeptides 31:537-563

Quirion R, Shults CW, Moody TW et al. (1983) Autoradiographic distribution of substance P receptors in rat central nervous system. Nature 303:714-716

Reiner A, Anderson KD (1990) The patterns of neurotransmitter and neuropeptide co-occurrence among striatal projection neurons: conclusions based on recent findings. Brain Res Rev 15:251-265

Reiner A, Medina L, Haber SN (1999) The distribution of dynorphinergic terminals in striatal target regions in comparison to the distribution of substance P-containing and enkephalinergic terminals in monkeys and humans. Neuroscience 88:775-793

Ribeiro-da-Silva A, Hökfelt T (2000) Neuroanatomical localisation of Substance P in the CNS and sensory neurons. Neuropeptides 34:256-271

Ricciardi KH, Blaustein JD (1994) Projections from ventrolateral hypothalamic neurons containing progestin receptor- and substance P-immunoreactivity to specific forebrain and midbrain areas in female guinea pigs. J Neuroendocrinol 6:135-144

Riche D, De Pommery J, Menetrey D (1990) Neuropeptides and catecholamines in efferent projections of the nuclei of the solitary tract in the rat. J Comp Neurol 293:399-424

Rieck RW, Nabors CC, Updyke BV et al. (1995) Organization of the basal forebrain in the cat: localization of L-enkephalin, substance P, and choline acetyltransferase immunoreactivity. Brain Res 672:237-250

Ronnekleiv OK, Kelly MJ, Eskay RL (1984) Distribution of immunoreactive substance P neurons in the hypothalamus and pituitary of the rhesus monkey. J Comp Neurol 224:51-59

Rothman RB, Herkenham M, Pert CB et al. (1984) Visualization of rat brain receptors for the neuropeptide, substance P. Brain Res 309:47-54.

Rustoni A, Weinberg RJ (1989) The somatosensory system. In Björklund A, Hökfelt T, Swanson LW (eds) Handbook of Chemical Neuroanatomy, Vol 7: Intergrated systems of the CNS, Part II: Central visual, auditory, somatosensory, gustatory. Elsevier: Amsterdam, pp 219-321

Saffroy M, Torrens Y, Glowinski J et al. (2003) Autoradiographic distribution of tachykinin NK_2 binding sites in the rat brain: comparison with NK_1 and NK_3 binding sites. Neuroscience 116:761-773

Sakamoto N, Takatsuji K, Shiosaka S et al. (1985) Evidence for the existence of substance P-like immunoreactive neurons in the human cerebral cortex: an immunohistochemical analysis. Brain Res 325:322-324

Sakanaka M (1992) Development of neuronal elements with substance P-like immunoreactivity in the central nervous system. In Björklund A, Hökfelt T, Tohyama M (eds) Handbook of Chemical Neuroanatomy, Vol 10: Ontogeny of Transmitters and Peptides in the CNS. Elsevier: Amsterdam, pp 197–255

Sakanaka M, Inagaki S, Shiosaka S et al. (1982a) Ontogeny of substance P-containing neuron system of the rat: immunohistochemical analysis. II. Lower brain stem. Neuroscience 7:1097–1126

Sakanaka M, Shiosaka S, Takatsuki K et al. (1982b) Origins of substance P-containing fibers in the lateral septal area of young rats: immunohistochemical analysis of experimental manipulations. J Comp Neurol 212:268–277

Sakanaka M, Shiosaka S, Takatsuki K et al. (1981) Evidence for the existence of a substance P-containing pathway from the nucleus laterodorsalis tegmenti (Castaldi) to the lateral septal area of the rat. Brain Res 230:351–355

Sakanaka M, Shiosaka S, Takatsuki K et al. (1983) Evidence for the existence of a substance P-containing pathway from the nucleus laterodorsalis tegmenti (Castaldi) to the medial frontal cortex of the rat. Brain Res 259:123–126

Sanides-Kohlrausch C, Wahle P (1991) Distribution and morphology of substance P-immunoreactive structures in the olfactory bulb and olfactory peduncle of the common marmoset (Callithrix jacchus), a primate species. Neurosci Lett 131:117–120

Sasai Y, Nakanishi S (1989) Molecular characterization of rat substance K receptor and its mRNAs. Biochem Biophys Res Commun 165:695–702

Schäfer MK, Varoqui H, Defamie N et al. (2002) Molecular cloning and functional identification of mouse vesicular glutamate transporter 3 and its expression in subsets of novel excitatory neurons. J Biol Chem 277:50734–50748

Senba E, Tohyama M (1985) Origin and fine structure of substance P-containing nerve terminals in the facial nucleus of the rat: an immunohistochemical study. Exp Brain Res 57:537–546

Seress L, Leranth C (1996) Distribution of substance P-immunoreactive neurons and fibers in the monkey hippocampal formation. Neuroscience 71:633–650

Sergeyev V, Hökfelt T, Hurd Y (1999) Serotonin and substance P co-exist in dorsal raphe neurons of the human brain. NeuroReport 10:3967–3970

Seroogy K, Tsuruo Y, Hökfelt T et al. (1988) Further analysis of presence of peptides in dopamine neurons. Cholecystokinin, peptide histidine-isoleucine/vasoactive intestinal polypeptide and substance P in rat supramammillary region and mesencephalon. Exp Brain Res 72:523–534

Severini C, Improta G, Falconieri-Erspamer G et al. (2002) The tachykinin peptide family. Pharmacol Rev 54:285–322

Shigemoto R, Nakaya Y, Nomura S et al. (1993) Immunocytochemical localization of rat substance P receptor in the striatum. Neurosci Lett 153:157–160

Shigemoto R, Yokota Y, Tsuchida K et al. (1990) Cloning and expression of a rat neuromedin K receptor cDNA. J Biol Chem 265:623–628

Shinoda K, Inagaki S, Shiosaka S et al. (1984) Experimental immunohistochemical studies on the substance P neuron system in the lateral habenular nucleus of the rat: distribution and origins. J Comp Neurol 222:578–588

Shughrue PJ, Lane MV, Merchenthaler I (1996) In situ hybridization analysis of the distribution of neurokinin-3 mRNA in the rat central nervous system. J Comp Neurol 372:395–414

Shults CW, Quirion R, Chronwall B et al. (1984) A comparison of the anatomical distribution of substance P and substance P receptors in the rat central nervous system. Peptides 5:1097–1128.

Silver R, Sookhoo AI, LeSauter J et al. (1999) Multiple regulatory elements result in regional specificity in circadian rhythms of neuropeptide expression in mouse SCN. NeuroReport 10:3165–3174

Sim LJ, Joseph SA (1992) Serotonin and substance P afferents to parafascicular and central medial nuclei. Peptides 13:171–176
Skirboll L, Hökfelt T, Dockray G et al. (1983) Evidence for periaqueductal cholecystokinin-substance P neurons projecting to the spinal cord. J Neurosci 3:1151–1157
Smith RL, Baker H, Greer CA (1993) Immunohistochemical analyses of the human olfactory bulb. J Comp Neurol 333, 519–530
Sternberger LA, Hardy PH, Cuculis JJ et al. (1970) The unlabelled antibody enzyme method of immunohistochemistry. Preparation and properties of soluble antigen-antibody complex (horseradish-peroxidase-anti-horseradish peroxidase) and its use in the identification of spirochetes. J Histochem Cytochem 18:315–333
Sternini C (1991) Tachykinin and calcitonin gene-related peptide immunoreactivities and mRNAs in the mammalian enteric nervous system and sensory ganglia. Adv Exp Med Biol 298:39–51
Sugimoto T, Takada M, Kaneko T et al. (1984) Substance P-positive thalamocaudate neurons in the center median-parafascicular complex in the cat. Brain Res 323:181–184
Sun Z, Del Mar N, Meade C et al. (2002) Differential changes in striatal projection neurons in R6/2 transgenic mice for Huntington's disease. Neurobiol Dis 11:369–385
Sutoo D, Akiyama K, Yabe K (2002) A novel technique for quantitative immunohistochemical imaging of various neurochemicals in a multiple-stained brain slice. J Neurosci Methods 118:41–50
Sutoo D, Yabe K, Akiyama K (1999) Quantitative imaging of substance P in the human brain using a brain mapping analyzer. Neurosci Res 35:339–346
Szeidemann Z, Jakab RL, Shanabrough M et al. (1995) Extrinsic and intrinsic substance P innervation of the rat lateral septal area calbindin cells. Neuroscience 69:1205–1221
Takatsuki K, Kawai Y, Sakanaka M et al. (1983a) Experimental and immunohistochemical studies concerning the major origins of the substance P-containing fibers in the lateral lemniscus and lateral parabrachial area of the rat, including the fiber pathways. Neuroscience 10:57–71
Takatsuki K, Sakanaka M, Takagi H et al. (1983b) Experimental immunohistochemical studies on the distribution and origins of substance P in the medial preoptic area of the rat. Exp Brain Res 53:183–192
Taquet H, Javoy-Agid F, Mauborgne A et al. (1988) Biochemical mapping of cholecystokinin-, substance P-, [Met]enkephalin-, [Leu]enkephalin- and dynorphin A (1–8)-like immunoreactivities in the human cerebral cortex. Neuroscience 27:871–883
Teichberg VI, Cohen S, Blumberg S (1981) Distinct classes of substance P receptors revealed by a comparison of the activities of substance P and some of its segments. Regul Pept 1:327–333
Todd AJ (2002) Anatomy of primary afferents and projection neurones in the rat spinal dorsal horn with particular emphasis on substance P and the neurokinin 1 receptor. Exp Physiol 87:245–249
Tregear GW, Niall HD, Potts JT et al. (1971) Synthesis of substance P. Nature New Biol 232:87–89
Triepel J, Weindl A, Kiemle I et al. (1985) Substance P-immunoreactive neurons in the brainstem of the cat related to cardiovascular centers. Cell Tissue Res 241:31–41
Truitt WA, Coolen LM (2002) Identification of a potential ejaculation generator in the spinal cord. Science 297:1566–1569
Tsuchida K, Shigemoto R, Yokota Y et al. (1990) Tissue distribution and quantitation of the mRNAs for three rat tachykinin receptors. Eur J Biochem 193:751–757
Turcotte JC, Blaustein JD (1997) Convergence of substance P and estrogen receptor immunoreactivity in the midbrain central gray of female guinea pigs. Neuroendocrinology 66:28–37
Ubink R, Calza L, Hökfelt T (2003) 'Neuro'-peptides in glia: focus on NPY and galanin. Trends Neurosci 26:604–609

Valentino KL, Tatemoto K, Hunter J et al. (1986) Distribution of neuropeptide K-immunoreactivity in the rat central nervous system. Peptides 7:1043–1059

Velasco A, De Leon M, Covenas R et al. (1993) Distribution of neurokinin A in the cat diencephalon: an immunocytochemical study. Brain Res Bull 31:279–285

Vigna SR, Bowden JJ, McDonald DM et al. (1994) Characterization of antibodies to the rat substance P (NK-1) receptor and to a chimeric substance P receptor expressed in mammalian cells. J Neurosci 14:834–845

Vincent SR, McGeer EG (1981) A substance P projection to the hippocampus. Brain Res 215:349–351

Vincent SR, Satoh K, Armstrong DM et al. (1983) Substance P in the ascending cholinergic reticular system. Nature 306:688–691

Vincent SR, Satoh K, Armstrong DM et al. (1986) Neuropeptides and NADPH-diaphorase activity in the ascending cholinergic reticular system of the rat. Neuroscience 17:167–182

von Euler US (1942) Herstellung und Eigenschaften von Substanz P. Acta Physiol Scand 4:373–375

von Euler US, Gaddum JH (1931) An unidentified depressor substance in certain tissue extracts. J Physiol 72:74–87

Wahle P, Meyer G (1986) The olfactory tubercle of the cat. II. Immunohistochemical compartmentation. Exp Brain Res 62:528–540

Wahle P, Sanides-Kohlrausch C, Meyer G et al. (1990) Substance P- and opioid-immunoreactive structures in olfactory centers of the cat: adult pattern and postnatal development. J Comp Neurol 302:349–369

Wahlestedt C, Yanaihara N, Håkanson R (1986) Evidence for different pre- and post-junctional receptors for neuropeptide Y and related peptides. Reg Pept 13:307–318

Walter A, Mai JK, Lanta L et al. (1991) Differential distribution of immunohistochemical markers in the bed nucleus of the stria terminalis in the human brain. J Chem Neuroanat 4:281–298

Warden MK, Young WS 3rd (1988) Distribution of cells containing mRNAs encoding substance P and neurokinin B in the rat central nervous system. J Comp Neurol 272:90–113

Yamano M, Hillyard CJ, Girgis S et al. (1988) Presence of a substance P-like immunoreactive neurone system from the parabrachial area to the central amygdaloid nucleus of the rat with reference to coexistence with calcitonin gene-related peptide. Brain Res 451:179–188

Yamano M, Inagaki S, Kito S et al. (1986) A substance P-containing pathway from the hypothalamic ventromedial nucleus to the medial preoptic area of the rat: an immunohistochemical analysis. Neuroscience 18:395–402

Yanagihara M, Niimi K (1989) Substance P-like immunoreactive projection to the hippocampal formation from the posterior hypothalamus in the cat. Brain Res Bull 22:689–694

Yao R, Rameshwar P, Donnelly RJ et al. (1999) Neurokinin-1 expression and co-localization with glutamate and GABA in the hypothalamus of the cat. Mol Brain Res 71:149–158

Yokota Y, Sasai Y, Tanaka K et al. (1989) Molecular characterization of a functional cDNA for rat substance P receptor. J Biol Chem 264:17649–17652

Zeng SL, Li YQ, Rao ZR et al. (1991) Projections from serotonin- and substance P-like immunoreactive neurons in the midbrain periaqueductal gray onto the nucleus reticularis gigantocellularis pars alpha in the rat. Neurosci Lett 131:205–209

Zetler G (1970) Distribution of peptidergic neurons in mammalian brain. In Bargmann W, Scharrer B (eds) Aspects of Neuroendocrinology. Springer: Berlin, pp 287–295

Zimmer A, Zimmer AM, Baffi J et al. (1998) Hypoalgesia in mice with a targeted deletion of the tachykinin 1 gene. Proc Natl Acad Sci USA 95:2630–2635

The Nomenclature of Tachykinin Receptors

R. Patacchini · C. A. Maggi

Pharmacology Department, Menarini Ricerche, Via Rismondo 12/A, 50131 Florence, Italy
e-mail: chimfarm@menarini-ricerche.it

1	Substance P and Amphibian Tachykinins: Unaware Members of a Common Family of Peptides	122
2	Evidence for the Existence of Multiple Tachykinin Receptors, and Early Classifications	125
2.1	SP-P and SP-E Receptor Classification	125
2.2	SP-P, SP-K and SP-E Receptor Classification	126
2.2.1	SP-P, SP-E and SP-N Receptor Classification	126
2.3	NK-P, NK-A and NK-B Receptor Classification	127
3	Nomenclature Committees at Work: The NK_1, NK_2 and NK_3 Receptor Classification	127
3.1	Weaknesses of the Present Nomenclature and Related Consequences	128
4	Evidence in Favor of and Against the Existence of Further Tachykinin Receptors/Receptor Subtypes	130
4.1	Variants of the Tachykinin NK_1 Receptor	131
4.2	Variants of the Tachykinin NK_2 Receptor	131
4.3	Variants of the Tachykinin NK_3 Receptor	133
5	Identification of New Mammalian Tachykinins: Hemokinin 1 and Endokinin A–D. Do They Stimulate Unknown Tachykinin Receptors?	133
6	Conclusions	135
	References	136

Abstract The first part of this chapter briefly reviews the studies leading to identification of mammalian and nonmammalian tachykinins, and of their multiple receptors. The second part deals with the issue of nomenclature. This was first addressed in 1984 during a IUPHAR Satellite Meeting held in Maidstone. The committee recommended the use of the name tachykinins to label the class of peptides, and the names NKA, NKB for the last discovered peptides. A second meeting held in Montreal in 1986 confirmed the nomenclature proposed in Maidstone, and the labels NK_1, NK_2 and NK_3 were introduced for the receptors. Unfortunately many authors interpreted the acronym NK as the abbreviation of the word neurokinin, and started using the terms neurokinins and neurokinin receptors (often denoted as neurokinin-1, neurokinin-2 and neurokinin-3) although these terms had not been approved in Montreal. A recent analysis of the literature shows that about half of the authors follow strictly the recommendations of the Montreal Nomenclature, whereas the rest deviate in various ways. None of the various proposals put forward to overcome the present confusing situation has yet received the necessary consensus by the experts of this field. Rather, in the last meeting on nomenclature held in La Grande Motte in 2000, the assembly agreed to avoid any change to the current nomenclature. Thus, the Montreal Nomenclature remains the only one voted and approved, to which all authors should be encouraged to adhere. Since the NK_1, NK_2 and NK_3 receptor classification was approved in 1986, several novel tachykinin receptor types/subtypes have been claimed to exist, but none of the proposed variants has since then gained the necessary status of scientific evidence to justify an update. In contrast, the number of mammalian tachykinins is growing by the introduction of the newly identified mouse, rat and human hemokinin 1 peptides and of the human endokinin A and B. Recently, a new era in tachykinin nomenclature was introduced by the Human Genome Organization Gene Nomenclature Committee which approved the names TAC1 and TAC3 for the gene encoding both substance P and NKA (preprotachykinin-A) and that encoding NKB (preprotachykinin-B), respectively, and named the genes for the three tachykinin receptors TACR1, TACR2 and TACR3.

Keywords Tachykinins · Tachykinin receptors · Nomenclature · Hemokinin 1 · Endokinin A,B,C,D

1
Substance P and Amphibian Tachykinins: Unaware Members of a Common Family of Peptides

The origin of the tachykinin peptide family dates back to 1931. At that time von Euler and Gaddum (1931) while studying the tissue distribution of acetylcholine in the horse, noted that the acid ethanol extract from the intestine possessed the unforeseen property to lower blood pressure in atropinized rabbits and contract the rabbit jejunum in the presence of atropine. The active component of the extract, whose pharmacological profile did not match any one possessed by the biologically active agents known at that time, was suspected to be a peptide and was named substance P (SP) (Gaddum and Schild 1934; von Euler 1942). Despite the fact that the chemical structure of SP was still unknown, and the purity of the biological extracts containing SP was variable, plenty of studies were undertaken aimed towards investigating tissue distribution and pharmacological activities of SP in mammalian and nonmammalian species. These studies showed that SP is present in both brain and gut of many species, and that it produces spasmogenic effects in various intestinal smooth muscles preparations, and stimulates peristalsis in vivo (for reviews see Pernow 1953, 1963; Lembeck and Zetler 1962; Pernow 1983; Maggio and Mantyh 1994). Moreover, an interesting observation was made by Lembeck (1953), i.e., that higher concentrations of SP-like material are contained in the dorsal than in the ventral half of the spinal cord, leading the author to speculate that SP might act as sensory neurotrans-

Table 1 Amino acid sequence of mammalian and certain nonmammalian tachykinins and of two tachykinin-gene related peptides

Tachykinin	Sequence
Mammalian tachykinins	
Substance P	Arg-Pro-Lys-Pro-Gln-Gln-**Phe**-Phe-**Gly-Leu-Met**-NH$_2$
Neurokinin A[a]	His-Lys-Thr-Asp-Ser-**Phe**-Val-**Gly-Leu-Met**-NH$_2$
Neurokinin B[b]	Asp-Met-His-Asp-Phe-**Phe**-Val-**Gly-Leu-Met**-NH$_2$
Mouse/rat hemokinin 1[c]	Arg-Ser-Arg-Thr-Arg-Gln-**Phe**-Tyr-**Gly-Leu-Met**-NH$_2$
Human hemokinin 1[c]	Thr-Gly-Lys-Ala-Ser-Gln-**Phe**-Phe-**Gly-Leu-Met**-NH$_2$
Human hemokinin 1 (4–11)[c]	Ala-Ser-Gln-**Phe**-Phe-**Gly-Leu-Met**-NH$_2$
Endokinin A/B C-terminal decapeptide[c,d]	Gly-Lys-Ala-Ser-Gln-**Phe**-Phe-**Gly-Leu-Met**-NH$_2$
Nonmammalian tachykinins	
Eledoisin	pGlu-Pro-Ser-Lys-Asp-Ala-**Phe**-Ile-**Gly-Leu-Met**-NH$_2$
Physalaemin	pGlu-Ala-Asp-Pro-Asn-Lys-**Phe**-Tyr-**Gly-Leu-Met**-NH$_2$
Kassinin	Asp-Val-Pro-Lys-Ser-Asp-Gln-**Phe**-Val-**Gly-Leu-Met**-NH$_2$
Tachykinin-gene related peptides	
Endokinin C[c]	Lys-Lys-Ala-Tyr-Gln-Leu-Glu-His-Thr-Phe-Gln-Gly-Leu-Leu-NH$_2$
Endokinin D[c]	Val-Gly-Ala-Tyr-Gln-Leu-Glu-His-Thr-Phe-Gln-Gly-Leu-Leu-NH$_2$

[a] Originally named substance K, neurokinin α, neuromedin L.

[b] Originally named neurokinin β, neuromedin K.

[c] At the present time only the corresponding mRNAs have been detected in various mammalian tissues: both the occurrence and physiological relevance of these peptides remains to be demonstrated.

[d] The common C-terminal decapeptidic fragment of endokinin A (47 amino acids in length) and endokinin B (41 amino acids in length) has been reported by Page et al. (2003). For other references, see text.

mitter. In retrospect, this finding represents the first experimental observation implicating a role of SP in pain transmission.

In a completely different framework Erspamer and coworkers, who were investigating the occurrence of biogenic amines in extracts obtained from the posterior salivary glands of the Mediterranean octopus *Eledone moscata*, found in 1947 a unidentified substance that proved to be hypotensive, sialogogic and spasmogenic in intestinal smooth muscle, as no other known substance was, except for SP (Erspamer 1949). The chemical structure of this substance (named moschatin at first, and eledoisin later) was established in 1962 to be an undecapeptide bearing an amidated C-terminal sequence (Anastasi and Erspamer 1962; Erspamer and Anastasi 1962) (Table 1). Now it can be stated that eledoisin represents *de facto* the first tachykinin whose sequence was determined. Comparison of eledoisin with SP revealed a wide qualitative overlap in the pharmacology of the two substances, although important quantitative differences were also reported (Erspamer and Falconieri-Erspamer 1962; Erspamer and Anastasi 1962). In 1962 Erspamer and coworkers found that extracts obtained from the skin of the frog *Physalaemus biligonigerus* were endowed with eledoisin-like pharmacological activity. The bioactive principle, named *physalaemin*, was recognized to be an undecapeptide exhibiting both a sequence and a pharmacological spectrum of activities similar to those possessed by eledoisin (Anastasi et al. 1964; Erspamer et al. 1964; Table 1).

Subsequently, many other eledoisin-related peptides were isolated from the skin, brain and gut of amphibians and other submammalian species. Currently their number exceeds 40 and altogether they represent the most abundant and varied family of biologically active peptides present in both vertebrate and invertebrate animals (see Severini et al. 2002 for an updated description of these peptides). Erspamer and coworkers named these peptides tachykinins, because of their fast spasmogenic activity exerted on intestinal preparations, as compared to the slower responses produced by bradykinins (Erspamer and Anastasi 1966; Bertaccini 1976). Importantly, it was noted that all these peptides bear a common C-terminal pentapeptide sequence: Phe-X-Gly-Leu-Met, and that this shared motif accounts for (most) biological effects produced by tachykinins (Bertaccini 1976). Considering this latter finding and the observed pharmacological similarities between the mammalian and amphibian tachykinins, Erspamer (1971) put forward the hypothesis (proven to be true soon afterwards) that SP should bear a Phe-X-Gly-Leu-Met C-terminal sequence. It happened that, 40 years after the isolation of SP from equine intestine, a sialogogic peptide isolated from bovine hypothalamus by Leeman and coworkers was recognized to correspond to SP (Chang and Leeman 1970). From the amino acid sequence, finally determined in 1971 (Chang et al. 1971), it was proven that SP bears the same C terminus common to all other nonmammalian tachykinins described by Erspamer and coworkers (Table 1), thereby representing the mammalian counterpart of the peptide compounds already isolated from amphibians.

The hypothesis that more than one tachykinin peptide could exist in mammals had been advanced since the 1940s by Erspamer (1949), and sustained later by others (e.g., Maggio et al. 1983). However, more than 10 years elapsed after the sequencing of SP before the number of tachykinins in mammals grew by the discovery of two novel peptides: neurokinin A (NKA) (originally named: substance K, neurokinin α, neuromedin L) and neurokinin B (NKB) (originally named: neurokinin β, neuromedin K) (Table 1). We owe the identification of these two novel tachykinins to the work of four independent groups: Kimura et al. (1983), Maggio et al. (1983), Nawa et al. (1983), Kangawa et al. (1983) and Minamino et al. (1984). Soon afterwards the group of Nakanishi (see Nakanishi 1991) identified two genes encoding the three tachykinin peptides, which were initially named preprotachykinin-A (PPT-A) and PPT-B, and more recently have been termed TAC1 (encoding SP and NKA) and TAC3 (encoding NKB) by the Human Genome Organization (HUGO) Gene Nomenclature Committee (http://www.gene.ucl.ac.uk/nomenclature/).

2
Evidence for the Existence of Multiple Tachykinin Receptors, and Early Classifications

The existence of multiple receptors for tachykinins in mammals was first postulated by the group of Erspamer, on the basis of the relative potencies shown by SP, eledoisin, physalaemin and by the newly discovered dodecapeptide kassinin (isolated from the African frog *Kassina senegalensis*) (Table 1) in various bioassays (Falconieri-Erspamer et al. 1980). In particular, SP and physalaemin, on the one hand, and eledoisin and kassinin, on the other, showed similar pharmacological profiles, suggesting the existence of at least two different receptors for tachykinins in mammals (Falconieri-Erspamer et al. 1980; Erspamer 1981). In the following years, accumulating evidence was collected for the existence of three different receptors for tachykinins in mammals by the introduction of the novel tachykinins identified in 1983–1984 (NKA, NKB), by the use of certain receptor-selective synthetic agonists and later by receptor-selective antagonists (see Maggi 1995 for review). The isolation of three distinct genes encoding three different receptor proteins for tachykinins has provided the final proof for the existence of (at least) three tachykinin receptors in mammals (see Nakanishi 1991 for a review). These genes have recently been termed tachykinin receptor 1, 2 and 3 (TACR1, TACR2, and TACR3) by the HUGO Gene Nomenclature Committee (http://www.gene.ucl.ac.uk/nomenclature/).

2.1
SP-P and SP-E Receptor Classification

The proposal of the existence of multiple receptors for tachykinins was further developed by Iversen and coworkers, who introduced a widely used nomenclature based on two different receptor types: the SP-P and the SP-E types (Lee et al. 1982). This classification stems from the observation that, whereas SP, physalaemin, eledoisin and kassinin are equipotent in the guinea-pig ileum, eledoisin and kassinin are at least 100-fold more potent than SP and physalaemin in the rat vas deferens (Lee et al. 1982). Thus, two classes of receptors were proposed to exist: the SP-P receptor (P for physalaemin) highly responsive to SP and physalaemin, and the SP-E receptor (E for eledoisin), highly responsive to eledoisin and kassinin (Table 2). Further support for this classification was given by the observation that SP and physalaemin show complete cross-desensitization in the guinea-pig ileum, whereas eledoisin and SP do not (Couture and Regoli 1982; Lee et al. 1982). Moreover, a synthetic analog of SP, SP methyl ester (SPOMe), was found almost inactive in the rat vas deferens while being equipotent to SP in the guinea-pig ileum (Watson et al. 1983).

Table 2 Early classifications of tachykinin receptors and their confluence into the Montreal Nomenclature

Reference	Receptors		
Lee et al. (1982)	SP-P	SP-E	
	(SP)	(NKA, NKB)	
Buck et al. (1984)	SP-P	SP-K	SP-E
	(SP)	(NKA)	(NKB)
Laufer et al. (1985)	SP-P	SP-E	SP-N
	(SP)	(NKA)	(NKB)
Regoli et al. (1985, 1987)	NK-P	NK-A	NK-B
	(SP)	(NKA)	(NKB)
Henry (1987)	NK-1	NK-2	NK-3
(Montreal Symposium)	(SP)	(NKA)	(NKB)

The preferred agonist for each receptor type is given in parentheses.

2.2
SP-P, SP-K and SP-E Receptor Classification

Evidence for the existence of a third receptor for tachykinins was provided by Buck et al. in 1984, by means of radioligand binding experiments. In this study, the interaction of radio-iodinated Bolton-Hunter eledoisin, SP and substance K (i.e., NKA) on cell membranes from several tissues was investigated. The rank order of inhibition of specific binding afforded by mammalian and nonmammalian tachykinins in cerebral cortex, gastrointestinal and bladder smooth muscle was used by Buck and coworkers (1984) to identify three different binding sites: two of them corresponding to the previously reported SP-P and SP-E, and a third, identified in the intestinal smooth muscle and bladder membranes, which they named SP-K (Table 2).

2.2.1
SP-P, SP-E and SP-N Receptor Classification

Evidence for the existence of a third receptor for tachykinins was also provided by Laufer and coworkers (1985), who distinguished between the muscular receptor (SP-P) mediating the contractile response to tachykinins in the guinea-pig ileum, and the neuronal receptor (that they named SP-N) mediating neurogenic contractile effects produced by tachykinins via acetylcholine released from intramural neurons. After SP-P receptors had been blocked (by desensitization with SPOMe) NKB was shown to be the most potent tachykinin at the novel SP-N receptor (Laufer et al. 1985) (Table 2). Further support for the existence of SP-N receptors was given by the use of senktide, a synthetic peptide agonist, that was proven to be as potent as NKB at SP-N receptors in the guinea-pig ileum, whereas it was almost inactive at SP-P and SP-E receptor sites (Wormser et al. 1986). It should be noted that in the classification of Laufer and coworkers

(1985) the SP-E receptor was thought to be the preferred site for NKA, whereas in Buck's classification SP-E was thought to be the preferred site for NKB.

2.3
NK-P, NK-A and NK-B Receptor Classification

This classification proposed by Regoli and coworkers was based on the rank order of potencies shown by mammalian and nonmammalian tachykinins, their fragments and several synthetic analogs, in isolated preparations from respiratory, gastrointestinal, cardiovascular and genitourinay systems of various species (Regoli et al. 1985, 1987; Dion et al. 1987). Three receptors were identified by Regoli's group: NK-P, NK-A and NK-B, on which the three mammalian tachykinins SP, NKA and NKB were the most potent agonists, respectively (Table 2).

3
Nomenclature Committees at Work:
The NK_1, NK_2 and NK_3 Receptor Classification

The different names used by different authors to label the 'novel' tachykinins NKA and NKB (Table 1) prompted efforts to establish a common nomenclature. This issue was first addressed during a IUPHAR Satellite Meeting entitled: 'Substance P: Metabolism and Biological Actions', held in Maidstone, UK, in 1984. The main recommendations made by the committee were to use the terms: neurokinin A and neurokinin B (and the corresponding acronyms NKA and NKB) in the place of their first names, and to use the term tachykinins to label the group of peptides defined by their C terminus (Jordan and Oehme 1985). The issue of nomenclature was further expanded in 1986 during the IUPS Satellite Symposium 'Substance P and Neurokinins—Montreal '86', which was summarized in a short preface to the proceedings of the meeting (Henry 1987). The so-called Montreal Nomenclature has remained the last published document in which a common position had been agreed on after joint discussion by experts of the field. The first agreed decision taken in Montreal was to confirm the nomenclature recommended at Maidstone 2 years before, including the use of the terms: SP, NKA, NKB and tachykinins (Henry 1987) (Table 3). Indeed, a proposal was advanced to use the term neurokinins in the place of the term tachykinins, but this proposal, which included the renaming of SP as neurokinin P, was opposed in view of the evidence that SP is expressed also in certain non-neuronal cell types.

In Montreal, the nomenclature of tachykinin receptors was also discussed. At that time the existence of three distinct receptors had became evident, accounting for the radioligand binding and functional data obtained by the use of natural and synthetic tachykinins. The three receptors were identified on the basis of various pharmacological criteria, the most notable of which was the preferential (albeit not exclusive) affinity of the three mammalian tachykinins (SP, NKA and NKB) for the three receptors. After a preliminary agreement for the names:

Table 3 The Montreal Nomenclature of tachykinins and tachykinin receptors, and the most frequent deviations appearing in the literature

Name of peptides approved in Montreal	Incorrect terms recurrent in literature
Substance P (SP)	–
Neurokinin A[a] (NKA)	–
Neurokinin B[b] (NKB)	–
Tachykinins[c] (TK)	Neurokinins
Name of receptors[d] approved in Montreal	Incorrect terms recurrent in literature
NK-1	Neurokinin-1 or substance P receptor
NK-2	Neurokinin-2
NK-3	Neurokinin-3
Tachykinin receptor(s)	Neurokinin receptor(s)

The Montreal Nomenclature is that approved in Montreal during the IUPS 'Satellite Symposium on Substance P and Neurokinins—Montreal '86' in the session on *Nomenclature* for Tachykinins and Tachykinin Receptors (Henry 1987).
[a] Previously known as neurokinin alpha, neuromedin L, substance K.
[b] Previously known as neurokinin beta, neuromedin K.
[c] Denotes the name of the group of peptides having the common C-terminal sequence: Phe-X-Gly-Leu-Met-NH$_2$.
[d] For previous classifications of tachykinin receptors see Table 2. The acronyms are given in parentheses.

SP-P, NK-A and NK-B, the three receptors for tachykinins were subsequently named by vote: TK-1 (preferential affinity for SP), TK-2 (preferential affinity for NKA) and TK-3 (preferential affinity for NKB) (Henry 1987). Once this vote had been taken, it was pointed out that this nomenclature was already present in the literature (Melchiorri and Negri 1984) for nonmammalian tachykinins. It was then decided to change the originally agreed label TK into NK and to name the three tachykinin receptors as NK-1, NK-2 and NK-3 (Henry 1987; Table 3).

3.1
Weaknesses of the Present Nomenclature and Related Consequences

As one of us has pointed out previously (Maggi 2000), a source of confusion in the Montreal Nomenclature was the choice of the label NK for designing the three receptors. The label NK lends itself to be interpreted as an abbreviation of the word neurokinin. Apparently, this was not the intention of the discussants of the Montreal symposium since the assembly voted for the use of the term tachykinin instead of neurokinin and the receptors were consequently designated as tachykinin receptors (Henry 1987). This decision is now backed by the HUGO Gene Nomenclature Committee (http://www.gene.ucl.ac.uk/nomenclature/) which bases the approved gene names on the term 'tachykinin'. Seen retrospectively, the choice of the label NK has had a very bad impact on the troubled nomenclature of this family of peptides. Thereafter, part of the authors adhered strictly to the Montreal Nomenclature, but many others have followed divergent terminologies, the most common of which can be summarized as follows:

a. Some authors use the terms neurokinins and neurokinin receptors and/or substance P receptor, although these were not approved in Montreal.
b. Some authors, by following the recommendations of some journals that have adopted the policy of avoiding the use of abbreviations in the title of articles, use the terms: neurokinin-1, neurokinin-2, neurokinin-3 receptors.
c. Other authors follow a mixed nomenclature by, e.g., using both the terms tachykinins and neurokinins interchangeably in the same article, or writing phrases such as substance P (NK-1) receptor (Table 3).

In order to ascertain the current adherence to the Montreal Nomenclature and the extent of deviations there from, we have performed a bibliographic search (by titles) of journal articles published over the year 2002 in this field. Of the 150 titles examined, 46% follows the Montreal Nomenclature by using the word tachykinin(s) for the peptides and/or the acronyms NK_1, NK_2 and NK_3 for receptors, whereas 44% of the titles contains deviations from the Montreal Nomenclature, and about 10% follow a mixed nomenclature, as reported before. These results are very similar to those reported 2 years ago (Maggi 2001).

The confusing terminology of tachykinin receptors is regarded as a real problem by most scientists in the field (Maggi 2001). This problem was reconsidered by experts in the field during a workshop on Nomenclature of Tachykinins and Tachykinin Receptors, held during the meeting Tachykinins 2000, La Grande Motte, France, in 2000, whose outcome was summarized by Maggi (2001). To solve the problem, the possibility was examined to shift from the Montreal Nomenclature to a new nomenclature. To this aim various proposals were discussed, including that to return to the original labels for receptors (i.e., TK-1, TK-2 and TK-3), as approved in Montreal at first, or to endorse the term neurokinins (and neurokinin receptors) for all peptides of this family, abandoning the term tachykinins (see Maggi 2000 for a discussion of these possibilities). However, the consensus reached during the workshop held in La Grande Motte was to avoid any changes/modifications to the Montreal Nomenclature, for several reasons: (a) despite its original weaknesses, the Montreal Nomenclature has been used since 1986, and it is estimated that about 50% of authors publishing in the field follows this nomenclature; (b) some terms derived from the word tachykinin (for example the terms preprotachykinin, to describe the gene/precursors of mature peptides and the term tachykininergic to describe the neurotransmitter coding of certain neurons) receive almost 100% consensus in the present literature. Thus, a shift to neurokinins would obviously introduce further confusion on this ground. Consequently, it appears unlikely that any proposal for changes/modifications leading to a new nomenclature could achieve the necessary consensus in the scientific community (Maggi 2001).

4
Evidence in Favor of and Against the Existence of Further Tachykinin Receptors/Receptor Subtypes

The introduction of the first potent and selective antagonists of the NK_1 and NK_2 receptors in the early 1990s has enabled the identification of significant pharmacological differences among tachykinin receptors belonging to different species. Later, a species-dependent pharmacological heterogeneity of the NK_3 receptor has also emerged from studies using certain receptor-selective agonists and antagonists. In general, it has been stated that the most relevant pharmacological heterogeneity exists among tachykinin NK_1, NK_2 and NK_3 receptors belonging to human, guinea-pig, rabbit and bovine species, on the one hand, and the corresponding receptors of rat and mouse, on the other. The issue of species-dependent heterogeneity of the tachykinin receptors has been reviewed extensively in the past (e.g., Maggi 1994, 1995). Furthermore, the species-dependent heterogeneity of the NK_1 receptor is briefly reported in the chapter on NK_1 receptor antagonists in this volume. It is worth noting that these species differences in receptor pharmacology have not led to a novel nomenclature, with the exception of the proposed NK_{2A} and NK_{2B} receptor subtypes (see Sect. 2).

The following sections discuss the claimed existence of further receptors or receptor subtypes for tachykinins, some of which have been marked by new labels and terms to distinguish them from the three known receptors. It is worth mentioning, however, that none of the proposed variants has gained the necessary status of scientific evidence to justify an update of the current NK_1, NK_2 and NK_3 receptor classification. The most relevant variants of tachykinin receptors suggested in the literature are summarized in Table 4.

Table 4 Most relevant claims for the existence of further tachykinin receptors/receptor subtypes from 1986 to 2003

New type/subtype	Evidence in favor	Evidence against
Septide-sensitive NK_1 subtype	Petitet et al. (1992)[a]	Pradier et al. (1994) Hastrup and Schwartz (1996)
NK_1 receptor short isoform	Fong et al. (1992) Kage et al. (1993)	Page and Bell (2002)
NK_4 (NK_2 variant)	McKnight et al. (1988a, 1988b)	Ireland et al. (1988, 1991) Devillier et al. (1988)
NK_{2A} and NK_{2B}	Maggi et al. (1990) Patacchini et al. (1991)	Maggi (1995)
NK_4 (NK_3 variant)	Donaldson et al. (1996) Krause et al. (1997)	Sarau et al. (2000) Page and Bell (2002)

[a] Further studies in favor of the existence of septide-sensitive NK_1 receptor subtype are discussed in the chapter on NK_1 receptor antagonists within this book.

4.1
Variants of the Tachykinin NK$_1$ Receptor

The most-investigated intraspecies heterogeneity of the tachykinin NK$_1$ receptor, originating from the work of Petitet et al. (1992) who proposed the existence of a so-called septide-sensitive subtype, is detailed in the chapter on tachykinin NK$_1$ receptor antagonists in this volume. It is worth mentioning here, however, that the existence of an independent tachykinin NK$_1$ receptor protein having the characteristics of the septide-sensitive subtype, as originally defined by Petitet and coworkers (1992), has thus far remained unproven. Rather, the pharmacology of septide and of other septide-like agonists can be convincingly explained by a model in which the NK$_1$ receptor exists in two different conformers: the general tachykinin conformer (recognized with high affinity by NKA, NKB, septide and related peptides) and the SP-preferring conformer (recognized with high affinity by SP and related peptides; for a review see Maggi and Schwartz 1997).

Another putative heterogeneity of the tachykinin NK$_1$ receptor arises from the original work of Fong and coworkers (1992) who reported the cloning of a shorter form of this receptor, in addition to the native longer form. It was shown that this C-terminally truncated NK$_1$ receptor isoform (311 instead of 407 amino acids), possibly generated through pre-mRNA alternative splicing, recognizes SP with tenfold lower affinity than the native form, and may have an inappropriate coupling with intracellular effector systems (Fong et al. 1992). Mantyh et al. (1996) have suggested that the short and long isoforms might have a different tissue distribution. Further biochemical evidence for the existence of a natural tail-less form of the NK$_1$ receptor has been provided by Kage et al. (1993) in the rat salivary glands. The issue of the physiological relevance of shorter isoforms of the NK$_1$ receptor has remained an open question for years. This issue has recently been addressed by Page and Bell (2002), who failed to detect the shorter variant of the tachykinin NK$_1$ receptor identified by Fong et al. (1992) in the human placenta and in 23 other human tissues. As the short isoform could be amplified only from human genomic DNA, Page and Bell (2002) were led to speak against a physiological role of truncated isoforms of the NK$_1$ receptor in humans.

4.2
Variants of the Tachykinin NK$_2$ Receptor

The first indication of a heterogeneity of the NK$_2$ receptor, since the introduction of the NK$_1$, NK$_2$ and NK$_3$ receptor classification in 1986, was put forward by McKnight and coworkers in 1988. These authors observed an anomalous rank order of potencies of certain agonists in the guinea-pig trachea (McKnight et al. 1988a) and, more importantly, they found the cyclic hexapeptide antagonist L659877 much more potent in the guinea-pig trachea (pA$_2$=8.0) than in the rat vas deferens (pA$_2$=5.3) (McKnight et al. 1988b). As the action of tachykinins in these two preparations was believed to be mediated by NK_2 receptors only, McKnight et al. (1988a) proposed the existence of a novel type of tachykinin receptor in the guinea pig trachea: the NK_4 receptor. However, (a) the demonstration that both NK_1 and NK_2 receptors mediate tachykinin-induced contraction in the guinea-pig trachea, while only NK_2 receptors are present in the rat vas deferens, (b) the use of an agonist (eledoisin) which can stimulate both NK_1 and NK_2 receptors (McKnight et al. 1988b) and (c) the recognition that a strong peptidase activity is present in guinea-pig airways (Ireland et al. 1988, 1991; Devillier et al. 1988) led to the rejection of the NK_4 receptor hypothesis. In retrospect, the results obtained by McKnight et al. (1988a, 1988b) may be interpreted as the first indication of a species-dependent pharmacological heterogeneity of NK_2 receptors.

In the early 1990s, we postulated the existence of two NK_2 receptor subtypes, which we named NK_{2A} and NK_{2B}, on the basis of our discovery that certain tachykinin NK_2 receptor-selective antagonists are 10–100-fold more potent in rabbit, human and guinea-pig than in hamster or rat tissues, whereas other antagonists display an opposite pattern of potencies (Maggi et al. 1990, 1991; Patacchini et al. 1991). Similar findings were obtained at the molecular level by van Giersbergen et al. (1991) who showed that two NK_2 receptor-selective antagonists (MEN10207 and L659877) bound with 100-fold different affinities to bovine vs. hamster NK_2 receptors. However, further investigations clearly demonstrated that these variants did not represent true receptor subtypes, but rather reflected the existence of species-related differences in the pharmacology of the NK_2 receptor. This understanding led us to suggest that the NK_{2A}/NK_{2B} labels should be abandoned (Maggi 1995). Our proposal was based on several observations. (a) No example of different preparations from the same species showing a divergent NK_{2A}/NK_{2B} pharmacology could be clearly identified. (b) Further data obtained with various ligands indicated subtle differences in the pharmacology of NK_2 receptors thought to belong to the same NK_{2A} or NK_{2B} category. (c) Only one copy of the NK_2 receptor gene could be detected in the genomes of the various species examined, leading to exclusion of intraspecies heterogeneity of this receptor (Maggi 1995). However, this latter assumption was challenged in 2002, when Candenas and coworkers provided evidence for the existence of a tachykinin NK_2 receptor splice variant. This variant receptor, resulting from skipping of exon 2 in the processing of tachykinin NK_2 receptor mRNA, was termed by Candenas et al. (2002) the $NK_2\beta$ receptor. The $NK_2\beta$ receptor isoform would be a truncated six-transmembrane domain receptor protein, lacking the transmembrane segment 4 and the second extracellular loop. These authors reported expression of $NK_2\beta$ receptor mRNA in the human and rat uterus and in the rat urinary bladder, colon, duodenum and stomach, in amounts from 15% to 109% relative to the wild-type NK_2 receptor mRNA detected in the same tissues (Candenas et al. 2002). The physiological relevance (if any) of the $NK_2\beta$ receptor isoform, and eventually the relationship between this receptor and the NK_{2A}/NK_{2B} receptor heterogeneity remain to be investigated.

4.3
Variants of the Tachykinin NK$_3$ Receptor

A novel human orphan receptor exhibiting a remarkable sequence similarity to the rat and human NK$_3$ receptor, and initially claimed to be an atypical opioid receptor, was cloned in 1992 by Xie and coworkers. Subsequently, Donaldson et al. (1996) showed that NKB was able to elicit potent concentration-dependent responses in *Xenopus* oocytes transfected with this orphan receptor, and by using a variety of tachykinin peptide agonists they found that this new receptor (which they termed NK$_4$) showed a pattern of responses similar to that of the human NK$_3$ receptor. Further pharmacological characterization of this new receptor type was performed by Krause and coworkers (1997) by means of arachidonic acid mobilization elicited in CHO receptor-transfected cells and radioligand binding experiments. Also these authors (Krause et al. 1997) observed very similar pharmacological profiles of the human NK$_3$ receptor and the new receptor type, which was therefore proposed to be an NK$_3$ receptor homolog or a novel human tachykinin receptor. The nearly identical pharmacological profile of the two receptors (NK$_3$ wild-type and NK$_4$) was further confirmed in the study of Sarau et al. (2000). However, these latter authors failed to demonstrate the presence of the novel NK$_4$ receptor in genomic DNA from human and other species, concluding that it was premature to extend the current tachykinin receptor classification (Sarau et al. 2000). More recently, Page and Bell (2002) failed to find expression of the putative NK$_4$ receptor in 24 different human tissues and to demonstrate its presence in the human genome. Rather, they observed that the sequence of the NK$_4$ receptor was 100% identical to that of the guinea-pig tachykinin NK$_3$ receptor, concluding that the NK$_4$ receptor is in fact the guinea-pig NK$_3$ receptor (Page and Bell 2002).

5
Identification of New Mammalian Tachykinins: Hemokinin 1 and Endokinin A–D. Do They Stimulate Unknown Tachykinin Receptors?

A breakthrough in the field of tachykinins has been the recent discovery of a third tachykinin gene in mammals, encoding previously unknown tachykinins that have a wide peripheral, non-neuronal distribution. In 2000, Zhang and coworkers reported the identification of a novel gene, originally named PPT-C, that was detected in lymphoid B hematopoietic cells of mouse bone marrow. This gene, renamed tachykinin 4 (TAC4) by the HUGO Gene Nomenclature Committee (http://www.gene.ucl.ac.uk/nomenclature/), encodes a 128-amino acid peptide that, by proteolytic cleavage, yields a predicted 11-amino acid tachykinin peptide, termed hemokinin 1 (HK-1), a peptide sharing a high sequence homology with SP (Table 1). Zhang et al. (2000) reported that HK-1 promotes survival and maturation of murine pre-B cells and suggested that this peptide may act as an autocrine factor in the hematopoietic system. The high homology with SP, along with the observation that treatment of mice with an

NK_1 receptor-selective antagonist blocked bone marrow B cell development, suggested that the biological effects of HK-1 could be mediated by tachykinin NK_1 receptors. Moreover, Zhang et al. (2000) reported that HK-1 induces certain typical effects of SP known to be mediated by NK_1 receptor stimulation, like plasma protein extravasation and mast cell degranulation. Nevertheless, Zhang et al. (2000) also observed that equimolar concentrations of SP were unable to reproduce the effects of HK-1 on murine pre-B cells, and that neither NK_1 nor NK_2 or NK_3 receptor mRNAs could be detected in murine bone marrow cells. On the basis of these latter observations they hypothesized that a novel tachykinin receptor may mediate the effects of HK-1 on murine B cells. This hypothesis was challenged by Morteau et al. (2001), who showed that HK-1 binds to the NK_1 receptor with an affinity identical to that of SP, and that it is equipotent to SP on cells transfected with the human NK_1 receptor. In a broad pharmacological characterization of HK-1, we traced the pharmacological profile of this novel tachykinin at NK_1, NK_2 and NK_3 receptors, by means of radioligand binding and functional in vitro and in vivo experiments (Bellucci et al. 2002). Our data indicated that HK-1 shows a SP-like behavior in vivo, while in vitro it is a full agonist at all three tachykinin receptors, with remarkable selectivity for NK_1 as compared to NK_2 and NK_3 receptors (Bellucci et al. 2002). No evidence arose from our study of HK-1-induced effects that could not be explained by interaction with tachykinin NK_1, NK_2 or NK_3 receptors. This profile of action of HK-1 was fully confirmed in the studies of Camarda et al. (2002) and Kurtz et al. (2002).

The identification of the PPT-C (TAC4) gene in the mouse genome prompted researchers to investigate human DNA in search of a possible equivalent gene transcript. The first group to reach this important goal was that of Kurtz and coworkers (2002). They reported the isolation and characterization of the human and rat orthologs of the mouse TAC4 gene. The rat TAC4 gene encodes a predicted undecapeptide (named rat HK-1) bearing a sequence identical to the mouse HK-1 (Table 1), whereas the human TAC4 gene encodes a different undecapeptide (named human HK-1) sharing 5 of 11 amino acids with the mouse and rat HK-1 (Table 1; Kurtz et al. 2002). All human, rat and mouse HK-1 peptides bear the common C-terminal sequence: Phe-X-Gly-Leu-Met (Table 1) and can therefore be included in the tachykinin peptide family. It is worth mentioning that neither of these HK-1 peptides has been isolated from tissues and sequenced, whereas TAC4 mRNA has been detected not only in hematopoietic cells but also in a variety of peripheral tissues such as skin, heart, stomach, lung, skeletal muscle and in the brain (Kurtz et al. 2002). Interestingly, Kurtz et al. (2002) observed that the human TAC4 gene has the potential to generate a truncated form of human HK-1: HK-1(4–11). As for the other HK-1 full length peptides, both tissue distribution and physiological relevance of human HK-1(4–11) remains to be investigated. The binding and functional analysis performed by Kurtz et al. (2002) indicated that both mouse/rat HK-1 and human HK-1 are nearly identical to SP in their overall pharmacological activity at NK_1, NK_2 and NK_3 receptors, with high selectivity for NK_1 receptors. It is worth noting, however, that human HK-1 and HK-1(4–11) bind to the human NK_1 receptor with 14- and 70-fold reduced affinities as compared to SP, respectively. As to whether this reduced affinity at the human NK_1 receptor indicates the existence of a more specific, still unknown, receptor for human HK-1 peptides is a question that awaits to be explored. Likewise, the overall physiological significance of these novel tachykinins needs to be proven.

More recently, another study has appeared in the literature in which Page et al. (2003) claim the identification of a novel tachykinin gene (named TAC4), encoding four different peptides that have been termed endokinin A, B, C, and D. Page and coworkers (2003) state that the human TAC4 gene identified in their study is the equivalent of the PPT-C (TAC4) gene identified by Zhang et al. (2000) in the mouse. Thus, TAC4 should correspond to the human PPT-C (TAC4) gene described by Kurtz et al. (2002), but this hypothesis awaits to be proven at the molecular level. Page et al. (2003) reported that the TAC4 gene is spliced into four different variants, corresponding to α, β, γ and δ mRNAs which encode the four different peptides. Endokinin A consists of 47 amino acids, while endokinin B is a 41-amino acid truncated form of endokinin A. Both endokinin A and B bear the common C-terminal sequence of tachykinins (Table 1). In contrast, the other two peptides, endokinin C and D, bear a slightly different terminal sequence (Table 1) that hampers these peptides to be included in the tachykinin peptide family. In our view, endokinin C and D could be regarded and denoted as tachykinin gene-related peptides. It should be noted that both endokinin A and B are de facto elongated forms of human HK-1 (see Table 1; Page et al. 2003; Kurtz et al. 2002). Page et al. (2003) characterized the pharmacological profile of the common decapeptide C-terminal fragment of endokinin A and B and that of endokinin C and D at tachykinin NK_1, NK_2 and NK_3 receptors in radioligand binding and functional in vitro and in vivo experiments. As observed for the various HK-1 peptides, the endokinin A and B decamers were equivalent to SP in their affinity for the three tachykinin receptors and were equipotent to SP in the functional bioassays. In contrast, endokinin C and D showed very low (if any) affinity for tachykinin receptors in competition binding experiments, and produced hemodynamic effects such as hypotension and tachycardia comparable to those of SP only at 100–1000 times higher doses. Interestingly, TAC4 expression was detected primarily in the adrenal gland and placenta, leading Page et al. (2003) to state that endokinin A and B could be the endocrine/paracrine agonists of peripheral tachykinin receptors. At present, the question whether the effects of endokinin C and D may be mediated by an as yet unidentified receptor remains to be addressed.

6
Conclusions

At the present stage, the following conclusions on the nomenclature of tachykinins and tachykinin receptors can be drawn.

a. The Montreal Nomenclature (Table 3) remains the last published classification that has been agreed on after joint discussion and vote by experts in the field. This nomenclature is followed by about half of the authors publishing in the tachykinin field
b. Since the introduction of the Montreal Nomenclature, several novel tachykinin receptor types/subtypes have been claimed to exist. However, none of the proposed variants has been endowed with scientific evidence sufficient to justify an update of the current NK_1, NK_2 and NK_3 receptor classification
c. In contrast, the number of mammalian tachykinins is going to be extended, to make room for the newly identified HK-1 peptides and endokinin A and B. A word of caution is yet necessary, since it has not yet been ascertained whether these novel tachykinins are actually released from the various cells having the possibility to produce them, and their eventual physiological relevance is unknown at present
d. As we have pointed out recently (Patacchini et al. 2004), in the light of the reported non-neuronal distribution of mouse, rat and human HK-1 peptides and endokinin A–D (Zhang et al. 2000; Kurtz et al. 2002; Page et al. 2003), the use of the term neurokinins (and neurokinin receptors) to label mammalian tachykinins should be definitely abandoned in favor of the term tachykinin. This would also be in agreement with the decision of the HUGO Gene Nomenclature Committee which bases the approved gene names on this latter term

Acknowledgements. The authors are grateful to Dr. Alessia Bernini, R&D Assistant, Menarini Ricerche SpA, Florence, Italy, for helpful technical assistance in the preparation of this article.

References

Anastasi A, Erspamer V (1962) Occurrence and some properties of eledoisin in extracts of posterior salivary glands of Eledone. Br J Pharmacol 19:326–336
Anastasi A, Erspamer V, Cei JM (1964) Isolation and amino acid sequence of physalaemin, the main active polypeptide of the skin of *Physalaemus fuscomaculatus*. Arch Biochem Biophys 108:341–348
Bellucci F, Carini F, Catalani C, Cucchi P, Lecci A, Meini S, Patacchini R, Quartara L, Ricci R, Tramontana M, Giuliani S, Maggi CA (2002) Pharmacological profile of the novel mammalian tachykinin, hemokinin 1. Br J Pharmacol 135:266–274
Bertaccini G (1976) Active peptides of nonmammalian origin. Pharmacol Rev 28:127–177
Buck SH, Burcher E, Shults CW, Lovenberg W, O'Donohue (1984) Novel pharmacology of substance K binding sites: a third type of tachykinin receptor. Science 226:987–989
Camarda V, Rizzi A, Calo G, Guerrini R, Salvadori S, Regoli D (2002) Pharmacological profile of hemokinin 1: a novel member of the tachykinin family. Life Sci 71:363–370
Candenas ML, Cintado CG, Pennefather JN, Pereda MT, Loizaga JM, Maggi CA, Pinto FM (2002) Identification of a tachykinin NK_2 receptor splice variant and its expression in human and rat tissues. Life Sci 72:269–277
Chang MM, Leeman SE (1970) Isolation of a sialogogic peptide from bovine hypothalamic tissue and its characterization as substance P. J Biol Chem 245:4784–4790

Chang MM, Leeman SE, Niall HD (1971) Aminoacid sequence of substance P. Nature New Biol 232:86–87

Couture R, Regoli D (1982) Mini review: smooth muscle pharmacology of substance P. Pharmacology 24:1–25

Devillier P, Advenier C, Drapeau G, Marsac J, Regoli D (1988) Comparison of the effects of epithelium removal and of an enkephalinase inhibitor on the neurokinin-induced contractions of guinea-pig isolated trachea. Br J Pharmacol 94:675–684

Dion S, D'Orleans-Juste P, Drapeau G, Rhaleb NE, Rouissi N, Tousignant C, Regoli D (1987) Characterization of neurokinin receptors in various isolated organs by the use of selective agonists. Life Sci 41:2269–2278

Donaldson LF, Haskell CA, Hanley MR (1996) Functional characterization by heterologous expression of a novel cloned tachykinin peptide receptor. Biochem J 320:1–5

Erspamer V (1949) Ricerche preliminari sulla moschatina. Experientia 5:79–81

Erspamer V (1971) Biogenic amines and active polypeptides of the amphibian skin. Ann Rev Pharmacol 11:327–350

Erspamer V (1981) The tachykinin peptide family. Trends Neurosci 4:267–269

Erspamer V, Anastasi A (1962) Structure and pharmacological actions of eledoisin, the active endecapeptide of the posterior salivary glands of Eledone. Experientia 18:58–59.

Erspamer V, Anastasi A (1966) Polypeptides active on plain muscle in the amphibian skin. In: Erdös EG, Back N, Sicuteri F, Wilde A (eds) Hypotensive peptides. Springer-Verlag, New York, pp 63–75

Erspamer V, Falconieri-Erspamer G (1962) Pharmacological actions of eledoisin on extravascular smooth muscle. Br J Pharmacol 19:337–354

Erspamer V, Anastasi A, Bertaccini G, Cei JM (1964) Structure and pharmacological actions of physalaemin, the main active polypeptide of the skin of *Physalaemus fuscomaculatus*. Experientia 20:489–490

Falconieri-Erspamer G, Erspamer V, Piccinelli D (1980) Parallel bioassay of physalaemin and kassinin, a tachykinin dodecapeptide from the skin of the African frog Kassina senegalensis. Naunyn-Schmiedeberg's Arch Pharmacol 311:61–65

Fong TM, Anderson S, Yu H, Huang RRC, Strader CD (1992) Differential activation of intracellular effector by two isoforms of human neurokinin 1 receptor. Mol Pharmacol 41:24–30

Gaddum JH, Schild H (1934) Depressor substances in extracts of intestine. J Physiol London 83:1–14

Hastrup, H, Schwartz, TW (1996) Septide and neurokinin A are high affinity ligands on the NK_1 receptor:evidence from homologus versus heterologus binding analysis. FEBS Lett 399:264–266

Henry JL (1987) Discussions of nomenclature for tachykinins and tachykinin receptors. In: Henry JL, Couture R, Cuello AC, Pelletier R, Quiron R, Regoli D (eds) Substance P and neurokinins. Springer Verlag, New York, pp xvii–xviii

Ireland SJ, Jordan CC, Stephens-Smith ML, Ward P (1988) Receptors mediating the contractile response to neurokinin agonists in the guinea-pig trachea. Regul Pept 22:93

Ireland SJ, Bailey F, Cook A, Hagan RM, Jordan CC, Stephens-Smith ML (1991) Receptors mediating tachykinin-induced contractile responses in guinea-pig trachea. Br J Pharmacol 103:1463–1469

Jordan CC, Oehme P (1985) Recommendations on nomenclature for tachykinins. In: Jordan CC, Oehme P (eds) Substance P: metabolism and biological actions. Taylor & Francis, London, Philadelphia, pp xi–xii

Kage R, Leeman SE, Boyd ND (1993) Biochemical characterization of two different forms of the substance P receptor in rat submaxillary gland. J Neurochem 60:347–351

Kanagawa K, Minamino N, Fukuda A, Matsuo H (1983) Neuromedin K: a novel mammalian tachykinin identified in porcine spinal cord. Biochem Biophys Res Comm 114:533–540

Kimura S, Okada M, Sugita Y, Kanazawa I, Munekata E (1983) Novel neuropeptides, neurokinin α and β, isolated from porcine spinal cord. Proc Jpn Acad Ser B Phys Biol Sci 59:101–104

Krause JE, Staveteig PT, Mentzer JN, Schmidt SK, Tucker JB, Brodbeck RM, Bu JY, Karpitskiy VV (1997) Functional expression of a novel human neurokinin-3 receptor homolog that binds [^{3}H]senktide and [^{125}I-MePhe^{7}]neurokinin B, and is responsive to tachykinin peptide agonists. Proc Natl Acad Sci USA 94:310–315

Kurtz MM, Wang R, Clements M, Cascieri M, Austin C, Cunningham B, Chicchi G, Liu Q (2002) Identification, localization and receptor characterization of novel mammalian substance P-like peptides. Gene 296:205–212

Laufer R, Wormser U, Friedman ZY, Gilon C, Chorev M, Selinger Z (1985) Neurokinin B is a preferred agonist for a neuronal substance P receptor and its action is antagonized by enkephalin. Proc Natl Acad Sci USA 82:7444–7448

Lee CM, Iversen LL, Hanley MR, Sandberg BEB (1982) The possible existence of multiple receptors for SP. Naunyn-Schmiedeberg's Arch Pharmacol 318:281–288

Lembeck F (1953) Zur frage der zentralen ubertragung afferenter impulse. III. Das vorkommen und die bedeutung der substanz P in den dorsalen wurzeln des ruckenmarks. Naunyn-Schmiedeberg's Arch Exp Pathol Pharmacol 219:197–213

Maggi C A (1994) Evidence for receptor subtypes/species variants of receptors. In: Buck SH (ed) The tachykinin receptors. Humana Press Inc, Totowa, NJ, pp 395–470

Maggi CA (1995) The mammalian tachykinin receptors. Gen Pharmacol 26:911–944

Maggi CA (2000) The troubled story of tachykinins and neurokinins. Trends Pharmacol Sci 21:173–175

Maggi CA (2001) The troubled story of tachykinins and neurokinins: an update. Trends Pharmacol Sci 22:16

Maggi CA, Schwartz TW (1997) The dual nature of the tachykinin NK_1 receptor. Trends Pharmacol Sci 18:351–355

Maggi CA, Patacchini R, Giuliani S, Rovero P, Dion S, Regoli D, Giachetti A, Meli A (1990) Competitive antagonists discriminate between NK_2 tachykinin receptor subtypes. Br J Pharmacol 100:588–592

Maggi CA, Patacchini R, Quartara L, Rovero P, Santicioli P (1991) Tachykinin receptors in the guinea-pig isolated bronchi. Eur J Pharmacol 197:167–174

Maggi CA, Patacchini R, Rovero P, Giachetti A (1993) Tachykinin receptors and tachykinin receptor antagonists. J Autonom Pharmacol 13:23–93

Maggio JE, Mantyh PW (1994) History of tachykinin peptides. In: Buck SH (ed) The tachykinin receptors. Humana Press Inc, Totowa NJ, pp 1–18

Maggio JE, Sandberg BEB, Bradley CV, Iversen LL, Santikarn S, Williams DH, Hunter JC, Hanley MR (1983) Substance K: a novel tachykinin in mammalian spinal cord. In: Skrabanek P, Powell D (eds) Substance P. Boole, Dublin, Ireland, pp 20–21

Mantyh PW, Rogers SD, Ghilardi JR, Maggio JE, Mantyh CR, Vigna SR (1996) Differential expression of two isoforms of the neurokinin-1 (substance P) receptor in vivo. Brain Res 719:8–13

McKnight AT, Maguire JJ, Varney, MA, Williams BJ, Iversen LL (1988a) Characterization of tachykinin receptors using selectivity of agonists and antagonists: evidence for an NK-4 type. Reg Peptides 22:126

McKnight AT, Maguire JJ, Williams BJ, Foster AC, Tridgett R, Iversen LL (1988b) Pharmacological specificity of synthetic peptides as antagonists at tachykinin receptors. Regul Pept 22:127

Melchiorri P, Negri L (1984) Evolutionary aspects of amphibian peptides. In: Faulkner et al. (eds) Evolution and tumor pathology of the neuroendocrine system. Elsevier, Amsterdam, pp 231–244

Minamino N, Kangawa K, Fukuda A, Matsuo H (1984) Neuromedin L: a novel mammalian tachykinin identified in porcine spinal cord. Neuropeptides 4:157–166

Morteau O, Lu B, Gerard C, Gerard NP (2001) Hemokinin 1 is a full agonist at the substance P receptor. Nature Immunol 2:1088
Nakanishi S (1991) Mammalian tachykinin receptors. Ann Rev Neurosci 14:123–136
Nawa H, Hirose T, Takashima H, Inayama S, Nakanishi S (1983) Nucleotide sequences of cloned cDNAs for two types of bovine substance P precursor. Nature 306:32–36
Page NM, Bell NJ (2002) The human tachykinin NK_1 (short form) and tachykinin NK_4 receptor: a reappraisal. Eur J Pharmacol 437:27–30
Page NM, Bell NJ, Gardiner SM, Manyonda IT, Brayley KJ, Strange PG, Lowry PJ (2003) Characterization of novel human peripheral substance P-like tachykinins with cardiovascular activity. Proc Natl Acad Sci USA 100:6245–6250
Patacchini R, Astolfi M, Quartara L, Rovero P, Giachetti A and Maggi CA (1991) Further evidence for the existence of NK_2 tachykinin receptor subtypes. Br J Pharmacol 104:91–96
Patacchini R, Lecci A, Holzer P, Maggi CA, (2004) Newly discovered tachykinins raise new questions about their peripheral roles and the tachykinin nomenclature. Trends Pharmacol Sci 25:1–3
Pernow B (1953) Studies on substance P: purification, occurrence and biological actions. Acta Physiol Scand Suppl 105:1–90
Pernow B (1963) Pharmacology of substance P. Ann NY Acad Sci 104:393–402
Pernow B (1983) Substance P. Pharmacol Rev 35:85–141
Petitet F, Saffroy M, Torrens Y, Lavielle S, Chassaing G, Loeuillet D, Glowinski J, Beaujouan JC (1992) Possible existence of a new tachykinin receptor subtype in the guinea-pig ileum. Peptides 13:383–388
Pradier L, Menager J, Le Guern J, Bock MD, Heuillet E, Fardin V, Garret C, Doble A, Mayaux JF (1994) Septide: an agonist for the NK_1 receptor acting at a site distinct from substance P. Mol Pharmacol 45:287–293
Regoli D, D'Orleans-Juste P, Drapeau G, Dion S, Escher E (1985) Pharmacological characterization of substance P antagonists. In: Hakanson R, Sundler F (eds) Tachykinin antagonists. Elsevier, Amsterdam, pp 277–287
Regoli D, Drapeau G, Dion S, D'Orleans-Juste P, (1987) Receptors for neurokinins in peripheral organs. In: Henry JL, Couture R, Cuello AC, Pelletier R, Quiron R, Regoli D (eds) Substance P and neurokinins. Springer Verlag, New York, pp 99–107
Sarau HM, Mooney JL, Schmidt DB, Foley JJ, Buckley PT, Giardina GAM, Wang DY, Lee JA, Hay DWP (2000) Evidence that the proposed novel human 'neurokinin-4' receptor is pharmacologically similar to the human neurokinin-3 receptor but is not of human origin. Mol Pharmacol 58:552–559
Van Giersbergen PLM, Shatzer SA, Henderson AK, Lai J, Nakanishi S, Yamamura HI, Buck SH (1991) Characterization of a tachykinin peptide NK-2 receptor transfected into murine fibroblast B82 cell. Proc Natl Acad Sci USA 88:1661–1665
Von Euler US (1942) Herstellung und Eigenschaften von Substanz P. Acta Physiol Scand 4:373–375
Von Euler US, Gaddum JH (1931) An unidentified depressor substance in certain tissue extracts. J Physiol London 72:74–87
Watson SP, Sandberg BE, Hanley MR, Iversen LL (1983) Tissue selectivity of substance P alkyl esters: suggesting multiple receptors. Eur J Pharmacol. 87:77–84
Wormser U, Laufer R, Hart Y, Chorev M, Gilon C, Selinger Z (1986) Highly selective agonists for substance P receptor subtypes. EMBO J 11:2805–2808
Xie GX, Miyajima A, Goldstein A (1992) Expression cloning of cDNA encoding a seven-helix receptor from human placenta with affinity for opioid ligands. Proc Natl Acad Sci USA 89:4124–4128
Zhang Y, Lu L, Furlonger C, Wu GE, Paige CJ (2000) Hemokinin is a hematopoietic-specific tachykinin that regulates B lymphopoiesis. Nature Immunol 1:392–397

The Mechanism and Function of Agonist-Induced Trafficking of Tachykinin Receptors

D. Roosterman · N. W. Bunnett

Departments of Surgery and Physiology, University of California, Room C317,
UCSF, 513 Parnassus Avenue, San Francisco, CA, 94143-0660 USA
e-mail: nigelb@itsa.ucsf.edu

1	Introduction	142
2	Desensitization of Tachykinin Receptors	143
2.1	General Mechanisms of Desensitization of GPCRs	143
2.2	Agonist-Induced Phosphorylation of Tachykinin Receptors	146
2.3	Mediated Desensitization of Tachykinin Receptors	147
2.4	Domains of the Tachykinin Receptors That Specify Desensitization.	148
2.5	Agonist Dependency of Desensitization of Tachykinin Receptors	148
3	Agonist-Induced Endocytosis of Tachykinin Receptors	149
3.1	General Mechanisms of Endocytosis of GPCRs	149
3.2	Molecular Mechanisms of Agonist-Induced Endocytosis of Tachykinin Receptors.	151
3.3	Endocytic Domains of Tachykinin Receptors	151
3.4	Interactions of Tachykinin Receptors with Arrestins	152
3.5	Trafficking of Tachykinin Receptors in the Nervous System	156
3.6	Endocytic Ablation of Neurons Expressing the Tachykinin NK_1 Receptor	159
4	Tachykinin Receptor Endocytosis and Mitogenic Signaling.	160
4.1	General Mechanisms of Arrestin-Dependent Mitogenic Signaling of GPCRs	160
4.2	Arrestin-Dependent Mitogenic Signaling of the Tachykinin NK_1 Receptor	161
5	Intracellular Trafficking of Tachykinin Receptors: Resensitization and Downregulation	163
5.1	General Mechanisms of Intracellular Trafficking of GPCRs	163
5.2	Recycling and Resensitization of Tachykinin Receptors	165
5.3	Downregulation of Tachykinin Receptors	166
6	Conclusions	166
References		167

Abstract Biological responses to agonists of G-protein coupled receptors are regulated by control of the interactions of receptors with signaling proteins and through control of the levels of receptors that are present at the cell surface. Agonist binding to receptors induces conformational changes that stabilize interaction between receptors and heterotrimeric G-proteins, which signal to a large number of pathways. This activation is often immediately followed by a series of events beginning with receptor phosphorylation that serve to uncouple receptors from heterotrimeric G-proteins and thereby terminate signaling. Another mechanism of control involves processes that regulate the trafficking of receptors to and from the cell surface. The capacity of cells to respond to agonists is critically dependent on the presence of receptors at the surface of cells, where they can bind to lipophobic agonists in the extracellular fluid and interact with heterotrimeric G-proteins and related signaling molecules that are often clustered at the plasma membrane or in microdomains within the cell. The presence of receptors at the plasma membrane is a balance between the removal of receptors by endocytosis, which is often regulated by binding to agonists, and replenishment by recycling of internalized receptors or by mobilization of stored or freshly synthesized receptors. This chapter focuses on our understanding of these mechanisms that control signaling of the neuropeptide SP and its neurokinin receptors.

Keywords Substance P · Neurokinin receptors · Endocytosis · Recycling · Desensitization · Downregulation · Resensitization · Trafficking

Abbreviations

β_2AR	β_2-Adrenergic receptor
ERK	Extracellular signal response kinases
GABA	Gamma-aminobutyric acid
GFP	Green fluorescent protein
GPCR	G-protein-coupled receptor
GRK	G-protein receptor kinase
MAP kinase	Mitogen-activated protein kinase
NKA	Neurokinin A
NKR	(NK_1R, NK_2R, NK_3R) tachykinin NK_1, NK_2 and NK_3 receptors
NMDA	N-methyl-D-aspartate
PKC	Protein kinase C
SP	Substance P

1
Introduction

Cells can regulate their capacity to respond to agonists of G-protein-coupled receptors (GPCRs) through control of the interactions between receptors and components of signaling cascades and through regulation of the levels of receptors that are present at the cell surface. The interaction of receptors with heterotrimeric G-proteins is regulated by agonist binding and by the process of desensitization. Thus, agonist binding to extracellular regions of GPCRs induces conformational changes in the receptor that, by undefined mechanisms, permit the intracellular portions to couple to heterotrimeric G-proteins, which signal to a large number of pathways. This activation is often immediately followed by a series of events beginning with receptor phosphorylation that serve to uncouple receptors from heterotrimeric G-proteins and thereby terminate signaling, a process known as desensitization. Another mechanism of control involves processes that regulate the trafficking of GPCRs between the cell surface and intracellular locations. The capacity of cells to respond to agonists is critically dependent on the presence of receptors at the surface of cells, where they can bind to lipophobic agonists in the extracellular fluid and interact with heterotrimeric G-proteins and related signaling molecules that are often clustered at the plasma membrane. Some components of signaling also require the translocation of receptors to microdomains within the cell where they can interact with particular signaling molecules. The presence of receptors at the plasma membrane is a balance between the removal of receptors by endocytosis, which is often regulated by binding to agonists, and replenishment by recycling of internalized receptors or by mobilization of stored or freshly synthesized receptors. These regulatory mechanisms are of critical importance since they determine the capacity of many receptors to signal and prevent uncontrolled signaling, that can cause disease.

This chapter focuses on the process of agonist-induced trafficking of tachykinin (NK_1, NK_2, NK_3) receptors (NKRs). Since trafficking of receptors and associated regulatory proteins is inextricably linked to desensitization, downregulation and mitogenic signaling, we will also discuss these events. Emphasis will be placed on the NK_1R, since comparatively little is known about trafficking of the NK_2R or NK_3R. The general approach will be to summarize briefly our current knowledge of regulation of GPCRs in general, and then to discuss regulation of the NKRs. There are several comprehensive reviews on general mechanisms of regulation of GPCRs (Bohm et al. 1997a; Lefkowitz 1998; Tsao et al. 2000; Luttrell et al. 2002).

2
Desensitization of Tachykinin Receptors

2.1
General Mechanisms of Desensitization of GPCRs

Desensitization is a rapid process that terminates signaling of GPCRs. There are two general mechanisms of desensitization: homologous desensitization refers to the process by which activation of one receptor prevents reactivation of that same receptor; heterologous desensitization is a process by which activation of one receptor induces desensitization of a different receptor.

The mechanisms of homologous desensitization have been investigated in detail for several GPCRs, in particular rhodopsin and the β_2-adrenergic receptor (β_2AR). Homologous desensitization is mediated by receptor phosphorylation by G-protein receptor kinases (GRK) followed by interaction of the phosphorylated receptor with β-arrestins, which uncouple receptors from G-proteins (Fig. 1A). Seven GRKs have been identified. GRK1 and GRK7 regulate photore-

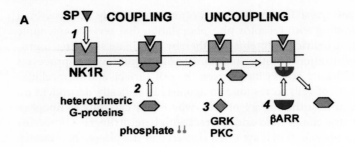

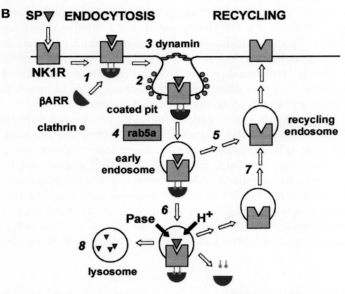

Fig. 1 A Potential mechanisms of desensitization of the tachykinin NK_1 receptor (NK_1R). (1) SP binding induces conformational changes in the NK_1R to facilitate coupling to heterotrimeric G-proteins (2) and signal transduction. (3) Second messenger kinases such as PKC and G-protein receptor kinases such as GRK2, 3 and 5 translocate to the plasma membrane and phosphorylate the NK_1R. (4) β-Arrestins (βARR) interact with GRK-phosphorylated receptors and disrupt interaction of the NK_1R with G-proteins, which desensitizes the NK_1R. **B** Potential mechanisms of endocytosis of the NK_1R. (1) β-Arrestins interact with the NK_1R and serve as adaptor proteins (2) that couple the NK_1R to clathrin and AP2 in coated pits. (3) The GTPase dynamin is required for the formation of vesicles from clathrin-coated pits. (4) Rab5a mediates the translocation of the NK_1R from superficial vesicles to early endosomes in a perinuclear region. (5) After brief stimulation with low concentrations of SP, the NK_1R rapidly recycles from superficially located early endosomes to the plasma membrane. (6) After longer stimulation with higher agonist concentrations, the NK_1R translocates to a perinuclear sorting compartment. Endosomal acidification promotes the dissociation of SP and endosomal phosphatases (*Pase*) promotes NK_1R dephosphorylation and dissociation of β-arrestins. (7) The NK_1R slowly recycles to the plasma membrane whereas SP is degraded in lysosomes (8)

ceptors: GRK1 is a rhodopsin kinase and GRK7 is a probable opsin kinase. The remaining kinases are widely distributed and mediate desensitization of GPCRs for hormones and neurotransmitters. Receptor activation triggers the translocation of GRKs from the cytosol to the plasma membrane, where they can interact with receptors. The mechanisms of membrane targeting differ between the GRKs, although this targeting depends on domains in the carboxyl tail. In the case of GRK1 and GRK7, a carboxyl CAAX motif is required for trafficking to the plasma membrane; light-induced translocation of GRK1 is facilitated by farnesylation of the CAAX domain. For GRK2 and GRK3 (formerly known as the β-adrenergic kinases 1 and 2), carboxyl terminal $G\beta$-subunit-binding and pleckstrin-homology domains are required for membrane trafficking; GRK2 and GRK3 translocate to the cell surface through interaction of these domains with free $G\beta$ subunits and inositol phospholipids. This is a precise mechanism of targeting since these binding partners are preferentially localized to activated receptors at the cell surface. For GRK5, membrane targeting involves electrostatic interaction of a basic carboxyl region with membrane phospholipids. For GRK4 and GRK6, palmitoylation of carboxyl cysteine residues results in constitutive membrane localization. GRKs are serine-threonine kinases that can phosphorylate agonist-occupied GPCRs within the third intracellular loop and carboxyl-terminus, although the precise residues that are phosphorylated after physiologically appropriate stimulation is not fully defined for most GPCRs. This phosphorylation alone is generally insufficient to induce desensitization.

Arrestins are a family of four proteins that are essential cofactors for GRK-induced desensitization of GPCRs. Visual arrestin (arrestin 1) and cone arrestin (arrestin 4) are expressed in the retina, where they interact with GRK1-phosphorylated rhodopsin and thus control the function of photoreceptors. The β-arrestins were discovered during attempts to study GRK2-mediated desensitization of β_2AR. The unexpected finding was that the ability of preparations of GRK2 to desensitize the β_2AR declined as the purity of the preparations increased, due to the removal of an essential cofactor during the purification. The observation that potency was restored by addition of visual arrestin led to the hypothesis that other arrestin-like proteins were required for GRK2 function, and subsequently β-arrestin 1 (arrestin 2) and β-arrestin 2 (arrestin 3) were isolated. It is now established that β-arrestins mediate desensitization of many GPCRs. In common with the GRKs, membrane translocation is essential for the function of arrestins. In unstimulated cells, arrestins are mostly present in the cytosol. However, they translocate to the plasma membrane within seconds of agonist binding and interact with activated GPCRs (Fig. 1A). This membrane translocation is caused by GRK-induced phosphorylation of the GPCR, which increases its affinity to interact with arrestins by many fold. The binding of β-arrestin to GRK-phosphorylated receptors serves to sterically interfere with interaction of GPCRs with heterotrimeric G-proteins and thereby to induce homologous desensitization.

Second-messenger-dependent protein kinases, including cyclic-AMP-dependent protein kinase (PKA) and protein kinase C (PKC), mediate heterologous desensitization of GPCRs. These kinases can phosphorylate serine and threonine residues within the cytoplasmic loops and carboxyl-terminal tail domains of many GPCRs. Notably, this phosphorylation does not induce interaction with β-arrestins but is sufficient to disrupt coupling of GPCRs to heterotrimeric G-proteins and thereby to terminate signal transduction. Although second messenger kinases are activated by receptor occupancy and could thus contribute to desensitization of the agonist-bound receptor (i.e., mediate homologous desensitization), these kinases are also activated by other receptors and can phosphorylate and thus inactivate different GPCRs to mediate heterologous desensitization.

2.2
Agonist-Induced Phosphorylation of Tachykinin Receptors

Responses to tachykinins that are mediated by all NKRs desensitize to repeated stimulation and fade in the continuous presence of agonist, indicative of homologous desensitization. However, there are some differences in the extent of desensitization between the NKRs. Thus, whereas NK_1R and NK_3R readily desensitize, the NK_2R desensitizes more slowly.

The phosphorylation of the NK_1R has been studied in detail using reconstituted systems and in NK_1R-transfected cell lines. The general consensus of these studies is that substance P (SP) stimulation induces rapid phosphorylation of the NK_1R through mechanisms that probably involve several GRKs; PKC can also phosphorylate the NK_1R but PKC does not mediate agonist-induced phosphorylation. The first study of NK_1R phosphorylation examined the rat NK_1R expressed in Sf9 insect cells and reconstituted in phospholipid vesicles (Kwatra et al. 1993). In this system, purified GRK2 and GRK3 were found to induce extensive phosphorylation in the presence of SP and Gq/11. GRK2 also phosphorylates the human NK_1R in expressed in Sf9 cell membranes in the presence of SP (Nishimura et al. 1998). Of interest, GRK2 strongly phosphorylates the NK_1R even in the absence of G$\beta\gamma$, which potently stimulates GRK2-catalyzed phosphorylation of the β_2AR. Purified GRK5 can also phosphorylate the human NK_1R in Sf9 membranes (Warabi et al. 2002). In the absence of SP, GRK2 and GRK5 only slowly phosphorylate the NK_1R to a small degree. However, SP promotes both GRK2-and GRK5-mediated phosphorylation of the NK_1R with similar potency and to a similar maximal stoichiometry of ~20 moles phosphate per mole of receptor. Studies in intact cells also confirm that SP induces phosphorylation of the NK_1R. SP stimulates a rapid, concentration dependent increase in phosphorylation of the human NK_1R expressed in Chinese hamster ovary (CHO) cells (Roush et al. 1999) and HEK293 cells (Barak et al. 1999). SP also induces phosphorylation of the rat NK_1R expressed in KNRK cells that correlates with rapid desensitization of generation of 1,4,5 inositol trisphosphate (Vigna 1999). The kinase(s) that are responsible for SP-induced phosphorylation of the NK_1R in these cells remain to be unequivocally identified, but candidates include GRK2 and GRK5. Thus, in HEK cells, SP-induced desensitization of the

NK_1R is accompanied by rapid membrane translocation of GRK2 tagged with green fluorescent protein (GFP; Barak et al. 1999). In myenteric neurons in primary culture, SP also triggers the rapid membrane translocation of GRK2/3 (McConalogue et al. 1998).

Second messenger kinase can also phosphorylate the NK_1R and induce desensitization (Roush et al. 1999; Vigna 1999; Dery et al. 2001). Thus, activation of PKC with phorbol 12-myristate 13-acetate (PMA) induces phosphorylation and desensitization of both receptors, whereas PKA does not contribute. However, several observations suggest that PKC does not mediate SP-induced phosphorylation. Firstly, SP-induced phosphorylation is unaffected by the general PKC inhibitors GF109203X and bisindolylmaleimide 1 (Roush et al. 1999; Vigna 1999; Dery et al. 2001). Secondly, whereas SP stimulates phosphorylation of the NK_1R at serine and threonine residues to a similar degree, PMA induces phosphorylation mostly at serine residues (Roush et al. 1999). Finally, although SP stimulates a very rapid membrane translocation of $PKC\beta II$ tagged with GFP (Barak et al. 1999), PMA-induced phosphorylation and desensitization of the NK_1R occurs with slower kinetics and to a lesser degree than SP (Vigna 1999).

2.3
Mediated Desensitization of Tachykinin Receptors

β-Arrestins, the essential cofactors for GRKs probably interact with GRK-phosphorylated NK_1R to uncouple the receptor from heterotrimeric G-proteins and mediate homologous desensitization (Fig. 1A). Thus, in transfected cell lines expressing the human and rat NK_1R, SP induces the translocation of β-arrestin 1 and β-arrestin 2 from the cytosol to the plasma membrane where they colocalize with the NK_1R (Barak et al. 1999; McConalogue et al. 1999). This translocation occurs within seconds of agonist binding and correlates with the membrane translocation of GRK2 in live cells. In a similar fashion, SP stimulates the trafficking of immunoreactive β-arrestin 1/2 from the cytosol to the plasma membrane of myenteric neurons in primary culture (McConalogue et al. 1998). β-Arrestins probably interact with domains in the carboxyl tail of the NK_1R that are sites of GRK phosphorylation. Thus, the microinjection of *Xenopus* oocytes expressing the rat NK_1R with inositol pentakisphosphate or inositol hexakisphosphate, which may block association of receptors with arrestins, reduces homologous desensitization of SP-induced currents (Sasakawa et al. 1994a). Moreover, in cells expressing a truncated NK_1R that lacks most of the carboxyl tail ($NK_1R\delta 325$), SP does not trigger the membrane translocation of β-arrestins (DeFea et al. 2000a), and this truncated receptor shows diminished desensitization (Li et al. 1997). A chimeric molecule comprised of β-arrestin 1 fused to the carboxyl terminus of the NK_1R is capable of binding agonists with high affinity but does not signal through the usual Gq/G11 and Gs pathways, suggesting that this molecule is completely uncoupled and thus permanently desensitized (Martini et al. 2002).

2.4
Domains of the Tachykinin Receptors That Specify Desensitization

The domains of the NK_1R that are required for desensitization are not fully characterized and there is some controversy in this area. Several observations suggest that domains in the carboxyl terminus are important for desensitization. Analysis of a series of truncation mutants of the rat NK_1R expressed in *Xenopus* oocytes indicated that a domain from residues 338 to 360 plays an essential role in agonist-induced desensitization, but is not necessary for functional activity of the receptor (Sasakawa et al. 1994b). A rat NK_1R lacking most of the carboxyl tail (truncated at residue 324), which may correspond to a naturally occurring isoform of the receptor, signals more robustly than the wild-type receptor and displays diminished desensitization to SP (Li et al. 1997). This truncation mutant is completely resistant to PKC-induced desensitization (Dery et al. 2001). Thus, whereas activation of PKC with PMA abolishes SP-induced mobilization of intracellular Ca^{2+} in cells expressing the wild-type rat NK_1R, it does not affect signaling of $NK_1R\delta 325$. In addition, homologous desensitization of SP-stimulated K^+ currents in bullfrog sympathetic neurons is selectively inhibited by intracellular injection of synthetic peptides corresponding to residues 325–360 and 361–375 of the carboxyl tail of the rat NK_1R. SP inhibits desensitization to SP, but not to the other agonists. A third peptide homologous to residues 376–407 and a peptide from the extracellular portion of the receptor (residues 168–179) did not affect desensitization. This suggests that the portion of the carboxyl tail of the NK_1R from amino acids 325–375 is involved in desensitization (Simmons et al. 1994).

In marked contrast, a truncation mutant of the human NK_1R ($NK_1R\delta 325$) desensitizes in a similar manner as the full length receptor, assessed by measurement of inositol trisphosphate accumulation or by transient translocation of PKCβII-GFP to the plasma membrane (Richardson et al. 2003). Moreover, in contrast to rat $NK_1R\delta 325$ that shows diminished interaction with β-arrestin 1 at the plasma membrane (DeFea et al. 2000a), translocation of β-arrestin 1 or 2 to the plasma membrane is similar in cells expressing human NK_1R or $NK_1R\delta 325$ (Richardson et al. 2003). Notably, $NK_1R\delta 325$ is resistant to agonist-induced phosphorylation. Together these results suggest that phosphorylation of the carboxyl tail is not required for interaction with arrestins and homologous desensitization of the human NK_1R. The discrepancy between these observations and those with the rat receptor could be related to structural differences or differences in the expression systems used to study the receptors.

2.5
Agonist Dependency of Desensitization of Tachykinin Receptors

Recent observations indicate that the characteristics of an agonist peptide affect the process of desensitization of the NK_1R. Although the carboxyl terminus of tachykinins is required for agonistic activity, regions in the amino terminus of

SP are required for receptor selectivity and for full potency, suggesting that amino-terminal domains are important for interactions between SP and the NK_1R. Thus, although a carboxyl-terminal SP fragment [SP(6–11)], the analog peptide and neurokinin A (NKA; which differs from SP in its amino terminus) are full agonists of the NK_1R, they do not desensitize the receptor to the same extent as SP (Vigna 2001). Neuropeptide K (NPK) and neuropeptide gamma (NPγ) contain NKA at their carboxyl terminals and are amino-terminally extended. In contrast to NKA, NPK and NPγ can induce full desensitization of the NK_1R (Vigna 2003). These findings suggest that domains of the amino terminus of SP are required for full desensitization. A similar study of the effectiveness of various tachykinins to regulate M-currents in sympathetic neurons from the bullfrog also found no relationship between the capacity of tachykinins to activate the receptor and to cause desensitization (Perrine et al. 2003). Although the mechanism of these effects is unknown, the findings that different forms of tachykinins desensitize the NK_1R to differing degrees could have important functional consequences for NK_1R signaling in vivo; the extent of receptor desensitization would depend on the available tachykinins. Thus, the same receptor would be expected to be desensitized to very different extents, depending on the available agonists.

3
Agonist-Induced Endocytosis of Tachykinin Receptors

3.1
General Mechanisms of Endocytosis of GPCRs

Many receptors undergo endocytosis, also referred to as internalization and sequestration, after binding agonists. In general, detectable endocytosis occurs more slowly than desensitization, although the early stages can be observed within seconds to minutes. In theory, agonist-induced endocytosis of GPCRs could contribute to desensitization, simply by removing receptors from the cell surface and thus making them inaccessible to agonists in the extracellular fluid. However, numerous studies show that rapid desensitization of signaling proceeds even when endocytosis is prevented, and thus endocytosis is not considered to contribute to desensitization. Rather, endocytosis plays a major role in resensitization and signaling of certain GPCRs, which will be discussed later. The mechanisms of receptor endocytosis differ between cells and for different receptors, but generally involve either clathrin-coated pits or caveolae. The mechanism of clathrin-mediated endocytosis has been examined in most detail for the GPCRs (Fig. 1B).

The role of clathrin for endocytosis of GPCRs is usually investigated by localizing receptors to clathrin-coated pits by microscopy and by treating cells with agents that disrupt clathrin-mediated endocytosis, such as hypertonic sucrose (causes abnormal clathrin polymerization) and monodansylcadaverine (disrupts invagination of clathrin-coated pits). These approaches have been used to show, for example, that the β_2AR (von Zastrow et al. 1992), gastrin-releasing peptide receptor (Grady et al. 1995b), and protease-activated receptors (Bohm et al. 1996; Paing et al. 2002) undergo clathrin-mediated endocytosis. Adaptor proteins couple receptors to clathrin and thereby play an essential role in clathrin-mediated endocytosis. In addition to their role in homologous desensitization, β-arrestins serve as clathrin adaptors for many GPCRs including the β_2AR (Ferguson et al. 1996; Goodman et al. 1996). Overexpression of β-arrestins promotes endocytosis of the β_2AR (Ferguson et al. 1996), whereas overexpression of dominant negative mutants of β-arrestins, such as β-arrestin$^{319-418}$, which comprises a clathrin binding domain, impairs endocytosis (Ferguson et al. 1996; Zhang et al. 1996; Krupnick et al. 1997b). GRK-mediated phosphorylation is also necessary for endocytosis of certain GPCRs, presumably by enhancing interactions of receptors with β-arrestins (Tsuga et al. 1994; Ferguson et al. 1995; Schlador et al. 1997). However, expression of dominant negative β-arrestin mutants has no effect on endocytosis of the angiotensin II type 1A receptor or the m1, m3 and m4 muscarinic cholinergic receptors (Zhang et al. 1996; Lee et al. 1998). Thus, β-arrestins do not mediate endocytosis of all GPCRs. β-Arrestins possess two motifs that allow them to function as clathrin adaptors. A LIEF sequence, which is located between residues 374 and 377 of β-arrestin 2, interacts with a domain between residues 89–100 within the amino terminus of the clathrin heavy chain (Krupnick et al. 1997a). An RxR sequence between residues 394 and 396 of β-arrestin binds to the b2 adaptin subunit of the heterotetrameric AP-2 adaptor complex (Laporte et al. 1999, 2000). The AP-2 complex plays an essential role in the formation of clathrin-coated pits by binding to clathrin, dynamin and EPS-15.

GTPases play an essential role in endocytosis and vesicular transport. Dynamin is a cytoplasmic GTPase that plays an essential role in endosome formation at sites of clathrin-coated pits and caveoli (McClure et al. 1996; McNiven et al. 2000). The role of dynamin is frequently investigated by expression of the GTPase-defective dominant-negative mutant, dynamin K44E. By this approach, dynamin has been shown to be required for constitutive endocytosis of the transferrin receptor, as well as agonist-induced endocytosis of numerous GPCRs, including the β_2-AR, δ-opioid, muscarinic m1, m3 and m4, and dopamine D_1 receptors (Herskovits et al. 1993; Zhang et al. 1996; Chu et al. 1997; Lee et al. 1998; Vogler et al. 1998; Iwata et al. 1999; Vickery et al. 1999). However, dynamin is not required for endocytosis of all GPCRs, since angiotensin II type 1A, muscarinic m2, dopamine D_2 and α_{2B}-adrenergic receptors internalize by a dynamin-independent mechanism (Zhang et al. 1996; Vogler et al. 1998; Schramm et al. 1999; Vickery et al. 1999). Ras-related GTPases (rabs) participate in multiple stages of vesicular transport (Olkkonen et al. 1997). Rab5a mediates the formation of early endosomes containing the transferrin receptor, and is required for endosomal translocation to a perinuclear region (Bucci et al. 1992; Roberts et al. 1999; Trischler et al. 1999). Internalized β_2AR colocalizes with rab5a in endosomes (Moore et al. 1995), and rab5a mediates endocytosis and vesicular trafficking of dopamine D2 receptors (Iwata et al. 1999) and protease-activated receptor 2 (Roosterman et al. 2003).

3.2
Molecular Mechanisms of Agonist-Induced Endocytosis of Tachykinin Receptors

Agonist-induced endocytosis of NKRs has been extensively examined in transfected cell lines by numerous experimental approaches, including localization of receptors using epitope tags, GFP, fluorescent ligands and receptor antibodies, and quantification of endocytosis using radiolabeled ligands and receptor antibodies with flow cytometry (Bowden et al. 1994; Garland et al. 1994; Grady et al. 1995a; Bohm et al. 1997a; McConalogue et al. 1999; Schmidlin et al. 2001, 2002, 2003). Experiments in transfected cell lines have allowed precise studies of the timing and pathway of agonist-stimulated trafficking of NKRs, and a detailed investigation of the molecular mechanisms involved (Fig. 1B). Within minutes of binding SP, the NK_1R redistributes to vesicles at or immediately beneath the plasma membrane (Bowden et al. 1994; Garland et al. 1994; Grady et al. 1995a). This process of endocytosis is inhibited by hypertonic sucrose and by phenylarsine oxide, and is thus dependent on clathrin-mediated mechanisms. In support of a role for clathrin, the pits and first-formed vesicles colocalize with immunoreactive clathrin. The endosomes contain SP and the NK_1R, and colocalize with transferrin receptors and with rab5a, and are thus early endosomes (Garland et al. 1994, 1995a; Schmidlin et al. 2001). β-Arrestins are required for endocytosis, since expression of a dominant negative mutant, β-arrestin[319–418] corresponding to the clathrin binding domain, strongly inhibits this trafficking (McConalogue et al. 1999). The GTPase-defective mutants dynamin K44E and rab5aS34 N also impede endocytosis, indicating a role for dynamin and rab5a (Schmidlin et al. 2001). Rab5a is probably involved in the more distal steps of endocytosis, since the main effect of rab5aS34 N is to impede the trafficking of the NK_1R from superficial vesicles to endosomes in a perinuclear location. Both the NK_2R and NK_3R also undergo agonist-stimulated endocytosis in transfected cell lines (Garland et al. 1996; Schmidlin et al. 2002, 2003). In the case of the NK_3R, this trafficking requires β-arrestins since it is inhibited by expression of β-arrestin[319–418]. Endocytosis does not mediate desensitization of the NK_1R because expression of dominant negative mutants of β-arrestins, dynamin and rab5a all impede endocytosis but do not affect homologous desensitization (McConalogue et al. 1999; Schmidlin et al. 2001).

3.3
Endocytic Domains of Tachykinin Receptors

For many receptors, particular domains are required for endocytosis. These domains may be sites of phosphorylation that interact with critically important adaptor proteins, such as β-arrestins. Endocytic domains of the rat NK_1R have been investigated by analysis of truncation mutants (Bohm et al. 1997b). Truncation of the carboxyl tail at residues 324, 342, and 354 reduces the rate of SP-induced endocytosis by up to 60%. However, these mutations do not prevent endocytosis, which still proceeds but at a slower rate. Although the explanation of this effect is not firmly established, it is likely that truncated receptors show impaired GRK-mediated phosphorylation and hence diminished interactions with β-arrestins (DeFea et al. 2000a). In contrast, the endocytosis of ^{125}I-SP proceeds at the same rate in cells expressing wild-type human NK_1R or NK_1R truncated at residue 325 (Richardson et al. 2003). SP induces membrane translocation of β-arrestin 1 and 2 in a similar manner in cells expressing full length or truncated NK_1R, although there are differences in the association of the truncated receptor with β-arrestins in endosomes.

Tyrosine-containing endocytic motifs have been identified for certain GPCRs, such as the β_2AR (Barak et al. 1994). The importance of such motifs for SP-induced endocytosis of the rat NK_1R has been examined by analysis of mutant receptors (Böhm et al. 1997). Mutation of Tyr-341 and Tyr-349 in potential tyrosine-containing endocytic motifs of the carboxyl tail of the rat NK_1R impairs, but does not prevent, SP-induced endocytosis. A Y305A mutant within the putative NPX2–3Y endocytic motif of the seventh transmembrane domain exhibits impaired signaling and is minimally expressed at the plasma membrane but instead found in cytoplasmic vesicles. In contrast, a Y305F mutant signals normally and is expressed at the plasma membrane but shows diminished SP-induced endocytosis. Thus, endocytosis of the NK_1R relies on several tyrosine-containing sequences in the carboxyl tail and seventh transmembrane domain.

PKC does not appear to play a role in SP-induced endocytosis of the rat NK_1R (Dery et al. 2001). Truncation of the NK_1R inhibits SP-induced endocytosis, indicating that carboxyl-terminal domains are necessary for trafficking. However, administration of the PKC inhibitor GF109203X does not affect NK_1R endocytosis, and endocytosis is also unaffected by mutations of potential PKC consensus sites within the carboxyl tail of the NK_1R.

3.4
Interactions of Tachykinin Receptors with Arrestins

There are distinct differences in the interaction of certain GPCRs with the two isoforms of β-arrestins that have important functional implications. Some receptors, designated 'class B' receptors, exhibit a high-affinity interaction with both isoforms of β-arrestins. These receptors interact with both isoforms of β-arrestins at the plasma membrane and usually colocalize with β-arrestins in endosomes for prolonged periods. For example, angiotensin II 1A, neurogenic 1, vasopressin V_2, and thyroid releasing hormone receptors interact with β-arrestin 1 and 2 with high affinity, and internalize with β-arrestins in endosomes (Oakley et al. 2000). The 'class A' GPCRs, including β_2- and β_{1B}-adrenergic, µ-opioid, endothelin A and dopamine D_{1A} receptors, transiently interact preferentially with β-arrestin 2 (Oakley et al. 2000). These receptors form low affinity, unstable interactions with β-arrestins, dissociate from β-arrestins near the plasma membrane, and are largely excluded from endosomes. Exchange of the carboxyl tails of class A and B receptors reverses their affinities for β-arrestins, suggesting that domains in the carboxyl tails specify interaction with β-arrestins (Oakley et al. 2000). The ability of β-arrestin 2 to remain associated with these receptors depends on the presence of a cluster of Ser and Thr residues in the carboxyl tail that may be phosphorylated and interact with arrestins (Oakley et al. 2001). Domains in the third intracellular loop of the α_{2B}-adrenergic receptor also determine its interaction with β-arrestin 2 (DeGraff et al. 2002). Interaction with β-arrestins determine the rate of receptor recycling and resensitization (Oakley et al. 1999). Thus, those receptors that form prolonged and high-affinity interactions with β-arrestins recycle and resensitize more slowly than receptors that rapidly dissociate from β-arrestins. The explanation for the slow recycling of class B receptors may be that they require dephosphorylation by endosomal phosphatases to dissociate from β-arrestins and recycle.

The NK_1R belongs to the class B of GPCRs. Thus, agonists of the NK_1R induce membrane translocation of both isoforms of β-arrestins, and the internalized receptor colocalizes with β-arrestins 1 and 2 for prolonged periods in endosomes, indicative of a stable and high-affinity interaction (McConalogue et al. 1998, 1999; Barak et al. 1999; Oakley et al. 2000; Schmidlin et al. 2003; Figs. 2A, 3A). In marked contrast, the NK_3R is a class A GPCR. Activation of the NK_3R induces prominent membrane translocation of only β-arrestin 2, and β-arrestin 2 rapidly dissociates from the NK_3R and returns to the cytosol (Schmidlin et al. 2003). Domains in the carboxyl tail and in intracellular loop 3 of the NK_1R and NK_3R specify this affinity of the interaction of these receptors with β-arrestins (Schmidlin et al. 2003). Thus, exchange of these domains of the NK_3R with equivalent domains of the NK_1R confers on the chimeric receptor the ability to form high-affinity interactions with both β-arrestin 1 and 2. In contrast, the interaction of the NK_1R with β-arrestin 1 is impeded by replacement of the carboxyl tail and third intracellular loop with those of the NK_3R. Interaction of the NK_1R and NK_3R with β-arrestins determines the rate of resensitization and probably recycling (Schmidlin et al. 2003). Thus, whereas the NK_1R interacts with high affinity with β-arrestins 1 and 2 and resensitizes slowly, the NK_3R, which interacts with low affinity only with β-arrestin 2, undergoes rapid resensitization. Resensitization of the NK_1R is accelerated by replacement of the carboxyl tail and loop III with those of the NK_3R, whereas NK_3R resensitization is delayed by substitution of these domains with those of the NK_1R (Schmidlin et al. 2003).

The affinity with which receptors interact with β-arrestins has potentially important implications for signaling in cells that co-express more than one receptor for the same agonists. A common theme of signaling by GPCRs is that cells often coexpress several distinct receptors for the same agonist and that several agonists can interact with the same receptor. This theme is well illustrated for the NKRs, which interact with several tachykinin peptides with varying affinity. The competition of different receptors for a common intracellular pool of β-arrestins can have important consequences for the control of signaling. NK_1R and

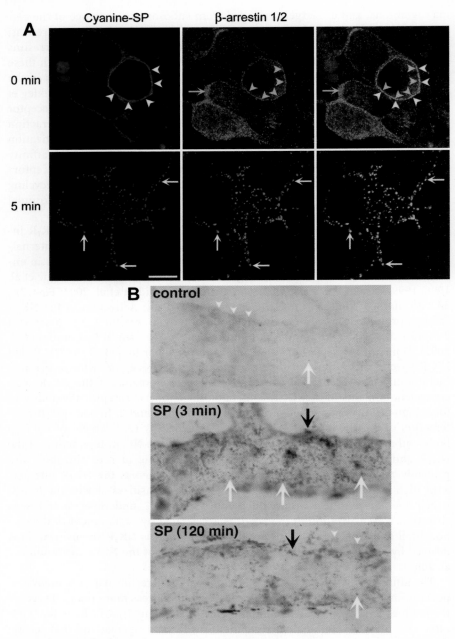

Fig. 2 A Agonist-induced trafficking of the tachykinin NK$_1$ receptor (*NK$_1$R*) and of β-arrestin 1 in myenteric neurons. Neurons were incubated with cyanine 3-labeled SP and β-arrestins were detected using a specific antibody. Note that SP and β-arrestins are found at the plasma membrane after incubation at 4°C and that they colocalize in endosomes of the soma and neurites after 5 min at 37°C. (Repro-

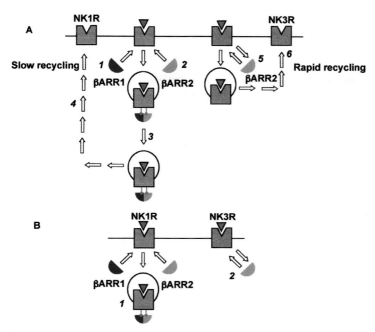

Fig. 3 A Contrasting regulation of tachykinin NK_1 and NK_3 receptors (NK_1R and NK_3R). Agonist binding to the NK_1R induces membrane translocation of β-arrestin 1 (*1*) and β-arrestin 2 (*2*), which form stable and high-affinity interactions with the NK_1R. Both isoforms of β-arrestins are sequestered with the NK_1R into endosomes (*3*) where they remain colocalized for prolonged periods. The NK_1R dissociates from β-arrestins and slowly recycles to the cell surface (*4*) for resensitization. In contrast, the NK_3R induces membrane translocation of β-arrestin 2 (*5*), which forms a low-affinity interaction with the NK_3R. Thus, β-arrestin 2 is not prominently colocalized with the NK_3R in endosomes but rapidly returns to the cytosol. In consequence the NK_3R rapidly recycles (*6*) and resensitizes. **B** Heterologous regulation of trafficking and signaling of NK_1R and NK_3R. The high-affinity interaction of the NK_1R with β-arrestin 1 and 2 (*1*) results in their sequestration with the NK_1R in endosomes, which depletes the cytosolic pools of β-arrestins and thereby impedes β-arrestin-mediated endocytosis and desensitization of the NK_3R. Conversely, the low-affinity interactions of the NK_3R with β-arrestin 2 (*2*) does not deplete the cytosolic pools and β-arrestins, so that NK_1R endocytosis and desensitization are unaffected

duced from McConalogue et al. 1998). **B** SP-induced endocytosis of the NK_1R in post-capillary venules. The NK_1R was detected by immunohistochemistry in endothelial cells of postcapillary venules at various times after intravenous injection of SP (*white arrows*). The *black arrows* show the accumulation of Monastral blue in the basement membrane, which is indicative of plasma extravasation. (Reproduced from Bowden et al. 1994)

NK_3R are co-expressed by certain cell types, such as neurons of the enteric nervous system (Grady et al. 1996a; Barak et al. 1997). NK_1R agonists induce a pronounced sequestration of β-arrestin 1 and 2 into endosomes, which depletes the cytosol (Schmidlin et al. 2002, 2003). Thus, in cells co-expressing the NK_1R and NK_3R, including transfected cell lines and enteric neurons in culture, selective activation of the NK_1R causes sequestration of β-arrestins into endosomes and thereby prevents β-arrestin-mediated endocytosis and homologous desensitization of the NK_3R (Fig. 3B). In contrast, NK_3R agonists induce a transient translocation of only β-arrestin 2 to the plasma membrane and endosomes; β-arrestin 2 rapidly returns to the cytosol and β-arrestin 1 is mostly unaffected. Thus, in cells expressing both receptors, selective activation of the NK_3R does not affect NK_1R endocytosis or homologous desensitization. This novel mechanism of regulation could have important functional implications in cells that naturally express these receptors. After activation of the NK_1R, the NK_1R would quickly desensitize and internalize; the NK_3R would remain at the cell surface and continue to signal due to impeded endocytosis and desensitization. Such regulation could permit cells to respond to tachykinins at a time when the NK_1R was desensitized and internalized. This state may persist until β-arrestins return to the cytosol, when the NK_3R could interact with β-arrestins, uncouple and internalize, and the NK_1R would have recycled. The rapid recycling of the NK_3R would allow cells to return to their steady state. This mechanism could be widespread since activation of the vasopressin V_2 receptor inhibits endocytosis of the β_2AR by a similar mechanism (Klein et al. 2001).

3.5
Trafficking of Tachykinin Receptors in the Nervous System

The use of transfected cells has allowed detailed investigation of the molecular mechanisms and potential functions of agonist-stimulated trafficking of the NKRs. However, the results from such studies can be influenced by the repertoire of regulatory proteins such as GRKs and β-arrestins that are expressed by the selected cell type, and by the overexpression of the receptor of interest. In view of these limitations, there has been great interest in studying trafficking of NKRs in cells that naturally express the receptors, in particular neurons, in primary cultures, tissues and in the intact animal. In particular, the first observation that an intravenous injection of SP into an intact rat induces a massive endocytosis of the NK_1R in post-capillary venules (Bowden et al. 1994; Fig. 2B) led to the realization that NK_1R endocytosis could be used to identify, with exquisite temporal and spatial resolution and a high degree of specificity and sensitivity, sites of release of endogenous SP in the intact animal under physiological circumstances. These studies have relied in large part on the generation of specific antibodies (Vigna et al. 1994; Grady et al. 1996a) and fluorescent peptides (Bunnett et al. 1995; Bennett et al. 2001), which can be used to localize receptors with a high degree of sensitivity and specificity.

In myenteric neurons from the guinea pig intestine in primary culture, exogenous SP as well as selective NK_1R agonists induce a rapid endocytosis of the NK_1R at sites of clathrin-coated pits by a mechanism that is inhibited by hypertonic sucrose and phenylarsine oxide and is thus clathrin mediated (Grady et al. 1996b). Endogenously released SP also causes endocytosis of the NK_1R in this system. SP induces the translocation of GRK2/3 and β-arrestins 1 and 2 to the plasma membrane, where they colocalize with the NK_1R (McConalogue et al. 1998). Within minutes of binding SP, the NK_1R internalizes into early endosomes in the soma and neurites that also contain β-arrestin 1/2, rab5a and the transferrin receptor (McConalogue et al. 1998; Schmidlin et al. 2001). It is therefore probable that SP-induced endocytosis in myenteric neurons requires GRK-induced phosphorylation and association with β-arrestins, although direct proof is lacking. After exposure to moderate concentrations of SP (10–100 nM, 60 min), the NK_1R slowly recycles to the plasma membrane by a mechanism that depends on acidification of endosomes (Grady et al. 1996b). SP-induced endocytosis of the NK_1R has similarly been examined in myenteric neurons using isolated segments of guinea pig (Southwell et al. 1996, 1998a) and rat ileum (Mann et al. 1999). In the rat ileum, there appears to be a sufficient release of endogenous SP from isolated tissues to trigger endocytosis of the NK_1R in intrinsic primary afferent neurons, implying that such neurons could be under the continuous influence from tachykinins in the normal intestine. Mechanical stimulation of the villi of the guinea pig ileum induces endocytosis of the NK_1R in a subpopulation of myenteric neurons due to reflex release of endogenous SP (Southwell et al. 1998b). In the myenteric plexus of rat ileum, SP and NKA induced endocytosis of NK_1R and NK_3R into distinct vesicles, and selective agonists of these receptors have similar effects (Jenkinson et al. 2000). Interestingly, whereas monensin treatment resulted in the accumulation of the NK_1R in endosomes, it had no effect on the NK_3R, suggesting distinct mechanisms of receptor recycling. SP also induces endocytosis of the NK_1R on interstitial cells of Cajal in the intestine (Lavin et al. 1998).

SP-induced trafficking of the NK_1R has also been extensively investigated in slice preparations of the spinal cord. Incubation of slices of rat spinal cord with SP triggers endocytosis of the NK_1R in most NK_1R-positive neurons in laminae I, IIo, and X and in half of the NK_1R-positive neurons in laminae III-V (Marvizon et al. 1997). Electrical stimulation of the dorsal root to activate the SP-containing C-fibers also stimulates NK_1R internalization in neurons in laminae I-IIo, indicating that endogenous SP can cause endocytosis of the NK_1R. Similarly, superfusion of the slices with capsaicin induces both the release of SP from primary afferent neurons that innervate the dorsal horn, and endocytosis of the NK_1R in lamina I-III (Marvizon et al. 2003). Since endocytosed receptors could be detected even in the absence of detectable SP release, endocytosis appears to be a highly sensitive index of SP secretion in the spinal cord. Endocytosis of the NK_1R in spinal cord slices has been used to examine the regulation of SP release in this tissue by glutamate receptors of the N-methyl-D-aspartate (NMDA) type and gamma-aminobutyric acid (GABA) receptors. Thus, the NMDA receptor antagonist, 2-amino-5-phosphonopentanoic acid, but not the AMPA and kainate receptor antagonist CNQX, promotes NK_1R endocytosis, whereas the NK_1R antagonist L-703606 abolishes endocytosis of the NK_1R induced by electrical stimulation or NMDA (Marvizon et al. 1997). NK_1R endocytosis in response to electrical stimulation is also inhibited by GABA and the $GABA_B$ receptor agonist R-baclofen, indicating that $GABA_B$ receptors inhibit SP release induced by dorsal root stimulation (Marvizon et al. 1997).

An intravenous injection of SP results in the rapid and marked endocytosis of the NK_1R in endothelial cells of post-capillary venules in intact rats (Bowden et al. 1994). The number of endosomes per endothelial cell increases from approximately ten under basal conditions to ~120 within 3 min SP injection, and this endocytosis correlates with extravasation of plasma proteins from the vasculature and desensitization of the NK_1R. In a similar manner, the striatal injection of SP in the anesthetized rat results in a massive endocytosis of the NK_1R within the cell body and dendrites of neurons within the zone of injection and a marked alteration in the morphology of dendrites, which change from structures of uniform diameter to ones characterized by large, swollen varicosities connected by thin fibers (Mantyh et al. 1995a). The findings that exogenous SP causes endocytosis in the intact animal led to the hypothesis that NK_1R endocytosis can be used to identify sites of release of endogenous SP in the intact animal. Thus, somatosensory stimuli that induce release of SP from the central projections of primary afferent neurons within the dorsal horn of the spinal cord also stimulate endocytosis of the NK_1R by spinal neurons and cause morphological alterations in the dendrites of these neurons (Mantyh et al. 1995b). These results suggest that endocytosis of the NK_1R correlates with structural changes in neurons within the central nervous system, which may be involved in neuronal plasticity. NK_1R endocytosis within spinal neurons has been used to monitor the release of SP within the spinal cord during states of acute and chronic inflammatory pain (Honoré et al. 1999).

Acute inflammatory pain, induced by unilateral injection of formalin into the paw, provokes rapid NK_1R endocytosis in lamina I neurons at 8 min and 60 min after formalin injection. In long-term inflammatory pain models, induced by intraplantar injection of complete Freund's adjuvant or induction of adjuvant-induced polyarthritis, there is sustained NK_1R endocytosis indicative of a tonic SP release that is accompanied by the upregulation of the NK_1R in lamina I neurons after 21 days. The extent of endocytosis of the NK_1R within the spinal cord that is induced by painful stimuli can be modified in the setting of chronic inflammation. Thus, in normal animals, noxious stimulation of the paw provokes release of SP and endocytosis of the NK_1R only in neurons with cell bodies and dendrites of lamina I (Abbadie et al. 1997). However, during chronic inflammation of the paw, the same stimuli provoke a greater internalization of the NK_1R in lamina I and additional internalization of the NK_1R in neurons of laminae III-VI. The pain induced by tooth extraction in rats stimulates internalization of the NK_1R in neurons of the trigeminal nucleus caudalis and spinal cord (Sabino et al. 2002). In general, the extent of the injury correlates with the degree of endocytosis, such that minor tissue injury (retraction of the gingiva) induces endocytosis in NK_1R neurons located in lamina I whereas more extensive and severe tissue injury (incisor or molar extraction) causes extensive endocytosis of the NK_1R in neurons located in both laminae I and III-V.

The release of endogenous SP in the peripheral nervous system also induces endocytosis of the NK_1R. Thus, stroking the villi of the guinea pig ileum provokes NK_1R endocytosis in the myenteric plexus, indicative of the reflex release of SP (Southwell et al. 1998b). Similarly, the injection of toxin A from *Clostridium difficile* into the rat ileum, which causes an acute inflammatory response that is dependent on release of SP and activation of the NK_1R, provokes marked endocytosis of the NK_1R in over 70% of enteric neurons, which is accompanied by the swelling of dendrites (Mantyh et al. 1996).

3.6
Endocytic Ablation of Neurons Expressing the Tachykinin NK_1 Receptor

The realization that SP causes a marked endocytosis of the NK_1R within the nervous system offered a novel means of ablating specific populations of neurons: if SP could be conjugated to a toxin in such a way that it retained biological activity, the toxin could be delivered to specific cell types by endocytosis of the NK_1R and would thus ablate these cells. Such an approach could be used to probe the function of particular populations of neurons that express the NK_1R. To test this concept, SP was fused to the ribosome-inactivating protein saporin, a toxin that exerts its effects of blocking protein synthesis only when it is taken up into cells (Mantyh et al. 1997). The SP–saporin conjugate retained biological activity since it was capable of binding to membrane preparations containing the NK_1R. SP–saporin also induced endocytosis of the NK_1R by spinal neurons in culture and, importantly, with 1 week there was destruction of 95% of the NK_1R-positive neurons. Infusion of SP–saporin into the spinal cord resulted in NK_1R endocytosis in neurons in lamina I of the dorsal horn and the selective ablation of 85% of NK_1R-positive neurons in lamina I after 28 days. This treatment markedly attenuated responses to highly noxious stimuli and to mechanical and thermal hyperalgesia, but left responses to mild noxious stimuli unchanged. These findings confirmed the essential role of NK_1R-positive neurons in the transmission of highly noxious stimuli and the maintenance of hyperalgesia. The technique has now been used to determine the function of NK_1R-positive neurons in other locations. For example, by ablating spinal neurons expressing the NK_1R using SP–saporin, it is apparent that NK_1R-expressing neurons located in the dorsal horn are essential for central sensitization and hyperalgesia (Khasabov et al. 2002). The function of neurons that drive ventilation has also been investigated using this approach. The normal breathing rhythm in mammals is hypothesized to be generated by NK_1R-expressing neurons in the pre-Botzinger complex, a medullary region proposed to contain the kernel of the circuits generating respiration. The role of these neurons in rats was assessed by using SP–saporin (Gray et al. 2001). Almost completely bilateral, but not unilateral, ablation of these neurons results in both an ataxic breathing pattern with markedly altered blood gases and pH, and pathological responses to challenges such as hyperoxia, hypoxia and anesthesia. Thus, these approximately 600 neurons seem necessary for the generation of normal breathing in rats.

4
Tachykinin Receptor Endocytosis and Mitogenic Signaling

4.1
General Mechanisms of Arrestin-Dependent Mitogenic Signaling of GPCRs

Many receptors couple to activation of mitogen activated protein kinases (MAP kinases). Several MAP kinase signaling pathways have been characterized in mammalian cells (reviewed in Widmann et al. 1999) with a common organization: a MAP kinase kinase kinase (a serine–threonine kinase) phosphorylates and activates MAP kinase kinase (threonine–tyrosine kinase), which in turn phosphorylates MAP kinase (serine–threonine kinase). Mammalian MAP kinase pathways include ERK1/2, JNK and p38, which regulate multiple substrates in the cytoplasm and nucleus and play major roles in multiple phenotypic responses. An emerging theme in MAP kinase signaling is that the components of different kinase cascades are organized into 'signaling modules' by interacting with molecular scaffold proteins. These scaffolds increase the efficiency of signaling between successive kinases in the phosphorylation cascade, minimize the cross-talk between various MAP kinase cascades and thereby promote the fidelity of signaling, and target MAP kinases to specific subcellular locations. In addition to their role in desensitization and endocytosis of GPCRs, β-arrestins are now known to be molecular scaffolds that recruit and organize various MAP kinases at the cell surface and in endosomes and which thereby play an important role in signal transduction.

The first evidence for a role of β-arrestins in MAP kinase signaling came from the observation that agonists of the β_2AR provoke the colocalization of the receptor with β-arrestins and Src kinases in clathrin-coated pits (Luttrell et al. 1999). Indeed, β-arrestins are the scaffolds for the assembly of a multiprotein complex comprised of β_2AR, β-arrestins and activated Src. There are two sites of interaction between Src kinase and β-arrestin 1. The first involves interaction between the Src homology or SH3 domain of the kinase and proline-rich PXXP motifs within the amino terminus of β-arrestin 1. The second involves the amino-terminal portion of the catalytic (SH1) domain of Src and domains within the amino-terminal 185 residues of β-arrestin 1 (Miller et al. 2000). Src kinase is necessary for the activation of the ERK1/2 MAP kinase pathway by several GPCRs (Luttrell et al. 1996), and for the β_2AR the interaction between Src and β-arrestins is required to activate ERK1/2 (Luttrell et al. 1999). Thus, expression of dominant negative mutants of β-arrestin that disrupt Src binding or block targeting of receptors to clathrin-coated pits inhibits β_2AR-mediated activation of ERK1/2. In a similar manner, agonists of protease activated receptor 2

(PAR2) stimulate the assembly of a MAP kinase signaling module, a high-molecular-weight complex containing PAR2, β-arrestins, raf-1 and activated ERK1/2, which forms at the plasma membrane or in early endosomes (DeFea et al. 2000b) The module retains activated ERK1/2 in the cytosol where they cannot induce proliferation. However, in cells expressing a mutant PAR2, which is unable to interact with β-arrestins, the module is unable to form. In these cells PAR2 agonists activate ERK1/2 by alternate pathways that allow nuclear translocation and stimulation of proliferation. Therefore, β-arrestins are molecular scaffolds that promote the formation of an ERK1/2 module at the plasma membrane or in endosomes. In the case of PAR2, the scaffold retains activated ERKs in these locations, where they may regulate cytosolic targets.

4.2
Arrestin-Dependent Mitogenic Signaling of the Tachykinin NK_1 Receptor

Arrestins also play an important role in SP-induced activation of ERK1/2 (Fig. 4A). In NK_1R-transfected KNRK cells and dermal endothelial cells that naturally express the NK_1R, SP strongly activates ERK1/2 (DeFea et al. 2000a). In these cells, the NK_1R couples to at least two pathways of ERK1/2 activation. One pathway depends on β-arrestin-mediated receptor endocytosis, since SP-stimulated ERK1/2 activation is inhibited by expression of dominant negative β-arrestin and there is diminished signaling in cells expressing an internalization-defective truncated mutant of the NK_1R ($NK_1R\delta325$). The second pathway is dependent on tyrosine kinases and transactivation of receptor tyrosine kinases. SP induces the formation of a multiprotein complex containing the NK_1R, β-arrestin, Src and ERK1/2, that can be detected by gel filtration analysis of cellular extracts, immunoprecipitation and Western blotting, and by confocal imaging. The formation of this complex is necessary for the nuclear translocation of ERK1/2 and thereby is essential for certain aspects of NK_1R signal transduction. Thus, inhibition of the assembly of this complex, for example by expression of dominant negative β-arrestin or in cells expressing truncated NK_1R, prevents both SP-induced endocytosis of the NK_1R and SP-induced ERK1/2 activation, which is required for the proliferative and anti-apoptotic effects of SP. Therefore, β-arrestins are required for nuclear targeting of activated ERK1/2 in the case of the NK_1R. In the case of other receptors, such as protease activated receptor 2, β arrestins serve to retain ERK1/2 in the Cytosol (DeFea 2000).

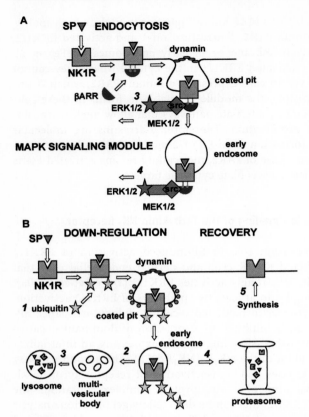

Fig. 4 A The role of β-arrestins in SP-induced MAP kinase activation. β-Arrestins interact with the tachykinin NK_1 receptor (NK_1R) (*1*) and with components of the MAP kinase pathway (*2*) in clathrin coated pits (*3*) or early endosomes (*4*). β-Arrestins are molecular scaffolds that interact with Src and thus couple the NK_1R to Src. This binding induces the formation of a multiprotein MAP kinase signaling module that comprises NK_1R, β-arrestins, Src, MEK1/2 and ERK1/2. The module determines the subcellular location and function of activated ERKs. **B** Potential mechanisms of downregulation of the NK_1R. (*1*) After prolonged exposure to SP, the NK_1R undergoes ubiquitination and subsequent degradation (N.W. Bunnett, unpublished observation). However, the E1, E2 and E3 ligases that mediate this ubiquitination are unknown, and it is not known whether ubiquitination occurs at the plasma membrane or in intracellular locations. Proteins that are monoubiquitinated are sorted through the multivesicular body (*2*) and are degraded in lysosomes (*3*). Proteins that are polyubiquitinated (i.e., chains of ubiquitin molecules are added to the protein) are targeted to the proteasome for degradation (*4*). Recovery from downregulation requires synthesis of new receptors (*5*)

5
Intracellular Trafficking of Tachykinin Receptors: Resensitization and Downregulation

5.1
General Mechanisms of Intracellular Trafficking of GPCRs

Once internalized, GPCRs have two general fates: recycling to the plasma membrane or trafficking to degradative organelles such as lysosomes or proteasomes. The fates vary between receptors and can differ for the same receptor depending on the nature of the stimulus. On the one hand, receptors for certain neurotransmitters such as the $\beta_2 AR$ (von Zastrow et al. 1992) and gastrin releasing peptide receptor (Grady et al. 1995b), which bind to agonists in a reversible manner and that can be reused by the cell, efficiently recycle to the plasma membrane. On the other hand, the protease-activated receptors, which are activated by irreversible proteolysis and that cannot be re-used by the cell, are mostly targeted to lysosomes for degradation (Hoxie et al. 1993; Bohm et al. 1996). However, all receptors are ultimately degraded, and even receptors that normally recycle can be targeted to lysosomes during continuous exposure of cells to agonists, a process known as downregulation. Thus, long-term (24 h) exposure of fibroblasts expressing the $\beta_2 AR$ to isoproterenol induces a loss of receptors due to degradation, and this is not observed in fibroblasts lacking β-arrestin 1 and 2, indicating the requirement for receptor endocytosis (Kohout et al. 2001).

Immediately after endocytosis, GPCRs enter the early endosomes, a tubulovesicular network that is frequently located towards the periphery of the cell. GPCRs may traffic from early endosomes into recycling endosomes, and thereby return to the cell surface, or can progress to the late endosomes that are usually more spherical than early endosomes and often are located near the nucleus. GPCRs can recycle from this compartment to the plasma membrane or can enter a subset of late endosomes that have a multivesicular appearance and which are referred to as multivesicular bodies. The multivesicular bodies play an essential role in protein sorting to lysosomes for degradation. Compared to endocytic mechanisms, that have been extensively investigated for many GPCRs, little is known about the molecular mechanisms of trafficking of internalized receptors to recycling or degradative pathways. However, emerging evidence indicates that ubiquitination may target some GPCRs to a degradative pathway. Ubiquitin is a conserved 76-amino acid protein that is covalently linked to proteins by the action of a cascade of three enzymes: an ubiquitin-activating enzyme (E1), an ubiquitin-conjugating enzyme (E2) and an ubiquitin ligase (E3). The cascade serves to form an isopeptide bond between the carboxyl terminus of ubiquitin and the amino group of lysine residues on target proteins. Monoubiquitination, which is addition of a single ubiquitin to a substrate, contributes to many regulatory processes, including endocytosis of receptors and trafficking of proteins to the multivesicular body and lysosomes. Polyubiquitination involves the addition of a series of ubiquitin molecules to the lysine residues of ubiquitin itself such that a chain of several ubiquitins can be attached to a single lysine residue in a protein substrate. Polyubiquitination targets proteins to be degraded by the 26S proteasome. Ubiquitin can target certain GPCRs to lysosomes. For example, the chemokine receptor CXCR4 is targeted to lysosomes, and mutation of ubiquitin-acceptor lysines in the cytoplasmic tail of this receptor prevents its degradation in lysosomes (Marchese et al. 2001).

However, not all receptors that are destined for degradation undergo ubiquitination and some interact with specific trafficking proteins that chaperone receptors to degradatory pathways. For example, the δ opioid receptor interacts with a novel protein named GTP-binding protein (G-protein)-coupled receptor-associated sorting protein or GASP (Whistler et al. 2002). Disruption of the receptor–GASP interaction by a dominant negative mutant of GASP inhibits receptor trafficking to lysosomes and promotes recycling. The GASP family of proteins may modulate lysosomal sorting and functional downregulation of other GPCRs, but this possibility remains to be investigated.

Lysosomal degradation of GPCRs serves to unequivocally terminate signaling, and recovery either requires the mobilization of intracellular stores of the receptors or the synthesis of new receptors. Thus, PAR2 is trafficked to lysosomes and resensitization of responses to agonists of this receptor is prevented by disruption of the prominent stores of the receptor in the Golgi apparatus by brefeldin or by inhibition of new protein synthesis with cycloheximide (Bohm et al. 1996).

Endocytosis and recycling are required for resensitization of other receptors. The requirement for endocytosis can be explained by the need to 'prepare' internalized receptors to be able to interact again with ligands and heterotrimeric G-proteins, and hence to signal. GPCRs in endosomes may be bound to agonists, extensively phosphorylated and associated with β-arrestins; resensitization would require removal of the ligand, dephosphorylation and dissociation of β-arrestins, and recycling to the plasma membrane. Thus, inhibition of endocytosis of the β_2AR by pharmacological means, such as treatment with concanavalin A or hypertonic sucrose, or by genetic strategies such as the study of mutant receptors that signal but do not internalize, indicates that endocytosis is required for resensitization (Sibley et al. 1986). Disruption of endocytosis of the β_2AR by expression of dominant negative mutants of β-arrestin or dynamin also impedes resensitization, whereas resensitization of a sequestration-impaired β_2AR mutant (Y326A) is reestablished by the overexpression of β-arrestin 1 (Pippig et al. 1993, 1995; Zhang et al. 1997). After activation, phosphorylated β_2AR can be detected in endosomes that are enriched with the GPCR-specific protein phosphatase PP2A (Pitcher et al. 1995).

5.2
Recycling and Resensitization of Tachykinin Receptors

Internalized NK_1Rs recycle back to the cell surface, but the timing and pathway of this recycling varies depending on the concentration of agonist and the duration of exposure (Fig. 1B). In KNRK cells expressing the rat NK_1R, exposure to moderate concentrations of SP (1–100 nM, 60 min) induces translocation of the NK_1R to endosomes in a perinuclear location (Garland et al. 1994; Grady et al. 1995a; McConalogue et al. 1999). SP and the NK_1R remain colocalized in early endosomes in a perinuclear region of the cell and are then sorted to distinct regions. SP is targeted to lysosomes and is degraded. In contrast, the NK_1R gradually recycles back to the plasma membrane, a process taking several hours, where it can again bind SP and signal. Acidification of endosomes is required for this recycling, since it is prevented by bafilomycin A1, an inhibitor of vacuolar ATPase, and by drugs such as ammonium chloride and monensin, which prevent endosomal acidification. However, after exposing cells to low concentrations of agonist (1 nM, 10 min), SP and the NK_1R traffic to endosomes that remain in a superficial location, and the NK_1R rapidly recycles to the cell surface by a process that does not require endosomal acidification (N.W. Bunnett, unpublished observation). The precise mechanism of this rapid recycling is unknown.

The NK_1R also recycles in neurons that naturally express the NK_1R. Thus, in myenteric neurons in culture and in intact tissues the reappearance of the NK_1R at the cell surface can be prevented by bafilomycin A and by inhibitors of endosomal acidification (Grady et al. 1996b; Southwell et al. 1998a). The NK_1R also recycles in neurons in the dorsal horn of the spinal cord (Wang et al. 2002). Similarly, in endothelial cells in the intact animal, the NK_1R recycles to the plasma membrane after exposure to SP (Bowden et al. 1994).

Endocytosis and recycling of the rat NK_1R are required for resensitization of SP signaling. In NK_1R transfected KNRK cells, expression of dominant negative mutants of dynamin and rab5a inhibits endocytosis of the NK_1R and also strongly inhibits resensitization of SP-induced Ca^{2+} mobilization (Schmidlin et al. 2001). These results suggest that the NK_1R must undergo endocytosis and appropriate sorting for resensitization to occur. This sorting may involve endosomal acidification, dissociation of SP, receptor dephosphorylation and dissociation of β-arrestins. In support of these possibilities, resensitization in CHO cells expressing the NK_1R is inhibited by bafilomycin (inhibitor of vacuolar ATPase) and okadaic acid (phosphatase inhibitor; Garland et al. 1996). In a similar manner, inhibition of endocytosis with concanavalin A and of recycling with monensin also inhibit resensitization of the NK_1R (Bennett et al. 2002). Similar mechanisms may account for resensitization of the NK_2R.

However, receptor recycling may not entirely explain resensitization of signaling. Thus, in many systems, including NK_1R transfected KNRK (Grady et al. 1995a; Bennett et al. 2002) and CHO cells (Garland et al. 1996), myenteric neurons in culture (Grady et al. 1996b; McConalogue et al. 1998) and in the intact animal (Bowden et al. 1994), resensitization of SP signaling occurs at a time when the NK_1R is still present in endosomes and before cycling is complete. The reason for this discrepancy is unknown. One possibility is that resensitization requires only a small number of recycled receptors and thus full resensitization precedes complete recycling.

5.3
Downregulation of Tachykinin Receptors

Prolonged exposure of KNRK cells expressing the NK_1R to high concentrations of SP (100 nM, 3 h) promotes the internalization and degradation of the NK_1R, indicative of downregulation (N.W. Bunnett, unpublished observation; Fig. 4B). This downregulation correlates with loss of SP signaling, and the recovery of signaling requires synthesis of new NK_1Rs. The mechanism of this downregulation is unknown, although prolonged treatment with SP promotes ubiquitination of the NK_1R by unknown mechanisms. Downregulation of the NK_1R could be of considerable interest during chronic inflammatory states where there may be a long-term tonic release of tachykinins.

6
Conclusions

Considerable progress has been made in understanding the molecular mechanisms of desensitization of the NK_1R and of the mechanism and function of agonist-induced trafficking of the NK_1R. Probably more than any other GPCR, the process of NK_1R endocytosis has been studied extensively in the nervous system of intact animals. However, there are several areas of limited understanding. Firstly, little is known about the regulation of desensitization and trafficking of the NK_2R and NK_3R. Secondly, the molecular mechanisms of receptor desensitization and resensitization are not fully understood. Thirdly, it is not known whether the mechanisms of NK_1R trafficking observed in transfected cell lines also apply to cells that naturally express the NK_1R, such as neurons. Finally, almost nothing is known about the processes and mechanisms of post-endocytic sorting and downregulation of the NK_1R. It will be important to address these mechanisms that regulate the NKRs given their important roles in inflammation and pain.

Acknowledgements. Research in the authors' laboratory in this area is supported by NIH grants DK43207, DK57489, DK39957, the RW Johnson Focused Giving Program (NWB) and the DFG (DR). The authors thank members of their laboratory for valuable discussion and advice and for critically reading this manuscript. This review is not exhaustive: some papers have been omitted due to space limitations.

References

Abbadie C, Trafton J, Liu H, Mantyh PW, Basbaum AI (1997) Inflammation increases the distribution of dorsal horn neurons that internalize the neurokinin-1 receptor in response to noxious and non-noxious stimulation. J Neurosci 17:8049–8060

Barak LS, Ferguson SS, Zhang J, Martenson C, Meyer T, Caron MG (1997) Internal trafficking and surface mobility of a functionally intact beta$_2$-adrenergic receptor-green fluorescent protein conjugate. Mol Pharmacol 51:177–184

Barak LS, Tiberi M, Freedman NJ, Kwatra MM, Lefkowitz RJ, Caron MG (1994) A highly conserved tyrosine residue in G protein-coupled receptors is required for agonist-mediated beta$_2$-adrenergic receptor sequestration. J Biol Chem 269:2790–2795

Barak LS, Warabi K, Feng X, Caron MG, Kwatra MM (1999) Real-time visualization of the cellular redistribution of G protein-coupled receptor kinase 2 and beta-arrestin 2 during homologous desensitization of the substance P receptor. J Biol Chem 274:7565–7569

Bennett VJ, Perrine SA, Simmons MA (2002) A novel mechanism of neurokinin-1 receptor resensitization. J Pharmacol Exp Ther 303:1155–1162

Bennett VJ, Simmons MA (2001) Analysis of fluorescently labeled substance P analogs: binding, imaging and receptor activation. BMC Chem Biol 1:1

Bohm SK, Grady EF, Bunnett NW (1997a) Regulatory mechanisms that modulate signalling by G-protein-coupled receptors. Biochem J 322:1–18.

Böhm SK, Khitin L, Smeekens SP, Grady EF, Payan DG, Bunnett NW (1997) Identification of potential tyrosine-containing endocytic motifs in the carboxyl-tail and seventh transmembrane domain domain of the neurokinin 1 receptor. J Biol Chem 272:2363–2372

Bohm SK, Khitin LM, Grady EF, Aponte G, Payan DG, Bunnett NW (1996) Mechanisms of desensitization and resensitization of proteinase- activated receptor-2. J Biol Chem 271:22003–22016

Bohm SK, Khitin LM, Smeekens SP, Grady EF, Payan DG, Bunnett NW (1997b) Identification of potential tyrosine-containing endocytic motifs in the carboxyl-tail and seventh transmembrane domain of the neurokinin 1 receptor. J Biol Chem 272:2363–2372

Bowden JJ, Garland AM, Baluk P, Lefevre P, Grady EF, Vigna SR, Bunnett NW, McDonald DM (1994) Direct observation of substance P-induced internalization of neurokinin 1 (NK$_1$) receptors at sites of inflammation. Proc Natl Acad Sci USA 91:8964–8968

Bucci C, Parton RG, Mather IH, Stunnenberg H, Simons K, Hoflack B, Zerial M (1992) The small GTPase rab5 functions as a regulatory factor in the early endocytic pathway. Cell 70:715–728

Bunnett NW, Dazin PF, Payan DG, Grady EF (1995) Characterization of receptors using cyanine 3-labeled neuropeptides. Peptides 16:733–740

Chu P, Murray S, Lissin D, von Zastrow M (1997) Delta and kappa opioid receptors are differentially regulated by dynamin-dependent endocytosis when activated by the same alkaloid agonist. J Biol Chem 272:27124–27130

DeFea KA, Vaughn ZD, O'Bryan EM, Nishijima D, Dery O, Bunnett NW (2000a) The proliferative and antiapoptotic effects of substance P are facilitated by formation of a beta-arrestin-dependent scaffolding complex. Proc Natl Acad Sci USA 97:11086–11091

DeFea KA, Zalevsky J, Thoma MS, Dery O, Mullins RD, Bunnett NW (2000b) Beta-arrestin-dependent endocytosis of proteinase-activated receptor 2 is required for intracellular targeting of activated ERK1/2. J Cell Biol 148:1267–1281

DeGraff JL, Gurevich VV, Benovic JL (2002) The third intracellular loop of alpha$_2$-adrenergic receptors determines subtype specificity of arrestin interaction. J Biol Chem 277:43247–43252

Dery O, Defea KA, Bunnett NW (2001) Protein kinase C-mediated desensitization of the neurokinin 1 receptor. Am J Physiol Cell Physiol 280: C1097–1106

Ferguson SS, Downey WE, Colapietro AM, Barak LS, Menard L, Caron MG (1996) Role of beta-arrestin in mediating agonist-promoted G protein-coupled receptor internalization. Science 271:363–366

Ferguson SS, Menard L, Barak LS, Koch WJ, Colapietro AM, Caron MG (1995) Role of phosphorylation in agonist-promoted beta$_2$-adrenergic receptor sequestration. Rescue of a sequestration-defective mutant receptor by beta ARK1. J Biol Chem 270:24782–24789

Garland AM, Grady EF, Lovett M, Vigna SR, Frucht MM, Krause JE, Bunnett NW (1996) Mechanisms of desensitization and resensitization of G protein-coupled neurokinin$_1$ and neurokinin$_2$ receptors. Mol Pharmacol 49:438–446

Garland AM, Grady EF, Payan DG, Vigna SR, Bunnett NW (1994) Agonist-induced internalization of the substance P (NK$_1$) receptor expressed in epithelial cells. Biochem J 303:177–186

Goodman OB, Jr., Krupnick JG, Santini F, Gurevich VV, Penn RB, Gagnon AW, Keen JH, Benovic JL (1996) Beta-arrestin acts as a clathrin adaptor in endocytosis of the beta$_2$-adrenergic receptor. Nature 383:447–450

Grady EF, Baluk P, Bohm S, Gamp PD, Wong H, Payan DG, Ansel J, Portbury AL, Furness JB, McDonald DM, Bunnett NW (1996a) Characterization of antisera specific to NK$_1$, NK$_2$, and NK$_3$ neurokinin receptors and their utilization to localize receptors in the rat gastrointestinal tract. J Neurosci 16:6975–6986

Grady EF, Gamp PD, Jones E, Baluk P, McDonald DM, Payan DG, Bunnett NW (1996b) Endocytosis and recycling of neurokinin 1 receptors in enteric neurons. Neuroscience 75:1239–1254

Grady EF, Garland AM, Gamp PD, Lovett M, Payan DG, Bunnett NW (1995a) Delineation of the endocytic pathway of substance P and its seven- transmembrane domain NK1 receptor. Mol Biol Cell 6:509–524

Grady EF, Slice LW, Brant WO, Walsh JH, Payan DG, Bunnett NW (1995b) Direct observation of endocytosis of gastrin releasing peptide and its receptor. J Biol Chem 270:4603–4611

Gray PA, Janczewski WA, Mellen N, McCrimmon DR, Feldman JL (2001) Normal breathing requires preBotzinger complex neurokinin-1 receptor-expressing neurons. Nat Neurosci 4:927–930

Herskovits JS, Burgess CC, Obar RA, Vallee RB (1993) Effects of mutant rat dynamin on endocytosis. J Cell Biol 122:565–578

Honoré P, Menning PM, Rogers SD, Nichols ML, Basbaum AI, Besson JM, Mantyh PW (1999) Spinal substance P receptor expression and internalization in acute, short-term, and long-term inflammatory pain states. J Neurosci 19:7670–7678

Hoxie JA, Ahuja M, Belmonte E, Pizarro S, Parton R, Brass LF (1993) Internalization and recycling of activated thrombin receptors. J Biol Chem 268:13756–13763

Iwata K, Ito K, Fukuzaki A, Inaki K, Haga T (1999) Dynamin and rab5 regulate GRK2-dependent internalization of dopamine D$_2$ receptors. Eur J Biochem 263:596–602

Jenkinson KM, Mann PT, Southwell BR, Furness JB (2000) Independent endocytosis of the NK$_1$ and NK$_3$ tachykinin receptors in neurons of the rat myenteric plexus. Neuroscience 100:191–199

Khasabov SG, Rogers SD, Ghilardi JR, Peters CM, Mantyh PW, Simone DA (2002) Spinal neurons that possess the substance P receptor are required for the development of central sensitization. J Neurosci 22:9086–9098

Klein U, Muller C, Chu P, Birnbaumer M, von Zastrow M (2001) Heterologous inhibition of G protein-coupled receptor endocytosis mediated by receptor-specific trafficking of beta-arrestins. J Biol Chem 276:17442–17447

Kohout TA, Lin FS, Perry SJ, Conner DA, Lefkowitz RJ (2001) Beta-arrestin 1 and 2 differentially regulate heptahelical receptor signaling and trafficking. Proc Natl Acad Sci USA 98:1601–1606

Krupnick JG, Goodman OB, Jr., Keen JH, Benovic JL (1997a) Arrestin/clathrin interaction. Localization of the clathrin binding domain of nonvisual arrestins to the carboxy terminus. J Biol Chem 272:15011–15016

Krupnick JG, Santini F, Gagnon AW, Keen JH, Benovic JL (1997b) Modulation of the arrestin-clathrin interaction in cells. Characterization of beta-arrestin dominant-negative mutants. J Biol Chem 272:32507–32512

Kwatra MM, Schwinn DA, Schreurs J, Blank JL, Kim CM, Benovic JL, Krause JE, Caron MG, Lefkowitz RJ (1993) The substance P receptor, which couples to Gq/11, is a substrate of beta-adrenergic receptor kinase 1 and 2. J Biol Chem 268:9161–9164

Laporte SA, Oakley RH, Holt JA, Barak LS, Caron MG (2000) The interaction of beta-arrestin with the AP-2 adaptor is required for the clustering of beta$_2$-adrenergic receptor into clathrin-coated pits. J Biol Chem 275:23120–23126

Laporte SA, Oakley RH, Zhang J, Holt JA, Ferguson SS, Caron MG, Barak LS (1999) The beta$_2$-adrenergic receptor/beta-arrestin complex recruits the clathrin adaptor AP-2 during endocytosis. Proc Natl Acad Sci USA 96:3712–3717

Lavin ST, Southwell BR, Murphy R, Jenkinson KM, Furness JB (1998) Activation of neurokinin 1 receptors on interstitial cells of Cajal of the guinea-pig small intestine by substance P. Histochem Cell Biol 110:263–271

Lee KB, Pals-Rylaarsdam R, Benovic JL, Hosey MM (1998) Arrestin-independent internalization of the m1, m3, and m4 subtypes of muscarinic cholinergic receptors. J Biol Chem 273:12967–12972

Lefkowitz RJ (1998) G protein-coupled receptors. III. New roles for receptor kinases and beta-arrestins in receptor signaling and desensitization. J Biol Chem 273:18677–18680

Li H, Leeman SE, Slack BE, Hauser G, Saltsman WS, Krause JE, Blusztajn JK, Boyd ND (1997) A substance P (neurokinin-1) receptor mutant carboxyl-terminally truncated to resemble a naturally occurring receptor isoform displays enhanced responsiveness and resistance to desensitization. Proc Natl Acad Sci USA 94:9475–9480

Luttrell LM, Ferguson SS, Daaka Y, Miller WE, Maudsley S, Della Rocca GJ, Lin F, Kawakatsu H, Owada K, Luttrell DK, Caron MG, Lefkowitz RJ (1999) Beta-arrestin-dependent formation of beta$_2$ adrenergic receptor-Src protein kinase complexes. Science 283:655–661

Luttrell LM, Hawes BE, van Biesen T, Luttrell DK, Lansing TJ, Lefkowitz RJ (1996) Role of c-Src tyrosine kinase in G protein-coupled receptor- and G$_{beta,gamma}$ subunit-mediated activation of mitogen-activated protein kinases. J Biol Chem 271:19443–19450

Luttrell LM, Lefkowitz RJ (2002) The role of beta-arrestins in the termination and transduction of G-protein-coupled receptor signals. J Cell Sci 115:455–465

Mann PT, Southwell BR, Furness JB (1999) Internalization of the neurokinin 1 receptor in rat myenteric neurons. Neuroscience 91:353–362

Mantyh CR, Pappas TN, Lapp JA, Washington MK, Neville LM, Ghilardi JR, Rogers SD, Mantyh PW, Vigna SR (1996) Substance P activation of enteric neurons in response to intraluminal Clostridium difficile toxin A in the rat ileum. Gastroenterology 111:1272–1280

Mantyh PW, Allen CJ, Ghilardi JR, Rogers SD, Mantyh CR, Liu H, Basbaum AI, Vigna SR, Maggio JE (1995a) Rapid endocytosis of a G protein-coupled receptor: substance P evoked internalization of its receptor in the rat striatum in vivo. Proc Natl Acad Sci USA 92:2622–2626

Mantyh PW, DeMaster E, Malhotra A, Ghilardi JR, Rogers SD, Mantyh CR, Liu H, Basbaum AI, Vigna SR, Maggio JE, Simone DA (1995b) Receptor endocytosis and dendrite reshaping in spinal neurons after somatosensory stimulation. Science 268:1629–1632

Mantyh PW, Rogers SD, Honore P, Allen BJ, Ghilardi JR, Li J, Daughters RS, Lappi DA, Wiley RG, Simone DA (1997) Inhibition of hyperalgesia by ablation of lamina I spinal neurons expressing the substance P receptor. Science 278:275–279

Marchese A, Benovic JL (2001) Agonist-promoted ubiquitination of the G protein-coupled receptor $CXCR_4$ mediates lysosomal sorting. J Biol Chem 276:45509–45512

Martini L, Hastrup H, Holst B, Fraile-Ramos A, Marsh M, Schwartz TW (2002). NK_1 receptor fused to beta-arrestin displays a single-component, high-affinity molecular phenotype. Mol Pharmacol 62:30–37

Marvizon JC, Martinez V, Grady EF, Bunnett NW, Mayer EA (1997) Neurokinin 1 receptor internalization in spinal cord slices induced by dorsal root stimulation is mediated by NMDA receptors. J Neurosci 17:8129–8136

Marvizon JC, Wang X, Matsuka Y, Neubert JK, Spigelman I (2003) Relationship between capsaicin-evoked substance P release and neurokinin 1 receptor internalization in the rat spinal cord. Neuroscience 118:535–545

McClure SJ, Robinson PJ (1996) Dynamin, endocytosis and intracellular signalling. Mol Membr Biol 13:189–215

McConalogue K, Corvera CU, Gamp PD, Grady EF, Bunnett NW (1998) Desensitization of the neurokinin-1 receptor (NK_1-R) in neurons: effects of substance P on the distribution of NK_1-R, $G_{alphaq/11}$, G- protein receptor kinase-2/3, and beta-arrestin-1/2. Mol Biol Cell 9:2305–2324

McConalogue K, Dery O, Lovett M, Wong H, Walsh JH, Grady EF, Bunnett NW (1999) Substance P-induced trafficking of beta-arrestins. The role of beta-arrestins in endocytosis of the neurokinin-1 receptor. J Biol Chem 274:16257–16268

McNiven MA, Cao H, Pitts KR, Yoon Y (2000) The dynamin family of mechanoenzymes: pinching in new places. Trends Biochem Sci 25:115–120

Miller WE, Maudsley S, Ahn S, Khan KD, Luttrell LM, Lefkowitz RJ (2000) Beta-arrestin1 interacts with the catalytic domain of the tyrosine kinase c-SRC. Role of beta-arrestin1-dependent targeting of c-SRC in receptor endocytosis. J Biol Chem 275:11312–11319

Moore RH, Sadovnikoff N, Hoffenberg S, Liu S, Woodford P, Angelides K, Trial JA, Carsrud ND, Dickey BF, Knoll BJ (1995) Ligand-stimulated $beta_2$-adrenergic receptor internalization via the constitutive endocytic pathway into rab5-containing endosomes. J Cell Sci 108:2983–2991

Nishimura K, Warabi K, Roush ED, Frederick J, Schwinn DA, Kwatra MM (1998) Characterization of GRK2-catalyzed phosphorylation of the human substance P receptor in Sf9 membranes. Biochemistry 37:1192–1198

Oakley RH, Laporte SA, Holt JA, Barak LS, Caron MG (1999) Association of beta-arrestin with G protein-coupled receptors during clathrin-mediated endocytosis dictates the profile of receptor resensitization. J Biol Chem 274:32248–32257

Oakley RH, Laporte SA, Holt JA, Barak LS, Caron MG (2001) Molecular determinants underlying the formation of stable intracellular g protein-coupled receptor-beta-arrestin complexes after receptor endocytosis. J Biol Chem 276:19452–19460

Oakley RH, Laporte SA, Holt JA, Caron MG, Barak LS (2000) Differential affinities of visual arrestin, beta-arrestin1, and beta-arrestin2 for G protein-coupled receptors delineate two major classes of receptors. J Biol Chem 275:17201–17210

Olkkonen VM, Stenmark H (1997) Role of Rab GTPases in membrane traffic. Int Rev Cytol 176:1–85

Paing MM, Stutts AB, Kohout TA, Lefkowitz RJ, Trejo J (2002) Beta-arrestins regulate protease-activated receptor-1 desensitization but not internalization or down-regulation. J Biol Chem 277:1292–1300

Perrine SA, Bennett VJ, Simmons MA (2003) Tachykinin peptide-induced activation and desensitization of neurokinin 1 receptors. Peptides 24:469–475

Pippig S, Andexinger S, Daniel K, Puzicha M, Caron MG, Lefkowitz RJ, Lohse MJ (1993) Overexpression of beta-arrestin and beta-adrenergic receptor kinase augment desensitization of beta$_2$-adrenergic receptors. J Biol Chem 268:3201–3208

Pippig S, Andexinger S, Lohse MJ (1995) Sequestration and recycling of beta$_2$-adrenergic receptors permit receptor resensitization. Mol Pharmacol 47:666–676

Pitcher JA, Payne ES, Csortos C, DePaoli-Roach AA, Lefkowitz RJ (1995) The G-protein-coupled receptor phosphatase: a protein phosphatase type 2A with a distinct subcellular distribution and substrate specificity. Proc Natl Acad Sci USA 92:8343–8347

Richardson MD, Balius AM, Yamaguchi K, Freilich ER, Barak LS, Kwatra MM (2003) Human substance P receptor lacking the C-terminal domain remains competent to desensitize and internalize. J Neurochem 84:854–863

Roberts RL, Barbieri MA, Pryse KM, Chua M, Morisaki JH, Stahl PD (1999) Endosome fusion in living cells overexpressing GFP-rab5. J Cell Sci 112:3667–3675

Roosterman D, Schmidlin F, Bunnett NW (2003) Rab5a and rab11a mediate agonist-induced trafficking of protease-activated receptor 2. Am J Physiol Cell Physiol 284:C1319–C1329

Roush ED, Warabi K, Kwatra MM (1999) Characterization of differences between rapid agonist-dependent phosphorylation and phorbol ester-mediated phosphorylation of human substance P receptor in intact cells. Mol Pharmacol 55:855–862

Sabino MA, Honore P, Rogers SD, Mach DB, Luger NM, Mantyh PW (2002) Tooth extraction-induced internalization of the substance P receptor in trigeminal nucleus and spinal cord neurons: imaging the neurochemistry of dental pain. Pain 95:175–186

Sasakawa N, Ferguson JE, Sharif M, Hanley MR (1994a) Attenuation of agonist-induced desensitization of the rat substance P receptor by microinjection of inositol pentakis-and hexakisphosphates in Xenopus laevis oocytes. Mol Pharmacol 46:380–385

Sasakawa N, Sharif M, Hanley MR (1994b) Attenuation of agonist-induced desensitization of the rat substance P receptor by progressive truncation of the C-terminus. FEBS Lett 347:181–184

Schlador ML, Nathanson NM (1997) Synergistic regulation of m2 muscarinic acetylcholine receptor desensitization and sequestration by G protein-coupled receptor kinase- 2 and beta-arrestin-1. J Biol Chem 272:18882–18890

Schmidlin F, Dery O, Bunnett NW, Grady EF (2002) Heterologous regulation of trafficking and signaling of G protein-coupled receptors: beta-arrestin-dependent interactions between neurokinin receptors. Proc Natl Acad Sci USA 99:3324–3329

Schmidlin F, Dery O, DeFea KO, Slice L, Patierno S, Sternini C, Grady EF, Bunnett NW (2001) Dynamin and Rab5a-dependent trafficking and signaling of the neurokinin 1 receptor. J Biol Chem 276:25427–25437

Schmidlin F, Roosterman D, Bunnett NW (2003) The third intracellular loop and carboxyl tail of the neurokinin 1 and 3 receptors determine interactions with beta-arrestins. Am J Cell Physiol 285: C945–958

Schramm NL, Limbird LE (1999) Stimulation of mitogen-activated protein kinase by G protein-coupled alpha$_2$-adrenergic receptors does not require agonist-elicited endocytosis. J Biol Chem 274:24935–24940

Sibley DR, Strausser RH, Benovic JA, Daniel K, Lefkowitz RJ (1986) Phosphorylation/dephosphorylation of the beta-adrenergic receptor regulates its functional coupling to adenylate cyclase and subcellular distribution. Proc Natl Acad Sci USA 83:9408–9412

Simmons MA, Schneider CR, Krause JE (1994) Regulation of the responses to gonadotropin-releasing hormone, muscarine and substance P in sympathetic neurons by changes in cellular constituents and intracellular application of peptide fragments of the substance P receptor. J Pharmacol Exp Ther 271:581–589

Southwell BR, Seybold VS, Woodman HL, Jenkinson KM, Furness JB (1998a) Quantitation of neurokinin 1 receptor internalization and recycling in guinea-pig myenteric neurons. Neuroscience 87:925–931

Southwell BR, Woodman HL, Murphy R, Royal SJ, Furness JB (1996) Characterisation of substance P-induced endocytosis of NK_1 receptors on enteric neurons. Histochem Cell Biol 106:563–571

Southwell BR, Woodman HL, Royal SJ, Furness JB (1998b) Movement of villi induces endocytosis of NK_1 receptors in myenteric neurons from guinea-pig ileum. Cell Tissue Res 292:37–45

Trischler M, Stoorvogel W, Ullrich O (1999) Biochemical analysis of distinct Rab5- and Rab11-positive endosomes along the transferrin pathway. J Cell Sci 112:4773–4783

Tsao P, von Zastrow M (2000) Downregulation of G protein-coupled receptors. Curr Opin Neurobiol 10:365–369

Tsuga H, Kameyama K, Haga T, Kurose H, Nagao T (1994) Sequestration of muscarinic acetylcholine receptor m2 subtypes. Facilitation by G protein-coupled receptor kinase (GRK2) and attenuation by a dominant-negative mutant of GRK2. J Biol Chem 269:32522–32527

Vickery RG, von Zastrow M (1999) Distinct dynamin-dependent and -independent mechanisms target structurally homologous dopamine receptors to different endocytic membranes. J Cell Biol 144:31–43

Vigna SR (1999) Phosphorylation and desensitization of neurokinin-1 receptor expressed in epithelial cells. J Neurochem 73:1925–1932

Vigna SR (2001) The N-terminal domain of substance P is required for complete homologous desensitization but not phosphorylation of the rat neurokinin-1 receptor. Neuropeptides 35:24–31

Vigna SR (2003) The role of the amino-terminal domain of tachykinins in neurokinin-1 receptor signaling and desensitization. Neuropeptides 37:30–35

Vigna SR, Bowden JJ, McDonald DM, Fisher J, Okamoto A, McVey DC, Payan DG, Bunnett NW (1994) Characterization of antibodies to the rat substance P (NK-1) receptor and to a chimeric substance P receptor expressed in mammalian cells. J Neurosci 14:834–845

Vogler O, Bogatkewitsch GS, Wriske C, Krummenerl P, Jakobs KH, van Koppen CJ (1998) Receptor subtype-specific regulation of muscarinic acetylcholine receptor sequestration by dynamin. Distinct sequestration of m2 receptors. J Biol Chem 273:12155–12160

von Zastrow M, Kobilka BK (1992) Ligand-regulated internalization and recycling of human $beta_2$-adrenergic receptors between the plasma membrane and endosomes containing transferrin receptors. J Biol Chem 267:3530–3538

Wang X, Marvizon JC (2002) Time-course of the internalization and recycling of neurokinin 1 receptors in rat dorsal horn neurons. Brain Res 944:239–247

Warabi K, Richardson MD, Barry WT, Yamaguchi K, Roush ED, Nishimura K, Kwatra MM (2002) Human substance P receptor undergoes agonist-dependent phosphorylation by G protein-coupled receptor kinase 5 in vitro. FEBS Lett 521:140–144

Whistler JL, Enquist J, Marley A, Fong J, Gladher F, Tsuruda P, Murray SR, von Zastrow M (2002) Modulation of postendocytic sorting of G protein-coupled receptors. Science 297:615–620

Widmann C, Gibson S, Jarpe MB, Johnson GL (1999) Mitogen-activated protein kinase: conservation of a three-kinase module from yeast to human. Physiol Rev 79:143–180

Zhang J, Barak LS, Winkler KE, Caron MG, Ferguson SS (1997) A central role for beta-arrestins and clathrin-coated vesicle-mediated endocytosis in $beta_2$-adrenergic receptor resensitization. Differential regulation of receptor resensitization in two distinct cell types. J Biol Chem 272:27005–27014

Zhang J, Ferguson SSG, Barak LS, Menard L, Caron MG (1996) Dynamin and beta-arrestin reveal distinct mechanisms for G protein-coupled receptor internalization. J Biol Chem 271:18302–18305

Tachykinin NK₁ Receptor Antagonists

R. Patacchini · C. A. Maggi

Pharmacology Department, Menarini Ricerche, Via Rismondo 12/A, 50131 Florence, Italy
e-mail: chimfarm@menarini-ricerche.it

1	Introduction	174
2	The Target Receptor	174
2.1	Heterogeneity of NK_1 Receptors Revealed by Selective Antagonists	175
2.1.1	Species-Dependent Heterogeneity of the NK_1 Receptor	175
2.1.2	Intraspecies Heterogeneity of the NK_1 Receptor	177
3	Peptide Antagonists	179
3.1	Early Peptide-Based SP Antagonists	179
3.2	NK_1 Receptor-Selective Peptide-Based Antagonists	181
4	Nonpeptide Antagonists	187
4.1	CP96345 and Related Compounds	187
4.2	RP67580 and Related Compounds	191
4.3	SR140333 and Related Compounds	192
4.4	GR203040 and Related Compounds	193
4.5	L754030 (MK869) and Related Compounds	196
4.6	LY 303870 and Related Compounds	198
4.7	CGP49823, SDZNKT343, NKP608	200
4.8	PD154075 (CI1021)	202
4.9	TAK637	203
4.10	Other Nonpeptide Compounds	204
5	Therapeutic Perspectives for NK_1 Receptor Antagonists	204
	References	206

Abstract This chapter is focused on the pharmacology of most relevant tachykinin NK_1 receptor-selective compounds, with emphasis on the progress of knowledge made possible by their use and on the therapeutic perspectives of these drugs. The first peptide antagonists of SP, synthesized about 20 years ago, were hampered by poor selectivity for NK_1 receptors and other serious side effects. Since then the number of new compounds, peptidic at first and nonpeptidic later, being endowed with increasing potency and selectivity toward the NK_1 receptor, has been growing continuously. A milestone in the field has been the introduction in 1991 of the first nonpeptide compound, CP96345, by Pfizer. CP96345 was instrumental, along with RP 67580 from Rhone-Poulenc, for recognition of the existence of species-related heterogeneity of NK_1 receptors. CP96345 has been used in a plenty of preclinical studies aiming at investigating the pathophysiological role of the NK_1 receptor, demonstrating the involvement of this receptor in pain perception and inflammation. The following compounds introduced in the field have been deprived of the undesired ion channel-blocking activity accompanying CP96345 and related compounds. By their use, the involvement of the NK_1 receptor in emesis and in mediating affective disorders such as depression and anxiety has been proven. Antiemetic and antidepressive activity in humans has been reported first by CP122721 from Pfizer and by L754030 (MK869) from Merck, respectively. In contrast, disappointing results have been obtained with tachykinin NK_1 antagonists tested as analgesics in patients affected by osteorathritis, neuropathic pain, dental pain and migraine, for reasons that are still not fully understood. Tachykinin NK_1 receptor-selective antagonists are expected to provide therapeutic effects in peripheral inflammatory diseases such as asthma and inflammatory bowel disease. However, the first trials with these compounds in asthmatic patients have been negative.

Keywords Tachykinins · Tachykinin NK_1 receptor · Tachykinin NK_1 receptor antagonists

1
Introduction

The appearance of the first peptidic antagonists of substance P (SP), at the beginning of the 1980s, triggered intense research programs undertaken by many pharmaceutical companies and which aimed at developing more potent and selective compounds to be used as drugs for treatment of painful/inflammatory conditions in which tachykinins were suspected to play a role. Since then, many potent antagonists of the NK_1 receptor have been developed, being devoid of the numerous drawbacks that characterized the earlier peptide and nonpeptide compounds. The very intense use of these tools in preclinical studies has enabled us to understand the contribution made by tachykinins acting via NK_1 NK_2 and NK_3 receptors to pathophysiological conditions in mammals. As a result of these efforts, several tachykinin NK_1 receptor antagonists are currently under clinical investigation for treatment of pain of different origin, inflammatory bowel disease (IBD), emesis, anxiety and depression. This chapter is focused on the pharmacology of the most relevant tachykinin NK_1 receptor-selective compounds, with emphasis on the progress of knowledge made possible by their use in the tachykinin field, and with appropriate hints at the therapeutic perspectives of these compounds.

2
The Target Receptor

The NK_1 receptor is widely distributed throughout the central and peripheral nervous system of mammals, and was originally defined as the mediator of the biological activities encoded by the C-terminal sequence of tachykinins for which SP is a more potent agonist than neurokinin A (NKA) or neurokinin B (NKB) (for reviews see Regoli et al. 1989; Guard and Watson 1991; Mussap et al. 1993; Maggi et al. 1993b; Maggi 1995). This concept remained unchanged until Schwartz and coworkers (Hastrup and Schwartz 1996) provided evidence, obtained by radioligand binding and functional experiments, that both NKA and NKB are as potent as SP at stimulating the tachykinin NK_1 receptor (see Sect. 2.1.2): this conclusion has recently been confirmed by in vivo experiments showing NKA to be a potent stimulator of salivary secretion in rats (Bremer et al. 2001). As reviewed elsewhere (Regoli et al. 1994; Maggi 1995; Quartara and Maggi 1997), the tachykinin NK_1 receptor has been cloned from several species, including man. It is worth mentioning that until now only one copy of the corresponding gene has been isolated in man and all other species examined. The NK_1 receptor belongs to the superfamily of rhodopsin-like G-protein-coupled receptors with seven transmembrane spanning segments. Stimulation of the NK_1 receptor leads to generation of at least three different second messenger pathways: phosphatidyl inositol breakdown (leading to elevation of intracellular calcium); arachidonic acid mobilization (leading to prostanoid generation) and cAMP accumulation (for a review see Quartara and Maggi 1997).

2.1
Heterogeneity of NK_1 Receptors Revealed by Selective Antagonists

2.1.1
Species-Dependent Heterogeneity of the NK_1 Receptor

The introduction of the first potent and selective nonpeptide antagonists of the NK_1 receptor, namely CP96345 and RP67580 (see Sects. 4.1 and 4.2), gave a boost to the research aimed at investigating the role of the NK_1 receptor as mediator of tachykinin-induced effects in mammals. From the many studies which were undertaken, a striking pharmacological heterogeneity among NK_1 receptors belonging to different species emerged. For example, CP96345 was found to be 30–100-fold more potent in displacing radiolabeled SP from human, guinea pig, bovine, hamster, gerbil and rabbit than from rat or mouse NK_1 receptors (Gitter et al. 1991; Beresford et al. 1991b). On the contrary, RP67580 displayed about 10–20-fold higher potency in blocking rat NK_1 receptors than guinea pig or human NK_1 receptors (Garret et al. 1991; Barr and Watson 1993; Beaujouan et al. 1993). The existence of pharmacological differences between NK_1 receptors present in rodents, on the one hand, and NK_1 receptors expressed in humans and guinea pigs, on the other, was corroborated with other NK_1 receptor antagonists which were subsequently introduced in this field (see Sect. 3.2), such as GR82334 (Meini et al. 1994) and FK888 (Fujii et al. 1992; Aramori et al. 1994). It should be noted that the ability to recognize with different affinities NK_1 receptors belonging to different species does not extend to all antagonist compounds thus far developed, as demonstrated by the selective antagonist SR140333 (see Sect. 4.3) which displaces ^{125}I-SP specific binding from rat cerebral cortex and from human IM9 cell membranes with identical affinities (0.02 and 0.01 nM, respectively) (Emonds-Alt et al. 1993). It is also worth mentioning that the reported species-related heterogeneity of NK_1 receptors is not recognizable by SP or other natural tachykinin agonists. The species-related pharmacological heterogeneity of the NK_1 receptor has been validated by point mutation studies on the human and rat NK_1 receptor. Fong and Strader (1992) and Sachais et al. (1993) have clarified that the diverging, species-related variable potencies of CP96345 and RP67580 in blocking the human and rat NK_1 receptor are determined by species-dependent variations at two discrete positions (116 and 290) of the amino acid sequence of the NK_1 receptor protein. In fact, the replacement of residues 116 and 290 of the human NK_1 receptor (Val and Ile) with the corresponding residues of the rat NK_1 receptor (Leu and Ser) conferred full sensitivity for RP67580 to the mutant receptor, while decreasing the binding affinity of CP96345. The opposite effect was observed when replacing residues 116 and 290 of the rat NK_1 receptor with those present in the human NK_1 receptor protein. Fong and Strader (1992) also provided evidence that residues 116 and 290 probably do not interact directly with antagonists but rather influence the overall conformation of the antagonist binding site(s) (Fong et al. 1994). Fong and Strader (1992) also noted that the exchange of these residues crucial for binding affinity of nonpeptide antagonists (like CP96345 and RP67580) did not affect the affinity of SP, indicating that they are not involved in the binding of the agonist. Further molecular studies have extended this finding, by showing that the residues of the NK_1 receptor involved in the binding of peptide agonists are mostly nonoverlapping with those interacting with nonpeptide antagonists (or low molecular weight peptide antagonists such as FK888) (for a review of these studies see Maggi 1995).

Looking at the results of these mutagenesis studies, the question arises as to how it is possible that CP96345, RP67580 and other nonpeptide compounds behave as competitive ligands in various binding and functional assays (e.g., Snider et al. 1991; Garret et al. 1991; Patacchini et al. 1992). This actually contrasts with the classical receptor theory, requiring that mutually exclusive binding sites are bound by both the agonist and the antagonist. Two main explanations have been advanced to account for this apparent discrepancy. Huang and coworkers (1994) proposed that "the competitive behavior can arise from a volume exclusion effect, in which the binding pockets for agonists and antagonists spatially overlap, even though agonists and antagonists do not utilize the same set of residues for their interactions with the NK_1 receptor". The second interpretation stems from the work of Schwartz's group, who elaborated a complex model, whereby the agonists stabilize the receptor in its active conformation, whereas the antagonists may not only stabilize the inactive conformation, but also may induce a structural change in the receptor that distorts the agonist binding site and "...drags the receptor equilibrium away from the active conformation that binds the agonist with high affinity" (Gether et al. 1995). Thus, agonists and antagonists would exclude each other from binding to the receptor, not by binding to the same sites, but through a strong negative cooperativity (Gether et al. 1995; Schwartz et al. 1995).

2.1.2
Intraspecies Heterogeneity of the NK_1 Receptor

The most studied and discussed intraspecies heterogeneity of the tachykinin NK_1 receptor originates from the work of Petitet et al. (1992) who showed the synthetic peptide septide to be as potent as SP in producing NK_1-receptor mediated contraction in the guinea pig ileum, while being a poor displacer (about 100-fold weaker than SP) of ^{3}H-SP specific binding of NK_1 receptors present on guinea pig ileum membranes. Two NK_1 receptor subtypes were proposed to exist: a classical one, sensitive to SP and other selective agonists (e.g., [Pro^9]SP, [Sar^9]SP sulfone), and a so-called septide-sensitive subtype, for which septide and some other analogs (e.g., SP methylester) have higher affinity than for the other subtype (Petitet et al. 1992). The main functional evidence supporting this conclusion was that a selective NK_1 receptor antagonist, GR71251 (see Sect. 3.2), blocked septide-mediated effects in the guinea pig ileum with approximately tenfold higher potency than [Pro^9]SP-mediated effects (Petitet et al. 1992). Subsequent investigations in which different NK_1 receptor-selective antagonists (such as CP96345, RP67580, GR82334, CP99994, LY303870 and others) were used, confirmed the early observation of Petitet et al. (1992), since all compounds tested were approximately tenfold more potent against septide (or septide-similar agonists) than against SP (or SP-related agonists) in various in vitro bioassays (Table 1). Similar results were obtained in in vivo investigations (e.g., Boni et al. 1994; Cellier et al. 1999). All these studies provided further evidence for the existence of pharmacological heterogeneity of the NK_1 receptor in the species examined. However, the early hypothesis of the existence of distinct NK_1 receptor subtypes declined quickly, for several reasons: (a) failure to isolate alternative gene(s) encoding receptor proteins bearing the characteristics of the septide-sensitive receptor; (b) demonstration that SP-preferring and septide-preferring binding sites could be described in one and the same receptor protein: the rat recombinant NK_1 receptor expressed in COS-1 cells (Pradier et al. 1994), and (c) demonstration that in homologous binding experiments radiolabeled NKA and septide bind with subnanomolar affinity to the NK_1 receptor, as expected from their functional activities (Hastrup and Schwartz 1996). On the basis of the former study, Maggi and Schwartz (1997) have elaborated a model accounting for all the reported pharmacological effects of septide, SP and other natural tachykinins; in this model the NK_1 receptor consists of two different conformers: the general tachykinin conformer (recognized with high affinity by NKA, NKB, septide and related peptides) and the SP-preferring conformer (recognized with high affinity by SP and related peptides). More recently, two residues (193 and 195) that are necessary for the binding of both NKA and septide but not of SP (Wijkhuisen et al. 1999) have been identified in the human sequence of the NK_1 receptor.

Table 1 Higher potency of tachykinin NK$_1$ receptor antagonists in blocking effects elicited by septide or septide-like agonists than in blocking effects elicited by SP or SP-like agonists in various in vitro bioassays from different species

Species	Tissue	Antagonist	Agonist 1 (pA$_2$)	Agonist 2 (pA$_2$)	Reference
Guinea-pig	Ileum	GR71251	Septide (7.9)	[Pro^9]SP (6.5)	Petitet et al. 1992
	Ileum	CP96345	Septide (9.2)	[Sar^9]SP sulfone (8.2)	Maggi et al. 1993a
	Ileum	CP99994	Septide (11.9)	SP (10.2)	Jenkinson et al. 1999
	Colon	GR82334	Septide (8.1)	[Sar^9]SP sulfone (7.5)	Maggi et al. 1994
	Colon	FK888	Septide (8.1)	[Sar^9]SP sulfone (7.1)	Maggi et al. 1994
	Bronchus	CP96345	Septide (7.8)	[Pro^9]SP (6.7)	Zeng and Burcher 1994
	Bronchus	GR82334	Septide (7.5)	[Pro^9]SP (6.3)	Zeng and Burcher 1994
	Bronchus	MEN10581	Septide (7.2)	[Sar^9]SP sulfone (6.3)	Patacchini et al. 1995
	Trachea	CP96345	Septide (8.4)	[Sar^9]SP sulfone (6.9)	Longmore et al. 1994
	Common bile duct	GR82334	Septide (7.5)	[Sar^9]SP sulfone (6.8)	Patacchini et al. 1997
Rat	Urinary bladder	RP67580	NKB (7.6)	[Sar^9]SP sulfone (7.1)	Meini et al. 1994
	Urinary bladder	GR82334	Septide (7.0)	[Sar^9]SP sulfone (5.9)	Meini et al. 1994
	Urinary bladder	CP96345	Septide (6.5)	[Sar^9]SP sulfone (5.7)	Montier et al. 1994
	Urinary bladder	GR82334	Septide (8.2)	[Pro^9]SP (6.6)	Torrens et al. 1995
Rabbit	Iris sphincter	CP96345	Septide (9.3)	SP (<7.0)	Hall et al. 1994
	Iris sphincter	RP67580	SPOMe[a] (7.0)	[Pro^9]SP (<6.0)	Hall et al. 1994
	Jugular vein	CP96345	Septide (9.0)	[Pro^9]SP (8.0)	Patacchini and Maggi 1995
Human	U373 astrocytoma cells	CP96345	Septide (9.7)[b]	SP (8.2)[b]	Palma et al. 1994
	U373 astrocytoma cells	GR82334	(7.1)[b]	(6.4)[b]	Palma et al. 1994

[a] Substance P methyl ester.
[b] Values are pIC$_{50}$.

3
Peptide Antagonists

3.1
Early Peptide-Based SP Antagonists

The first antagonists of SP became available at the beginning of the 1980s. These compounds were close analogs of SP with two or more D-amino acids replacing natural amino acids in the SP sequence (e.g., [DPro^2, DTrp^{7,9}]SP), and they were useful tools to prove that tachykinins play a neurotransmitter role in the peripheral nervous system (e.g., Leander et al. 1981). One of the most potent and widely used compounds of this series, spantide, is a peptide bearing two DTrp residues in position 7 and 9 of the SP backbone and a DArg in the N-terminal position to improve metabolic stability (Table 2). Spantide was described to block NK$_1$ receptors with a potency (pA$_2$) exceeding 7.0 in the guinea pig taenia coli (Folkers et al. 1984), and greater than 6.0 in the guinea pig ileum, whereas it was found to be tenfold weaker at NK$_2$ or NK$_3$ receptors (Table 3). However, spantide suffered from several drawbacks (shared by other DTrp-substituted analogs of SP, and summarized in Table 4), including neurotoxicity and ability to degranulate mast cells and release histamine (Folkers et al. 1984; Hoover 1991). To overcome these 'side effects' and obtain a more potent and selective antagonist, Folkers and coworkers (1990) synthesized a linear peptide compound bearing only 4 out of 11 amino acids of the SP sequence: spantide II (Table 2). Spantide II was devoid of neurotoxicity and showed about tenfold higher potency than spantide at NK$_1$ receptors of the guinea pig (Table 3). Other examples of early SP antagonists are represented by a series of N-terminal truncated analogs of SP

Table 2 Amino acid sequence of substance P (SP) and peptide-based NK$_1$ receptor antagonists (for references, see text)

Compound	Sequence
Substance P	Arg-Pro-Lys-Pro-Gln-Gln-Phe-Phe-Gly-Leu-Met-NH$_2$
Spantide	[dArg^1, dTrp^{7,9}, Leu^{11}]SP
Spantide II	dLys(Nic)-Pro-Pal(3)-Pro-dPhe(Cl$_2$)-Asn-dTrp-Phe-dTrp-Leu-Nle-NH$_2$
L668169	(cyclo[Gln-dTrp(Me)-Phe-(R)Gly[ANC-2]Leu-Met]^2)
GR71251	[dPro^9 [spiro-γ-lactam]Leu^{10}, Trp^{11}]SP(1–11)
GR82334	[DPro^9 [spiro-γ-lactam]Leu^{10}, Trp^{11}]physalaemin (1–11)
FR113680	Ac-Thr-dTrp(for)-Phe-NMeBzl
FK888	(2-(N-Me)indolyl)-CO-Hyp-Nal-NMeBzl
S18523	{1-[4-(H-tetrazol-5-yl)butyl]indol-3-yl}CO-Hyp-Nal-NMeBzl
MEN10930	Nα{N-[(1H)indol-3-yl-carbonyl]1-amino-cyclohexane-1-carbonyl}L-3-(2-naphthyl)alanine N-(benzyl) N methyl amide
MEN11467	(1R,2S)-2-N[1(H)indol-3-yl-carbonyl]-1-N-{N$^\alpha$ (p-tolylacetyl)-N$^\alpha$ (methyl)-d-3-(2-naphthyl)alanyl}diaminocycloexane
Cam2819	[2-Fl]Z-α-aMe-Trp-NH(S)-CH-Me-Ph
Sendide	[Tyr^6, dTrp^7, dHis^9]SP(6–11)

Table 3 Potencies of tachykinin NK_1 receptor antagonists at tachykinin NK_1, NK_2 and NK_3 receptors measured on isolated smooth muscle preparations

Compound	NK_1 bioassays (pK_B)	NK_2 bioassays (pK_B)	NK_3 bioassays (pK_B)
Spantide	GPI (6.2)[a]	RPA (5.3)[a] HT (4.9)[a]	RPV (<5.0)a
Spantide II	GPI (7.1)[a,b] RJV (6.8)[b] RUB (<5.0)[b]	RPA (5.4)[a,b] HT (6.0)[a,b]	RPV (<5.0)[a,b]
L668169	GPI (7.0)[c] (6.4)[b] RJV (6.3)[b] RUB (<5.0)[b] HI (6.3)[d]	RPA (<5.0)[b] HT (6.2)[b] HUB (6.2)[c]	RPV (<5.0)[b]
GR71251	GPI (7.7)[e]	RCMM (<5.0)[e]	RPV (<5.0)[e]
GR82334	GPI (7.6)[f]	RCMM (<5.0)[f] RPA (5.1)[un]	RPV (<5.0)[f]
FR113680	GPI (7.5)g (6.6)[b]	RPA (5.4)[b] RVD (<5.0)[g]	RPV (<5.0)[b,g]
FK888	GPI (9.3)h (7.5–8.3)[i] RIS (7.1)[j]	RPA (<5.0)[h] RVD (<5.0)[h]	RPV (<5.0)[h,un]
S18523	RVC (9.6)[k]	RPA (5.6)[k]	RPV (5.0)[k]
Cam 2819	GPI (8.4)[l] RJV (8.9)[l]	–	–
CP96345	DCA (8.7)[m] GPI (8.1)[b] RJV (8.3)[b] RUB (6.2)[b]	RPA (<5.0)[b]	RPV (<5.0)[b]
RP67580	GPI (7.2–7.6)[n]	RPA (<6.0)[n]	RPV (<6.0)[n]
SR 140333	GPI (~9.7)[o]	RPA (<6.0)[o]	RPV (<6.0)[o]
GR 203040	GPI (~11.9)[p] DCA (~11.2)[p]	–	–
GR 205171	DCA (11.4)[q]	–	–
LY 303870	RVC (9.4)[r]	RPA (4.7)[r]	RPV (4.7)[r]
CGP 49823	RVC (7.7)[s]	RA (<6.0)[s]	–
PD 154075	GPI (9.5)[t]	–	–
TAK 637	GPC (8.3–8.7)[u]	GPC (<7.0)[u]	GPTC (<7.0)[u]

DCA, Dog cerebral arteries; GPI, longitudinal muscle strip of guinea pig ileum; GPTC, guinea pig taenia coli; HI, human ileum; HT, hamster trachea; HUB, hamster urinary bladder; RA, rabbit aorta; RCMM, rat colon muscularis mucosae; RIS, rabbit iris sphincter; RJV, rabbit jugular vein; RPA, rabbit pulmonary artery (endothelium-denuded); RPV, rat portal vein; RUB, rat urinary bladder; RVD, rat vas deferens; RVC, rabbit vena cava.

References: [a] Maggi et al. 1991; [b] Patacchini et al. 1992; [c] Williams et al. 1988; [d] Maggi et al. 1992; [e] Hagan et al. 1990; [f] Hagan et al. 1991; [g] Morimoto et al. 1992; [h] Fujii et al. 1992; [i] Maggi et al. 1994; [j] Wang et al. 1994; [k] Bonnet et al. 1996; [l] McKnight et al. 1994; [m] Snider et al. 1991; [n] Garret et al 1991; [o] Emonds-Alt et al. 1993; [p] Beattie et al. 1995; [q] Gardner et al. 1996; [r] Gitter et al. 1995; [s] Hauser et al. 1994; [t] Boyle et al. 1994; [u] Venkova et al. 2002; [un] Patacchini et al. unpublished observations.

Table 4 Drawbacks and side effects shown by several tachykinin NK_1 receptor antagonists

Compound	Drawbacks and/or side effects	References
Linear SP analogs bearing dTrp (e.g., spantide)	Low potency; poor selectivity between NK_1 and NK_2 receptors; neurotoxicity; local anesthetic activity; mast cell degranulation activity; blockade of tachykinin-unrelated receptors	Folkers et al. 1984 Hoover 1991 Hökfelt et al. 1981 Maggi et al. 1993b Regoli et al. 1994
GR71251	Mast cell degranulation activity; poor metabolic stability	Hagan et al. 1990, 1991
FR113680	Poor bioavailability and solubility	Hagiwara et al. 1993
Sendide	Inability to penetrate the CNS; interaction with tachykinin-unrelated receptors; antagonist activity restricted to rodents	Minami et al. 1998 R. Patacchini, unpublished observations
CP96345	Depression of smooth muscle contractility in various isolated tissues; nonspecific inhibitory effects on neurotransmission; local anesthetic-like effects; depression of blood pressure and +/− effects on heart rate	Patacchini et al. 1991 Wang et al. 1994a Karlsson et al. 1994 Tamura et al. 1993 Lembeck et al. 1992 Constantine et al. 1991
CP99994	Nonspecific reduction of formalin-induced nociceptive responses in gerbils and rats	Smith et al. 1994 Rupniak et al. 1995
RP67580	Nonspecific inhibitory effects on neurotransmission; insufficient penetration of the blood–brain barrier; nonspecific inhibitory effects	Wang et al. 1994a Holzer-Petsche and Rodorf-Nikolic 1995 Rupniak et al. 1993
FK888 SR140333 CGP49823 LY303870	Poor brain penetration, as shown by their low potency (effective doses: 3–10 mg/kg) in preventing foot tapping induced by a SP analog in gerbils, and by their ineffectiveness in preventing cisplatin-induced acute retching in ferrets (at the same doses)	Rupniak et al. 1997
CP122721 GR205171	Brief duration of action at central NK_1 receptors after oral administration	Hale et al. 1998

containing two or three dTrp residues, such as [dPro^4, DTrp^{7,9,10}]SP (Regoli et al. 1984; see Regoli et al. 1994 for review of these analogs).

3.2
NK_1 Receptor-Selective Peptide-Based Antagonists

Historically, the first compound reported to be selective for NK_1 vs. NK_2 and NK_3 receptors was the cyclic antagonist L668169, developed at the end of the 1980s (Williams et al. 1988). L668169, a dodecapeptide bearing a lactam constraint between Gly and Leu and a dTrp residue (Table 2), showed at least tenfold higher potency at guinea pig or rabbit NK_1 receptors as compared to rabbit NK_2 or rat NK_3 receptors, whereas it was apparently inactive at rat NK_1 receptors (Table 3). L668169 was also able to produce weak but selective blockade of contractions produced by SP in the human isolated ileum (Maggi et al. 1992; Table 3).

Soon after the introduction of L668169 a group of researchers at Glaxo (Hagan et al. 1990) presented GR71251 (Table 2), an undecapeptide analog of SP

Table 5 Affinities of tachykinin NK_1 receptor antagonists for NK_1, NK_2 and NK_3 receptors measured in radioligand binding experiments

Compound	NK_1 receptor bioassay (pKi or pIC_{50})	NK_2 receptor bioassay (pKi or pIC_{50})	NK_3 receptor bioassay (pKi or pIC_{50})
FR113680	GPL (pIC_{50}=7.8)[a] RCC (pIC_{50}<5.0)[a]	RD (pIC_{50}<5.0)[a]	RCC (pIC_{50}<5.0)[a]
FK888	GPL (pKi=9.2)[b] RCC (pKi=6.3)[b]	–	–
S18523	IM9 (pKi=9.1)[c]	CHO (pKi=5.5)[c]	–
MEN10930	IM9 (pKi=9.0)[d] U 373 MG (pKi=8.6)[d] RUB (pKi<5.0)[d]	HUB (pKi=5.8)[d]	RCC (pKi<5.0)[d]
MEN11467	IM9 (pKi=9.4)[e]	CHO (pKi=6.4)[e]	RCC (pKi<5.0)[e]
Cam2819	IM9 (pIC_{50}=8.3)[f]	–	–
Sendide	MSC (pKi=11.4)[g]	–	–
CP96345	BCM (pIC_{50}=8.5)[h] RFM (pIC_{50}=6.6)[h] IM9 (pKi=9.6)[i]	HUB (pIC_{50}<5.0)[i]	GPC (pIC_{50}<5.0)[i]
CP99994	IM9 (pKi=9.6)[i]	HUB (pIC_{50}<5.0)[i]	GPC (pIC_{50}<5.0)[i]
CP122721	IM9 (pIC_{50}=9.8)[j]	CHO (pIC_{50}<5.0)[j]	GPC (pIC_{50}<5.0)[j]
CJ11974	IM9 (pKi=9.7)[k]	CHO (pKi<6.0)[k]	GPC (pKi<6.0)[k]
RP67580	RB (pKi=8.4)[l]	RD (pIC_{50}<5.0)[l]	GPC (pIC_{50}<5.0)[l]
RPR100893	IM9 (pKi=7.9)[m]	–	–
SR140333	IM9 (pKi=10.7)[n] U373MG (pKi=9.2)[n] RB (pKi=10.6)[n]	RD (pKi<6.0)[n]	RCC (pKi<6.0)[n]
SSR240600	IM9 (pKi=11.2)[o] U373MG (pKi=10.0)[o] RI (pKi=9.0)[o]	CHO (pKi=7.6)[o]	CHO (pKi=6.7)[o]
GR203040	CHO (pKi=10.3)[p] U373MG (pKi=10.5)[p] RCC (pKi=8.6)[p]	CHO (pKi<5.0)[p]	GPC (pKi<6.0)[p]
GR205171	CHO (pKi=10.6)[q] RCC (pKi=9.5)[q] FCC (pKi=9.8)[q]	RC (pKi<5.0)[q]	GPC (pKi<5.0)[q]
L733060	CHO (pIC_{50}=9.0)[r]	–	–
L742694	CHO (pIC_{50}=10.0)[s]	CHO (pIC_{50}=5.2)[s]	CHO (pIC_{50}=6.8)[s]
L754030	CHO (pIC_{50}=10.0)[t]	CHO (pIC_{50}<6.0)[t]	CHO (pIC_{50}<7.0)[t]
L732138	CHO (pIC_{50}=8.6)[u]	CHO (pIC_{50}<6.0)[u]	CHO (pIC_{50}<6.0)[u]
L737488	CHO (pIC_{50}=9.8)[v]	CHO (pIC_{50}<6.0)[v]	CHO (pIC_{50}<6.0)[v]
LY303870	IM9 (pKi=9.8)[w] H Caud. (pKi=10.0)[w]	CHO (pKi<6.0)[w]	CPC (pKi<6.0)[w]
CGP49823	BR (pIC_{50}=7.9)[x]	BB (pIC_{50}=5.6)[x]	GC (pIC_{50}=5.6)[x]
SDZNKT343	COS-7 (pKi=9.2)[y]	COS-7 (pKi=6.3)[y]	COS-7 (pKi=5.5)[y]
NKP608	BR (pIC_{50}=8.6)[z]	BB (pIC_{50}=5.7)[z]	GC (pIC_{50}=5.8)[z]
PD154075	IM9 (pKi=9.5)[aa]	HUB (<5.0)[bb]	CHO (<6.0)[bb]
TAK637	IM9 (pIC_{50}=9.3)[cc]	–	GPC (pIC_{50}<6.0)[cc]
R116301	CHO (pKi=9.3)[dd]	CHO (pKi=6.1)[dd]	CHO (pKi=7.0)[dd]

BB, Bovine bladder membranes; BCM, bovine caudate membranes; BR, bovine retina membranes; CHO, Chinese hamster ovary cells bearing transfected human NK_1, NK_2 or NK_3 receptors; COS-7, cell line bearing transfected human NK_1, NK_2 or NK_3 receptors; FCC, ferret cerebral cortex; GC, gerbil cortical mem-

incorporating a bicyclic conformational constraint designed to eliminate NK_1 receptor-activating conformations. GR71251 was endowed with good potency at NK_1 receptors and selectivity vs. NK_2 and NK_3 receptors (Table 3), but it was shown to release histamine from rat peritoneal mast cells (Hagan et al. 1990). This side effect, along with rapid metabolic degradation (Hagan et al. 1991), limited the use of GR71251 (Table 4). These problems were overcome with the synthesis of GR82334, an undecapeptide compound bearing the same substitutions introduced in GR71251, but in this case on the backbone of the nonmammalian tachykinin physalaemin (Hagan et al. 1991; Table 2). GR82334 retained the same potency and selectivity for NK_1 receptors as GR71251 (Table 3) and, being devoid of important side effects and being resistant to metabolic degradation, became a very useful tool for both in vitro (e.g., Guo et al. 1995) and in vivo (e.g., Hagan et al. 1991; Beresford et al. 1991a) studies (see also Table 1).

A systematic approach was undertaken by researchers at Fujisawa in the early 1990s to identify the minimal sequence contained in the octapeptide SP antagonist [$DPro^4$ $DTrp^{7,9,10}$ Phe^{11}]SP(4–11) that was capable of binding with sufficient affinity to tachykinin receptors. After a subsequent lead optimization, low-molecular weight peptide fragments endowed with high affinity and selectivity for NK_1 receptors were identified, the most interesting being the tripeptide FR113680 (Morimoto et al. 1992; Table 2). FR113680 displayed a fairly good affinity for guinea pig NK_1 receptors, whereas it was inactive/much less active at rat NK_1 receptors and at NK_2 and NK_3 receptors (Tables 3, 5). In in vivo experiments, FR113680 (1–10 mg/kg intravenously) fully prevented both bronchoconstriction and airway edema induced by SP, and only partially the effects produced by NKA or capsaicin, thought to be mediated by both NK_2 and NK_1 receptors (Murai et al. 1992). Nevertheless, FR113680 was not developed as antiasthmatic drug by Fujisawa, because of its unfavorable chemico-physical characteristics (Hagiwara et al. 1993; Table 4). Rather, a new series of short peptides was synthesized that had reduced weight and lipophilicity: the most promising compound of this series was FK888 (Table 2; Fujii et al. 1992). FK888 showed

◀───

branes; GPC, guinea pig cortex membranes; GPL, guinea pig lung membranes; Hcaud, human caudate homogenate; HUB, hamster urinary bladder cell membranes; IM9, human lymphoblastoma cell line membranes; MSC, mouse spinal cord membranes; NE, not effective; RB, rat brain membranes; RCC, rat cerebral cortex membranes; RD, rat duodenum smooth muscle membranes; RFM, rat forebrain membranes; RI, rat ileum; RUB, rat urinary bladder cell membranes; U373MG, human astrocytoma cell line membranes.
References: [a] Morimoto et al. 1992; [b] Fujii et al. 1992; [c] Bonnet et al. 1996; [d] Astolfi et al. 1997; [e] Cirillo et al. 2001; [f] McKnight et al. 1994; [g] Sakurada et al. 1992; [h] Snider et al. 1991; [i] McLean et al. 1993); [j] McLean et al. 1996; [k] Tsuchiya et al. 2002; [l] Garret et al. 1991; [m] Fardin et al. 1994; [n] Emonds-Alt et al. 1993; [o] Emonds-Alt et al. 2002; [p] Beattie et al. 1995; [q] Gardner et al. 1996; [r] Harrison et al. 1994; [s] Hale et al. 1996; [t] Hale et al. 1998; [u] Cascieri et al. 1994; [v] MacLeod et al. 1995; [w] Gitter et al. 1995; [x] Hauser et al. 1994; [y] Walpole et al. 1998a; [z] Vassout et al. 2000; [aa] Boyle et al. 1994; [bb] Singh et al. 1997; [cc] Natsugari et al. 1999; [dd] Megens et al. 2002.

high affinity for (guinea pig) NK_1 receptors and remarkable selectivity vs. rat NK_1, NK_2 and NK_3 receptors (Tables 3, 5). Furthermore, FK888 was found to be active in blocking airway edema, plasma protein extravasation and other responses produced by exogenous and endogenous tachykinins in in vivo bioassays, even after oral administration (Table 6; Fujii et al. 1992; Murai et al. 1993; Hirayama et al. 1993; Wang et al. 1994b). Owing to its potency and lack of nonspecific effects, FK888 was regarded as promising candidate for the treatment of asthma in clinical trials. However, FK888 failed to afford beneficial effects on exercise-induced bronchoconstriction in asthmatics, even though it significantly shortened the recovery times, suggesting a possible role of tachykinin NK_1 receptors in the late bronchoconstrictor response to exercise (Ichinose et al. 1996).

A strict analog of FK888 is represented by the Servier compound S18523: a dipeptide bearing a tetrazolyl-butyl group on the indole nitrogen of FK888 (Table 2; Bonnet et al. 1996). S18523 shows highly improved water solubility compared to FK888 and slightly higher affinity for human and rabbit NK_1 receptors (Tables 3, 5; Bonnet et al. 1996). In vivo, S18523 exerts potent antinociceptive effects in classical pain tests (hot-plate in mice, phenylbenzoquinone-induced writhing in mice and formalin test in rats), and blocks SP- or capsaicin-induced plasma protein extravasation in guinea pigs, also after oral administration.

Another example of a short peptide endowed with high potency and selectivity for the NK_1 receptor is represented by MEN10930, a tripeptide analog of FK888 synthesized at Menarini Laboratories (Table 2; Astolfi et al. 1997). MEN10930 potently and selectively bound to human transfected NK_1 receptors, whereas it showed approximately 10,000-fold lower affinity for the rat NK_1 receptor subtype (Table 5). Further modifications of MEN10930 led to discovery of a partially retro-inverse peptidomimetic antagonist: MEN11467 (Table 2). MEN11467 is characterized by the presence of a diaminocyclohexane moiety and of a (2-naphthyl)alanine residue, inserted in a reversed direction starting from the C terminus. These modifications preserve MEN11467 from enzymatic degradation, while the high affinity for the NK_1 receptor is conserved (Cirillo et al. 2001). MEN11467 showed high affinity and selectivity for human NK_1 receptors in binding experiments, with clearly insurmountable kinetics of action (Cirillo et al. 2001; Table 5). In vivo, oral administration of MEN11467 produced a long-lasting inhibition of both bronchoconstriction and plasma protein extravasation induced in anesthetized guinea pigs by a selective NK_1 receptor agonist (Cirillo et al. 2001; Table 6). MEN11467 was also able to prevent albumin-induced mucus secretion in sensitized ferrets (Khan et al. 2001) and was therefore proposed for development as an antisecretory drug in allergic asthma. In contrast to its peripheral activity, MEN11467 showed poor ability to penetrate into the central nervous system (CNS) and block central NK_1 receptors (Cirillo et al. 2001).

An approach similar to that of Fujisawa–that is identification of a short peptide sequence bearing appreciable affinity for tachykinin receptors followed by optimization–was followed by a group of researchers at Parke-Davis, leading to

Table 6 Preclinical in vivo effects of tachykinin NK_1 receptor antagonists being of possible therapeutic relevance

Compound	Effect	Dosage	Therapeutic perspective
FK888	Blockade of SP-induced airway edema[a]	0.011 mg/kg i.v. 4.2 mg/kg p.o.	Antiasthmatic
	Blockade of SP-induced plasm exudation in guinea pig lower trachea and bronchi[b]	0.1 µmol/kg i.v.	Antiasthmatic
MEN1467	Blockade of [Sar^9] SPsulfone-induced bronchoconstriction in anesthetized guinea pig[c]	29 µg/kg i.v.	Antiasthmatic
	Blockade of [Sar^9] SPsulfone-induced plasma protein extravasation in guinea pig bronchi[c]	6.7 mg/kg p.o.	Antiasthmatic
CP96345	Mild thermal analgesia in mice (hot plate test)[d]	30 mg/kg i.p.	Analgesic
	Blockade of SP- or noxious heat-induced tail flick in rats[e]	5 mg/kg s.c.	Analgesic
	Blockade of plasma protein extravasation in the rat hind paw induced by various irritants[f]	3–9 µmol/kg i.v.	Antiinflammatory
	Blockade of plasma protein extravasation induced by vagus nerve stimulation or capsaicin in guinea pigs[g] (see text for further examples)	100 nmol/kg i.v.	Antiinflammatory
CP99994	Blockade of plasma protein extravasation induced by capsaicin in guinea pigs[h]	3–30 mg/kg p.o.	Antiinflammatory
	Prevention of tracheal vascular permeability induced by hypertonic saline or capsaicin in rats[i]	1–4 mg/kg i.v.	Antiinflammatory
	Prevention of foot tapping in gerbils elicited by i.t. GR73632[j]	01–1 mg/kg s.c.	Analgesic
	Attenuation of cisplatin-induced emesis in the ferret[k]	0.3–3 mg/kg i.v.	Antiemetic
	Reduction of vomiting response to $CuSO_4$ and apomorphine in the dog[l] (see text for further examples)	40 µg/kg bolus plus 0.3 mg/kg/h i.v.	Antiemetic
RP67580	Reduction of writhings in phenylbenzo-quinone-injected mice[m]	70 µg/kg s.c.	Analgesic
	Reduction of paw lickings in formalin-injected mice[m] (see text for further examples)	3.7 mg/kg s.c.	Analgesic
RPR100893	Blockade of plasma protein extravasation within the dura mater elicited by electrical stimulation of the trigeminal ganglion in guinea pigs[n]	0.01–100 µg/kg p.o.	Antimigraine
SR140333	Reduction of fecal mass excretion and diarrhea in castor oil-pretreated rats[o]	0.02–20 µg/kg s.c.	Antidiarrheal
	Blockade of secretory responses evoked by alphaIgE or capsaicin in the human colonic mucosa (note: in vitro study)[p]	–	Antidiarrheal in food allergy and inflammatory bowel disease
GR203040	Antagonism of SPOMe-induced reduction of carotid vascular resistance in rabbits[q]	1–100 µg/kg i.v.	Antimigraine
	Prolonged antiemetic activity against radiation, cisplatin and other stimuli in the ferret[r]	0.03–0.3 mg/kg (s.c. or p.o.)	Antiemetic
GR205171	Antiemetic activity against X-radiation in the ferret[s]	0.1 mg/kg p.o.	Antiemetic
	Antiemetic activity against ipecacuanha in dogs[s]	0.2 mg/kg p.o.	Antiemetic
L733060	Abolition of vocalizations elicited by i.c.v. infusion of GR73632 in guinea pigs[t]	3 mg/kg s.c.	Antidepressant
	Abolition of maternal separation-induced vocalizations in guinea pig pups[t]	3 mg/kg s.c.	Antidepressant

Table 6 (continued)

Compound	Effect	Dosage	Therapeutic perspective
LY303870	Reduction of the severity of inflammatory bowel disease (IBD) in mice and facilitation of partial healing of lesions in mice with pre-existing IBD[u]	–	Antinflammatory in inflammatory bowel disease
CGP49823	Increase in active social time and reduction of the immobility time of rats in the swim test[v]	3–30 mg/kg p.o.	Anxiolytic antidepressant
NKP608	Increase in the time spent by rats in social contact, in a social interaction test	0.01–1 mg/kg p.o.	Anxiolytic
	Increase in the time spent by the intruder rat in social contact with the resident rat in a social exploration test[w]	0.03–3 mg/kg p.o.	Anxiolytic
PD154075	Prevention of both thermal and mechanical hypersensitivity induced by surgery in rats[x]	3–100 mg/kg s.c.	Antihyperalgesic in postoperative pain
	Blockade of hypersensitivity induced by chronic constriction injury (sciatic nerve ligation) in both rats and guinea pigs[y]	10–100 mg/kg s.c. 0.1 mg/kg p.o.	Analgesic in neuropathic pain
TAK637	Increase in the volume threshold required to elicit micturition reflex in awake guinea pigs[z]	0.01 mg/kg p.o.	Control of urinary incontinence
	Decrease in the number of rhythmic bladder contractions elicited by infusion of saline[aa]	1 mg/kg i.v.	
	Reduction of capsaicin-induced micturition reflex in guinea pigs[aa]	0.03–0.3 mg/kg i.v.	
	Reduction of restraint stress-stimulated fecal pellet output in gerbils[bb]	0.1–1 mg/kg p.o.	Control of irritable bowel syndrome
	Reduction of the number of distension-induced abdominal contractions in rabbits previously subjected to colonic irritation[cc]	0.1–3 mg/kg i.d.	

ic.v., Intracerebroventricular; i.d., intraduodenal; i.p., intraperitoneal; i.t., intrathecal; i.v., intravenous; s.c., subcutaneous; p.o., per os.

References: [a] Fujii et al. 1992; [b] Hirayama et al. 1993; [c] Cirillo et al. 2001; [d] Lecci et al. 1991; [e] Yashpal et al. 1993; [f] Lembeck et al. 1992; [g] Lei et al. 1992; [h] McLean et al. 1993; [i] Piedimonte et al. 1993; [j] Rupniak et al. 1995; [k] Tattersall et al. 1993; [l] Watson et al. 1995; [m] Garret et al. 1991; [n] Lee et al. 1994; [o] Croci et al. 1997; [p] Moriarty et al. 2001; [q] Beattie et al. 1995; [r] Gardner et al. 1995; [s] Gardner et al. 1996; [t] Kramer et al. 1998; [u] Sonea et al. 2002; [v] Vassout et al. 1994; [w] Vassout et al. 2000; [x] Gonzalez et al. 1998; [y] Gonzalez et al. 2000; [z] Doi et al. 1999; [aa] Doi et al. 2000; [bb] Okano et al. 2001; [cc] Okano et al. 2002.

development of a number of peptoid compounds, the most potent of which was Cam2819 (Table 2). Cam2819 showed nanomolar affinity for guinea pig, rabbit and human NK_1 receptors (Tables 3, 5; McKnight et al. 1994).

In 1992 Sakurada and coworkers reported the identification (from random screening) of an extremely potent antagonist of SP: sendide (or selective NK_1 receptor D-amino acid-containing peptide; Table 2). Sendide showed a subnanomolar affinity for NK_1 receptors in the mouse spinal cord (Table 6). Selectivity of sendide for NK_1 over NK_2 and NK_3 receptors was claimed on the basis of an in vivo experiment in which sendide (intrathecally administered) was at least 100-fold more potent in preventing SP from inducing scratching, biting and licking in the mouse than in preventing NKA or NKB from eliciting these responses (Sakurada et al. 1992). Subsequently, sendide was shown to produce antinociceptive effects in the mouse paw formalin test (Sakurada et al. 1995) and antiemetic effects in the cisplatin model of emesis in the ferret (Minami et al. 1998). However, the utility of sendide as NK_1 receptor antagonist was hampered by several factors, including its poor activity at NK_1 receptors present in species different from the mouse (Table 4).

4
Nonpeptide Antagonists

4.1
CP96345 and Related Compounds

The era of nonpeptide antagonists for the NK_1 receptor of tachykinins began in 1991, when the first one of this series, CP96345, was reported by Pfizer (Snider et al. 1991). The introduction of CP96345 gave rise to plenty of structure–activity studies all over the world, leading to discovery of new nonpeptide compounds, and to plenty of studies aiming at investigating the pathophysiological role of the NK_1 receptor by the use of this new potent antagonist. CP96345 (Fig. 1) is a quinuclidine derivative, discovered by the screening of a chemical collection followed by lead optimization (Lowe et al. 1992). CP96345, in its active form [the (2S,3S)-enantiomer], displays high affinity for NK_1 receptors present in human, guinea pig, rabbit and other human-related species, and remarkable selectivity over NK_2 and NK_3 receptors. In contrast, it recognizes with much less affinity NK_1 receptors of rodents (Tables 3 and 5). These observations were instrumental for recognition of the existence of species-related heterogeneity of the NK_1 receptor (see Sect. 2.1.1). From the many in vivo studies in which CP96345 was used, the role played by NK_1 receptors in nociception received strong support (Table 6). For example, (±)-CP96345 selectively blocked the aversive behavior induced by intrathecal SP in mice, and also produced mild analgesic effects in the hot plate test (Lecci et al. 1991). CP96345 prevented excitation of dorsal horn neurons in cat spinal cord following application of noxious heat to the appropriate receptive field in the hind limb (Radhakrishan and Henry 1991). In other studies, CP96345 was proven to inhibit carrageenin-induced mechanical hyperalgesia in rats (Birch et al. 1992), paw licking induced by formalin in mice (Sakurada et al. 1993), abdominal stretching induced by intracolonic instillation of acetic acid in mice (Nagahisa et al. 1992), tail flick responses induced by noxious thermal and chemical stimuli in the rat (Yashpal et al. 1993) and vocalization induced by mechanical stimuli of mononeuropathic and diabetic rats (Coudoré-Civiale et al. 1998).

Unequivocal evidence for the involvement of NK_1 receptors in mediating inflammatory responses was provided by means of CP96345 (Table 6). For example, CP96345 was found effective in preventing plasma protein extravasation

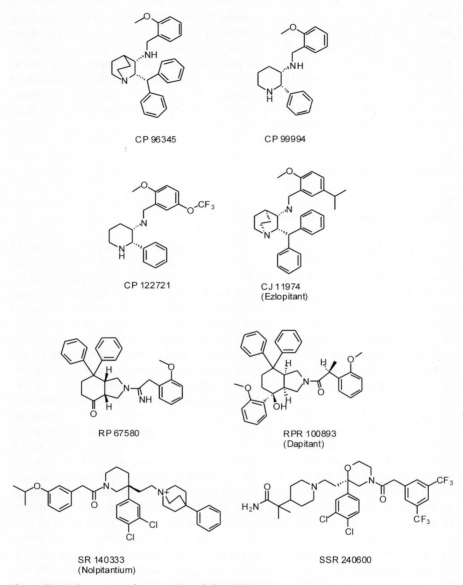

Fig. 1 Chemical structures of nonpeptide tachykinin NK_1 receptor antagonists: I

and/or edema caused by several pro-inflammatory stimuli such as capsaicin in the rat urinary bladder (Eglezos et al. 1991), SP, mustard oil and stimulation of the saphenous nerve in the rat hind paw skin (Lembeck et al. 1992) as well as cigarette smoke in the rat trachea (Delay-Goyet and Lundberg 1991; Delay-Goyet et al. 1992). Likewise CP96345 prevented neurogenic plasma exudation evoked by electrical stimulation of the cervical vagus and by intravenous (i.v.) capsaicin in guinea pigs (Lei et al. 1992). CP96345 (administered chronically) has recently been shown to attenuate significantly colonic inflammation and oxidative stress produced by dextran sulfate in rats: an animal disease model resembling chronic ulcerative colitis in humans (Stucchi et al. 2000). Although not described in the original reports on CP96345 pharmacology, several nonspecific effects produced by this compound, both in vitro and in vivo, were reported (Table 4). Subsequent investigations of these nonspecific effects showed that they are due to interaction of CP96345 with several ion channels, including L- and N-type calcium channels (Schmidt et al. 1992; Guard et al. 1993), and voltage-dependent sodium channels (Caesar et al. 1993). Two relevant observations dealing with nonspecific activities of CP96345 are worth mentioning: (a) these effects are equally exerted by either the (+) (named: CP96344) or (−) isomer of CP96345, while the NK_1 receptor affinity is present in the (−) isomer only, and (b) they are evident at micromolar concentrations. It follows that, due to the reduced affinity of CP96345 for rat and mouse NK_1 receptors (compared to that for NK_1 receptors of other species), the window of selectivity between doses of CP96345 producing specific (i.e., NK_1 receptor-mediated) vs. nonspecific (i.e., membrane channel-mediated) effects is narrow in the rodent species. Thus, the introduction of the (+) isomer (CP96344) as control drug in studies in which CP96345 is used, is necessary to avoid misinterpretation of the results.

In an attempt to overcome the many drawbacks of CP96345, Pfizer researchers synthesized CP99994, a compound in which the quinuclidine ring of CP96345 was replaced by a piperidine, and the benzhydryl moiety by a benzyl group (Fig. 1; McLean et al. 1993). Only the (2S, 3S) isomer of CP99994 (like CP96345) shows considerable affinity for NK_1 receptors, whereas the (2R, 3R) enantiomer (named CP100263) is more than 1000-fold weaker at NK_1 receptors. CP99994 shows high affinity for human NK_1 receptors comparable to that of CP96345 (Table 5), but a 100-fold reduced affinity ($IC_{50}=3$ μM) for L-type calcium channels relative to CP96345 ($IC_{50}=27$ nM) (McLean et al. 1993). Similarly to CP96345, CP99994 showed marked antinociceptive activity and was a potent inhibitor of plasma protein extravasation in the airways (Table 6). However, some nonspecific effects of CP99994 (i.e., effects shared by its inactive enantiomer CP100263) in antinociceptive tests in the rat, mouse and gerbil have been reported (Table 4; Smith et al. 1994; Rupniak et al. 1995). In this regard it is worth to mention the study of Lombet and Spedding (1994) who showed CP99994 to interact with phenylalkylamine calcium channels in rat skeletal muscle membranes, with quite the same potency of CP96345 ($K_i=0.17$ vs. 0.94 μM, respectively), indicating that even CP99994 may have the potential of producing 'nonspecific' effects under certain circumstances. On the basis of its preclinical profile, CP99994 has been used in clinical trials to prove its analgesic and anti-asthmatic properties, obtaining conflicting results. Dionne et al. (1998) reported that CP99994 reduced postoperative pain in patients undergoing dental extraction, although the nonsteroidal antinflammatory agent, ibuprofen, was more effective. On the other hand, CP99994 failed to produce significant analgesic activity in patients suffering from peripheral neuropathy (Suarez et al. 1994), and was proven unable to prevent bronchoconstriction and cough induced by hypertonic saline in mild asthmatic patients (Fahy et al. 1995).

CP99994 was the first NK_1 receptor antagonist to be used for demonstrating that blockade of NK_1 receptors produces antiemetic effects against a broad variety of emetogenic stimuli (Table 6). In 1993 Bountra et al. and Tattersall et al. (1993) reported that CP99994 is effective in the ferret in attenuating emesis induced by cisplatin, morphine, ipecacuana, copper sulfate, radiation and other cytotoxic drugs. Subsequently the antiemetic activity of CP99994 was proven in the dog (Watson et al. 1995), *Suncus murinus* (Tattersall et al. 1995) and cat (Lucot et al. 1997) against a variety of central, peripheral and mixed central and peripheral emetic stimuli. An interesting follow-up of CP99994 was CP122721 (Fig. 1), an analog bearing an additional trifluoromethoxy group in the benzyl ring of the molecule. CP122721 shows similar high affinity for human NK_1 receptors as CP99994 (Table 5) and 1000-fold reduced ability to interact with calcium channels than its parent compounds, CP96345 and CP99994 (McLean et al. 1996). The analysis of the kinetics of interaction of CP122721 with the NK_1 receptor indicated a noncompetitive/insurmountable behavior of this antagonist (McLean et al. 1996). CP122721 was consistently more potent than CP99994 in blocking capsaicin-induced plasma protein extravasation in guinea pig ureter and lung: the greater potency of CP122721 was attributed to its improved bioavailability and to its insurmountable mechanism of action (McLean et al. 1996). Like CP99994, CP122721 was shown to be a potent inhibitor of both retching and vomiting elicited by a broad spectrum of emetic agents in ferrets (Gonsalves et al. 1996). The efficacy of CP122721 has been confirmed in clinical trials in which this compound significantly attenuated delayed emesis due to cisplatin chemotherapy (Kris et al. 1997) and postoperative nausea and vomiting (Gesztesi et al. 1998, 2000).

A further development of Pfizer compounds is represented by CJ11974, or ezlopitant, a quinuclidine analog of CP96354 bearing an additional isopropyl group in the benzyl ring (Fig. 1). CJ11974 is endowed with high affinity and selectivity for human NK_1 receptors, and oral efficacy in preventing cisplatin-induced emesis in ferrets (Tsuchiya et al. 2002; Table 5). The therapeutic potential of this compound has been documented in Phase II clinical trials (Evangelista 2001). CJ11974 proved to be effective in controlling chemotherapy-induced emesis (Hesketh et al. 1999), but as it was less effective for nausea it was discontinued for this indication (Evangelista 2001). Subsequently, CJ11974 has been shown to reduce emotional responses to rectosigmoid distension and related symptoms in patients affected by irritable bowel syndrome (IBS) (Lee et al. 2000), and consequently it is currently being developed for this indication (Evangelista 2001).

4.2
RP67580 and Related Compounds

Soon after the introduction of the first nonpeptide NK_1 receptor antagonist, CP96345, another interesting compound was described by researchers at Rhone-Poulenc: RP67580 (Garret et al. 1991). As discussed before (see Sect. 2.1.1) RP67580 has been instrumental in unraveling the species-dependent variations in the pharmacology of NK_1 receptor antagonists, being significantly more potent at rat or mouse than guinea pig or human NK_1 receptors. RP67580 is a perhydroisoindolone derivative (Fig. 1), whose affinity for NK_1 receptors (Table 5) is confined to the (3aR, 7aR) enantiomer, whereas the (3aS, 7aS) enantiomer (or RP 68651) is inactive up to 10 μM (Garret et al. 1991). RP67580 showed good selectivity relative to NK_2 and NK_3 tachykinin receptors (Tables 3 and 5) and was found to be as potent as morphine in two classical tests for analgesia: phenylbenzoquinone-induced writhing and formalin test (Garret et al. 1991; Table 6). Further in vivo studies confirmed the ability of RP67580 to produce stereoselective analgesic/antiinflammatory effects against various stimuli. As an example, RP67580 attenuated chronic hyperalgesia in streptozotocin-induced diabetic rats (Courtiex et al. 1993), prevented capsaicin-induced plasma protein extravasation in the dura mater of guinea pigs (Moussaoui et al. 1993) and plasma protein extravasation in the dura mater of rats elicited by electrical stimulation of the trigeminal ganglion (Shepheard et al. 1993). Holzer-Petsche and Rordorf-Nikolic (1995) noticed that intrathecal RP67580 was able to inhibit reflex changes of mean arterial pressure and intragastric pressure in response to systemic capsaicin administration in anesthetized rats, whereas intravenous administration of RP67580 was not, concluding that this could be due to insufficient penetration of this antagonist through the blood–brain barrier (Table 4). A further drawback of RP67580 is its ability to interact with various calcium channels at submicromolar to micromolar concentrations (Lombet and Spedding 1994; Rupniak et al. 1993). Although it remains to be clarified whether this calcium blocking activity of RP67580 resides equally in the two enantiomers, it has been argued that this mechanism could be responsible for acute antinociceptive effects produced by RP67580 in several animal models of pain. The former conclusion is supported by the observation that the channel blockers nifedipine and verapamil are as effective as RP67580 in one of these models (formalin paw test in gerbils; Rupniak et al. 1993).

The recognition of the existence of species-dependent variations of the NK_1 receptor, and the observation that RP67580 was more potent on the rodent type rather than on the human type, prompted Rhone-Poulenc researchers to undertake structure–activity studies in order to improve the selectivity of their compound toward the human type NK_1 receptor. The most interesting outcome of this search was the nonpeptide RPR100893 (Fardin et al. 1994), a product bearing an additional aromatic ring in the RP67580 structure: a modification leading to increasing structural complexity of the molecule (i.e., two additional stereogenic centers) (Fig. 1). RPR100893 has 100-fold higher affinity for human NK_1

receptors than for rat or mouse receptors (Fardin et al. 1994) (Table 5), showing time-dependent kinetics in its binding to the human receptor, and insurmountable antagonism of SP-induced inositol phosphate production in cultured human U373MG astrocytoma cells (Fardin et al. 1994). In vivo, RPR 100893 showed good oral bioavailability and efficacy in preventing plasma protein extravasation induced by either tachykinins or capsaicin in guinea pig trachea, and in reducing formalin-induced nociceptive behavior in guinea pigs (Moussaoui et al. 1994). In particular, Lee et al. (1994) demonstrated the ability of RPR100893 to block plasma protein extravasation within the dura mater and conjunctiva of guinea pigs induced by capsaicin or electrical stimulation of the trigeminal ganglion, and suggested that this compound (or others of the same series) could prove to be useful for treating migraine and cluster headaches (Table 6). This latter hypothesis was further supported by Cutrer et al. (1995) who showed RPR100893 capable of selectively reducing c-*fos* expression in the trigeminal nucleus caudalis in guinea pigs, elicited by intracisternal injection of capsaicin. However, subsequent clinical trials with RPR100893 have failed to show any efficacy against migraine pain (Rupniak and Kramer 1999).

4.3
SR140333 and Related Compounds

In 1993 Emonds-Alt and coworkers at Sanofi discovered the piperidine-based compound SR140333 (Fig. 1), one of the most potent and selective NK_1 receptor antagonists since then developed (Table 5). SR140333 was reported to possess subnanomolar affinity for NK_1 receptors of various species, including human, guinea pig and rat: thereby showing no preference for any of the two NK_1 receptor 'families' previously identified by the use of other antagonists (see Sect. 2.1.1; Emonds-Alt et al. 1993). As for other nonpeptide antagonists containing stereogenic centers in the chemical structure, the affinity of SR140333 for NK_1 receptors was (much) higher in one enantiomer, [the (S) enantiomer], than in the corresponding (R) enantiomer (or SR140603). SR140333 was proven to be a highly potent antagonist at NK_1 receptors present in classical isolated smooth muscle preparations such as the guinea pig ileum and the rabbit pulmonary artery. However, in these bioassays SR140333 produced insurmountable effects, probably due to the slow rate of reversibility of its receptor interaction (Emonds-Alt et al. 1993; Table 3). In vivo, SR140333 potently [50% effective dose (ED_{50}) ranging from 3 to 42 µg/kg i.v.] antagonized hypotension, bronchoconstriction and plasma protein extravasation in the skin elicited by selective NK_1 receptor agonists in dogs, guinea pigs and rats, respectively (Emonds-Alt et al. 1993). Subsequent investigations aiming at clarifying the role of NK_1 receptors in nociception and neurogenic inflammation demonstrated the efficacy of SR140333 in various animal models of pain and inflammation. As an example, SR140333 reduced the mouse ear edema provoked by topical capsaicin application (Inoue et al. 1996), abolished the tail-flick reflex facilitation induced by noxious heat in rats (Jung et al. 1994), abolished the cutaneous edema induced by electrical stimulation of the rat saphenous nerve (Towler and Brain 1998), reduced colonic inflammatory responses induced by trinitrobenzene sulfonic acid in the rat (Mazelin et al. 1998) and produced a long-lasting inhibition of mustard oil-induced plasma protein extravasation in the dorsal skin of the rat hind paw (Amann et al. 1995). However, Amann and coworkers (1995) also reported that SR140333, at doses that caused inhibition of neurogenic inflammation, was unable to prevent the acute behavioral responses elicited in conscious rats by chemical irritants (capsaicin, PGE_2) or noxious heat. In an in vivo model of chemically induced diarrhea, Croci et al. (1997) have shown that systemic pretreatment with SR140333 may (partially) reduce both fecal mass excretion and the rate of diarrhea in castor-oil treated rats, without producing constipation in normal, untreated rats. More recently, the potential use of SR140333 as an antidiarrheal drug for food allergy or IBS has been supported by the ability of SR140333 to reduce the secretory responses evoked in the human colonic mucosa by either mast cells or by primary afferent neuron stimulation (Moriarty et al. 2001; Table 6).

A follow-up compound of SR140333 is SSR240600 (Emonds-Alt et al. 2002; Fig. 1). SSR240600 maintains a very high affinity for the human NK_1 receptor present in different human cell lines (Table 5), comparable to that shown by the parent compound SR140333. However SSR240600, unlike SR140333, can distinguish among species, being consistently less potent on rat and gerbil NK_1 receptors than on human or guinea pig receptors (Emonds-Alt et al. 2002). Like SR140333, SSR240600 has been reported to produce insurmountable antagonism of NK_1-receptor mediated responses obtained in isolated preparations, and long-lasting inhibitory effects against hypotension, bronchoconstriction and plasma protein extravasation elicited by selective NK-1 receptor agonists in dogs and guinea pigs, respectively (Emonds-Alt et al. 2002). Moreover, SSR240600 was able to prevent citric acid-induced cough in guinea pigs (whereas SR140333 was not): an effect claimed to be due to its easy penetration into the brain (Emonds-Alt et al. 2002). Concomitantly, Steinberg et al. (2002) have reported that SSR240600 inhibits distress vocalization induced by maternal separation in guinea pig pups and counteracts the increase in body temperature induced by isolation stress: two effects that suggest a potential antidepressant-like activity of this compound, as shown previously for certain Merck compounds found active in similar animal models (see Sect. 4.5).

4.4
GR203040 and Related Compounds

Structure–activity studies on the backbone of the Pfizer compound CP99994 were undertaken by researchers at Glaxo, aiming at improving oral bioavailability, metabolic stability and specificity of action of this molecule. Their efforts lead to the discovery of a tetrazole-substituted analog of CP99994, named GR203040 (Fig. 2; Ward et al. 1995; Beattie et al. 1995). GR203040 showed slightly higher affinity for human NK_1 receptors expressed by different cell lines than

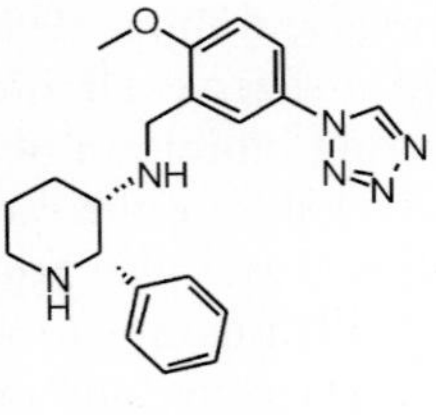

GR 203040

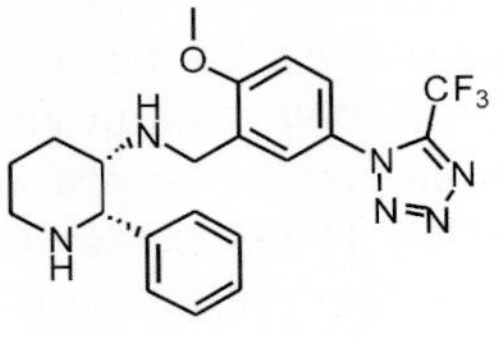

GR 205171
(Vofopitant)

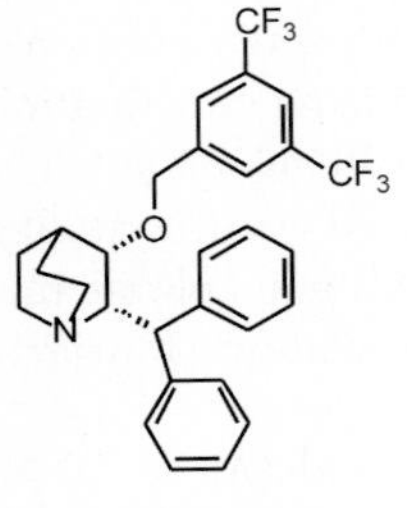

L 709210

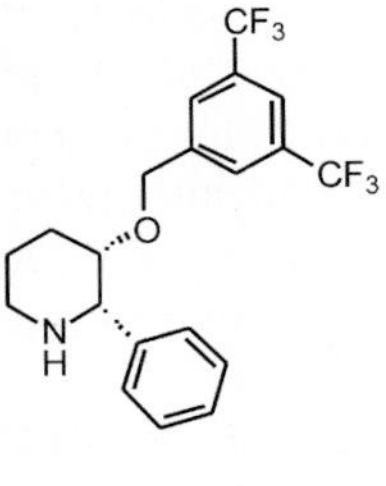

L 733060

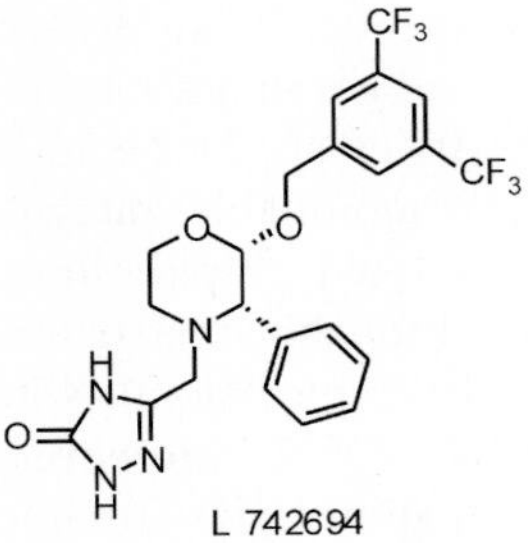

L 742694

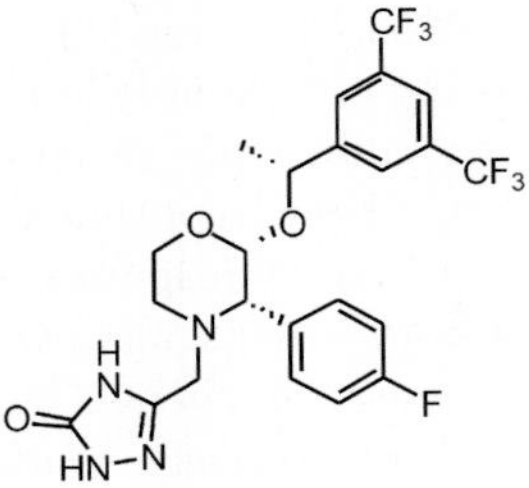

L 754030
(Aprepitant, MK-869)

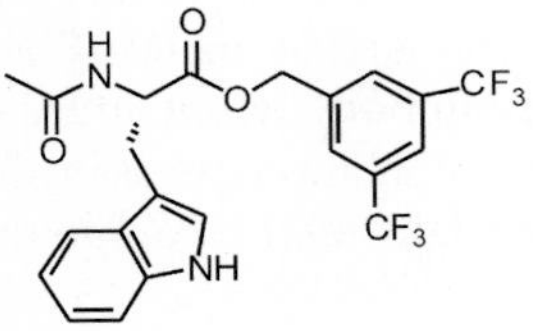

L 732138

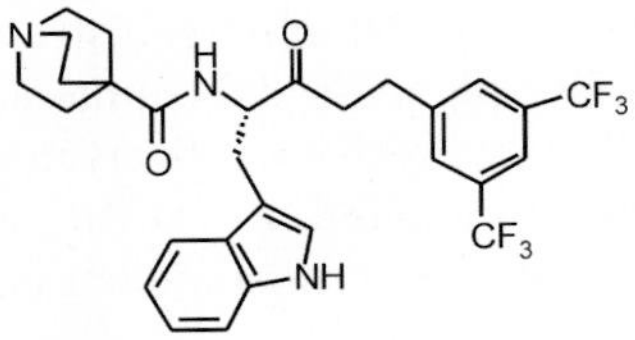

L 737488

Fig. 2 Chemical structures of nonpeptide tachykinin NK_1 receptor antagonists: II

the parent compounds CP99994 and CP96345, whereas at rat NK_1 receptors GR203040 was approximately 50-fold less potent than at human receptors, as observed for CP99994 and CP96345 (see Table 5; Beattie et al. 1995). As for the Pfizer compounds, the affinity for NK_1 receptors was almost completely confined in the (2*S*, 3*S*) enantiomer (i.e., GR203040), while the corresponding enantiomer (2R, 3R) (or GR205608) was about 10,000-fold weaker at the human-type NK_1 receptor (Beattie et al. 1995). A significant improvement in the pharmacology of GR203040, over CP96345 and CP99994, was the reduced/lack of affinity of the Glaxo compound for various Na^+ or Ca^{2+} ion channels; a side-effect that had seriously hampered the use of both Pfizer compounds (Beattie et al. 1995). On isolated tissues GR203040 potently, but insurmountably, antagonized NK_1 receptor-mediated responses, probably due to its slow dissociation rate from the receptor (Table 3). In vivo, GR203040 was found to gain rapid access to the CNS (in the gerbil) and to reduce carotid vascular resistance (induced by local administration of SP methyl ester; Table 6); this latter effect was thought to be representative of NK_1 receptor-mediated craniovascular dilatation. Moreover, GR203040 antagonized NK_1 receptor-mediated cranial vasodilatation in vitro (in dog cerebral arteries) and plasma protein extravasation in the rat dura mater; these results prompted Beattie and coworkers (1995) to propose GR203040 as a candidate therapeutic agent in the clinical management of migraine. In addition, GR203040 was shown to possess potent (about 30-fold more than CP99994), prolonged and broad-spectrum antiemetic activity in the ferret, dog and *Suncus murinus*, after intravenous or oral administration (Table 6), suggesting that it could prove useful in the control of emesis associated with cancer chemotherapy and possibly with other diseases (Ward et al. 1995; Gardner et al. 1995).

Further structure–activity studies on close analogs of GR203040, aimed at the optimization of the antiemetic properties of this structural type of molecule, led to the identification of GR205171, a trifluoromethyl-substituted compound at the C-1 position of the tetrazole ring present in the GR203040 structure (Gardner et al. 1996; Fig. 2). GR205171 was shown to possess a pattern of affinities for tachykinin and nontachykinin receptors very similar to that shown by GR203040, with a potency at various ion channels being at least 1000-fold lower than that at human NK_1 receptors (Gardner et al. 1996; Table 5). GR205171 was reported to produce long-lasting antiemetic activity in the ferret, dog and *Suncus murinus*, against a variety of emetic stimuli, including stimuli (like morphine, copper sulfate and motion) that produce emetic responses refractory to treatment with 5-HT_3 receptor antagonists, at doses approximately threefold lower than those required with GR203040 (Gardner et al. 1996). In particular GR205171 was fully active in both ferret and dog following oral administration at doses only threefold higher than the minimum fully effective parenteral dose (Gardner et al. 1996; Table 6). Subsequent investigations have suggested a potential usefulness of GR205171 in counteracting drug-induced conditioned aversive behavior and nausea, as shown by McAllister and Pratt (1998) who found GR205171 capable of preventing conditioned taste aversions in rats provoked by administration of apomorphine or amphetamine. In clinical trials, GR205171 has been proven effective in reducing the rate of postoperative nausea and vomiting in patients undergoing major gynecological surgery, being well tolerated and producing no major adverse effect (Diemunsch et al. 1999). In contrast, GR205171 was found ineffective in preventing motion-induced nausea in 16 healthy subjects, even in combination with the 5-HT$_3$ receptor antagonist ondansetron (Reid et al. 2000).

4.5
L754030 (MK869) and Related Compounds

Structure–activity studies aimed at obtaining highly potent and selective NK$_1$ receptor antagonists, devoid of the side effects accompanying the first Pfizer compounds, were undertaken at Merck, based on the backbone of CP96345 first, and of CP99994 later. A first class of Merck compounds derived in such way were quinuclidine benzylether derivatives; one of the most important representative was a bis-trifluoromethyl analog, named L709210 (Fig. 2) whose affinity for human NK$_1$ receptors ranged from 0.7 nM to 1.3 nM, depending on the diastereoisomer under examination (Swain et al. 1993). Another series of early Merck compounds is that of benzylamino piperidines. The first example of these structures is L733060 (Fig. 2), which bears a 3,5-bistrifluoromethyl benzylether moiety in the place of the 2-methoxy benzylamine moiety present in CP99994 (Harrison et al. 1994). L733060 was shown to be a stereoselective (2S, 3S being the active enantiomer) highly potent ligand of human NK$_1$ receptors in vitro (Harrison et al. 1994; Seabrook et al. 1996; Table 5) and a potent antagonist of NK$_1$ receptor-mediated neurogenic plasma protein extravasation in rats (Seabrook et al. 1996). L733060 also reduces the late phase of the nociceptive response (paw licking) to formalin injection in gerbils [50% inhibitory dose IC$_{50}$) of about 0.2 mg/kg i.v.] (Rupniak et al. 1996). L 733060 has been instrumental in demonstrating an antidepressant-like activity of the tachykinin NK$_1$ receptor antagonists. In preclinical assays, Kramer and coworkers (1998) showed that L733060 (but not its less active enantiomer L733061) abolished vocalizations elicited in guinea pigs by intracerebroventricular (i.c.v.) injection of an NK$_1$ receptor-selective agonist, an effect shared by the antidepressants imipramine and fluoxetine. Furthermore, L733060 was able to prevent vocalizations evoked in guinea pig pups by transient maternal separation, as did certain antidepressant or anxiolytic drugs (Kramer et al. 1998). This study provided the first preclinical evidence that blockade of central NK$_1$ receptors could inhibit a psychological stress response, and led Merck researchers to a undertake clinical trials to evaluate MK869 (see below) in patients suffering from depression associated with anxiety.

Subsequent structure–activity efforts on this series of compounds were devoted to reducing the affinity–typical of such structures–for Ca^{2+} channels and improving oral bioavailability (for a chemical review of these structures see Quartara and Maggi 1997). One of the most promising outcomes was L742694 (Fig. 2), a compound bearing morpholine at the place of piperidine in the precursor L733060 and a triazolone structure attached to the morpholine nitrogen (Hale et al. 1996). As compared to CP99994, L742694 gained affinity at human NK$_1$ receptors (Table 5) while showing threefold lower affinity than CP99994 for L-type Ca^{2+} channels (Hale et al. 1996). Further investigation into the molecular mechanism of interaction of L742694 with the human NK_1 receptor, in which the radiolabeled [^{3}H]L742694 was used (Cascieri et al. 1997), demonstrated that it behaves as a competitive antagonist but, due to its slow rate of dissociation from the receptor, can act as a pseudoirreversible antagonist under particular experimental conditions. In vivo, L742694, given orally, inhibited SP-induced plasma protein extravasation in guinea pig skin with about 170-fold higher potency than CP99994 (ID_{50}=0.009 vs. 1.6 mg/kg) (Hale et al. 1996), and blocked acute retching induced by cisplatin in ferrets (Rupniak et al. 1997). Structural modifications of L742694 were introduced to reduce metabolic degradation of this morpholine acetal derivative. The result of these efforts was the discovery of L754030 (also known as aprepitant or MK869; Fig. 2) a fluorine derivative of L742694, showing conserved high affinity for the human NK_1 receptor (Table 5) and high oral efficacy in preventing vascular leakage induced by resiniferatoxin in guinea pigs or foot tapping behavior induced by central infusion of an NK_1 selective agonist (GR73632) in gerbils (Hale et al. 1998). In particular, antagonist administration 24 h prior to resiniferatoxin challenge or GR73632 infusion revealed a three- to tenfold higher potency of L754030 than L742694, whereas antagonists like CP122721 (Sect. 4.1) or GR205171 (Sect. 4.4), although being equipotent/more potent than L754030 within 1h of administration, were ineffective or tenfold less effective than L754030 24 h post-treatment, demonstrating a very long duration of action of L754030 relative to piperidine-based compounds (Hale et al. 1998). In addition, L754030 proved to be more potent than L742694 in preventing cisplatin-, morphine- or apomorphine-induced emesis in ferrets, and was selected by Merck as candidate for treatment of various pathological conditions in which NK_1 receptors are thought to play a role (Hale et al. 1998). As a matter of fact, L754030 (300 mg/day) was the first tachykinin NK_1 receptor antagonist to be shown efficacious in producing antidepressant and anxiolytic activity in outpatients with major depressive disorder and high anxiety levels, in a randomized, double-blind, placebo-controlled study (Kramer et al. 1998). Concomitant biochemical investigations in gerbils led Kramer and coworkers (1998) to claim that L754030 works differently from established antidepressant drugs, with no augmentation of norepinephrine or serotonin function provoked by L754030 administration. Subsequently, L754030 and its water-soluble prodrug, L758298 (Hale et al. 2000) were found effective in reducing delayed emesis induced by cisplatin treatment in patients with malignant disease (Navari et al. 1999; Van Belle et al. 2002). L754030 (given as dual therapy with dexamethasone) was also effective in reducing acute emesis, although dual therapy with ondansetron and dexamethasone was superior in this early phase (van Belle et al. 2002). A higher efficacy of L754030 in controlling delayed vs. acute emesis was further supported by a study performed on 53 cisplatin-naive patients who received 60–100 mg of the L754030 prodrug, L758298 (Cocquyt et al. 2001). Owing to its effectiveness in preventing chemotherapy-induced nausea/emesis, L754030 (Emend^R) has recently been approved in the USA for the treatment of this pathological condition. Thus, L754030 is the first compound arising from the research efforts in the tachykinin field to become a drug for the management of human diseases. L754030 underwent further clinical trials to assess its efficacy in the control of painful conditions such as dental pain or osteoarthritic neuropathic pain. However, doses [300 mg per os (p.o.)] of L754030 found effective in the treatment of depression or emesis failed to be analgesic in both painful conditions (Rupniak and Kramer 1999).

Several chemical analogs of L754030 have been synthesized by Merck, and proven effective as antidepressant/antiemetic agents in preclinical animal models. One of the most relevant is L760735 which is able to prevent vocalizations evoked in guinea pig pups by transient maternal separation (Kramer et al. 1998) and to increase the time spent by gerbils in social interaction (Cheeta et al. 2001); this latter study suggests that this compound could be useful in anxiolytic therapy.

A different class of NK_1 receptor antagonists, derived from the screening of a Merck sample collection, is represented by N-ethyl-L-tryptophan benzyl esters, whose most representative and active compound is L732138 (Fig. 2; MacLeod et al. 1994). L732138 possesses nanomolar affinity for human NK_1 receptors (Table 5) and a 200-fold reduced affinity for the rat-type NK_1 receptor (Cascieri et al. 1994). Intensive work was undertaken by Merck chemists on the L732138 backbone, in order to minimize its rapid metabolic degradation and improve both solubility and bioavailability. Their efforts led to the synthesis of L737488 (Fig. 2), a compound showing higher affinity than L732138 at human NK_1 receptors (Table 5) along with good oral activity in reducing SP-induced plasma protein extravasation in guinea pigs (ID_{50}=1.8 mg/kg, p.o.) and improved solubility in water (MacLeod et al. 1995).

4.6
LY 303870 and Related Compounds

A new class of NK_1 receptor antagonists was developed at Lilly, by optimization of N-acetylated tryptophan amides and esters (Hipskind et al. 1996). The most interesting compound of this series is LY 303870 (lanepitant), in which the carboxy group of tryptophan is reduced and functionalized with an *N*-acetyl tertiary amide (Fig. 3). LY 303870 showed subnanomolar affinity for human NK_1 peripheral and central receptors, approximately 50-fold lower affinity for rat or mouse NK_1 receptors, and very low affinity for tachykinin NK_2 or NK_3 receptors, T, L and N-type calcium channels (pKi values lower than 6.0 at these latter receptors) (Table 5; Gitter et al. 1995). The ability of LY303870 to interact with NK_1 receptors was highly stereoselective, as demonstrated by its (*S*)-enantiomer (named LY306155) which bound to human or animal NK_1 receptors with 1,000–15,000-fold lower affinities (Gitter et al. 1995). LY303870 proved to be a very potent antagonist of NK_1-receptor mediated effects in both the rabbit isolated vena cava (Table 3) and in vivo experiments, in which it potently inhibited both bronchoconstriction (ID_{50}=13µg/kg i.v.) and pulmonary microvascular leakage (ID_{50}=8µg/kg i.v.) induced by [Sar^9]SP sulfone in the guinea pig (Gitter et al. 1995). Subsequently, the antinociceptive potential of LY303870 was examined in

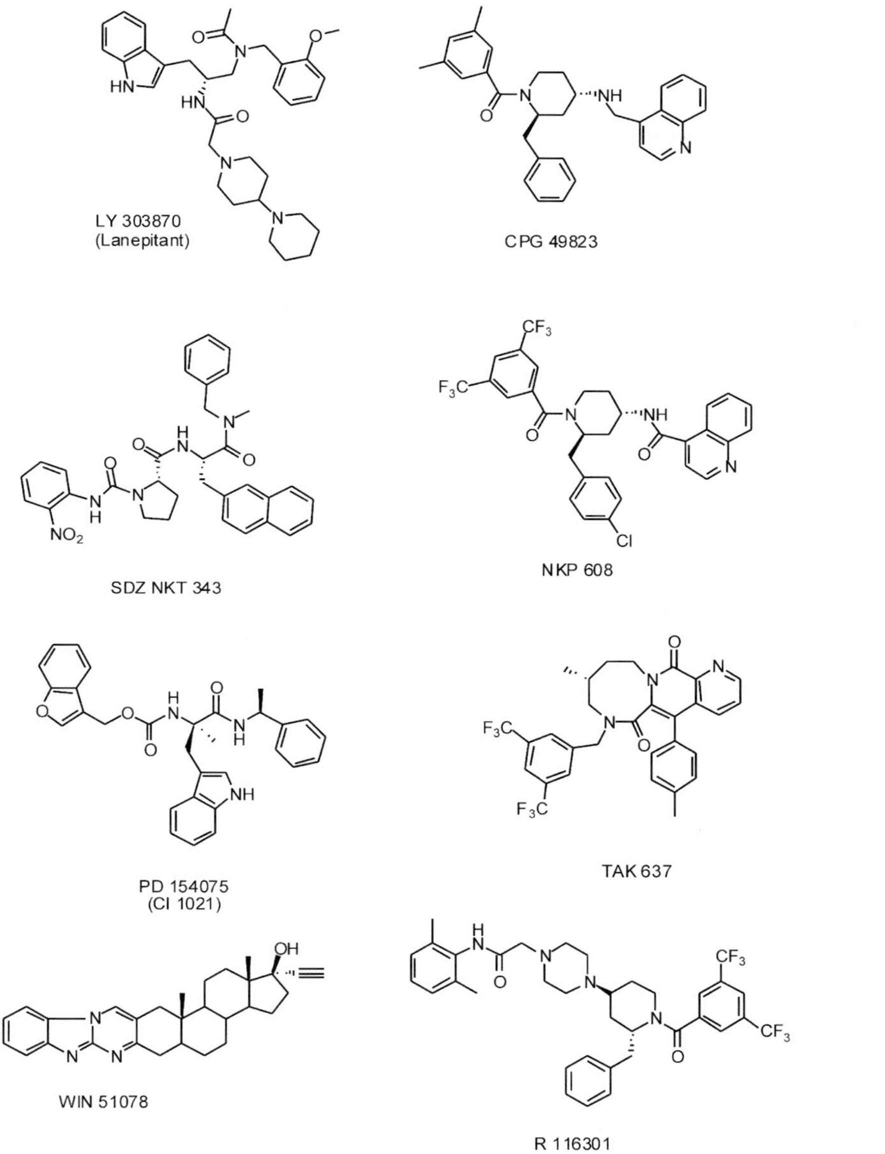

LY 303870
(Lanepitant)

CPG 49823

SDZ NKT 343

NKP 608

PD 154075
(CI 1021)

TAK 637

WIN 51078

R 116301

Fig. 3 Chemical structures of nonpeptide tachykinin NK$_1$ receptor antagonists: III

typical tests of pain such as the tail flick latency test and the formalin test in rats (Iyengar et al. 1997). In both bioassays LY303870 was shown to produce long-lasting antinociceptive effects after systemic or oral administration, in a range of quite high doses: 1–30 mg/kg (Iyengar et al. 1997). Lower doses of LY303870

(1–100μg/kg) were required to prevent plasma protein extravasation induced by electrical stimulation of the trigeminal ganglion in guinea pigs, a model thought to be representative of migraine of neurogenic origin (Phebus et al. 1997). Nevertheless, subsequent clinical trials with LY303870 provided disappointing results, as it was found ineffective in relieving osteoarthritic pain (Goldstein et al. 2000), in preventing migraine (Goldstein et al. 2001a) and in relieving pain of diabetic neuropathy (Goldstein et al. 2001b). The failure of LY303870 as an analgesic in the clinical trials was first attributed to its inadequate penetration of the blood–brain barrier (Goldstein et al. 2000), a possibility that would explain its low potency in preventing foot-tapping in gerbils and cisplatin-induced emesis in ferrets (Table 4; Rupniak et al. 1997). In this regard, Urban and Fox (2000) have argued that the failure of LY303870 to exert analgesic effects is not surprising, since this compound is poorly effective even in inflammatory and neuropathic models of the guinea pig, a species endowed with a human-like NK_1 receptor, and for this reason more predictive than other models of pain in rodents. Whatever the explanation, it is worth mentioning that other trials with compounds better at penetrating the CNS (such as L754030, see Sect. 4.5) did not provide pain relief either. More recently, the effectiveness of LY303870 in a mouse model of inflammatory bowel disease has been reported, both in reducing the severity of IBD and in allowing the partial healing of intestinal lesions in mice with preexisting IBD (Sonea et al. 2002; Table 6).

4.7
CGP49823, SDZNKT343, NKP608

A new class of piperidine-based antagonists was obtained at Ciba-Geigy, through the modification of a lead derived from the screening of a chemical collection. One of the most interesting compounds of this series derived from lead optimization is CGP49823 (Fig. 3; Hauser et al. 1994). CGP49823 showed good affinity for bovine NK_1 receptors (Table 5) and antagonized with similar affinity (IC_{50}=13 nM) SP-induced inositol monophosphate production in human U-373MG cells and SP-induced contractions in the rabbit vena cava (Table 3), whereas it was approximately 100-fold weaker at NK_2 or NK_3 receptors (Table 5; Hauser et al. 1994). In vivo, CGP49823 increased active social time in rats and reduced the immobility time of rats in the swim test, after both single or subchronic oral administration of high doses (3–30 mg/kg) (Vassout et al. 1994). These results led the authors to indicate a potential anxiolytic and antidepressant effect of CGP49823 (Table 6). High doses of CGP49823 were also required to inhibit thumping behavior induced by i.c.v. administration of SP methylester in gerbils (ED_{50}=50 mg/kg p.o.) (Vassout et al. 1994). This result was confirmed by Rupniak and coworkers (1997) who, in addition, found CGP49823 to be ineffective in preventing cisplatin-induced acute retching in ferrets and concluded that this compound does not readily cross the blood–brain barrier. On this basis, Rupniak and Kramer (1999) questioned whether the reported antidepressant effects produced by CGP49823 in rats are attributable to antagonism of NK_1 receptor rather than another pharmacological effect.

Another series of compounds discovered at Novartis was obtained by combination of a series of 2-halo-substituted benzylthioureas with aromatic amino acid amides leading to a series of 2-NO_2 phenylthiourea analogs, whose most active derivative was SDZNKT343 (Fig. 3; Walpole et al. 1998b). SDZNKT343 showed subnanomolar affinity for human NK_1 receptors (Table 5) and at least 500-fold lower affinity for the rat NK_1 receptor, whereas its less active (R,R)-enantiomer was about 1,000 times less active at all receptors (Walpole et al. 1998a). The mechanism of interaction of SDZNKT343 with human NK_1 receptors was clearly noncompetitive, as was the antagonism by SDZNKT343 of [Sar^9]SP sulfone-induced contractions in the guinea pig ileum (Walpole et al. 1998a). SDZNKT343 is an extremely selective ligand of human NK_1 receptors relative to a wide array of binding sites: however, at $10\mu M$ it inhibited voltage-activated Ca^{2+} and Na^+ currents in guinea pig dorsal root ganglion neurons, as did CP96345 and CP99994 (Walpole et al. 1998a). In vivo, SDZNKT343 (30 mg/kg p.o.) reduced mechanical hyperalgesia elicited by intraplantar carrageenan in guinea pigs, being significantly more potent than other NK_1 receptor antagonists like RPR100893 or SR140333, whereas it was less efficacious in reducing thermal hyperalgesia in carrageenan-pretreated guinea pigs (Campbell et al. 2000). In addition, SDZNKT343 (30 mg/kg p.o.) reversed by 60% the plasma protein extravasation induced by Freund's complete adjuvant in the guinea pig knee joint. On the basis of these results there is great expectation that SDZNKT343 could exert analgesic activity in man, as proposed by Urban and Fox (2000) who outlined the high reliability of the guinea pig models of neuropathic pain and inflammation over similar models in rodents.

More recently Novartis has presented a novel 4-amino piperidine derivative: NKP608 (Fig. 3; Vassout et al. 2000). In vitro, NKP608 showed nanomolar affinity for bovine NK_1 receptors (Table 5), and blocked with similar affinity (IC_{50}=2.6 nM) SP-induced inositol monophosphate production in human U-373MG cells, while it bound with only tenfold reduced affinity to the rat NK_1 receptor (Vassout et al. 2000). In in vivo behavioral tests, such as the social interaction test, oral doses of NKP608 specifically increased the time spent by rats in social contact, and in a social exploration test similar doses of NKP608 increased the time spent by the intruder rat in social contact with the resident rat. Both effects of NKP608 resembled those exerted by the benzodiazepine drug chlordiazepoxide, and suggested a possible anxiolytic use of NKP608 in humans (Table 6; Vassout et al. 2000). In the hind foot thumping model in gerbils, NKP608 was a highly effective and long-lasting antagonist following oral administration of doses comparable to those found effective in the rat: this finding provided an indirect demonstration of a good ability of NKP608 to cross the blood–brain barrier (Vassout et al. 2000). Anxiolytic-like effects of NKP608 have recently been reported also in gerbils, a species bearing a NK_1 receptor more similar to the human type than the rat. In the gerbil model, low doses of NKP608 given orally were reported to be as effective as the anxiolytic drug chlordiazepoxide (Gentsch et al. 2002). In addition to its anxiolytic activity, NKP608 has been reported to produce antidepressant-like effects in a chronic mild stress model of depression in rats (Papp et al. 2000). In this latter model NKP608 (0.03–0.1 mg/kg p.o.) produced effects comparable to those of imipramine, but the onset of action was faster with NKP608 than with the tricyclic antidepressant. Interestingly, the dose–response curves for both the anxiolytic (Vassout et al. 2000) and antidepressant (Papp et al. 2000) effects of NKP608 in rats were bell-shaped: i.e., the positive effect seen at relatively low doses was fading out at higher doses. The cause of this behavior of NKP608 is not fully understood (see Vassout et al. 2000, for tentative explanations).

4.8
PD154075 (CI1021)

A new series of compounds containing a tryptophan motif was developed at Parke-Davis in the mid-1990s, starting from a dipeptide library, from which a lead peptide compound (Z-Trp-Phe-NH$_2$) showing micromolar affinity at tachykinin receptors was obtained (Boyle et al. 1994). Further modifications of this lead yielded PD154075 (or CI1021), a compound bearing a methyl-tryptophan group and a 2-benzofuran moiety (Fig. 3). PD154075 showed high affinity for NK$_1$ receptors of man, guinea pig and other human-related species in binding and functional assays (Tables 3 and 5), whereas it bound with approximately 300-fold lower affinity to rat or mouse NK$_1$ receptors (Boyle et al. 1994; Singh et al. 1997). In vivo, PD154075 potently (ID$_{50}$=0.02 mg/kg i.v.) blocked SP methylester-induced plasma protein extravasation in the guinea pig bladder (Boyle et al. 1994) and dose-dependently [1–100 mg/kg subcutaneously (s.c.)] antagonized [Sar^9]SP sulfone-induced foot tapping in the gerbil, this latter effect supporting a central mechanism of action for this compound (Singh et al. 1997). Brain penetration by PD154075 after oral administration in rats was confirmed by extraction and HPLC assay: its absolute oral bioavailability in this species was 49±15% (Singh et al. 1997). An antiemetic potential of PD154075 was shown in the ferret: in this species PD154075 (10 mg/kg s.c. three times a day for 3 days) almost completely blocked both the acute and delayed emetic response to cisplatin (Singh et al. 1997). In a model of postoperative pain in the rat, PD154075–given before, not after surgery–selectively blocked both mechanical and thermal hypersensitivity of the operated rats and, unlike morphine, did not modify the length of anesthetic-induced sleep (Gonzalez et al. 1998; Table 6). The analgesic activity of PD154075 has further been shown in a chronic constrictive injury model (sciatic nerve ligation), in which rats developed thermal and mechanical hyperalgesia along with cold, dynamic and static allodynia (Gonzalez et al. 2000). PD154075 blocked all these responses (except dynamic allodynia), without inducing tolerance during a 10-day long treatment with 100 mg/kg/day s.c. Importantly, PD154075 blocked hypersensitivity in guinea pigs induced by sciatic nerve ligation, with a minimum effective dose of 0.1 mg/kg p.o.; on the basis of these results it has been proposed as a possible therapeutic agent in inflammatory and neuropathic pain (Gonzalez et al. 2000).

4.9
TAK637

In 1995 Natsugari and coworkers at Takeda presented a new series of N-benzylcarboxyamide compounds, that were the result of optimization of a benzodiazepine-based cholecystokinin antagonist chosen as the lead compound. The most active compound of this series showed subnanomolar affinity for the human NK_1 receptor along with high oral potency in preventing capsaicin-induced plasma protein extravasation in guinea pigs (Natsugari et al. 1995). Further optimization of this series of compounds yielded TAK637 (Fig. 3), a very potent antagonist of the human NK_1 receptor, with >100-fold reduced potency at the rat NK_1 receptor (Natsugari et al. 1999; Table 5). Moreover, TAK637 has been claimed to be at least 2000 times weaker at NK_2 or NK_3 than human NK_1 receptors (Okano et al. 2002). However, it should be noted that while the affinity of TAK637 for NK_3 receptors (pK_i <6.0) has been checked in a binding assay on guinea pig cerebral cortex membranes, that for NK_2 receptors (pA_2=6.0) has been evaluated in functional experiments on the guinea pig ileum, using NKA as the agonist (Natsugari et al. 1999). Thus, both the presence of additional NK_1 and NK_3 receptors in the tissue, and the poor selectivity of the agonist used (NKA) make the outcome of the latter experiments questionable. More recently, Venkova and coworkers (2002) have established that in the guinea pig isolated colon the ratio between antagonist potencies of TAK637 at NK_1 vs. NK_2 receptors is at least 20–50:1; however concentrations of TAK637 higher than $0.1\mu M$ were not tested in that study, so that determination of an exact potency ratio was hampered (Table 3). In vivo, systemic TAK637 increased the volume threshold of saline required to elicit the micturition reflex in anesthetized guinea pigs without decreasing the voiding pressure; an effect that was reproduced in unanesthetized animals after oral administration of low doses of TAK637 (Doi et al. 1999). Furthermore, TAK637 was able to decrease the number, but not amplitude, of rhythmic bladder contractions elicited by urinary bladder wall distension (by saline), and also dose-dependently reduced the number of animals responding with micturition to topical application of capsaicin onto the surface of the bladder dome (Doi et al. 2000). These results suggest TK637 as possible drug candidate for the control of urinary incontinence due to detrusor overactivity (Doi et al. 1999, 2000; Table 6)

In a model of intestinal transit in the gerbil, TK637 selectively reduced the increase of fecal pellet output provoked by both administration of an NK_1 receptor-selective agonist or by restraint stress, without modifying spontaneous excretion of fecal pellets in control animals (Okano et al. 2001). In a model of visceral pain, intraduodenal TK637 stereoselectively reduced the number of abdominal contractions provoked by colorectal distension in rabbits previously subjected to colonic irritation (Okano et al. 2002). Intrathecal administration of

TK637 was also effective in this model (Okano et al. 2002), thus showing that central (spinal cord located) receptors are the most probable target of the action of TK637. These results led Okano and coworkers (2001, 2002) to suggest that TK637 may be useful in treating functional bowel disorders such as IBS.

4.10
Other Nonpeptide Compounds

WIN51078 (Fig. 3) is one of the most interesting heterosteroid compounds of a series obtained through the screening of natural products, being endowed with appreciable affinity for the rat NK_1 receptor (IC_{50}=50 nM in rat forebrain membranes) (Venepalli et al. 1992). However, WIN51078 possesses a dramatically lower (about 400-fold) affinity for the human NK_1 receptor as compared to the rat type (Sachais and Krause 1994). This characteristic of action, while reinforcing the concept that important species-related differences exist among tachykinin receptors, has represented a serious drawback that hampered the development of WIN51078 and related compounds into drugs to be used in human diseases.

R116301 (Fig. 3) has recently been introduced by Johnson & Johnson as a potent and selective piperidine-based NK_1 receptor antagonist (Megens et al. 2002). R116301 is endowed with subnanomolar affinity for human NK_1 receptors and over 200-fold selectivity relative to human NK_2, NK_3 and rat NK_1 receptors (Table 5). In vivo, R116301 has been shown to potently antagonize both peripheral (SP-induced plasma protein extravasation) and central (thumping in gerbils) NK_1 receptor-mediated effects. In addition, R116301 is able to reduce/prevent emesis caused by various emetic stimuli in ferrets, cats and dogs, also after oral pretreatment (Megens et al. 2002). The ratio of oral vs. parenteral activity ranges from 0.2 to 2.7 in the species examined, and the action lasts from 6.5 to 16 h. An oral dose of 300 mg R116301 was found effective in reducing SP-induced dilation of the precontracted hand vein in human volunteers (Romerio et al. 1999). The available results obtained with R116301 make it a promising tool to be exploited in clinical trials for various human diseases involving a role of NK_1 receptors.

5
Therapeutic Perspectives for NK_1 Receptor Antagonists

Historically, the first indication for which a tachykinin NK_1 receptor antagonist was expected to be efficacious was pain, as the early evidence for an involvement of SP in pain transmission and perception had been collected since the 1950s, and subsequently this concept had been widely supported by a plethora of preclinical studies (for a review see Quartara and Maggi 1998). Thus, it was not surprising to see the first potent and selective nonpeptide antagonists such as CP96345 (Sect. 4.1), RP67580 (Sect. 4.2) and SR140333 (Sect. 4.3) to produce antinociceptive and antinflammatory effects in classical animal models for testing analgesic drugs, thereby providing further support for a role of NK_1 receptors in nociception and inflammation (Table 6). Nevertheless, after promising results had been obtained with CP99994 in patients suffering with postoperative dental pain (Sect. 4.1), all subsequent clinical trials with CP99994 itself and several other nonpeptide compounds such as LY303870 (Sect. 4.6), L754030 (or MK869; Sect. 4.5) and RPR100893 (Sect. 4.2) were disappointing, as they failed to demonstrate analgesic effects in patients affected by osteoarthritis, neuropathic pain, dental pain and migraine. Several explanations have been put forward in an attempt to explain this discrepancy between preclinical and clinical results (Rupniak and Kramer 1999; Hill 2000a, 2000b; Urban and Fox 2000). One of the criticisms was that certain antagonists (like LY303870 and others) are ineffective as analgesics in clinical trials because of an inadequate penetration of the blood–brain barrier. However, it is worth noting that better CNS penetrating compounds (like L754030) did not provide pain relief either (Rupniak and Kramer 1999).

At present, there are two main pathological conditions in which tachykinin NK_1 receptor-selective antagonists have been confirmed effective in humans: emesis, and affective disorders such as depression and anxiety. The first evidence that an NK_1 antagonist could be useful in the control of emesis provoked by cisplatin and various other stimuli, was collected 10 years ago with CP99994 in the ferret (Sect. 4.1). Subsequently, CP99994 and other antagonist compounds like GR203040, GR205171 (Sect. 4.4), PD154075 (Sect. 4.8) and others have proved effective in blocking the effects of various emetogens in animal species such as *Suncus murinus*, dog and cat. In man, CP122721 has been the first compound reported successful in reducing delayed emesis in patients under chemotherapy (Sect. 4.1). Since it was less effective in the control of nausea it was discontinued for this indication (Evangelista 2001). Another compound shown to be effective in the control of chemotherapy-induced emesis in humans is L754030 (aprepitant, Emend^R; Sect. 4.5) that has recently been approved in the USA for the combination treatment of this pathological condition. Thus, L754030 is the first compound arising from research in the field of tachykinins to become a drug for the care of human diseases. GR205171 has also been proven efficacious in reducing postoperative nausea and emesis, but unable to reduce motion-induced nausea in healthy volunteers (Sect. 4.4).

CGP49823 (Sect. 4.7) was shown in 1994 to be active in the social interaction test and swimming test in rats, and on the basis of these results it was claimed to possess a potential anxiolytic/antidepressant property. The efficacy of CGP49823 in these tests has been questioned afterwards, because of the poor ability shown by this compound to cross the blood–brain barrier (Rupniak and Kramer 1999). Demonstration of antidepressant and possibly anxiolytic activity afforded by an NK_1 receptor antagonist has been furnished by Kramer and coworkers (1998), in both animals and humans, by the use of L754030 (MK869) and related compounds (Sect. 4.5). NKP608 is another promising compound in this area, having been shown effective in affording anxiolytic-like effects in rats and gerbils and antidepressant effects in rats (Sect. 4.7; Table 6). Tachykinin

NK_1 receptor-selective antagonists are also expected to provide therapeutic effects in peripheral inflammatory diseases in which the role of both tachykinins and NK_1 receptors have been clearly documented in preclinical investigations. The main peripheral indications are: asthma/bronchial hyperreactivity and inflammatory bowel disease. A possible antiasthmatic activity of these compounds was envisaged with the first peptidic antagonists such as FK888 or MEN11467 (Sect. 3.2) which were found effective in preventing both bronchoconstriction and plasma protein extravasation in the airways, induced by tachykinins and/or other stimuli. However, the clinical trials performed with FK888 and CP99994 (Sect. 4.1) have provided negative results. Now there is general expectation that mixed NK_1/NK_2 receptor antagonists might afford beneficial effects in asthma, relative to NK_1 or NK_2 receptor-selective compounds. Tachykinins are thought to be important mediators involved in the genesis/maintenance of various inflammatory gastrointestinal diseases, a concept that is based on an increasing number of studies that have been reviewed extensively elsewhere (e.g., Holzer and Holzer-Petsche 1997; Holzer 1998; Bueno 2000). Various NK_1 receptor antagonists have been found effective in reducing the severity of experimental IBD in animals, including SR140333 (Sect. 4.3), LY303870 (Sect. 4.6) and TAK637 (Sect. 4.9; Table 6). This latter compound has been found effective in a model of visceral pain in rabbits and on this basis it has been proposed for the treatment of IBS. However, its selectivity for NK_1 over NK_2 receptors is not well documented, and the possibility that reduction of visceral pain by TAK637 stems from blockade of NK_2 receptors remains to be investigated. Nevertheless, the effectiveness of a very selective NK_1 receptor antagonist, such as CJ11974 (ezlopitant), in patients affected by IBShas been documented recently (Sect. 4.1).

Acknowledgements. The authors are grateful to Dr. Laura Quartara, Chemical Department, Menarini Ricerche, Florence, Italy, for drawing the chemical structures of the nonpeptide compounds and for critically reading the text.

References

Amann R, Schuligoi R, Holzer P, Donnerer J (1995) The non-peptide NK_1 receptor antagonist SR140333 produces long-lasting inhibition of neurogenic inflammation, but does not influence acute chemo- or thermonociception in rats. Naunyn Schmiedebergs Arch Pharmacol 352:201–205

Aramori I, Morikawa N, Zenkoh J, O'Donnell N, Iwami M, Kojo H, Notsu Y, Okuhara M, Ono S, Nakanishi S (1994) Subtype- and species-selectivity of a tachykinin receptor antagonist, FK888, for cloned rat and humantachykinin receptors. Eur J Pharmacol 269:277–281

Astolfi M, Patacchini R, Maggi M, Manzini S (1997) Improved discriminatory properties between human and murine tachykinin NK_1 receptors of MEN 10930: a new potent and competitive antagonist. Neuropeptides 31:373–379

Barr AJ, Watson SP (1993) Non-peptide antagonists CP 96,345 and RP 67,580, distinguish species variants in tachykinin NK-1 receptors. Br J Pharmacol 108:223–227

Beattie DT, Beresford IJ, Connor HE, Marshall FH, Hawcock AB, Hagan RM, Bowers J, Birch PJ, Ward P (1995) The pharmacology of GR203040, a novel, potent and selective non-peptide tachykinin NK_1 receptor antagonist. Br J Pharmacol 116:3149–3157

Beaujouan JC, Heuillet E, Petitet F, Saffroy M, Torrens Y, Glowinski J (1993) Higher potency of RP 67580, in the mouse and the rat compared with other nonpeptide and peptide tachykinin NK_1 antagonists. Br J Pharmacol 108:793–800

Beresford IJ, Birch PJ, Basalt M-C, Rogers H, Fernandez L, Hagan RM (1991a) Effect of the spirolactam NK-1 receptor antagonist GR 82334 on neurokinin-1 and electrical stimulation-induced oedema in the rat. Br J Pharmacol 102:360P

Beresford IJ, Birch PJ, Hagan RM, Ireland SJ (1991b) Investigation into species variants in tachykinin NK_1 receptors by use of the non-peptide antagonist, CP-96,345. Br J Pharmacol 104:292–293

Birch PJ, Harrison SM, Hayes AG, Rogers H, Tyers MB (1992) The non-peptide NK_1 receptor antagonist, (+/-)-CP-96,345, produces antinociceptive and anti-oedema effects in the rat. Br J Pharmacol 105:508–510

Boni P, Maggi CA, Evangelista S (1994) In vivo evidence for the activation of a septide-sensitive tachykinin receptor in guinea-pig bronchoconstriction. Life Sci 54:PL327–PL332

Bonnet J, Kucharczyk N, Robineau P, Lonchampt M, Dacquet C, Regoli D, Fauchere JL, Canet E A (1996) Water-soluble, stable dipeptide NK_1 receptor-selective neurokinin receptor antagonist with potent in vivo pharmacological effects: S18523. Eur J Pharmacol 310:37–46

Bountra C, Bunce K, Dale T, Gardner C, Jordan C, Twissell D, Ward P (1993) Anti-emetic profile of a non-peptide neurokinin NK_1 receptor antagonist, CP-99,994, in ferrets. Eur J Pharmacol 249:R3–R4

Boyle S, Guard S, Higginbottom M, Horwell DC, Howson W, McKnight AT, Martin K, Pritchard MC, O'Toole J, Raphy J (1994) Rational design of high affinity tachykinin NK_1 receptor antagonists. Bioorg Med Chem 2:357–370

Bremer AA, Tansky FM, Wu M, Boyd ND, Leeman S (2001) Direct evidence for the interaction of neurokinin A with the tachykinin NK_1 receptor in tissue. Eur J Pharmacol 423:143–147

Bueno L, Fioramonti J, Garcia-Villar R (2000) Pathobiology of visceral pain: molecular mechanisms and therapeutic implications III Visceral afferent pathways: a source of new therapeutic targets for abdominal pain. Am J Physiol 278:G670–G676

Caesar M, Seabrook GR, Kemp JA (1993) Block of voltage-dependent sodium currents by the substance P receptor antagonist (±)CP 96,345 in neurons cultured from the rat cortex. J Physiol 459:397P

Campbell EA, Gentry C, Patel S, Kidd B, Cruwys S, Fox AJ, Urban L (2000) Oral anti-hyperalgesic and anti-inflammatory activity of NK_1 receptor antagonists in models of inflammatory hyperalgesia of the guinea-pig. Pain 87:253–263

Cascieri MA, Macleod AM, Underwood D, Shiao LL, Ber E, Sadowski S, Yu H, Merchant KJ, Swain CJ, Strader CD, Fong TM (1994) Characterization of the interaction of N-acyl-L-tryptophan benzyl ester neurokinin antagonists with the human neurokinin-1 receptor. J Biol Chem 269:6587–6591

Cascieri MA, Ber E, Fong TM, Hale JJ, Tang F, Shiao LL, Mills SG, MacCoss M, Sadowski S, Tota MR, Strader CD (1997) Characterization of the binding and activity of a high affinity, pseudoirreversible morpholino tachykinin NK_1 receptor antagonist. Eur J Pharmacol 325:253–261

Cellier E, Barbot L, Iyengar S, Couture R (1999) Characterization of central and peripheral effects of septide with the use of five tachykinin NK_1 receptor antagonists in the rat. Br J Pharmacol 127:717–728

Cheeta S, Tucci S, Sandhu J, Williams AR, Rupniak NM, File SE (2001) Anxiolytic actions of the substance P (NK_1) receptor antagonist L-760735 and the 5-HT1A agonist 8-OH-DPAT in the social interaction test in gerbils. Brain Res 915:170–175

Cirillo R, Astolfi M, Conte B, Lopez G, Parlani M, Sacco G, Terracciano R, Fincham CI, Sisto A, Evangelista S, Maggi CA, Manzini S (2001) Pharmacology of MEN 11467: a potent new selective and orally- effective peptidomimetic tachykinin NK_1 receptor antagonist. Neuropeptides 35:137–147

Cocquyt V, Van Belle S, Reinhardt RR, Decramer ML, O'Brien M, Schellens JH, Borms M, Verbeke L, Van Aelst F, De Smet M, Carides AD, Eldridge K, Gertz BJ (2001) Comparison of L-758,298, a prodrug for the selective neurokinin-1 antagonist, L-754,030, with ondansetron for the prevention of cisplatin-induced emesis. Eur J Cancer 37:835–842

Constantine JW, Lebel WS, Woody HA (1991) Inhibition of tachykinin-induced hypotension in dogs by CP-96,345, a selective blocker of NK-1 receptors. Naunyn Schmiedeberg's Arch Pharmacol 344:471–477

Coudore'-Civiale MA, Courteix C, Eschalier A and Fialip J (1998) Effect of tachykinin receptor antagonists in experimental neuropathic pain. Eur J Pharmacol 361:175–184

Courteix C, Lavarenne J, Eschalier A (1993) RP-67580, a specific tachykinin NK_1 receptor antagonist, relieves chronic hyperalgesia in diabetic rats. Eur J Pharmacol 241:267–270

Croci T, Landi M, Emonds-Alt X, Le Fur G, Maffrand JP, Manara L (1997) Role of tachykinins in castor oil diarrhoea in rats. Br J Pharmacol 121:375–380

Cutrer FM, Moussaoui S, Garret C, Moskowitz MA (1995) The non-peptide neurokinin-1 antagonist, RPR 100893, decreases c-fos expression in trigeminal nucleus caudalis following noxious chemical meningeal stimulation. Neuroscience 64:741–750

Delay-Goyet P, Lundberg JM (1991) Cigarette smoke-induced airway oedema is blocked by the NK-1 antagonist, CP 96,345. Eur J Pharmacol 203:157–158

Delay-Goyet P, Franco-Cereceda A, Gonsalves SF, Clingan CA, Lowe JA 3rd, Lundberg JM (1992) CP-96,345 antagonism of NK_1 receptors and smoke-induced protein extravasation in relation to its cardiovascular effects. Eur J Pharmacol 222:213–218

Diemunsch P, Schoeffler P, Bryssine B, Cheli-Muller LE, Lees J, McQuade BA, Spraggs CF (1999) Antiemetic activity of the NK_1 receptor antagonist GR205171 in the treatment of established postoperative nausea and vomiting after major gynaecological surgery. Br J Anaesth 82:274–276

Dionne RA, Max MB, Gordon SM, Parada S, Sang C, Gracely RH, Sethna NF, MacLean DB (1998) The substance P receptor antagonist CP-99,994 reduces acute postoperative pain. Clin Pharmacol Ther 64:562–568

Doi T, Kamo I, Imai S, Okanishi S, Ishimaru T, Ikeura Y, Natsugari H (1999) Effects of TAK-637, a tachykinin receptor antagonist, on lower urinary tract function in the guinea-pig. Eur J Pharmacol 383:297–303

Doi T, Kamo I, Imai S, Okanishi S, Ikeura Y, Natsugari H (2000) Effects of TAK-637, a tachykinin receptor antagonist, on the micturition reflex in guinea-pigs. Eur J Pharmacol 395:241–246

Eglezos A, Giuliani S, Viti G, Maggi CA (1991) Direct evidence that capsaicin-induced plasma protein extravasation is mediated through tachykinin NK_1 receptors. Eur J Pharmacol 209:277–279

Emonds-Alt X, Doutremepuich J, Heaulme M, Neliat G, Santucci V, Steinberg R, Vilain P, Bichon D, Ducoux J, Proietto V, Van Broeck D, Soubrié P, Le Fur G and Breliere, J (1993) In vitro and in vivo biological activities of SR140333, a novel potent non-peptide NK_1 receptor antagonist. Eur J Pharmacol 250:403–413

Emonds-Alt X, Proietto V, Steinberg R, Oury-Donat F, Vige X, Vilain P, Naline E, Daoui S, Advenier C, Le Fur G, Maffrand JP, Soubrie P, Pascal M (2002) SSR240600 [(R)-2-(1-[2-[4-[2-[3,5-bis(trifluoromethyl)phenyl]acetyl]-2-(3,4-dichlorophenyl)-2-morpholinyl]ethyl]-4-piperidinyl)-2-methylpropanamide], a centrally active nonpeptide antagonist of the tachykinin neurokinin-1 receptor: I biochemical and pharmacological characterization. J Pharmacol Exp Ther 303:1171–1179

Evangelista S (2001) Eziopitant. Curr Opin Invest Drugs 2:1441–1443

Fahy JV, Wong HH, Geppetti P, Reis JM, Harris SC, Maclean DB, Nadel JA Boushey HA (1995) Effect of an NK_1 receptor antagonist (CP-99,994) on hypertonic saline-induced bronchoconstriction and cough in male asthmatic subjects. Am J Resp Crit Care Med 152:879–884

Fardin V, Carruette A, Menager J, Bock M, Flamand O, Foucault F, Heuillet E, Moussaoui SM, Tabart M, Peyronel JF, Garret C (1994) In vitro pharmacological properties of RPR 100,893, a novel nonpeptide antagonist of the human NK_1 receptor. Neuropeptides 26:34

Folkers K, Hakanson R, Horig J, Xu JC, Leander S (1984) Biological evaluation of substance P antagonists. Br J Pharmacol 83:449–456

Folkers K, Feng DM, Asano N, Hakanson R, Weisenfeld-Hallin Z, Leander S (1990) Spantide II, an effective tachykinin antagonist having high potency and negligible neurotoxicity. Proc Natl Acad Sci USA 87:4833–4835

Fong TM, Yu H, Strader C (1992) Molecular bases for the species selectivity of the NK_1 receptor antagonists CP 96,345 and RP 67,580. J Biol Chem 267:25668–25671

Fong TM, Yu H, Cascieri MA, Underwood D, Swian CJ, Strader CD (1994)The role of Histidine 265 in antagonist binding to the NK_1 receptor. J Biol Chem 269:2728–2732

Fujii T, Murai M, Morimoto H, Maeda Y, Yamaoka M, Hagiwara D, Miyake H, Ikari N, Matsuo M (1992) Pharmacological profile of a high affinity dipeptide NK_1 receptor antagonist, FK888. Br J Pharmacol 107:785–789

Gardner CJ, Twissell DJ, Dale TJ, Gale JD, Jordan CC, Kilpatrick GJ, Bountra C, Ward P (1995) The broad-spectrum anti-emetic activity of the novel non-peptide tachykinin NK_1 receptor antagonist GR203040. Br J Pharmacol 116:3158–3163

Gardner CJ, Armour DR, Beattie DT, Gale JD, Hawcock AB, Kilpatrick GJ, Twissell DJ, Ward P (1996) GR205171: a novel antagonist with high affinity for the tachykinin NK_1 receptor, and potent broad-spectrum anti-emetic activity. Regul Pept 65:45–53

Garret C, Carruette A, Fardin V, Moussaoui S, Peyronel JF, Blanchard JC, Laduron PM (1991) Pharmacological properties of a potent and selective nonpeptide SP antagonist. Proc Natl Acad Sci USA 88:10208–10211

Gentsch C, Cutler M, Vassout A, Veenstra S, Brugger F (2002) Anxiolytic effect of NKP608, a NK_1-receptor antagonist, in the social investigation test in gerbils. Behav Brain Res 133:363–368

Gesztesi Z, Song D, White PF (1998) Comparison of a new NK-1 antagonist (CP-122,721) to ondansetron in the prevention of post-operative nausea and vomiting. Anesth Anal 86:S32

Gesztesi Z, Scuderi PE, White PF, Wright W, Wender RH, D'Angelo R, Black LS, Dalby PL, MacLean D (2000) Substance P (Neurokinin-1) antagonist prevents postoperative vomiting after abdominal hysterectomy procedures. Anesthesiology 93:931–937

Gether U, Lowe JA 3rd, Schwartz TW (1995) Tachykinin non-peptide antagonists: binding domain and molecular mode of action. Biochem Soc Trans 23:96–102

Gitter BD, Waters DC, Bruns RF, Mason NR Nixon JA, Howbert JJ (1991) Species differences in affinities of nonpeptide antagonists for SP receptors. Eur J Pharmacol 197:237–238

Gitter BD, Bruns RF, Howbert JJ, Waters DC, Threlkeld PG, Cox LM, Nixon JA, Lobb KL, Mason NR, Stengel PW, Cockerham SL, Silbaugh SA, Gehlert DR, Schober DA, Iyengar S, Calligaro DO, Regoli D, Hipskind PA (1995) Pharmacological characterization of LY303870: a novel, potent and selective nonpeptide substance P (neurokinin-1) receptor antagonist. J Pharmacol Exp Ther 275:737–744

Goldstein DJ, Wang O, Todd LE, Gitter BD, DeBrota DJ, Iyengar S (2000) Study of the analgesic effect of lanepitant in patients with osteoarthritis pain. Clin Pharmacol Ther 67:419–426

Goldstein DJ, Offen WW, Klein EG, Phebus LA, Hipskind P, Johnson KW, Ryan RE Jr (2001a) Lanepitant, an NK-1 antagonist, in migraine prevention. Cephalalgia 21:102–106

Goldstein DJ, Wang O, Gitter BD, Iyengar S (2001b) Dose-response study of the analgesic effect of lanepitant in patients with painful diabetic neuropathy. Clin Neuropharmacol 24:16–22

Gonsalves S, Watson J, Ashton C (1996) Broad spectrum antiemetic effects of CP-122,721, a tachykinin NK_1 receptor antagonist, in ferrets. Eur J Pharmacol 305:181–185

Gonzalez MI, Field MJ, Holloman EF, Hughes J, Oles RJ, Singh L (1998) Evaluation of PD 154075, a tachykinin NK_1 receptor antagonist, in a rat model of postoperative pain. Eur J Pharmacol 344:115–120

Gonzalez MI, Field MJ, Hughes J, Singh L (2000) Evaluation of selective NK_1 receptor antagonist CI-1021 in animal models of inflammatory and neuropathic pain. J Pharmacol Exp Ther 294:444–450

Guard S, Watson SP (1991) Tachykinin receptor types: classification and membrane signalling mechanisms. Neurochem Int 18:149–165

Guard S, Boyle SJ, Tang KW, Watling KJ, McKnight AT, Woodruff GN (1993) The interaction of the NK_1 receptor antagonist CP-96,345 with L-type calcium channels and its functional consequences. Br J Pharmacol 110:385–391

Guo JZ, Yoshioka K, Zhao FY, Hosoki R, Maehara T, Yanagisawa M, Hagan RM, Otsuka M (1995) Pharmacological characterization of GR82334, a tachykinin NK_1 receptor antagonist, in the isolated spinal cord of the neonatal rat. Eur J Pharmacol 281:49–54

Hagan RM, Ireland SJ, Jordan CC, Beresford IJM, Stephens-Smith ML, Ewan G, Ward P (1990) GR 71251, a novel, potent and highly selective antagonist at neurokinin NK-1 receptors. Br J Pharmacol 99:62P

Hagan RM, Ireland SJ Bailey F, Mcbride C, Jordan CC, Ward, P (1991) A spirolactam conformationally-constrained analogue of physalaemin which is a peptidase-resistant selective neurokinin NK_1 receptor antagonist. Br J Pharmacol 102:168P

Hagiwara D, Miyake H, Igari N, Murano K, Morimoto H, Murai M, Fujii T, Matsuo M (1993) Design of a novel dipeptide substance P antagonist FK888. Regul Pept 46:332–334

Hale JJ, Mills SG, MacCoss M, Shah SK, Qi H, Mathre DJ, Cascieri MA, Sadowski S, Strader CD, MacIntyre DE, Metzger JM (1996) 2(S)-((3,5-bis(trifluoromethyl)benzyl)-oxy)-3(S)-phenyl-4- ((3-oxo-1,2,4-triazol-5-yl)methyl)morpholine (1): a potent, orally active, morpholine-based human neurokinin-1 receptor antagonist. J Med Chem 39:1760–1762

Hale JJ, Mills SG, MacCoss M, Finke PE, Cascieri MA, Sadowski S, Ber E, Chicchi GG, Kurtz M, Metzger J, Eiermann G, Tsou NN, Tattersall FD, Rupniak NM, Williams AR, Rycroft W, Hargreaves R, MacIntyre DE (1998) Structural optimization affording 2-(R)-(1-(R)-3, 5-bis(trifluoromethyl)phenylethoxy)-3-(S)-(4-fluoro)phenyl-4-(3-oxo-1,2,4-triazol-5-yl)methylmorpholine, a potent, orally active, long-acting morpholine acetal human NK-1 receptor antagonist. J Med Chem 41:4607–4614

Hale JJ, Mills SG, MacCoss M, Dorn CP, Finke PE, Budhu RJ, Reamer RA, Huskey SE, Luffer-Atlas D, Dean BJ, McGowan EM, Feeney WP, Chiu SH, Cascieri MA, Chicchi GG, Kurtz MM, Sadowski S, Ber E, Tattersall FD, Rupniak NM, Williams AR, Rycroft W, Hargreaves R, Metzger JM, MacIntyre DE (2000) Phosphorylated morpholine acetal human neurokinin-1 receptor antagonists as water-soluble prodrugs. J Med Chem 43:1234–1241

Hall JM, Mitchell D, Morton IK (1994) Typical and atypical NK_1 tachykinin receptor characteristics in the rabbit isolated iris sphincter. Br J Pharmacol 112:985–991

Harrison T, Williams BJ, Swain CJ, Ball RG (1994) Piperidine-ether based hNK_1 antagonists 1: determination of the relative and absolute stereochemical requirements. Biomed Chem Lett 4:2545–2550

Hastrup H, Schwartz TW (1996) Septide and neurokinin A are high affinity ligands on the NK_1 receptor:evidence from homologus versus heterologus binding analysis. FEBS Lett 399:264–266

Hauser K, Heid J, Criscione L, Brugger F, Ofner S, Veenstra S, Schilling W (1994) CGRP 49,823, a novel, nonpeptidic NK_1 receptor antagonist : in vitro pharmacology. Neuropeptides 26:37

Hesketh PJ, Gralla RJ, Webb RT, Ueno W, DelPrete S, Bachinsky ME, Dirlam NL, Stack CB, Silberman SL (1999) Randomized phase II study of the neurokinin 1 receptor antagonist CJ-11,974 in the control of cisplatin-induced emesis. J Clin Oncol 17:338–343

Hill R (2000a) NK_1 (substance P) receptor antagonists—why are they not analgesic in humans? Trends Pharmacol Sci 21:244–246

Hill R (2000b) Reply: will changing the testing paradigms show that NK_1 receptor antagonists are analgesic in humans? Trends Pharmacol Sci 21:465

Hipskind PA, Howbert JJ, Bruns RF, Cho SS, Crowell TA, Foreman MM, Gehlert DR, Iyengar S, Johnson KW, Krushinski JH, Li DL, Lobb KL, Mason NR, Muehl BS, Nixon JA, Phebus LA, Regoli D, Simmons RM, Threlkeld PG, Waters DC, Gitter BD (1996) 3-Aryl-1,2-diacetamidopropane derivatives as novel and potent NK-1 receptor antagonists. J Med Chem 39:736–748

Hirayama Y, Lei YH, Barnes PJ, Rogers DF (1993) Effects of two novel tachykinin antagonists, FK224 and FK888, on neurogenic airway plasma exudation, bronchoconstriction and systemic hypotension in guinea-pigs in vivo. Br J Pharmacol 108:844–851

Hökfelt T, Vincent S, Hellsten L, Rosell S, Folkers K, Markey K, Goldstein M, Cuello C (1981) Immunohistochemical evidence for a 'neurotoxic' action of (D-Pro^2, D-Trp^{7,9})-substance P, an analogue with substance P antagonistic activity. Acta Physiol Scand 113:571–573

Holzer P (1998) Tachykinins as targets of gastroenterological pharmacotherapy. Drugs News Perspect 11:394–401

Holzer P, Holzer-Petsche U (1997) Tachykinins in the gut Part II Roles in neural excitation, secretion and inflammation. Pharmacol Ther 73:219–263

Holzer-Petsche U, Rordorf-Nikolic T (1995) Central versus peripheral site of action of the tachykinin NK_1-antagonist RP 67580 in inhibiting chemonociception. Br J Pharmacol 115:486–490

Hoover DB (1991) Effects of spantide on guinea-pig coronary resistance vessels. Peptides 12:983–988

Huang R-RC, Yu H, Strader CD, Fong TM (1994) Interaction of substance P with the second and seventh transmembrane domain of the NK_1 receptor. Biochemistry 33:3007–3013

Ichinose M, Miura M, Yamauchi H, Kageyama N, Tomaki M, Oyake T, Ohuchi Y, Hida W, Miki H, Tamura G, Shirato K (1996) A neurokinin 1-receptor antagonist improves exercise-induced airway narrowing in asthmatic patients. Am J Resp Crit Care Med 153:936–941

Inoue H, Nagata N, Koshihara Y (1996) Effect of the tachykinin receptor antagonists, SR 140333, FK 888, and SR 142801, on capsaicin-induced mouse ear oedema. Inflamm Res 45:303–307

Iyengar S, Hipskind PA, Gehlert DR, Schober D, Lobb KL, Nixon JA, Helton DR, Kallman MJ, Boucher S, Couture R, Li DL, Simmons RM (1997) LY303870, a centrally active neurokinin-1 antagonist with a long duration of action. J Pharmacol Exp Ther 280:774–785

Jenkinson KM, Southwell BR, Furness JB (1999) Two affinities for a single antagonist at the neuronal NK_1 tachykinin receptor: evidence from quantitation of receptor endocytosis. Br J Pharmacol 126:131–136

Jung M, Calassi R, Maruani J, Barnouin MC, Souilhac J, Poncelet M, Gueudet C, Emonds-Alt X, Soubrie P, Breliere JC (1994) Neuropharmacological characterization of SR 140333, a non peptide antagonist of NK_1 receptors. Neuropharmacology 33:167–179

Karlsson U, Nasstrom J, Berge OG (1994) (+/-)-CP-96,345, an NK_1 receptor antagonist, has local anaesthetic-like effects in a mammalian sciatic nerve preparation. Regul Pept 52:39-46

Khan S, Liu YC, Khawaja AM, Manzini S, Rogers DF (2001) Effect of the long-acting tachykinin NK_1 receptor antagonist MEN 11467 on tracheal mucus secretion in allergic ferrets. Br J Pharmacol 132:189-196

Kramer MS, Cutler N, Feighner J, Shrivastava R, Carman J, Sramek JJ, Reines SA, Liu G, Snavely D, Wyatt-Knowles E, Hale JJ, Mills SG, MacCoss M, Swain CJ, Harrison T, Hill RG, Hefti F, Scolnick EM, Cascieri MA, Chicchi GG, Sadowski S, Williams AR, Hewson L, Smith D, Carlson EJ, Hargreaves RJ, Rupniak NM (1998) Distinct mechanism for antidepressant activity by blockade of central substance P receptors. Science 281:1640-1645

Kris MG, Radford JE, Pizzo BA, Inabinet R, Hesketh A, Hesketh PJ (1997) Use of an NK_1 receptor antagonist to prevent delayed emesis after cisplatin. J Natl Cancer Inst 89:817-818

Lei YH, Barnes PJ, Rogers DF (1992) Inhibition of neurogenic plasma exudation in guinea-pig airways by CP-96,345, a new non-peptide NK_1 receptor antagonist. Br J Pharmacol 105:261-262

Leander S, Hakanson R, Rosell S, Folkers K, Sundler F, Tornqvist K (1981) A specific SP antagonist blocks smooth muscle contractions induced by noncholinergic nonadrenergic nerve stimulation. Nature 294:467-469

Lecci A, Giuliani S, Patacchini R, Viti G, Maggi CA (1991) Role of NK_1 tachykinin receptors in thermonociception: effect of (±)- CP 96,345, a non peptide substance P antagonist on the hot plate test in mice. Neurosci Letters 129:299-302

Lee OY, Munakata J, Naliboff, BD, Chang L, Mayer E (2000) A double blind parallel group pilot study of the effects of CJ-11,974 and placebo on perceptual and emotional responses to rectosigmoid distension in IBS patients. Gastroenterology 118:4439

Lee WS, Moussaoui SM, Moskowitz MA (1994) Blockade by oral or parenteral RPR 100893 (a non-peptide NK_1 receptor antagonist) of neurogenic plasma protein extravasation within guinea-pig dura mater and conjunctiva. Br J Pharmacol 112:920-924

Lembeck F, Donnerer J, Tsuchiya M, Nagahisa A (1992) The non-peptide tachykinin antagonist, CP-96,345, is a potent inhibitor of neurogenic inflammation. Br J Pharmacol 105:527-530

Lombet A, Spedding M (1994) Differential effects of nonpeptidic tachykinin receptor antagonists on Ca2+ channels. Eur J Pharmacol 267:113-115

Longmore J, Razzaque Z, Shaw D, Hill RG (1994) Differences in the effects of NK_1-receptor antagonists, (+/-)-CP 96,345 and CP 99,994, on agonist-induced responses in guinea-pig trachea. Br J Pharmacol 112:176-178

Lowe JA 3rd, Drozda SE, Snider RM, Longo KP, Zorn SH, Morrone J, Jackson ER, McLean S, Bryce DK, Bordner J (1992) The discovery of (2S,3S)-cis-2-(diphenylmethyl)-N-[(2-methoxyphenyl)methyl]-1- azabicyclo[222]-octan-3-amine as a novel, nonpeptide substance P antagonist. J Med Chem 35:2591-2600

Lucot JB, Obach RS, McLean S, Watson JW (1997) The effect of CP-99994 on the responses to provocative motion in the cat. Br J Pharmacol 120:116-120

MacLeod AM, Merchant KJ, Brookfield F, Kelleher F, Stevenson G, Owens AP, Swain CJ, Casiceri MA, Sadowski S, Ber E (1994) Identification of L-tryptophan derivatives with potent and selective antagonist activity at the NK_1 receptor. J Med Chem 37:1269-1274

MacLeod AM, Cascieri MA, Merchant KJ, Sadowski S, Hardwicke S, Lewis RT, MacIntyre DE, Metzger JM, Fong TM, Shepheard S (1995) Synthesis and biological evaluation of NK_1 antagonists derived from L-tryptophan. J Med Chem 38:934-941

Maggi CA (1995) The mammalian tachykinin receptors. Gen Pharmacol 26:911-944

Maggi CA, Schwartz TW (1997) The dual nature of the tachykinin NK_1 receptor. Trends Pharmacol Sci 18:351-355

Maggi CA, Patacchini R, Feng DM, Folkers K (1991) Activity of spantide I and II at various tachykinin receptors and NK-2 tachykinin receptor subtypes. Eur J Pharmacol 199:127–129

Maggi CA, Giuliani S, Patacchini R, Santicioli P, Theodorsson E, Barbanti G, Turini D, Giachetti A (1992) Tachykinin antagonists inhibit nerve-mediated contractions in the circular muscle of the human ileum. Gastroenterology 102:88–96

Maggi CA, Patacchini R, Meini S, Giuliani S (1993a) Evidence for the presence of a septide-sensitive tachykinin receptor in the circular muscle of the guinea-pig ileum. Eur J Pharmacol 235:309–311

Maggi CA, Patacchini R, Rovero P, Giachetti A (1993b) Tachykinin receptors and tachykinin receptor subtypes. J Autonom Pharmacol 13:23–93

Maggi CA, Patacchini R, Meini S, Quartara L, Sisto A, Potier E, Giuliani S, Giachetti A (1994) Comparison of tachykinin NK_1 and NK_2 receptors in the circular muscle of the guinea-pig ileum and proximal colon. Br J Pharmacol 112:150–160

Mazelin L, Theodorou V, More J, Emonds-Alt X, Fioramonti J and Bueno L (1998) Comparative effects onf nonpeptide tachykinin receptor antagonists in experimental gut inflammation in rats and guinea-pigs. Life Sci 63:293–304

McAllister KH, Pratt JA (1998) GR205171 blocks apomorphine and amphetamine-induced conditioned taste aversions Eur J Pharmacol 353:141–148

McKnight AT, Boyle SJ, Tang K-W, Hill DR, Sauman-Chauhan N, Woodruff GN (1994) Functional activity at NK_1 receptors of novel peptoid antagonists. Br J Pharmacol 111:49P

McLean S, Ganong A, Seymour PA, Snider RM, Desai MC, Rosen T, Bryce DK, Longo KP, Reynolds LS, Robinson G, Schmidt AW, Siok C, Heym J (1993) Pharmacology of CP-99,994; a nonpeptide antagonist of the tachykinin neurokinin-1 receptor. J Pharmacol Exp Ther 267:472–479

McLean S, Ganong A, Seymour PA, Bryce DK, Crawford RT, Morrone J, Reynolds LS, Schmidt AW, Zorn S, Watson J, Fossa A, DePasquale M, Rosen T, Nagahisa A, Tsuchiya M, Heym J (1996) Characterization of CP-122,721; a nonpeptide antagonist of the neurokinin NK_1 receptor. J Pharmacol Exp Ther 277:900–908

Megens AA, Ashton D, Vermeire JC, Vermote PC, Hens KA, Hillen LC, Fransen JF, Mahieu M, Heylen L, Leysen JE, Jurzak MR, Janssens F (2002) Pharmacological profile of (2R-trans)-4-[1-[3,5-bis(trifluoromethyl)benzoyl]-2-(phenylmethyl)-4-piperidinyl]-N-(2,6-dimethylphenyl)-1-acetamide (S)-Hydroxybutanedioate (R116301), an orally and centrally active neurokinin-1 receptor antagonist. J Pharmacol Exp Ther 302:696–709

Meini S, Patacchini R, Maggi CA (1994) Tachykinin NK_1 receptor subtypes in the rat urinary bladder. Br J Pharmacol 111:739–746

Minami M, Endo T, Kikuchi K, Ihira E, Hirafuji M, Hamaue N, Monma Y, Sakurada T, Tan-no K, Kisara K (1998) Antiemetic effects of sendide, a peptide tachykinin NK_1 receptor antagonist, in the ferret. Eur J Pharmacol 363:49–55

Montier F, Carruette A, Moussaoui S, Boccio D, Garret C (1994) Antagonism of substance P and related peptides by RP 67580 and CP-96,345, at tachykinin NK_1 receptor sites, in the rat urinary bladder. Eur J Pharmacol 251:9–14

Moriarty D, Goldhill J, Selve N, O'Donoghue DP, Baird AW (2001) Human colonic antisecretory activity of the potent NK_1 antagonist, SR140333: assessment of potential anti-diarrhoeal activity in food allergy and inflammatory bowel disease. Br J Pharmacol 133:1346–1354

Morimoto H, Murai M, Maeda Y, Hagiwara D, Miyake H, Matsuo M, Fujii T (1992) FR 113680: a novel tripeptide substance P antagonist with NK_1 receptor selectivity. Br J Pharmacol 106:123–126

Moussaoui SM, Montier F, Carruette A, Blanchard JC, Laduron PM, Garret C A (1993) non-peptide NK_1-receptor antagonist, RP 67580, inhibits neurogenic inflammation postsynaptically. Br J Pharmacol 109:259–264

Moussaoui SM, Montier F, Carruette A, Fardin V, Floch A, Garret C (1994) In vivo pharmacological properties of RPR 100893, a novel non-peptide antagonist of the human NK_1 receptor. Neuropeptides 26:35

Murai M, Morimoto H, Maeda Y, Fujii T (1992) Effects of the tripeptide substance P antagonist, FR113680, on airway constriction and airway edema induced by neurokinins in guinea-pigs. Eur J Pharmacol 217:23–29

Murai M, Maeda Y, Hagiwara D, Miyake H, Ikari N, Matsuo M, Fujii T (1993) Effects of an NK_1 receptor antagonist, FK888, on constriction and plasma extravasation induced in guinea-pig airway by neurokinins and capsaicin. Eur J Pharmacol 236:7–13

Mussap CJ, Geraghty DP, Burcher E (1993) Tachykinin receptors: a radioligand binding perspective. J Neurochem 6:1987–2009

Nagahisa A, Kanai Y, Suga O, Taniguchi K, Tsuchiya M, Lowe JA 3rd, Hess HJ (1992) Antiinflammatory and analgesic activity of a non-peptide substance P receptor antagonist. Eur J Pharmacol 217:191–195

Natsugari H, Ikeura Y, Kiyota Y, Ishichi Y, Ishimaru T, Saga O, Shirafuji H, Tanaka T, Kamo I, Doi T (1995) Novel, potent, and orally active substance P antagonists: synthesis and antagonist activity of N-benzylcarboxamide derivatives of pyrido[3,4-b]pyridine. J Med Chem 38:3106–3120

Natsugari H, Ikeura Y, Kamo I, Ishimaru T, Ishichi Y, Fujishima A, Tanaka T, Kasahara F, Kawada M, Doi T (1999) Axially chiral 1,7-naphthyridine-6-carboxamide derivatives as orally active tachykinin NK_1 receptor antagonists: synthesis, antagonistic activity, and effects on bladder functions. J Med Chem 42:3982–3993

Navari RM, Reinhardt RR, Gralla RJ, Kris MG, Hesketh PJ, Khojasteh A, Kindler H, Grote TH, Pendergrass K, Grunberg SM, Carides AD, Gertz BJ (1999) Reduction of cisplatin-induced emesis by a selective neurokinin-1-receptor antagonist. New Engl J Med 340:190–195

Okano S, Nagaya H, Ikeura Y, Natsugari H, Inatomi N (2001) Effects of TAK-637, a novel neurokinin-1 receptor antagonist, on colonic function in vivo. J Pharmacol Exp Ther 298:559–564

Okano S, Ikeura Y, Inatomi N (2002) Effects of tachykinin NK_1 receptor antagonists on the viscerosensory response caused by colorectal distention in rabbits. J Pharmacol Exp Ther 300:925–931

Palma C, Goso C, Manzini S (1994) Different susceptibility to neurokinin 1 receptor antagonists of substance P and septide-induced interleukin-6 release from U373 MG human astrocytoma cell line. Neurosci Lett 171:221–224

Papp M, Vassout A, Gentsch C (2000) The NK_1-receptor antagonist NKP608 has an antidepressant-like effect in the chronic mild stress model of depression in rats. Behav Brain Res 115:19–23

Patacchini R, Maggi CA (1995) Tachykinin NK_1 receptors mediate both vasoconstrictor and vasodilator responses in the rabbit isolated jugular vein Eur J Pharmacol 283:233–240

Patacchini R, Santicioli P Astolfi M Rovero P, Viti G & Maggi CA (1992) Activity of peptide and non-peptide antagonists at peripheral NK-1 tachykinin receptors. Eur J Pharmacol 215:93–98

Patacchini R, Quartara L, Astolfi M, Goso C, Giachetti A, Maggi CA (1995) Activity of cyclic pseudopeptide antagonists at peripheral tachykinin receptors. J Pharmacol Exp Ther 272:1082–1087

Patacchini R, Bartho L, Maggi CA (1997) Characterization of receptors mediating contraction induced by tachykinins in the guinea-pig isolated common bile duct. Br J Pharmacol 122:1633–1638

Petitet F, Saffroy M, Torrens Y, Lavielle S, Chassaing G, Loeuillet D, Glowinski J & Beaujouan JC (1992) Possible existence of a new tachykinin receptor subtype in the guinea-pig ileum. Peptides 13:383–388

Phebus LA, Johnson KW, Stengel PW, Lobb KL, Nixon JA, Hipskind PA (1997) The non-peptide NK-1 receptor antagonist LY303870 inhibits neurogenic dural inflammation in guinea-pigs. Life Sci 60:1553–1561

Piedimonte G, Bertrand C, Geppetti P, Snider RM, Desai MC, Nadel JA (1993) A new NK_1 receptor antagonist (CP-99,994) prevents the increase in tracheal vascular permeability produced by hypertonic saline. J Pharmacol Exp Ther 266:270–273

Pradier L, Menager J, Le Guern J, Bock MD, Heuillet E, Fardin V, Garret C, Doble A, Mayaux JF (1994) Septide: an agonist for the NK_1 receptor acting at a site distinct from substance P. Mol Pharmacol 45:287–93

Quartara L, Maggi CA (1997) The tachykinin NK-1 receptor. Part I: ligands and mechanisms of cellular activation. Neuropeptides 31:537–553

Quartara L, Maggi CA (1998) The tachykinin NK-1 receptor. Part II: distribution and pathopysiological roles. Neuropeptides 32:1–49

Radhakrishnan V, Henry JL (1991) Novel substance P antagonist, CP-96,345, blocks responses of cat spinal dorsal horn neurons to noxious cutaneous stimulation and to substance P. Neurosci Lett 132:39–43

Regoli D, Escher E, Drapeau G, D'Orleans-Juste P, Mizrahi J (1984) Receptors for substance P III Classification by competitive antagonists. Eur J Pharmacol 97:179–189

Regoli D, Drapeau G, Dion S, D'Orleans-Juste P (1989) Receptors for substance P and related neurokinins. Pharmacology 38:1–15

Regoli D, Boudon A and Fauchere J (1994) Receptors and antagonists for substance P and related peptides. Pharmacol Rev 46:551–599

Reid K, Palmer JL, Wright RJ, Clemes SA, Troakes C, Somal HS, House F, Stott JR (2000) Comparison of the neurokinin-1 antagonist GR205171, alone and in combination with the 5-HT3 antagonist ondansetron, hyoscine and placebo in the prevention of motion-induced nausea in man. Br J Clin Pharmacol 50:61–64

Romerio SC, Linder L, Haefeli WE (1999) Neurokinin-1 receptor antagonist R116301 inhibits substance P-induced venodilation. Clin Pharmacol Ther 66:522–527

Rupniak NMJ, Kramer MS (1999) Discovery of the anti-depressant and anti-emetic efficacy of substance P receptor (NK_1) antagonists. Trends Pharmacol Sci 20:485–489

Rupniak NM, Boyce S, Williams AR, Cook G, Longmore J, Seabrook GR, Caeser M, Iversen SD, Hill RG (1993) Antinociceptive activity of NK_1 receptor antagonists: nonspecific effects of racemic RP67580. Br J Pharmacol 110:1607–1613

Rupniak NM, Webb JK, Williams AR, Carlson E, Boyce S, Hill RG (1995) Antinociceptive activity of the tachykinin NK_1 receptor antagonist, CP-99,994, in conscious gerbils. Br J Pharmacol 116:1937–1943

Rupniak NM, Carlson E, Boyce S, Webb JK, Hill RG (1996) Enantioselective inhibition of the formalin paw late phase by the NK_1 receptor antagonist L-733,060 in gerbils. Pain 67:189–195

Rupniak NM, Tattersall FD, Williams AR, Rycroft W, Carlson EJ, Cascieri MA, Sadowski S, Ber E, Hale JJ, Mills SG, MacCoss M, Seward E, Huscroft I, Owen S, Swain CJ, Hill RG, Hargreaves RJ (1997) In vitro and in vivo predictors of the anti-emetic activity of tachykinin NK_1 receptor antagonists. Eur J Pharmacol 326:201–209

Sachais BS, Krause JE (1994) Both extracellular and transmembrane residues contribute to the species selectivity of the neurokinin-1receptor antagonist WIN 51708. Mol Pharmacol 46:122–128

Sachais BS, Snider RM, Lowe JA 3rd, Krause JE (1993) Molecular basis for the species selectivity of the substance P antagonist CP-96,345. J Biol Chem 268:2319–23

Sakurada T, Manome Y, Tan-No K, Sakurada S, Kisara K, Ohba M, Terenius L A (1992) Selective and extremely potent antagonist of the neurokinin-1 receptor. Brain Res 593:319–322

Sakurada T, Katsumata K, Yogo H, Tan-No K, Sakurada S, Kisara K (1993) Antinociception induced by CP 96,345, a non-peptide NK-1 receptor antagonist, in the mouse formalin and capsaicin tests. Neurosci Lett 151:142–145

Sakurada T, Katsumata K, Yogo H, Tan-No K, Sakurada S, Ohba M, Kisara K (1995) The neurokinin-1 receptor antagonist, sendide, exhibits antinociceptive activity in the formalin test. Pain 60:175–180

Schmidt AW, McLean S, Heym J (1992) The substance P receptor antagonist CP-96,345 interacts with Ca^{2+} channels. Eur J Pharmacol 219:491–492

Schwartz TW, Gether U, Schambye HT, Hjorth SA (1995) Molecular mechanisms of action of nonpeptide ligands for peptide receptors. Curr Pharmac Design 1:325–342

Seabrook GR, Shepheard SL, Williamson DJ, Tyrer P, Rigby M, Cascieri MA, Harrison T, Hargreaves RJ, Hill RG (1996) L-733,060, a novel tachykinin NK_1 receptor antagonist; effects in [Ca2+]i mobilisation, cardiovascular and dural extravasation assays. Eur J Pharmacol 317:129–135

Shepheard SL, Williamson DJ, Hill RG, Hargreaves RJ (1993) The non-peptide neurokinin1 receptor antagonist, RP 67580, blocks neurogenic plasma extravasation in the dura mater of rats. Br J Pharmacol 108:11–12

Singh L, Field MJ, Hughes J, Kuo BS, Suman-Chauhan N, Tuladhar BR, Wright DS, Naylor RJ (1997) The tachykinin NK_1 receptor antagonist PD 154075 blocks cisplatin-induced delayed emesis in the ferret. Eur J Pharmacol 321:209–216

Smith G, Harrison S, Bowers J, Wiseman J, Birch P (1994) Non-specific effects of the tachykinin NK_1 receptor antagonist, CP-99,994, in antinociceptive tests in rat, mouse and gerbil. Eur J Pharmacol 271:481–487

Snider RM, Constantine JW, Lowe III JA, Longo KP, Lebel WS, Woody HA, Drozda SE, Desai MC, Vinick FJ, Spencer RW Hess HJ (1991) A potent nonpeptide antagonist of the SP (NK-1) receptor. Science 251:435–437

Sonea IM, Palmer MV, Akili D, Harp JA (2002) Treatment with neurokinin-1 receptor antagonist reduces severity of inflammatory bowel disease induced by Cryptosporidium parvum. Clin Diagn Lab Immunol 9:333–340

Steinberg R, Alonso R, Rouquier L, Desvignes C, Michaud JC, Cudennec A, Jung M, Simiand J, Griebel G, Emonds-Alt X, Le Fur G, Soubrie P (2002) SSR240600 [(R)-2-(1-[2-[4-[2-[3,5-bis(trifluoromethyl)phenyl]acetyl]-2-(3,4-dichlorophenyl)-2-morpholinyl]ethyl]-4-piperidinyl)-2-methylpropanamide], a centrally active nonpeptide antagonist of the tachykinin neurokinin 1 receptor: II Neurochemical and behavioral characterization. J Pharmacol Exp Ther 303:1180–1188

Stucchi AF, Shofer S, Leeman S, Materne O, Beer E, McClung J, Shebani K, Moore F, O'Brien M, Becker JM (2000) NK-1 antagonist reduces colonic inflammation and oxidative stress in dextran sulfate-induced colitis in rats. Am J Physiol 279:G1298–G1306

Suarez GA, Opfer-Gehrking TL, McLean DB (1994) Double-blind, placebo-controlled study of the efficacy of a substance P (NK_1) receptor antagonist in painful peripheral neuropathy. Neurology 44:A220

Swain CJ, Seward EM, Sabin V, Cascieri MA (1993) Quinuclidine based NK-1 antagonists 2: determination of the absolute stereochemical requirements. Biomed Chem Lett 3:1703–1706

Tamura K, Mutabagani K, Wood JD (1993) Analysis of a nonpeptide antagonist for substance P on myenteric neurons of guinea-pig small intestine. Eur J Pharmacol 232:235–239

Tattersall FD, Rycroft W, Hargreaves RJ, Hill RG (1993) The tachykinin NK_1 receptor antagonist CP-99,994 attenuates cisplatin induced emesis in the ferret. Eur J Pharmacol 250:R5–R6

Tattersall FD, Rycroft W, Marmont N, Cascieri M, Hill RG, Hargreaves RJ (1995) Enantiospecific inhibition of emesis induced by nicotine in the house musk shrew (Suncus murinus) by the neurokinin 1 (NK_1) receptor antagonist CP-99,994. Neuropharmacology 34:1697–1699

Torrens Y, Beaujouan JC, Saffroy M, Glowinski J (1995) Involvement of septide-sensitive tachykinin receptors in inositol phospholipid hydrolysis in the rat urinary bladder. Peptides 16:587–594

Towler PK, Brain SD (1998) Activity of tachykinin NK_1 and bradykinin B_2 receptor antagonists, and an opioid ligand at different stimulation parameters in neurogenic inflammation in the rat. Neurosci Lett 257:5–8

Tsuchiya M, Fujiwara Y, Kanai Y, Mizutani M, Shimada K, Suga O, Ueda S, Watson JW, Nagahisa A (2002) Anti-emetic activity of the novel nonpeptide tachykinin NK_1 receptor antagonist ezlopitant (CJ-11,974) against acute and delayed cisplatin-induced emesis in the ferret. Pharmacology 66:144–152

Urban LA, Fox AJ NK_1 receptor antagonists—are they really without effect in the pain clinic? (2000) Trends Pharmacol Sci 21:462–464

Van Belle S, Lichinitser MR, Navari RM, Garin AM, Decramer ML, Riviere A, Thant M, Brestan E, Bui B, Eldridge K, De Smet M, Michiels N, Reinhardt RR, Carides AD, Evans JK, Gertz BJ (2002) Prevention of cisplatin-induced acute and delayed emesis by the selective neurokinin-1 antagonists, L-758,298 and MK-869. Cancer 94:3032–3041

Vassout A, Schaub M, Gentsch C, Ofner S, Schilling W, Veenstra S (1994) CGRP 49,823, a novel NK_1 receptor antagonist: behavioural effects. Neuropeptides 26:37

Vassout A, Veenstra S, Hauser K, Ofner S, Brugger F, Schilling W, Gentsch C (2000) NKP608: a selective NK-1 receptor antagonist with anxiolytic-like effects in the social interaction and social exploration test in rats. Regul Pept 96:7–16

Venepalli BR, Aimone LD, Appell KC, Bell MR, Dority JA, Goswami R, Hall PL, Kumar V, Lawrence KB, Logan ME (1992) Synthesis and substance P receptor binding activity of androstano[3,2-b]pyrimido[1,2-a]benzimidazoles. J Med Chem 35:374–378

Venkova K, Sutkowski-Markmann DM, Greenwood-Van Meerveld B (2002) Peripheral activity of a new NK_1 receptor antagonist TAK-637 in the gastrointestinal tract. J Pharmacol Exp Ther 300:1046–1052

Walpole CS, Brown MC, James IF, Campbell EA, McIntyre P, Docherty R, Ko S, Hedley L, Ewan S, Buchheit KH, Urban LA (1998a) Comparative, general pharmacology of SDZ NKT 343, a novel, selective NK_1 receptor antagonist. Br J Pharmacol 124:83–92

Walpole C, Ko SY, Brown M, Beattie D, Campbell E, Dickenson F, Ewan S, Hughes GA, Lemaire M, Lerpiniere J, Patel S, Urban L (1998b) 2-Nitrophenylcarbamoyl-(S)-prolyl-(S)-3-(2-naphthyl)alanyl-N-benzyl-N-methylamide (SDZ NKT 343), a potent human NK_1 tachykinin receptor antagonist with good oral analgesic activity in chronic pain models. J Med Chem 41:3159–3173

Wang ZY, Tung SR, Strichartz GR, Hakanson R (1994a) Nonspecific actions of the nonpeptide tachykinin receptor antagonists, CP 96,345, RP 67580 and SR 48968, on neurotransmission. Br J Pharmacol 111:179–184

Wang ZY, Tung SR, Strichartz GR, Hakanson R (1994b) Investigation of the specificity of FK 888 as a tachykinin NK_1 receptor antagonist. Br J Pharmacol 111:1342–1346

Ward P, Armour DR, Bays DE, Evans B, Giblin GM, Heron N, Hubbard T, Liang K, Middlemiss D, Mordaunt J, Pegg NA, Vinader MV, Watson SP, Heron N, Liang K, Bountra C, Evans DC (1995) Discovery of an orally bioavailable NK_1 receptor antagonist, (2S,3S)-(2-methoxy-5-tetrazol-1-ylbenzyl)(2-phenylpiperidin-3-yl)amine (GR203040), with potent antiemetic activity. J Med Chem 38:4985–4992

Watson JW, Gonsalves SF, Fossa AA, McLean S, Seeger T, Obach S, Andrews PL (1995) The anti-emetic effects of CP-99,994 in the ferret and the dog: role of the NK_1 receptor. Br J Pharmacol 115:84–94

Wijkhuisen A, Sagot MA, Frobert Y, Creminon C, Grassi J, Boquet D, Couraud JY (1999) Identification in the NK_1 tachykinin receptor of a domain involved in recognition of neurokinin A and septide but not of substance P. FEBS Lett 447:155–159

Williams, BJ, NR Curtis, AT McKnight, J Maguire, A Foster and R Tridgett, (1988) Development of NK_2 selective antagonists. Reg Peptides 22:189

Yashpal K, Radhakrishnan V, Coderre TJ, Henry JL (1993) CP-96,345, but not its stereo-isomer, CP-96,344, blocks the nociceptive responses to intrathecally administered substance P and to noxious thermal and chemical stimuli in the rat. Neuroscience 52:1039–1047

Zeng XP, Burcher E (1994) Use of selective antagonists for further characterization of tachykinin NK-2, NK-1 and possible 'septide-selective' receptors in guinea-pig bronchus. J Pharmacol Exp Ther 270:1295–1300

Tachykinin NK$_2$ Receptor Antagonists

X. Emonds-Alt

Sanofi-Synthelabo Recherche, 371 Rue du Professeur J. Blayac,
34184 Montpellier, Cedex 04, France
e-mail: xavier.emonds-alt@sanofi-synthelabo.com

1	Introduction	220
2	Peptide-Based Tachykinin NK$_2$ Receptor Antagonists	220
2.1	Linear Peptides	220
2.2	Cyclic Peptides	222
2.2.1	Nepadutant (MEN11420)	223
2.2.2	Cyclic Pseudopeptides	226
2.3	Other Peptides	227
3	Nonpeptide Tachykinin NK$_2$ Receptor Antagonists	227
3.1	Saredutant (SR48968)	227
3.2	ZM253270	231
3.3	GR159897	232
3.4	SR144190	232
3.5	Other Compounds	233
4	Conclusion	234
	References	235

Abstract The tachykinin NK$_2$ receptor is widely distributed in the peripheral and central nervous system. It mediates numerous pharmacological effects of neurokinin A, suggesting that the tachykinin NK$_2$ receptor may be possible target for new therapeutics. Many efforts to identify and characterize selective antagonists of this receptor have been undertaken and several antagonists of both peptide and nonpeptide nature have been disclosed. Among these compounds, nepadutant, a peptide-based antagonist, and saredutant, a nonpeptide antagonist, are certainly among the most potent compounds. These two antagonists combine high affinity for the human tachykinin NK$_2$ receptor with strong antagonist activities in a large variety of in vitro and in vivo pharmacological models. Contrary to nepadutant, saredutant has been shown to be orally active. In addition to their potent activities in the airways they show potent inhibitory activities in visceral nociception suggesting that a tachykinin NK$_2$ receptor antagonist may be useful for the treatment of functional gastrointestinal disorders such as irritable bowel syndrome. With the use of saredutant, major progress has also been made in the understanding of the role of brain tachykinin NK$_2$ receptors.

Contrary to nepadutant, saredutant is centrally active and shows antidepressant/anxiolytic activity.

Keywords Neurokinin A · Tachykinin NK_2 receptor · Antagonist · SR48968 · MEN11420 · Saredutant · Nepadutant

1
Introduction

The tachykinin NK_2 receptor which preferentially binds neurokinin A is widely distributed in the periphery, especially in the airways, gastrointestinal and urinary tracts (Regoli et al. 1994; Maggi 1995). The tachykinin NK_2 receptor is also present in the rat and human brain (Hagan et al. 1993; Steinberg et al. 1998a; Bensaid et al. 2001; Saffroy et al. 2001) and in the rat spinal cord (Yashpal et al. 1990) but in much lower quantity than in the periphery. The large variety of effects produced by neurokinin A (Regoli et al. 1994; Advenier 1995; Maggi 1995; Holzer and Holzer-Petsche 1997a, 1997b; Patacchini and Maggi 2001) has stimulated interest for the tachykinin NK_2 receptor as a possible target for new therapeutics and, for about 15 years, many efforts to identify and characterize tachykinin NK_2 receptor antagonists have been undertaken. Peptide-derived antagonists have been designed from short ligands for tachykinin receptors combining amino acid substitutions in these templates to increase potency/selectivity, deletion–optimization strategies to reduce the molecular weight, introduction of structural constraints to favor an active conformation and modifications of the peptide backbone to increase metabolic stability and solubility. However, the usefulness of peptide-based antagonists as drugs is likely to be limited by low oral bioavailability, brain penetration and metabolic stability. These problems might be overcome by nonpeptide antagonists and many efforts to identify such antagonists have been undertaken by screening of chemical libraries and further optimization of lead compounds.

2
Peptide-Based Tachykinin NK_2 Receptor Antagonists

2.1
Linear Peptides

A first class of tachykinin NK_2 receptor antagonists was characterized by a linear peptide structure. MEN10207 (H-Asp-Tyr-(DTrp)-Val-(DTrp)-(DTrp)-Arg-NH$_2$) (Rovero et al. 1990), MEN10376 (H-Asp-Tyr-(DTrp)-Val-(DTrp)-(DTrp)-Lys-NH$_2$) (Maggi et al. 1991; Quartara et al. 1995) and MEN10456 (H-Asp-Tyr-(DTrp)-Val-(DTrp)-(DTrp)-Lys-OH) (Patacchini et al. 1993; Quartara et al. 1995) were obtained from the C-terminal heptapeptide of neurokinin A by classical amino acid substitutions. They are characterized by three DTrp residues which were shown to be crucial for both affinity and selectivity (Rovero et al. 1990;

Quartara et al.1995). Among these compounds, MEN10376 antagonized tachykinin NK_2 receptor-mediated contractions of different human (pA_2=6.93–7.34), rat (pA_2=6.20–6.67), guinea pig (pA_2=6.77–7.41), rabbit (pA_2=7.87–8.19) and hamster (pA_2=5.64–6.07) isolated tissues (Maggi et al. 1993a) (pA_2 or apparent affinity as defined by Arunlakshana and Schild 1959). In vivo, intravenous (i.v.) administration of MEN10376 (1–3 µmol/kg) inhibited urinary bladder contraction in rats and bronchoconstriction in guinea pigs provoked by i.v. treatment with a specific tachykinin NK_2 receptor agonist. It did not show any inhibitory effect on different in vivo pharmacological responses provoked by a specific tachykinin NK_1 receptor agonist (Maggi et al. 1991).

Short ligands for tachykinin NK_1 receptors have also been used as templates to obtain selective antagonists of the tachykinin NK_2 receptor. Amino acid substitutions in a weakly active and nonselective tachykinin receptor antagonist previously obtained by modification of the C-terminal heptapeptide of substance P provided a first selective tachykinin NK_2 receptor antagonist, GR83074 (tert-butyloxycarbonyl-Arg-Ala-DTrp-Phe-DPro-Pro-Nle-NH_2), with a pK_B of 8.20 on rat receptors (pK_B or apparent affinity according to Kenakin 1993). This compound (0.3 µmol/kg i.v.) inhibited bronchoconstriction provoked by a specific tachykinin NK_2 receptor agonist in guinea pigs but its inhibitory effect seemed to be short-lived (McElroy et al. 1992). Further optimization by increasing lipophilicity of the N terminus of GR83074 yielded a ten times more potent compound, GR94800 [Benzoyl-Ala-Ala-(DTrp)-Phe-(DPro)-Pro-Nle-NH_2] (pK_B= 9.60 on rat receptors) (McElroy et al. 1992). This last compound was then modified by a deletion–optimization strategy to obtain the tetrapeptide GR100679 [Benzoyl-Gly-Ala-(DTrp)-Phe-N(CH_3)$_2$] which was shown in vitro to be as potent as GR94800 (pK_B=9.05 on rat receptors) (Smith et al. 1993). Further deletion and introduction of an imidazole group at the N terminus led to a low molecular weight derivative, GR112000 [Phenyl-imidazole-CO-Ala-(DTrp)-N(CH_3)-(CH_2)$_2$-CH(CH_3)$_2$], with an about ten times lower in vitro potency (pK_B=7.64 on rat receptors). In vivo, inhibition of bronchoconstriction provoked by a specific tachykinin NK_2 receptor agonist in guinea pigs was obtained after administration of GR112000 (5 µmol/kg i.v.) and GR100679 (0.1 µmol/kg i.v.) with a potency correlating with their respective in vitro affinity for the tachykinin NK_2 receptor. However, the reduced peptidic character of GR112000 allowed a much longer biological half-life to be obtained than with GR100679 (Smith et al. 1993).

Another approach has been to use as template a known linear peptide with weak antagonist activity at tachykinin NK_2 receptors, L659874 (Ac-Leu-Met-Gln-Trp-Phe-Gly-NH_2) (McKnight et al. 1988; Williams et al. 1988). A selective and potent antagonist of the tachykinin NK_2 receptor, the hexapeptide R396 (Ac-Leu-Asp-Gln-Trp-Phe-Gly-NH_2), was obtained directly by a crucial amino acid substitution (replacement of Met by Asp) in L659874 (Dion et al. 1990; Maggi et al. 1992; Rovero et al. 1992; Regoli et al. 1994). R396 antagonized tachykinin NK_2 receptor-mediated contractions of different human (pA_2=5.41–6.10), rat (pA_2=6.20–6.53), guinea pig (pA_2=5.26–6.43), rabbit (pA_2=5.42–5.56) and hamster (pA_2=7.46–7.63) isolated tissues (Maggi et al. 1993a). Two other selective tachykinin NK_2 receptor antagonists with reduced peptidic character have been obtained using L659874 as starting template. Firstly, PD147714 ([2,3di-CH_3O-Phenyl]-CH2-O-CO-[(S)-Trp]-[(S)-α-MePHe]-Gly-NH_2) was produced after identification of two amino acids which mainly contribute to the tachykinin NK_2 receptor binding affinity of L659874. The resulting dipeptide with micromolar affinity was then conformationally restricted and optimized by N- and C- terminus modifications. PD147714 showed nanomolar affinity for the hamster tachykinin NK_2 receptor in radioligand binding experiments and in vitro antagonist activity at rat tachykinin NK_2 receptors (pK_B=7.9) (Boyle et al. 1994). Secondly, a classical deletion of amino acids in L659874 followed by N- and C-terminus modifications of the resulting short peptide yielded TAC 363 (α-[tert-butylcarbamoyl]-Gln-Trp-α-aza-Phe-[2-benzyloxyethylamide]) (Higashide et al. 1996). This compound (0.1–1 mg/kg i.v.) was shown to be a very potent and selective tachykinin NK_2 receptor antagonist with high efficacy to block capsaicin-, citric acid- and neurokinin A-induced bronchoconstriction as well as capsaicin- and antigen-induced airway hyperresponsiveness to acetylcholine in guinea pigs (Higashide et al. 1996, 1997).

2.2
Cyclic Peptides

A second class of tachykinin NK_2 receptor antagonists was characterized by a cyclic peptide structure. The first example of cyclic peptides with relatively high potency and selectivity at tachykinin NK_2 receptors has been the cyclic hexapeptide L659877 (cyclo[Leu-Met-Gln-Trp-Phe-Gly]). L659877 was obtained by removing the lactam constraint in another less active compound, L659837 (cyclo[Gln-Trp-Phe-{(R)-Gly(ANC-2)Leu}-Met]). This last compound is a cyclic derivative of a poor ligand for tachykinin NK_2 receptors, which was obtained by introduction of a lactam conformational constraint in a modified C-terminal hexapeptide of substance P (Williams et al. 1988, 1993; McKnight et al. 1988, 1991). In tachykinin receptor radioligand binding experiments, L659877 showed nanomolar affinity for the hamster tachykinin NK_2 receptor but a ten times lower affinity for the rat receptor. It had almost no affinity for tachykinin NK_1 and NK_3 receptors (Maggi et al. 1994). In vitro, L659877 antagonized tachykinin NK_2 receptor-mediated contractions of different human (pA_2=6.64–7.0), rat (pA_2=7.75–7.90), guinea pig (pA_2=6.29–7.15), rabbit (pA_2=6.72–6.97) and hamster (pA_2=7.59–8.24) isolated tissues and had negligible activity in functional bioassays for tachykinin NK_1 and NK_3 receptors (Maggi et al. 1993a, 1993b, 1994). In vivo, L659877 (0.03–1 μmol/kg i.v.) was shown to inhibit urinary bladder contraction provoked by i.v. treatment with a specific tachykinin NK_2 receptor agonist in rats but its inhibitory effect was short-lived (Maggi et al. 1993b).

2.2.1
Nepadutant (MEN11420)

Increase of conformational constraint and rigidity in the cyclic hexapeptide L659877 by introduction of a second ring has yielded a much more potent peptide, MEN10627 (cyclo[(Met-Asp-Trp-Phe-Dap-Leu)cyclo(2β-5β)]; Fig. 1A). Different conformational studies of L659877 showed that the cyclic peptide retained a sufficient degree of conformational freedom to adopt various conformations not necessarily optimal for high affinity interaction with the tachykinin NK_2 receptor (Wollborn et al. 1993; Quartara et al. 1995). Addition of a further constraint in the cyclic structure of L659877 was then suspected as a good approach to favor a better conformation for potent and selective interactions with tachykinin NK_2 receptors. After substitution of the Gly and Gln residues of L659877 by a L-2,3-diaminopropionic acid (Dap) and an Asp residue, respectively, the side chains of these two amino acids were linked by a lactam bridge (Maggi et al. 1994; Pavone et al. 1995a, 1995b; Quartara et al. 1995, 1996). The resultant bicyclic hexapeptide MEN10627 was certainly the first selective peptide-based antagonist of the tachykinin NK_2 receptor, combining high efficacy both in vitro and in vivo. In tachykinin receptor radioligand binding experiments it displayed subnanomolar and nanomolar affinity for the hamster and rat tachykinin NK_2 receptor, respectively. It had only micromolar affinity for the guinea pig NK_3 receptor and no affinity for the rat NK_1 receptor (Maggi et al. 1994; Quartara et al.

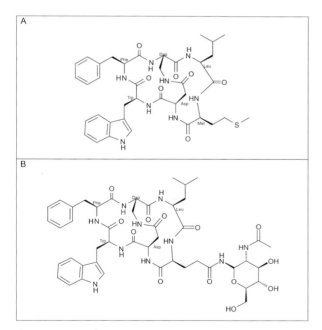

Fig. 1 Structure of the bicyclic peptide NK_2 receptor antagonists MEN10627 (**A**) and nepadutant (**B**)

1996). Moreover, it showed subnanomolar affinity for the human tachykin NK_2 receptor (Quartara et al. 1996; Catalioto et al. 1998). In vitro, MEN10627 potently antagonized tachykinin NK_2 receptor-mediated contractions of human (pK_B=8.4–8.8), rat (pK_B=8.8), guinea pig (pK_B=8.2–8.6), rabbit (pK_B=8.1), hamster (pK_B=9.7–10.1) and mouse (pK_B=10.3) isolated tissues (Maggi et al. 1994; Giuliani et al. 1996; Patacchini et al. 1997b; Catalioto et al. 1998; Tramontana et al. 1998a, 1998b). On the other hand, its activity in functional bioassays for tachykinin NK_1 and NK_3 receptors remained very low and its specificity was shown by its inactivity at 1 µM on contractile responses of different isolated tissues to various nontachykinin spasmogens (Maggi et al. 1994; Catalioto et al. 1998). In vivo, contrary to L659877, MEN10627 (10–100 nmol/kg i.v.) produced a long-lasting inhibition of contraction of the rat urinary bladder and duodenum provoked by a specific tachykinin NK_2 receptor agonist. Inhibition of tachykinin NK_2 receptor-mediated urinary bladder contraction in rats was also achieved after intranasal and intraduodenal (i.d.) administration (Maggi et al. 1994). Likewise, tachykinin NK_2 receptor-mediated bronchoconstriction was inhibited by MEN10627 (30–1000 nmol/kg i.v.) with a long duration of action in guinea pigs (Tramontana et al. 1998a). In addition, MEN10627 (30–100 nmol/kg i.v.) reduced antigen-induced bronchoconstriction and platelet-activating factor (PAF)-induced airway hyperresponsiveness to histamine in guinea pigs (Perretti et al. 1995).

Unfortunately, despite its remarkable activity profile, MEN10627 was characterized by a very low hydrophilicity and this property notably reduced its interest as drug candidate. This problem has been overcome by the replacement of the Met residue by a (2-cetylamino-2-deoxy-β-D-glucopyranosyl)-L-asparaginyl residue in MEN10627. The resultant bicyclic peptide MEN11420 or nepadutant (cyclo[(Asn{β-D-(2-acetylamino-2-deoxy-glucopyranosyl)}-Asp-Trp-Phe-Dap-Leu) cyclo(2β-5β)]; Fig. 1B) showed an 80-fold increase in water solubility as compared to the parent compound MEN10627 while maintaining very high and selective antagonist activity at tachykinin NK_2 receptors (Catalioto et al. 1998). Furthermore, nepadutant displayed improved metabolic and pharmacokinetic characteristics. Compared to MEN10627, it was shown to be more resistant to hydrolytic and oxidative metabolism, at least after incubation with rat tissue homogenates and liver microsomes (Catalioto et al. 1998). After i.v. administration, nepadutant showed a longer plasma half-life and a greater area under the plasma concentration–time curve (AUC) than MEN10627 in rats (Lippi et al. 1998) and guinea pigs (Tramontana et al. 1998a). However, the absolute bioavailability of nepadutant in rats remained limited after intrarectal (i.r.) administration and very low, if any, after sublingual or oral (p.o.) administration even if it was complete after intranasal or intraperitoneal (i.p.) administration (Lippi et al. 1998). Nevertheless, its bioavailability after i.d. or p.o. administration was shown to increase after intestinal inflammation (Lecci et al. 2001b).

In pharmacological studies, nepadutant has been shown to be a potent and selective antagonist of the tachykinin NK_2 receptor with improved in vivo pharmacological activities as compared to MEN10627. Nepadutant displayed nanomolar affinity for the human tachykinin NK_2 receptor with negligible affinities for tachykinin NK_1 and NK_3 receptors. At 1 μM, it was devoid of affinity for other neurotransmitter and neuropeptide receptors as well as for ion channels (Catalioto et al. 1998; Giuliani et al. 2001). A tritium-labeled version of nepadutant bound to the human NK_2 receptor with a dissociation constant (K_d) of 2.1 nM (Renzetti et al. 1998). In in vitro functional studies, nepadutant potently antagonized tachykinin NK_2 receptor-mediated contractions of different human (pK_B=8.1–8.5), rat (pK_B=9–9.3), guinea pig (pK_B=8.1–8.4), rabbit (pK_B=8.6), hamster (pK_B=9.3–10.2) and mouse (pK_B=9.8) isolated tissues (Santicioli et al. 1997; Catalioto et al. 1998; Tramontana et al. 1998a, 1998b; Patacchini et al. 1997a, 1998b, 2000, 2001). It inhibited the tachykinin NK_2 receptor-mediated component involved in the contractile response of different isolated tissues to endogenous tachykinins released upon application of electrical field stimulation (Santicioli et al. 1997; Patacchini et al. 1998a, 1998b). The selectivity of nepadutant was demonstrated by its very low activity in functional bioassays for tachykinin NK_1 and NK_3 receptors (pK_B<6) (Santicioli et al. 1997; Catalioto et al. 1998).

In in vivo pharmacological studies, nepadutant inhibited various pharmacological effects induced by i.v. administration of neurokinin A or a specific tachykinin NK_2 receptor agonist with a long duration of action: bronchoconstriction in guinea pigs (30–100 nmol/kg i.v.) (Tramontana et al. 1998a), urinary bladder contractions in rats (1–10 nmol/kg, 3–10 nmol/kg, 30–100 nmol/kg and 100–300 nmol/kg after i.v., intranasal, i.r. and i.d. administration, respectively) (Catalioto et al. 1998) and hamster (3–10 nmol/kg i.v.) (Tramontana et al. 1998b) as well as colonic contractions in rats (10–1000 nmol/kg i.v.) (Lecci et al. 1997b; Carini et al. 2001), guinea pigs (10–100 nmol/kg i.v.) (Santicioli et al. 1997) and dogs (100 nmol/kg i.v.) (Giuliani et al. 2001). Furthermore, nepadutant was always more potent and had a longer duration of action than MEN10627 (Lecci et al. 1997b; Catalioto et al. 1998; Tramontana et al. 1998a, 1998b). On the other hand, nepadutant never displayed any activity on the pharmacological effects provoked by i.v. treatment with a specific tachykinin NK_1 receptor agonist (Lecci et al. 1997a, 1997b; Santicioli et al. 1997; Catalioto et al. 1998).

The potent in vivo activity of nepadutant has been further characterized in pharmacological models where endogenous neurokinin A acting on NK_2 receptors plays a crucial role. Nepadutant inhibited the tachykinin NK_2 receptor-mediated component of the bronchial responses provoked by capsaicin (3–30 nmol/kg i.v.), vagal electrical stimulation (10-100 nmol/kg i.v.) or sephadex (0.1-1 μmol/kg i.v.) in guinea pigs (Tramontana et al. 1998a, 2002). In the rat urinary system, nepadutant (100 nmol/kg i.v.) reduced the bladder motor responses and the urodynamic modifications provoked by intravesical administration of capsaicin. It decreased the bladder tone at high degree of bladder distension but did not modify the micturition reflex evoked by bladder distension (Lecci et al. 1997a, 1998a, 2001a). At the same dose, it reduced detrusor hyperreflexia induced by intravesical treatment with bacterial toxin (Lecci et al. 1998b) or spinal cord injury (Abdel-Gawad et al. 2001).

In the gastrointestinal system, the enhanced colonic contractions associated with colon irritation were reduced by nepadutant (100 nmol/kg, i.v. or i.r.) in rats (Carini et al. 2001; Lecci et al. 2001b). The compound (10–300 nmol/kg i.v.) also produced a long-lasting inhibition of noncholinergic colonic contractions provoked by distension in guinea pigs (Santicioli et al. 1997). Subcutaneous (s.c.) administration of nepadutant (0.3–10 mg/kg) exerted an anti-inflammatory activity in the acute rectocolitis induced by acetic acid in guinea pigs (Cutrufo et al. 2000). Furthermore, nepadutant showed antinociceptive activity in the rat gut. The gastric sensitivity to distension (Toulouse et al. 2001) and the gastric hypersensitivity after exposure to a food contaminant (Anton et al. 2001) was reduced by nepadutant (100 µg/kg i.v.). Furthermore, it blocked basal sensitivity (5–100 µg/kg i.v.) and the hypersensitivity to rectal distension after gut inflammation (5–100 µg/kg i.v.) and stress (100–200 µg/kg i.v.) (Toulouse et al. 2000). Moreover, nepadutant (ID_{50}=402 µg/kg i.v.) inhibited the enhanced responses of nociceptive spinal neurons to distension of the inflamed colon but had no effect on neuronal responses to distension of the normal colon (Laird et al. 2001). In addition, nepadutant (200 nmol/kg i.v.) reduced the enhanced c-*fos* proto-oncogene expression in spinal cord neurons and c-*jun* proto-oncogene expression in dorsal root ganglion neurons after inflammation of the colon but had no effect on proto-oncogene expression after a non-noxious colorectal distension (Birder et al. 2003).

2.2.2
Cyclic Pseudopeptides

Among the tachykinin NK_2 receptor antagonists with a cyclic peptide structure, there is a group of derivatives which conserve the cyclic peptide structure but contain pseudopeptide bounds. Replacement of the Leu–Met peptide segment in the cyclic peptide L659877 with the Leuψ(CH_2NH)Leu reduced dipeptide moiety and substitution of the secondary amine of the reduced amide bound with a methyl group gave MDL29913 (cyclo[Leuψ[CH_2NCH_3]Leu-Gln-Trp-Phe-Gly]) (Harbeson and Rovero 1994). Similarly, the reduction of the amide bound in the Leu–Met peptide segment in L659877 and replacement of the Met residue with another residue with higher lipophilicity led to MEN10573 (cyclo[Leuψ[CH2NH]Asp(*O*-benzyl)-Gln-Trp-Phe-βAla]) (Maggi et al. 1993b; Giuliani et al. 1996; Patacchini et al. 1997), MEN10612 (cyclo[Leuψ[CH_2NH](β-cyclohexylalanine)-Gln-Trp-Phe-βAla]) (Maggi et al. 1993b; Quartara et al. 1995; Giuliani et al. 1996; Patacchini et al. 1997b) and MEN10677 (cyclo[Leuψ[CH2NH]Asp(NH-benzyl)-Gln-Trp-Phe-βAla]) (Quartara et al. 1994, 1995; Giuliani et al. 1996; Patacchini et al. 1997b). All these cyclic pseudopeptides showed in vitro antagonist activity at human, rat, rabbit and hamster tachykinin NK_2 receptors but, as observed with the other peptide-based antagonists, their potency remained markedly species-dependent with pA_2 or pK_B values ranging from 7.31 to 9.26 (Maggi et al. 1993b; Quartara et al. 1995; Giuliani et al. 1996; Patacchini et al. 1997b). In vivo, MEN10612 (0.03–0.1 µmol/kg i.v.) and, to a lesser extent,

MDL29913 (0.03–0.1 μmol/kg i.v.) produced a more long-lasting blockade of rat urinary bladder contractions provoked by i.v. application of a specific tachykinin NK_2 receptor agonist than the cyclic hexapeptide L659877 (0.03–1 μmol/kg i.v.) (Maggi et al. 1993b). The duration of action of MEN10612 was comparable to that reported for the bicyclic hexapeptide MEN10627 (Maggi et al. 1994). Finally, another family of cyclic pseudopeptides with high affinity for the human tachykinin NK_2 receptor has been derived from the bicyclic hexapeptide MEN10627 and its analogue MEN11420. These compounds have a reduced molecular weight and retained in vitro functional activity at tachykinin NK_2 receptors (Giannotti et al. 2000; Giolitti et al. 2002; Harmat et al. 2002).

2.3
Other Peptides

A family of natural peptides were isolated from the fermentation of a taxonomically unidentified fungus and were shown to be selective antagonists of the human tachykinin NK_2 receptor in radioligand binding and cellular functional assays (Chu et al. 2000; Hedge et al. 2001). Two bicyclic peptides closely related to MEN10627 and MEN11420, Neuronorm (cyclo[(Cys{β-D-galactopyranosyl}-Asp-Trp-Phe-Dap-Leu)cyclo(2β-5β)]) (Lombardi et al. 1998a) and a derivative (Lombardi et al. 1998b) were also described as antagonists of the rat and guinea pig tachykinin NK_2 receptors in in vitro bioassays.

3
Nonpeptide Tachykinin NK_2 Receptor Antagonists

3.1
Saredutant (SR48968)

In 1992 the first nonpeptide antagonist of the tachykinin NK_2 receptor, SR48968 or saredutant ([S]-N-methyl-N[4-(4-acetylamino-4-phenylpiperidino)-2-(3,4-dichlorophenyl)butyl]benzamide; Fig. 2A) was described (Emonds-Alt et al. 1992, 1993d). This compound was obtained by optimizing a lead structure discovered by random screening of diverse chemical libraries involving several thousand compounds.

Saredutant has been shown to be a selective, potent and orally active tachykinin NK_2 receptor antagonist, which efficiently penetrates into the brain. In radioligand binding assays, saredutant showed subnanomolar affinity for the human, rat and guinea pig tachykinin NK_2 receptors. However, it had a lower affinity for the hamster tachykinin NK_2 receptor with an inhibition constant (K_i) of 2.3 nM (Emonds-Alt et al. 1992, 1993a, 1995a). A tritium-labeled version of saredutant bound with high affinity to human, rat and guinea pig tachykinin NK_2 receptors with dissociation constants (K_d) in the range 0.4–2.9 nM, depending on the species (Emonds-Alt et al. 1993c; Bhogal et al. 1994). In vitro, saredutant potently antagonized tachykinin NK_2 receptor-mediated contractions of differ-

ent human (pA_2=9.36–9.59), rat (pA_2=9.41–9.58), guinea pig (pA_2=10.51), rabbit (pA_2=9.78–10.34) and hamster (pA_2=7.5–8.66) isolated tissues (Advenier et al. 1992a, 1992b; Emonds-Alt et al. 1992, 1993a; Croci et al. 1995). It inhibited the tachykinin NK_2 receptor-mediated component involved in the contractile response of the isolated guinea pig main bronchus to endogenous tachykinins released upon application of electrical field stimulation with a 50% inhibitory concentration (IC_{50}) of 0.056 nM (Martin et al. 1992). The selectivity and specificity of saredutant for tachykinin NK_2 receptors was demonstrated in radioligand binding and in vitro functional studies (Advenier et al. 1992b; Emonds-Alt et al. 1992, 1993a, 1995a; Martin et al. 1993; Vilain et al. 1993).

In vivo, saredutant potently and specifically inhibited several responses provoked by the i.v. administration of a specific agonist of the tachykinin NK_2 receptor or neurokinin A: bronchoconstriction [50% inhibitory dose (ID_{50}) of 37 µg/kg i.v. and 350 µg/kg i.d.] (Emonds-Alt et al. 1992, 1993a) and increase of airway microvascular leakage (0.1–0.3 mg/kg i.v.) (Qian et al. 1993) in guinea pigs, fecal excretion in rats (ID_{50} 0.3 µg/kg s.c. and 0.4 µg/kg p.o.) (Emonds-Alt et al. 1993a; Croci et al. 1994) and urinary bladder contractions in rats (ID_{50} 31 µg/kg i.v.) (Maggi et al. 1993c, 1994; Emonds-Alt et al. 1997). In addition, in the central nervous system, the turning behavior provoked by intrastriatal injection of a specific tachykinin NK_2 receptor agonist in mice was antagonized by saredutant (ID_{50} 0.15 mg/kg i.p. and 0.19 mg/kg p.o.) (Poncelet et al. 1993). Moreover, it specifically inhibited the hyperalgesic effect of intrathecally injected neurokinin A in a nociceptive spinal reflex test after systemic (1 mg/kg) and intrathecal (6.5–65 nmol) administration in rats (Picard et al. 1993; Yashpal et al. 1996). Furthermore, these antagonist activities of saredutant were long lasting, in particular, after oral administration (Emonds-Alt et al. 1993d; Poncelet et al. 1993; Croci et al. 1994).

The potent activity of saredutant has been further demonstrated in different pharmacological models where endogenous neurokinin A acting via the tachykinin NK_2 receptor plays a crucial role.

In the guinea pig airways, saredutant inhibited coughs induced by inhaled citric acid in conscious animals (ID_{50} 0.1 mg/kg i.p.) (Advenier et al. 1993; Girard et al. 1995; Advenier and Emonds-Alt 1996). Furthermore, it (1 mg/kg i.p.) prevented airway hyperresponsiveness to acetylcholine induced either by aerosolized citric acid (Girard et al. 1996) and substance P (Boichot et al. 1996) or after an ovalbumin challenge in sensitized animals (Boichot et al. 1995). At 1 mg/kg i.p., saredutant reduced the enhanced in vitro reactivity of alveolar macrophages from animals exposed to aerosolized substance P (Boichot et al.

Fig. 2 Structure of nonpeptide antagonists of the tachykinin NK_2 receptor: (**A**) saredutant, (**B**) ZM253270, (**C**) GR159897, (**D**) SR144190, (**E**) SB414240, (**F**) YM38336, (**G**) NK5807, (**H**) ZD7944 and (**I**) UK224671

1998). In addition, saredutant specifically inhibited the tachykinin NK_2 receptor-mediated component of the bronchoconstriction provoked by several bronchoconstrictor agents such as capsaicin, resiniferatoxin, citric acid, sodium metabisulfite and ovalbumin challenge (Advenier 1995).

In the rat gut, saredutant prevented castor oil-induced diarrhea (0.02 to 20 µg/kg s.c. and p.o.) (Croci et al. 1997) and interleukin 1β-induced colonic hypersecretion (5 mg/kg i.p.) (Eutamène et al. 1995). Saredutant (1–250 µg/kg s.c.) reduced the stimulation of fecal excretion evoked by idazoxan and salmonella endotoxin (lipopolysaccharide) but not by serotonin, carbachol, or PAF (Croci et al. 1994). The trinitro-benzenesulfonic acid (TNBS)-induced gut inflammation was effectively attenuated by saredutant given at the dose of 5 mg/kg i.p. for 4 days (Mazelin et al. 1998). Furthermore, the viscerosensitive responses provoked by rectal distension (Julia et al. 1994) and acute inflammation induced by acetic acid (Julia and Buéno 1997) were reduced by saredutant after i.p. and intracerebroventricular (i.c.v.) injection (5–10 mg/kg i.p. and 0.4–0.8 mg/kg i.c.v.), respectively. The visceral allodynia following intestinal inflammation and the visceral hyperalgesia induced by restraint stress was effectively blocked by the antagonist (1 mg/kg i.p.) while colonic transit remained unaltered (Gaultier et al. 2002). In addition, the visceral hyperalgesia following intestinal nematode (*Nippostrongylus brasilensis*) infection was likewise inhibited by saredutant (ID_{50} 0.1 mg/kg i.p.) (McLean et al. 1997).

Saredutant has been shown to be a very potent antagonist at brain tachykinin NK_2 receptors, and the results obtained suggest that saredutant is endowed with antidepressant/anxiolytic-like activity. The central activity of saredutant has been evaluated by combining behavioral, neurochemical and electrophysiological approaches. Saredutant reduced the immobility time in the forced swimming tests in mice (2–8 mg/kg i.p.) and rats (0.3–3 mg/kg p.o) (Steinberg et al. 2001). Maternal separation-induced vocalizations in guinea pig pups were also inhibited by saredutant (0.3–10 mg/kg i.p.). In addition, it (3 mg/kg i.p.) reduced the tachykinin NK_1 receptor internalization in the amygdala due to maternal separation (Steinberg et al. 2001). In all these different behavioral models, the magnitude of the effect of saredutant was comparable to that of the antidepressant drug fluoxetine. As observed with antidepressants such as fluoxetine, daily injection of saredutant (1 mg/kg i.p.) for 3 weeks, but not acute treatment, was shown to increase expression of mRNA for the cAMP response element binding protein (Steinberg et al. 2001) and the peripheral benzodiazepine receptor-associated protein in the rat hippocampus (Chardenot et al. 2002). Moreover, saredutant displayed a potent inhibitory action on anxiety-like behaviors following aversive conditions or upon forced and unavoidable contact with a threatening stimulus. Its activity was observed in the rat elevated plus-maze test (3 mg/kg i.p.), the mouse defense test battery (0.03–1 mg/kg i.p.), the staircase test following cat exposure in rats (3 mg/kg i.p.) and the free-exploration test following cat exposure in mice (0.1–0.3 mg/kg i.p., twice a day for 5 days) (Griebel et al. 2001a, 2001b). Its anxiolytic-like action was further evidenced in the mouse light–dark box test (0.0005–50 µg/kg s.c.) (Stratton et al. 1993), the marmoset human intruder test (0.2–50 μg/kg s.c.) (Walsh et al. 1995) as well as the mouse elevated plus-maze test (1–100 pmol i.c.v.) (Teixeira et al.1996).

The activity of saredutant in several regions of the central nervous system has been evaluated in in vivo microdialysis studies in rats. In the striatum, saredutant (1–3 mg/kg i.p.) reduced the acetylcholine release evoked by neurokinin A (Steinberg et al. 1995) and by stimulation of dopamine D_1 receptors by [+]-SKF38393 (a dopamine D_1 agonist) (Steinberg et al. 1998b). In the septohippocampal cholinergic system, saredutant (1 mg/kg i.p.) prevented the increase of acetylcholine release in the hippocampus induced by intraseptal application of neurokinin A (Steinberg et al. 1998a) or i.c.v. injection of corticotropin releasing factor (CRF) (Desvignes et al. 2003). In addition, it (1 mg/kg i.p.) blocked the increase of acetylcholine release in the hippocampus evoked by sensory stimulation in conscious animals (Steinberg et al. 1998a). Moreover, in the prefrontal cortex, saredutant (0.3–1 mg/kg i.p.) inhibited the increase of norepinephrine release both in conscious animals exposed to tail pinch and in anesthetized animals injected i.c.v. with CRF (Steinberg et al. 2001). Similarly, it (1 mg/kg i.p.) blocked the cortical release of norepinephrine induced by infusion of neurokinin A into the locus coeruleus (Steinberg et al. 2001). In electrophysiological experiments in rats, saredutant (1 mg/kg i.p.) totally suppressed the enhancing effect of i.c.v. injection of CRF on the discharge rate of noradrenergic neurons in the locus coeruleus (Steinberg et al. 2001).

3.2
ZM253270

Another new class of selective nonpeptide antagonists of the tachykinin NK_2 receptor was obtained by optimizing lead compounds discovered by random screening of chemical libraries. A series of substituted 2,4-diaminoquinazolines and 2,4-diamino-8-alkylpurines has been discovered by this way (Jacobs et al. 1995) and among these compounds, ZM253270 (4-[(4-chlorophenyl)amino]-5,6-dihydro-2-(4-(2-(hydroxyimino)1-oxopropyl)-1-piperazyinyl)-6-(1-methylethyl) 7H-pyrrolo(3,4-d)pyrimidin-7-one; Fig. 2B) has been identified as the most potent compound (Aharony et al. 1995).

In radioligand binding assays for tachykinin NK_2 receptors, ZM253270 showed nanomolar affinity for the hamster tachykinin NK_2 receptor while it and all its analogs displayed a very low affinity for the human (K_i=105 nM) and guinea pig (K_i=9 μM) tachykinin NK_2 receptor. The poor affinity of ZM253270 for human tachykinin NK_2 receptors was confirmed by its low potency to antagonize neurokinin A-induced contractions of the human isolated bronchus with a pK_B of 5.4 (Aharony et al. 1995).

3.3
GR159897

Another successful approach to obtain potent nonpeptide antagonists of the tachykinin NK_2 receptor has been based on the identification of some important structural requirements for activity in peptide-based tachykinin NK_2 receptor antagonists, followed by a directed screening for nonpeptide compounds sharing the same structural features in chemical libraries (Cooper et al. 1994). Different small peptides were previously identified as potent antagonists of the tachykinin NK_2 receptor (McElroy et al. 1992; Smith et al. 1993). All these peptide-based NK_2 receptor antagonists shared common structural features; they contained an indole ring and at least another aromatic group. A screening for compounds containing these two structural features in chemical libraries allowed the identification of lead nonpeptide compounds having low affinity for the tachykinin NK_2 receptor. The optimization of these lead compounds led to the discovery of GR159897 ([R]-1-[2-(5-fluoro-1H-indol-3-yl)ethyl]-4-methoxy-4-[(phenylsulfinyl)methyl]piperidine; Fig. 2C), a very potent and selective antagonist of the tachykinin NK_2 receptor (Cooper et al. 1994; Beresford et al. 1995). A series of spiropiperidines closely related to GR159897 have been subsequently disclosed as potent tachykinin NK_2 receptor antagonists with subnanomolar affinity (Smith et al. 1995).

In radioligand binding assays, GR159897 displayed subnanomolar affinity for the rat and human tachykinin NK_2 receptors, with negligible affinity for the tachykinin NK_1 or NK_3 receptor (Beresford et al. 1995). It exhibited high antagonist potencies at tachykinin NK_2 receptors in functional bioassays using guinea pig, cynomolgus monkey and human isolated airway tissues, with pA_2 values between 8.2 and 8.7 (Ball et al. 1994; Beresford et al. 1995; Rizzo et al. 1999; Sheldrick et al. 1995). In vivo, in guinea pigs, GR159897 (120 µg/kg i.v.) efficiently inhibited bronchoconstriction provoked by a specific agonist for tachykinin NK_2 receptors with a long duration of action (Beresford et al. 1995). It blocked the tachykinin NK_2 receptor-mediated contractions of the urinary bladder in rats (ID_{50}=20 µg/kg i.v.) (Chopin et al. 2002). Furthermore, GR159897 has been shown to be centrally active. Like saredutant, GR159897 was shown to have a potent disinhibitory action on suppressed behaviors induced by aversive conditions in the mouse light–dark box test (0.0005–50 µg/kg s.c.) and in the marmoset human intruder test (0.2–50 µg/kg s.c.) (Walsh et al. 1995).

3.4
SR144190

Modification of the chemical structure of known potent nonpeptide antagonists for the different tachykinin receptors has been another approach to searching for selective nonpeptide antagonists of the tachykinin NK_2 receptor. This approach led to the discovery of SR144190 ([R]-3-{1-[2-(4-benzoyl-2-(3,4-difluorophenyl)-morpholin-2-yl)-ethyl]-4-phenylpiperidin-4-yl}-1-dimethylurea;

Fig. 2D) (Emonds-Alt et al. 1997), based on the structure–activity information gained during the discovery of saredutant and related selective nonpeptide antagonists of the tachykinin NK_1 and NK_3 receptor (Emonds-Alt et al. 1993b, 1995b).

SR144190 has been shown to be a selective, potent and orally active tachykinin NK_2 receptor antagonist with a biochemical and pharmacological profile comparable to that of saredutant. In radioligand binding assays, SR144190 showed subnanomolar affinity for the human, rat and guinea pig tachykinin NK_2 receptors. However, it had a much lower affinity for the hamster tachykinin NK_2 receptor (K_i=12 nM) (Emonds-Alt et al. 1997). In vitro, SR144190 potently antagonized tachykinin NK_2 receptor-mediated contractions of human (pA_2= 9.42–9.86), rat (pA_2=9.5), guinea pig (pA_2=10.1) and rabbit (pA_2=9.64) isolated tissues (Emonds-Alt et al. 1997; Croci et al. 1998). Moreover, the selectivity and specificity of SR144190 for tachykinin NK_2 receptors was demonstrated in radioligand binding and in vitro functional studies (Emonds-Alt et al. 1997).

In vivo, antagonist activity of SR144190 was demonstrated on the tachykinin NK_2 receptor-mediated bronchoconstriction in guinea pigs (ID_{50}=21 µg/kg i.v. and 250 µg/kg i.d.) and urinary bladder contraction in rats (ID_{50}=11 µg/kg i.v. and 190 µg/kg i.d.). Its activity has been further demonstrated in animal models where endogenous neurokinin A acting via the NK_2 receptor plays an important role. SR144190 prevented citric acid-induced cough and airway hyperresponsiveness to acetylcholine in guinea pigs (1 mg/kg i.p.) as well as castor oil-induced diarrhea in rats (0.01–10 µg/kg s.c. or p.o.) (Emonds-Alt et al. 1997). Finally, SR144190 has been shown to be centrally active. The turning behavior induced by intrastriatal injections of a specific tachykinin NK_2 receptor agonist in mice was blocked by SR144190 with a potency higher than that of saredutant (ID_{50}=3 µg/kg i.v. and 16 µg/kg p.o.). At 0.1 mg/kg p.o., SR144190 showed a long-lasting effect (Emonds-Alt et al. 1997). Its efficient antagonism at brain tachykinin NK_2 receptors has been supported by different in vivo microdialysis studies in rats. SR144190 inhibited the release of acetycholine in the hippocampus after either septal application of neurokinin A or sensory stimulation (0.1–0.3 mg/kg i.p.) (Steinberg et al. 1998a). In addition, it blocked the release of acetycholine in the striatum provoked by striatal application of a dopamine D_1 receptor agonist ([+]-SKF-38393) (0.3–1 mg/kg i.p.) or neurotensin under dopamine D_2 receptor blockade (1 mg/kg i.p.) (Steinberg et al. 1998b).

3.5
Other Compounds

SB414240 is another example of a tachykinin NK_2 receptor antagonist that has been obtained by chemical modification of known antagonists for other tachykinin receptors. Stepwise chemical modification of a series of highly potent nonpeptide tachykinin NK_3 receptor antagonists (Sarau et al. 1997, 2000; Hay et al. 2002) led to the identification of a novel, selective compound, SB414240 ([S]-(+)-N-(1,2,2-trimethylpropyl)-3-{(4-piperidin-1-yl)piperidin-1-yl}methyl-2-

phenylquinoline-4-carboxamide; Fig. 2E), with nanomolar affinity for the human tachykinin NK_2 receptor (Blaney et al. 2001).

On the other hand, following an approach rather close to that used for the discovery of GR159897, novel nonpeptide compounds with nanomolar affinities for the human tachykinin NK_2 receptor and equivalent functional activities were obtained from a structural element essential for activity in a series of peptide-based tachykinin NK_2 receptor antagonists (Altamura et al. 2002).

Finally, since the discovery of saredutant numerous compounds closely derived from saredutant have been reported as selective tachykinin NK_2 receptor antagonists. Different series of compounds obtained in such chemical programs using saredutant as template were described (Jacobs et al. 1998; MacKenzie et al. 2002; Selway and Terrett 1996). Among the compounds closely related to saredutant, YM38336 ([+/−]-1'-4-[4-(N-benzoyl-N-methylamino)-3-(3,4-dichlorophenyl)butyl]spiro[benzo[c]thiophene-1(3H), 4'-piperidine]; Fig. 2F) (Kubota et al. 1998a, 1998b), NK5807 ([−]-10-acetylamino-2-[4-(N-benzoyl-N-methyl)amino-3-(3,4-dichlorophenyl)butyl]-1,2,3,4-tetrahydro-benzo[b][1,6]naphthyridine; Fig. 2G) (Iida et al. 2001; Matsumoto et al. 2001), ZD7944 ([S]-3-{1-(3,4-dichlorophenyl)-3-[4−2(methanesulfonyl)-piperidin-1-yl]-propyl}-2-ethyl-2,3dehydro-insoindol-1-one; Fig. 2H) (Bernstein et al. 2001) and UK224671 ([S]-(+)-4-(1-{2-[1-cyclopropylmethyl)-3-(3,4-dichlorophenyl)-6-oxo-3-piperidyl]ethyl}azetidin-3-yl)-1-piperazine sulfonamide; Fig. 2I) (Beaumont et al. 2000a, 2000b; MacKenzie et al. 2002; Middleton 2002) have been reported as functional tachykinin NK_2 receptor antagonists in preliminary in vitro or in vivo pharmacological studies.

4
Conclusion

During the last 15 years, several selective antagonists of the tachykinin NK_2 receptor have been disclosed but, in spite of a large structural diversity of these compounds, only a very limited number of compounds shows a real potential as drug. Among the peptide-based antagonists, nepadutant is certainly the most potent molecule whereas saredutant is the leading compound in the class of nonpeptide antagonists.

Both saredutant and nepadutant have in common a high affinity for the human NK_2 receptor although saredutant has in fact a ten times higher affinity for the human receptor than nepadutant. The affinity for the human tachykinin NK_2 receptor is a critical point in the search for antagonists. As also observed with antagonists for tachykinin NK_1 and NK_3 receptors, both peptide-based and nonpeptide tachykinin NK_2 receptor antagonists can show important differences in affinity for the receptor expressed in different species. This probably reflects species-dependent variations at the binding site in the amino acid sequence of the tachykinin NK_2 receptor protein (Maggi et al. 1993c; Maggi 1995; Patacchini and Maggi 1995; Quartara et al. 1995; Catalioto et al. 1998).

On the other hand, with regard to tachykinin NK_2 receptors located in the peripheral nervous system, nepadutant and saredutant have a comparable pharmacological profile in terms of the magnitude and duration of their action. They potently and selectively antagonize various effects provoked by both exogenous and endogenous neurokinin A in animals. However, two major characteristics discriminate between these two compounds. Saredutant is orally active while the pharmacological activity of nepadutant by the oral route has not been proved. Moreover, in contrast with nepadutant, saredutant is centrally active. Importantly, all the central activities of saredutant observed in behavioral, neurochemical and electrophysiological experiments suggest that a brain-penetrant tachykinin NK_2 receptor antagonist may provide a new therapeutic approach in affective disorders. In addition to their potent activities in the airways, nepadutant and saredutant also show interesting activities in visceral nociception by reducing both the reflex responses to acute painful stimuli (rectal distension, irritation) and the gut hypersensitivity seen after inflammation, infection and stress. Therefore, a tachykinin NK_2 receptor antagonist may be useful for the treatment of functional gastrointestinal disorders such as irritable bowel syndrome (IBS). Moreover, an antidepressant/anxiolytic-like activity as displayed by saredutant may be an advantage as psychosocial stressors are considered to be an important trigger in IBS (Mayer and Collins 2002).

Up to now, only a few clinical results have been reported with tachykinin NK_2 receptor antagonists, and these results mainly show that the antagonist activity of saredutant and nepadutant at tachykinin NK_2 receptors can also be observed in humans. Administration of saredutant p.o. and nepadutant i.v. have been shown to reduce the bronchoconstrictor effect of neurokinin A in mild asthmatic patients (Van Schoor et al. 1998; Joos and Powels 2001) but saredutant did not reduce airway hyperresponsiveness to adenosine in allergic asthmatic patients (Kraan et al. 2001). In healthy male volunteers, i.v. administration of nepadutant has been reported to inhibit the motility-stimulating effects of neurokinin A in the small intestine (Lördal et al. 2001). There is now a clear need for more clinical studies to prove whether the tachykinin NK_2 receptor is a possible target for new therapeutics.

References

Abdel-Gawad M, Dion SB, Elhilali MM (2001) Evidence of a peripheral role of neurokinins in detrusor hyperreflexia: a further study of selective tachykinin antagonists in chronic spinal injured rats. J Urol 165:1739–1744

Advenier C (1995) Tachykinin NK_2 receptor further characterized in the lung with nonpeptide receptor antagonists. Can J Physiol Pharmacol 73:878–884

Advenier C, Emonds-Alt X (1996) Tachykinin receptor antagonists and cough. Pulmon Pharmacol 9:329–333

Advenier C, Naline E, Toty L, Bakdach H, Emonds-Alt X, Vilain P, Brelière JC, Le Fur G (1992) Effects on the isolated human bronchus of SR48968, a potent and selective nonpeptide antagonist of the neurokinin A (NK_2) receptors. Am Rev Respir Dis 146:1177–1181

Advenier C, Rouissi N, Nguyen QT, Emonds-Alt X, Brelière JC, Neliat G, Naline E, Regoli D (1992b) Neurokinin A (NK_2) receptor revisited with SR48968, a potent non-peptide antagonist. Biochem Biophys Res Commun 184:1418–1424

Advenier C, Girard V, Naline E, Vilain P, Emonds-Alt X (1993) Antitussive effect of SR48968, a non-peptide tachykinin NK_2 receptor antagonist. Eur J Pharmacol 250:169–171

Aharony D, Buckner CK, Ellis JL, Ghanekar SV, Graham A, Kays JS, Little J, Neeker S, Miller SC, Undem BJ, Waldron IR (1995) Pharmacological characterization of a new class of non-peptide neurokinin A antagonists that demonstrate species selectivity. J Pharmacol Exp Ther 274:1216–1221

Altamura M, Canfarini F, Catalioto RM, Guidi A, Pasqui F, Renzetti AR, Triolo A, Maggi CA (2002) Successful bridging from a peptide to a nonpeptide antagonist at the human tachykinin NK-2 receptor. Bioorg Med Chem Lett 12:2945–2948

Anton PM, Theodorou V, Fioramonti J, Buéno L (2001) Chronic low-level administration of diquat increases the nociceptive response to gastric distension in rats: role of mast cells and tachykinin receptor activation. Pain 92:219–227

Arunlakshana O, Schild HO (1959) Some quantitative uses of drug anatgonists. Br J Pharmacol 14:48–58

Ball DI, Beresford IJM, Wren GPA, Pendry YD, Sheldrick RLG, Walsh DM, Turpin MP, Hagan RM, Coleman RA (1994) In vitro and in vivo pharmacology of the non-peptide antagonist at tachykinin NK_2-receptor, GR159897. Br J Pharmacol 112 (Suppl):48P

Beaumont K, Harper A, Smith DA, Abel S (2000a) Pharmacokinetics and metabolism of a sulphamide NK_2 antagonist in rat and human. Xenobiotica 30:627–642

Beaumont K, Harper A, Smith DA, Bennett J (2000b) The role of P-glycoprotein in determining the oral absorption and clearance of the NK_2 antagonist, UK-224,671. Eur J Pharm Sci 12:41–50

Bensaid M, Faucheux BA, Hirsch E, Agid Y, Soubrié P, Oury-Donat F (2001) Expression of tachykinin NK_2 receptor mRNA in human brain. Neurosci Lett 303:25–28

Beresford IJM, Sheldrick RLG, Ball DI, Turpin MP, Walsh DM, Hawcock AB, Coleman RA, Hagan RM, Tyers MB (1995) GR 159897, a potent non-peptide antagonist at tachykinin NK_2 receptors.. Eur J Pharmacol. 272:241–248

Bernstein PR, Aharony D, Albert JS Andisik D, Barthlow HG, Bialecki R, Davenport T, Dedinas BT, Koether G, Kosmider BJ, Kirkland K, Ohnmacht CJ, Potts, W, Rumsey WL, Shen L, Shenvi A, Sherwood S, Stollman D, Russell K (2001) Discovery of novel, orally active dual NK_1/NK_2 antagonists. Bioorg Med Chem Lett 11:2769–2773

Bhogal N, Donnelly D, Findlay JB (1994) The ligand binding site of the neurokinin-2 receptor. Site-directed mutagenesis and identification of neurokinin A binding residues in the human neurokinin-2 receptor. J Biol Chem 269:27269–27274

Birder LA, Kiss S, DE Groat WC, Lecci A, Maggi CA (2003) Effect of nepadutant, a neurokinin-2 Tachykinin receptor antagonist, on immediate-early gene expression after trinitrobenzenzsulfonic acid-induced colitis in the rat. J Pharmacol Exp Ther 304:272–276

Blaney FE, Raveglia LF, Artico M, Cavagnera S, Dartois C, Farina C, Grugni M, Gagliardi S, Luttmann MA, Martinelli M, Nadler GMMG, Parini C, Petrillo P, Sarau HM, Scheideler MA, Hay DWP, Giardina GAM (2001) Stepwise modulation of neurokinin-3 and neurokinin-2 receptor affinity and selectivity in quinoline tachykinin receptor antagonists. J Med Chem 44:1675–1689

Boichot E, Germain N, Lagente V, Advenier C (1995) Prevention by the tachykinin NK_2 receptor antagonist, SR48968, of antigen-induced airway hyperresponsiveness in sensitized guinea pigs. Br J Pharamcol 114:259–261

Boichot E, Biyah K, Germain N, Emonds-Alt X, Lagente V, Advenier C (1996) Involvement of tachykinin NK_1 and NK_2 receptors in substance P-induced microvascular leakage hypersensitivity and airway hyperesponsiveness in guinea pigs. Eur Respir J 9:1445–1450

Boichot E, Germain N, Emonds-Alt X, Advenier C, Lagente V (1998) Effects of SR140333 and SR48968 on antigen and substance P-induced activation of guinea-pig alveolar macrophages. Clin Exp Allergy 28:1299–1305

Boyle S, Guard S, Hodgson J, Horwell DC, Howson W, Hughes J, McKnight AT, Martin K, Pritchard MC, Watling KJ, Woodruff GN (1994) Rational design of high affinity tachykinin NK_2 receptor antagonists. Bioorg Med Chem 2:101–113

Carini F, Lecci A, Tramontana M, Giuliani S, Maggi CA (2001) Tachykinin NK_2 receptors and enhancement of cholinergic transmission in the inflamed rat colon: an in vivo motility study. Br J Pharmacol 133:1107–1113

Catalioto RM, Criscuoli M, Cucchi P, Giachetti A, Giannotti D, Giuliani S, Lecci A, Lippi A, Patacchini R, Quartara L, Renzetti AR, Tramontana M, Arcamone F, Maggi CA (1998) MEN 11420 (Nepadutant), a novel glycosylated bicyclic peptide tachykinin NK_2 receptor antagonist. Br J Pharmacol 123:81–91

Chardenot P, Roubert C, Galiègue S, Casellas P, Le Fur G, Soubrié P, Oury-Donat F (2002) Expression profile and up-regulation of PRAX-1 mRNA by antidepressant treatment in the rat brain. Mol Pharmacol 62:1314–1320

Chopin A, Groke G, Bringas A, Stepan G, Dillon MP (2002) Effect of YM-44781, YM-44778 and YM-49598, novel tachykinin antagonists, in a drug-induced bladder contraction model. Pharmacology 65:96–102

Chu M, Chan TM, Das P, Mierzwa R, Patel M, Puar MS (2000) Structure of SCH 218157, a cyclodepsipeptide with neurokinin A activity. J Antibiot 53:736–738

Cooper AWJ, Adams HS, Bell R, gore PM, McElroy AB, Pritchard JM, Smith PW, Ward P (1994) GR159897 and related analogues as highly potent, orally active non-peptide neurokinin NK_2 receptor antagonists. Bioorg Med Chem Lett 4:1951–1956

Croci T, Emonds-Alt X, Manara L (1994) SR48968 selectively prevents faecal excretion following activation of tachykinin NK2 receptors in rats. J Pharm Pharmacol 46:383–385

Croci T, Emonds-Alt X, Le Fur G, Manara L (1995) In vitro characterization of the non-peptide tachykinin NK_1 and NK_2 receptor antagonists, SR140333 and SR48968, in different rat and guinea-pig intestinal segments. Life Sci 56:267–275

Croci T, Landi M, Emonds-Alt X, Le Fur G, Maffrand JP, Manara L (1997) Role of tachykinin in castor oil diarrhoea in rats. Br J Pharmacol 121:375–380

Croci T, Aureggi G, Manara L, Emonds-Alt X, Le Fur G, Maffrand JP, Mukenge S, Ferla G (1998) In vitro characterization of tachykinin NK-2 receptors modulating motor responses of human colonic muscle strips. Br J Pharmacol 124:1321–1327

Cutrufo C, Evangelista S, Cirillo R, Ciucci A, Conte B, Lopez G, Manzini S, Maggi CA (2000) Protective effect of the tachykinin NK_2 receptor antagonist nepadutant in acute rectocolitis induced by diluted acetic acid in guinea pigs. Neuropeptides 34:355–359

Desvignes C, Rouquier L, Souilhac J, Mons G, Rodier S, Soubrié P, Steinberg R (2003) Control by tachykinin NK_2 receptor of CRF_1 receptor-mediated activation of hippocampal acetylcholine release in the rat and guinea pig. Neuropeptides 37:89–97

Dion S, Rouissi N, Nantel F, Jukic D, Rhaleb NE, Tousignant C, Telemaque S, Drapeau G, Regoli D, Naline E, Advenier C, Rovero P, Maggi CA (1990) Structure-activity study of neurokinins: antagonists for the neurokinin-2 receptor. Pharmacology 41:184–194

Emonds-Alt X, Vilain P, Goulaouic P, Proietto V, Van Broeck D, Advenier C, Naline E, Neliat G, Le Fur G, Brelière JC (1992) A potent and selective non-peptide antagonist of the neurokinin A (NK_2) receptor. Life Sci 50:PL101–PL106

Emonds-Alt X, Advenier C, Croci T, Manara L, Neliat G, Poncelet M, Proietto V, Santucci V, Soubrié P, Van Broeck D, Vilain P, Le Fur G, Brelière JC (1993a) SR 48968, a neurokinin A (NK_2) receptor antagonist. Regul Pept 46:31–36

Emonds-Alt X, Doutremepuich JD, Healme M, Neliat G, Santucci V, Steinberg R, Vilain P, Bichon D, Ducoux JP, Proietto V, Van Broeck D, Soubrié P, Le Fur G, Brelière JC (1993b) In vitro and in vivo biological activities of SR 140333, a novel potent non-peptide tachykinin NK_1 receptor antagonist. Eur J Pharmacol 250:403–413

Emonds-Alt X, Golliot F, Pointeau P, Le Fur G, Brelière JC (1993c) Characterization of the binding sites of [^{3}H]SR 48968, a potent nonpeptide radioligand antagonist of the neurokinin-2 receptor. Biochem Biophys Res Commun 191:1172–1177

Emonds-Alt X, Proietto V, Van Broeck D, Vilain P, Advenier C, Neliat G, Le Fur G, Brelière JC (1993d) Pharmacological profile and chemical synthesis of SR 48968, a non-peptide antagonist of the neurokinin A (NK_2) receptor. Bioorg Med Chem Lett 3:925–930

Emonds-Alt X, Advenier C, Soubrié P, Le Fur G, Brelière JC (1995a) SR 48968: nonpeptide antagonist of the tachykinin NK_2 receptor. Drugs of the Future 20:701–707

Emonds-Alt X, Bichon D, Ducoux JP, Heaulme M, Miloux B, Poncelet M, Proietto V, Van Brock D, Vilain P, Neliat G, Soubrié P, Le Fur G, Brelière JC (1995b) SR 142801, the first potent non-peptide antagonist of the tachykinin NK_3 receptor. Life Sci 56:PL27–PL32

Emonds-Alt X, Advenier C, Cognon C, Croci T, Daoui S, Ducoux JP, Landi M, Naline E, Neliat G, Poncelet M, Proietto V, Van Broeck D, Vilain P, Soubrié P, Le Fur G, Maffrand JP, Brelière JC (1997) Biochemical and pharmacological activities of SR 144190, a new potent non-peptide tachykinin NK_2 receptor antagonist. Neuropeptides 31:449–458

Eutamène H, Theodorou V, Fioramonti J, Buéno L (1995) Implication of NK_1 and NK_2 receptors in rat colonic hypersecretion induced by interleukin 1β: role of nitric oxide. Gastroenterology 109:483–489

Gaultier E, Emonds-Alt X, Fioramonti J, Buéno L (2002) Neurokinin 2 and neurokinin 3 receptor antagonists, SR48968 and SR142801, reduce colonic hypersensitivity induced by inflammation and stress in rats, but do not modify colonic transit time. Gastroenterology 122:A-526

Giannotti D, Perrotta E, Di Bugno C, Nannicini R, Harmat JS, Giolitti A, Patacchini R, Renzetti AR, Rotondaro L, Giuliani S, Altamura M, Maggi CA (2000) Discovery of cyclic pseudopeptide human tachykinin NK-2 receptor antagonists. J Med Chem 43:4041–4044

Giolitti A, Altamura M, Bellucci F, Giannotti D, Meini S, Patacchini R, Rotondara L, Zappitelli S, Maggi CA (2002) Monocyclic human tachykinin NK-2 receptor antagonists as evolution of a potent bicyclic antagonist: QSAR and site-directed mutagenesis studies. J Med Chem 45:3418–3429

Girard V, Naline E, Emonds-Alt X, Advenier C (1995) Effect of two tachykinin antagonists, SR48968 and SR140333, on cough induced by citric acid in the unanaesthetized guinea pig. Eur Respir J 8:1110–1114

Girard V, Yavo JC, Emonds-Alt X, Advenier C (1996) The tachykinin NK_2 receptor antagonist SR48968 inhibits citric acid-induced airway hyperresponsiveness in guinea pigs. Am J Respir Crit Care Med 153:1496–1502

Giuliani S, Patacchini R, Lazzeri M, Nenaim G, Turini D, Quartara L, Maggi Ca (1996) Effect of several bicyclic and cyclic pseudopeptide tachykinin NK_2 receptor antagonists in the human isolated urinary bladder. J Auton Pharmacol 16:251–259

Giuliani S, Guelfi M, Toulouse M, Buéno L, Lecci A, Tramontana M, Criscuoli M, Maggi CA (2001) Effect of a tachykinin NK_2 receptor antagonist, nepadutant, on cardiovascular and gastrointestinal function in rats and dogs. Eur J Pharmacol 415:61–71

Griebel G, Moindrot N, Alaiga C, Simiand J, Soubrié P (2001a) Characterization of the profile of neurokinin-2 and neurotensin receptor antagonists in the mouse defense test battery. Neurosci Biobehav Rev 25:619–626

Griebel G, Perrault G, Soubrié P (2001b) Effects of SR48968, a selective non-peptide NK_2 receptor antagonist on emotional processes in rodents. Psychopharmacology 158:241–251

Hagan, RM, Beresford IJ, Stables J, Dupere J, Stubbs CM, Elliott PJ, Sheldrick RL, Chollet A, Kawashima E, McElroy AB, Ward P (1993) Characterization, CNS distribution and function of NK_2 receptors studied using potent NK_2 receptor antagonists. Regul Pept 46:9-19

Harbeson SL, Rovero P (1994) Structure-activity relationships of agonist and antagonist ligands. In: Buck SH (ed) The tachykinin receptor. Humana Press, Totowa, New Jersey, pp 329-365

Harmat NJS, Giannotti D, Nannicini R, Perrotta E, Criscuoli M, Patacchini R, Renzetti AR, Giuliani S, Altamura M, Maggi CA (2002) Insertion of 2-carboxysuccinate and tricarballylic acid fragments into cyclo-pseudopeptides: new antagonists for the human tachykinin NK-2 receptor. Bioorg Med Chem Lett 12:693-696

Hay DWP, Giardina GAM, Griswold DE, Underwood DC, Kotzer CJ, Bush B, Potts W, Sandhu P, Lundberg D, Foley JJ, Schmidt DB, Martin LD, Kilian D, Legos JJ, Barone FC, Luttmann MA, Grugni M, Raveglia LF, Sarau HM (2002) Nonpeptide tachykinin receptor antagonists. III. SB 235375, a low central nervous system-penetrant, potent and selective neurokinin-3 receptor antagonist, inhibits citric acid-induced cough and airways hyper-reactivity in guinea pigs. J Pharmacol Exp Ther 300:314-323

Hedge VR, Puar MS, Dai P, Pu H, Patel M, Anthes JC, Richard C, Terracciano J, Das PR, Gullo V (2001) A family of depsi-peptide fungal metabolites, as selective and competitive human tachykinin receptor (NK_2) antagonists: fermentation, isolation, physico-chemical properties, and biological activities. J Antibiot 54:125-135

Higashide Y, Yatabe Y, Arai Y, Nakajima Y, Shibata M, Yamaura T (1996) Pharmacological profile of a novel tachykinin NK_2 receptor antagonist TAC-363. Yakugaku Zasshi 116:884-891

Higashide Y, Yatabe Y, Arai Y, Shibata M, Yamaura T (1997) Effect of a novel tachykinin NK_2 receptor antagonist, TAC-363, on bronchoconstriction and airway hyperresponsiveness in guinea pigs. Nippon Yakurigaku Zasshi 109:19-29

Holzer P, Holzer-Petsche U (1997a) Tachykinins in the gut. Part I. Expression, relaese and motor function. Pharmacol Ther 73:173-217

Holzer P, Holzer-Petsche U (1997b) Tachykinins in the gut. Part II. Role in neural excitation, secretion and inflammation. Pharmacol Ther 73:219-263

Iida M, Matsumoto T, Ishida, K, Sakitama K, Nishikawa K (2001) NK5807, a novel NK_2 receptor antagonist, on animal model of asthma. Abstracts of the 74th Annual Meeting of the Japanese Pharmacological Society P-551

Jacobs RT, Shenvi AB, Mauger RC, Ulatowski TG, Aharony D, Buckner CK (1995) Substituted 2,4-diaminoquinazolines and 2,4-diamino-8-alkylpurines as antagonists of the neurokinin-2 (NK_2) receptor. Bioorg Med Chem Lett 5:2879-2884

Jacobs RT, Shenvi AB, Mauger RC, Ulatowski TG, Aharony D, Buckner CK (1998) 4-alkyl-piperidines related to SR-48968: potent antagonists of the neurokinin-2 (NK_2) receptor. Bioorg Med Chem Lett 8:1935-1940

Joos GF, Pauwels RA (2001) Tachykinin receptor antagonists: potential in airways diseases. Curr Opin Pharmacol 1:235-241

Julia V, Buéno L (1997) Tachykinergic mediation of viscerosensitive responses to acute inflammation in rats: role of CGRP. Am J Physiol 272:G141-G146

Julia V, Morteau O, Buéno L (1994) Involvement of neurokinin a nad neurokinin 2 receptors in viceronsensitive response to rectal distension in rats. Gastroenterology 107:94-102

Kenakin T (1993) Pharmacologic analysis of drug-receptor interaction. 2nd edn. Raven Press, New York

Kraan J, Vink-Klooster H, Potsma DS (2001) The NK-2 receptor anatgonist SR48968C does not improve adenosine hyperresponsiveness and airawy obstruction in allergic asthma. Clin Exp Allergy 31:274-278

Kubota H, Fujii M, Ikeda K, Takeuchi M, Shibanuma T, Isomura Y (1998a) Spiro-substituted piperidines as neurokinin receptor antagonists. I. Design and synthesis of

(+/−)-N-[2-(3,4-dichlorophenyl)-4-(spiro[isobenzofuran-1(3H),4'piperidin]-1'-yl)butyl]-N-methylbenzamide, YM-35375, as a new lead compound for novel neurokinin receptor antagonists. Chem Pharm Bull 46:351–354

Kubota H, Kakefuda A, Nagaoka H, Yamamoto O, Ikeda K, Takeuchi M, Shibanuma T, Isomura Y (1998b) Spiro-substituted piperidines as neurokinin receptor antagonists. II. Synthesis and NK_2 receptor–antagonistic activities of N-[2-aryl-4-(spiro-substituted piperidin-1'-yl)butyl]carboxamides. Chem Pharm Bull 46:242–254

Laird JMA, Olivar T, Lopez-Garcia JA, Maggi CA, Cervero F (2001) Responses of rat spinal neurons to distension of inflamed colon: role of tachykinin NK_2 receptors. Neuropharmacology 40:696–701

Lecci A, Giuliani S, Tramontana M, Criscuoli M, Maggi CA (1997a) MEN 11420, a peptide tachykinin NK_2 receptor antagonist, reduces motor responses induced by the intravesical administration of capsaicin in vivo. Naunyn-Schmiedeberg's Arch Pharmacol 356:182–188

Lecci A, Tramontana M, Giuliani S, Maggi CA (1997b) Role of tachykinin NK_1 and NK_2 receptors on colonic motility in anesthetized rats: effects of agonists. Can J Physiol Pharmacol 74:582–586

Lecci A, Giuliani S, Tramontana M, Santicioli P, Criscuoli M, Dion S, Maggi CA (1998a) Bladder distension and activation of the efferent function of sensory fibres: similarities with the effect of capsaicin. Br J Pharmacol 124:259–266

Lecci A, Tramontana M, Giuliani S, Criscuoli M, Maggi CA (1998b) Effect of tachykinin NK_2 receptor blockade on detrusor hyperreflexia induced by bacterial toxin in rats. J Urol 160:206–209

Lecci A, Carini F, Tramontana M, Birder LA, de Groat WC, Saznticioli P, Giuliani S, Maggi CA (2001a) Urodynamic effects induced by intravesical capsaicin in rats and hamsters. Auton Neurosci 91:37–46

Lecci A, Carini F, Tramontana M, D'Aranno V, Marinoni E, Crea A, Buéno L, Fioramonti J, Criscuoli M, Giuliani S, Maggi CA (2001b) Nepadutant pharmacokinetics and dose-effect relationships as tachykinin NK_2 receptor antagonist are altered by intestinal inflammation in rodent models. J Pharmacol Exp Ther 299:247–254

Lippi A, Criscuoli M, Guelfi M, Santicioli P, Maggi CA (1998) Pharmacokinetics of the bicyclic peptide tachykinin NK_2 receptor anatgonist MEN 11420 (NEPADUTANT) in rats. Drug Metab Dispos 26:1077–1081

Lombardi A, D'Agostino B, Filippelli A, Pedone C, Matera MG, Falciani M, De Rosa M, Rossi F, Pavone V (1998a) Neuronorm is apotent and water soluble neurokinin A receptor antagonist. Bioorg Med Chem Lett 8:1735–1740

Lombardi A, D'Agostino B, Nastri F, D'Andrea LD, Filippelli A, Falciani M, Rossi F, Pavone V (1998b) A novel super-potent neurokinin A receptor antagonist containing dehydroalanine. Bioorg Med Chem Lett 8:1153–1156

Lördal M, Navalesi G, Theodorsson E, Maggi CA, Hellstrom PM (2001) A novel tachykinin NK_2 receptor antagonist prevents motility-stimulating effects of neurokinin A in small intestine. Br J Pharmacol 134:215–223

MacKenzie AR, Marchington AP, Middleton DS, Newman SD, Jones BC (2002) Structure-activity relationships of 1-alkyl-5-(3,4-dichlorophenyl)-5-[2-[(substituted)-1-azetidinyl]ethyl]-2-piperidones. 1. Selective antagonists of neurokinin-2 receptor. J Med Chem 45:5365–5377

Maggi CA (1995) Tachykinin receptors.In: Ruffolo RR, Hollinger MA (eds) G-protein coupled transmembrane signaling mechanisms. CRC Press, Boca Raton, Ann Arbor, London, Tokyo, pp 95–151

Maggi CA, Giuliani S, Ballati L, Lecci A, Manzini S, Patacchini R, Renzetti AR, Rovero P, Quartara L, Giachetti A (1991) In vivo evidence for tachykinergic transmission using a new NK-2 receptor selective antagonist, MEN 10,376. J Pharmacol Exp Ther 257:1172–1178

Maggi CA, Patacchini R, Astolfi M, Rovero P, Giachetti A, Van Giersbergen PL (1992) Affinity of R 396, an NK_2 tachykinin receptor antagonist, for NK_2 receptors in preparation from different species. Neuropeptides 22:93–98

Maggi CA, Patacchini R, Rovero P, Giachetti A (1993a) Tachykinin receptors and tachykinin receptor antagonists. J Auton Pharmacol 13:23–93

Maggi CA, Quartara L, Patacchini R, Giuliani S, Barbanti G, Turini D, Giachetti A (1993b) MEN 10,573 and MEN 10,612, novel cyclic pseudopeptides which are potent tachykinin NK-2 receptor antagonists. Regul Pept 47:151–158

Maggi CA, Patacchini R, Giuliani S, Giachetti A (1993c) In vivo and in vitro pharmacology of SR48968, a non-peptide tachykinin NK_2 receptor antagonist. Eur J Pharmacol 234:83–90

Maggi CA, Astolfi M, Giuliani S, Goso C, Manzini S, Meini S, Patacchini R, Pavone V, Pedone C, Quartara L, Renzetti AR, Giachetti A (1994) MEN 10,627 a novel polycyclic peptide antagonist of tachykinin NK_2 receptors. J Pharmacol Exp Ther 271:1489–1500

Martin CAE, Naline E, Emonds-Alt X, Advenier C (1992) Influence of (±)-CP-96,345 and SR48968 on electrical field stimulation of the isolated guinea-pig main bronchus. Eur J Pharmacol 224:137–143

Martin CAE, Emonds-Alt X, Advenier C (1993) Inhibition of cholinergic neurotransmission in isolated guinea-pig main bronchi by SR48968. Eur J Pharmacol 243:309–312

Matsumoto T, Iida M, Ishida, K, Sakitama K, Nishikawa K (2001) NK5807, a novel nonpeptide antagonist of tachykinin NK2 receptor. Abstracts of the 74th Annual Meeting of the Japanese Pharmacological Society. P-561

Mayer EA, Collins SM (2002) Evolving pathophysiologic models of functional gastrointestinal disordres. Gastroenterology 122:2032–2048

Mazelin L, Theodorou V, More J, Emonds-Alt X, Fioramonti J, Buéno L (1998) Comparative effects of nonpeptide tachykinin receptor antagonists on experimental gut inflammation in rats and guinea pigs. Life Sci 63:293–304

McElroy AB, Clegg SP, Deal MJ, Ewan GB, Hagan RM, Ireland SJ, Jordan CC, Porter B, Ross BC, Ward P, Whittington AR (1992) Highly potent and selective heptapeptide antagonists of the neurokinin NK-2 receptor. J Med Chem 35:2582–2591

McKnight AT, Maguire JJ, Williams BJ, Foster AC, Tridgett R, Iversen LL (1988) Pharmacological specificity of synthetic peptides as antagonists at tachykinin receptors. Regul Pept 22:127

McKnight AT, Maguire JJ, Elliott NJ, Fletcher AE, Foster AC, Tridgett R, Williams BJ, Longmore J, Iversen LL (1991) Pharmacological specificity of novel, synthetic, cyclic peptides as antagonists at tachykinin receptors. Br J Pharmacol 104:355–360

McLean PG, Picard C, Garcia-Villar R, More J, Fioramonti J, Buéno L (1997) Effects of nematode infection on sensitivity to intestinal distension: role of tachykinin NK_2 receptors. Eur J Pharmacol 337:279–282

Middleton D (2002) Balancing potency and oral bioavailability. The design and profile of neurokinin-2 antagonists for the treatment of urinary urge incontinence. Abstracts of the 223rd National Meeting of the American Chemical Society MEDI-137

Patacchini R, Maggi CA (1995) Tachykinin receptors and receptor subtypes. Arch Int Pharmacodyn 329:161–184

Patacchini R, Maggi CA (2001) Peripheral tachykinin receptors as targets for new drugs. Eur J Pharmacol 429:13–21

Patacchini R, Quartara L, Rovero P, Goso C, Maggi CA (1993) Role of C-terminal amidation on the biological activity of neurokinin A derivatives with agonist and antagonist properties. J Pharmacol Exp Ther 264:17–21

Patatacchini R, Bartho L, Maggi Ca (1997a) Characterization of receptors mediating contraction induced by tachykinins in the guinea-pig isolated common bile duct. Br J Pharmacol 122:1633–1638

Patacchini R, Giuliani S, Lazzeri M, Turini A, Quartara L, Maggi CA (1997b) Effect of several bicyclic and cyclic pseudopeptide tachykinin NK_2 receptor antagonists in the human isolated ileum and colon. Neuropeptides 31:71–77

Patacchini R, De Giorgio R, Bartho L, Barbara G, Corinaldesi R, Maggi CA (1998a) Evidence that tachykinins are the main NANC excitatory neurotransmitters in the guinea-pig common bile duct. Br J Pharmacol 124:1703–1711

Patacchini R, Santicioli P, Zagorodnyuk V, Lazzeri M, Turini D, Maggi CA (1998b) Excitatory motor and electrical effects produced by tachykinins in the human and guinea-pig isolated ureter and guinea-pig renal pelvis. Br J Pharmacol 125:987–996

Patacchini R, Cox HM, Stahl S, Tough IR, Maggi CA (2001) Tachykinin NK_2 receptor mediates contraction and ion transport in rat colon by different mechanisms. Eur J Pharmacol 415:277–283

Pavone V, Lombardi A, Maggi CA, Quartara L, Pedone C (1995a) Conformational rigidity versus flexibility in a novel peptidic neurokinin A receptor antagonist. J Pept Sci 1:236–240

Pavone V, Lombardi A, Nastri F, Saviano M, Maglio O, D'Auria G, Quartara L, Maggi CA, Pedone C (1995b) Design and structure of a novel neurokinin A receptor anatgonist cyclo(Met-Asp-Trp-Phe-Dap-Leu)cyclo(2β-5β). Perkin Trans 2:987–993

Perretti F, Ballati L, Manzini S, Maggi CA, Evangelista S (1995) Antibronchopastic activity of MEN10,627, a novel tachykinin NK_2 receptor antagonist, in guinea-pig airways. Eur J Pharmacol 273:129–135

Picard P, Boucher S, Regoli D, Gitter BD, Howbert JJ, Couture R (1993) Unse of non-peptide receptor antagonists to substantiate the involvement of NK_1 and NK_2 receptors in a spinal nociceptive reflex in the rat. Eur J Pharmacol 232:255–261

Poncelet M, Gueudet C, Emonds-Alt X, Brelière JC, Le Fur G, Soubrié P (1993) Turning behavior induced in mice by a neurokinin A receptor agonist: stereoselective blockade by SR48968, a non-peptide receptor antagonist. Neursci Lett 149:40–42

Qian Y, Emonds-Alt X, Advenier C (1993) Effects of capsaicin, (±)-CP-96,345 and SR48968 on the bradykinin-induced airways microvascular leakage in guinea pigs. Pulm Pharmacol 6:63–67

Quartara L, Fabbri G, Ricci R, Patacchini R, Pestellini V, Maggi CA, Pavone V, Giachetti A, Arcamone F (1994) Influence of lipophilicity on the biological activity of cyclic pseudopeptide NK-2 receptor antagonists. J Med Chem 37:3630–3638

Quartara L, Rovero P, Maggi CA (1995) Peptide-based tachykinin NK_2 receptor antagonists. Med Res Rev 15:139–155

Quartara L, Pavone V, Pedone C, Lombardi A, Renzetti AR, Maggi CA (1996) A review of the design, synthesis and biological activity of the bicyclic hexapeptide tachykinin NK_2 antagonist MEN 10627. Regul Pept 65:55–59

Regoli D, Boudon A, Fauchère JL (1994) Receptors and antagonists for substance P and related peptides. Pharmacol Rev 46:551–599

Renzetti AR, Catalioto RM, Criscuoli M, Cucchi P, Lippi A, Guelfi M, Quartara L, Maggi CA (1998) Characterization of [^{3}H]MEN 11420, a novel glycosylated peptide antagonist radioligand of the tachykinin NK_2 receptor. Biochem Biophys Res Commun 248:78–82

Rizzo CA, Valentine AF, Egan RW, Kreutner W, Hey Ja (1999) NK_2-receptor mediated contraction in monkey, guinea-pig and human aiway smooth muscle. Neuropeptides 33:27–34

Rovero P, Pestellini V, Maggi CA, Patacchini R, Regoli D, Giachetti A (1990) A highly potent NK-2 tachykinin receptor antagonist containing D-tryptophan. Eur J Pharmacol 175:113–115

Rovero P, Astolfi M, Manzini S, Jukic D, Rouissi N, Maggi CA, Regoli D (1992) Structure-activity relationship study of R 396, an NK_2 tachykinin antagonist selective for the NK_{2B} receptor subtype. Neuropeptides 23:143–145

Saffroy M, Torrens Y, Glowinski J, Beaujouan JC (2001) Presence of NK_2 binding sites in the rat brain. J Neurochem 79:985–996

Santicioli P, Giuliani S, Patacchini R, Tramontana M, Criscuoli M, Maggi CA (1997) MEN 11420, a potent and selective tachykinin NK_2 receptor antagonist in the guinea-pig and human colon. Naunyn-Schmiedeberg's Arch Pharmacol 356:678–688

Sarau HM, Griswold DE, Potts W, Foley JJ, Schmidt DB, Webb EF, Martin LD, Brawner ME, Elshourbagy NA, Medhurst AD, Giardina GAM, Hay DWP (1997) Nonpeptide tachykinin receptor antagonists. I. Pharmacological and pharmacokinetic characterization of SB 223412, a novel, potent and selective neurokinin-3 receptor antagonist. J Pharmacol Exp Ther 281:1303–1311

Sarau HM, Griswold DE, Bush B, Potts W, Sandhu P, Lundberg D, Foley JJ, Schmidt DB, Webb EF, Martin LD, Legos JJ, Whitmore RG, Barone FC, Medhurst AD, Luttmann MA, Giardina GAM, Hay DWP (2000) Nonpeptide tachykinin receptor antagonists. II. Pharmacological and pharmacokinetic profile of SB-222200, a central nervous system penetrant, potent and selective NK-3 receptor antagonist. J Pharmacol Exp Ther 295:373–381

Selway CN, Terrett NK (1996) Parallel-compound synthesis: methodology for accelerating drug discovery. Bioorg Med Chem 4:645–654

Sheldrick RL, Rabe KF, Fischer A, Magnussen H, Coleman RA (1995) Further evidence that tachykinin-induced contraction of human isolated bronchus is mediated only by NK_2 receptors. Neuropeptides 29:281–292

Smith PW, McElroy AB, Pritchard JM, Deal MJ, Ewan GB, Hagan RM, Ireland SJ, Ball D, Beresford I, Sheldrick R, Jordan CC, Ward P (1993) Low molecular weight neurokinin NK_2 antagonists Bioorg Med Chem Lett 3:931–936

Smith PW, Cooper AWJ, Bell R, Beresford IJM, Gore PM, McElroy AB, Pritchard JM, Saez V, Taylor NR, Sheldrick RLG, Ward P (1995) New spiropiperidines as potent and selective non-peptide tachykinin NK_2 receptor antagonists. J Med Chem 38:3772–3779

Steinberg R, Rodier D, Souilhac J, Bougault I, Emonds-Alt X, Soubrié P, Le Fur G (1995) Pharmacological characterization of tachykinin receptors controlling acetylcholine release from rat striatum. An in vivo microdyalysis study. J Neurochem 65:2543–2548

Steinberg R, Marco N, Voutsinos B, Bensaid M, Rodier D, Souilhac J, Alonso R, Oury-Donat F, Le Fur G, Soubrié P (1998a) Expression and presence of septal neurokinin-2 receptors controlling hippocampal acetylcholine release during sensory stimulation in rat. Eur J Neurosci 10:2337–2345

Steinberg R, Souilhac J, Rodier D, Alonso R, Emonds-Alt X, Le Fur G, Soubrié P (1998b) Facilitation of striatal acetylcholine release by dopamine D_1 receptor stimulation: involvement of enhanced nitric oxide production via neurokinin-2 receptor activation. Neuroscience 84:511–518

Steinberg R, Alonso R, Griebel G, Bert L, Jung M, Oury-Donat F, Poncelet M, Gueudet C, Desvignes C, Le Fur G, Soubrié P (2001) Selective blockade of neurokinin-2 receptor produces antidepressant-like effects associated with reduced corticotropin-releasing factor function. J Pharmacol Exp Ther 299:449–458

Stratton SC, Beresford IJ, Harvey FJ, Turpin MP, Hagan RM, Tyers MB (1993) Anxiolytic activity of tachykinin NK_2 receptor antagonists in mouse light-dark box. Eur J Pharmacol 250:R11–R12

Teixeira RM, Santos ARS, Ribeiro SJ, Calixto JB, Rae GA, De Lima TCM (1996) Effects of central administration of tachykinin receptor agonists and antagonists on plus-maze behavior in mice. Eur J Pharmacol 311:7–14

Toulouse M, Coelho AM, Fioramonti J, Lecci A, Maggi CA, Buéno L (2000) Role of tachykinin NK_2 receptors in normal and altered rectal sensitivity in rats. Br J Pharmacol 129:193–199

Toulouse M, Fioramonti J, Maggi CA, Buéno L (2001) Role of NK_2 receptors in gastric barosensitivity and in experimental ileus in rats. Neurogastroenterol Motil 13:45–53

Tramontana M, Patacchini R, Giuliani S, Lippi A, Lecci A, Santicioli P, Criscuoli M, Maggi CA (1998a) Characterization of the antibronchoconstrictor activity of M 11420, a tachykinin NK_2 receptor antagonist, in guinea pigs. Eur J Pharmacol 352:279–288

Tramontana M, Patacchini R, Lecci A, Giuliani S, Maggi CA (1998b) Tachykinin NK_2 receptors in the hamster urinary bladder: in vitro and in vivo characterization. Naunyn-Schmiedeberg's Arch Pharmacol 358:293–300

Tramontana M, Santicioli P, Giuliani S, Catalioto RM, Lecci A, Carini F, Maggi CA (2002) Role of tachykinins in sephadex-indiced airway hyperreactivity and inflammation in guinea pigs. Eur J Pharmacol 439:149–158

Van Schoor J, Joos GF, Chasson BL, Brouard RJ, Pauwels RA (1998) The effect of the NK_2 tachykinin receptor antagonist SR48968 (saredutant) on beurokinin A-induced bronchoconstriction in asthmatics. Eur Respir J 12:17–23

Vilain P, Emonds-Alt X, Brelière JC (1993) Characterization of SR48968. Specificity for neurokinin A (NK_2) receptor. Neuropeptides 24(Suppl):236

Walsh DM, Stratton SC, Harvey IJ, Hagan RM (1995) The anxiolytic-like activity of GR159897, a non-peptide NK_2 receptor antagonist, in rodent and primate models of anxiety. Psychopharmacology 121:186–191

Williams BJ, Curtis NR, McKnight AT, Maguire J, Foster A, Tridgett R (1988) Development of NK-2 selective antagonists. Regul Pept 22:189

Williams BJ, Curtis NR, McKnight AT, Maguire JJ, Young SC, Veber DF, Baker R (1993) Cyclic peptides as selective tachykinin antagonists. J Med Chem 36:2–10

Wollborn U, Brunne RM, Harting J, Holzemann G, Leibfritz D (1993) Comparative conformational analysis and in vitro pharmacological evaluation of three cyclic hexapeptide NK_2 antagonists. Int J Peptide Protein Res 41:376–384

Yashpal K, Dam TV, Quirion R (1990) Quantitative autoradiographic distribution of multiple neurokinin binding sites in rat spinal bord. Brain Res 506:259–266

Yashpal K, Hui-Chan CW, Henry JL (1996) SR48968 specifically depresses neurokinin A-vs. substance P-induced hyperalgesia in a nociceptive withdrawal reflex. Eur J Pharmacol 308:41–48

Tachykinin NK₃ Receptor Antagonists

S. B. Mazzone[1] · B. J. Canning[2]

[1] Department of Neurobiology, Howard Florey Institute, University of Melbourne, VIC, Australia 3010
[2] The Johns Hopkins Asthma and Allergy Center, 5501 Hopkins Bayview Circle, Baltimore, MD, 21224 USA
e-mail: bjc@jhmi.edu

1   Introduction   246

2   NK₃ Receptor Antagonists   248
2.1   Selective Nonpeptide Antagonists   252
2.2   Nonselective Antagonists   255

3   Clinical Applications for NK₃ Receptor Antagonists   256
3.1   Gastrointestinal Disorders   257
3.2   Obstructive Airways Disease   257
3.3   Cardiovascular Diseases   259
3.4   Pre-eclampsia   259
3.5   Somatic and Visceral Pain   260
3.6   CNS Disorders   261

4   Concluding Remarks   262

References   263

Abstract A growing body of evidence suggests that tachykinin NK₃ receptors at both central and peripheral sites may contribute to the pathogenesis of a variety of diseases. Although the discovery of potent and selective nonpeptide NK₃ receptor antagonists initially lagged behind the development of comparable molecules for NK₁ and NK₂ receptors, work in this field has rapidly progressed. Both selective and nonselective antagonists displaying nanomolar potencies for the human NK₃ receptor are now available. The results from preclinical animal trials are encouraging, and suggest clinical applications for NK₃ receptor antagonists in the treatment of human diseases. However, the therapeutic potential for NK₃ receptor antagonists in humans awaits controlled clinical trials.

Keywords NK3 receptor antagonists · Selective nonpeptide antagonists · Nonselective nonpeptide antagonists · Clinical applications

1
Introduction

Prior to the development of selective and potent NK_3 receptor antagonists and the molecular characterization of the NK_3 receptor, initial insights into the distribution and function of this tachykinin receptor relied on in vivo and in vitro functional and radioligand binding experiments (Fig. 1, Table 1). Such experi-

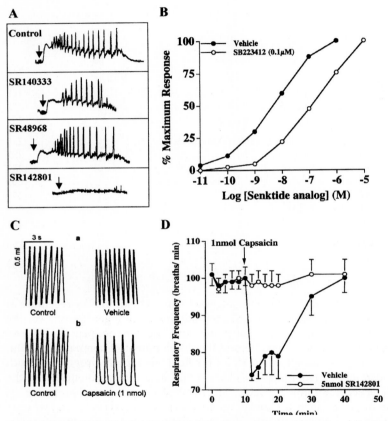

Fig. 1A–D Functional assays for NK_3 receptors. **A, B** NK_3 receptor-dependent contractions of the longitudinal muscle from isolated guinea pig ileum. **A** Phasic and tonic contractions of guinea pig ileum evoked by senktide ([Asp[6], Me-Phe[8]]-substance P(6–11)) are prevented by pretreatment with the selective NK_3 receptor antagonist SR142801, but not by the NK_1 or NK_2 receptor antagonists SR140333 and SR48968, respectively. (Modified with permission from Vilain et al. 1997). **B** Senktide analog ([Asp[5,6],Me-Phe[8]]-substance P(5–11)) evoked contractions of guinea pig ileum are competitively inhibited by the NK_3 receptor antagonist SB223412. (Data provided by Bradley J. Undem, Johns Hopkins Asthma and Allergy Center, Baltimore, MD, USA). **C, D** NK_3 receptor-dependent reductions in rat respiratory rate. Representative traces (**C**) and mean data (**D**) demonstrating that microinjection of capsaicin into the nucleus of the solitary tract of the rat evokes a reduction in respiratory rate. In this model, respiratory slowing can be mimicked by the NK_3 receptor agonists, NKB and senktide, and is prevented by the selective NK_3 receptor antagonist SR142801 (see Mazzone and Geraghty 1999, 2000b for further details)

Table 1 Functional and binding assays for studying the pharmacology of NK_3 receptor antagonists

Tissue/cell type	Species	Assay details[a]	References
In vitro			
Ileum	GP	Agonist-evoked contraction	Jacoby et al. 1986; Patacchini et al. 1995
	GP	Radioligand binding	Guard et al. 1990
	GP	Agonist-evoked IP_3 hydrolysis	Guard et al. 1988; Petitet et al. 1993
Portal vein	Rat	Agonist-evoked contraction	Mastrangelo et al. 1987
Iris sphincter muscle	Rbt	Agonist-evoked contraction	Medhurst et al. 1997
CHO/HEK (cloned NK_3R)	Hm, Ms	Agonist-evoked Ca^{2+} mobilization	Oury-Danat et al. 1995; Sarau et al. 2000
	Hm, Rat	Agonist-evoked IP_3 hydrolysis	Chung et al. 1994
	Hm, Rat, Ms	Radioligand binding	Chung et al. 1994; Suman-Chauhan et al. 1994; Sarau et al. 2000
Habenula neurons	GP	Agonist-evoked spontaneous firing	Boden and Woodruf 1994
Cortex	GP, Rat	Radioligand binding	Guard et al. 1990; Langlois et al. 2001
In vivo			
Brain	GP	i.c.v. agonist-evoked c-fos expression	Ding et al. 2000; Yip and Chahl 1997
	GP, Ms	i.c.v. agonist-evoked behavioral responses	Cellier et al. 1997; Yip and Chahl 1997; Sarau et al. 2000
Brainstem (nTS)	Rat	Agonist and capsaicin-evoked respiratory slowing	Mazzone and Geraghty 1999, 2000b

CHO, Chinese hamster ovary cells; GP, guinea-pig; HEK, human embryonic kidney cells; Hm, human; i.c.v., intracerebroventricular; IP_3, inositol trisphosphate; Ms, mouse; NK_3R, tachykinin NK_3 receptor; nTS, nucleus of the solitary tract (nucleus tractus solitarii); Rbt, rabbit.

[a] See references for a complete description of the methods for each assay.

ments used the endogenous tachykinin, neurokinin B (NKB), and selective NK_3 receptor agonists, such as [MePhe^7]NKB and senktide to localize and characterize NK_3 receptors in several tissues from a variety of mammalian species (Guard and Watson 1987; Guard et al. 1988, 1990; Nguyen et al. 1994; Petitet et al. 1993b; Pinnock et al. 1994; Regoli et al. 1994; Sadowski et al. 1993; Saffroy et al. 1988; Suman-Chauhan et al. 1994; Wang and Hakanson 1993; Weinrich et al. 1989). Although these studies indicated the presence of NK_3 receptors in both central and peripheral sites, such approaches often provided little detail as to the specific cell types expressing NK_3 receptors.

The NK_3 receptor has now been cloned and sequenced from a variety of species, including rat, mouse, rabbit and human (Buell et al. 1992; Huang et al. 1992; Medhurst et al. 1999; Ohkubo and Nakanishi 1991; Sarau et al. 2001; Shigemoto et al. 1990). In addition, potent and selective NK_3 receptor antagonists are available and an NK_3 receptor knockout mouse has also been described (Kung et al. 2002). By using a combination of immunohistochemical, molecular biological and electrophysiological approaches, coupled with conventional functional and binding studies, it has been possible to describe better and identify the distribution and function of NK_3 receptors. In both the peripheral and central nervous systems, NK_3 receptors are found primarily on neurons. In particular, peripheral neurons within the myenteric, cardiac, bronchial, celiac and submandibular ganglia as well as retinal and gallbladder neurons express functional NK_3 receptors (Canning et al. 2002; Hoover et al. 2000; Jenkinson et al. 1999; Mann et al. 1997; Mawe 1995; Oyamada et al. 1999; Soejima et al. 1999; Zhao et al. 1995). NK_3 receptor activation at these sites is believed to play an important role in the modulation of neuronal excitability. Other cell types including blood vessels, iris sphincter and uterine smooth muscle, epithelium and glands may also express NK_3 receptors (Barr et al. 1991; Mastrangelo et al. 1987; Medhurst et al. 1997; Phillips et al. 2003).

Despite the wide distribution of NK_3 receptors in the central nervous system (CNS; including both the brain and spinal cord), the role of NK_3 receptors in central neuronal functioning is somewhat poorly described. NK_3 activation has been shown to participate in somatic and visceral sensory integration, cardiorespiratory regulation, water and electrolyte balance, and is probably important for learning, memory and behavioral responses (Cellier et al. 1997; Flynn 1999; Kamp et al. 2001; Massi et al. 2000; Mazzone and Geraghty 2000a; Santos and Calixto 1997; Yip and Chahl 1997; Yuan and Couture 1997).

The aim of this chapter is to describe the evolution of nonpeptide NK_3 receptor antagonists, and highlight their therapeutic potential in a variety of human diseases.

2
NK_3 Receptor Antagonists

The study of tachykinin receptors was greatly advanced in the early 1990s following the development of potent and selective nonpeptide NK_1 and NK_2 receptor antagonists (Garret et al. 1991; Emonds-Alt et al. 1992; Snider et al. 1991). However, comparable compounds with activity at the NK_3 receptor (Figs. 2, 3, Tables 2, 3) proved more difficult to obtain. Modifications of the peptide backbones of neurokinin A (NKA) and NKB yielded several peptide antagonists with moderate selectivity and affinity for the NK_3 receptor. For example, Drapeau et al. (1990) described a modified NKA peptide, [Trp^7,beta-Ala^8]-NKA(4–10), which inhibited NK_3 receptors in rat portal vein. However this compound was shown also to possess agonist activity (Canning and Undem 1994). A conformationally constrained analog of NKB (named GR138676) was reported to inhibit NK_3 receptor-mediated effects in various cell systems and tissues (pK_b 8.2–8.3). However, GR138676 was not specific for NK_3 receptors as NK_1 receptor-mediated increases in intracellular calcium in human astrocytoma cells were also prevented by this molecule with a comparable potency (Stables et al. 1994).

In 1993 it was recognized that a newly developed nonpeptide NK_2 receptor antagonist, SR48968, also displayed affinity for NK_3 receptors, and several studies successfully took advantage of the NK_3 receptor blocking properties of high concentrations of SR48968 to study NK_3 receptor function (Canning and Undem

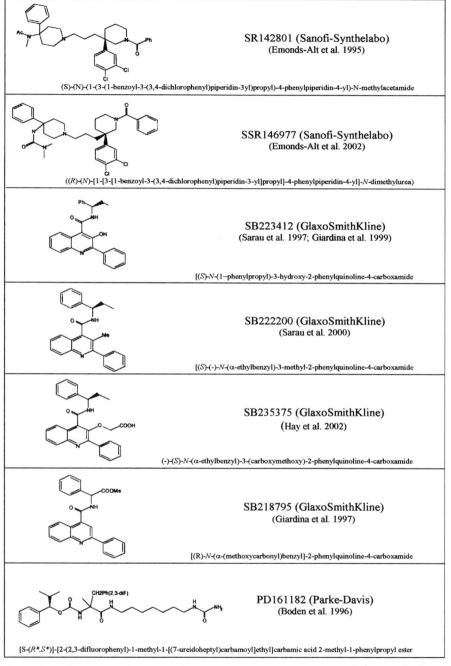

Fig. 2 Chemical structures of reported selective NK$_3$ receptor antagonists. See references and Tables 2 and 3 for further details

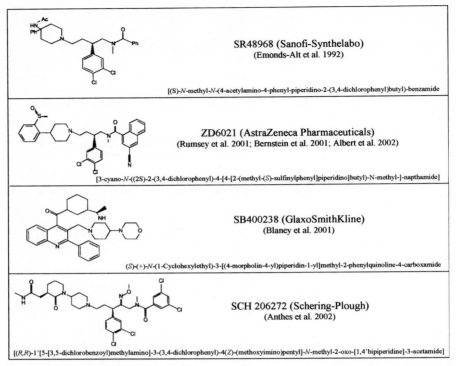

Fig. 3 Chemical structures of reported nonselective NK$_3$ receptor antagonists. See references and Tables 2 and 3 for further details

1994; Chung et al. 1994; Petitet et al. 1993a). The nonspecificity of SR48968 appeared to be species dependent. Thus, SR48968 was reported to block NK$_3$ receptors at submicromolar concentrations in guinea pigs and humans, but not rats (Chung et al. 1994; Petitet et al. 1993a). The SR48968 binding site on the NK$_3$ receptor was shown to involve five amino acids in the first and second transmembrane spanning domains (Wu et al. 1994). Mutation of two of these residues in the rat NK$_3$ receptor yielded a domain with a binding affinity for SR48968 comparable to that in humans and guinea pigs. Parke-Davis also reported the development of several peptide-based NK$_3$ receptor antagonists, PD154740 and PD157672, which like SR48968, revealed species differences in NK$_3$ receptor binding (Chung et al. 1995). Thus, PD154740 and PD157672 bound to human NK$_3$ receptors with greater affinity than to NK$_3$ receptors from rat. Interestingly, although these compounds were structurally distinct from SR48968, both classes of molecules appeared to interact with the same binding domain on the NK$_3$ receptor (Chung et al. 1995). Substitution of the two amino acid residues in the first and second transmembrane spanning domains of the rat NK$_3$ receptor that conferred species selectivity for SR48968, also enhanced the affinity of both PD154740 and PD157672.

Table 2 Antagonist affinities (K_i) estimated from radioligand binding experiments using [^{3}H]-senktide or [^{125}I]-(MePhe^7)NKB

	Receptor binding affinity, K_i (nM)					
	Human (cloned)[a]			GP (brain)	Rat (brain)	References
	NK$_1$	NK$_2$	NK$_3$	NK$_3$	NK$_3$	
Selective antagonists						
SR142801	744	40	0.21	0.11	15	Emonds-Alt et al. 1995; Beaujouan et al. 1997; Oury-Donat et al. 1995
SSR146977	175[b]	19.3	0.26	0.09	7.2	Emonds-Alt et al. 2002
SB223412	>100,000	144	1.0	0.79	30.1	Sarau et al. 1997; Giardina et al. 1999
SB222200	>100,000	250	4.4	3.0	88	Sarau et al. 2000
SB235375	>100,000	209	2.2	1.6	17.9	Hay et al. 2002
SB218795	>100,000	1221	13.3	NR	NR	Giardina et al. 1997
PD161182	3000[c]	NR	7.0	4.0	30	Boden et al. 1996
Nonselective antagonists						
SR48968	454	0.23	350	205	>10,000[d]	Chung et al. 1994; Petitet et al. 1993; Chung et al. 1995
ZD6021	0.12[e]	0.61[e]	74[e]	NR	NR	Rumsey et al. 2001; Bernstein et al. 2001; Albert et al. 2002
SB400238	NR	0.8	0.8	NR	NR	Giardina et al. 2001
SCH206272	1.3	0.4	0.3	NR	NR	Anthes et al. 2002; Reichard et al. 2002

NR, Not reported.
[a] Unless otherwise noted, Chinese hamster ovary (CHO) cells were transfected with the cloned tachykinin receptors.
[b] Human astrocytoma cell line.
[c] Human lymphoma cell line.
[d] Rat cloned NK$_3$ receptors.
[e] Murine erythroleukemia (MEL) cells.

Another important step towards the development of NK$_3$ receptor antagonists was the observation that nonpeptide angiotensin (AT) 1 receptor antagonists (such as losartan), but not AT2 receptor antagonists, interacted with NK$_3$ receptors in guinea pigs and rats. For example, in guinea pig and rat brain membranes, AT1 receptor antagonists displaced [^{3}H]-senktide binding with 50% inhibitory concentrations (IC$_{50}$) ranging between 18 and 50 μM (Chretien et al. 1994). Similarly, in functional studies, losartan inhibited cardiovascular and behavioral responses to centrally administered [MePhe^7]NKB in conscious rats (Picard et al. 1995). Although none of the molecules described above displayed sufficient selectivity or potency at NK$_3$ receptors to adequately study NK$_3$ receptor function, important insights into the structural determinants of potential nonpeptide NK$_3$ receptor antagonists were gained from such observations.

Table 3 Potency of NK$_3$ receptor antagonists in functional assays

	IC$_{50}$ (nM)[a]	Cell line	K$_b$ (nM)[b]	Assay	References
Selective antagonists					
SR142801	6.1	CHO	0.54	GPI	Emonds-Alt et al. 1995; Patacchini et al. 1995
SSR146977	10.0	CHO	0.85	GPI	Emonds-Alt et al. 2002
SB223412	16.6	HEK	1.6, 1.1	RISM, GPI	Sarau et al. 1997; Giardina et al. 1999; Goldhill et al. 1999
SB222200	18.4	HEK	3.3	RISM	Sarau et al. 2000
SB235375	81.9	HEK	8.1, 8.3	RISM, GPI	Hay et al. 2002
SB218795	NR	–	43.0	RISM	Giardina et al. 1997
PD161182	0.9	CHO	6.0	GP Habenula	Boden et al. 1996
Nonselective antagonists					
SR48968	1000	CHO	83	GPI	Pinnock et al. 1994; Petitet et al. 1993
ZD6021	NR	–	Noncompetitive[c]	GPI	Rumsey et al. 2001
SB400238	NR	–	NR	–	Giardina et al. 2001
SCH206272	12.0	CHO	Noncompetitive[c]	GPI	Anthes et al. 2002; Reichard et al. 2002

CHO, Chinese hamster ovary cells; HEK, human embryonic kidney cells; NR, not reported.
[a] Concentration of antagonist needed to inhibit by 50% calcium mobilization in cells expressing the cloned human NK$_3$ receptor.
[b] Antagonist dissociation constants (K$_b$) were estimated in vitro from NK$_3$ agonist-evoked: contractions of guinea-pig ileum (GPI); contractions of rabbit iris sphincter muscle (RISM); spontaneous discharge in guinea-pig habenula neurons.
[c] Not calculated as the antagonist inhibited maximal attainable response in these assays.

2.1
Selective Nonpeptide Antagonists

In 1995 Sanofi Recherche (now Sanofi-Synthelabo Recherche) described the first potent and selective nonpeptide NK$_3$ receptor antagonist (SR142801, named osanetant; Emonds-Alt et al. 1995). SR142801 is a constrained analog of SR48968 that inhibits NK$_3$-mediated responses in a variety of species, including humans. In binding studies, SR142801 bound with high affinity to membrane preparations from guinea pig and gerbil brain (0.11 and 0.42 nM, respectively), but displayed lower affinity in rat. In functional studies, SR142801 inhibited NK$_3$-dependent contractions of isolated guinea pig ileum, and dose-dependently prevented behavioral responses in gerbils evoked by intrastriatal injections of senktide, indicating CNS penetration when given orally. Patacchini and colleagues (1995) reported that SR142801 produced insurmountable antagonism of NK$_3$ receptors in guinea pig ileum (pK$_b$=9.27), although the potency was decidedly less in rat portal vein preparations (pK$_b$=7.49). The activity of SR142801 was enhanced by increasing the incubation time from 15 min to 120 min, but the potency of SR142801 remained lower in rat preparations. The species selectivity

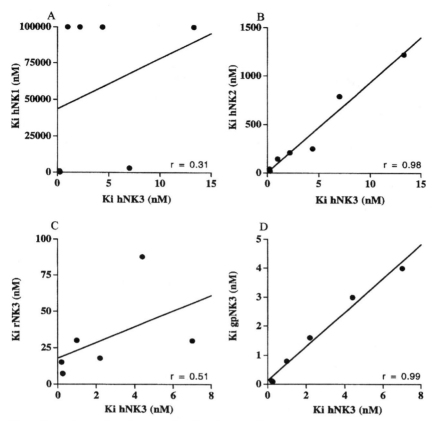

Fig. 4 Regression analysis demonstrating the relationship between the binding affinity of selective NK$_3$ receptor antagonists at human NK$_3$ (*hNK$_3$*) receptors and (**A**) hNK$_1$ receptors, (**B**) hNK$_2$ receptors, (**C**) rat NK$_3$ (*rNK$_3$*) receptors and (**D**) guinea pig NK$_3$ (*gpNK$_3$*) receptors. In general, hNK$_3$ receptor binding affinity appears to be highly correlated with the affinity of antagonists at hNK$_2$. Furthermore, the analysis suggests that the binding affinity of currently available NK$_3$ receptor antagonists at hNK$_3$ receptors is also highly conserved at gpNK$_3$ receptors, but not rNK$_3$ receptors. Analysis of the functional data contained in Table 3 demonstrates a similar correlation (r=0.998) between the activity of NK$_3$ receptor antagonists in human and guinea pig preparations (analysis not shown)

(Figs. 2, 4, Tables 2, 3), slow onset of action and the seemingly irreversible nature of SR142801 were further demonstrated in binding and functional studies using guinea pig and rat tissues (Beaujouan et al. 1997; Nguyen-Le et al. 1996). In cells expressing the human NK$_3$ receptor, Oury-Donat et al. (1995) showed that SR142801 displaced radiolabeled [MePhe^7]NKB with an estimated binding affinity (K_i) of 0.21 nM. In functional assays (Table 3), SR142801 blocked human NK$_3$ receptor-dependent responses with a potency comparable to that in guinea pigs (Oury-Donat et al. 1995).

Preclinical studies have shown SR142801 to be effective in animal models of pain, inflammation, obstructive airways disease and various CNS disorders (see below). SR142801 is also reportedly undergoing clinical trials in humans for the treatment of schizophrenia (Kamali 2001). In addition, Sanofi-Synthelabo Recherche has recently reported the development of another selective NK_3 receptor antagonist, SSR146977 (Emonds-Alt et al. 2002). This compound is structurally similar to SR142801, and displays a potency at the NK_3 receptor similar to that of the parent molecule. Like SR142801, SSR146977 can cross the blood–brain barrier, and is less potent in rodent preparations. There are no independent studies published using this molecule.

Although SR142801 remains the best characterized nonpeptide NK_3 receptor antagonist, a number of other pharmaceutical companies have disclosed nonpeptide compounds with activity at NK_3 receptors (Fig. 2, Tables 2, 3). In 1996, Parke-Davis reported the development of PD161182 (Boden et al. 1996). This molecule was developed using a peptide modification strategy that initially identified a urea derivative of a dipeptide (PD157672) with high affinity (7 nM) for the cloned human NK_3 receptor. Further modifications of the dipeptide lead to the development of the nonpeptide antagonist, PD161182, which displayed low nanomolar affinity for the cloned human NK_3 receptor. PD161182, like SR142801, also displayed activity at guinea pig NK_3 receptors, and was moderately potent in rat preparations. Also in 1996, SmithKlineBeecham (now GlaxoSmithKline) reported the development of several 4-quinolinecarboxamide based molecules which were unrelated to the piperidine derivatives SR142801 and SR48968, but structurally related to the angiotensin receptor antagonists that displayed some affinity for the NK_3 receptor (Giardina et al. 1996). SB218795 was disclosed in the following year and incorporated the 4-quinolinecarboxamide framework (Giardina et al. 1997). This antagonist displayed approximately 90-fold selectivity for the human NK_3 receptor versus the human NK_2 receptor and over 7,000-fold selectivity for the human NK_3 receptor versus the human NK_1 receptor. In binding studies, SB218795 bound to cells expressing the human NK_3 receptor with a K_i of 13 nM. Furthermore, SB218795 blocked senktide-evoked contractions of the rabbit iris sphincter muscle with a K_b of 43 nM (Giardina et al. 1997).

SmithKlineBeecham continued their search for novel nonpeptide NK_3 receptor antagonists, and in 1997 reported the development of the highly selective NK_3 receptor antagonist, SB223412 (also named talnetant; Sarau et al. 1997; Giardina et al. 1999b). This molecule was again based on the 4-quinolinecarboxamide framework of SB218795, but possessed modified chemical side chains to reduce biological instability and minimize binding affinity for the NK_2 receptor. As a result, SB223412 displayed a potency at human NK_3 receptors that was comparable to SR142801, but significantly lower affinity for NK_2 and NK_1 receptors. Furthermore, SB223412 showed oral bioavailability, low plasma clearance in most species (with the exception of guinea pig), reversibility in binding studies and moderate CNS penetration (Sarau et al. 1997; Giardina et al. 1999b, D.W. Hay, personal communication 2003). Attempts to replace the 4-quinolinecarboxamide back bone led to less active compounds (Giardina et al. 1999a). In recent years, GlaxoSmithKline has disclosed two additional selective NK_3 receptor antagonists based on the 4-quinolinecarboxamide framework. SB222200 and SB235375 are both highly selective for NK_3 receptors, display oral bioavailability and are stable in plasma (Sarau et al. 2000; Hay et al. 2002). SB222200 and SB235375, however, differ substantially in their ability to cross the blood–brain barrier. Studies in mice show that oral administration of SB222200, but not SB235375, inhibits behavioral effects evoked by centrally administered senktide. There are no published reports of any of the GlaxoSmithKline molecules undergoing clinical trials for the treatment of human diseases.

Merck has reported the development of several molecules that display nanomolar affinity at human NK_3 receptors in binding and functional studies (Harrison et al. 1998). These compounds were synthesized from a common structural template by transposing derivatives on the piperidine ring of SR142801. The most potent of these molecules (an amide series) displayed IC_{50} values at the human NK_3 receptor in the nanomolar range. Finally, a number of other companies have patented molecules that are claimed to be NK_3 receptor antagonists (for review see Rogers 2001). Amongst these are derivatives of piperidines (Pfizer, Sankyo and Warner-Lambert), oximes, hydrazones and olefins (Schering Plough); however, there are no published data detailing the pharmacology of these molecules.

2.2
Nonselective Antagonists

Endogenous tachykinins are able to interact, albeit with differing affinities, with each of the three tachykinin receptors. Furthermore, multiple receptors often contribute to a single biological effect of tachykinins in vivo. Redundancies in tachykinin-mediated responses have lead to proposals that antagonists with activity at multiple tachykinin receptors may provide greater clinical effectiveness than those that simply target a single receptor subtype. Whilst the majority of nonselective nonpeptide tachykinin receptor antagonists display the greatest affinity for NK_1 and NK_2 receptors, a number of molecules have been synthesized which also possess activity at NK_3 receptors.

Given its lack of selectivity in humans and guinea pigs, SR48968 served as a starting molecule for many nonselective tachykinin receptor antagonists. A number of pharmaceutical companies have now reported nonselective antagonists with activity at NK_3 receptors. Hoechst Marion Roussel (now Aventis) reported a dual NK_1/NK_2 receptor antagonist (HMR2091) with an NK_3 receptor binding affinity of 8.2 nM (Mauser et al. 2000). Sankyo has also disclosed the development of several morpholine analogs that display high affinity for all three tachykinin receptors (Nishi et al. 2000). The most potent of these compounds (Sankyo compound 12 in Nishi et al. 2000) has IC_{50} values of 21 nM, 3.1 nM and 0.72 nM at guinea pig NK_1, NK_2 and NK_3 receptors, respectively. Merck and Marion Merell Dow disclosed nonselective tachykinin receptor antagonists with activity at NK_3 receptors (reviewed in Leroy et al. 2000). L743986 (Merck) is a potent dual NK_1/NK_2 receptor antagonist with submicromolar affinity for the NK_3 receptor. MDL105212 (Marion Merell Dow) displays IC_{50} values of 3.1 nM, 8.4 nM and 21 nM at human NK_1, NK_2 and NK_3 receptors, respectively. There are no reports of any of these compounds undergoing extensive pharmacological characterization by independent groups.

AstraZeneca has disclosed an orally active dual NK_1/NK_2 receptor antagonist (ZD6021) with some affinity for human and guinea pig NK_3 receptors (Albert et al. 2002; Bernstein et al. 2001; Rumsey et al. 2001). In binding studies, ZD6021 inhibited radiolabeled [MePhe^7]NKB binding to the human NK_3 receptor with a K_i of 74 nM. ZD6021 (10 nM) also substantially inhibited senktide-evoked contractions of isolated guinea pig ileum, but displayed signs of noncompetitiveness in this assay. There were no attempts to determine whether ZD6021 could cross the blood–brain barrier, although independent studies would suggest that ZD6021 can block central synapses when given intravenously (Canning et al. 2001; Mazzone and Canning 2002). GlaxoSmithKline has reported several NK_2/NK_3 receptor antagonists that are based on the quinoline structure of their selective NK_3 receptor antagonists described above (Blaney et al. 2001). The most potent of these molecules, SB400238, displays a binding affinity at both NK_3 and NK_2 receptors of 0.8 nM.

Finally, a recent report from Schering-Plough has described an oxime derivative as a new potent nonselective, orally active, tachykinin receptor antagonist, SCH206272 (Anthes et al. 2002). In binding studies, SCH206272 displays high (and almost equal) affinity for human NK_1 (1.3 nM), NK_2 (0.4 nM) and NK_3 (0.3 nM) receptors. However, in functional studies, SCH206272 showed superior activity at human NK_2 and NK_3 receptors compared to human NK_1 receptors. In vivo studies in guinea pigs and dogs confirm the ability of SCH206272 to block multiple tachykinin receptors. There are no reports describing whether SCH206272 can cross the blood–brain barrier. Interestingly, Schering has now gone on to design additional nonselective tachykinin receptor antagonists that are devoid of any appreciable NK_3 receptor activity (Reichard et al. 2002). The decision to limit the activity of their compounds at NK_3 receptors came from recent suggestions that NK_3 receptors play a critical role in the regulation of the hypothalamic–pituitary axis and the production and secretion of androgens and gonadotrophins, which may result in unwanted side effects of any potential therapeutic compounds.

3
Clinical Applications for NK_3 Receptor Antagonists

There are currently only limited published data describing the use of NK_3 receptor antagonists in the treatment of human diseases. Consequently, the discussion that follows describes human conditions in which preclinical animal studies have suggested that an NK_3-mediated mechanism may underlie some or all of the symptoms of the disease, and highlights the potential for safe and effective NK_3 receptor antagonists in the treatment of such conditions.

3.1
Gastrointestinal Disorders

The role of tachykinins in gastrointestinal function has been studied in depth, and is reviewed elsewhere in detail (Holzer 1998; Holzer and Holzer-Petsche 1997; Holzer and Holzer-Petsche 2001). Throughout the gastrointestinal tract, NK_3 receptors are primarily distributed on subsets of neurons in the myenteric and submucosal plexuses. In addition, there are several reports of NK_3 receptor-dependent responses in submandibular and gallbladder neurons, pancreatic acinar cells and mesenteric endothelial cells (Linari et al. 2002; Mawe 1995; Mizuta et al. 1995; Soejima et al. 1999). In general, the NK_3 receptor is thought to play an important role in the generation of slow excitatory postsynaptic potentials in enteric neurons (Alex et al. 2001; Johnson et al. 1998; Manning and Mawe 2001; Mawe 1995; Thornton and Bornstein 2002). Activation of NK_3 receptors on excitatory and inhibitory motor pathways therefore results in either the facilitation of intestinal motility (Fig. 1A,B) or, in some species, depression of motility, secondary to the release of nitric oxide (Maggi et al. 1994). In addition, NK_3 receptor activation may play a role in the regulation of secretion in the gastrointestinal tract and accessory organs, and has been shown to be involved in gut-related nociceptive processing in regions of the CNS (Julia et al. 1999; Michl et al. 2001).

Under physiological conditions, tachykinins contribute minimally to intestinal motility, although qualitative and quantitative alterations in tachykinins and their receptors have been reported in the diseased intestinal tract (Evangelista 2001; Holzer 1998; Holzer and Holzer-Petsche 2001). As a result, a great deal of attention has been ascribed to a potential role of tachykinins in gastrointestinal disorders, including inflammatory bowel disease, dysmotility disorders, secretory diarrhea and intestinal pain. However, the majority of these investigations have been directed towards the role of NK_1 and NK_2 receptors in such disorders. In general, with the exception of intestinal pain (see below), there is little evidence to suggest an involvement of NK_3 receptors in intestinal dysfunction. Indeed, it has been suggested that a reduction in the activity of NK_3 receptors may precipitate chronic constipation (Mitolo-Chieppa et al. 2001). Whether NK_3 receptor antagonists are a viable therapeutic approach for the treatment of intestinal diseases, possibly in combination with NK_1 and/ or NK_2 receptor antagonists, awaits further investigation.

3.2
Obstructive Airways Disease

Obstructive airways disease is a term encompassing two distinct diseases (asthma and chronic obstructive pulmonary disease) that are characterized by varying degrees and types of airways inflammation. Symptoms of obstructive airways disease differ depending on the defining etiology, but include airways obstruction (which may or may not be reversible), excessive mucous secretion, mucosal edema, airways hyperreactivity and cough. Preclinical animal studies (particularly in guinea pigs) have provided compelling evidence for a role of tachykinins and their receptors in the symptoms of airways disease. Consistent with this, all three tachykinin receptors are present in the airways and are associated with a variety of different tissue types, including vascular endothelium, smooth muscle, glands, epithelium and nerves.

With respect to NK_3 receptors, aerosolized treatment of guinea pigs with [MePhe^7]NKB and senktide leads to airways hyperreactivity, demonstrated by exaggerated bronchoconstrictor responses to acetylcholine 24 h later (Daoui et al. 2000). Hyperresponsivess to acetylcholine occurs in the absence of any direct bronchoconstriction evoked by the NK_3 receptor agonists. Furthermore, pretreatment with the NK_3 receptor antagonist, SR142801, prevents the hyperresponsiveness evoked by [MePhe^7]NKB, supporting a direct role of NK_3 receptors in this response (Daoui et al. 2000). Similar effects of SR142801 and/or SB223412 have been demonstrated against hyperresponsiveness and microvascular hyperpermeability in the airways of guinea pigs and rabbits evoked by aerosolized tachykinin exposure or intraesophageal challenges with acid (Daoui et al. 1997, 2001, 2002). In isolated guinea pig tracheal preparations, prolonged exposure to the β_2-adrenoceptor agonist, fenoterol, induces smooth muscle hyperresponsiveness to neurokinin A, a response that is also prevented by SR142801 (Pinto et al. 2002). NK_3 receptor antagonists are antitussive in standard cough models in guinea pigs, modulate cholinergic nerve-mediated submucosal secretions in pigs and inhibit inflammatory cell recruitment to the airways in antigen challenged mice (Daoui et al. 1998; Mazzone and Canning 2003; Moreaux et al. 2000; Nenan et al. 2001; Phillips et al. 2003). NK_3 receptors also play a critical role in the central reflex regulation of respiration (Fig. 1C,D), and may therefore contribute to respiratory abnormalities (such as dyspnea) in airways disease (Mazzone and Geraghty 1999, 2000b).

The exact mechanisms underlying the beneficial effects of NK_3 receptor antagonists in animal models of obstructive airways disease are not well defined. However, NK_3 receptors are present in airway parasympathetic ganglia, and NK_3 receptor activation modulates nerve-mediated responses at these sites (Canning et al. 1998, 2002; Myers and Undem 1993). In addition, blockade of NK_3 receptors in the brain stem may contribute to the beneficial in vivo effects of NK_3 receptor antagonists (Canning et al. 2001; Mazzone and Canning 2002, 2003; Mazzone and Geraghty 1999, 2000b).

There are no reported clinical trials in humans assessing the potential for NK_3 receptor antagonists in obstructive airways disease.

3.3
Cardiovascular Diseases

NK_3 receptors in the cardiovascular system, like the gastrointestinal and pulmonary systems, are distributed primarily on subsets of autonomic neurons. However, a distinct role for cardiac NK_3 receptors in the manifestation of heart disease has not been described. Not surprisingly, NK_3 receptor stimulation modulates the excitability of intrinsic parasympathetic neurons in the heart, perhaps producing spontaneous action potential formation and therefore evoking changes in heart rate and coronary vascular tone (Chang et al. 2000; Hardwick et al. 1997; Hoover et al. 2000; Thompson et al. 1998). In addition, NK_3 receptors may be distributed on sympathetic nerves and in sympathetic ganglia (Jobling et al. 2001; Roccon et al. 1996; Seabrook et al. 1992). Accordingly, intravenous injections of senktide and other NK_3 receptor agonists evoke a pressor response that is blocked by SR142801 and the α-adrenoceptor antagonist prazosin, suggesting an involvement of noradrenergic mechanisms (Roccon et al. 1996). This pressor response following peripheral NK_3 receptor activation may have important implications in hypertensive disorders, such as pre-eclampsia, which are thought to manifest as a result of excessive circulating levels of NKB (see below).

Several studies have also implicated NK_3 receptors at supraspinal sites in the regulation of cardiovascular activity, although mixed responses have been reported depending on the dose of agonist used. For example, Cellier et al. (1997) reported that low doses of senktide injected directly into the cerebral ventricles produces a fall in mean arterial pressure, whereas higher doses produce hypertension. Both responses appear to be NK_3 receptor dependent since they were prevented by NK_3 receptor blockade. Although the site of action of senktide was not determined in the study by Cellier and colleagues, NK_3 receptors in the hypothalamus and substantia nigra may be involved in the effects of senktide (Couture et al. 1995; Lessard and Couture 2001). More recently, NK_3 receptors in the substantia nigra have also been shown to regulate cardiovascular activity in spontaneously hypertensive rats (Lessard et al. 2003). Despite the growing evidence that central NK_3 receptors may play a role in cardiovascular control, there is little data to suggest that central NK_3 receptor blockade may be of therapeutic benefit in cardiovascular disease.

3.4
Pre-eclampsia

Pre-eclampsia is a multisystem disorder characterized by maternal hypertension and proteinuria that become apparent during the second half of pregnancy. Although pre-eclampsia is typically associated with pregnancy, pre-eclamptic conditions can occur in the absence of fetal tissue (e.g., hydatidiform mole) and resolves following the removal of the placenta. This suggests a causative link between the placenta and the manifestation of the disorder (reviewed by Page et al. 2001). Although the exact cause of pre-eclampsia remains the subject of debate, it has been suggested that it results from the release of a vasoactive substance from placental tissue which is poorly perfused and/or ischemic.

Page and colleagues have identified high levels of NKB in placental tissue from pregnant women, and have shown that plasma levels of NKB are grossly elevated in pregnancy-induced hypertension and pre-eclampsia (Page et al. 2000). They believe that placental secretion of NKB plays an important role in improving utero-placental blood supply, and is directly responsible for the shunting of maternal blood to the uterus. In support of this assertion, administration of NKB to female rats evokes a pressor response and increases placental mass by up to 30% (Page et al. 2000). In addition, NK_3 receptors are expressed in the placenta of rats and humans (Cintado et al. 2001; Crane et al. 2002; Page and Bell 2002). In contrast, recent data suggest that NKB fails to constrict (but rather dose-dependently dilates) human myometrial or omental blood vessels, casting doubt on the exact role of NKB in pre-eclampsia (Wareing et al. 2003). Moreover, it has yet to be demonstrated whether there is a direct role for NK_3 receptors (or other tachykinin receptors) in pre-eclampsia, or if there will be any potential therapeutic effects of NK_3 receptor antagonists in the treatment of this disorder.

3.5
Somatic and Visceral Pain

NK_3 receptors have been localized to both spinal and supraspinal sites, and are believed to play a key role in pain (or nociception). Using a combination of immunohistochemistry and electron microscopy, Zerari and colleagues described the presence of NK_3 receptors on both pre- and postsynaptic neurons in the rat spinal cord, suggesting that NK_3 receptor activation may regulate both primary afferent nerve activity and spinal dorsal horn neuron excitability (Zerari et al. 1997). Consistent with this, cultured dorsal root ganglia neurons have been shown to express functional NK_3 receptors which potentiate neuronal activity (Brechenmacher et al. 1998; Wang et al. 2001). Furthermore, NK_3 receptor activation enhances the release of substance P from rat spinal cord synaptosomes, and induces neuronal bursting in neonatal rat spinal cords in vitro (Guo et al. 1998; Marchetti et al. 2001; Schmid et al. 1998). Spinal neurons in the superficial dorsal horn expressing NK_3 receptors can be further subclassified into two distinct phenotypes, those that also express either the μ-opioid receptor or nitric oxide synthase (Ding et al. 2002). At supraspinal sites (i.e., brain stem), NK_3 receptor expressing neurons in the rat nucleus of the solitary tract demonstrate early gene activation (c-*fos*) following noxious chemical stimulation of the peritoneum (Chen et al. 1997). In addition, the NK_3 receptor antagonist SB222200 reduces c-*fos* expression in the area postrema following noxious acidification of the gastric mucosa (Michl et al. 2001). Thus, multiple populations of NK_3 receptor expressing neurons may be important in sensory signaling in mammals.

In animal models of inflammatory pain, spinal NK_3 receptors have been shown to undergo quantitative changes in expression and function. For example, McCarson and Krause (1994) demonstrated an increase in the expression of NK_3 mRNA in the rat spinal cord following induction of nociception. The NK_3 receptor antagonist, SR142801, depresses the characteristic 'wind-up' in spinal cord neurons following repetitive dorsal root stimulation in vitro, and has been shown to prevent capsaicin and formalin-evoked pain responses in mice in vivo (Barbieri and Nistri 2001). Similar results have been obtained using SB223412 (Zeratin et al. 2000). In models of visceral hyperalgesia, NK_3 receptor antagonists modify responses to noxious visceral stimulation, possibly via both central and peripheral mechanisms (Julia et al. 1999). Consistent with receptor localization studies, spinal NK_3 receptor-dependent modulation of nociception in functional studies has been shown to involve both opioid and nitric oxide-mediated mechanisms (Linden et al. 1999; Linden and Seybold 1999). Of particular interest, in addition to preventing nociceptive-like responses in animal models, NK_3 receptor blockade can reverse the central changes in spinal neuron excitability (Houghton et al. 2000). Thus, NK_3 receptors may participate in both the induction and the ongoing maintenance of nociceptive responses, making NK_3 receptors an ideal therapeutic target in a variety of pain states.

Although a number of clinical trials have assessed the therapeutic potential of NK_1 receptor antagonists in inflammatory pain in humans, there have been no reported clinical trials using compounds with activity at NK_3 receptors.

3.6
CNS Disorders

NK_3 receptors are widely distributed throughout the CNS. Autoradiography (using [^{3}H]-senktide and/or [^{3}H]-SR142801) and immunohistochemical studies in guinea pigs, rats and humans indicate the presence of NK_3 receptors on neurons and/or glial cells in many central structures, including the cortex, medial habenula nucleus, amygdaloid complex, superior colliculus, interpeduncular nucleus, thalamus, hypothalamus, periaqueductal gray, area postrema and the nucleus of the solitary tract (Langlois et al. 2001; Mileusnic et al. 1999; Yip and Chahl 2001). Although there is significant overlap of the distribution of NK_3 receptors across species, some species differences are also noted.

The widespread distribution of NK_3 receptors in the mammalian brain suggests an involvement of NK_3 receptors in many central processes. NK_3 receptor activation in vitro has been shown to modify neural signaling in the striatum, basal ganglia, substantia nigra, medial habenula nucleus and the cortex (Kemel et al. 2002; Keegan et al. 1992; Nalivaiko et al. 1997; Norris et al. 1993; Stacey et al. 2002; Tremblay et al. 1992). In vivo, central NK_3 receptor activation in animal models evokes a variety of motor and behavioral responses, modifies cardiorespiratory activity, alters fluid and electrolyte balance and produces anxiolytic and antidepressant responses (Cellier et al. 1997; Flynn 1999; Mazzone and Geraghty 1999, 2000b; Panocka et al. 2001; Piot et al. 1995; Ribeiro et al. 1999).

The only reported clinical trial involving an NK_3 receptor antagonist in humans is currently assessing the potential therapeutic benefit of SR142801 (osanetant) in the treatment of schizophrenia (reviewed by Kamali 2001). Sanofi-Synthelabo has disclosed that they are performing phase II clinical trials comparing SR142801 to three other compounds (a 5-HT$_2$ receptor antagonist, a neurotensin receptor antagonist and a cannabinoid receptor antagonist). Although the current pathophysiological mechanisms underlying schizophrenia are still not fully understood, one of the most widely accepted hypotheses relates to the over-activity of dopaminergic neurons in the brain. In preclinical studies, NK$_3$ receptors are involved in: (a) senktide-evoked [$_3$H]-dopamine release from cultured mesencephalon neurons; (b) spontaneous and senktide-evoked firing of locus coeruleus neurons in guinea pigs; (c) nigrostriatal dopamine release in rat and guinea pig; and (d) senktide evoked dopamine release from neurons in the pars compacta and ventral tegmental area in guinea pigs (Alonso et al. 1996; Bert et al. 2002; Gueudet et al. 1999; Humpel and Saria 1993; Keegan et al. 1992; Marco et al. 1998; Michaud et al. 1995; Nalivaiko et al. 1997; Seabrook et al. 1995). These data strongly support the hypothesis that central NK$_3$ receptors play an important role in central dopaminergic circuits, and that NK$_3$ receptor antagonists may provide therapeutic benefit in schizophrenia. However, at present there are no data available detailing the outcomes of the Sanofi-Synthelabo clinical trials.

4
Concluding Remarks

The availability of potent NK$_3$ receptor antagonists has shed light on the multiple roles of NK$_3$ receptors in the central and peripheral nervous systems. However, despite the tremendous advances made in the development of these compounds, there are essentially no reports from any independent groups or pharmaceutical companies testing currently available NK$_3$ receptor antagonists in human diseases. This may reflect, in some cases, a lack of preclinical data to support a potential benefit of NK$_3$ receptor antagonists in disease. Alternatively, it may relate to recent suggestions that widespread blockade of NK$_3$ receptors may evoke undesirable side effects, since NK$_3$ receptor activation is reportedly anxiolytic and is involved in normal hypothalamic–pituitary functioning. Regardless of the reasons, preclinical studies would suggest that NK$_3$ receptor antagonists may provide therapeutic benefit in a variety of gastrointestinal, respiratory, and cardiac diseases, and may prove useful in the management of somatic and visceral pain in humans. Furthermore, the clinical manifestations of other diseases in which the pathophysiology shares similarities with those described above (e.g., gastroesophageal reflux, rhinitis and urinary bladder dysfunction) may also benefit from NK$_3$ receptor antagonists, although there are limited preclinical data available to support this assertion. Clearly, more research in this area of tachykinin neurobiology is required to adequately address the potential therapeutic benefit of NK$_3$ receptor antagonists in human disease.

References

Albert JS, Aharony D, Andisik D, Barthlow H, Bernstein PR, Bialecki RA, Dedinas R, Dembofsky BT, Hill D, Kirkland K, Koether GM, Kosmider BJ, Ohnmacht C, Palmer W, Potts W, Rumsey W, Shen L, Shenvi A, Sherwood S, Warwick PJ, Russell K (2002) Design, synthesis, and SAR of tachykinin antagonists: modulation of balance in NK_1/NK_2 receptor antagonist activity. J Med Chem 45:3972–3983

Alex G, Kunze WA, Furness JB, Clerc N (2001) Comparison of the effects of neurokinin-3 receptor blockade on two forms of slow synaptic transmission in myenteric AH neurons. Neuroscience 104:263–269

Alonso R, Fournier M, Carayon P, Petitpretre G, Le Fur G, Soubrie P (1996) Evidence for modulation of dopamine-neuronal function by tachykinin NK_3 receptor stimulation in gerbil mesencephalic cell cultures. Eur J Neurosci 8:801–808

Anthes JC, Chapman RW, Richard C, Eckel S, Corboz M, Hey JA, Fernandez X, Greenfeder S, McLeod R, Sehring S, Rizzo C, Crawley Y, Shih NY, Piwinski J, Reichard G, Ting P, Carruthers N, Cuss FM, Billah M, Kreutner W, Egan RW (2002) SCH 206272: a potent, orally active tachykinin NK_1, NK_2, and NK_3 receptor antagonist. Eur J Pharmacol 450:191–202

Barbieri M, Nistri A (2001) Depression of windup of spinal neurons in the neonatal rat spinal cord in vitro by an NK_3 tachykinin receptor antagonist. J Neurophysiol 85:1502–1511

Barr AJ, Watson SP, Bernal AL, Nimmo AJ (1991) The presence of NK_3 tachykinin receptors on rat uterus. Eur J Pharmacol 203:287–290

Beaujouan JC, Saffroy M, Torrens Y, Glowinski J (1997) Potency and selectivity of the tachykinin NK_3 receptor antagonist SR 142801. Eur J Pharmacol 319:307–316

Bernstein PR, Aharony D, Albert JS, Andisik D, Barthlow HG, Bialecki R, Davenport T, Dedinas RF, Dembofsky BT, Koether G, Kosmider BJ, Kirkland K, Ohnmacht CJ, Potts W, Rumsey WL, Shen L, Shenvi A, Sherwood S, Stollman D, Russell K (2001) Discovery of novel, orally active dual NK_1/NK_2 antagonists. Bioorg Med Chem Lett 11:2769–2773

Bert L, Rodier D, Bougault I, Allouard N, Le Fur G, Soubrie P, Steinberg R (2002) Permissive role of neurokinin NK_3 receptors in NK_1 receptor-mediated activation of the locus coeruleus revealed by SR142801. Synapse 43:62–69

Blaney FE, Raveglia LF, Artico M, Cavagnera S, Dartois C, Farina C, Grugni M, Gagliardi S, Luttmann MA, Martinelli M, Nadler GM, Parini C, Petrillo P, Sarau HM, Scheideler MA, Hay DW, Giardina GA (2001) Stepwise modulation of neurokinin-3 and neurokinin-2 receptor affinity and selectivity in quinoline tachykinin receptor antagonists. J Med Chem 44:1675–1689

Boden P, Woodruff GN (1994) Presence of NK_3-sensitive neurons in different proportions in the medial habenula of guinea pig, rat and gerbil. Br J Pharmacol 112:717–719

Boden P, Eden JM, Hodgson J, Horwell DC, Hughes J, McKnight AT, Lewthwaite RA, Pritchard MC, Raphy J, Meecham K, Ratcliffe GS, Suman-Chauhan N, Woodruff GN (1996) Use of a dipeptide chemical library in the development of non-peptide tachykinin NK_3 receptor selective antagonists. J Med Chem 39:1664–1675

Brechenmacher C, Larmet Y, Feltz P, Rodeau JL (1998) Cultured rat sensory neurones express functional tachykinin receptor subtypes 1, 2 and 3. Neurosci Lett 241:159–162

Buell G, Schulz MF, Arkinstall SJ, Maury K, Missotten M, Adami N, Talabot F, Kawashima E (1992) Molecular characterisation, expression and localisation of human neurokinin-3 receptor. FEBS Lett 299:90–95

Canning BJ, Undem, BJ (1994) Evidence that antidromically stimulated vagal afferents activate inhibitory neurones innervating guinea-pig trachealis. J Physiol 480:613–625

Canning BJ, Fischer A, Undem BJ (1998) Pharmacological analysis of the tachykinin receptors that mediate activation of nonadrenergic, noncholinergic relaxant nerves that innervate guinea pig trachealis. J Pharmacol Exp Ther 284:370–377

Canning BJ, Reynolds SM, Mazzone SB (2001) Multiple mechanisms of reflex bronchospasm in guinea pigs. J Appl Physiol 91:2642–2653

Canning BJ, Reynolds SM, Anukwu LU, Kajekar R, Myers AC (2002) Endogenous neurokinins facilitate synaptic transmission in guinea pig airway parasympathetic ganglia. Am J Physiol 283:R320–R330

Cellier E, Barbot L, Regoli D, Couture R (1997) Cardiovascular and behavioural effects of intracerebroventricularly administered tachykinin NK_3 receptor antagonists in the conscious rat. Br J Pharmacol 122:643–654

Chang Y, Hoover DB, Hancock JC, Smith FM (2000) Tachykinin receptor subtypes in the isolated guinea pig heart and their role in mediating responses to neurokinin A. J Pharmacol Exp Ther 294:147–154

Chen LW, Guan ZL, Ding YQ (1997) Neurokinin B receptor (NK_3)-positive neurons expressing c-fos after chemical noxious stimulation on the peritoneum: a double staining study in the nucleus tractus solitarius of the rat. J Hirnforsch 38:363–367

Chung FZ, Wu LH, Vartanian MA, Watling KJ, Guard S, Woodruff GN, Oxender DL (1994) The non-peptide tachykinin NK_2 receptor antagonist SR 48968 interacts with human, but not rat, cloned tachykinin NK_3 receptors. Biochem Biophys Res Commun 198:967–972

Chung FZ, Wu LH, Tian Y, Vartanian MA, Lee H, Bikker J, Humblet C, Pritchard MC, Raphy J, Suman-Chauhan N, Horwell DC, Lalwani ND, Oxender DL (1995) Two classes of structurally different antagonists display similar species preference for the human tachykinin neurokinin$_3$ receptor. Mol Pharmacol 48:711–716

Chretien L, Guillemette G, Regoli D (1994) Non-peptide angiotensin receptor antagonists bind to tachykinin NK_3 receptors of rat and guinea pig brain. Eur J Pharmacol 256:73–78

Cintado CG, Pinto FM, Devillier P, Merida A, Candenas ML (2001) Increase in neurokinin B expression and in tachykinin NK_3 receptor-mediated response and expression in the rat uterus with age. J Pharmacol Exp Ther 299:934–938

Couture R, Picard P, Poulat P, Prat A (1995) Characterization of the tachykinin receptors involved in spinal and supraspinal cardiovascular regulation. Can J Physiol Pharmacol 73:892–902

Crane LH, Williams MJ, Nimmo AJ, Hamlin GP (2002) Estrogen-dependent regulation of neurokinin 3 receptor-mediated uterine contractility in the rat. Biol Reprod 67:1480–1487

Daoui S, Cui YY, Lagente V, Emonds-Alt X, Advenier C (1997) A tachykinin NK_3 receptor antagonist, SR 142801 (osanetant), prevents substance P-induced bronchial hyperreactivity in guinea-pigs. Pulm Pharmacol Ther 10:261–270

Daoui S, Cognon C, Naline E, Emonds-Alt X, Advenier C (1998) Involvement of tachykinin NK_3 receptors in citric acid-induced cough and bronchial responses in guinea pigs. Am J Respir Crit Care Med 158:42–48

Daoui S, Naline E, Lagente V, Emonds-Alt X, Advenier C (2000) Neurokinin B- and specific tachykinin NK_3 receptor agonists-induced airway hyperresponsiveness in the guinea-pig. Br J Pharmacol 130:49–56

Daoui S, Ahnaou A, Naline E, Emonds-Alt X, Lagente V, Advenier C (2001) Tachykinin NK_3 receptor agonists induced microvascular leakage hypersensitivity in the guinea-pig airways. Eur J Pharmacol 433:199–207

Daoui S, D'Agostino B, Gallelli L, Alt XE, Rossi F, Advenier C (2002) Tachykinins and airway microvascular leakage induced by HCl intra-oesophageal instillation. Eur Respir J 20:268–273

Ding YD, Shi J, Su LY, Xu JQ, Su CJ, Guo XE, Ju G (2000) Intracerebroventricular injection of senktide-induced Fos expression in vasopressin-containing hypothalamic neurons in the rat. Brain Res 882:95–102

Ding YQ, Lu CR, Wang H, Su CJ, Chen LW, Zhang YQ, Ju G (2002) Two major distinct subpopulations of neurokinin-3 receptor-expressing neurons in the superficial dorsal horn of the rat spinal cord. Eur J Neurosci 16:551–556

Drapeau G, Rouissi N, Nantel F, Rhaleb NE, Tousignant C, Regoli D (1990) Antagonists for the neurokinin NK-3 receptor evaluated in selective receptor systems. Regul Pept 31:125–135

Emonds-Alt X, Vilain P, Goulaouic P, Proietto V, Van Broeck D, Advenier C, Naline E, Neliat G, Le Fur G, Breliere JC (1992) A potent and selective non-peptide antagonist of the neurokinin A (NK_2) receptor. Life Sci 50:PL101–PL106

Emonds-Alt X, Bichon D, Ducoux JP, Heaulme M, Miloux B, Poncelet M, Proietto V, Van Broeck D, Vilain P, Neliat G, Soubrie P, Le Fur G, Breliere JC (1995) SR 142801, the first potent non-peptide antagonist of the tachykinin NK_3 receptor. Life Sci 56:PL27–PL32

Emonds-Alt X, Proietto V, Steinberg R, Advenier C, Daoui S, Naline E, Gueudet C, Michaud JC, Oury-Donat F, Poncelet M, Vilain P, Le Fur G, Maffrand JP, Soubrie P, Pascal M (2002) Biochemical and pharmacological activities of SSR 146977, a new potent nonpeptide tachykinin NK_3 receptor antagonist. Can J Physiol Pharmacol 80:482–488

Evangelista S (2001) Involvement of tachykinins in intestinal inflammation. Curr Pharm Des 7:19–30

Flynn FW (1999) Brain tachykinins and the regulation of salt intake. Ann N Y Acad Sci 897:173–181

Garret C, Carruette A, Fardin V, Moussaoui S, Peyronel JF, Blanchard JC, Laduron PM (1991) Pharmacological properties of a potent and selective nonpeptide substance P antagonist. Proc Natl Acad Sci USA 88:10208–10212

Giardina GA, Sarau HM, Farina C, Medhurst AD, Grugni M, Foley JJ, Raveglia LF, Schmidt DB, Rigolio R, Vassallo M, Vecchietti V, Hay DW (1996) 2-Phenyl-4-quinolinecarboxamides: a novel class of potent and selective non-peptide competitive antagonists for the human neurokinin-3 receptor. J Med Chem 39:2281–2284

Giardina GA, Sarau HM, Farina C, Medhurst AD, Grugni M, Raveglia LF, Schmidt DB, Rigolio R, Luttmann M, Vecchietti V, Hay DW (1997) Discovery of a novel class of selective non-peptide antagonists for the human neurokinin-3 receptor. 1. Identification of the 4-quinolinecarboxamide framework. J Med Chem 40:1794–1807

Giardina GA, Artico M, Cavagnera S, Cerri A, Consolandi E, Gagliardi S, Graziani D, Grugni M, Hay DW, Luttmann MA, Mena R, Raveglia LF, Rigolio R, Sarau HM, Schmidt DB, Zanoni G, Farina C (1999a) Replacement of the quinoline system in 2-phenyl-4-quinolinecarboxamide NK-3 receptor antagonists. Farmaco 54:364–374

Giardina GA, Raveglia LF, Grugni M, Sarau HM, Farina C, Medhurst AD, Graziani D, Schmidt DB, Rigolio R, Luttmann M, Cavagnera S, Foley JJ, Vecchietti V, Hay DW (1999b) Discovery of a novel class of selective non-peptide antagonists for the human neurokinin-3 receptor. 2. Identification of (S)-N-(1-phenylpropyl)-3-hydroxy-2-phenylquinoline-4-carboxamide (SB223412). J Med Chem 42:1051–1065

Goldhill J, Porquet MF, Selve N (1999) Antisecretory and relaxatory effects of tachykinin antagonists in the guinea pig intestinal tract. J Pharm Pharmacol 51:1041–1048

Guard S, Watson SP (1987) Evidence for neurokinin-3 receptor-mediated tachykinin release in the guinea-pig ileum. Eur J Pharmacol 144:404–412

Guard S, Watling KJ, Watson SP (1988) Neurokinin 3-receptors are linked to inositol phospholipid hydrolysis in the guinea-pig ileum longitudinal muscle-myenteric plexus preparation. Br J Pharmacol 94:141–154

Guard S, Watling KJ, Watson SP (1990) Pharmacological analysis of [^{3}H]-senktide binding to NK$_3$ tachykinin receptors in guinea-pig ileum longitudinal muscle-myenteric plexus and cerebral cortex membranes. Br J Pharmacol 99:767–773

Gueudet C, Santucci V, Soubrie P, Le Fur G (1999) Blockade of neurokinin-3 receptors antagonizes drug-induced population response and depolarization block of midbrain dopamine neurons in guinea pigs. Synapse 33:779

Guo JZ, Yoshioka K, Otsuka M (1998) Effects of a tachykinin NK$_3$ receptor antagonist, SR 142801, studied in isolated neonatal rat spinal cord. Neuropeptides 32:535–542

Hardwick JC, Mawe GM, Parsons RL (1997) Tachykinin-induced activation of non-specific cation conductance via NK$_3$ neurokinin receptors in guinea-pig intracardiac neurones. J Physiol 504:674

Harrison T, Korsgaard MP, Swain CJ, Cascieri MA, Sadowski S, Seabrook GR (1998) High affinity, selective neurokinin 2 and neurokinin 3 receptor antagonists from a common structural template. Bioorg Med Chem Lett 8:1341–1348

Hay DW, Giardina GA, Griswold DE, Underwood DC, Kotzer CJ, Bush B, Potts W, Sandhu P, Lundberg D, Foley JJ, Schmidt DB, Martin LD, Kilian D, Legos JJ, Barone FC, Luttmann MA, Grugni M, Raveglia LF, Sarau HM (2002) Nonpeptide tachykinin receptor antagonists. III. SB235375, a low central nervous system-penetrant, potent and selective neurokinin-3 receptor antagonist, inhibits citric acid-induced cough and airways hyper-reactivity in guinea pigs. J Pharmacol Exp Ther 300:313–323

Holzer P (1998) Implications of tachykinins and calcitonin gene-related peptide in inflammatory bowel disease. Digestion 59:262–283

Holzer P, Holzer-Petsche U (1997) Tachykinins in the gut. Part I. Expression, release and motor function. Pharmacol Ther 73:172–178

Holzer P, Holzer-Petsche U (2001) Tachykinin receptors in the gut: physiological and pathological implications. Curr Opin Pharmacol 1:585–590

Hoover DB, Chang Y, Hancock JC, Zhang L (2000) Actions of tachykinins within the heart and their relevance to cardiovascular disease. Jpn J Pharmacol 84:363–373

Houghton AK, Ogilvie J, Clarke RW (2000) The involvement of tachykinin NK$_2$ and NK$_3$ receptors in central sensitization of a spinal withdrawal reflex in the decerebrated, spinalized rabbit. Neuropharmacology 39:131–140

Huang RR, Cheung AH, Mazina KE, Strader CD, Fong TM (1992) cDNA sequence and heterologous expression of the human neurokinin-3 receptor. Biochem Biophys Res Commun 184:969–972

Humpel C, Saria A (1993) Intranigral injection of selective neurokinin-1 and neurokinin-3 but not neurokinin-2 receptor agonists biphasically modulate striatal dopamine metabolism but not striatal preprotachykinin-A mRNA in the rat. Neurosci Lett 157:222–226

Jacoby HI, Lopez I, Wright D, Vaught JL (1986) Differentiation of multiple neurokinin receptors in the guinea pig ileum. Life Sci 39:1995–2003

Jenkinson KM, Morgan JM, Furness JB, Southwell BR (1999) Neurons bearing NK$_3$ tachykinin receptors in the guinea-pig ileum revealed by specific binding of fluorescently labelled agonists. Histochem Cell Biol 112:233–246

Jobling P, Messenger JP, Gibbins IL (2001) Differential expression of functionally identified and immunohistochemically identified NK$_1$ receptors on sympathetic neurons. J Neurophysiol 85:1888–1898

Johnson PJ, Bornstein JC, Burcher E (1998) Roles of neuronal NK$_1$ and NK$_3$ receptors in synaptic transmission during motility reflexes in the guinea-pig ileum. Br J Pharmacol 14:1375–1384

Julia V, Su X, Bueno L, Gebhart GF (1999) Role of neurokinin 3 receptors on responses to colorectal distention in the rat: electrophysiological and behavioral studies. Gastroenterology 116:1124–1131

Kamali F (2001) Osanetant Sanofi-Synthelabo. Curr Opin Invest Drugs 2:950–956

Kamp EH, Beck DR, Gebhart GF (2001) Combinations of neurokinin receptor antagonists reduce visceral hyperalgesia. J Pharmacol Exp Ther 299:105–113

Keegan KD, Woodruff GN, Pinnock RD (1992) The selective NK_3 receptor agonist senktide excites a subpopulation of dopamine-sensitive neurones in the rat substantia nigra pars compacta in vitro. Br J Pharmacol:3–5

Kemel ML, Perez S, Godeheu G, Soubrie P, Glowinski J (2002) Facilitation by endogenous tachykinins of the NMDA-evoked release of acetylcholine after acute and chronic suppression of dopaminergic transmission in the matrix of the rat striatum. J Neurosci 22:1929–1936

Kung T, Crawley Y, Jones H, Greenfeder S, Anthes J, Lira S, Wiekowski M, Cook D, Hey J, Egan RW, Chapman RW (2002) Development of allergic pulmonary inflammation and airway hyperresponsiveness in NK_3 deficient mice. Am J Resp Crit Care Med 165:A311

Langlois X, Wintmolders C, te Riele P, Leysen JE, Jurzak M (2001) Detailed distribution of Neurokinin 3 receptors in the rat, guinea pig and gerbil brain: a comparative autoradiographic study. Neuropharmacology 40:242–253

Leroy V, Mauser P, Gao Z, Peet NP (2000) Neurokinin receptor antagonists. Expert Opin Investig Drugs 9:735–746

Lessard A, Couture R (2001) Modulation of cardiac activity by tachykinins in the rat substantia nigra. Br J Pharmacol 134:1749–1759

Lessard A, Campos MM, Neugebauer W, Couture R (2003) Implication of nigral tachykinin NK_3 receptors in the maintenance of hypertension in spontaneously hypertensive rats: a pharmacologic and autoradiographic study. Br J Pharmacol 138: 554–563

Linari G, Broccardo M, Nucerito V, Improta G (2002) Selective tachykinin NK_3-receptor agonists stimulate in vitro exocrine pancreatic secretion in the guinea pig. Peptides 23:947–953

Linden DR, Jia YP, Seybold VS (1999) Spinal neurokinin$_3$ receptors facilitate the nociceptive flexor reflex via a pathway involving nitric oxide. Pain 80:301–308

Linden DR, Seybold VS (1999) Spinal neurokinin$_3$ receptors mediate thermal but not mechanical hyperalgesia via nitric oxide. Pain 80:309–317

Maggi CA, Zagorodnyuk V, Giuliani S (1994) Tachykinin NK_3 receptor mediates NANC hyperpolarization and relaxation via nitric oxide release in the circular muscle of the guinea-pig colon. Regul Pept 53:259–274

Mann PT, Southwell BR, Ding YQ, Shigemoto R, Mizuno N, Furness JB (1997) Localisation of neurokinin 3 (NK_3) receptor immunoreactivity in the rat gastrointestinal tract. Cell Tissue Res 289:1–9

Manning BP, Mawe GM (2001) Tachykinins mediate slow excitatory postsynaptic transmission in guinea pig sphincter of Oddi ganglia. Am J Physiol 281:G357–G364

Marchetti C, Nistri A (2001) Neuronal bursting induced by NK3 receptor activation in the neonatal rat spinal cord in vitro. J Neurophysiol. 86:2939–2950.

Marco N, Thirion A, Mons G, Bougault I, Le Fur G, Soubrie P, Steinberg R (1998) Activation of dopaminergic and cholinergic neurotransmission by tachykinin NK_3 receptor stimulation: an in vivo microdialysis approach in guinea pig. Neuropeptides 32:481–488

Massi M, Panocka I, de Caro G (2000) The psychopharmacology of tachykinin NK-3 receptors in laboratory animals. Peptides 21:1597–1609

Mastrangelo D, Mathison R, Huggel HJ, Dion S, D'Orleans-Juste P, Rhaleb NE, Drapeau G, Rovero P, Regoli D (1987) The rat isolated portal vein: a preparation sensitive to neurokinins, particularly neurokinin B. Eur J Pharmacol 134:321–326

Mauser PJ, Quinn S, Selig W, Alvarez A, Marcus, R, Brooks K, Wasserman MA, (2000) HMR 2091: A dual NK_1/NK_2 neurokinin receptor antagonist. Am J Resp Crit Care Med 161:A437

Mawe GM (1995) Tachykinins as mediators of slow EPSPs in guinea-pig gall-bladder ganglia: involvement of neurokinin-3 receptors. J Physiol 485:513–524

Mazzone SB, Geraghty DP (1999) Respiratory action of capsaicin microinjected into the nucleus of the solitary tract: involvement of vanilloid and tachykinin receptors. Br J Pharmacol 127:473–481

Mazzone SB, Geraghty DP (2000a) Characterization and regulation of tachykinin receptors in the nucleus tractus solitarius. Clin Exp Pharmacol Physiol 27:939–942

Mazzone SB, Geraghty DP (2000b) Respiratory actions of tachykinins in the nucleus of the solitary tract: characterization of receptors using selective agonists and antagonists. Br J Pharmacol 129:1121–1231

Mazzone SB, Canning BJ (2002) Synergistic interactions between airway afferent nerve subtypes mediating reflex bronchospasm in guinea pigs. Am J Physiol 283:R86–R98

Mazzone SB, Canning BJ (2003) Interactions between airway afferent nerve subtypes mediating cough. Am J Resp Crit Care Med 167:A146

Mazzone SB (2003) Targeting tachykinins for the treatment of obstructive airways disease. Am J Resp Med in press

Medhurst AD, Parsons AA, Roberts JC, Hay DW (1997) Characterization of NK_3 receptors in rabbit isolated iris sphincter muscle. Br J Pharmacol 120:93–101

Medhurst AD, Hirst WD, Jerman JC, Meakin J, Roberts JC, Testa T, Smart D (1999) Molecular and pharmacological characterization of a functional tachykinin NK_3 receptor cloned from the rabbit iris sphincter muscle. Br J Pharmacol 128:627–636

McCarson KE, Krause JE (1994) NK-1 and NK-3 type tachykinin receptor mRNA expression in the rat spinal cord dorsal horn is increased during adjuvant or formalin-induced nociception. J Neurosci 14:712–720

Michaud JC, Soubrie P, Le Fur G (1995) Antagonism by SR142801, a non-peptide NK_3 receptor antagonist, of senktide-evoked increase in firing rate in guinea-pig locus coeruleus slices. Proc Br Pharmacol Soc P59

Michl T, Jocic M, Schuligoi R, Holzer P (2001) Role of tachykinin receptors in the central processing of afferent input from the acid-threatened rat stomach. Regul Pept 102:119–126

Mileusnic D, Lee JM, Magnuson DJ, Hejna MJ, Krause JE, Lorens JB, Lorens SA (1999) Neurokinin-3 receptor distribution in rat and human brain: an immunohistochemical study. Neuroscience 89:1269–1290

Mitolo-Chieppa D, Mansi G, Nacci C, De Salvia MA, Montagnani M, Potenza MA, Rinaldi R, Lerro G, Siro-Brigiani G, Mitolo CI, Rinaldi M, Altomare DF, Memeo V (2001) Idiopathic chronic constipation: tachykinins as cotransmitters in colonic contraction. Eur J Clin Invest 31:349–355

Mizuta A, Takano Y, Honda K, Saito R, Matsumoto T, Kamiya H (1995) Nitric oxide is a mediator of tachykinin NK_3 receptor-induced relaxation in rat mesenteric artery. Br J Pharmacol 116:2919–2922

Moreaux B, Nemmar A, Vincke G, Halloy D, Beerens D, Advenier C, Gustin P (2000) Role of substance P and tachykinin receptor antagonists in citric acid-induced cough in pigs. Eur J Pharmacol 408:305–312

Myers AC, Undem BJ (1993) Electrophysiological effects of tachykinins and capsaicin on guinea-pig bronchial parasympathetic ganglion neurones. J Physiol 470:665–679

Nalivaiko E, Michaud JC, Soubrie P, Le Fur G, Feltz P (1997) Tachykinin neurokinin-1 and neurokinin-3 receptor-mediated responses in guinea-pig substantia nigra: an in vitro electrophysiological study. Neuroscience 78:745–757

Nenan S, Germain N, Lagente V, Emonds-Alt X, Advenier C, Boichot E (2001) Inhibition of inflammatory cell recruitment by the tachykinin NK_3 receptor antagonist, SR 142801, in a murine model of asthma. Eur J Pharmacol 421:201–205

Nguyen QT, Jukic D, Chretien L, Gobeil F, Boussougou M, Regoli D (1994) Two NK-3 receptor subtypes: demonstration by biological and binding assays. Neuropeptides 27:157–161

Nguyen-Le XK, Nguyen QT, Gobeil F, Pheng LH, Emonds-Alt X, Breliere JC, Regoli D (1996) Pharmacological characterization of SR 142801: a new non-peptide antagonist of the neurokinin NK-3 receptor. Pharmacology 52:283–291

Nishi T, Ishibashi K, Takemoto T, Nakajima K, Fukazawa T, Iio Y, Itoh K, Mukaiyama O, Yamaguchi T (2000) Combined tachykinin receptor antagonist: Synthesis and stereochemical structure-activity relationships of novel morpholine analogues. Bioorg Med Chem Lett 10:1665–1668

Norris SK, Boden PR, Woodruff GN (1993) Agonists selective for tachykinin NK_1 and NK_3 receptors excite subpopulations of neurons in the rat medial habenula nucleus in vitro. Eur J Pharmacol 234:223–228

Ohkubo H, Nakanishi S (1991) Molecular characterization of the three tachykinin receptors. Ann N Y Acad Sci 632:53–62

Oury-Donat F, Carayon P, Thurneyssen O, Pailhon V, Emonds-Alt X, Soubrie P, Le Fur G (1995) Functional characterization of the nonpeptide neurokinin3 (NK_3) receptor antagonist, SR142801 on the human NK_3 receptor expressed in Chinese hamster ovary cells. J Pharmacol Exp Ther 274:148–154

Oyamada H, Takatsuji K, Senba E, Mantyh PW, Tohyama M (1999) Postnatal development of NK_1, NK_2, and NK_3 neurokinin receptors expression in the rat retina. Brain Res Dev Brain Res 117:59–70

Page NM, Woods RJ, Gardiner SM, Lomthaisong K, Gladwell RT, Butlin DJ, Manyonda IT, Lowry PJ (2000) Excessive placental secretion of neurokinin B during the third trimester causes pre-eclampsia. Nature 405:797–800

Page NM, Woods RJ, Lowry PJ (2001) A regulatory role for neurokinin B in placental physiology and pre-eclampsia. Regul Pept 98:97–104

Page NM, Bell NJ (2002) The human tachykinin NK_1 (short form) and tachykinin NK_4 receptor: a reappraisal. Eur J Pharmacol 437:27–30

Panocka I, Massi M, Lapo I, Swiderski T, Kowalczyk M, Sadowski B (2001) Antidepressant-type effect of the NK_3 tachykinin receptor agonist aminosenktide in mouse lines differing in endogenous opioid system activity. Peptides 22:1037–1042

Patacchini R, Bartho L, Holzer P, Maggi CA (1995) Activity of SR 142801 at peripheral tachykinin receptors. Eur J Pharmacol 278:17–25

Petitet F, Beaujouan JC, Saffroy M, Torrens Y, Glowinski J (1993a) The nonpeptide NK-2 antagonist SR 48968 is also a NK-3 antagonist in the guinea but not in the rat. Biochem Biophys Res Commun 191:180–187

Petitet F, Saffroy M, Torrens Y, Glowinski J, Beaujouan JC (1993b) A new selective bioassay for tachykinin NK_3 receptors based on inositol monophosphate accumulation in the guinea pig ileum. Eur J Pharmacol 247:185–191

Phillips JE, Hey JA, Corboz MR (2003) Tachykinin NK_3 and NK_1 receptor activation elicits secretion from porcine airway submucosal glands. Br J Pharmacol 138:254–260

Picard P, Chretien L, Couture R (1995) Functional interaction between losartan and central tachykinin NK_3 receptors in the conscious rat. Br J Pharmacol 114:1563–1570

Pinnock RD, Suman-Chauhan N, Chung FZ, Webdale L, Madden Z, Hill DR, Woodruff GN (1994) Characterization of tachykinin mediated increases in $[Ca^{2+}]_i$ in Chinese hamster ovary cells expressing human tachykinin NK_3 receptors. Eur J Pharmacol 269:73–78

Pinto FM, Saulnier JP, Faisy C, Naline E, Molimard M, Prieto L, Martin JD, Emonds-Alt X, Advenier C, Candenas ML (2002) SR 142801, a tachykinin NK_3 receptor antagonist, prevents beta$_2$-adrenoceptor agonist-induced hyperresponsiveness to neurokinin A in guinea-pig isolated trachea. Life Sci 72:307–320

Piot O, Betschart J, Grall I, Ravard S, Garret C, Blanchard JC (1995) Comparative behavioural profile of centrally administered tachykinin NK_1, NK_2 and NK_3 receptor agonists in the guinea-pig. Br J Pharmacol 116:2496–2502

Ribeiro SJ, Teixeira RM, Calixto JB, De Lima TC (1999) Tachykinin NK_3 receptor involvement in anxiety. Neuropeptides 33:181–188

Roccon A, Marchionni D, Nisato D (1996) Study of SR 142801, a new potent non-peptide NK$_3$ receptor antagonist on cardiovascular responses in conscious guinea-pig. Br J Pharmacol 118:1095–1102

Regoli D, Nguyen QT, Jukic D (1994) Neurokinin receptor subtypes characterized by biological assays. Life Sci 54:2035–2047

Reichard GA, Grice CA, Shih NY, Spitler J, Majmundar S, Wang SD, Paliwal S, Anthes JC, Piwinski JJ (2002) Preparation of oxime dual NK$_1$/NK$_2$ antagonists with reduced NK$_3$ affinity. Bioorg Med Chem Lett 12:2355–2358

Rogers DF (2001) Tachykinin receptor antagonists for asthma and COPD. Expert Opin Ther Patents 11:1097–1121

Rumsey WL, Aharony D, Bialecki RA, Abbott BM, Barthlow HG, Caccese R, Ghanekar S, Lengel D, McCarthy M, Wenrich B, Undem B, Ohnmacht C, Shenvi A, Albert JS, Brown F, Bernstein PR, Russell K (2001) Pharmacological characterization of ZD6021: a novel, orally active antagonist of the tachykinin receptors. J Pharmacol Exp Ther 298:307–315

Sadowski S, Huang RR, Fong TM, Marko O, Cascieri MA (1993) Characterization of the binding of [^{125}I-iodo-histidyl, methyl-Phe7] neurokinin B to the neurokinin-3 receptor. Neuropeptides 24:317–319

Saffroy M, Beaujouan JC, Torrens Y, Besseyre J, Bergstrom L, Glowinski J (1988) Localization of tachykinin binding sites (NK$_1$, NK$_2$, NK$_3$ ligands) in the rat brain. Peptides 9:227–241

Santos AR, Calixto JB (1997) Further evidence for the involvement of tachykinin receptor subtypes in formalin and capsaicin models of pain in mice. Neuropeptides 31:381–389

Sarau HM, Griswold DE, Potts W, Foley JJ, Schmidt DB, Webb EF, Martin LD, Brawner ME, Elshourbagy NA, Medhurst AD, Giardina GA, Hay DW (1997) Nonpeptide tachykinin receptor antagonists: I. Pharmacological and pharmacokinetic characterization of SB 223412, a novel, potent and selective neurokinin-3 receptor antagonist. J Pharmacol Exp Ther 281:1303–1311

Sarau HM, Feild JA, Ames RS, Foley JJ, Nuthulaganti P, Schmidt DB, Buckley PT, Elshourbagy NA, Brawner ME, Luttmann MA, Giardina GA, Hay DW (2001) Molecular and pharmacological characterization of the murine tachykinin NK$_3$ receptor. Eur J Pharmacol 413:143–150

Sarau HM, Griswold DE, Bush B, Potts W, Sandhu P, Lundberg D, Foley JJ, Schmidt DB, Webb EF, Martin LD, Legos JJ, Whitmore RG, Barone FC, Medhurst AD, Luttmann MA, Giardina GA, Hay DW (2000) Nonpeptide tachykinin receptor antagonists. II. Pharmacological and pharmacokinetic profile of SB-222200, a central nervous system penetrant, potent and selective NK-3 receptor antagonist. J Pharmacol Exp Ther 295:373–381

Schmid G, Carita F, Bonanno G, Raiteri M (1998) NK-3 receptors mediate enhancement of substance P release from capsaicin-sensitive spinal cord afferent terminals. Br J Pharmacol 125:621–626

Seabrook GR, Main M, Bowery B, Wood N, Hill RG (1992) Differences in neurokinin receptor pharmacology between rat and guinea pig superior cervical ganglia. Br J Pharmacol.105:925–928

Seabrook GR, Bowery BJ, Hill RG (1995) Pharmacology of tachykinin receptors on neurons in the ventral tegmental area of rat brain slices. Eur J Pharmacol 273:113–119

Shigemoto R, Yokota Y, Tsuchida K, Nakanishi S (1990) Cloning and expression of a rat neuromedin K receptor cDNA. J Biol Chem 265:623–628

Snider RM, Constantine JW, Lowe JA, Longo KP, Lebel WS, Woody HA, Drozda SE, Desai MC, Vinick FJ, Spencer RW, Hess HJ (1991) A potent nonpeptide antagonist of the substance P (NK$_1$) receptor. Science 251(4992):435–437

Soejima T, Endoh T, Suzuki T (1999) Tachykinin-induced responses via neurokinin-1 and -3 receptors in hamster submandibular ganglion neurones. Arch Oral Biol 44:455–463

Stables JM, Beresford IJ, Arkinstall S, Ireland SJ, Walsh DM, Seale PW, Ward P, Hagan RM (1994) GR138676, a novel peptidic tachykinin antagonist which is potent at NK_3 receptors. Neuropeptides 27:333–341

Stacey AE, Woodhall GL, Jones RS (2002) Neurokinin-receptor-mediated depolarization of cortical neurons elicits an increase in glutamate release at excitatory synapses. Eur J Neurosci 16:1896–1906

Suman-Chauhan N, Grimson P, Guard S, Madden Z, Chung FZ, Watling K, Pinnock R, Woodruff G (1994) Characterisation of [^{125}I][MePhe7]neurokinin B binding to tachykinin NK_3 receptors: evidence for interspecies variance. Eur J Pharmacol 269:65–72

Thompson GW, Hoover DB, Ardell JL, Armour JA (1998) Canine intrinsic cardiac neurons involved in cardiac regulation possess NK_1, NK_2, and NK_3 receptors. Am J Physiol. 275:R1683–R1689

Thornton PD, Bornstein JC (2002) Slow excitatory synaptic potentials evoked by distension in myenteric descending interneurones of guinea-pig ileum. J Physiol 539:589–602

Tremblay L, Kemel ML, Desban M, Gauchy C, Glowinski J (1992) Distinct presynaptic control of dopamine release in striosomal- and matrix-enriched areas of the rat striatum by selective agonists of NK_1, NK_2, and NK_3 tachykinin receptors. Proc Natl Acad Sci USA 89:11214–11218

Vilain P, Emonds-Alt X, Le Fur G, Breliere J-C (1997) Tachykinin-induced contractions of the guinea pig ileum longitudinal smooth muscle: tonic and phasic muscular activities. Can J Physiol Pharmacol 75:587–590

Wang MJ, Xiong SH, Li ZW (2001) Neurokinin B potentiates ATP-activated currents in rat DRG neurons. Brain Res 923:157–162

Wang ZY, Hakanson R (1993) The rabbit iris sphincter contains NK_1 and NK_3 but not NK_2 receptors: a study with selective agonists and antagonists. Regul Pept 44:269–275

Wareing M, Bhatti H, O'Hara M, Kenny L, Warren AY, Taggart MJ, Baker PN (2003) Vasoactive effects of neurokinin B on human blood vessels. Am J Obstet Gynecol 188:196–202

Wienrich M, Reuss K, Harting J (1989) Effects of receptor-selective neurokinin agonists and a neurokinin antagonist on the electrical activity of spinal cord neurons in culture. Br J Pharmacol 98:914–920

Wu LH, Vartanian MA, Oxender DL, Chung FZ (1994) Identification of methionine134 and alanine146 in the second transmembrane segment of the human tachykinin NK_3 receptor as reduces involved in species-selective binding to SR 48968. Biochem Biophys Res Commun 198:961–966

Yip J, Chahl LA (1997) Localization of Fos-like immunoreactivity induced by the NK_3 tachykinin receptor agonist, senktide, in the guinea-pig brain. Br J Pharmacol 122:715–725

Yip J, Chahl LA (2001) Localization of NK_1 and NK_3 receptors in guinea-pig brain. Regul Pept 98:55–62

Yuan YD, Couture R (1997) Renal effects of intracerebroventricularly injected tachykinins in the conscious saline-loaded rat: receptor characterization. Br J Pharmacol 120:785–796

Zaratin P, Angelici O, Clarke GD, Schmid G, Raiteri M, Carita F, Bonanno G (2000) NK_3 receptor blockade prevents hyperalgesia and the associated spinal cord substance P release in monoarthritic rats. Neuropharmacology 39:141–149

Zerari F, Karpitskiy V, Krause J, Descarries L, Couture R (1997) Immunoelectron microscopic localization of NK-3 receptor in the rat spinal cord. NeuroReport 8:2661–2664

Zhao FY, Saito K, Yoshioka K, Guo JZ, Murakoshi T, Konishi S, Otsuka M (1995) Subtypes of tachykinin receptors on tonic and phasic neurones in celiac ganglion of the guinea-pig. Br J Pharmacol 115:25–30

Combined Tachykinin NK$_1$, NK$_2$, and NK$_3$ Receptor Antagonists

W. L. Rumsey · J. K. Kerns

Respiratory and Inflammation Center of Excellence for Drug Discovery, GlaxoSmithKline, 709 Swedeland Road, P.O. Box 1539, King of Prussia, PA, 19406-0939 USA
e-mail: Bill.2.Rumsey@gsk.com

1	Introduction	274
2	Tachykinins and the Lung	275
3	Synergy of Combined Tachykinin Receptor Blockade in Animals	278
4	Medicinal Chemistry Overview of Combined Antagonists	280
5	Pharmacology of Combined Antagonists	286
5.1	Nonpeptide Antagonists	286
5.2	Pseudopeptide Antagonists	288
5.3	Clinical Trials	289
6	Summary	289
	References	290

Abstract The tachykinins substance P, neurokinin A, and neurokinin B bind to their preferred receptors, termed NK$_1$, NK$_2$, and NK$_3$, respectively, and with appreciable affinity to each of the other tachykinin receptors, e.g., neurokinin A to the NK$_1$ receptor. In some tissues, substance P and neurokinin A are colocalized, and upon release from nerve terminals, produce a robust response such as bronchoconstriction. In many instances, pharmacological blockade of more than one of the tachykinin receptors results in far greater protection than is afforded by antagonism of a single receptor. There has been considerable effort to discover potent antagonists with combined affinity for two or three of the tachykinin receptors, termed dual or pan-tachykinin antagonists. These compounds have mostly derived from a single molecular template and the key substituents that drive pan-antagonism have been identified. As yet, there have been few reports of clinical trials to evaluate the efficacy of these compounds for human disease.

Keywords Dual · Tachykinin · Pan-tachykinin · Airway · Respiratory · Synergism

1
Introduction

In mammalian tissues, the tachykinins (substance P, SP; neurokinin A, NKA; and neurokinin B, NKB) represent a small group of structurally related peptides that are widely distributed in the central and peripheral nervous systems. Neuronal cells are the primary source of the tachykinins where they serve as neurotransmitters and/or modulators of neurotransmission to influence a broad array of biological actions such as contraction of smooth muscle within a number of tissues; cardiovascular, respiratory, digestive, reproductive, and musculoskeletal (Otsuka and Yoshioka 1993). Cellular sources of SP other than neurons are limited but include immune cells (Maggi 1997).

The mammalian tachykinins are encoded by two different genes, termed preprotachykinin-I and preprotachykinin-II (also known as PPT-A and PPT-B). The former contains the blueprint for SP and NKA while the latter contains that for NKB. The amounts of the precursor mRNAs are regulated in a tissue-specific manner and this determines, in part, the local levels of the peptides (Otsuka and Yoshioka 1993). Not surprisingly, SP and NKA are co-expressed in several tissues; examples include skin (Schulze et al. 1997), bladder (Smet et al. 1997) and lung (Kummer et al. 1992). SP and NKB (or their mRNAs) are colocalized in enteric nerves of the rat ileum (Yunker et al. 1999) and medial habenula (Burgunder and Young 1989).

Biological activity of SP, NKA, and NKB is conferred via binding to three distinct receptors, termed NK_1R, NK_2R, and NK_3R respectively, which are members of the G-protein-coupled receptor superfamily. Distribution of these tachykinin receptors is species- and tissue-specific. Each of the tachykinin peptides is capable of binding with appreciable affinity to its preferred receptor, e.g., SP to the NK_1R, and to another of the tachykinin receptors, SP to the NK_2R (Maggi and Schwartz 1997). In guinea pig lung, both SP and NKA are released from nerve terminals upon exposure to the C-fiber irritant, capsaicin, which provokes a profound decline in airway function (Saria et al. 1988). Capsaicin-mediated bronchoconstriction is attenuated by blocking the NK_2R, but airway resistance is markedly improved by concomitant antagonism of the NK_1R (Buckner et al. 1993; Rumsey et al. 2001). Although several selective tachykinin receptor antagonists have been discovered, a greater understanding of tachykinin-mediated processes may be afforded by discovery of compounds with affinity for more than one of the tachykinin receptors. This chapter focuses on dual (NK_1R plus NK_2R) and combined NK_1R, NK_2R, and NK_3R (termed pan-tachykinin) antagonists. Although such molecules may be applied for study of extra-pulmonary tissues and diseases, specific attention, albeit limited, is given to tachykinins and their actions in the lung since much of the pharmacology of these antagonists has been determined with respiratory tissues and models. A broader review of tachykinins and airway disease is provided in the chapter by G.F.Joos in this volume.

2
Tachykinins and the Lung

The principal tachykinins in the lung are SP and NKA whereas NKB is not found in significant quantities (Hua et al. 1985). These peptides are present in vagal, typically C-fiber, afferent neurons, and are synthesized in neuronal cell bodies. They are carried via axonal transport to central and peripheral terminals for release and subsequent interaction with their respective tachykinin receptors. SP and NKA are colocalized in the sensory neurons that essentially innervate all compartments of the airway wall, from trachea to bronchioles (Kummer et al. 1992; Advenier et al. 1997). SP-immunoreactive (SP-IR) nerves are found within four main locations of the guinea pig and human lung; the epithelium, around blood vessels, within the bronchial smooth muscle layer, and around local tracheo-bronchial ganglion cells (Lundberg et al. 1984; Myers et al. 1996; Olerenshaw et al. 1991). In rat, SP-IR cells possess mostly single axons which contain varicosities, and are associated with the epithelium, mucosal arterioles, post-capillary venules, and smooth muscle of the trachea and bronchi (Baluk et al. 1992). The distribution of SP-IR nerves in the lungs of rodents as compared to large mammals and humans has been open to debate (see for example, Laitinen et al. 1983). Lamb and Sparrow (2002) found that the arrangement of SP-IR nerves to be very similar in pig and human samples. Three-dimensional maps produced with confocal microscopy revealed varicose SP-IR nerves in nerve trunks adjacent to airway smooth muscle and bundles traveling to the epithelium. The nerves formed lateral branches that were near arterioles and mucous glands. Those nerves that reached the epithelium (94% stained for SP in pig but modestly in human) passed through the basement membrane and spread out to form a network at the base of the epithelium sending fibers toward the lumen and around the circumference of goblet cells. Unlike those found in rat (Baluk et al. 1996), few SP-IR fibers were observed near postcapillary venules of porcine and human lung (Lamb and Sparrow 2002).

Autoradiographic mapping of NK_1R distribution in the guinea pig lung using either [^{125}I]-Bolton-Hunter-SP or the selective NK_1R antagonist, [^{3}H]CP96345, showed specific labeling on airway smooth muscle from central to peripheral airways, submucosal glands, nerve fibers of the trachea, and pulmonary blood vessels (Carstairs and Barnes 1986; Zhang et al. 1995). As the size of the airway decreased, the labeling of epithelium progressively increased. Although similar findings were obtained in human lung (Carstairs and Barnes 1986), Walsh and colleagues (1994) did not find NK_1R on smooth muscle of bronchus, trachea or pulmonary artery. Rather, these receptors appeared limited to microvascular endothelium. Binding affinity of iodinated SP, however, may be influenced by G-protein coupling or peptidase-induced degradation (Coats and Gerard 1989), possibly contributing to discrepancies in the different studies. Immunohistochemical detection showed NK_1R on endothelial cells of capillaries and postcapillary venules in the tracheo-bronchial circulation, epithelial cells and glands of rats, mice and guinea pigs (Baluk et al. 1996; Bowden et al. 1996). Until recently,

NK$_2$R appeared to be restricted to airway smooth muscle on mostly large airways (Baluk et al. 1996; Mak et al. 1996; Strigas and Burcher 1996). Positive immunostaining for both NK$_1$R and NK$_2$R was obtained in the bronchial smooth muscle layer, myoepithelial cells of bronchial glands, smooth muscle layer of bronchial vessels, and in the endothelium and smooth muscle layer of pulmonary arteries of human lung (Mapp et al. 2000).

Neurons containing NKB are primarily in the central nervous system (for review see Severini et al. 2002) where they are found in many brain areas (Merchenthaler et al. 1992) and in the spinal cord, mostly the dorsal horn (Ogawa et al. 1985). SP is also present in the dorsal horn but is not colocalized with NKB (McLeod et al. 2000). NKB has not been identified in airway nerves (Kummer et al. 1992). Nonetheless, the NK$_3$R is detectable in human airway ganglia (A. Myers, personal communication) and is broadly distributed in the brain, including the nucleus of the solitary tract (Ding et al. 1996).

SP, NKA, NKB or their analogues stimulate contraction of isolated human bronchus. Independent of the specific agonist, the primary contractile response comes about by activation of the NK$_2$R (Ellis et al. 1993, 1997; Sheldrick et al. 1995). The increased tone resulting from the SP analog, Ac-[Arg^6,Sar^9,Met(O$_2$)^{11}]SP(6–11) (or ASMSP) was weak (<50% of the maximal response) relative to that elicited by the NK$_2$R agonist, β-Ala^8-NKA(4–10), and in some cases, was prevented by selective NK$_2$R antagonists (Ellis et al. 1993). However, there are reports of these weak contractions, in particular with peripheral bronchi, being blocked by selective NK$_1$R antagonists and unaffected by NK$_2$R blockade (Naline et al. 1996; Amadesi et al. 2001). Specific immunohistochemical staining identified the presence of NK$_1$R in the bronchial smooth muscle layer of these tissue segments (Amadesi et al. 2001). Thus, tachykinins produce contraction of isolated human bronchus either directly via activation of their preferred receptors or indirectly through binding to another tachykinin receptor, i.e., SP to the NK$_2$R. Although NK$_1$R activation cannot be dismissed as a contributor to tachykinin-mediated bronchoconstriction, the more pronounced contractile effect is brought about by the interaction of NKA with the NK$_2$R.

Activation of the tachykinin receptors in the human lung in vivo via inhalation of SP or NKA alters airway function (Fuller et al. 1987; Joos et al. 1987; Crimi et al. 1988, 1990; Cheung et al. 1992; 1993, 1994; Van Rensen et al. 2002). In normal persons and in asthmatics, NKA induces bronchoconstriction that is enhanced by inhibiting neutral endopeptidase, a principal route of tachykinin degradation (Cheung et al. 1992, 1993). Inhalation of SP weakly stimulates cholinergic neurons (Crimi et al. 1990) and increases responsiveness to methacholine in asthmatics (Cheung et al. 1994). The effect of inhaled SP is not limited to changes of airway caliber in asthmatics. SP, but not NKA, significantly elevates markers of microvascular leakage in induced sputum, i.e., α_2-macroglobulin, ceruplasmin, and albumin (Van Ressen et al. 2002). Some workers suggest that pulmonary responses to exogenous tachykinins are restricted to patients with respiratory diseases, with SP or NKA having little influence on pulmonary function in normal individuals (Joos et al. 1987).

Neural regulation of airway smooth muscle tone and airway secretions occurs in large part via excitatory parasympathetic and nonadrenergic, noncholinergic innervation. The vagal autonomic efferent pathway is the dominant bronchoconstrictor limb in man and other species (Goldie et al. 2001). Sensory afferent signals initiate or influence efferent vagal impulses to the airway wall (Goldie et al. 2001). Activation of NK_1R and NK_3R in airway ganglia by endogenous tachykinins contributes to parasympathetic nerve trafficking (Canning et al. 2002; Myers and Undem 1993; Myers et al. 1996; Watson et al. 1993). Airway ganglion neurons express PPT-A and NK_1R mRNAs and their protein products may serve in an autocrine or paracrine manner to modify neural input to the lung (Perez Fontan et al. 2000).

In the rat (Springall et al. 1987) and guinea pig (Kummer et al. 1992), afferent nerve fibers innervating the trachea arise from the superior (jugular) and inferior (nodose) vagal sensory ganglia. The lung also receives sensory innervation from the dorsal root ganglion. Tachykinin-immunoreactive fibers innervating the airways originate in the jugular ganglion and dorsal root ganglion but not the nodose ganglion (Saria et al. 1985; Kummer et al. 1992; Riccio et al. 1996). Normally, the tachykinin-expressing fibers innervating the guinea pig airways are limited to nociceptive C-fibers. However, during pathological conditions, i.e., in response to allergen exposure or respiratory infection, a phenotypic switch occurs and SP- and NKA-immunoreactivity are found in myelinated low threshold mechanosensitive fibers (Myers et al. 2002; Carr et al. 2002). This induction of tachykinins in nodose neurons appears to be secondary to an elevation in the transcription of PPT-A (Fischer et al. 1996).

Afferent neurons which affect respiratory reflexes terminate in the nucleus of the solitary tract (Jordan 2001). Activation of secondary afferent neurons in this region elicits a variety of reflexes resulting in changes in breathing pattern, cough, bronchoconstriction and mucus secretion (Carr and Ellis 2002). Tachykinins released from nerve terminals within the brain stem also facilitate cholinergic reflexes, thereby modulating these airway functions (Mazzone and Canning 2002).

Tachykinins may contribute to airway dysfunction and other diseases. For example, SP was elevated in samples of induced sputum obtained from patients with asthma and chronic obstructive pulmonary disease (Tomaki et al. 1995). Tachykinin receptor mRNA expression was also increased in these patient populations (Adcock et al. 1993; Bai et al. 1995). Patients with rhinitis, asthma, chronic obstructive pulmonary disease, and interstitial lung disease, as well as those suffering from gastroesophageal reflux disease experience increased sensitivity to capsaicin inhalation (Higgenbottam 2002; Doherty et al. 2000). Capsaicin, the pungent substance extracted from red peppers, is a strong irritant that excites sensory C-fibers in the lung and releases both SP and NKA (Saria et al. 1988). Chronic cough may arise from an aberrant and enhanced sensitivity of the afferent limb of the cough reflex (Higgenbottam 2002). In animals, administration of SP or NKA induces an inflammation-like response, termed neurogenic inflammation, and the effects include hyperresponsiveness, vasodilatation, plasma extravasation, leukocyte adhesion and migration, and increased mucus secretion (Advenier et al. 1997). These responses are consistent with those found in some airway diseases.

3
Synergy of Combined Tachykinin Receptor Blockade in Animals

SP and NKA cause profound pulmonary effects in animals. Treatment with either selective NK_1R or NK_2R antagonists provides partial protection, however, preservation of airway function following tachykinin exposure has been optimally achieved by blocking both receptors (e.g., Kusner et al. 1992, Bertrand et al. 1993, Buckner et al. 1993). In many cases, a clear synergistic action can be obtained when these selective antagonists are used in combination.

Bertrand and coworkers (1993) showed that bronchoconstriction in response to NKA was reduced by 82% in guinea pigs by pretreatment with the selective NK_2R antagonist, SR48968 [0.3 µmol/kg intravenous (i.v.)]. Bronchospasm was completely blocked when SR48968 was combined with the selective NK_1R antagonist, CP96345 (2 µmol/kg i.v.). The dose of the NK_2R agonist, β-Ala^8-NKA (4–10), required to elicit a 50% lowering of airway conductance was increased in a dose-dependent manner by i.v. injection of the animals with SR48968 (Buckner et al. 1993). This beneficial change was also augmented by administration of CP96345 (9 µmol/kg i.v.). Although this dose of CP96345 produced a tenfold rightward shift of the dose-response curve to NK_1R agonists, i.e., SP and ASMSP, it was without effect when given alone against the NK_2R agonists, NKA or β-Ala^8-NKA(4–10). Foulon and colleagues (1993) obtained a far greater effect using the combination of CP99994, a selective NK_1R antagonist, and SR48968: the dose-response curve to NKA-mediated bronchoconstriction was displaced to the right with SR48968 (at 1 mg/kg i.v., no effect against NK_1R agonists) only by fivefold but together with CP99994 (1 mg/kg i.v.), the displacement was more than 300-fold. Treatment of guinea pigs with CP99994 alone provided modest protection against the SP-induced decline in airway conductance but produced significantly greater inhibition when given with SR48968. Thus, tachykinin receptor agonists seem to activate more than one tachykinin receptor to decrease airflow. The magnitude of protection afforded by treatment with both NK_1R and NK_2R antagonists appears to be much larger than what would have been predicted from the data obtained from each antagonist alone.

Electrical stimulation of the vagus nerve produces microvascular leakage and extravasation of plasma proteins in lung in addition to narrowing the airways of guinea pigs. The extravasation response was nearly prevented by administration of CP99994 but it was nominal following administration of this compound plus SR48968 (Savoie et al. 1995). In addition, the negative change in airway mechanical activity was maximally inhibited by the combined therapy.

When capsaicin served as the airway spasmogen, the findings were analogous to those described above. In anesthetized animals, the rise in airway resistance elicited by capsaicin was decreased by 64% by treatment with SR48968 alone and completely abolished by the combination of SR48968 and CP96345 (Bertrand et al. 1993). CP96345 alone did not improve changes in airway resistance. In conscious guinea pigs, Kudlacz and coworkers (1993) demonstrated that the combination of SR48698 and CP96345 decreased the incidence of capsaicin-induced dyspnea by 40%; neither compound given alone was beneficial although significant blockade of NKA- or SP-mediated dysfunction was evident. Rumsey and coworkers (2001) showed that capsaicin-mediated bronchoconstriction was unaffected by low doses of selective NK_1R (SR140333, 1 µmol/kg i.v.) or NK_2R (ZD7944, 0.3 µmol/kg i.v.) antagonists despite significant inhibition of elevated airway resistance mediated by exogenous delivery of tachykinin analogues. In combination, SR140333 and ZD7944 resulted in a marked rightward displacement, 29-fold, of the capsaicin dose-response curve. Resiniferatoxin, like capsaicin, releases sensory neuropeptides (Szallasi and Blumberg 1989). It too markedly constricted guinea pig airways but was completely blocked by treatment with atropine (partial antagonism) plus CP99994 and SR48968 (Foulon et al. 1993). Individually, neither tachykinin receptor antagonist showed any improvement.

In models of airway inflammation, combination therapy has had mixed results. Turner and coworkers (1996) reported that airway hyperreactivity to methacholine and eosinophilia in *Ascaris suum*-sensitized nonhuman primates was completely eliminated by giving both CP99994 [10 mg/kg per os (p.o.)] and SR48968 (10 mg/kg p.o.). Given separately, neither compound influenced these two parameters or acute challenge with antigen. The authors postulated that in atopic monkeys, the combined receptor inhibition of eosinophil influx indirectly afforded protection against airway hyperresponsiveness. By contrast, ovalbumin-induced bronchoconstriction in ovalbumin-sensitized guinea pigs was unaffected by pretreatment with CP99994 or SR48968, alone or in combination (Foulon et al. 1993).

Inhalation of acidic substances, such as citric acid, provoke cough, bronchoconstriction, and increased bronchial responsiveness in asthmatics (Boyle et al. 1985). Gastroesophageal reflux disease may promote bronchial hyperresponsiveness via microaspiration of esophageal substances (Ricciardolo 2001; Kiljander et al. 2002). In guinea pigs, bronchoconstriction elicited by inhaled citric acid decreased by 50% following treatment with SR48968 (0.3 µmol/kg i.v.; Ricciardolo et al. 1999). No protection was obtained with CP99994 (8 µmol/kg i.v.). When given together, citric acid-induced bronchconstriction was essentially abolished. Similarly, SR48968 was an effective antitussive against aerosolized citric acid and its effects were enhanced by SR140333 (Girard et al. 1995). By itself and at the same doses, SR140333 [(0.1–1 mg/kg intraperitoneally (i.p.)] did not provide any antitussive effect. Daoui and colleagues (1998) evaluated the effects of selective antagonists, including one for the NK_3R (SR142801) on citric acid-induced bronchospasm. While SR140333, SR48968 or SR142801 alone had little or no impact on bronchoconstriction, the combinations of SR140333 and SR48968, SR48968 and SR142801, or SR140333, SR48968 plus SR142801, significantly decreased airway spasm. Maximal inhibition resulted from blockade of the NK_1R and NK_2R whereas antagonism of NK_1R and NK_3R was not beneficial.

The studies above exemplify the additive and often synergistic effects of combined blockade, mostly with selective NK_1R and NK_2R antagonists. The mechanism underlying this phenomenon has not been investigated; however, a few observations are noteworthy. First, most of the studies used guinea pigs. As such, synergism may be limited to this species, although the work of Turner and coworkers (1996) using nonhuman primates argues against this suggestion. Second, the findings were obtained with an assortment of chemically distinct compounds for blocking the NK_1R (CP96345, CP99994, or SR140333) and the NK_2R (SR48698 or ZD4974). Although selective NK_3R antagonists have been effective in models of cough and airway hyperresponsiveness (Daoui et al. 1998), their use in combination with other selective agents has not been thoroughly characterized. Lastly, synergy was observed whether the tachykinins were administered exogenously or were released from endogenous stores. One plausible explanation for these findings is that synergy resulted from either a pharmacokinetic or pharmacodynamic effect of the two compounds in circulation. For example, one antagonist may interfere with hepatic or renal clearance of the other, thereby increasing the plasma concentrations of the more relevant compound. An alternate possibility is that dual receptor blockade of airway smooth muscle impacts the intracellular generation of a mediator common to both receptors. The NK_1R and NK_2R are both linked via G-proteins to the phospholipase C and adenylate cyclase signal transduction pathways (Nakanishi et al. 1993). Dual receptor blockade could ultimately converge to modulate intracellular calcium concentrations and abolish tachykinin-mediated changes in airway resistance. Whatever the mechanism(s), a more complete understanding of this synergistic action seems warranted and may aid discovery of therapeutic molecules.

4
Medicinal Chemistry Overview of Combined Antagonists

Given the tissue colocalization, the cross-reactivity of the natural ligands with the tachykinin receptors, and the additivity or apparent synergy provided by combined receptor blockade, a therapeutic advantage might be gained by constructing a single molecular entity capable of antagonizing two or possibly three tachykinin receptors. To this end, the molecular template exhibited by SR48968 (Table 1, entry 1) has been explored extensively. The predisposition of this template toward NKR antagonism is evidenced by its ability to produce selective, dual, and pan-antagonists. While the available data largely address selective and dual (NK_1R and NK_2R) antagonists, a basic understanding of the factors necessary for pan-tachykinin antagonism can be gleaned. This section aims to present an understanding of the factors revealed in the general literature which drive nonpeptide, dual and pan-tachykinin antagonists. A general overview will be provided followed by specific examples.

Table 1 Potency of selected dual and pan-tachykinin receptor antagonists

Entry	Structure	Potency (nM)	Reference
1[a,b] SR 48968		hNK$_1$=593 hNK$_2$=0.44 NK$_3$=208	Burkholder et al. 1996
2[a] MDL 103,220		hNK$_1$=161 hNK$_2$=2.25 hNK$_3$=N.D.	Burkholder et al. 1996
3[a,b] MDL 105,212		hNK$_1$=3.1 hNK$_2$=8.4 NK$_3$=21.0	Burkholder et al. 1996
4[c] DNK 333		hNK$_1$=4.8 hNK$_2$=5.5 hNK$_3$=N.D.	Gerspacher et al. 2001
5[d] ZD6021		hNK$_1$=0.12 hNK$_2$=0.61 hNK$_3$=74	Rumsey et al. 2001
6[d] ZD4974		hNK$_1$=0.17 hNK$_2$=N.D. hNK$_3$=220	Albert et al. 2002

Table 1 (continued)

Entry	Structure	Potency (nM)	Reference
7[d]		hNK_1=0.27 hNK_2=35 hNK_3=9.4	Albert et al. 2002
8[e] SCH 206272		hNK_1=1.3 hNK_2=0.4 hNK_3=0.3	Anthes et al. 2002
9[e]		hNK_1=4 hNK_2=9 hNK_3=N.D.	Shih et al. 2002
10[e]		hNK_1=1.9 hNK_2=4.1 hNK_3=945	Reichard et al. 2002
11[f] R113281		hNK_1=2.2 hNK_2=0.51 hNK_3=0.95	Ito et al. 2001
12[f]		hNK_1=40 hNK_2=6.8 hNK_3=2.7	Nishi et al. 2000

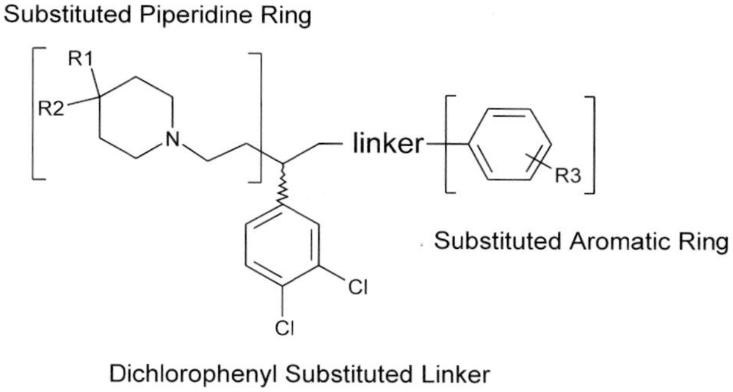

Substituted Piperidine Ring

Fig. 1 General chemical template

Figure 1 illustrates the general template from which a variety of selective, dual and pan-tachykinin antagonists have arisen. Key features of this template include: (1) a substituted piperidine ring connected to a (2) substituted aromatic ring via a (3) dichlorophenyl-substituted linker. The structure–activity relationships described to date have focused largely on substitution of the piperidine and aromatic rings and on modifications to the linker region.

Distal substitution of the piperidine ring producing mono-, di-, and spiro-substitution patterns has been explored by several investigators (Albert et al. 2002; Bernstein et al. 2001; Burkholder et al. 1997; Gerspacher et al. 2001; Mah et al. 2002; Nishi et al. 2000; Reichard et al. 2002a; Ting et al. 2001). In general, monosubstitution altered potency equally at the NK_1R and NK_2R, while 4,4-disubstitution (including spiro) improved it only at the NK_2R. Potency against the NK_3R could also be modulated by substitution at this position.

The aromatic ring also offers a handle to alter the selectivity profile with the NK_1R being more sensitive relative to the NK_2R. Substitution of the aromatic

---

Structure and potency of selected dual and pan-tachykinin receptor antagonists.
[a] Data expressed as IC_{50} against NK_1R from human IM-9 cells using [^{125}I]-Bolton Hunter labeled SP and NK_2R in HSKR-1 cells using [^{125}I]-Iodohistidyl NKA.
[b] Data expressed as IC_{50} against guinea pig cerebral cortex NK_3R using [^{125}I]-Bolton Hunter labeled eledoisin.
[c] Data expressed as Ki against membranes from recombinant NK_2R expressed in CHO cells using [^{125}I]-NKA.
[d] Data expressed as Ki. Membranes prepared from recombinant NKR expressed in MEL cells using [^{3}H]-SP, [^{3}H]-NKA, and [^{125}I]-MePhe^{7}NKB.
[e] Data expressed as Ki. Membranes prepared from recombinant NKR expressed in CHO cells using [^{3}H][Sar^9, Met(O$_2$)^{11}]-SP, [^{3}H]-NKA, and [^{125}I]-MePhe^{7}NKB.
[f] Data expressed as Ki. NKR were obtained from human cells expressing NKR: NK_1R using [^{3}H]-SP; NK_2R using [^{3}H]-SR 48968; and NK_3R using [^{3}H]-senktide.
N.D. = Not Determined

ring appears necessary for robust NK_1R affinity and pan-tachykinin antagonism (Albert et al. 2002; Burkholder et al. 1996; Bernstein et al. 2001; Shih et al. 2002; Mah et al. 2002; Nishi et al. 2000). Tolerated substituents are dependent on the nature of the linker (vide infra), however, both electron-rich and electron-poor substituents are useful (e.g., 3,5-bis-trifluoromethylphenyl and 3,4,5-trimethoxyphenyl).

Acting in cooperation with the other substituents, the linker provides further opportunity to adjust the selectivity profile of the antagonist. Modifications of the linker include; chain length, variation of substitution patterns, and the incorporation of heteroatoms and heterocycles (Burkholder et al. 1996; Mah et al. 2002; Nishi et al. 2000; Reichard et al. 2000, 2002a, 2002b; Shih et al. 2002). The methyl amide has been most commonly used (Table 1, entries 4–10), often in conjunction with the 3,5-bis-trifluoromethylphenyl aromatic. Those antagonists employing a cyclic linker have relied on the 3,4,5-trimethoxyphenyl group (Table 1, entries 2, 3 and 11, 12). To maintain a dual or pan-tachykinin profile, variations in the linker have necessitated optimization of the other substituents including the absolute configuration of the 3,4-dichlorophenyl ring (vide infra).

The derivation of MDL105212 ([(*R*)-1-[2-[3-(3,4-dichlorophenyl)-1-(3,4,5-trimethoxy benzoyl)-pyrrolidin-3-yl]-4-phenylpiperidine-4-carboxamide] was driven principally via molecular modeling and analog preparation (Burkholder et al. 1996). Conformational analysis of SR48968 suggested a cyclic linker resulting in the preparation of MDL103220 (Table 1, entry 2). This possesses excellent affinity for the NK_2R and a four- to fivefold improvement at the NK_1R. Comparison to CP96345 suggested modifications aimed at enhancing NK_1R activity. Varying the substitution on the aromatic ring, MDL103220 gave rise to MDL103392, a mixture of *R* and *S* enantiomers at the 3,4-dichlorophenyl position with good potency at the NK_1R and the NK_2R. Resolution of the mixture identified the *R* enantiomer, MDL105212, as the most potent stereoisomer (Table 1, entry 3). Replacements for the 3,4-dichlorophenyl group of MDL105212 were not as effective (Burkholder et al. 1997).

Studies designed to elucidate the binding mode of MDL103392 using site-directed mutagenesis revealed that mutation of residues Leu203, Ile204, Phe264 and His265 markedly decreased NK_1R affinity (Greenfeder et al. 1999). Homologous mutations at the NK_2R, however, demonstrated that the Leu and Ile residues were not key for binding. Mutation of Tyr289 within the NK_2R resulted in a nearly 300-fold loss of potency while the corresponding mutation of the NK_1R had no effect. The authors suggested that the binding mode of MDL103392 differed for each receptor and that the number of conformational degrees of freedom was a principal factor in acquiring dual and pan-tachykinin antagonism.

Recently, a set of dual (NK_1R and NK_2R) antagonists was identified that are distinguishable from the general template described above. DNK333 or *N*-[(*R,R*)-(E)-1-(3,4-dichlorobenzyl)-3-(2-oxoazepan-3-yl)carbamoyl]allyl-N-methyl-3,5-bis(trifluoromethyl)benzamide (Table 1, entry 4) possesses the 3,4-dichlorophenyl-methyl amide linker, and a 3,5-bis-trifluoromethylphenyl aromatic (Gerspacher et al. 2001). The piperidine ring, however, was replaced by a caprolactam moiety connected to the linker region through an E-enamide. To eliminate the chiral centers, a series of biaryl replacements for the dichlorophenyl group and replacements for the caprolactam were produced (Mah et al. 2002). While many of these compounds exhibit dual activity, none are as potent as DNK333.

Starting with ZD7944, an isoindolone derivative of SR48968, Bernstein and co-workers (2001) applied an array approach to identify naphthalene as a viable aromatic group. Further work culminated in the preparation of ZD6021, [3-cyano-N-((2S)-2-(3,4-dichlorophenyl)-4-[4-[2-(methyl-(S)-sulfinylphenyl)piperidino] butyl)-N-methyl-]-naphthamide] (Table 1, entry 5), an orally active pan-tachykinin antagonist (Bernstein et al. 2001; Rumsey et al. 2001). Substitution of the naphthalene ring was undertaken to reduce metabolic turnover (Albert et al. 2002). While modifications at the 4- and 6- positions decreased NK_1R potency, a 2-methoxy group improved NK_1R affinity with a corresponding loss at the NK_2R and NK_3R. The 2-methoxy group also formed a set of atropisomers around the linker region (Table 1, entry 6). This is likely to have implications for the relative affinities of each conformer for each receptor. By preparing analogs of the piperidine ring, potency at the NK_3R was markedly improved with only modest change at the NK_2R (Table 1, entry 7).

Reichard and co-workers (2000) combined the piperidine and 3,4-dichlorophenyl substituents of SR48968 and the 3,5-bis-trifluoromethylphenyl portion of a selective NK_1R antagonist (Williams et al. 1994). This combination strategy culminated in the pan-tachykinin antagonist SCH206272 [(R,R)-1'[5-[3,5-dichlorobenzoyl]methylamino]-3-(3,4-dichlorophenyl)-4(Z)-(methoxyimino)-pentyl]- N-methyl-2-oxo-[1,4'bipiperidine]-3-acetamide] (Table 1, entry 8; Anthes et al. 2002; Reichard et al. 2002b). By incorporating an oxime into the linker, a reversal of the stereochemical preference of the 3,4-dichlorophenyl substituent was necessary for better potency (Table 1, entries 8–10; Reichard et al. 2000). Oxime geometry and substitution were important for avid binding to the NK_2R (Reichard et al. 2000, 2002b). In addition to the oxime, inclusion of a methyl amide in the linker provided flexibility in the choice of aromatic groups while maintaining potency at both the NK_1R and NK_2R (Shih et al. 2002). Selectivity over the NK_3R was preferable because of its potential involvement in the hypothalamic-pituitary-gonadal axis (Debeljuk and Lasaga 1999) and was accomplished through substitution of the piperidine ring and alkylation of the amide in the linker region (Table 1, entry 10; Reichard et al. 2002a).

Nishi and co-workers (1999) recently reported the optimization of a series of antagonists with an oxazolidine as the linker. Optimal activity for the NK_1R and NK_2R was obtained with a chiral sulfoxide on the piperidine ring, and the 3,4-dichlorophenyl and 3,4,5-trimethoxyphenyl substituents. This report was followed by the description of the synthesis of the pan-tachykinin antagonist R113281 (Table 1, entry 11), however, only a few key elements of the structure–activity response were disclosed (Nishi et al. 2000). R-113281 contains a 4,4-disubstituted piperidine with chiral sulfoxide, a 3,4,5-trimethoxyphenyl, the 3,4-dichlorophenyl group, and a cyclic morpholine linker. The absolute configuration of the 3,4-dichlorophenyl (R) group and the sulfoxide (S) was key for pan-tachykinin activity. Replacement of the chiral sulfoxide with a carbinol (Table 1, entry 12) resulted in pan-tachykinin antagonism albeit with reduced affinity relative to the sulfoxide.

While the construction of dual and pan-tachykinin receptor antagonists remains a challenge, the molecular template represented by SR48968 has proven versatile in preparing nonpeptide antagonists with a variety of selectivity profiles. The potential therapeutic benefit gained from blockade of more than one tachykinin receptor continues to be elucidated through pre-clinical and ultimately clinical studies.

5 Pharmacology of Combined Antagonists

This section details the pharmacology of dual (NK_1R and NK_2R) and pan-tachykinin receptor antagonists, beginning with the nonpeptide type. Most workers in the field have focused their investigations on this class of compounds. In depth reports of their efficacy in human studies have not yet been published. The pharmacology of two pseudopeptide dual receptor antagonists is also described; one example, FK224, is notable for its progression to early clinical trials.

5.1 Nonpeptide Antagonists

The first nonpeptide, orally available NK_1R and NK_2R antagonist was MDL105212 (Kudlacz et al. 1996a, 1996b). Potency values for the NK_1R and NK_2R were: 50% inhibitory concentration (IC_{50})=3.1 nM and 8.4 nM (IM-9 and HSKR-1 cells) and pA_2=8.2 and 8.7 ([3H]-inositol phosphate accumulation in monoreceptor cell lines), respectively. Selectivity over the NK_3R was modest, IC_{50}=21 nM (guinea pig cortical membranes). As such, MDL105212 might be considered a pan-tachykinin antagonist, although its activity at the human NK_3R has not been established. Capsaicin-induced dyspnea, cough, gasps, and plasma protein extravasation in conscious guinea pigs were dose-dependently blocked by administration of MDL105212 [50% effective dose (ED_{50})=50 mg/kg p.o.; Kudlacz et al. 1996b]. In addition, airway hyperresponsiveness to methacholine resolved in ovalbumin-sensitized guinea pigs given this antagonist (Kudlacz et al. 1996a). Interestingly, pretreatment of normal, nonsensitized animals with MDL105212 at doses sufficient to significantly block capsaicin-induced respiratory events had no effect on methacholine-mediated bronchoconstriction. Thus, blockade of both NK_1R and NK_2R per se with MDL105212A would not produce nonspecific bronchodilation.

An analogue of MDL105212, termed MDL105172A ((R)-1-[3-(3,4-dichlorophenyl)-1-(3,4,5-trimethoxybenzoyl)-3-pyrrolidinyl]-4-phenylpiperidine-4-morpholinecarboxamide) avidly bound to all three receptors: IC_{50}=4.3 nM (guinea pig lung), 2.1 nM (HSKR cells), and 2.5 nM (guinea pig cortex) for the NK_1R,

NK$_2$R and NK$_3$R, respectively (Kudlacz et al. 1997). In conscious guinea pigs it blocked capsaicin-elicited dyspnea: ED$_{50}$=20 mg/kg p.o.

Other workers sought to construct dual NK$_1$R and NK$_2$R antagonists beginning with the potent, selective NK$_1$R antagonist, CGP49823, as a synthetic template (Gerspacher et al. 2000). One compound, N-[(R,R)-(E)-1-(4-chlorobenzyl)-3-(2-oxo-azepan-3-ylcarbamoyl)-allyl]-N-methyl-3,5-bis-trifluoromethylbenzamide, potently displaced tachykinins from the cloned human NK$_1$R and NK$_2$R: IC$_{50}$=0.5 nM and 24 nM, respectively, and blocked tachykinin-mediated responses in guinea pigs: ED$_{50}$=0.04 mg/kg and 0.9 mg/kg p.o., respectively. Further refinements of this compound led to the discovery of DNK333 (Gerspacher et al. 2001) which has essentially equivalent inhibition of ligand binding at the NK$_1$R (4.8 nM) and NK$_2$R (IC$_{50}$=5.5 nM).

The first pan-tachykinin receptor antagonist that avidly bound to all three cloned human tachykinin receptors and displayed pharmacological activity in animal tissues and models was ZD6021 (Rumsey et al. 2001). The inhibition constant (K_i) was highest at the NK$_1$R (0.12 nM), followed by the NK$_2$R (0.61 nM) and NK$_3$R (74 nM). For the NK$_1$R and NK$_2$R, selective agonist concentration–response curves were displaced rightward in a concentration-dependent manner using tissues isolated from rabbit and human: rabbit pulmonary artery, pA_2=8.7 and 8.5, human pulmonary artery and bronchus, pK_B=8.9 and 7.5 (at 0.1 μM), respectively. Concentration-response curves of isolated guinea pig ileum stimulated by the selective NK$_3$R agonist, senktide, were also shifted to the right by ZD6021 in a concentration-dependent fashion with substantial displacement at 10 nM. The maximal response to senktide was reduced by 50% by ZD6021 and was independent of drug concentration. These data suggested that antagonism of the NK$_3$R was not purely competitive. In guinea pigs, half-maximal protection (ED$_{50}$) against ASMSP-induced tracheal plasma protein extravasation and β-Ala^8-NKA(4–10)-mediated bronchoconstriction resulted from oral administration of ZD6021 at 0.8 μmol/kg (or 500 μg/kg) and 20 μmol/kg (or 13 mg/kg) for the NK$_1$R and NK$_2$R, respectively. Capsaicin-mediated airway narrowing was also significantly blocked by ZD6021 (10 μmol/kg i.v.). The response was comparable to that obtained when using selective NK$_1$R and NK$_2$R antagonists in combination (see previous section on synergy).

Anthes and colleagues (Anthes et al. 2002; Reichard et al. 2000, 2002b; Ting et al. 2001, 2002; Shih et al. 2002) showed that SCH206272 was a potent, orally active pan-tachykinin antagonist with activity for the human receptors. This molecule yielded K_i values of 1.3 nM, 0.4 nM, and 0.3 nM for the NK$_1$R, NK$_2$R and NK$_3$R, respectively. Tachykinin-mediated changes of contractile tone of human pulmonary artery or bronchus were potently blocked by SCH206272: pK_B=7.7 and 8.2 for NK$_1$R and NK$_2$R, respectively. Inhibition of senktide-induced contraction of isolated guinea pig ileum was obtained at 10 nM (63%) and 100 nM (92%) and displayed noncompetitive features. Treatment of guinea pigs with SCH206272 dose-dependently prevented tachykinin-mediated airway plasma protein extravasation and bronchoconstriction; half-maximal inhibition was achieved at 1 and 10 mg/kg p.o. for both pharmacological assays. When administered orally to dogs at 1 mg/kg, SCH206272 significantly reduced the SP-mediated fall in arterial blood pressure and NKA-induced bronchoconstriction.

A series of pan-tachykinin antagonists has been described that was derived from oxazolidine analogues (Nishi et al. 1999). Replacement of the oxazolidine with morpholine resulted in compounds with high binding affinity for the tachykinin receptors (Nishi et al. 2000). One example, R113281, demonstrated strict stereochemical requirements (S,R) and potency at the nanomolar level for the guinea pig receptors; IC_{50} values were 21 nM, 3.1 nM and 7.2 nM for NK_1R, NK_2R and NK_3R, respectively. Binding affinity for the human receptors was also high: Ki=2.2 nM, 0.51 nM, and 0.95 nM (Ito et al. 2001). In guinea pigs, it inhibited SP-induced airway plasma protein extravasation (ED_{50}=5.1 mg/kg p.o.) and NKA- or NKB-induced bronchoconstriction (ED_{50}=1.3 mg/kg and 0.45 mg/kg p.o., respectively). Moreover, it significantly prevented airway narrowing in response to tobacco smoke.

5.2
Pseudopeptide Antagonists

Rational design of low molecular weight peptides derived from the C terminus of SP (Kucharczyk et al. 1993) resulted in the partially cyclized, pseudo-tetrapeptide NK_1R and NK_2 R antagonist, S16474 (cyclo-[Abo-Asp(D-Trp(Suc)Na)-Phe-N-(Me)Bzl)]; Robineau et al. 1995). This water-soluble cyclopeptide potently inhibited ligand binding for the cloned human NK_1R and NK_2R: IC_{50}=85 nM and 129 nM, respectively, and was selective over the NK_3R, IC_{50}=3 μM. Pharmacological activity in vitro was markedly greater at the NK_1R than the NK_2R using tissues isolated from rabbit, pA_2=7.0 and 5.6, respectively. When the compound was tested in guinea pig in vivo, potency for the two receptors was comparable: ID_{50}=4 μmol/kg i.v. and 7 μmol/kg i.v. vs. SP- or NKA-induced bronchoconstriction, respectively. Capsaicin-mediated bronchospasm was also inhibited by 60% at 20 μmol/kg i.v.

The dual NK_1R and NK_2R antagonist, FK224 (N-[N2-[N-[N-[N-[2,3-didehydro-N-methyl-N-[N-[3-(2-pentylphenyl)-proionyl]-L-threonyl]tyrosyl-L-leucynyl]-D-phenylalanyl]-L-allo-threonyl]-L-asparaginyl]-L-serine-lactone), was produced by the catalytic hydrogenation of a fermentation product that was isolated from *Streptomyces violaceoniger* (Hayashi et al. 1992; Hashimoto et al. 1992). The inhibition constant (Ki vs. 3[H]-SP) of FK224 for guinea pig lung membranes was 67 M. Using rat cortical or duodenal membranes, the IC_{50} values were 1.7 μM and 1.9 μM for NK_1R and NK_2R, respectively (Morimoto et al. 1992). There was essentially no activity against the NK_3R. Electrical field stimulation of isolated guinea pig trachea produced contractions that were prevented by FK224 (IC_{50}=3.5 μM, this value being similar to those obtained with either SP or NKA as the contractile stimulus; Murai et al. 1992). When FK224 was administered intravenously to guinea pigs, it blocked SP-, NKA- or capsaicin-mediated bronchoconstriction with ED_{50} values of 0.39 mg/kg, 0.36 mg/kg and 1.1 mg/kg, respectively. Airway edema was similarly prevented by FK224.

5.3
Clinical Trials

To date, there have been remarkably few clinical studies evaluating dual or pan-tachykinin receptor antagonists. There are potentially several reasons for the lack of trials, e.g., adverse side effects in animals, but the basis for not advancing compounds to testing in humans is not generally reported. Thus far, the only reports in the literature are those describing the effects of FK224 in asthmatics. Ichinose and colleagues (1992) initially described that inhalation of FK224 (4 mg) by ten patients significantly attenuated bronchoconstriction caused by inhaled bradykinin which was used to provoke asthma symptoms (Fuller et al. 1987). In this double-blind, placebo-controlled crossover trial, pretreatment with the NK_1R and NK_2R antagonist resulted in a significant increase in the concentration of bradykinin required to provoke a 35% fall in specific airway conductance from 5.3 $\mu g/ml$ to 40 $\mu g/ml$. Three of the patients coughed on exposure to bradykinin but this response was also suppressed by FK224. Subsequently, a lower inhaled dose (2 mg) offered little protection to bradykinin challenge (Schmidt et al. 1996). In a third study, inhalation of FK224 (4 mg) did not prevent bronchoconstriction provoked by inhaled NKA (Joos et al. 1996) yet it significantly inhibited NKA-induced contraction of isolated human bronchial tissue (Honda et al. 1997).

6
Summary

Tachykinins are involved in a diverse array of cellular processes and are implicated in a broad spectrum of diseases including those within the respiratory, psychiatric and gastrointestinal areas. To aid in the determination of how these neuropeptides function in healthy and diseased tissue, both selective and nonselective receptor antagonists are now available in the pharmacological toolbox. The preparation of dual and pan-tachykinin antagonists remains a challenge, however, significant progress has been made. The template represented by SR48968 has been explored extensively and is unique in lending itself to the preparation of either selective, dual, or pan-antagonists via judicious choice of substituents. The hunt for additional templates is likely to reveal molecular entities with varying degrees of complexity and selectivity and remains an exciting area of research. Rationale for the design of these molecules draws, in part, from the additive or synergistic protection that is achieved by blocking both the NK_1R and NK_2R in several animal models of airway function and pathology. Inhibiting the activation of the NK_3R may also be important; however, the benefits of its blockade in combination with either the NK_1R or NK_2R (or both) are less documented. The location of synergistic action for the NK_1R and NK_2R antagonists may reside at postsynaptic sites, e.g., smooth muscle cells, while that obtained with NK_3R antagonists is likely to include the airway ganglia or brain stem. Synergy has been repeatedly demonstrated. The cellular mechanism(s) responsible for this phenomenon and the potential advantage it may offer for disease therapy await further elucidation. The advent of dual and pan-tachykinin antagonists may advance this area of tachykinin biology.

The application of tachykinin receptor blockade to pulmonary function is central to this chapter, in part because the pharmacological testing of dual or pan-tachykinin antagonists has been mostly performed in animal models of respiratory disease and, in a few cases, in humans with asthma. A few comments about tachykinins and pulmonary function are notable from these studies. Although blockade of the NK_1R and NK_2R, either by combining selective antagonists (Turner et al. 1996) or with a dual antagonist (Kudlacz et al. 1996b), restores hyperresponsive airways to their normal state, it does little to affect airway resistance of ordinary animals normally or when challenged with nonspecific stimuli. Superseding airway contractile tone established by cholinergic input may not result from blockade of a single or multiple tachykinin receptors. Rather, attenuation of aberrant or heightened responses that exist in inflammatory diseases may be better served by tachykinin receptor antagonists, particularly those with combined receptor affinity. Further evaluation of these compounds, whenever possible in human trials, would enhance our understanding of tachykinin processes in health and disease.

References

Adcock IM, Peters M, Gelder C, Shirasaki H, Brown CR, Barnes PJ (1993) Increased tachykinin receptor gene expression in asthmatic lung and its modulation by steroids. J Mol Endocrinol 11:1–7

Advenier C, Lagente V, Boichot E (1997) The role of tachykinin receptor antagonists in the prevention of bronchial hyperresponsiveness, airway inflammation and cough. Eur Respir J 10:1892–1906

Albert JS, Aharony D, Andisik D, Barthlow H, Bernstein PR, Bialecki RA, Dedinas R, Dembofsky BT, Hill D, Kirkland K, Koether GM, Kosmider BJ, Ohnmacht C, Plamer W, Potts W, Rumsey W, Shen L, Shenvi A, Sherwood S, Warwick PJ, Russell K (2002) Design, synthesis, and SAR of tachykinin antagonists: modulation of balance in NK_1/NK_2 receptor antagonist activity. J Med Chem 45:3972–3983

Amadesi S, Moreau J, Tognetto M, Springer J, Trevisani M, Naline E, Advenier C, Fisher A, Vinci D, Mapp C, Miotto D, Cavallesco G, Geppetti P (2001) NK_1 receptor stimulation causes contraction and inositol phosphate increase in medium-size human isolated bronchi. Am J Respir Crit Care Med 163:1206–1211

Anthes JC, Chapman RW, Richard C, Eckel S, Corboz M, Hey JA, Fernandez X, Greenfeder S, McLeod R, Sehring S, Rizzo C, Crawley Y, Shih NY, Piwinski J, Reichard G, Ting P, Carruthers N, Cuss FM, Billah M, Kreutner W, Egan RW (2002) SCH 206272: a potent, orally active tachykinin NK_1, NK_2, and NK_3 receptor antagonist. Eur J Pharmacol 450:191–202

Bai TR, Zhou D, Weir T, Walker B, Hegele R, Hayashi S, McKay K, Bondy GP, Fong T (1995) Substance P (NK-1)- and neurokinin A (NK-2)-receptor gene expression in inflammatory airway diseases. Am J Physiol 269:L310–L317

Baluk P, Bunnett NW, McDonald DM (1996) Localization of tachykinin NK-1, NK-2, and NK-3 receptors in airways by immunohistochemistry. Am J Respir Crit Care Med 153:A161

Baluk P, Nadel JA, McDonald DM (1992) Substance P-immunoreactive sensory axons in the rat respiratory tract: a quantitative study of their distribution and role in neurogenic inflammation. J Comp Neurol 319:586–598

Bernstein PR, Aharony D, Albert JS, Andisik D, Barthlow HG, Bialecki R, Davenport T, Dedinas RF, Dembofsky BT, Koether G, Kosmider BJ, Kirkland K, Ohnmacht CJ, Potts W, Rumsey WL, Shen L, Shenvi A, Sherwood S, Stollman D, Russell K (2001) Discovery of novel, orally active dual NK_1/NK_2 antagonists. Bioorg Med Chem Lett 11:2769–2773.

Bertrand C, Nadel JA, Graf PD, Geppetti P (1993) Capsaicin increases airflow resistance in guinea pigs in vivo by activating both NK_2 and NK_1 tachykinin receptors. Am Rev Respir Dis 148:909–914

Bowden JJ, Baluk P, Lefevre PM, Vigna SR, McDonald DM (1996) Substance P (NK_1) receptor immunoreactivity on endothelial cells of the rat tracheal mucosa. Am J Physiol 270:L404–L414

Boyle JT, Tuchman DN, Altschuler SM, Nixon TE, Pack AI, Cohen S (1985) Mechanisms for the association of gastroesophageal reflux and bronchospasm. Am Rev Respir Dis 131:S16–S20

Buckner CK, Liberati N, Dea D, Lengel D, Stinson-Fisher C, Campbell J, Miller S, Shenvi A, Krell RD (1993) Differential blockade by tachykinin NK_1 and NK_2 receptor antagonists of bronchoconstriction induced by direct-acting agonists and the indirect-acting mimetics capsaicin, serotonin and 2-methyl-serotonin in the anesthetized guinea pig. J Pharmacol Exp Ther 267:1168–1175

Burkholder TP, Kudlacz EM, Le TB, Knippenber RW, Shatzer SA, Maynard GD, Webster ME, Horgan SW (1996) Identification and chemical synthesis of MDL105,212, a nonpeptide tachykinin antagonist with high affinity for NK_1 and NK_2 receptors. Bioorg Med Chem Lett 6:951–956

Burkholder TP, Kudlacz EM, Maynard GD, Liu XG, Le TB, Webster ME, Horgan SW, Wenstrup DL, Freund DW, Boyer F, Bratton L, Gross RS, Knippenberg RW, Logan DE, Jones BK, Chen TM (1997) Synthesis and structure-activity relationships for a series of substituted pyrrolidine NK_1/NK_2 receptor antagonists. Bioorg Med Chem Lett 7:2531–2536

Burgunder JM, Young WS 3rd (1989) Neurokinin B and substance P genes are co-expressed in a subset of neurons in the rat habenula. Neuropeptides 13:165–169

Canning BJ, Reynolds SM, Anukwu LU, Kajekar R, Myers AC (2002) Endogenous neurokinins facilitate synaptic transmission in guinea pig airway parasympathetic ganglia. Am J Physiol 283:R320–R330

Carr MJ, Ellis JL (2002) The study of afferent neuron excitability. Curr Opin Pharmacol 2:216–219

Carr MJ, Hunter DD, Jacoby DB, Undem BJ (2002) Expression of tachykinins in nonnociceptive vagal afferent neurons during respiratory viral infection in guinea pigs. Am J Respir Crit Care Med 165:1071–1075

Carstairs JL, Barnes PJ (1986) Autoradiographic mapping of substance P receptors in lung. Eur J Pharmacol 127:295–296

Cheung D, Bel EH, Den Hartigh J, Dijkman JH, Sterk PJ (1992) The effect of an inhaled neutral endopeptidase inhibitor, thiorphan, on airway responses to neurokinin A in normal humans in vivo. Am Rev Respir Dis 145:1275–1280

Cheung D, Timmers MC, Zwinderman AH, den Hartigh J, Dijkman JH, Sterk PJ (1993) Neutral endopeptidase activity and airway hyperresponsiveness to neurokinin A in asthmatic subjects in vivo. Am Rev Respir Dis 148:1467–1473

Cheung D, van der Veen H, den Hartigh J, Dijkman JH, Sterk PJ (1994) Effects of inhaled substance P on airway responsiveness to methacholine in asthmatic subjects in vivo. J Appl Physiol 77:1325–1332

Coats SR, Gerard NP (1989) Characterization of the substance P receptor in guinea pig lung tissues. Am J Respir Cell Mol Biol 1:269–275

Crimi N, Palermo F, Oliveri R, Palermo B, Vancheri C, Polosa R, Mistretta A (1988) Effect of nedocromil on bronchospasm induced by inhalation of substance P in asthmatic subjects. Clin Allergy 18:375–382

Crimi N, Palermo F, Oliveri R, Palermo B, Vancheri C, Polosa R, Mistretta A (1990) Influence of antihistamine (astemizole) and anticholinergic drugs (ipratropium bromide) on bronchoconstriction induced by substance P. Ann Allergy 65:115–120

Daoui S, Cognon C, Naline E, Emods-Alt X, Advenier C (1998) Involvment of tachykinin NK_3 receptors in citric acid-induced cough and bronchial responsesin guinea pigs. Am J Respir Crit Care Med 158:42–48

Debeljuk L, Lasaga M (1999) Modulation of the hypothalamo-pituitary-gonadal axis and the pineal gland by neurokinin A, neuropeptide K and neuropeptide gamma. Peptides 20:285–299

Ding YQ, Shigemoto R, Takada M, Ohishi H, Nakanishi S, Mizuno N (1996) Localization of the neuromedin K receptor (NK_3) in the central nervous system of the rat. J Comp Neurol 364:290–310

Doherty MJ, Mister R, Pearson MG, Calverley PM (2000) Capsaicin responsiveness and cough in asthma and chronic obstructive pulmonary disease. Thorax 55:643–649

Ellis JL, Undem BJ, Kays JS, Ghanekar SV, Barthlow HG, Buckner CK (1993) Pharmacological examination of receptors mediating contractile responses to tachykinins in airways isolated from human, guinea pig and hamster. J Pharmacol Exp Ther 267:95–101

Ellis JL, Sham JSK, Undem BJ (1997) Tachykinin-independent effects of capsaicin on smooth muscle in human isolated bronchi. Am J Respir Crit Care Med 155:751–755

Fischer A, McGregor GP, Saria A, Philippin B, Kummer W (1996) Induction of tachykinin gene and peptide expression in guinea pig nodose primary afferent neurons by allergic airway inflammation. J Clin Invest 98:2284–2291

Foulon DM, Champion E, Masson P, Rodger IW, Jones TR (1993) NK_1 and NK_2 receptors mediate tachykinin and resiniferatoxin-induced bronchospasm in guinea pigs. Am Rev Respir Dis 148:915–921

Fuller RW, Dixon CM, Cuss FM, Barnes PJ (1987) Bradykinin-induced bronchoconstriction in humans. Mode of action. Am Rev Respir Dis 135:176–180

Fuller RW, Maxwell DL, Dixon CM, McGregor GP, Barnes VF, Bloom SR, Barnes PJ (1987) Effect of substance P on cardiovascular and respiratory function in subjects. J Appl Physiol 62:1473–1479

Gerspacher M, La Vecchia L, Mah R, von Sprecher A, Anderson GP, Subramanian N, Hauser K, Bammerlin H, Kimmel S, Pawelzik V, Ryffel K, Ball HA (2001) Dual neurokinin NK_1/NK_2 antagonists: N-[(R,R)-(E)-1-arylmethyl-3-(2-oxo-azepan-3-yl)carbamoyl]allyl-N-methyl-3,5-bis(trifluoromethyl)benzamides and 3-[N'-3,5-bis(trifluoromethyl)benzoyl-N-arylmethyl-N'-methylhydrazino]-N-[(R)-2-oxo-azepan-3-yl]propionamides. Bioorg Med Chem Lett 11:3081–3084

Gerspacher M, von Sprecher A, Mah R, Anderson GP, Bertrand C, Subramanian N, Hauser K, Ball HA (2000) N-[(R,R)-(E)-1-(4-chloro-benzyl)-3-(2-oxo-azepan-3-ylcarbamoyl)-allyl]-N-methyl-3,5-bis-trifluoromethyl-benzamide: an orally active neurokinin NK_1/NK_2 antagonist. Bioorg Med Chem Lett 10:1467–1470

Girard V, Naline E, Vilain P, Emonds-Alt X, Advenier C (1995) Effect of the two tachykinin antagonists, SR48968 and SR140333, on cough induced by citric acid in the unanaesthetized guinea pig. Eur Respir J 8:1110–1114

Goldie RG, Rigby PJ, Fernandes LB, Henry PJ (2001) The impact of inflammation on bronchial neuronal networks. Pulm Pharmacol Ther 14:177–182

Greenfeder S, Cheewatrakoolpong B, Billah M, Egan RW, Keene E, Murgolo NJ, Anthes JC (1999) The neurokinin-1 and neurokinin-2 receptor binding sites of MDL103,392 differ. Bioorg Med Chem 7:2867–2876

Hashimoto M, Hayashi K, Murai M, Fujii T, Nishikawa M, Kiyoto S, Okuhara M, Kohsaka M, Imanaka H (1992) WS9326A, a novel tachykinin antagonist isolated from Streptomyces violaceusniger no. 9326. II. Biological and pharmacological properties of WS9326A and tetrahydro-WS9326A (FK224). J Antibiot (Tokyo) 45:1064–1070

Hayashi K, Hashimoto M, Shigematsu N, Nishikawa M, Ezaki M, Yamashita M, Kiyoto S, Okuhara M, Kohsaka M, Imanaka H (1992) WS9326A, a novel tachykinin antagonist isolated from Streptomyces violaceusniger no. 9326. I. Taxonomy, fermentation, isolation, physico-chemical properties and biological activities. J Antibiot (Tokyo) 45:1055–1063

Higenbottam T (2002) Chronic cough and the cough reflex in common lung diseases. Pulm Pharmacol Ther 15:241–247

Honda I, Kohrogi H, Yamaguchi T, Hamamoto J, Hirata N, Iwagoe H, Fujii K, Goto E, Ando M (1997) Tachykinin antagonist FK224 inhibits neurokinin A-, substance P- and capsaicin-induced human bronchial contraction. Fundam Clin Pharmacol 11:260–266

Hua XY, Theodorsson-Norheim E, Brodin E, Lundberg JM, Hokfelt T (1985) Multiple tachykinins (neurokinin A, neuropeptide K and substance P) in capsaicin-sensitive sensory neurons in the guinea-pig. Regul Pept 13:1–19

Ichinose M, Nakajima N, Takahashi T, Yamauchi H, Inoue H, Takishima T (1992) Protection against bradykinin-induced bronchoconstriction in asthmatic patients by neurokinin receptor antagonist. Lancet 340:1248–1251

Ito K, Watanabe S, Mukaiyama O, Nosaka E, Satoh Y, Nishi T, Yamaguchi Y (2001) Pharmacological activities of R-113281 a triple $NK_1/NK_2/NK_3$ neurokinin receptor antagonist Eur Respir J 18:P1787

Joos G, Pauwels R, van der Straeten M (1987) Effect of inhaled substance P and neurokinin A on the airways of normal and asthmatic subjects. Thorax 42:779–783

Joos GF, Van Schoor J, Kips JC, Pauwels RA (1996) The effect of inhaled FK224, a tachykinin NK-1 and NK-2 receptor antagonist, on neurokinin A-induced bronchoconstriction in asthmatics. Am J Respir Crit Care Med 153:1781–1784

Jordan D (2001) Central nervous pathways and control of the airways. Resp Physiol 125:67–81

Kiljander TO, Salomaa ER, Hietanen EK, Ovaska J, Helenius H, Liippo K (2002) Gastroesophageal reflux and bronchial responsiveness: correlation and the effect of fundoplication. Respiration 69:434–439

Kucharczyk N, Thurieau C, Paladino J, Morris AD, Bonnet J, Canet E, Krause JE, Regoli D, Couture R, Fauchere JL (1993) Tetrapeptide tachykinin antagonists: synthesis and modulation of the physicochemical and pharmacological properties of a new series of partially cyclic analogs. J Med Chem 36:1654–1661

Kudlacz EM, Logan DE, Dhatzer SA, Farrell AM, Baugh LE (1993) Tachykinin-mediated respiratory effects in conscious guinea pigs: modulation by NK_1 and NK_2 receptor antagonists. Eur J Pharmacol 241:17–25

Kudlacz EM, Knippenberg RW, Logan DE, Burkholder TP (1996a) Effect of MDL 105,212, a nonpeptide NK-1/NK-2 receptor antagonist in an allergic guinea pig model. J Pharmacol Exp Ther 279:732–739

Kudlacz EM, Shatzer SA, Knippenberg RW, Logan DE, Poirot M, van Giersbergen PL, Burkholder TP (1996b) In vitro and in vivo characterization of MDL 105,212A, a nonpeptide NK-1/NK-2 tachykinin receptor antagonist. J Pharmacol Exp Ther 277:840–851

Kudlacz EM, Knippenberg RW, Shatzer SA, Kehne JH, McCloskey TC, Burkholder TP (1997) The peripheral NK-1/NK-2 receptor antagonist MDL 105,172A inhibits tachykinin-mediated respiratory effects in guinea-pigs. J Auton Pharmacol 17:109–119

Kummer W, Fischer A, Kurkowski R, Heym (1992) The sensory and sympathetic innervation of guinea-pig lung and trachea as studied by retrograde neuronal tracing and double-labeling immunohistochemistry. Neuroscience 49:715–737

Kusner EJ, Buckner CK, DeHaas CJ, Lengel DJ, Marks RL, Krell RD (1992) Tachykinin-induced dyspnea in conscious guinea pigs. Eur J Pharmacol 210:299–306

Laitinen LA, Laitinen A, Panula PA, Partanen M, Tervo K, Tervo T (1983) Immunohistochemical demonstration of substance P in the lower respiratory tract of the rabbit and not of man. Thorax 38:531–536

Lamb JP, Sparrow MP (2002) Three-dimensional mapping of sensory innervation with substance P in porcine bronchial mucosa: comparison with human airways. Am J Respir Crit Care Med 166:1269–1281

Lundberg JM, Hokfelt T, Martling CR, Saria A, Cuello C (1984) Substance P-immunoreactive sensory nerves in the lower respiratory tract of various mammals including man. Cell Tissue Res 235:251–261

Maggi CA (1997) The effects of tachykinins on inflammatory and immune cells. Regul Pept 70:75–90

Maggi CA, Schwartz TW (1997) The dual nature of the tachykinin NK_1 receptor. Trends Pharmacol Sci 18:351–355

Mah R, Gerspacher M, von Sprecher A, Stutz S, Tschinke V, Anderson GP, Bertrand C, Subramanian N, Ball HA (2002) Biphenyl derivatives as novel dual NK_1/NK_2-receptor antagonists. Bioorg Med Chem Lett 12:2065–2068

Mak JC, Astolfi M, Zhang XL, Evangelista S, Manzini S, Barnes PJ (1996) Autoradiographic mapping of pulmonary NK_1 and NK_2 tachykinin receptors and changes after repeated antigen challenge in guinea pigs. Peptides 17:1389–1395

Mapp CE, Miotto D, Braccioni F, Saetta M, Turato G, Maestrelli P, Krause JE, Karpitskiy V, Boyd N, Geppetti P, Fabbri LM (2000) The distribution of neurokinin-1 and neurokinin-2 receptors in human central airways. Am J Respir Crit Care Med 161:207–215

Mazzone SB, Canning BJ (2002) Central nervous system control of the airways: pharmacological implications. Curr Opin Pharmacol 2:220–228

McLeod AL, Krause JE, Ribeiro-Da-Silva A (2000) Immunocytochemical localization of neurokinin B in the rat spinal dorsal horn and its association with substance P and GABA: an electron microscopic study. J Comp Neurol 420:349–362

Merchenthaler I, Maderdrut JL, O'Harte F, Conlon JM (1992) Localization of neurokinin B in the central nervous system of the rat. Peptides 13:815–829

Morimoto H, Murai M, Maeda Y, Yamaoka M, Nishikawa M, Kiyotoh S, Fujii T (1992) FK 224, a novel cyclopeptide substance P antagonist with NK_1 and NK_2 receptor selectivity. J Pharmacol Exp Ther 262:398–402

Murai M, Morimoto H, Maeda Y, Kiyotoh S, Nishikawa M, Fujii T (1992) Effects of FK224, a novel compound NK_1 and NK_2 receptor antagonist, on airway constriction and airway edema induced by neurokinins and sensory nerve stimulation in guinea pigs. J Pharmacol Exp Ther 262:403–408

Myers AC, Undem BJ (1993) Electrophysiological effects of tachykinins and capsaicin on guinea-pig bronchial parasympathetic ganglion neurones. J Physiol 470:665–679

Myers AC, Kajekar R, Undem BJ (2002) Allergic inflammation-induced neuropeptide production in rapidly adapting afferent nerves in guinea pig airways. Am J Physiol 282:L775–L781

Myers A, Undem B, Kummer W (1996) Anatomical and electrophysiological comparison of the sensory innervation of bronchial and tracheal parasympathetic ganglion neurons. J Auton Nerv Sys 61:162–168

Nakanishi S, Nakajima Y, Yokota Y (1993) Signal transduction and ligand-binding domains of the tachykinin receptors. Regul Pept 46:37–42

Naline E, Molimard M, Regoli D, Emonds-Alt X, Bellamy JF, Advenier C (1996) Evidence for functional tachykinin NK_1 receptors on human isolated small bronchi. Am J Physiol 271:L763–L767

Nishi T, Fukazawa T, Ishibashi K, Nakajima K, Sugioka Y, Iio Y, Kurata H, Itoh K, Mukaiyama O, Satoh Y, Yamaguchi T (1999) Combined NK_1 and NK_2 tachykinin receptor antagonists: synthesis and structure-activity relationships of novel oxazolidine analogues. Bioorg Med Chem Lett 9:875–880

Nishi T, Ishibashi K, Takemoto T, Nakajima K, Fukazawa T, Iio Y, Itoh K, Mukaiyama O, Yamaguchi T (2000) Combined tachykinin receptor antagonist: synthesis and stereochemical structure-activity relationships of novel morpholine analogues. Bioorg Med Chem Lett 10:1665–1668

Ogawa T, Kanazawa I, Kimura S (1985) Regional distribution of substance P, neurokinin alpha and neurokinin beta in rat spinal cord, nerve roots and dorsal root ganglia, and the effects of dorsal root section or spinal transection. Brain Res 359:152–157

Ollerenshaw SL, Jarvis D, Sullivan CE, Woolcock AJ (1991) Substance P immunoreactive nerves in airways from asthmatics and nonasthmatics. Eur Respir J 4:673–682

Otsuka M, Yoshioka K (1993) Neurotransmitter functions of mammalian tachykinins. Physiol Rev 73:229–308

Perez Fontan JJ, Cortright DN, Krause JE, Velloff CR, Karpitskyi VV, Carver Jr TW, Shapiro SD, Mora BN (2000) Substance P and neurokinin-1 receptor expression by intrinsic airway neurons in the rat. Am J Physiol 278:L344–L355

Reichard GA, Ball ZT, Aslanian R, Anthes JC, Shih NY, Piwinski JJ. (2000) The design and synthesis of novel NK_1/NK_2 dual antagonists. Bioorg Med Chem Lett 10:2329–2332

Reichard GA, Grice CA, Shih NY, Spitler J, Majmundar S, Wang SD, Paliwal S, Anthes JC, Piwinski JJ (2002a) Preparation of oxime dual NK_1/NK_2 antagonists with reduced NK_3 affinity. Bioorg Med Chem Lett 12:2355–2358

Reichard GA, Spitler J, Aslanian R, Mutahi M, Shih NY, Lin L, Ting PC, Anthes JC, Piwinski JJ (2002b) Structure-activity relationships of oxime neurokinin antagonists: oxime modifications. Bioorg Med Chem Lett 12:833–836

Ricciardolo FLM, (2001) Mechanisms of citric acid-induced bronchoconstriction. Am J Med 111:18S–24S

Ricciardolo FL, Rado V, Fabbri LM, Sterk PJ, Di Maria GU, Geppetti P (1999) Bronchoconstriction induced by citric acid inhalation in guinea pigs: role of tachykinins, bradykinin, and nitric oxide. Am J Respir Crit Care Med 159:557–562

Riccio MM, Myers AC, Undem BJ (1996) Immunomodulation of afferent neurons in guinea-pig isolated airway. J Physiol 491:499–509

Robineau P, Lonchampt M, Kucharczyk N, Krause JE, Regoli D, Fauchere JL, Prost JF, Canet E (1995) In vitro and in vivo pharmacology of S 16474, a novel dual tachykinin NK_1 and NK_2 receptor antagonist. Eur J Pharmacol 294:677–684

Rumsey WL, Aharony D, Bialecki RA, Abbott BM, Barthlow HG, Caccese R, Ghanekar S, Lengel D, McCarthy M, Wenrich B, Undem B, Ohnmacht C, Shenvi A, Albert JS, Brown F, Bernstein PR, Russell K (2001) Pharmacological characterization of ZD6021: a novel, orally active antagonist of the tachykinin receptors. J Pharmacol Exp Ther 298:307–315

Saria A, Martling CR, Dalsgaard CJ, Lundberg JM (1985) Evidence for substance P-immunoreactive spinal afferents that mediate bronchoconstriction. Acta Physiol Scand 125:407–414

Saria A, Martling CR, Yan Z, Theodorsson-Norheim E, Gamse R, Lundberg JM (1988) Release of multiple tachykinins from capsaicin-sensitive sensory nerves in the lung by bradykinin, histamine, dimethylphenyl piperazinium, and vagal nerve stimulation. Am Rev Respir Dis 137:1330–1335

Savoie C, Tousignant C, Rodger IW, Chan CC (1995) Involvement of NK_1 and NK_2 receptors in pulmonary responses elicited by non-adrenergic, non-cholinergic vagal stimulation in guinea-pigs. J Pharm Pharmacol 47:914–920

Schmidt D, Jorres RA, Rabe KF, Magnussen H (1996) Reproducibility of airway response to inhaled bradykinin and effect of the neurokinin receptor antagonist FK-224 in asthmatic subjects. Eur J Clin Pharmacol 50:269–273

Schulze E, Witt M, Fink T, Hofer A, Funk RH (1997) Immunohistochemical detection of human skin nerve fibers. Acta Histochem 99:301–309

Severini C, Improta G, Falconieri-Erspamer G, Salvadori S, Erspamer V (2002) The tachykinin peptide family. Pharmacol Rev 54:285–322

Sheldrick RL, Rabe KF, Fischer A, Magnussen H, Coleman RA (1995) Further evidence that tachykinin-induced contraction of human isolated bronchus is mediated only by NK_2-receptors. Neuropeptides 29:281–292

Shih NY, Albanese M, Anthes JC, Carruthers NI, Grice CA, Lin L, Mangiaracina P, Reichard GA, Schwerdt J, Seidl V, Wong SC, Piwinski JJ (2002) Synthesis and structure-activity relationships of oxime neurokinin antagonists: discovery of potent arylamides. Bioorg Med Chem Lett 12:141–145

Smet PJ, Moore KH, Jonavicius J (1997) Distribution and co-localization of calcitonin gene-related peptide, tachykinins, and vasoactive intestinal peptide in normal and idiopathic unstable human urinary bladder. Lab Invest 77:37–49

Springall DR, Cadieux A, Oliveira H, Su H, Royston D, Polak JM (1987) Retrograde tracing shows that CGRP-immunoreactive nerves of rat trachea and lung originate from vagal and dorsal root ganglia. J Auton Nerv Syst 20:155–166

Strigas J, Burcher E (1996) Autoradiographic localization of tachykinin NK_2 and NK_1 receptors in the guinea-pig lung, using selective radioligands. Eur J Pharmacol 311(2–3):177–186

Szallasi A, Blumberg PM (1989) Resiniferatoxin, a phorbol-related diterpene, acts as an ultrapotent analog of capsaicin, the irritant constituent in red pepper. Neuroscience 30:515–520

Ting PC, Lee JF, Anthes JC, Shih NY, Piwinski JJ (2001) Synthesis of substituted 4(Z)-(methoxyimino)pentyl-1-piperidines as dual NK_1/NK_2 inhibitors. Bioorg Med Chem Lett 11:491–494

Ting PC, Lee JF, Shih NY, Piwinski JJ, Anthes JC, Chapman RW, Rizzo CA, Hey JA, Ng K, Nomeir AA (2002) Identification of a novel 1'-[5-((3,5-dichlorobenzoyl)methylamino)-3-(3,4-dichlorophenyl)-4-(methoxyimino)pentyl]-2-oxo-(1,4'-bipiperidine) as a dual NK_1/NK_2 antagonist. Bioorg Med Chem Lett 12:2125–2128

Tomaki M, Ichinose M, Miura M, Hirayama Y, Yamauchi H, Nakajima N, Shirato K (1995) Elevated substance P content in induced sputum from patients with asthma and patients with chronic bronchitis. Am J Respir Crit Care Med 151:613–617

Turner CR, Andresen CJ, Patterson DK, Keir RF, Obach S, Lee P, Watson JW (1996) Dual antagonism of NK_1 and NK_2 receptors by CP-99,994 and SR48968 prevents airway hyperresponsiveness in primates Am J Respir Crit Care Med 153:A160

Van Rensen EL, Hiemstra PS, Rabe KF, Sterk PJ (2002) Assessment of microvascular leakage via sputum induction: the role of substance P and neurokinin A in patients with asthma. Am J Respir Crit Care Med 165:1275–1279

Walsh DA, Salmon M, Featherstone R, Wharton J, Church MK, Polak JM (1994) Differences in the distribution and characteristics of tachykinin NK_1 binding sites between human and guinea pig lung. Br J Pharmacol 113:1407–1415

Watson N, Maclagan J, Barnes PJ (1993) Endogenous tachykinins facilitate transmission through parasympathetic ganglia in guinea-pig trachea. Br J Pharmacol 109:751–759

Williams BJ, Teall M, McKenna J, Harrison T, Swain CJ, Cascieri MA, Sadowski S, Strader C, Baker R (1994) Acyclic NK-1 Antagonists: 2-Benzhydryl-2-Aminoethyl Ethers. Bioorg Med Chem Lett 4: 1903–1908.

Yunker AM, Krause JE, Roth KA (1999) Neurokinin B- and substance P-like immunoreactivity are co-localized in enteric nerves of the rat ileum. Regul Pept 80:67–74

Zhang, Z-L, Mak JC, Barnes PJ (1995) Characterization and autoradiographic mapping of [3H]CP96,345 a nonpeptide selective NK_1 receptor antagonist in guinea pig lung. Peptides 16:867–872

Pre-protachykinin and Tachykinin Receptor Knockout Mice

C. A. Gadd · M. Sukumaran · S. P. Hunt

Department of Anatomy and Developmental Biology, University College London, Medawar Building, Gower Street, London, WC1E 6BT, UK
e-mail: hunt@ucl.ac.uk

1	Introduction	298
2	Distribution and Actions of Substance P	298
2.1	Central Nervous System	298
2.2	Peripheral Nervous System	299
3	The Tachykinin NK_1 Receptor	300
4	Loss-of-Function Alleles of Tachykinins and Tachykinin Receptors	301
4.1	Pharmacogenetics: Strategies, Cautions and Caveats	303
5	The Effects of *PPT-A* or NK_1 Receptor Gene Knockout	305
5.1	Nociception	305
5.2	Affective Behaviours	309
5.2.1	Depression and Anxiety	309
5.2.2	Stress	313
5.2.3	Reward and Addiction	314
5.3	Learning and Memory	317
5.4	Inflammation and Response to Pathogens	317
5.5	Epilepsy and Other Phenotypes	319
6	Conclusion	326
	References	326

Abstract This chapter reviews the current contribution of genetics to tachykinin research. The discovery of substance P, related tachykinins and their receptors is reviewed, and the phenotypes of mice lacking functional tachykinin NK_1 receptors are described in the context of the biological activities of substance P, its receptor antagonists and the tissue distribution of the expression of the ligand and receptor. We compare and contrast the phenotypes of different null alleles of the NK_1 receptor with the pre-protachykinin-1 gene that encodes substance P. Loss-of-function alleles of the NK_1 receptor and the pre-protachykinin-1 gene show remarkably similar phenotypes with deficits in nociceptive processing, attenuated responses to pathogenic and inflammatory challenge and altered responses in affective behaviours relating to anxiety, depression, stress and reward.

Keywords Substance P · Genetics · Pre-protachykinin-1 gene

1
Introduction

The tachykinins are a family of peptides which share the common C-terminal sequence Phe-Xaa-Gly-Leu-Met-NH$_2$ and act at three clearly established receptor types, NK$_1$, NK$_2$ and NK$_3$ (Maggi et al. 1993). Since the discovery of tachykinins a large body of data has accumulated establishing the varied physiological roles of these peptides but it is relatively recently that molecular genetics has been used to complement these observations. To date only the pre-protachykinin A (*PPT-A*) or *TAC1* gene and the NK$_1$ receptor gene have been deleted using homologous recombination in mouse embryonic stem cells. However a large number of new and complementary physiological and behavioural observations have been made using these mice. This chapter will review these findings.

2
Distribution and Actions of Substance P

2.1
Central Nervous System

Although suggested in the 1950s, the excitatory nature of substance P (SP) on central neurons was not demonstrated until 1972, in frog spinal motor neurons (Otsuka et al. 1972). This, along with the observation that SP's concentration was around 20 times higher in dorsal than ventral root extracts from bovine spinal cord (Lembeck et al. 1953; Takahashi et al. 1974), led to the suggestion that SP was located in primary afferent neurons. This was supported by observations that a dramatic decrease in SP concentrations was seen in the superficial spinal cord following rhizotomy or nerve section (Takahashi et al. 1975; Jessell et al. 1979), and that SP immunoreactivity and PPT-A mRNA were observed in the cell bodies of largely nociceptive primary afferent cells in the dorsal root ganglia, which possessed unmyelinated or thinly myelinated axons (Hökfelt et al. 1975, 1976; Ljungdahl et al. 1978).

The localization of SP to unmyelinated C-fibres was confirmed using the neurotoxin capsaicin. In adult animals, chronic capsaicin administration results in an insensitivity to painful stimuli, in part due to the reversible depletion of SP from the terminals of primary afferent C-fibres (Jancsó et al. 1977; Jessell et al. 1978; Gamse et al. 1979). Furthermore, the release of SP from primary afferent nerve terminals is induced by sciatic nerve stimulation, but only at stimulus intensities high enough to activate C-fibres (Gamse et al. 1979; Theriault et al. 1979; Yaksh et al. 1980). These findings have subsequently been verified in ultrastructural studies, demonstrating the presence of SP-containing vesicles within nerve terminals located in the superficial dorsal horn (Chan-Palay and Palay 1977; Barber et al. 1979; Ribeiro-da-Silva et al. 1989). SP application brings about a slow, long-lasting depolarization of dorsal horn neurons both in vivo and in culture, which is excitatory and often accompanied by a fast excitatory post-synaptic potential (Konishi and Otsuka 1974; Krnjevic and Morris 1974; Henry 1976; Randic and Miletic 1977; Zieglgänsberger and Tulloch 1979; Nowak and Macdonald 1982).

In the brain, SP is found in a wide range of structures. Immunohistochemical studies have shown that the area with the highest levels of SP is the substantia nigra, where SP terminals derive from the ipsilateral striatum (Davies and Dray 1976). SP release in this brain region causes a long-lasting excitation of the nigrostriatal dopaminergic neurones that are under inhibitory control by dopamine and γ-aminobutyric acid (GABA) although paradoxically expression of the NK_1 receptor is extremely low in this brain region.

In the striatum, SP is found in the dynorphin-positive population of medium-sized GABAergic spiny neurons of the caudate putamen (Haber and Nauta 1983; Penny et al. 1986; Anderson and Reiner 1990; Besson et al. 1990; Napier et al. 1995). These neurons also possess SP-positive collaterals that project back onto the large cholinergic interneurons of the striatum (Bolam et al. 1983, 1986). The release of SP in this region increases the firing rate of these cholinergic interneurons (Le Gal La Salle and Ben Ari 1977), and leads to the release of dopamine and acetylcholine within the striatum (Starr 1978; Petit and Glowinski 1986; Arenas et al. 1991; Boix et al. 1992; Anderson et al. 1993; Steinberg et al. 1995; Tang et al. 1998; Galarraga et al. 1999).

Elsewhere in the brain, SP is found in low levels in the cortex and hippocampus whilst in the hypothalamus, it is thought to be involved in the regulation of hormone release from the pituitary gland (Aronin et al. 1986). In the amygdala, SP fibres have been observed primarily in the medial and central subnuclei, where it is involved in the control of affective behaviour (Emson et al. 1978; Kramer et al. 1998; Steinberg et al. 2001). SP is also found in the parabrachial nucleus and in the locus coeruleus, where SP terminals contact noradrenergic neurons, whose firing rate is increased by SP application (Guyenet and Aghajanian 1977; Bert et al. 2002). The nucleus tractus solitarius possesses high levels of SP, where it plays a role in respiratory, cardiovascular and emetic reflexes (Gillis et al. 1980; Tattersall et al. 1996), whilst the distribution and action of SP in the trigeminal nucleus (Henry et al. 1980; Yonehara et al. 1986) correspond to that found in the dorsal horn of the spinal cord.

2.2
Peripheral Nervous System

In the periphery, SP is found in most tissues, where it is usually derived from the peripheral endings of primary afferent neurons. Additionally, in the eye, SP is found in the ganglion cell layer of the retina, where it acts as a neurotransmitter (Brecha et al. 1987), and in the innervation of the iris, where it causes contraction (Tornqvist et al. 1982; Ueda et al. 1986; Andersson 1987). SP is also a potent stimulator of salivation, being found at high levels in the parasympathetic innervation of the salivary glands (Ekström et al. 1988). In the urinary bladder and the respiratory system, SP mediates smooth muscle contraction, resulting in micturition or bronchoconstriction respectively (Falconieri-Erspamer et al. 1973; Sjögren et al. 1982; Carruette et al. 1992). Furthermore, exposure to irritants in the airways leads to the release of SP and other tachykinins resulting in coughing, sneezing, vasodilation, plasma extravasation and the secretion of mucus: SP release may therefore be an important factor in asthma (Barnes 1987; Barnes et al. 1988, 1990; Delay-Goyet and Lundberg 1991; Frossard and Advenier 1991; Lei et al. 1992; Bertrand and Geppetti 1996). In the circulatory system, SP induces vasodilation and hypotension via a nitric oxide-dependent mechanism (Pernow and Rosell 1975; Whittle et al. 1989; Fiscus et al. 1992).

Within the gastrointestinal tract, SP is found both in primary afferent terminals and in intrinsic neurons (Franco et al. 1979; Sternini et al. 1995). SP causes smooth muscle contraction in the gut, coupled with an increase in motility and secretion of water and electrolytes, mediated both by the direct action of SP on muscle cells, and by a cholinergic mechanism (Bauer and Kuriyama 1982; Holzer and Lembeck 1980; Huidobro-Toro et al. 1982; Kachur et al. 1982; Holzer and Petsche 1983).

3
The Tachykinin NK_1 Receptor

The tachykinin receptors, including the NK_1 receptor, were discovered in the early 1980s. Following comparison of SP and other non-mammalian tachykinins in various bioassays, it was found that they displayed dramatic differences in potencies in different tissues (Falconieri-Erspamer et al. 1980; Erspamer 1981; Lee et al. 1982; Watson et al. 1983; Laufer et al. 1985). It was noted that all three mammalian tachykinins were able to act as full agonists at each of the three receptor types, but with different affinities (Regoli et al. 1987; Saffroy et al. 1988; Mantyh et al. 1989; Ingi et al. 1991). In 1986, tachykinin receptor classification was standardized at the Substance P and Neurokinin Meeting in Montreal, Canada, and the three tachykinin receptors named NK_1, NK_2 and NK_3. (Henry 1987). These three receptors were cloned from the cow and the rat in the late 1980s, and were found to be very similar in amino acid sequence and in structure, as guanine nucleotide binding protein (G-Protein)-coupled receptors (Masu et al. 1987; Yokota et al. 1989; Hershey and Krause 1990; Shigemoto et al. 1990; Nakanishi 1991). The three receptors differ primarily in the third intracellular loop and the C terminus, which may confer their differential coupling to G-proteins (Nakanishi 1991). The human NK_1 receptor was cloned in 1991, and was found to be 94.8% homologous to the rat receptor (Gerard et al. 1991).

The C-terminal hexapeptide of SP is essential for binding to and activation of the NK_1 receptor (Bury and Mashford 1976; Cascieri et al. 1992). Activation of the NK_1 receptor results in the activation of G-proteins (Mantyh et al. 1984). The local increase in inositol 1,4,5-triphosphate results in the release of calcium ions from intracellular stores and the subsequent activation of calcium–calmodulin-dependent kinases, as well as the influx of extracellular calcium, whereas diacyl glycerol activates protein kinase C (Merritt and Rink 1987; Sugiya et al. 1987; Womack et al. 1988; Seabrook and Fong 1993). NK_1 receptor activation also stimulates arachidonic acid mobilization via phospholipase A_2 and cyclic adenosine monophosphate (Mitsuhashi et al. 1992; Nakajima et al. 1992). These effects on second messenger pathways are followed by a slow excitatory response mediated by increases in membrane sodium conductance and decreases in potassium conductance (Nicoll 1978; Yamaguchi et al. 1990; Shen and North 1992).

The distribution of the NK_1 receptor in the brain is extensive, being heavily expressed in the olfactory bulbs, the striatum, the amygdala, the habenula, the periaqueductal grey, the superior colliculus, the parabrachial nucleus and the locus coeruleus. Elsewhere in the central nervous system, it is found in high levels in the superficial laminae of the spinal cord, where it is expressed on the neurons that are activated by SP released from nociceptive C-fibres (Nakaya et al. 1994). Peripherally, the expression of the NK_1 receptor is similar to that of SP, particularly in the salivary glands and in the small and large intestines (Dray and Pinnock 1982; Sternini et al. 1995, Andoh et al. 1996; Li and Zhao 1998).

4
Loss-of-Function Alleles of Tachykinins and Tachykinin Receptors

The tachykinin family of ligands and their receptors are phylogenetically conserved and expressed tissue-specifically (Nakanishi 1987, 1991). The NK_1 receptor locus is located on chromosome 6 of the mouse genome, spanning ~180 kb (Yokota et al. 1989; Hershey and Krause 1990; Hershey et al. 1991; Sundelin et al. 1992). The gene comprises five exons interrupted by relatively large first and second introns followed by smaller third and fourth introns (Fig. 1A).

Substance P is generated from the proteolytic processing of the product of the *PPT-A* gene (Nakanishi 1987). In addition to SP, proteolytic processing can also generate neurokinin A (NKA) and the N-terminal variants of NKA, neuropeptide K (NPK) and neuropeptide γ (NPγ) (Carter and Krause 1990). The segments of the gene encoding SP, NPK and NKA are illustrated in Fig. 1A. The *PPT-A* locus is composed of seven exons and six introns spanning 7.6 kb of the mouse genome also on chromosome 6. Differential alternative splicing and alternate exon use generate α-, β- and γ-PPT-A primary transcripts, from which different combinations of the three peptides are synthesized in a tissue-dependent manner (Nakanishi 1987; Carter and Krause 1990). α-PPT-A mRNA encodes SP only and is produced primarily within the central nervous system; β- and γ-PPT-A mRNAs are found both in peripheral and central nervous system tissues and encode NKA, NPK and NPγ in addition to SP. Substance P and NKA are always produced and released together from neurons; NKB, a lower affinity agonist of the NK_1 receptor, is encoded by the *PPT-B* locus, located on mouse chromosome 10. *PPT-B* and *PPT-A* have a similar structural organization suggestive of evolution from a common ancestral gene (Nakanishi 1987).

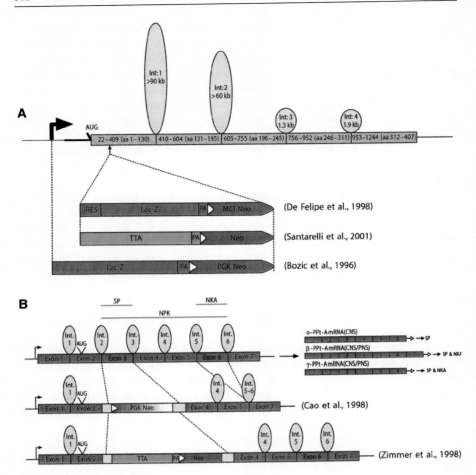

Fig. 1 A Structural organization of the murine NK$_1$ receptor locus and loss-of-function alleles (De Felipe et al. 1998; Santarelli et al. 2001; Bozic et al. 1996). **B** Schematic of the structural organization and loss-of-function alleles of the murine pre-protachykinin-A locus (Cao et al. 1998; Zimmer et al. 1998). *AUG*, translation initiating methionine, 22–1244 (amino acids 1–407); *cDNA*, nucleotide sequence and corresponding amino acid sequence of receptor; *Int*, intron; *IRES*, intra-ribosomal entry site; *LacZ*, gene encoding β-galactosidase; *TTA*, gene encoding tetracycline transactivator; *Neo*, neomycin resistance gene and promoters (PGK or MCI); *PA*, polyadenylation site; *SP*, substance P; *NKA*, neurokinin A; *NPK*, neuropeptide K, a N-terminal splice variant of NKA; *CNS*, central nervous system; *PNS*, peripheral nervous system

The two mutant alleles of the *PPT-A* locus involved replacing the SP coding exon 3 of the locus with the neomycin resistance gene (Cao et al. 1998; Zimmer et al. 1998). Cao et al. (1998) replaced exon 3 and also deleted exon 6 sequences that encode NKA (Fig. 1B). This ensured that alternative differential splicing does not produce NKA- or NPγ-encoding mRNA. Zimmer et al. (1998) introduced a tTA/Neo cassette between exons 2 and 4, replacing the SP-encoding exon and retained the remainder of the locus intact as illustrated. Both these mutant strains have been shown to be null for both SP and NKA.

Recently generated loss-of-function alleles of the NK_1 (Bozic et al. 1996; De Felipe et al. 1998; Santarelli et al. 2001) and *PPT-A* (Cao et al. 1998; Zimmer et al. 1998) loci have highlighted the functions of NK_1 receptor-mediated tachykinin signalling. These mutant strains have deficits in sensory processing and affective behaviours, and altered responses to pathogenic and inflammatory challenge. The different genetic strategies used by various independent research groups to generate these mutant strains are illustrated in Fig. 1B.

Three mutant alleles of the NK_1 receptor locus have been generated (Bozic et al. 1996; De Felipe et al. 1998; Santarelli et al. 2001). All involved targeted disruption of exon 1 coding sequences by the insertion of the neomycin resistance gene linked with the LacZ reporter gene (Bozic et al. 1996; De Felipe et al. 1998) or the tetracycline transactivator (Santarelli et al. 2001; Gross et al. 2002). De Felipe and colleagues (1998) used the IRES-LacZ cassette as a reporter gene. In this allele, LacZ-encoded β-galactosidsase comes under the transcriptional control of the NK_1 receptor locus and is translated under the influence of an intraribosomal entry site (Houdebine and Attal 1999). Similarly, Bozic et al. (1996) used a LacZ reporter gene but avoided the use of IRES sequences by inserting the gene in the 5' untranslated region.

4.1
Pharmacogenetics: Strategies, Cautions and Caveats

Molecular genetic experiments start with the premise of studying gene function by examining the phenotype of loss-of-function or gain-of-function alleles. However, several caveats need to be considered before a causal link between the engineered molecular change and the observed phenotype can be made (Gerlai 1996). This is of particular importance when considering the effects of the mutation on complex polygenic traits such as anxiety, depression and pain perception that have cognitive and emotional dimensions (Clement et al. 2002). First, wild-type congenic strains of mice of different genetic backgrounds behave differently in tests designed to assess cognitive and emotional functions (Crawley et al. 1997). It has been suggested that to study the effect of a mutation on a particular trait, the mutant allele has to be transferred into a background that is most sensitive and responsive for the parameter assayed. However, certain mutations can have different phenotypes in different genetic backgrounds (Threadgill et al. 1995). Therefore the genetic determinants of these traits are multi-factorial and pleiotropic, i.e. alterations in one component can result in compensatory modulation by parallel pathways, and in addition there may also be an epistatic element, i.e. complex interactions between the mutant allele, background genes and environmental factors can also influence these traits (Gerlai 1996). Therefore, to identify the molecular determinants of complex behavioural traits requires careful experimental design to decipher the consequential effects of the mutation from the modifying functions of background genes and the environment (Crawley et al. 1997). This necessitates analysis of the effects of the mutation on different congenic backgrounds by comparative analysis of mutants with wild-type littermate controls of similar genetic background (Banbury Conference Report 1997).

As embryonic stem (ES) cells derived from the 129 strain of mice proved to be the most robust for experimental manipulations, most targeting experiments use ES cells derived from inbred 129 substrains of undefined genetic backgrounds (Simpson et al. 1997). Consequently, after homologous recombination, the mutant allele is presented in this complex genetic background in the recombinant ES cells. When chimeras generated from blastocysts injected with recombinant ES cells are crossed with the C57BL/6 strain, germ line transmission of the mutation generates F1 progeny heterozygous for the mutation in a 129 and C57BL/6 hybrid background. A commonly used strategy to generate homozygous mutants for experimentation is from F1 interbreeding and by maintaining the colony by subsequent homozygous inbreeding. Although this approach is cost effective, less time consuming and the apparent total background genetic contribution from the hybrid strains remains constant in each generation, random segregations of alleles in successive generations produce variability in the genotype of littermate controls. In addition, selection of background genes which modify the deleterious effects of the mutation can result in progressive changes in the genotype of the hybrid mutant strain (Banbury Conference Report 1997). To avoid these problems the mutant allele should be backcrossed into defined genetic backgrounds and maintained as congenic lines. This is accomplished by successive backcrossing of heterozygous progeny with wild-types of defined inbred strains (Banbury Conference Report 1997). Simultaneous derivation of congenic mutant strains in different inbred lines permits analysis of the phenotype in different backgrounds and also in the F1 hybrids of these strains. In many behavioural investigations it has also been thought advisable to breed knockout and wild-type mice from the same heterozygotic parents further avoiding differences in genetic background. This is a hugely expensive and time consuming exercise but may be advisable when subtle changes in phenotype have been observed (Morcuende et al. 2003)

Apart from the issues arising from strain differences and genetic backgrounds, other less confounding problems associated with the knockout approach relate to the differential tissue specific functions of certain genes and compensatory developmental changes that result from deletion of the gene throughout development. For example, in NK_1 knockout mice there is a down-regulation of 5-hydroxytryptamine (5-HT) $5-HT_{1A}$ receptor function resulting in increased stimulation-induced release of 5-HT (Froger et al. 2001; Santarelli et al. 2002) These problems may be circumvented through the use of Cre/Lox mediated conditional and inducible gene ablation strategies (Picciotto et al. 1998; Sauer 1998; Lewandowski 2001; Nagy et al. 2003). For these experiments to be successful it is important to characterize the *cis*-regulatory elements that control the cell-type specificity of the gene. Transgenic mice lines have been generated from engineered bacterial artificial chromosomes allow for the identification of regulatory elements (Copeland et al. 2001; Heintz et al. 2001). These types of conditional knockouts and indeed the generation of inducible deletion of tachykinin related gene knockouts has yet to be achieved but will undoubtedly be necessary to clarify the role of NK_1 and SP in different areas of the brain.

5
The Effects of *PPT-A* or NK_1 Receptor Gene Knockout

5.1
Nociception

The most widely studied physiological role of SP is that of a transmitter of nociceptive information (Table 1). Since painful stimuli bring about the release of SP onto dorsal horn neurons (Randic and Miletic 1977), SP was hypothesized to play an important role in pain transmission. Behavioural studies supported this hypothesis, since intrathecal and intravenous administration of SP elicit scratching and biting of the abdomen (Hylden and Wilcox 1981; Piercey et al. 1981; Gamse and Saria 1986), behaviours which are blocked by an NK_1 receptor antagonist (Yamamoto and Yaksh 1991; Yasphal et al. 1993) and by systemic morphine injections (Hylden and Wilcox 1981). Opiates inhibit the release of SP in the trigeminal nucleus induced by a high potassium concentration and in the dorsal horn after stimulation of primary afferents (Jessell and Iversen 1977; Yaksh et al. 1980). Noxious stimulation also brings about an increase in PPT-A and NK_1 receptor mRNA expression in the dorsal root ganglion (Noguchi et al. 1988; McCarson 1999), further supporting SP's involvement in the detection of pain.

However, early conclusions that SP was the 'primary' pain transmitter have not been upheld by more recent findings (Hill 2000). Despite the anatomical evidence for a role of SP and the NK_1 receptor in pain, they seem to have little influence on acute pain sensation, although the responses to intense or inflammatory pain seem to engage NK_1 receptor-dependent processes. Hence, in the mouse, rat and guinea pig, NK_1 receptor antagonists have been shown to reduce SP-induced plasma extravasation and hyperalgesia in models of chronic inflammatory and neuropathic pain, but with relatively few effects on acute tests (Garret et al. 1991; Birch et al. 1992; Nagahisa et al. 1992; Chapman and Dickenson 1993; Laird et al. 1993; Yasphal et al. 1993; Ma and Woolf 1995, 1997; Campbell et al. 1998; Lecci et al. 1998). Based on findings that NK_1 receptor antagonists can attenuate the increase in pain sensitivity that follows tissue or nerve injury, NK_1 receptors have been proposed to participate in the generation of central excitability (Baranauskas and Nistri 1998). In support of this, NK_1 receptor immunoreactivity increases following persistent pain states (Abbadie et al. 1996) and administration of NK_1 receptor antagonists produces antinociception in studies that include acute pain models (Birch et al. 1992; Dougherty et al. 1994; Santos and Calixto 1997; Sluka et al. 1997; Campbell et al. 1998; Cumberbatch et al. 1998; Field et al. 1998; Henry et al. 1999; Coudore-Civiale et al. 2000). However, data from pharmacological studies using NK_1 receptor antago-

Table 1 Summary of observed phenotypes in *PPT*-A and NK$_1$ receptor knockout mice: pain

PPT-A$^{-/-}$		NK$_1$$^{-/-}$	
Cao et al. (1998) Strain: Cd-1×129	Zimmer et al. (1998) Strain: C57/6 J×129	Bozic et al. (1996) Strain: C57/6 J×129/Sv	De Felipe et al. (1998) Strain: C57/BL6×129/Sv	Santarelli et al. (2001) Strain: 129/SvEv pure

Acute pain				
Reduced response to moderate–intense thermal pain (Cao et al. 1998). Drink more capsaicin solution, but do show concentration-dependent aversion (Simons et al. 2001)	No pain response following formalin injection (1st and 2nd phases), increased pain threshold on hotplate, normal behaviour in tail flick and acetic acid-induced writhing (Zimmer et al. 1998)	Normal responses to acute thermal/mechanical stimulation, except at high temperature (Mansikka et al. 1999). Responses of L4–L5 neurones in deep dorsal horn to acute noxious stimuli unchanged (Weng et al. 2001)	Normal acute pain, intensity coding absent (De Felipe et al. 1998). Normal responses to visceral mechanical stimuli (acetylcholine, hypertonic saline, colon distension) but failure to encode intensity, reduced response to acute chemical visceral stimulation (Laird et al. 2000). Normal responses to mechanical stimuli and cooling (Martinez-Caro and Laird 2000). Normal acute mechanical thresholds (Laird et al. 2001)	Reduced response to high intensity heat/mechanical stimulation (King et al. 2000)

Wind-up/hyperalgesia				
		Capsaicin-induced paw-licking and mechanical / thermal hyperalgesia reduced (Mansikka et al. 1999). Sensitization of L4–L5 neurones in deep dorsal horn to repeated C-fibre stimulation reduced (Weng et al. 2001)	Wind-up absent, reduced 2nd phase of formalin test, but normal hyperalgesia to CFA (De Felipe et al. 1998). Inflammatory hyperalgesia following CFA normal up to 4d (Cao et al. 1998; De Felipe et al. 1998) then decreases (Kidd et al. 2003), SP and dynorphin up-regulation in sensory neurones and spinal cord normal; (Palmer et al. 1999) hyperalgesia to decreased capsaicin, but normal response to mustard oil (Laird et al. 2000). Shorter duration of paw-licking after capsaicin at lower dose, no allodynia or hyperalgesia following capsaicin—mainly due to central NK$_1$ receptors/changes in descending inhibition and excitation (Laird et al. 2001). Response to CFA normal for first 4 days then reduced (Kidd et al. 2003)	Reduced 2nd phase of formalin test (King et al. 2000)

Table 1 (continued)

	PPT-A⁻/⁻		NK₁⁻/⁻	
	Cao et al. (1998) Strain: Cd-1×129	Zimmer et al. (1998) Strain: C57/6 J×129	Bozic et al. (1996) Strain: C57/6 J×129/Sv	De Felipe et al. (1998) Strain: C57/BL6×129/Sv	Santarelli et al. (2001) Strain: 129/SvEv pure
Neuropathic pain/nerve injury hyperalgesia			No increases in sensitivity to mechanical stimuli after L5 spinal nerve ligation, but normal increases to thermal stimuli (Mansikka et al. 2000)	Normal responses to mechanical stimuli and cooling before and after partial sciatic nerve ligation (Martinez-Caro and Laird 2000)	
Descending inhibition				Stimulation of hindpaw and forepaw causes increased c-Fos expression in lumbar laminae I-II, but reduced response in raphe magnus and pallidus: NK_1 receptor essential for noxious descending inhibition (Bester et al. 2001; Suzuki et al. 2002)	
Development					Normal response to heat, mechanical or chemical stimuli at 3 days, but reduced responses to high intensity heat/mechanical stimulation and 2nd phase of formalin test at 21 d (King et al. 2000)

CFA, complete Freund's adjuvant.

nists have been difficult to interpret due to conflicting reports that may arise from differences in the selectivity of the compounds, as well as their route of administration, dose and modalities tested. In humans NK_1 receptor antagonists have been ineffective in the majority of clinical trials for analgesia (Hill 2000) although a single trial showed that an NK_1 receptor antagonist reduced postoperative pain after dental extraction (Dionne et al. 1998). Some success has also been reported for patients with fibromyalgia (Russell 2002). It is possible that this apparent difference between humans and preclinical species may reflect a difference in NK_1 receptor distribution (Moussaoui et al. 1993; O'Shaughnessy and Connor 1993; Shepheard et al. 1995; Binder et al. 1999; Beattie et al. 2002) or targeting of inappropriate pain states.

Mice with genetic disruption of the NK_1 receptor or *PPT-A* gene have also been reported to have normal acute pain behaviour, apart from some difficulty in coding the intensity of intermediate to very noxious thermal stimuli (see Table 1 for references; Mansikka et al. 2000). The hyperalgesia evoked by sciatic mononeuropathy in NK_1 receptor knockout mice was similar to that in wild-type mice (Martinez-Caro and Laird 2000). In contrast, secondary visceral nociception (Laird et al. 2000), the second phase of the formalin response (De Felipe et al. 1998), and hyperalgesia induced by capsaicin (Mansikka et al. 1999; Laird et al. 2000) or inflammation (Kidd et al. 2003) (Fig. 2) were similarly depressed. Electrophysiological studies have revealed that although acute mechanical and thermal nociception remained unaltered in $NK_1^{-/-}$ mice, there is a reduction in central sensitization following mustard oil application, as well as attenuated 'wind-up' of spinal neurons (Weng et al. 2001). However, a second study reported that $NK_1^{-/-}$ mice do display marked deficits in mechanical and thermal coding in the noxious range of stimuli, as well as exhibiting attenuated wind-up, thereby implicating NK_1 receptors in the accurate intensity coding of suprathreshold stimuli (Suzuki et al. 2002).

The failure to alter acute pain processing but the observation that many aspects of secondary changes in pain behaviour are reduced is consistent with what we are beginning to understand about pain processing: many of these observations can be explained by deficits in descending facilitation of spinal processing through pathways originating in the brain stem. Substance P is released and activates the NK_1 receptor on neurons mainly in lamina I of the dorsal horn (Allen et al. 1997). Using SP conjugated to the cytotoxin saporin to selectively ablate neurons expressing the NK_1 receptor (Mantyh et al. 1997), a marked reduction of hyperalgesia was demonstrated following tissue and peripheral nerve injury, with no effect on acute behavioural nociceptive thresholds (Nichols et al. 1999). This suggested that the NK_1 receptor on lamina I projection neurons was crucial for the development of hyperalgesia. Destruction of lamina I neurons also resulted in the loss of central sensitization, as revealed by diminished wind-up and decreased formalin second phase response in deeper lying lamina V neurons. Many of these changes were due to the disruption of a serotonergic spinal-bulbo-spinal facilitatory loop driven by lamina I NK_1 expressing neurons (Suzuki et al. 2002).

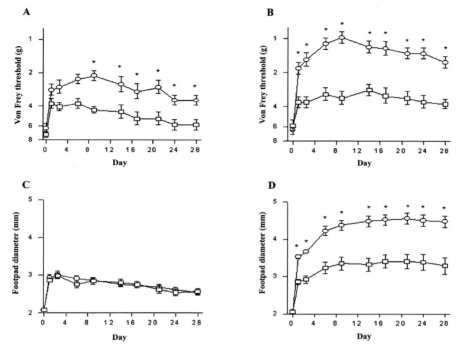

Fig. 2A–D Late, but not early, stage hyperalgesia is influenced by NK_1R knockout. Outcome measures of complete Freund's adjuvant (*CFA*)-induced inflammation in wild-type (*circles*) and NK_1 receptor knockout (*squares*) groups. Mechanical hyperalgesia following (**A**) 1 mg/ml and (**B**) 10 mg/ml CFA-induced inflammation. Footpad diameter following (**C**) 1 mg/ml and (**D**) 10 mg/ml CFA-induced inflammation. *Asterisks* indicate significant differences ($P<0.05$). (From Kidd et al. 2003)

In summary, loss of the NK_1 receptor results in changes in pain behaviour that are seen primarily in the secondary responses to noxious stimuli such as inflammatory hyperalgesia. It seems likely that the NK_1 receptor may be important in the high intensity noxious signalling of acute peripheral (mechanical or thermal) stimuli. However this may also result from the lack of wind-up or disruption of spinal-bulbo-spinal loops.

5.2
Affective Behaviours

5.2.1
Depression and Anxiety

In humans, the NK_1 receptor antagonist MK869 has been shown to alleviate major depressive disorder with moderately high anxiety (Kramer et al. 1998). NK_1 receptor antagonists also bring about increases in the firing rate of dorsal raphe serotonergic neurons, in common with the effects observed with established antidepressant drugs (Haddjeri and Blier 2001; Conley et al. 2002).

NK_1 receptor antagonists also show an anxiolytic profile in preclinical tests. MK869 has been found to inhibit maternal separation-induced vocalizations in guinea pig pups (Kramer et al. 1998), as do the antagonists L733060 and GR205171 in an enantioselective manner (Rupniak et al. 2000). Furthermore, SP is released in the amygdala following aversive stimuli such as immobilization stress or maternal separation, and blockade of NK_1 receptors in the amygdala prevents the vocalizations associated with these stimuli (Kramer et al. 1998; Smith et al. 1999; Boyce et al. 2001; Steinberg et al. 2002). NK_1 receptor antagonists reduce anxiety in the social interaction test in the rat (File 1997, 2000; Vassout et al. 2000) and the gerbil (Cheeta et al. 2001; Gentsch et al. 2002). In mice, intracerebroventricular administration of FK888 brought about an anxiolytic profile on the elevated plus maze test (Teixeira et al. 1996), an effect that was also seen with oral administration of a range of NK_1 antagonists in the gerbil (Varty et al. 2002). NK_1 receptor antagonists have also been shown to reduce anxiety-related foot-tapping behaviour in the gerbil (Ballard et al. 2001; Duffy et al. 2002). Intracerebroventricular SP administration or microinjection into the dorsal periaqueductal grey brings about an increase in anxiety-related behaviour in mice and rats (Aguiar and Brandao 1996; Teixeira et al. 1996; De Araujo et al. 1999), but it has an anxiolytic effect when injected systemically or into the nucleus basalis magnocellularis (Hasenöhrl et al. 1998; Nikolaus et al. 2000). SP signalling is therefore of importance in the mediation of anxiety-related behaviour, in a region-specific manner.

$NK_1^{-/-}$ mice have been shown to exhibit alterations in brain function similar to those observed after chronic antidepressant treatment, notably down-regulation and desensitization of inhibitory presynaptic $5\text{-}HT_{1A}$ receptors in the dorsal raphe nucleus (Table 2; Froger et al. 2001; Santarelli et al. 2001). Behaviourally, $NK_1^{-/-}$ mouse pups emit fewer isolation-induced vocalizations than their wild-type littermates (Rupniak et al. 2000; Santarelli et al. 2001). They also exhibit more struggling in the forced swim and tail suspension tests than wild-type littermates (Rupniak et al. 2001), as do mice lacking the *PPT-A* gene (Bilkei-Gorzo et al. 2002), providing further support for a role of NK_1 receptor blockade in the effective treatment of depressive illness (Santarelli et al. 2001). The findings of Bilkei-Gorzo et al. (2002) with *PPT-A$^{-/-}$* mice have been replicated in the strain of $NK_1^{-/-}$ mouse produced by Hunt and colleagues (De Felipe et al. 1998) using the light–dark shuttle box (Herpfer, Hunt and Stanford, unpublished results).

It has previously been shown that chronic treatment with antidepressant drugs increases neurogenesis and levels of brain-derived neurotrophic factor (BDNF) in the hippocampus. These changes have been correlated with changes in learning and long-term potentiation and may contribute to the therapeutic efficacy of antidepressant drug treatment (Nibuya et al. 1995; Chen et al. 2000; Malberg et al. 2000; Manev et al. 2001). Recently, it has been demonstrated that there is a significant elevation of neurogenesis but not cell survival in the hippo-

Table 2 Summary of observed phenotypes in *PPT-A* and NK$_1$ receptor knockout mice: affective behaviours

	PPT-A$^{-/-}$		NK$_1$$^{-/-}$	
	Cao et al. (1998) Zimmer et al. (1998) Strain: C57/6 J × 129		De Felipe et al. (1998) Strain: C57/BL6×129/Sv Bozic et al. (1996)	Santarelli et al. (2001) Strain: 129/SvEv pure
Depression/anxiety/stress	More active in Porsolt test and tail suspension test, not hyperactive after bulbectomy, anxiolytic profile in open field, more social interactions, more active in open areas of elevated O-maze, reduced latency to feed in Thatcher-Britton paradigm (Bilkei-Gorzo et al. 2002)		Impaired stress-induced analgesia and aggressive response to territorial challenge (De Felipe et al. 1998). Neonates emit fewer ultrasonic vocalizations (Rupniak et al. 2000). Similar on resident-intruder and forced swim tests to fluoxetine and NK$_1$ antagonism, reduced struggling in tail suspension (not seen with NK$_1$ antagonist—developmental peculiarity?), no change on elevated plus maze (Rupniak et al. 2001; Murtra et al. 2000b). Dorsal raphe 5-HT$_{1A}$ receptors, 5-HT$_{1A}$-dependent [^{35}S]GTP-γ-S binding and [5-HT] in anterior raphe reduced, decrease in function of agonist and reduced hypothermic response, but fluoxetine causes more cortical 5-HT overflow, i.e., down-regulation/desensitization of 5-HT$_{1A}$ receptors (Froger et al. 2001). Increased hippocampal neurogenesis: not further increased by citalopram or desipramine (Morcuende et al. 2003)	Reduction in anxiety and stress-related behaviours, increase in firing rate of 5-HT neurones in dorsal raphe and desensitization of 5-HT$_{1A}$ receptors (Santarelli et al. 2001)
Reward/addiction			Absence of conditioned place preference to morphine and amphetamine, but not cocaine or food, physical response and aversion to opiate withdrawal reduced (Murtra et al. 2000a, 2000b). Absence of self-administration and behavioural sensitization to morphine, but not cocaine (Ripley et al. 2002)	

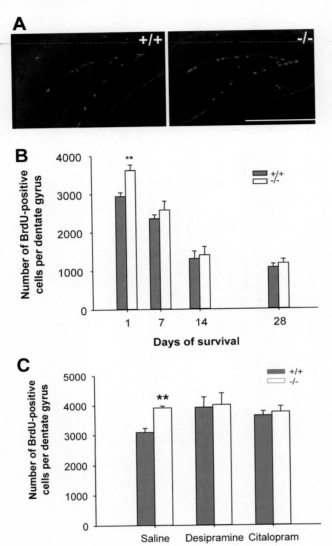

Fig. 3A–C Increase of neurogenesis in the dentate gyrus of $NK_1R^{-/-}$ mice (Morcuende et al. 2003). **A** Representative photomicrographs from the dentate gyrus of the wild-type and $NK_1R^{-/-}$ mice 1 day after BrdU injections. *Scale bar*, 500 μm. **B** The analysis of the BrdU-labelled cells in the dentate gyrus 1, 7, 14 and 28 days after BrdU injections showed that proliferation was enhanced in the hippocampus of the $NK_1R^{-/-}$ mice compared to the wild-type animals but that the number of BrdU-positive cells in both genotypes decreased to a similar level after 28 days. **C** Antidepressants increase neurogenesis in wild-type, but not $NK_1R^{-/-}$ mice. Chronic treatment with antidepressants (desipramine or citalopram) increased proliferation in wild-type mice, but failed to increase the number of newborn cells in the $NK_1R^{-/-}$ animals compared to the control animals chronically injected with saline. $**P<0.01$ (vs. +/+)

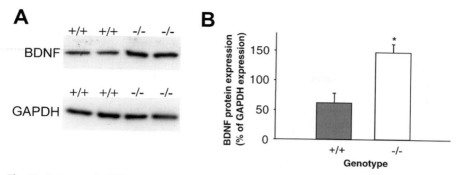

Fig. 4A, B Increase in BDNF in the hippocampus, but not cortex, of $NK_1R^{-/-}$ mice. **A** Western blot from hippocampus of both genotypes showed an increased basal level of BDNF protein expression in $NK_1R^{-/-}$ mice compared to wild-type mice. **B** Histograms representing the increased BDNF levels expressed as a percentage of GAPDH expression in terms of relative grey-scale values after quantitative densitometry of enhanced chemiluminescent films *P<0.05 (vs. +/+). GAPDH = Glyceraldehyde-3-phosphate dehydrogenase. (From Morcuende et al. 2003)

campus of NK_1 receptor knockout mice (Fig. 3; Morcuende et al. 2003). Neurogenesis was also increased in wild-type, but not NK_1 receptor knockout mice by chronic treatment with antidepressant drugs that preferentially target noradrenergic (desipramine) and serotonergic pathways (paroxetine).

Hippocampal levels of BDNF were twofold higher in $NK_1^{-/-}$ mice (Fig. 4), whereas cortical levels were similar in knockout and wild-type mice (Morcuende et al. 2003). This was of some interest as BDNF has recently been shown to have antidepressant activity when injected directly into the hippocampus (Shirayama et al. 2002).

These data support the hypothesis that increased neurogenesis, perhaps accompanied by higher levels of BDNF, may contribute to the efficacy of antidepressant drug therapy. In this respect it is interesting to note that levels of the NK_1 receptor although low are greater in the ventral than dorsal hippocampus. This is in keeping with the recent observation that the ventral hippocampus is more closely involved with anxiety behaviours while the dorsal hippocampus has more obvious learning and memory functions (Bannerman et al. 2003).

5.2.2
Stress

Central injection of NK_1 receptor agonists produces a range of defensive behavioural and cardiovascular responses, which are consistent with reactions to stressful stimuli (Elliott 1988; Krase et al. 1994; Aguiar and Brandao 1996). Conversely, NK_1 receptor antagonists reduce the stress-induced activation of ascending neural pathways originating in the locus coeruleus (McLean et al. 1993; Hahn and Bannon 1999), as well as the cardiovascular and behavioural responses to the stress associated with chronic formalin injection (Culman et al. 1997). The facilitation of defensive rage behaviours in the cat induced by stimulation of the amygdala is also reduced with NK_1 receptor antagonists (Shaikh et al. 1993), suggesting an important role for the NK_1 receptor in modulating the responses to stressful stimuli. This is supported by findings in $NK_1^{-/-}$ mice, which have reduced stress-induced analgesia and aggressive responses to territorial challenge (Table 2; De Felipe et al. 1998). The NK_1 receptor therefore seems to be necessary for the orchestrated response to a variety of stressors, including pain, injury, invasion of territory or psychological stress, which may partly explain their antidepressant efficacy.

5.2.3
Reward and Addiction

A potential role for SP and the NK_1 receptor in reward processes was suggested by the observations that systemic injections of SP, as well as focal injections into the nucleus basalis magnocellularis, induce a place preference in rats (Holzhäuer-Oitzl et al. 1988; Hasenöhrl et al. 1998), which is dependent upon NK_1 receptor activation (Nikolaus et al. 1999). These focal injections are coupled with an increase in dopamine turnover in the contralateral nucleus accumbens (NAcc), a brain region known to be of importance in the execution of reward-related behaviours (Boix et al. 1995). Furthermore, administration of SP intracerebroventricularly or into the VTA, NAcc or ventral pallidum brings about increases in locomotor behaviour in a dopamine-dependent manner, in a way similar to rewarding stimuli such as drugs of abuse (Iversen 1982; Elliott and Iversen 1986; Elliott et al. 1992; Kalivas and Alesdatter 1993; Napier et al. 1995; Piot et al. 1995). Rats will self-administer SP into the ventromedial caudate-putamen, although repeated injections can have aversive properties (Krappmann et al. 1994). Furthermore, the cocaine-induced increases in dopamine release in the striatum are reduced with local infusion of NK_1 receptor antagonists (Noailles and Angulo 2002), and self-administration of cocaine causes up-regulation of *PPT-A* mRNA in the shell region of the NAcc (Arroyo et al. 2000). A further link between drugs of abuse and the NK_1 receptor was provided by work demonstrating that some of the behavioural responses to morphine withdrawal were reduced in rats following intracerebroventricular injection of the NK_1 receptor antagonist RP67580 (Maldonado et al. 1993), and following systemic injection of the SP amino terminal metabolite SP_{1-7} (Kreeger and Larson 1996), suggesting that NK_1 receptors may also be involved in the aversive aspects of drug administration.

However, the clearest indication of a role for the NK_1 receptor in drug reward was the demonstration that $NK_1^{-/-}$ mice do not show a conditioned place preference (CPP) to morphine (Murtra et al. 2002b). $NK_1^{-/-}$ mice also failed to show CPP to amphetamine (Murtra et al. 2000a). They are therefore insensitive to the pleasurable, or rewarding aspects of morphine and amphetamine. The loss of CPP to morphine but not cocaine was also seen when specific populations of

NK_1 receptor-expressing neurons in the amygdala were ablated, indicating a crucial role for the amygdala in this task (Gadd et al. 2003). Such locomotor responses to drugs of abuse are believed to be mediated by the same neural mechanisms as reward (Wise 1987; Wise and Bozarth 1987). Importantly, morphine metabolism and binding characteristics of the μ-opioid receptor were found to be normal in $NK_1^{-/-}$ mice (Murtra et al. 2000b).

$NK_1^{-/-}$ mice did not show a general disruption of reward-related behaviours, since CPP to food (a natural reinforcer) and to the psychostimulant cocaine were unaffected by the genetic manipulation. In addition to a loss of the rewarding response to morphine, the mice did not show many of the physical withdrawal signs following naloxone-precipitated withdrawal from chronic morphine, and did not demonstrate a conditioned place aversion to naloxone, suggesting a loss of both some of the physical as well as the motivational aspects of opiate withdrawal (Murtra et al. 200b). Subsequent work has extended these observations showing that the self-administration of opiates and sensitization to morphine (Fig. 5) but not cocaine are selectively lost in NK_1 −/− knockout mice (Ripley et al. 2002).

De Felipe et al. (1998) found that the analgesic properties of morphine were not impaired in $NK_1^{-/-}$ mice in the hot-plate test, indicating that the SP/NK_1 receptor system is not critically involved in opiate analgesia. The NK_1 receptor therefore plays an essential and specific role in the motivational, but not analgesic, properties of opiates (Nestler and Aghajanian 1987; Nestler et al. 1993, 1999). Although the noradrenergic system is one plausible candidate for the behavioural changes observed in NK_1 receptor knockout mice, other systems could be involved. The loss of NK_1 receptors in the VTA, or on the cholinergic interneurons of the NAcc could perturb dopaminergic transmission between these nuclei and bring about changes in behaviour as observed in this experiment. An additional candidate is 5-HT (Froger et al. 2001; Haddjeri and Blier 2001). Although there is no change in the basal concentrations of extracellular 5-HT in the forebrain, sensitization to morphine can be blocked by pre-treatment with a selective serotonin reuptake inhibitor (Wennemer and Kornetsky 1999). The present results clearly demonstrate a functional distinction in the participation of NK_1 receptors in opiate and psychostimulant motivational behaviours. The failure of NK_1 −/− mice to self-administer morphine or to sensitize to its locomotor stimulant effects suggests that manipulation of the SP and the NK_1 receptor may be of use in preventing the development of opiate addiction or reducing craving and relapse. NK_1 receptor antagonists may therefore offer a powerful new approach to the management of opiate addiction, in addition to their predicted effectiveness in alleviating stress, anxiety and depression.

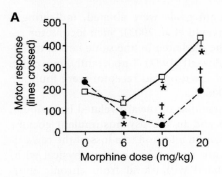

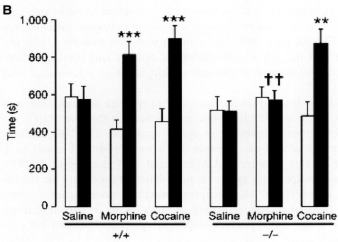

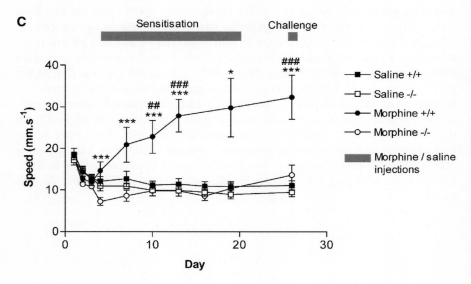

5.3
Learning and Memory

There has been a large number of studies examining the effects of SP infusion, either systemically (Tomaz and Huston 1986; Hasenöhrl et al. 1990a, 1990b; Tomaz and Nogueira 1997) or into various brain structures (Kafetzopoulos et al. 1986; Gerhardt et al. 1992; Hasenöhrl et al. 1998), on learning and memory tasks, the majority of which have demonstrated that SP brings about a facilitation of avoidance learning and can prevent the age-induced deficits in spatial learning ability (Hasenöhrl et al. 1990a, 1994) or the deficits in spontaneous alternation behaviour (Wolff et al. 2003). These effects seem to be dependent on the brain region into which the injections are made: infusions of SP into the nucleus basalis magnocellularis or the medial septum (Kafetzopoulos et al. 1986; Gerhardt et al. 1992; Hasenöhrl et al. 1998) bring about a facilitation of memory, whereas injections into the substantia nigra or amygdala can produce retrograde amnesia (Huston and Stäubli 1978, 1979). However in an extensive series of experiments we were unable to detect differences in contextual or trace fear conditioning or in the Morris water maze between wild-type and $NK_1^{-/-}$ mice (Figs. 6; Morcuende et al. 2003) despite the observed changes in BDNF and neurogenesis seen in the hippocampus of $NK_1^{-/-}$ mice.

5.4
Inflammation and Response to Pathogens

Apart from neuronal stimulation, tachykinins have been shown to have other biological actions in endothelium-dependent vasodilatation, smooth muscle contraction, plasma protein extravasation, mast cell degranulation, the recruitment or stimulation of inflammatory cells and the stimulation of secretions (Tables 3, 4; Maggi et al. 1993).

Recent research with NK_1 −/− mice has confirmed that early oedema formation following injury is NK_1 receptor-dependent, usually via NK_1 receptors situated on postcapillary endothelial cells (Cao et al. 2000). In chronic inflamma-

Fig. 5A–C Lack of increase in locomotor activity, conditioned place preference and sensitization to morphine in $NK_1^{-/-}$ mice. **A** Locomotor behaviour does not increase in $NK_1^{-/-}$ mice after morphine administration. Motor response to morphine administration in wild-type (*squares*) and $NK_1^{-/-}$ (*circles*) mice *$P<0.05$ (vs. 0 mg/kg); †$P<0.05$ (vs. +/+). **B** Lack of conditioned place preference induced by morphine (3 mg/kg) or cocaine (5 mg/kg) in $NK_1^{-/-}$ mice. Time spent in the drug-associated compartment during the preconditioning (*white bars*) and the testing phases (*black bars*) **$P<0.01$; ***$P<0.001$ (vs. preconditioning); †$P<0.01$ (vs. +/+). **C** $NK_1^{-/-}$ mice do not show sensitization to morphine. Locomotor responses to morphine. Mean ± standard error speeds moved by $NK_1^{-/-}$ and wild-type mice during habituation to the activity boxes (days 1–3), chronic morphine [(15 mg/kg, intraperitoneally (i.p.)] or saline administration (days 4–19; two injections per day), and after a challenge dose of morphine (15 mg/kg, i.p.) or saline on day 26. (From Murtra et al. 2000b; Ripley et al. 2002)

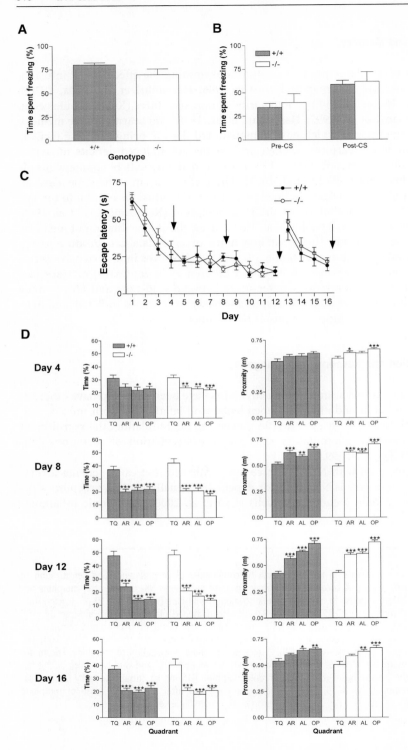

tion, such as the 6-day mouse air pouch model, the inflammation seen 5 h after cutaneous injection of carrageenan or the immune complex-mediated inflammatory model in lung, neutrophil accumulation is significantly reduced in NK_1 receptor knockout mice implying a role for locally released SP on neutrophil activation. However, the effect of SP on neutrophils may not be direct and mediated by other plasma constituents of the inflammatory response (Ahluwalia et al. 1998; Cao et al. 2000). The reduced inflammatory response in *PPT-A* gene-deleted mice has also been implicated in the reduced lung impairment in acute pancreatitis as well as the attenuation of pancreatic damage itself (Bhatia et al. 2003). Similarly, a lessening of the inflammatory response in NK_1 receptor gene knockout mice has been suggested to reduce the clinical impairment in a mouse model of post-treatment reactive encephalopathy but to increase the severity of the neuroinflammatory component of the disease (Kennedy et al. 2003). The increase in severity was explained by possible compensatory changes in NK_2 and NK_3 receptor usage although it remains unclear what changes in tachykinin signalling may have occurred in this model.

5.5
Epilepsy and Other Phenotypes

Mice with a deletion of the *PPT-A* gene are resistant to hippocampal cell death caused by intraperitoneal kainate injection and show a reduced duration and severity of epileptiform seizures (Table 5; Liu et al. 1999). This may correlate with the observations that NK_1 receptor expression is entirely on hippocampal interneurons and that selective ablation of these neurons with SP–saporin conjugates results in 'epileptic' pathophysiology. Thus, elimination of the NK_1 receptor or *PPT-A* gene would remove excitatory drive to interneurons that normally keep hippocampal activity in check (Martin and Sloviter 2001).

Fig. 6A–D Normal behaviour of $NK_1^{-/-}$ mice in tests of fear conditioning and learning and memory. **A, B** Fear conditioning is unchanged in $NK_1R^{-/-}$ mice. **A** Freezing behaviour in first minute of re-exposure to shock-associated context 15 days after training. **B** Freezing behaviour in the minute before and the minute after conditioned stimulus (*CS*) presentation following trace fear conditioning. Data are means ± standard error. **C, D** Spatial learning behaviour is essentially unchanged in $NK_1R^{-/-}$ mice in the Morris water maze. **C** Escape latencies during training (days 1–12) and reversal learning (days 13–16). *Arrows* indicate probe trial positions. **D** Probe trial behaviour (days 4, 8, 12 and 16). *Left column* shows time spent searching in each quadrant during probe trials (*TQ*, target quadrant; *AR*, adjacent right; *AL*, adjacent left; *OP*, opposite). *Right column* shows distance of mice from target position relative to each quadrant. (From Morcuende et al. 2003)

Table 3 Summary of observed phenotypes in *PPT-A* and NK_1 receptor knockout mice: inflammation/oedema

	PPT-A$^{-/-}$ Cao et al. (1998) Strain: Cd-1×129	PPT-A$^{-/-}$ Zimmer et al. (1998)	$NK_1^{-/-}$ Bozic et al. (1996) Strain: C57/6 J x 129/Sv	$NK_1^{-/-}$ De Felipe et al. (1998) Strain: C57/BL6×129/Sv	$NK_1^{-/-}$ Santarelli et al. (2001)
Skin	Almost no neurogenic inflammation (Cao et al. 1998)		No oedema following intradermal SP, septide or GR-73632, oedema greater following mast cell-degranulating agent: NK_1 receptor-mediated mechanism includes a mast cell-dependent component (Cao et al. 1999). SP does not induce oedema in skin, IL-1β-induced neutrophil accumulation normal, but enhanced in wild-type mice by NK_1 agonists (which have no effect alone), $NK_1R^{-/-}$ mice have reduced neutrophil accumulation following carrageenan but not zymosan: carrageenan accumulation in wild-type mice reduced with bradykinin B_1 and B_2 antagonists, but no further reduction in $NK_1^{-/-}$ mice (Cao et al. 2000). Oedema absent following capsaicin, but no differences in wound healing (Cao et al. 2001). Scald-induced oedema reduced at 10 min and 30 min but not 60 min, no reduction in neutrophil accumulation at 4 h (Rawlingson et al. 2001). No plasma extravasation to NKB in skin (Grant et al. 2002b). Oedema response to capsaicin absent in ear, blood flow potentiated and abolished by CGRP receptor antagonist (Grant et al. 2002a)	Neutrophil accumulation following IL-1β or SP reduced in $+/-$ and $-/-$ mice, normal response to cytokine-induced neutrophil chemoattractant (Ahluwalia et al. 1998)	
Lungs	Protected against injury from immune complex and stretch, not rescued by wild-type bone marrow: synergy between sensory nerve fibres and haemopoietic cells (Chavolla-Calderon et al. 2003)		Immune complex inflammation attenuated (Bozic et al. 1996). Normal plasma extravasation to NKB in lung (and liver), unaffected by NK_2 antagonist, NK_3 antagonist or COX inhibitor: but partially inhibited by eNOS inhibitor: tachykinin-independent pathway (Grant et al. 2002a)		

Table 3 (continued)

	PPTA−/−		NK$_1$−/−		
	Cao et al. (1998) Strain: Cd-1×129	Zimmer et al. (1998)	Bozic et al. (1996) Strain: C57/6 J × 129/Sv	De Felipe et al. (1998) Strain: C57/BL6×129/Sv	Santarelli et al. (2001)
Cystitis	Reduced effect of lower urinary tract intravesical acetic acid-evoked stimulation to induce c-Fos in spinal cord, unaffected by NK$_2$ antagonist, more mice exhibiting urinary retention and overflow incontinence (Kiss et al. 2001)		More mast cells in bladder but without inflammation, sensitizing antigen produces normal mast cell degranulation but without inflammatory response (Saban et al. 2000). Absence of bladder inflammation and different profile of gene up-regulation, involving neprilysin and AP-1 (Saban et al. 2002; Dozmorov et al. 2003)	Impaired responses to neurogenic inflammatory stimuli: cyclophosphamide cystitis (Laird et al. 2000)	
Pancreatitis	Protected against acute pancreatitis-associated lung injury and partially protected against local pancreatic damage (Bhatia et al. 2003)		Increased survival and reduced pancreatitis following choline-deficient and ethionine-supplemented diet (Maa et al. 2000). SP- and cerulein-induced plasma extravasation reduced (Grady et al. 2000). Reduced cerulein-induced hyperamylasaemia, hyperlipasaemia, pancreatic neutrophil sequestration and acinar cell necrosis, and pancreatitis-associated lung injury (Bhatia et al. 1998)		
Viscera				No oedema to capsaicin, but normal response to mustard oil, impaired responses to neurogenic inflammatory stimuli: acetic acid (Laird et al. 2000)	
Clostridium difficile-induced enteritis			Reduced response to toxin A (inflammation, secretion and epithelial cell damage), reduced levels of TNF-α and myeloperoxidase (Castagliuolo et al. 1998)		

CGRP, Calcitonin gene-related peptide; COX, cyclooxygenase; eNOS, endothelial nitric oxide synthase; IL, interleukin; TNF, tumour necrosis factor.

Table 4 Summary of observed phenotypes in *PPT-A* and NK_1 receptor knockout mice: responses to pathogens

	PPT-A$^{-/-}$		$NK_1^{-/-}$		
	Cao et al. (1998) Strain: Cd-1×129	Zimmer et al. (1998)	Bozic et al. (1996) Strain: C57/6 J × 129/Sv	De Felipe et al. (1998) Strain: C57/BL6×129/Sv	Santarelli et al. (2001) Strain: 129/SvEv pure
Herpes virus infection:	higher viral titres in lungs, more rapid influx of inflammatory cells and slower clearance of latently infected cells from spleen (Payne et al. 2001)		Increased viral load due to reduced CTL response and IL-12 expression (Elsawa et al. 2003)		
Schistosomiasis			Smaller liver granulomas following infection, splenocytes and granulomas produce less IFN-γ, IgG2a and IgE, but normal amounts of Th2 cytokines (Blum et al. 1999). T cells from $NK_1^{-/-}$ mice cause abnormal liver granulomas following infection with schistosomiasis: smaller size and less production of IFN-γ and IgG2a (Blum et al. 2003)		
African trypanosomiasis				Post-treatment reactive encephalitis reduced, but neuroinflammatory response increased: this, but not clinical score, decreased with combined NK_2/NK_3 antagonist treatment (Kennedy et al. 2003)	

CTL, cytotoxic T lymphocyte; IL, interleukin; IFN, interferon.

Table 5 Summary of observed phenotypes in *PPT-A* and NK$_1$ receptor knockout mice: other phenotypes

Epilepsy				
PPT-A$^{-/-}$			NK$_1$$^{-/-}$	
Cao et al. (1998) Strain: Cd-1×129	Zimmer et al. (1998)		Bozic et al. (1996)	De Felipe et al. (1998)	Santarelli et al. (2001)
Resistant to kainate excitotoxicity, reduced duration and severity of kainate or pentylenetetrazole-induced seizures, have necrosis and apoptosis of hippocampal neurones, kainate-induced up-regulation of bax and caspase 3 in hippocampus reduced (Liu et al. 1999)				
Respiration				
PPT-A$^{-/-}$			NK$_1$$^{-/-}$	
Cao et al. (1998)	Zimmer et al. (1998)		Bozic et al. (1996)	De Felipe et al. (1998)	Santarelli et al. (2001)
Medullary respiratory network				Normal respiratory frequency, rhythm variability and amplitude of motor output, but SP-induced increases in respiratory frequency or inspiratory motor output absent (Ptak et al. 2000). Pressor response-induced baroreceptor reflex-mediated bradycardia greater in −/− (but not +/−) mice, normal falls in heart rate, arterial pressure and respiratory depression with pulmonary chemoreflex (Butcher et al. 1998)	

Table 5 (continued)

	$PPT\text{-}A^{-/-}$		$NK_1^{-/-}$	
Respiration	Cao et al. (1998)	Zimmer et al. (1998)	Bozic et al. (1996)	De Felipe et al. (1998) Santarelli et al. (2001)
Hypoxic chemoreception				Facilitation of ventilation by hypoxia weaker in adults, but no differences in prenatal development of respiratory network (Ptak et al. 2002). No differences in activity or catecholamine content of carotid body neurones in response to hypoxia (Rigual et al. 2002)
Tracheal contraction				Reduced contraction by electrical field stimulation and less acetylcholine release: NK_1 receptor augments cholinergic transmission (Tournoy et al. 2003)
Gastrointestinal motility	Cao et al. (1998)	Zimmer et al. (1998)	Bozic et al. (1996)	De Felipe et al. (1998) Santarelli et al. (2001)
			Ileal segments from –/– mice require higher concentrations of SP to undergo contraction, have increased inhibitory non-adrenergic non-cholinergic transmission, excitatory component of this transmission absent (Saban et al. 1999)	

Table 5 (continued)

Itch			
$PPT\text{-}A^{-/-}$		$NK_1{-}/{-}$	
Cao et al. (1998) Strain: Cd-1×129	Zimmer et al. (1998)	Bozic et al. (1996) De Felipe et al. (1998)	Santarelli et al. (2001)
Normal number and duration of intradermal 5-HT-induced scratching bouts, but more spontaneous and 5-HT-evoked scratching (Cuellar et al. 2003)			

6
Conclusion

Gene knockout mice have become an invaluable resource in determining the contribution of tachykinins to physiology and behaviour. In most cases, the results have confirmed and extended previous research using specific antagonists at the NK_1 receptor. This was important as many of these antagonists had reduced specificity in rats and mice compared to other species and man. Future research will need to concentrate on specific deletions of tachykinins or their receptors from specific areas of the central nervous system and to understand the molecular compensations that occur in other neurochemical networks such as the monoaminergic pathways.

References

Abbadie C, Brown JL, Mantyh PW, Basbaum AI (1996) Spinal cord substance P receptor immunoreactivity increases in both inflammatory and nerve injury models of persistent pain. Neuroscience 70:201–209

Aguiar MS, Brandão ML (1996) Effects of microinjections of the neuropeptide substance P in the dorsal periaqueductal gray on the behaviour of rats in the plus-maze test. Physiol Behav 60:1183–1186

Ahluwalia A, De Felipe C, O'Brien J, Hunt SP, Perretti M (1998) Impaired IL-1α-Induced Neutrophil Accumulation in Tachykinin NK_1 Receptor Knockout Mice. Br J Pharmacol 124:1013–1015

Allen BJ, Rogers SD, Ghilardi JR, Menning PM, Kuskowski MA, Basbaum AI, Simone DA, Mantyh PW (1997) Noxious cutaneous thermal stimuli induce a graded release of endogenous substance P in the spinal cord: imaging peptide action in vivo. J Neurosci 17:5921–5927

Anderson JJ, Chase TN, Engber TM (1993) Substance P increases release of acetylcholine in the dorsal striatum of freely moving rats. Brain Res 623:189–194

Anderson KD, Reiner A (1990) Extensive co-occurrence of substance P and dynorphin in striatal projection neurons: an evolutionarily conserved feature of basal ganglia organization. J Comp Neurol 295:339–369

Andersson SE (1987) Responses to antidromic trigeminal nerve stimulation, substance P, NKA, CGRP and capsaicin in the rat eye. Acta Physiol Scand 131:371–376

Andoh T, Nagasawa T, Kuraishi Y (1996) Expression of tachykinin NK_1 receptor mRNA in dorsal root ganglia of the mouse. Mol Brain Res 35:329–332

Arenas E, Alberch J, Perez-Navarro E, Solsona C, Marsal J (1991) Neurokinin receptors differentially mediate endogenous acetylcholine release evoked by tachykinins in the neostriatum. J Neurosci 11:2332–2338

Aronin N, Coslovsky R, Leeman SE (1986) Substance P and neurotensin: their roles in the regulation of anterior pituitary function. Annu Rev Physiol 48:537–549

Arroyo M, Baker WA, Everitt BJ (2000) Cocaine self-administration in rats differentially alters mRNA levels of the monoamine transporters and striatal neuropeptides. Mol Brain Res 83:107–120

Ballard TM, Sänger S, Higgins GA (2001) Inhibition of shock-induced foot tapping behaviour in the gerbil by a tachykinin NK_1 receptor antagonist. Eur J Pharmacol 412:255–264

Banbury Conference Report (1997) Mutant mice and neuroscience: recommendations concerning genetic background. Banbury Conference on Genetic Background in Mice. Neuron 19:755–759

Bannerman DM, Grubb M, Deacon RM, Yee BK, Feldon J, Rawlins JN (2003) Ventral hippocampal lesions affect anxiety but not spatial learning. Behav Brain Res 139:197–213

Baranauskas G, Nistri A (1998) Sensitization of pain pathways in the spinal cord: cellular mechanisms. Prog Neurobiol 54:349–365

Barber RP, Vaughn JE, Slemmon JR, Salvaterra PM, Roberts E, Leeman SE (1979) The origin, distribution and synaptic relationships of substance P axons in rat spinal cord. J Comp Neurol 184:331–351

Barnes PJ (1987) Airway neuropeptides and asthma. Trends Pharmacol Sci 8:24–27

Barnes PJ, Belvisi MG, Rogers DF (1990) Modulation of neurogenic inflammation: novel approaches to inflammatory disease. Trends Pharmacol Sci 11:185–189

Barnes PJ, Chung KF, Page CP (1988) Inflammatory mediators and asthma. Pharmacol Rev 40:49–84

Bauer V, Kuriyama H (1982) The nature of non-cholinergic, non-adrenergic transmission in longitudinal and circular muscles of the guinea-pig ileum. J Physiol 332:375–391

Beattie DT, Stubbs CM, Connor HE, Feniuk W (1993) Neurokinin-induced changes in pial artery diameter in the anaesthetized guinea-pig. Br J Pharmacol 108:146–149

Bertrand C, Geppetti P (1996) Tachykinin and kinin receptor antagonists: therapeutic perspectives in allergic airway disease. Trends Pharmacol Sci 17:255–259

Bert L, Rodier D, Bougault I, Allouard N, Le-Fur G, Soubrié P, Steinberg R (2002) Permissive Role of neurokinin NK_3 receptors in NK_1 receptor-mediated activation of the locus coeruleus revealed by SR 142801. Synapse 43:62–69

Besson MJ, Graybiel AM, Quinn B (1990) Co-expression of neuropeptides in the cat's striatum: an immunohistochemical study of substance P, dynorphin B and enkephalin. Neuroscience 39:33–58

Bester H, De Felipe C, Hunt SP (2001) The NK_1 receptor is essential for the full expression of noxious inhibitory controls in the mouse. J Neurosci 21:1039–1046

Bhatia M, Saluja AK, Hofbauer B, Frossard JL, Lee HS, Castagliuolo I, Wang CC, Gerard N, Pothoulakis C, Steer ML (1998) Role of substance P and the neurokinin 1 receptor in acute pancreatitis and pancreatitis-associated lung injury. Proc Natl Acad Sci USA 95:4760–4765

Bhatia M, Slavin J, Cao Y, Basbaum AI, Neoptolemos JP (2003) Preprotachykinin-A gene deletion protects mice against acute pancreatitis and associated lung injury. Am J Physiol 284:G830–G836

Bilkei-Gorzo A, Racz I, Michel K, Zimmer A (2002) Diminished Anxiety- and depression-related behaviors in mice with selective deletion of the *Tac1* gene. J Neurosci 22:10046–10052

Binder W, Scott C, Walker JS (1999) Involvement of substance P in the anti-inflammatory effects of the peripherally selective kappa-opioid asimadoline and the NK_1 antagonist GR205171. Eur J Neurosci 11:2065–2072

Birch PJ, Harrison SM, Hayes AG, Rogers H, Tyers MB (1992) The non-peptide NK_1 receptor antagonist, (±)-CP-96,345, produces antinociceptive and anti-oedema effects in the rat. Br J Pharmacol 105:508–510

Blum AM, Metwali A, Elliott DE, Weinstock JV (2003) T cell substance P receptor governs antigen-elicited IFN-γ Production. Am J Physiol 284:G197–G204

Blum AM, Metwali A, Kim-Miller M, Li J, Qadir K, Elliott DE, Lu B, Fabry Z, Gerard N, Weinstock JV (1999) The substance P receptor is necessary for a normal granulomatous response in murine Schistosomiasis mansoni. J Immunol 162:6080–6085

Boix F, Huston JP, Schwarting RK (1992) The C-terminal fragment of substance P enhances dopamine release in nucleus accumbens but not in neostriatum in freely moving rats. Brain Res 592:181–186

Boix F, Sandor P, Nogueira PJ, Huston JP, Schwarting RK (1995) Relationship between dopamine release in nucleus accumbens and place preference induced by substance P injected into the nucleus basalis magnocellularis region. Neuroscience 64:1045–1055

Bolam JP, Ingham CA, Izzo PN, Levey AI, Rye DB, Smith AD, Wainer BH (1986) Substance P-containing terminals in synaptic contact with cholinergic neurons in the neostriatum and basal forebrain: a double immunocytochemical study in the rat. Brain Res 397:279–289

Bolam JP, Somogyi P, Takagi H, Fodor I, Smith AD (1983) Localization of substance P-like immunoreactivity in neurons and nerve terminals in the neostriatum of the rat: a correlated light and electron microscopic study. J Neurocytol 12:325–344

Boyce S, Smith D, Carlson E, Hewson L, Rigby M, O'Donnell R, Harrison T, Rupniak NM (2001) Intra-amygdala injection of the substance P (NK_1 Receptor) antagonist L-760735 inhibits neonatal vocalisations in guinea-pigs. Neuropharmacology 41:130–137

Bozic CR, Lu B, Hopken UE, Gerard C, Gerard NP (1996) Neurogenic amplification of immune complex inflammation. Science 273:1722–1725

Brecha N, Johnson D, Bolz J, Sharma S, Parnavelas JG, Lieberman AR (1987) Substance P-immunoreactive retinal ganglion cells and their central axon terminals in the rabbit. Nature 327:155–158

Bury RW, Mashford ML (1976) Biological activity of C-terminal partial sequences of substance P. J Med Chem 19:854–856

Butcher JW, De Felipe C, Smith AJH, Hunt SP, Paton JFR (1998) Comparison of cardiorespiratory reflexes in NK_1 receptor knockout, heterozygous and wild-type mice in vivo. J Auton Nerv Syst 69:89–95

Campbell EA, Gentry CT, Patel S, Panesar MS, Walpole CSJ, Urban L (1998) Selective neurokinin-1 receptor antagonists are anti-hyperalgesic in a model of neuropathic pain in the guinea-pig. Neuroscience 87:527–532

Cao T, Gerard NP, Brain SD (1999) Use of NK_1 knockout mice to analyze substance P-induced edema formation. Am J Physiol 277:R476–R481

Cao T, Grant AD, Gerard NP, Brain SD (2001) Lack of a significant effect of deletion of the tachykinin neurokinin-1 receptor on wound healing in mouse skin. Neuroscience 108:695–700

Cao T, Pintér E, Al Rashed S, Gerard N, Hoult JR, Brain SD (2000) Neurokinin-1 receptor agonists are involved in mediating neutrophil accumulation in the inflamed, but not normal, cutaneous microvasculature: an in vivo study using neurokinin-1 receptor knockout mice. J Immunol 164:5424–5429

Cao YQ, Mantyh PW, Carlson EJ, Gillespie AM, Epstein CJ, Basbaum AI (1998) Primary afferent tachykinins are required to experience moderate to intense pain. Nature 392:390–394

Carruette A, Moussaoui SM, Champion A, Cottez D, Goniot P, Garret C (1992) Comparison in different tissue preparations of the in vitro pharmacological profile of RP 67580, a new non-peptide substance P antagonist. Neuropeptides 23:245–250

Carter MS, Krause JE (1990) Structure, expression, and some regulatory mechanisms of the rat preprotachykinin gene encoding substance P, neurokinin A, neuropeptide K, and neuropeptide gamma. J Neurosci 10:2203–2214

Cascieri MA, Huang RRC, Fong TM, Cheung AH, Sadowski S, Ber E, Strader CD (1992) Determination of the amino acid residues in substance P conferring selectivity and specificity for the rat neurokinin receptors. Mol Pharmacol 41:1096–1099

Chan-Palay V, Palay SL (1977) Ultrastructural identification of substance P cells and their processes in rat sensory ganglia and their terminals in the spinal cord by immunocytochemistry. Proc Natl Acad Sci USA 74:4050–4054

Chapman V, Dickenson AH (1993) The effect of intrathecal administration of RP67580, a potent neurokinin 1 antagonist on nociceptive transmission in the rat spinal cord. Neurosci Lett 157:149–152

Chavolla-Calderón M, Bayer MK, Peréz Fontán JJ (2003) Bone marrow transplantation reveals an essential synergy between neuronal and hemopoietic cell neurokinin production in pulmonary inflammation. J Clin Invest 111:973–980

Cheeta S, Tucci S, Sandhu J, Williams AR, Rupniak NMJ, File SE (2001) Anxiolytic actions of the substance P (NK_1) receptor antagonist L-760735 and the 5-HT1A agonist 8-OH-DPAT in the social interaction test in gerbils. Brain Res 915:170–175

Chen LW, Wei LC, Liu HL, Rao ZR (2000) Noradrenergic neurons expressing substance P receptor (NK_1) in the locus coeruleus complex: a double immunofluorescence study in the rat. Brain Res 873:155–159

Clement Y, Calatayud F, Belzung C (2002) Genetic basis of anxiety-like behaviour: a critical review. Brain Res Bull 57:57–71

Conley RK, Cumberbatch MJ, Mason GS, Williamson DJ, Harrison T, Locker K, Swain C, Maubach K, O'Donnell R, Rigby M, Hewson L, Smith D, Rupniak NM (2002) Substance P (neurokinin 1) receptor antagonists enhance dorsal raphe neuronal activity. J Neurosci 22:7730–7736

Copeland NG, Jenkins NA, Court DL (2001) Recombineering: a powerful new tool for mouse functional genomics. Nat Rev Genet 2:769–779

Coudore-Civiale M, Courteix C, Boucher M, Fialip J, Eschalier A (2000) Evidence for an involvement of tachykinins in allodynia in streptozocin-induced diabetic rats. Eur J Pharmacol 401:47–53

Crawley JN, Belknap JK, Collins A, Crabbe JC, Frankel W, Henderson N, Hitzemann RJ, Maxson SC, Miner LL, Silva AJ, Wehner JM, Wynshaw-Boris A, Paylor R (1997) Behavioral phenotypes of inbred mouse strains: implications and recommendations for molecular studies. Psychopharmacology (Berlin) 132:107–124

Cuellar JM, Jinks SL, Simons CT, Carstens E (2003) Deletion of the preprotachykinin A gene in mice does not reduce scratching behavior elicited by intradermal serotonin. Neurosci Lett 339:72–76

Culman J, Klee S, Ohlendorf C, Unger T (1997) Effect of tachykinin receptor inhibition in the brain on cardiovascular and behavioral responses to stress. J Pharmacol Exp Ther 280:238–246

Cumberbatch MJ, Carlson E, Wyatt A, Boyce S, Hill RG, Rupniak NM (1998) Reversal of behavioural and electrophysiological correlates of experimental peripheral neuropathy by the NK_1 receptor antagonist GR205171 in rats. Neuropharmacology 37:1535–1543

Davies J, Dray A (1976) Substance P in the substantia nigra. Brain Res 107:623–627

De Araújo JE, Silva RC, Huston JP, Brandão ML (1999) Anxiogenic Effects of substance P and Its 7–11 C terminal, but not the 1–7 N terminal, injected into the dorsal periaqueductal gray. Peptides 20:1437–1443

De Felipe C, Herrero JF, O'Brien JA, Palmer JA, Doyle CA, Smith AJ, Laird JM, Belmonte C, Cervero F, Hunt SP (1998) Altered nociception, analgesia and aggression in mice lacking the receptor for substance P. Nature 392:394–397

Delay-Goyet P, Lundberg JM (1991) Cigarette smoke-induced airway oedema is blocked by the NK_1 antagonist, CP-96,345. Eur J Pharmacol 203:157–158

Dionne RA, Max MB, Gordon SM, Parada S, Sang C, Gracely RH, Sethna NF, MacLean DB (1998)The substance P receptor antagonist CP-99,994 reduces acute postoperative pain. Clin Pharmacol Ther 64:562–568

Dougherty PM, Palecek J, Paleckova V, Willis WD (1994) Neurokinin 1 and 2 antagonists attenuate the responses and NK_1 antagonists prevent the sensitization of primate spinothalamic tract neurons after intradermal capsaicin. J Neurophysiol 72:1464–1475

Dozmorov I, Saban MR, Gerard NP, Lu B, Nguyen NB, Centola M, Saban R (2003) Neurokinin 1 receptors and neprilysin modulation of mouse bladder gene regulation. Physiol Genomics 12:239–250

Dray A, Pinnock RD (1982) Effects of substance P on adult rat sensory ganglion neurones in vitro. Neurosci Lett 33:61–66

Duffy RA, Varty GB, Morgan CA, Lachowicz JE (2002) Correlation of neurokinin (NK) 1 receptor occupancy in gerbil striatum with behavioral effects of NK_1 antagonists. J Pharmacol Exp Ther 301:536–542

Ekström J, Ekman R, Håkanson R, Sjögren S, Sundler F (1988) Calcitonin gene-related peptide in rat salivary glands: neuronal localization, depletion upon nerve stimulation, and effects on salivation in relation to substance P. Neuroscience 26:933–949

Elliott PJ (1988) Place aversion induced by the substance P analogue, dimethyl-C7, is not state dependent: implication of substance P in aversion. Exp Brain Res 73:354–356

Elliott PJ, Iversen SD (1986) Behavioural effects of tachykinins and related peptides. Brain Res 381:68–76

Elliott PJ, Mason GS, Graham EA, Turpin MP, Hagan RM (1992) Modulation of the rat mesolimbic dopamine pathway by neurokinins. Behav Brain Res 51:77–82

Elsawa SF, Taylor W, Petty CC, Marriott I, Weinstock JV, Bost KL (2003) Reduced CTL response and increased viral burden in substance P receptor-deficient mice infected with murine gamma-herpesvirus 68. J Immunol 170:2605–2612

Emson PC, Cuello AC, Paxinos G, Jessell T, Iversen LL (1977) The origin of substance P and acetylcholine projections to the ventral tegmental area and interpeduncular nucleus in the rat. Acta Physiol Scand 452(Suppl):43–46

Emson PC, Jessell T, Paxinos G, Cuello AC (1978) Substance P in the amygdaloid complex, bed nucleus and stria terminalis of the rat brain. Brain Res 149:97–105

Erspamer V (1981) The tachykinin peptide family. Trends Neurosci 4:267–269

Falconieri-Erspamer G, Erspamer V, Piccinelli D (1980) Parallel bioassay of physalaemin and kassinin, a tachykinin dodecapeptide from the skin of the african frog *Kassina senegalensis*. Naunyn-Schmiedeberg's Arch Pharmacol 311:61–65

Falconieri-Erspamer G, Negri L, Piccinelli D (1973) The use of preparations of urinary bladder smooth muscle for bioassays of and discrimination between polypeptides. Naunyn-Schmiedeberg's Arch Pharmacol 279:61–74

Field MJ, McCleary S, Boden P, Suman-Chauhan N, Hughes J, Singh L (1998) Involvement of the central tachykinin NK_1 receptor during maintenance of mechanical hypersensitivity induced by diabetes in the rat. J Pharmacol Exp Ther 285:1226–1232

File SE (1997) Anxiolytic action of a neurokinin1 receptor antagonist in the social interaction test. Pharmacol Biochem Behav 58:747–752

File SE (2000) NKP608, an NK_1 Receptor antagonist, has an anxiolytic action in the social interaction test in rats. Psychopharmacology (Berlin) 152:105–109

Fiscus RR, Gross DR, Hao H, Wang X, Arden WA, Maley RH, Salley RKN (1992) Omega-nitro-L-arginine blocks the second phase but not the first phase of the endothelium-dependent relaxations induced by substance P in isolated rings of pig carotid artery. J Cardiovasc Pharmacol 20(Suppl 12):S105–108

Franco R, Costa M, Furness JB (1979) Evidence that axons containing substance P in the guinea pig ileum are of intrinsic origin. Naunyn-Schmiedeberg's Arch Pharmacol 326:111–115

Froger N, Gardier AM, Moratalla R, Alberti I, Lena I, Boni C, De Felipe C, Rupniak NM, Hunt SP, Jacquot C, Hamon M, Lanfumey L (2001) 5-Hydroxytryptamine 5-HT_{1A} autoreceptor adaptive changes in substance P (neurokinin 1) receptor knock-out mice mimic antidepressant-induced desensitization. J Neurosci 21:8188–8197

Frossard N, Advenier C (1991) Tachykinin receptors and the airways. Life Sci 49:1941–1953

Gadd CA, Murtra P, De Felipe C, Hunt SP (2003) Neurokinin-1 receptor-expressing neurons in the amygdala modulate morphine reward and anxiety behaviors in the mouse. J Neurosci 23:8271–8280

Galarraga E, Hernández-López S, Tapia D, Reyes A, Bargas J (1999) Action of substance P (neurokinin-1) receptor activation on rat neostriatal projection neurons. Synapse 33:26–35

Gamse R, Molnar A, Lembeck F (1979) Substance P release from spinal cord slices by capsaicin. Life Sci 25:629–636

Gamse R, Saria A (1986) Nociceptive behavior after intrathecal injections of substance P, neurokinin A and calcitonin gene-related peptide in mice. Neurosci Lett 70:143–147

Garret C, Carruette A, Fardin V, Moussaoui S, Peyronel JF, Blanchard JC, Laduron PM (1991) Pharmacological properties of a potent and selective nonpeptide substance P antagonist. Proc Natl Acad Sci USA 88:10208–10212

Gentsch C, Cutler M, Vassout A, Veenstra S, Brugger F (2002) Anxiolytic effect of NKP608, a NK1-receptor antagonist, in the social investigation test in gerbils. Behav Brain Res 133:363–368

Gerard NP, Garraway LA, Eddy RL, Shows TB, Iijima H, Paquet JL, Gerard C (1991) Human substance P receptor (NK-1): organization of the gene, chromosome localization, and functional expression of CDNA clones. Biochemistry 30:10640–10646

Gerhardt P, Hasenöhrl RU, Huston JP (1992) Enhanced learning produced by injection of neurokinin substance P into the region of the nucleus basalis magnocellularis: mediation by the N-terminal sequence. Exp Neurol 118:302–308

Gerlai R (1996) Gene-targeting studies of mammalian behavior: is it the mutation or the background genotype? Trends Neurosci 19:177–181

Gillis RA, Helke CJ, Hamilton BL, Norman WP, Jacobowitz DM (1980) Evidence that substance P Is a neurotransmitter of baro- and chemoreceptor afferents in nucleus tractus solitarius. Brain Res 181:476–481

Grady EF, Yoshimi SK, Maa J, Valeroso D, Vartanian RK, Rahim S, Kim EH, Gerard C, Gerard N, Bunnett NW, Kirkwood KS (2000) Substance P mediates inflammatory oedema in acute pancreatitis *via* activation of the neurokinin-1 receptor in rats and mice. Br J Pharmacol 130:505–512

Grant AD, Akhtar R, Gerard NP, Brain SD (2002a) Neurokinin B induces oedema formation in mouse lung via tachykinin receptor-independent mechanisms. J Physiol 543:1007–1014

Grant AD, Gerard NP, Brain SD (2002b) Evidence of a role for NK_1 and CGRP receptors in mediating neurogenic vasodilatation in the mouse ear. Br J Pharmacol 135:356–362

Gross C, Zhuang X, Stark K, Ramboz S, Oosting R, Kirby L, Santarelli L, Beck S, Hen R (2002) Serotonin$_{1A}$ receptor acts during development to establish normal anxiety-like behaviour in the adult. Nature 416:396–400

Guyenet PG, Aghajanian GK (1977) Excitation of neurons in the nucleus locus coeruleus by substance P and related peptides. Brain Res 136:178–184

Haber SN, Nauta WJ (1983) Ramifications of the globus pallidus in the rat as indicated by patterns of immunohistochemistry. Neuroscience 9:245–260

Haddjeri N, Blier P (2001) Sustained blockade of neurokinin-1 receptors enhances serotonin neurotransmission. Biol Psychiatry 50:191–199

Hahn MK, Bannon MJ (1999) Stress-induced c-fos expression in the rat locus coeruleus is dependent on neurokinin 1 receptor activation. Neuroscience 94:1183–1188

Hasenöhrl RU, Frisch C, Nikolaus S, Huston JP (1994) Chronic administration of neurokinin SP improves maze performance in aged Rattus norvegicus. Behav Neural Biol 62:110–120

Hasenöhrl RU, Gerhardt P, Huston JP (1990a) Substance P enhancement of inhibitory avoidance learning: mediation by the N-terminal sequence. Peptides 11:163–167

Hasenöhrl RU, Huston JP, Schuurman T (1990b) Neuropeptide substance P improves water maze performance in aged rats. Psychopharmacology (Berlin) 101:23–26

Hasenöhrl RU, Jentjens O, Souza Silva MA, Tomaz C, Huston JP (1998) Anxiolytic-like action of neurokinin substance P administered systemically or into the nucleus basalis magnocellularis region. Eur J Pharmacol 354:123–133

Heintz N (2001) Bac to the future: the use of Bac transgenic mice for neuroscience research. Nat Rev Neurosci 2:861–870

Henry JL (1976) Effects of substance P on functionally identified units in cat spinal cord. Brain Res 114:439–451

Henry JL (1987) Discussions of nomenclature for tachykinins and tachykinin receptors. In: Henry JL, Couture R, Cuello AC, Pelletier G, Quirion R, Regoli D (eds) Substance P and Neurokinins. New York: Springer, pp xxvii–xxviii

Henry JL, Sessle BJ, Lucier GE, Hu JW (1980) Effects of substance P on nociceptive and non-nociceptive trigeminal brain stem neurons. Pain 8:33–45

Henry JL, Yashpal K, Pitcher GM, Chabot J, Coderre TJ (1999) Evidence for tonic activation of NK-1 receptors during the second phase of the formalin test in the rat. J Neurosci 19:6588–6598

Hershey AD, Dykema PE, Krause JE (1991) Organization, structure, and expression of the gene encoding the rat substance P receptor. J Biol Chem 266:4366–4374

Hershey AD, Krause JE (1990) Molecular Characterization of a functional cDNA encoding the rat substance P Receptor. Science 247:958–962

Hill R (2000) NK_1 (Substance P) Receptor antagonists—why are they not analgesic in humans? Trends Pharmacol Sci 21:244–246

Hökfelt T, Elde R, Johansson O, Luft R, Nilsson G, Arimura A (1976) Immunohistochemical evidence for separate populations of somatostatin-containing and substance P-Containing primary afferent neurons in the rat. Neuroscience 1:131–136

Hökfelt T, Kellerth JO, Nilsson G, Pernow B (1975) Substance P: localization in the central nervous system and in some primary sensory neurons. Science 190:889–890

Holzhauer-Oitzl MS, Hasenöhrl R, Huston JP (1988) Reinforcing properties of substance P in the region of the nucleus basalis magnocellularis in rats. Neuropharmacology 27:749–756

Holzer P, Lembeck F (1980) Neurally mediated contraction of ileal longitudinal muscle by substance P. Neurosci Lett 17:101–105

Holzer P, Petsche U (1983) On the mechanism of contraction and desensitization induced by substance P in the intestinal muscle of the guinea-pig. J Physiol 342:549–568

Houdebine LM, Attal J (199) Internal ribosome entry sites (IRESs): reality and use. Transgenic Res 8:157–177

Huidobro-Toro JP, Chelala CA, Bahouth S, Nodar R, Musacchio JM (1982) Fading and tachyphylaxis to the contractile effects of substance P in the guinea-pig ileum. Eur J Pharmacol 81:21–34

Huston JP, Stäubli U (1978) Retrograde amnesia produced by post-trial injection of substance P into substantia nigra. Brain Res 159:468–472

Huston JP, Stäubli U (1979) Post-trial injection of substance P into lateral hypothalamus and amygdala, respectively, facilitates and impairs learning. Behav Neural Biol 27:244–248

Hylden JLK, Wilcox GL (1981) Intrathecal substance P elicits a caudally-directed biting and scratching behavior in mice. Brain Res 217:212–215

Ingi T, Kitajima Y, Minamitake Y, Nakanishi S (1991) Characterization of ligand-binding properties and selectivities of three rat tachykinin receptors by transfection and functional expression of their cloned cDNAs in mammalian cells. J Pharmacol Exp Ther 259:968–975

Iversen SD (1982) Behavioural effects of substance P through dopaminergic pathways in the brain. Ciba Found Symp 91:307–324

Jancsó G, Kiraly E, Jancsó-Gábor A (1977) Pharmacologically induced selective degeneration of chemosensitive primary sensory neurones. Nature 270:741–743

Jessell TM, Iversen LL (1977) Opiate analgesics inhibit substance P release from rat trigeminal nucleus. Nature 268:549-551
Jessell TM, Iversen LL, Cuello AC (1978) Capsaicin-induced depletion of substance P from primary sensory neurones. Brain Res 152:183-188
Jessell T, Tsunoo A, Kanazawa I, Otsuka M (1979) Substance P: depletion in the dorsal horn of rat spinal cord after section of the peripheral processes of primary sensory neurons. Brain Res 168:247-259
Kachur JF, Miller RJ, Field M, Rivier J (1982) Neurohumoral control of ileal electrolyte transport. II. Neurotensin and substance P. J Pharmacol Exp Ther 220:456-463
Kafetzopoulos E, Holzhauer MS, Huston JP (1986) Substance P injected into the region of the nucleus basalis magnocellularis facilitates performance of an inhibitory avoidance task. Psychopharmacology (Berlin) 90:281-283
Kalivas PW, Alesdatter JE (1993) Involvement of N-methyl-D-aspartate receptor stimulation in the ventral tegmental area and amygdala in behavioral sensitization to cocaine. J Pharmacol Exp Ther 267:486-495
Kennedy PGE, Rodgers J, Bradley B, Hunt SP, Gettinby G, Leeman SE, De Felipe C, Murray M (2003) Clinical and neuroinflammatory responses to meningoencephalitis in substance P receptor knockout mice. Brain 126:1-8
Kidd BL, Inglis J, Hood VC, De Felipe C, Bester H, Hunt SP, Cruwys SC (2003) Inhibition of inflammation and hyperalgesia in NK-1 receptor knock-out mice. NeuroReport 17:2189-2192
King TE, Heath MJS, Debs P, Davis MBE, Hen R, Barr GA (2000) The development of the nociceptive responses in neurokinin-1 receptor knockout mice. NeuroReport 11:587-591
Kiss S, Yoshiyama M, Cao YQ, Basbaum AI, de Groat WC, Lecci A, Maggi CA, Birder LA (2001) Impaired response to chemical irritation of the urinary tract in mice with disruption of the preprotachykinin gene. Neurosci Lett 313:57-60
Konishi S, Otsuka M (1974) excitatory action of hypothalamic substance p on spinal motoneurones of newborn rats. Nature 252:734-735
Kramer MS, Cutler N, Feighner J, Shrivastava R, Carman J, Sramek JJ, Reines SA, Liu G, Snavely D, Wyatt-Knowles E, Hale JJ, Mills SG, MacCoss M, Swain CJ, Harrison T, Hill RG, Hefti F, Scolnick EM, Cascieri MA, Chicchi GG, Sadowski S, Williams AR, Hewson L, Smith D, Rupniak NM (1998) Distinct mechanism for antidepressant activity by blockade of central substance P receptors. Science 281:1640-1645
Krappmann P, Hasenöhrl RU, Frisch C, Huston JP (1994) Self-administration of neurokinin substance P into the ventromedial caudate-putamen in rats. Neuroscience 62:1093-1101
Krase W, Koch M, Schnitzler HU (1994) Substance P is involved in the sensitization of the acoustic startle response by footshocks in rats. Behav Brain Res 63:81-88
Kreeger JS, Larson AA (1996) The substance P amino-terminal metabolite substance P(1-7), administered peripherally, prevents the development of acute morphine tolerance and attenuates the expression of withdrawal in mice. J Pharmacol Exp Ther 279:662-667
Krnjevic K, Morris ME (1974) An excitatory action of substance P on cuneate neurones. Can J Physiol Pharmacol 52:736-744
Laird JMA, Hargreaves RJ, Hill RG (1993) Effect of RP 67580, a non-peptide neurokinin1 receptor antagonist, on facilitation of a nociceptive spinal flexion reflex in the rat. Br J Pharmacol 109:713-718
Laird JMA, Olivar T, Roza C, De Felipe C, Hunt SP, Cervero F (2000) Deficits in visceral pain and hyperalgesia of mice with a disruption of the tachykinin NK_1 receptor gene. Neuroscience 98:345-352
Laird JMA, Roza C, De Felipe C, Hunt SP, Cervero F (2001) Role of central and peripheral tachykinin NK_1 receptors in capsaicin-induced pain and hyperalgesia in mice. Pain 90:97-103

Laufer R, Wormser U, Friedman ZY, Gilon C, Chorev M, Selinger Z (1985) Neurokinin B is a preferred agonist for a neuronal substance P receptor and its action is antagonized by enkephalin. Proc Natl Acad Sci USA 82:7444–7448

Lecci A, Giuliani S, Patacchini R, Viti G, Maggi CA (1991) Role of NK_1 tachykinin receptors in thermonociception: effect of (±)-CP 96,345, a non-peptide substance P antagonist, on the hot plate test in mice. Neurosci Lett 129:299–302

Lee CM, Iversen LL, Hanley MR, Sandberg BEB (1982) The possible existence of multiple receptors for substance P. Naunyn-Schmiedeberg's Arch Pharmacol 318:281–287

Le Gal La Salle G, Ben Ari Y (1977) Microiontophoretic effects of substance P on neurons of the medial amygdala and putamen of the rat. Brain Res 135:174–179

Lei YH, Barnes PJ, Rogers DF (1992) Inhibition of neurogenic plasma exudation in guinea-pig airways by CP-96,345, a new non-peptide NK_1 receptor antagonist. Br J Pharmacol 105:261–262

Lembeck F (1953) Zur Frage der zentralen Übertragung afferenter Impulse. III. Mitteilung. Das Vorkommen und die Bedeutung der Substanz P in den dorsalen Wurzeln des Rückenmarks. Naunyn-Schmiedeberg's Arch Exp Pathol Pharmacol 219:197–213

Lewandoski M (2001) Conditional control of gene expression in the mouse. Nat Rev Genet 2:743–755

Li HS, Zhao ZQ Small sensory neurons in the rat dorsal root ganglia express functional NK-1 tachykinin receptor. Eur J Neurosci 10:1292–1299

Liu H, Cao Y, Basbaum AI, Mazarati AM, Sankar R, Wasterlain CG (1999) Resistance to excitotoxin-induced seizures and neuronal death in mice lacking the preprotachykinin A gene. Proc Natl Acad Sci USA 96:12096–12101

Ljungdahl Å, Hökfelt T, Nilsson G (1978) Distribution of substance P-like immunoreactivity in the central nervous system of the rat I. Cell bodies and nerve terminals. Neuroscience 3:861–943

Ma QP, Woolf CJ (1995) Involvement of neurokinin receptors in the induction but not the maintenance of mechanical allodynia in rat flexor motoneurones. J Physiol 486:769–777

Ma QP, Woolf CJ (1997) Tachykinin NK_1 Receptor antagonist RP67580 attenuates progressive hypersensitivity of flexor reflex during experimental inflammation in rats. Eur J Pharmacol 322:165–171

Maa J, Grady EF, Yoshimi SK, Drasin TE, Kim EH, Hutter MM, Bunnett NW, Kirkwood KS (2000) Substance P is a determinant of lethality in diet-induced hemorrhagic pancreatitis in mice. Surgery 128:232–239

Maggi CA, Patacchini R, Rovero P, Giachetti A (1993) Tachykinin receptors and tachykinin receptor antagonists. J Auton Pharmacol 13:23–93

Malberg JE, Eisch AJ, Nestler EJ, Duman RS (2000) Chronic antidepressant treatment increases neurogenesis in adult rat hippocampus. J Neurosci 20:9104–9110

Maldonado R, Girdlestone D, Roques BP (1993) RP 67580, a selective antagonist of neurokinin-1 receptors, modifies some of the naloxone-precipitated morphine withdrawal signs in rats. Neurosci Lett 156:135–140

Manev H, Uz T, Smalheiser NR, Manev R (2001) Antidepressants alter cell proliferation in the adult brain in vivo and in neural cultures in vitro. Eur J Pharmacol 411:67–70

Mansikka H, Sheth RN, DeVries C, Lee H, Winchurch R, Raja SN (2000) Nerve injury-induced mechanical but not thermal hyperalgesia is attenuated in neurokinin-1 receptor knockout mice. Exp Neurol 162:343–349

Mansikka H, Shiotani M, Winchurch R, Raja SN (1999) Neurokinin-1 receptors are involved in behavioral responses to high-intensity heat stimuli and capsaicin-induced hyperalgesia in mice. Anesthesiology 90:1643–1649

Mantyh PW, Gates T, Mantyh CR, Maggio JE (1989) Autoradiographic localization and characterization of tachykinin receptor binding sites in the rat brain and peripheral tissues. J Neurosci 9:258–279

Mantyh PW, Pinnock RD, Downes CP, Goedert M, Hunt SP (1984) Correlation between inositol phospholipid hydrolysis and substance P receptors in rat CNS. Nature 309:795–797

Mantyh PW, Rogers SD, Honore P, Allen BJ, Ghilardi JR, Li J, Daughters RS, Lappi DA, Wiley RG, Simone DA (1997) Inhibition of hyperalgesia by ablation of lamina I Spinal neurons expressing the substance P receptor. Science 278:275–279

Martin JL, Sloviter RS (2001) Focal inhibitory interneuron loss and principal cell hyperexcitability in the rat hippocampus after microinjection of a neurotoxic conjugate of saporin and a peptidase-resistant analog of substance P. J Comp Neurol 436:127–152

Martinez-Caro L, Laird JM (2000) Allodynia and hyperalgesia evoked by sciatic mononeuropathy in NK1 receptor knockout mice. NeuroReport 11:1213–1217

Masu Y, Nakayama K, Tamaki H, Harada Y, Kuno M, Nakanishi S (1987) CDNA cloning of bovine substance-K receptor through oocyte expression system. Nature 329:836–838

McCarson KE (1999) Central and peripheral expression of neurokinin-1 and neurokinin-3 receptor and substance P-encoding messenger RNAs: peripheral regulation during formalin-induced inflammation and lack of neurokinin receptor expression in primary afferent sensory neurons. Neuroscience 93:361–370

McLean S, Ganong A, Seymour PA, Snider RM, Desai MC, Rosen T, Bryce DK, Longo KP, Reynolds LS, Robinson G, Schmidt AW, Siok C, Heym J (1993) Pharmacology of CP-99,994; a nonpeptide antagonist of the tachykinin neurokinin-1 receptor. J Pharmacol Exp Ther 267:472–479

Merritt JE, Rink TJ (1987) The effects of substance P and carbachol on inositol tris- and tetrakisphosphate formation and cytosolic free calcium in rat parotid acinar cells. A correlation between inositol phosphate levels and calcium entry. J Biol Chem 262:14912–14916

Mitsuhashi M, Ohashi Y, Shichijo S, Christian C, Sudduth-Klinger J, Harrowe G, Payan DG (1992) Multiple intracellular signaling pathways of the neuropeptide substance P receptor. J Neurosci Res 32:437–443

Morcuende S, Gadd CA, Peters M, Moss A, Harris EA, Sheasby A, Fisher AS, De Felipe C, Mantyh PW, Rupniak NJ, Giese PK, Hunt SP (2003) Increased neurogenesis and brain-derived neurotrophic factor in neurokinin-1 receptor gene knockout mice. Eur J Neurosci 18:1828–1836

Moussaoui SM, Philippe L, Le Prado N, Garret C (1993) Inhibition of neurogenic inflammation in the meninges by a non-peptide NK_1 receptor antagonist, RP 67580. Eur J Pharmacol 238:421–424

Murtra P, Sheasby AM, Hunt SP, De Felipe C (2000a) Loss of rewarding effects of morphine and amphetamine, but not cocaine, in mice with disruption of the substance P receptor (NK_1) gene. Soc Neurosci Abstr 26:1769

Murtra P, Sheasby AM, Hunt SP, De Felipe C (2000b) Rewarding effects of opiates are absent in mice lacking the receptor for substance P. Nature 405:180–183

Nagahisa A, Kanai Y, Suga O, Taniguchi K, Tsuchiya M, Lowe JA, Hess HJ (1992) Antiinflammatory and analgesic activity of a non-peptide substance P receptor antagonist. Eur J Pharmacol 217:191–195

Nagy A, Perrimon N, Sandmeyer S, Plasterk R (2003) Tailoring the genome: the power of genetic approaches. Nat Genet 33(Suppl):276–284

Nakajima Y, Tsuchida K, Negishi M, Ito S, Nakanishi S (1992) Direct linkage of three tachykinin receptors to stimulation of both phosphatidylinositol hydrolysis and cyclic AMP cascades in transfected Chinese hamster ovary cells. J Biol Chem 267:2437–2442

Nakanishi S (1987) Substance P precursor and kininogen: their structures, gene organizations, and regulation. Physiol Rev 67:1117–1142

Nakanishi S (1991) Mammalian tachykinin receptors. Annu Rev Neurosci 14:123–136

Nakaya Y, Kaneko T, Shigemoto R, Nakanishi S, Mizuno N (1994) Immunohistochemical localization of substance P Receptor in the central nervous system of the adult rat. J Comp Neurol 347:249–274

Napier TC, Mitrovic I, Churchill L, Klitenick MA, Lu XY, Kalivas PW (1995) Substance P in the ventral pallidum: projection from the ventral striatum, and electrophysiological and behavioral consequences of pallidal substance P. Neuroscience 69:59–70

Nestler EJ, Aghajanian GK (1997) Molecular and cellular basis of addiction. Science 278:58–63

Nestler EJ, Alreja M, Aghajanian GK (1999) Molecular control of locus coeruleus neurotransmission. Biol Psychiatry 46:1131–1139

Nestler EJ, Hope BT, Widnell KL (1993) Drug addiction: a model for the molecular basis of neural plasticity. Neuron 11:995–1006

Nibuya M, Morinobu S, Duman RS (1995) Regulation of BDNF and TrkB mRNA in rat brain by chronic electroconvulsive seizure and antidepressant drug treatments. J Neurosci 15:7539–7547

Nichols ML, Allen BJ, Rogers SD, Ghilardi JR, Honore P, Luger NM, Finke MP, Li J, Lappi DA, Simone DA, Mantyh PW (1999) Transmission of chronic nociception by spinal neurons expressing the substance P receptor. Science 286:1558–1561

Nicoll RA (1978) The Action of thyrotropin-releasing hormone, substance P and related peptides on frog spinal motoneurons. J Pharmacol Exp Ther 207:817–824

Nikolaus S, Huston JP, Hasenöhrl RU (1999) Reinforcing effects of neurokinin substance P in the ventral pallidum: mediation by the tachykinin NK1 receptor. Eur J Pharmacol 370:93–99

Nikolaus S, Huston JP, Hasenöhrl RU (2000) Anxiolytic-like effects in rats produced by ventral pallidal injection of both N- and C-terminal fragments of substance P. Neurosci Lett 283:37–40

Noailles PA, Angulo JA (2002) Neurokinin receptors modulate the neurochemical actions of cocaine. Ann New York Acad Sci 965:267–273

Noguchi K, Morita Y, Kiyama H, Ono K, Tohyama M (1988) A noxious stimulus induces the preprotachykinin-a gene expression in the rat dorsal root ganglion: a quantitative study using in situ hybridization histochemistry. Brain Res 464:31–35

Nowak LM, Macdonald RL (1982) Substance P: ionic basis for depolarizing responses of mouse spinal cord neurons in cell culture. J Neurosci 2:1119–1128

O'Shaughnessy CT, Connor HE (1993) Neurokinin NK_1 receptors mediate plasma protein extravasation in guinea-pig dura. Eur J Pharmacol 236:319–321

Otsuka MS, Konishi S, Takahashi T (1972) The presence of a motoneuron depolarizing peptide in bovine dorsal roots of spinal nerves. Proc Jpn Acad 48:342–346

Palmer JA, De Felipe C, O'Brien JA, Hunt SP (1999) Disruption of the substance P receptor (neurokinin-1) gene does not prevent upregulation of preprotachykinin-A mRNA in the spinal cord of mice following peripheral inflammation. Eur J Neurosci 11:3531–3538

Payne CM, Heggie CJ, Brownstein DG, Stewart JP, Quinn JP (2001) Role of tachykinins in the host response to murine gammaherpesvirus infection. J Virol 75:10467–10471

Penny GR, Afsharpour S, Kitai ST (1986) The glutamate decarboxylase-, leucine enkephalin-, methionine enkephalin- and substance P-immunoreactive neurons in the neostriatum of the rat and cat: evidence for partial population overlap. Neuroscience 17:1011–1045

Pernow B, Rosell S (1975) Effect of substance P on blood flow in canine adipose tissue and skeletal muscle. Acta Physiol Scand 93:139–141

Petit F, Glowinski J (1986) Stimulatory effect of substance P on the spontaneous release of newly synthesized [^{3}H]dopamine from rat striatal slices: a tetrodotoxin-sensitive process. Neuropharmacology 25:1015–1021

Picciotto MR, Wickman K (1998) Using knockout and transgenic mice to study neurophysiology and behavior. Physiol Rev 78:1131–1163

Piercey MF, Dobry PJK, Schroeder LA, Einspahr FJ (1981) Behavioral evidence that substance p may be a spinal cord sensory neurotransmitter. Brain Res 210:407–412

Piot O, Betschart J, Grall I, Ravard S, Garret C, Blanchard JC (1995) Comparative behavioural profile of centrally administered tachykinin NK_1, NK_2 and NK_3 receptor agonists in the guinea-pig. Br J Pharmacol 116:2496–2502

Ptak K, Burnet H, Blanchi B, Sieweke M, De Felipe C, Hunt SP, Monteau R, Hilaire G (2002) The murine neurokinin NK_1 receptor gene contributes to the adult hypoxic facilitation of ventilation. Eur J Neurosci 16:2245–2252

Ptak K, Hunt SP, Monteau R (2000) Substance P and central respiratory activity: a comparative in vitro study in NK_1 Receptor knockout and wild-type mice. Pflügers Arch 440:446–451

Randic M, Miletic V (1977) Effect of substance P in cat dorsal horn neurones activated by noxious stimuli. Brain Res 128:164–169

Rawlingson A, Gerard NP, Brain SD (2001) Interactive contribution of NK_1 and kinin receptors to the acute inflammatory oedema observed in response to noxious heat stimulation: studies in NK_1 receptor knockout mice. Br J Pharmacol 134:1805–1813

Regoli D, Drapeau G, Dion S, D'Orleans-Juste P (1987) Pharmacological receptors for substance P and neurokinins. Life Sci 40:109–117

Ribeiro-da-Silva A, Tagari P, Cuello AC (1989) Morphological characterization of substance P-like immunoreactive glomeruli in the superficial dorsal horn of the rat spinal cord and trigeminal subnucleus caudalis: a quantitative study. J Comp Neurol 281:497–515

Rigual R, Rico AJ, Prieto-Lloret J, De Felipe C, González C, Donnelly DF (2002) Chemoreceptor activity is normal in mice lacking the NK_1 receptor. Eur J Neurosci 16:2078–2084

Ripley TL, Gadd CA, De Felipe C, Hunt SP, Stephens DN (2002) Lack of self-administration and behavioural sensitisation to morphine, but not cocaine, in mice lacking NK1 receptors. Neuropharmacology 43:1258–1268

Rupniak NM, Carlson EC, Harrison T, Oates B, Seward E, Owen S, De Felipe C, Hunt S, Wheeldon A (2000) Pharmacological blockade or genetic deletion of substance P (NK_1) receptors attenuates neonatal vocalisation in guinea-pigs and mice. Neuropharmacology 39:1413–1421

Rupniak NMJ, Carlson EJ, Webb JK, Harrison T, Porsolt RD, Roux S, De Felipe C, Hunt SP, Oates B, Wheeldon A (2001) Comparison of the phenotype of $NK1R^{-/-}$ mice with pharmacological blockade of the substance P (NK_1) Receptor in assays for antidepressant and anxiolytic drugs. Behav Pharmacol 12:497–508

Russell IJ (2002) The promise of substance P inhibitors in fibromyalgia. Rheum Dis Clin North Am 28:329–342

Saban R, Gerard NP, Saban MR, Nguyen NB, DeBoer DJ, Wershil BK (2002) Mast cells mediate substance P-induced bladder inflammation through an NK_1 receptor-independent mechanism. Am J Physiol 283:F616–F629

Saban R, Nguyen N, Saban MR, Gerard NP, Pasricha PJ (1999) Nerve-mediated motility of ileal segments isolated from NK_1 receptor knockout mice. Am J Physiol 277:G1173–G1179

Saban R, Saban MR, Nguyen NB, Lu B, Gerard C, Gerard NP, Hammond TG (2000) Neurokinin-1 (NK-1) receptor is required in antigen-induced cystitis. Am J Pathol 156:775–780

Saffroy M, Beaujouan JC, Torrens Y, Besseyre J, Bergstrom L, Glowinski J (1988) Localization of tachykinin binding sites (NK_1, NK_2, NK_3 ligands) in the rat brain. Peptides 9:227–241

Santarelli L, Gobbi G, Blier P, Hen R (2002) Behavioral and physiologic effects of genetic or pharmacologic inactivation of the substance p receptor (NK_1). J Clin Psychiatry 63(Suppl 11):11–17

Santarelli L, Gobbi G, Debs PC, Sibille EL, Blier P, Hen R, Heath MJ (2001) Genetic and pharmacological disruption of neurokinin 1 receptor function decreases anxiety-related behaviors and increases serotonergic function. Proc Natl Acad Sci USA 98:1912–1917

Santos AR, Calixto JB (1997) Further evidence for the involvement of tachykinin receptor subtypes in formalin and capsaicin models of pain in mice. Neuropeptides 31:381–389

Sauer B (1998) Inducible Gene targeting in mice using the Cre/Lox system. Methods 14:381–392

Seabrook GR, Fong TM (1993) thapsigargin blocks the mobilisation of intracellular calcium caused by activation of human NK_1 (long) receptors expressed in Chinese hamster ovary cells. Neurosci Lett 152:9–12

Shaikh MB, Steinberg A, Siegel A (1993) Evidence that substance P Is utilized in medial amygdaloid facilitation of defensive rage behavior in the cat. Brain Res 625:283–294

Shen KZ, North RA (1992) Substance P opens cation channels and closes potassium channels in rat locus coeruleus neurons. Neuroscience 50:345–353

Shepheard SL, Williamson DJ, Williams J, Hill RG, Hargreaves RJ (1995) Comparison of the effects of sumatriptan and the NK_1 antagonist CP-99,994 on plasma extravasation in dura mater and c-Fos mRNA expression in trigeminal nucleus caudalis of rats. Neuropharmacology 34:255–261

Shigemoto R, Yokota Y, Tsuchida K, Nakanishi S (1990) Cloning and expression of a rat neuromedin K receptor cDNA. J Biol Chem 265:623–628

Shirayama Y, Chen ACH, Nakagawa S, Russell DS, Duman RS (2002) Brain-derived neurotrophic factor produces antidepressant effects in behavioral models of depression. J Neurosci 22:3251–3261

Simons CT, Dessirier JM, Jinks SL, Carstens E (2001) An animal model to assess aversion to intra-oral capsaicin: increased threshold in mice lacking substance P. Chem Senses 26:491–497

Simpson EM, Linder CC, Sargent EE, Davisson MT, Mobraaten LE, Sharp JJ (1997) Genetic variation among 129 substrains and its importance for targeted mutagenesis in mice. Nat Genet 16:19–27

Sjögren S, Andersson KE, Husted S (1982) Contractile effects of some polypeptides on the isolated urinary bladder of guinea pig, rabbit, and rat. Acta Pharmacol Toxicol 50:175–184

Sluka KA, Milton MA, Willis WD, Westlund KN (1997) Differential roles of neurokinin 1 and neurokinin 2 receptors in the development and maintenance of heat hyperalgesia induced by acute inflammation. Br J Pharmacol 120:1263–1273

Smith DW, Hewson L, Fuller P, Williams AR, Wheeldon A, Rupniak NMJ (1999) The substance P antagonist L-760,735 inhibits stress-induced NK_1 receptor internalisation in the basolateral amygdala. Brain Res 848:90–95

Starr MS (1978) Investigation of possible interactions between substance P and transmitter mechanisms in the substantia nigra and corpus striatum of the rat. J Pharm Pharmacol 30:359–363

Steinberg R, Alonso R, Griebel G, Bert L, Jung M, Oury-Donat F, Poncelet M, Gueudet C, Desvignes C, Le Fur G, Soubrie P (2001) Selective blockade of neurokinin-2 receptors produces antidepressant-like effects associated with reduced corticotropin-releasing factor function. J Pharmacol Exp Ther 299:449–458

Steinberg R, Alonso R, Rouquier L, Desvignes C, Michaud JC, Cudennec A, Jung M, Simiand J, Griebel G, Emonds-Alt X, Le Fur G, Soubrié P (2002) SSR240600 [(R)-2-(1-{2-[4-{2-[3,5-Bis(Trifluoromethyl)Phenyl]Acetyl}-2-(3,4-dichloro phenyl)-2-morpholinyl]Ethyl}-4-piperidinyl)-2-methylpropanamide], a centrally active nonpeptide antagonist of the tachykinin neurokinin 1 receptor. II. Neurochemical and behavioral characterization. J Pharmacol Exp Ther 303:1180–1188

Steinberg R, Rodier D, Souiclhac J, Bougault I, Emonds-Alt X, Soubrie P, Le Fur G (1995) Pharmacological characterization of tachykinin receptors controlling acetylcholine release from rat striatum: an in vivo microdialysis study. J Neurochem 65:2543–2548

Sternini C, Su D, Gamp PD, Bunnett NW (1995) Cellular sites of expression of the neurokinin-1 receptor in the rat gastrointestinal tract. J Comp Neurol 358:531–540

Sugiya H, Tennes KA, Putney JW (1987) Homologous desensitization of substance-P-induced inositol polyphosphate formation in rat parotid acinar cells. Biochem J 244:647–653

Sundelin JB, Provvedini DM, Wahlestedt CR, Laurell H, Pohl JS, Peterson PA (1992) Molecular cloning of the murine substance K and substance P receptor genes. Eur J Biochem 203:625–631

Suzuki R, Morcuende S, Webber M, Hunt SP, Dickenson AH (2002) Superficial NK_1-expressing neurons control spinal excitability through activation of descending pathways. Nat Neurosci 5:1319–1326

Takahashi K, Konishi S, Powell D, Leeman SE, Otsuka M (1974) Identification of the motoneuron-depolarizing peptide in bovine dorsal root as hypothalamic substance P. Brain Res 73:59–69

Takahashi T, Otsuka M (1975) Regional distribution of substance P in the spinal cord and nerve roots of the cat and the effect of dorsal root section. Brain Res 87:1–11

Tang FI, Chiu TH, Wang Y (1998) Electrochemical studies of the effects of substance P on dopamine terminals in the rat striatum. Exp Neurol 152:41–49

Tattersall FD, Rycroft W, Francis B, Pearce D, Merchant K, MacLeod AM, Ladduwahetty T, Keown L, Swain C, Baker R, Cascieri M, Ber E, Metzger J, MacIntyre DE, Hill RG, Hargreaves RJ (1996) Tachykinin NK_1 receptor antagonists act centrally to inhibit emesis induced by the chemotherapeutic agent cisplatin in ferrets. Neuropharmacology 35:1121–1129

Teixeira RM, Santos ARS, Ribeiro SJ, Calixto JB, Rae GA, De Lima TCM (1996) Effects of central administration of tachykinin receptor agonists and antagonists on plus-maze behavior in mice. Eur J Pharmacol 311:7–14

Theriault E, Otsuka M, Jessell T (1979) Capsaicin-evoked release of substance P From primary sensory neurons. Brain Res 170:209–213

Threadgill DW, Dlugosz AA, Hansen LA, Tennenbaum T, Lichti U, Yee D, LaMantia C, Mourton T, Herrup K, Harris RC et al. (1995) Targeted disruption of mouse EGF receptor: effect of genetic background on mutant phenotype. Science 269:230–234

Tomaz C, Huston JP (1986) Facilitation of conditioned inhibitory avoidance by post-trial peripheral injection of substance P. Pharmacol Biochem Behav 25:469–472

Tomaz C, Nogueira PJC (1997) Facilitation of memory by peripheral administration of substance P. Behav Brain Res 83:143–145

Tornqvist K, Mandahl A, Leander S, Loren I, Håkanson R, Sundler F (1982) Substance P-immunoreactive nerve fibres in the anterior segment of the rabbit eye. Distribution and possible physiological significance. Cell Tissue Res 222:467–477

Tournoy KG, De Swert KO, Leclere PG, Lefebvre RA, Pauwels RA, Joos GF (2003) Modulatory role of tachykinin NK_1 receptor in cholinergic contraction of mouse trachea. Eur Respir J 21:3–10

Ueda N, Muramatsu I, Taniguchi T, Nakanishi S, Fujiwara M (1986) Effects of neurokinin A, substance P and electrical stimulation on the rabbit iris sphincter muscle. J Pharmacol Exp Ther 239:494–499

Varty GB, Cohen-Williams ME, Morgan CA, Pylak U, Duffy RA, Lachowicz JE, Carey GJ, Coffin VL (2002) The gerbil elevated plus-maze II: anxiolytic-like effects of selective neurokinin NK_1 receptor antagonists. Neuropsychopharmacology 27:371–379

Vassout A, Veenstra S, Hauser K, Ofner S, Brugger F, Schilling W, Gentsch C (2000) NKP608: a selective NK-1 receptor antagonist with anxiolytic-like effects in the social interaction and social exploration test in rats. Regul Pept 96:7–16

Watson SP, Sandberg BEB, Hanley MR, Iversen LL (1983) Tissue selectivity of substance P alkyl esters: suggesting multiple receptors. Eur J Pharmacol 87:77–84

Weng HR, Mansikka H, Winchurch R, Raja SN, Dougherty PM (2001) Sensory processing in the deep spinal dorsal horn of neurokinin-1 receptor knockout mice. Anesthesiology 94:1105–1112

Wennemer HK, Kornetsky C (1999) fluoxetine blocks expression but not development of sensitization to morphine-induced oral stereotypy in rats. Psychopharmacology (Berlin) 146:19–23

Whittle BJ, Lopez-Belmonte J, Rees DD (1989) Modulation of the vasodepressor actions of acetylcholine, bradykinin, substance P and endothelin in the rat by a specific inhibitor of nitric oxide formation. Br J Pharmacol 98:646–652

Wise RA (1987) The role of reward pathways in the development of drug dependence. Pharmacol Ther 35:227–263

Wise RA, Bozarth MAA (1987) Psychomotor stimulant theory of addiction. Psychol Rev 94:469–492

Wolff M, Benhassine N, Costet P, Hen R, Segu L, Buhot MC (2003) Delay-dependent working memory impairment in young-adult and aged 5-HT1BKO mice as assessed in a radial-arm water maze. Learn Mem 10:401–409

Womack MD, MacDermott AB, Jessell TM (1988) Sensory transmitters regulate intracellular calcium in dorsal horn neurons. Nature 334:351–353

Yaksh TL, Jessell TM, Gamse R, Mudge AW, Leeman SE (1980) Intrathecal morphine inhibits substance P release from mammalian spinal cord in vivo. Nature 286:155–157

Yamaguchi K, Nakajima Y, Nakajima S, Stanfield PR (1990) Modulation of inwardly rectifying channels by substance P in cholinergic neurones from rat brain in culture. J Physiol 426:499–520

Yamamoto T, Yaksh TL (1991) Stereospecific effects of a nonpeptidic NK_1 selective antagonist, CP-96,345: antinociception in the absence of motor dysfunction. Life Sci 49:1955–1963

Yashpal K, Radhakrishnan V, Coderre TJ, Henry JL (1993) CP-96,345, but not its stereoisomer, CP-96,344, blocks the nociceptive responses to intrathecally administered substance P and to noxious thermal and chemical stimuli in the rat. Neuroscience 52:1039–1047

Yokota Y, Sasai Y, Tanaka K, Fujiwara T, Tsuchida K, Shigemoto R, Kakizuka A, Ohkubo H, Nakanishi S (1989) Molecular characterization of a functional cDNA for rat substance P receptor. J Biol Chem 264:17649–17652

Yonehara N, Shibutani T, Tsai HY, Inoki R (1986) Effects of Opioids And Opioid Peptide On The Release Of Substance P-Like material induced by tooth pulp stimulation in the trigeminal nucleus caudalis of the rabbit. Eur J Pharmacol 129:209–216

Zieglgänsberger W, Tulloch IF (1979) Effects of substance P on neurones in the dorsal horn of the spinal cord of the cat. Brain Res 166:273–282

Zimmer A, Zimmer AM, Baffi J, Usdin T, Reynolds K, König M, Palkovits M, Mezey É (1998) Hypoalgesia in mice with a targeted deletion of the tachykinin 1 gene. Proc Natl Acad Sci USA 95:2630–2635

Therapeutic Potential of Tachykinin Receptor Antagonists in Depression and Anxiety Disorders

N. M. J. Rupniak

Clinical Neuroscience, Merck Research Laboratories, BL2-5, West Point, PA 19486, USA
e-mail: nadiarupniak@concast.net

1 Introduction 342
2 Substance P (NK_1) Receptor Antagonists 342
2.1 Clinical Trials with NK_1RAs in Depression and Anxiety Disorders 343
2.2 Mechanism of Action of NK_1RAs in Depression. 345
2.2.1 Selectivity of NK_1RAs for NK_1 Receptors over Monoamine Reuptake Sites . . . 346
2.2.2 Substance P Levels in CSF and Plasma of Depressed Patients 346
2.2.3 Possible Sites of Action of NK_1RAs in the CNS 347
2.3 Anxiolytic-Like Activity of NK_1RAs in Animal Models 350

3 NK_2 Receptor Antagonists 351

4 NK_3 Receptor Antagonists 352

5 Conclusions 352

References 353

Abstract Neuropeptides, including the tachykinins substance P, neurokinin A and neurokinin B, are potentially important new targets for the development of psychotropic drugs. NK_1, NK_2 and NK_3 tachykinin receptor antagonists are currently undergoing clinical trials to investigate their therapeutic applications in central nervous system (CNS) disorders. Substance P (NK_1) receptor antagonists (NK_1RAs) have been clinically validated as a novel approach to treat major depression. Confirmation of the antidepressant efficacy of NK_1RAs has been obtained with three separate compounds in double-blind, placebo-controlled clinical trials. NK_1RAs are active in a range of preclinical assays that detect clinically used antidepressant and anxiolytic drugs, but they have a profile of activity that is distinct from established drugs. There is preliminary evidence that substance P function may be altered in depressed patients, suggesting a possible link between substance P and depressive pathophysiology. Preclinical studies indicate that the psychotherapeutic effects of NK_1RAs may be mediated by direct blockade of NK_1 receptors in the amygdala and its associated output projections, through stimulation of hippocampal neurogenesis, and also via interactions with monoamines. Clinical assessment of NK_2 and NK_3 receptor antagonists in psychiatric disorders is currently in progress.

Keywords Substance P · NK$_1$ receptor · NK$_2$ receptor · NK$_3$ receptor · Stress · Amygdala

1
Introduction

The tachykinins substance P, neurokinin A (NKA) and neurokinin B (NKB) belong to a family of neuropeptides that share a common C-terminal sequence, Phe-X-Gly-Leu-Met-NH$_2$ (where X is Phe, Tyr, Val or Ile). The biological actions of tachykinins are mediated through G-protein coupled receptors designated NK$_1$ for substance P, NK$_2$ for NKA, and NK$_3$ for NKB (Cascieri et al. 1992a). During decades of research to map the distribution and functions of tachykinins, there has been much speculation concerning their physiological functions, and potential therapeutic applications. Only since the development of the first highly selective, nonpeptide tachykinin receptor antagonists in the last decade has it been possible to test these hypotheses. These studies have been complicated by marked species differences in tachykinin receptor pharmacology and localization. Small–molecule NK$_1$, NK$_2$ and NK$_3$ receptor antagonists that are able to cross the blood-brain barrier are currently undergoing clinical trials to investigate their potential therapeutic applications in central nervous system (CNS) disorders. The vast majority of these are substance P (NK$_1$) receptor antagonists (NK$_1$RAs), and consequently these are much better characterized than other tachykinin antagonists. However, recently NK$_2$ and NK$_3$ receptor antagonists have also entered clinical development for psychiatric disorders. This review will provide an update on the current status of clinical development of these compounds and our understanding of their effects on CNS function.

Despite the availability of selective antagonists for tachykinin receptors, investigation of their actions in experimental models has been greatly complicated by inter-species differences in tachykinin receptor pharmacology and CNS distribution. Thus, whilst rats are the preferred species in preclinical models of CNS disorders, most NK$_1$ receptor antagonists have only low affinity for the rat variant NK$_1$ receptor, so that onerous experiments were needed to establish models in alternative species (Rupniak et al. 2003a). Similar species variants in NK$_2$ and NK$_3$ receptor pharmacology have also been reported (Aharony et al. 1995; Beaujouan et al. 1997). Difficulties in interpreting preclinical studies have also arisen because of marked inter-species differences in the CNS distribution of tachykinin receptors. This is particularly striking for NK$_3$ receptors, which are relatively abundant in the brains of rodents, but in extremely low abundance in human brain (Dietl and Palacios 1991).

2
Substance P (NK$_1$) Receptor Antagonists

Research into the biological functions of substance P has a long history. Since its discovery in 1931 by von Euler and Gaddum, substance P has arguably been the most extensively studied of all neuropeptides. Aprepitant (MK0869; Emend) recently became the first selective NK_1RA to be approved as a medicinal product for the relief of cancer chemotherapy-induced nausea and vomiting (FDA news release, March 2003). The approval of Emend is a significant milestone in neuropeptide research, being only the second neuropeptide antagonist (after the opioid antagonist, naloxone) ever to be approved for clinical use. Proof of the clinical utility of NK_1RAs in a human CNS disorder is an important spur to explore other therapeutic utilities for the drug class, and to pursue other neuropeptide targets. Aprepitant is currently in Phase III clinical trials for depression, and a number of other NK_1RAs are progressing in clinical development for this and other indications[1]. However, it is sobering to note that realization of the clinical potential of NK_1RAs has taken over 70 years since the peptide was first discovered (Wahlestedt 1998). This painfully slow progress is attributable to the absence of small molecule NK_1RAs prior to the 1990s. Initial efforts to understand the role of substance P attempted to use peptide antagonists, but these were difficult to deliver to the CNS, exhibited partial agonist activity, or were neurotoxic. In contrast, clinical development has proceeded swiftly and aggressively during the last decade following publication of the identity of the first nonpeptide NK_1 receptor antagonist, CP96345 (Snider et al. 1991). The accelerated development of NK_1RAs in recent years owes much to technological innovations in genomics, high throughput screening, molecular modeling and neuroimaging (Cascieri et al. 1992a, 1992b; Hargreaves et al. 2002).

2.1
Clinical Trials with NK₁RAs in Depression and Anxiety Disorders

The discovery of the antidepressant efficacy of NK_1RAs provides a revealing insight into the many challenges of bringing molecules with novel pharmacologies into clinical development, especially in psychiatry. For NK_1RAs, a vexing question has been how to prioritize the plethora of speculative therapeutic uses that encompass all of the major organ systems of the body (including the CNS, gastrointestinal, respiratory, cardiovascular, urinary and immune systems; Quartara and Maggi 1998). The decision to conduct clinical trials in patients with major depression was part of a focused effort to explore the potential CNS applications of NK_1RAs in pain, emesis and psychiatric disorders (Rupniak and Kramer 1999). Clinical trials in these patient populations were hypothesis-driven, underpinned in each case by evidence from preclinical models, and human pathological specimens that suggested an abnormality in substance P function in the disease. Based on positive findings in these exploratory studies, full-scale clinical programs were launched for the depression and emesis indications. Trials conducted in patient populations with pain or schizophrenia were negative, and so these indications were not pursued further.

---

[1] In November 2003, Merck announced that it had terminated the development of aprepitant for depression due to insufficient efficacy at the doses tested.

The safety and efficacy of aprepitant was evaluated in outpatients with major depressive disorder and moderately high anxiety levels. Aprepitant was chosen to test the concept clinically because of its high affinity, selectivity, brain penetrance, duration and oral bioavailability that permitted a once daily oral dosing regimen. Importantly, a clinically active dose in humans was already known from emesis studies conducted in patients undergoing cancer chemotherapy (Navari et al. 1999). Positron emission tomography studies established that more than 90% of NK_1 receptors in the human brain were blocked at this dose (Hargreaves et al. 2002), thereby providing an adequate test of whether aprepitant also possessed psychotherapeutic properties. The trial confirmed that aprepitant was an efficacious antidepressant and also demonstrated significant anxiolytic activity in this population of depressed patients (Kramer et al. 1998). It is sometimes suggested that the decision to examine aprepitant in depressed patients was based on spontaneous reports of acute 'mood altering' effects of the drug in other patient populations. However, this has not been observed in normal volunteers or any of the patient populations in which this drug has been studied to date. The tolerability of aprepitant was excellent and generally similar to placebo, except for mild and transient headaches, somnolence, nausea and fatigue. Strikingly, the incidence of sexual dysfunction, a common side effect of established antidepressants, resembled that for placebo. The manuscript was a landmark publication of the first new mechanistic approach to treat depression for four decades.

Following the intense interest and high expectations generated by the expedited publication of the first trial, the time taken for clinical replication and delivery of a long awaited new therapy has seemed frustrating. During this period, the clinical development of aprepitant, and also a second NK_1RA, L759274 ('Compound A'), proceeded, but was impeded by a problem that often afflicts the clinical evaluation of antidepressant drugs: failed trials. Failure to differentiate even clinically used antidepressant drugs from placebo occurs in around half of all clinical trials in this patient population, and reflects shortcomings in the approved trial methodology, rather than a lack of efficacy of the experimental drug. As has been observed with other antidepressant drugs, when aprepitant was examined in a subsequent large, dose-finding study, the data were not definitive due to a high placebo response rate, although the data suggested that there was antidepressant activity for at least one of the doses tested (Kramer 2000). Meanwhile, the principle that selectively blocking the NK_1 receptor can alleviate major depression and accompanying anxiety has been replicated using a second compound in a placebo-controlled, blinded study using L759274 published recently (Kramer et al. 2004). This study provided a stringent test of antidepressant efficacy, as the patients who benefited from L759274 suffered from severe depression with melancholic features, a condition that is often difficult to treat. The overall tolerability of L759274 was generally similar to that observed in patients treated with placebo, concurring with the excellent side-effect profile seen previously with aprepitant. Confirmatory clinical efficacy data were also presented for a third compound, CP122721 (Chappell et al., Association for European Psychiatry congress, Stockholm, May 2002). As an important contribution to the clinical literature in this newly emerging field, publication of these findings is eagerly awaited.

NK_1RAs therefore remain the best validated and clinically most advanced novel therapy for depression, and a number of compounds are currently progressing in clinical development. It is likely that the first NK_1RAs will be launched for use in depression in the next few years. Further studies will be required to fully characterize the clinical profile of NK_1RAs in depressed patients (speed of onset, efficacy across the severity spectrum of depression, compliance, relapse rates, and combination with established antidepressants). In addition, aprepitant and L759274 exhibited anxiolytic activity in the depressed patient populations in which they were examined. Established antidepressants are being used increasingly in place of benzodiazepines to treat a variety of anxiety disorders. It is a logical extension to enquire whether NK_1RAs might also have beneficial effects in patients with social phobia, or generalized anxiety, panic or post-traumatic stress disorder. Clinical trials of the NK_1RA NKP608 have been conducted in patients with social phobia (Ameringen et al. 2000), but the outcome of these studies has not been published. This will be an important area for future clinical investigation, especially in light of the growing preclinical evidence for anxiolytic-like effects of NK_1RAs (see Sect. 3).

2.2
Mechanism of Action of NK_1RAs in Depression

Speculation about the antidepressant and anxiolytic potential of NK_1RAs had its origins in the neuroanatomical localization of substance P and its receptor. Their presence in brain circuits that regulate stress responses and autonomic function (amygdala, hippocampus, hypothalamus, periaqueductal gray, raphe nuclei and locus coeruleus; Mantyh et al. 1984; Hokfelt et al. 1987) suggested an important role of substance P as a stress neurotransmitter. The content of substance P in these regions is altered by acute exposure to psychological stress and noxious stimuli (Siegel et al. 1987; Brodin et al. 1994). Activation of these pathways by central injection of substance P agonists elicits defensive cardiovascular (Unger et al. 1988) and behavioral changes (Aguiar and Brandao 1996; Teixeira et al. 1996; Kramer et al. 1998), suggesting that the release of endogenous substance P might contribute to the clinical manifestations of stress disorders. Finally, it was found that NK_1RAs were able to attenuate a stress response in preclinical species in a manner that resembled the effect of clinically used antidepressant drugs (Kramer et al. 1998; Rupniak et al. 2000). These preclinical findings provided the rationale for the first clinical study using an NK_1RA in depressed patients.

Publication of the clinical proof of concept study stimulated a flood of experiments to more fully understand the antidepressant mechanism of action of NK_1RAs. These have led to identification of some key sites of action of NK_1RAs in the brain that most likely confer their psychotropic effects. There is also preliminary evidence for abnormally high levels of substance P in plasma and cerebrospinal fluid (CSF) from depressed patients that suggests a possible link between this peptide and the pathophysiology of depression.

2.2.1
Selectivity of NK$_1$RAs for NK$_1$ Receptors over Monoamine Reuptake Sites

Nearly all of the currently used antidepressant drugs are selective serotonin (5-hydroxytryptamine; 5-HT) and norepinephrine reuptake inhibitors (SSRIs and SNRIs, respectively). Therefore, in considering the mechanism of aprepitant's antidepressant efficacy, it is essential to first rule out trivial explanations, such as direct actions on monoamine transporters, receptors or metabolic enzymes. Aprepitant, and its major des-triazolone metabolite, exhibit high (IC$_{50}$ 0.1–1 nM) affinity for NK$_1$ receptors, but are inactive (IC$_{50}$ >3 µM) in a comprehensive panel of in vitro counterscreens, including monoamine oxidase A and B, norepinephrine and serotonin reuptake sites, and monoamine receptors (Kramer et al. 1998). Conversely, SSRIs and SNRIs, respectively exhibit the opposite in vitro selectivity, namely nanomolar affinity for 5-HT and norepinephrine reuptake sites (Beique et al. 1998), and low (>10 µM) affinity for NK$_1$ receptors (Kramer et al. 1998). Moreover, simple markers of monoamine function were not altered after acute administration of an NK$_1$RA in vivo (i.e., NK$_1$RAs did not reverse reserpine-induced hypothermia, potentiate 5-HT-mediated behavioral syndrome, or increase extracellular 5-HT efflux), indicating that the profile of NK$_1$RAs is clearly distinct from that observed with SSRIs and SNRIs (Kramer et al. 1998). Hence, the in vitro and in vivo profile of NK$_1$RAs indicates that their antidepressant efficacy is mediated through a pharmacological site of action that is distinct from that of established therapies. This is also strongly suggested by the distinct clinical side effect profiles of the NK$_1$RAs aprepitant and L759274 as compared with the SSRI paroxetine (Kramer et al. 1998; Kramer et al. 2004).

2.2.2
Substance P Levels in CSF and Plasma of Depressed Patients

The simplest interpretation of how NK$_1$RAs are able to alleviate depression is that the pathophysiology of depression involves over-activity of substance P pathways. At present, only preliminary data are available regarding this possibility. However, there are reports of an increase in the level of substance P in the CSF of patients with major depression (Rimon et al. 1984; Heikkila et al. 1990). These findings were not replicated by Berrettini et al. (1995), who suggested that the findings of Rimon's earlier study might be an artifact, and until recently there was no interest in exploring this field further. A number of studies examining CSF samples from patients with a range of symptom severities and neuropsychiatric diagnoses are now in progress and the results are awaited with interest.

In addition to its putative role as a stress neurotransmitter in the CNS, there is also evidence that substance P may act as a stress hormone that is secreted by the adrenal glands in response to noxious stimulation (Vaupel et al. 1988). Two published studies have reported elevated serum substance P levels associated with severe stressors in humans: war time missile attacks (Weiss et al. 1996), and parachute jumping (Schedlowski et al. 1995). Recently, preliminary evidence has been obtained that circulating levels of substance P are also elevated in patients with major depression (Bondy et al. 2003). The source of origin of the elevated circulating peptide in these conditions is not known. The consequences are also not fully understood, although some intriguing possibilities have emerged. Intravenous injection of substance P has been shown to stimulate the secretion of adrenocorticotrophic hormone and corticosterone in humans (Coiro et al. 1992; Lieb et al. 2002), raising the possibility that elevated levels of circulating substance P might be a contributing factor in hypercortisolemia, a common endocrine abnormality in depressed patients. Intravenous injection of substance P in normal volunteers was also reported to cause sleep and mood disturbance (Lieb et al. 2002), also common features of depression. The question of whether abnormally high levels of substance P in the CNS and periphery contribute to the symptoms of depression is a fascinating area for future investigation.

2.2.3
Possible Sites of Action of NK_1RAs in the CNS

A large number of recent studies in preclinical species have identified key brain regions that most likely mediate the psychotropic actions of NK_1RAs. Our understanding of the role of NK_1 receptors in these pathways continues to evolve, but three brain regions that have emerged as being of particular interest are the amygdala, hippocampus and brainstem.

Amygdala. The amygdala plays a critical role in integrating stress responses through its connections with other brain regions, including the locus coeruleus, hypothalamus, periaqueductal gray, septum and thalamus. It is noteworthy that substance P is expressed in the pathways that connect these nuclei (Li and Ku 2002). Patients with depression have impaired function in amygdala-dependent tasks, such as tests that require the discrimination of facial expressions of emotion (Gur et al. 1992; Drevets 1998; Sheline et al. 2001).

There is evidence that substance P affects behavioral markers of emotionality in animals through direct actions in the amygdala and its output connections. Maternal separation or immobilization stress causes internalization of NK_1 receptors in the amygdala (Kramer et al. 1998; Smith et al. 1999), a marker of local release of substance P. Conversely, focal injection of an NK_1RA into the amygdala attenuated distress vocalizations in guinea pig pups (Boyce et al. 2001). Release of substance P in the amygdala is also implicated in a conditioned fear behavior in gerbils. In this species, foot drumming is used as an alarm signal and can be elicited by placing animals in an environment in which they had previously received foot shocks (Ballard et al. 2001; Rupniak et al. 2003b). It is striking that this behavior is also exhibited vigorously in gerbils following central injection of substance P agonists (Graham et al. 1993; Bristow and Young 1994). Foot drumming elicited by fear conditioning was blocked by NK$_1$RAs, and also by amygdalectomy (Ballard et al. 2001; Rupniak et al. 2003b).

Recently, a highly sensitive method for measuring extracellular efflux of substance P in the brain in vivo has been developed using microdialysis and micropush–pull superfusion techniques (Ebner et al. 2004). This innovation enabled the first ever real-time measurement of substance P release in the amygdala in conscious rats subjected to immobilization stress. Immobilization induced a long-lasting increase in substance P release in the medial amygdala, and subsequently decreased the time spent on the open arms of an elevated plus maze, a measure of anxiety. This anxiogenic-like effect was also produced by infusion of exogenous substance P into the medial amygdala. These anxiogenic-like effects of immobilization stress or substance P infusion were both blocked by intra-amygdala infusion of an NK$_1$RA. These findings provide further evidence that the amygdala is a critical brain area through which substance P mediates stress responses, and is a likely site of action for the psychotropic effects of NK$_1$RAs.

Other studies have demonstrated that local blockade of NK$_1$ receptors in regions receiving projections from the amygdala also inhibits stress responses in animals. These regions include the hypothalamus (Shaikh et al. 1993), reticulopontine nucleus (Krase et al. 1994), and periaqueductal gray (Teixeira et al. 1996).

Hippocampus. Like the amygdala, the hippocampus plays important roles in emotionality, cognition and endocrine function. A reduction in the volume of the hippocampus has been consistently observed in depressed patients (Sheline et al. 1996; Shah et al. 1998). An emerging new hypothesis to explain the lag period of several weeks before clinical improvement can be detected after commencing antidepressant therapy is that they might act to reverse impairments in the neural plasticity of the hippocampus that are induced by chronic stress (Jacobs et al. 2000).

A well-characterized change in neural plasticity is a reversible reduction in nerve cell proliferation (neurogenesis) in the hippocampal dentate gyrus caused by chronic stress. Chronic treatment with antidepressant drugs, and electroconvulsive shock (but not antipsychotic or anxiolytic drugs) reverses stress-induced impairments in cell proliferation, and increases levels of brain derived neurotrophic factor (BDNF; Madsen et al. 2000; Malberg et al. 2000). A chronic stress paradigm that has been used to study neurogenesis is psychosocial stress in tree shrews, induced by housing subordinate shrews in olfactory and visual contact with dominant animals. This produces disturbances of behavior, sleep, and hypercortisolemia that are reminiscent of human depression. These disturbances can be reversed by chronic administration of antidepressant drugs (van Kampen et al. 2002). These animals also exhibit stress-induced reductions in hippocampal neurogenesis and volume that were reversed by chronic administration of a tricyclic antidepressant drug (Czéh et al. 2001). Recently, it was demonstrated that chronic administration of an NK$_1$RA also increased hippocampal neurogenesis in stressed tree shrews (van der Hart et al. 2002). An increase in hippocampal neurogenesis has also been observed in NK$_1$ receptor null mutant (NK$_1$R$^{-/-}$) mice. Moreover, hippocampal levels of BDNF, which are elevated by antidepressant treatment, were twofold higher in NK$_1$R$^{-/-}$ mice (Morcuende et al. 2003).

A direct association between impaired hippocampal neurogenesis and reduced hippocampal volume in depressed patients has not been established and so this intriguing explanation of antidepressant drug action remains somewhat speculative at present. However, the ability of an NK$_1$RA to promote neurogenesis lends support to this proposed generic action of antidepressant therapy. The increase in hippocampal BDNF immunoreactivity in NK$_1$R$^{-/-}$ mice also warrants further investigation since evidence that this may be a clinically relevant marker is now emerging. BDNF immunoreactivity was increased in postmortem hippocampal brain tissue from subjects with major depression treated with antidepressant medications at the time of death, compared with untreated subjects (Chen et al. 2001).

Brain Stem. Since augmentation of 5-HT and norepinephrine systems is thought to underlie the antidepressant effects of clinically used drugs, there has been much interest in the effect of NK$_1$RAs on monoaminergic cell bodies located in the dorsal raphe nucleus and locus coeruleus. An intriguing aspect of the neuroanatomical localization of substance P is its intimate association with dorsal raphe 5-HT neurons in the human brain, approximately 50% of which co-express substance P with 5-HT (Sergeyev et al. 1999).

Several studies have now clearly established that NK$_1$RAs alter 5-HT neuronal function in a manner that is distinct from that seen with SSRIs. One of the most striking differences between these drug classes is that, unlike SSRIs, NK$_1$RAs do not increase extracellular 5-HT efflux (Kramer et al. 1998; Millan et al. 2001). Another difference is that NK$_1$RAs cause a marked increase (doubling) in the firing rate of dorsal raphe neurons (Santarelli et al. 2001; Conley et al. 2002), a change that is also seen in NK$_1$R$^{-/-}$ mice (Santarelli et al. 2001). A feature of the overall increase in firing rate was the induction of a burst firing pattern (Conley et al. 2002), which is thought to increase the probability of synaptic release of the neurotransmitter. Given this change in firing rate and pattern, it is curious that basal 5-HT efflux is not also increased. A likely explanation for this is that any increase in synaptic release of 5-HT can be sequestered through the intact 5-HT reuptake mechanism without overflowing into the extracellular space. This interpretation is suggested by an experiment using NK$_1$R$^{-/-}$ mice, in which basal 5-HT efflux was similar to that of wild-type mice, but was dramatically increased in the null mutants by administration of the SSRI fluoxetine (Froger et al. 2001). These studies provide evidence of alterations in 5-HT function by NK$_1$RAs that are likely to contribute to their antidepressant efficacy, but differ from those produced by SSRIs. The observation that NK$_1$RAs do not increase extracellular 5-HT efflux may explain the low incidence of SSRI-like side effects (nausea, gastrointestinal disturbance, sweating, sexual dysfunction) seen with apreiptant and L759274 in depressed patients (Kramer et al. 1998, 2003). These preclinical studies also suggest that a synergistic interaction might be seen after combined administration of NK$_1$RAs with established drugs.

In addition to their effects on 5-HT neuronal function, NK$_1$RAs also alter the firing characteristics of norepinephrine neurons originating in the locus coeruleus. These neurons are innervated by substance P-containing fibers and express NK$_1$ receptors on their cell bodies (Hahn and Bannon 1999). There is preliminary evidence that NK$_1$RAs may increase norepinephrine efflux in the hippocampus and frontal cortex (Millan et al. 2001) and alter the firing rate of locus coeruleus neurons (Millan et al. 2001; Maubach et al. 2002). There is also evidence that stress-induced activation of locus coeruleus neurons can be inhibited by NK$_1$RAs (Hahn and Bannon 1999).

As is the case for other antidepressant drugs, the precise mechanism that underlies the therapeutic efficacy of NK$_1$RAs in depression is not yet established. Preclinical studies provide strong evidence that their psychotherapeutic effects may be mediated through direct blockade of substance P receptors in the amygdala and its associated output projections, by promoting hippocampal neurogenesis, and by augmenting monoamine function.

2.3
Anxiolytic-Like Activity of NK$_1$RAs in Animal Models

In addition to their antidepressant activity, an expanding preclinical literature strongly suggests that NK$_1$RAs have anxiolytic properties. NK$_1$RAs have been shown to have anxiolytic-like effects in an impressively long list of ethological models of anxiety. These include: neonatal vocalization (Kramer et al. 1998; Rupniak et al. 2000), social interaction (File 1997; File et al. 2000; Cheeta et al. 2001; Gentsch et al. 2002), elevated plus maze (Teixeira et al. 1996; Santarelli et al. 2001; Varty et al. 2002a; Ebner et al. 2003), 4-plate test (Rupniak et al. 2003b), and fear conditioning (Ballard et al. 2001; Rupniak et al. 2003b). An interesting finding in those studies that incorporated SSRIs, SNRIs or tricyclic antidepressants as comparator drugs was that the effect of the NK$_1$RA on behavior was markedly different, and in some cases opposite, to that seen with monoamine reuptake inhibitors. In the elevated plus maze, the effect of acute administration of apreiptant and other NK$_1$RAs was comparable to that observed for the benzodiazepine anxiolytic diazepam, whereas fluoxetine and imipramine were ineffective (Varty et al. 2002a, 2002b). In the social interaction and fear conditioning tasks, acute administration of SSRIs and SNRIs was anxiogenic, in marked contrast to the anxiolytic-like effects of the NK$_1$RAs (Cheeta et al. 2001; Ballard et al. 2001; Rupniak et al. 2003). Acute worsening of anxiety on commencing therapy with conventional antidepressant drugs in patients can be problematic, and often requires gradual titration of the dose. These preclinical studies suggest that this may not be seen with NK_1RAs, although clinical studies that are specifically designed to address this have not yet been performed. The preclinical studies do, however, provide further proof of the distinct mechanism of action of NK_1RAs compared with other classes of antidepressant drugs.

3
NK_2 Receptor Antagonists

There has been interest in NK_2 receptor antagonists as anxiolytic and antidepressant agents for a number of years. This is based entirely on empirical observations of the effects of these compounds on behavior and brain physiology in animal models. However, the rationale for progressing this class of compound into clinical development is greatly weakened by the failure to establish that NK_2 receptors are expressed in the adult mammalian brain, including human (Saffroy et al. 1987; Dietl and Palacios 1991). Even using the improved nonpeptide ligands [^{3}H]GR100679 and [^{3}H]SR48968 for autoradiography studies, only 10%–30% of binding was specific in the adult rat brain (Hagan et al. 1993; Stratton et al. 1996). Until there is a clear demonstration of NK_2 receptor expression in the human brain, it should not be expected that selective antagonists at this receptor will have clinically useful psychotropic effects.

The selective NK_2 receptor antagonists SR48968, GR100679 and GR159897 have been reported to exhibit anxiolytic-like effects across a range of rodent and primate behavioral assays (mouse light–dark box, rat social interaction test, rat elevated plus maze and marmoset threat test). These compounds were found to be extremely potent [0.0005–200 µg/kg subcutaneously (s.c.); Stratton et al. 1993, 1994; Walsh et al. 1995]. Importantly, it was not established that the doses that were administered were able to block NK_2 receptors in the CNS. This is an important omission since the putative anxiolytic dose range (µg/kg) is considerably lower than that required to block NK_2 agonist-mediated effects in peripheral tissues (0.6–1.2 mg/kg; Tousignant et al. 1993; Beresford et al. 1995).

More recently, NK_2 receptor antagonists have been revisited as a putative antidepressant drug target. Using reverse transcription–polymerase chain reaction analysis, Steinberg et al. (1998) showed the presence of NK_2 receptor mRNA in the septal area of the rat brain, and detected septal NK_2 binding sites using a fluorescent-tagged NKA derivative. In vivo microdialysis suggested a functional role of NK_2 receptors regulating the release of acetylcholine in the hippocampus. Hippocampal acetylcholine release was prevented by administration of the selective NK_2 receptor antagonists SR144190 (0.03–0.3 mg/kg) and saredutant (SR48968; 0.3 and 1 mg/kg). NK_2 receptors were also found to control acetylcholine release in vitro using striatal brain slices from mice (Preston et al. 2000). These functional studies were extended to show that saredutant (0.3–10 mg/kg) exhibited antidepressant-like activity in two animal models (forced swim test, and neonatal vocalization). Saredutant also inhibited a stress-induced increase in locus coeruleus neuronal firing and cortical norepinephrine release. These findings suggested that NK_2 receptor blockade may constitute a novel mechanism in the treatment of depression and corticotropin-releasing factor-related disorders (Steinberg et al. 2001). Saredutant is currently in Phase IIb clinical trials for depression/anxiety.

4
NK$_3$ Receptor Antagonists

Unlike NK$_2$ receptors, it has been established that NK$_3$ receptors are expressed, albeit in low abundance, in the human brain (Buell et al. 1992). Caution is required in extrapolating possible clinical implications from functional assays using rodents because of the marked phylogenetic differences in the expression of NK$_3$ receptors in the CNS, which are more abundant in the rat than the human brain (Dietl and Palacios. 1991). In the human brain, NK$_3$ receptor immunoreactivity has been found in the paraventricular and perifornical nuclei of the hypothalamus, a localization that resembles that seen in rat brain (Koutcherov et al. 2000). In rats, hypothalamic NK$_3$ receptors are involved in the neural circuitry regulating blood pressure (Polidori et al. 1989) and these receptors may serve a similar function in humans. Central injection of selective NK$_3$ receptor agonists such as senktide produces a distinctive, 5-HT-like, behavioral syndrome of 'wet dog shakes' in rodents (Itoi et al. 1992; Cellier et al. 1997). This is blocked by 5-HT receptor antagonism, indicating that NK$_3$ receptor agonists cause release of 5-HT in rodent brain (Stoessl et al. 1988). This might be expected to confer activity of NK$_3$ receptor agonists in animal models of depression that are sensitive to SSRIs, but there are no published reports that this has been examined. However, it has been reported that senktide possesses anxiolytic-like activity in the elevated plus maze test in rats (Ribiero et al. 1999). These studies suggest that agonists, rather than antagonists, at the NK$_3$ receptor might possess antidepressant and/or anxiolytic-like effects, at least in rodents. There are no NK$_3$ receptor agonists in clinical development. Osanetant (SR142801) is a selective NK$_3$ receptor antagonist that was originally investigated as a potential treatment for anxiety and psychosis and is currently in Phase IIb clinical trials as an antipsychotic (Kamali 2001).

5
Conclusions

During the many decades of neuropeptide research, there has been much speculation about the physiological roles of tachykinins, but it is only since selective antagonists of NK$_1$, NK$_2$ and NK$_3$ receptors have been available that these hypotheses could be tested. Clinical data confirm the ability of NK$_1$RAs to alleviate symptoms of depression and anxiety in patients with major depression, and they may also have utility in other patient populations, such as social phobia. The utility of NK$_2$ and NK$_3$ receptor antagonists is less certain as these compounds are less advanced in clinical development, and species differences in the expression of these receptors in the CNS complicate extrapolation based on their effects in preclinical models. Ligands for many other neuropeptide receptors are also progressing in clinical development and, if successful, may launch a new era in psychotherapeutics.

References

Aguiar MS, Brandao ML (1996). Effects of microinjections of the neuropeptide substance P in the dorsal periaqueductal gray on the behaviour of rats in the plus-maze test. Physiol Behav 60:1183–1186

Aharony D, Buckner CK, Ellis JL, Ghanekar SV, Graham A, Kays JS, Little J, Meeker S, Miller SC, Undem BJ (1995) Pharmacological characterization of a new class of non-peptide neurokinin A antagonists that demonstrate species selectivity. J Pharmacol Exp Ther 274:1216–1221

Ballard TM, Sanger S, Higgins GA (2001) Inhibition of shock-induced foot tapping behaviour in the gerbil by a tachykinin NK_1 receptor antagonist. Eur J Pharmacol 412:255–264

Beaujouan JC, Saffroy M, Torrens Y, Glowinski J (1997) Potency and selectivity of the tachykinin NK_3 receptor antagonist SR 142801. Eur J Pharmacol 319:307–316

Beique JC, Lavoie N, de Montigny C, Debonnel G (1998) Affinities of venlafaxine and various reuptake inhibitors for the serotonin and norepinephrine transporters. Eur J Pharmacol 349:129–132

Beresford IJ, Sheldrick RL, Ball DI, Turpin MP, Walsh DM, Hawcock AB, Coleman RA, Hagan RM, Tyers MB (1995) GR159897, a potent non-peptide antagonist at tachykinin NK_2 receptors. Eur J Pharmacol 272:241–248

Berrettini WH, Rubinow DR, Nurnberger JI, Simmons-Alling S, Post RM, Gershon ES (1985) CSF substance P immunoreactivity in affective disorders. Biol Psychiat 20:965–970

Boyce S, Smith D, Carlson E, Hewson L, Rigby M, O'Donnell R, Harrison T, Rupniak NMJ (2001) Intra-amygdala injection of the substance P [NK_1 receptor] antagonist L-760735 inhibits neonatal vocalisations in guinea-pigs. Neuropharmacology 41:130–137

Bondy B, Baghai TC, Minov C, Schule C, Schwarz MJ, Zwanzger P, Rupprecht R, Moller HJ (2003) Substance P serum levels are increased in major depression: preliminary results. Biol Psychiat 53:538–542

Bristow LJ, Young L (1994) Chromodacryorrhoea and repetitive hind paw tapping: models of peripheral and central tachykinin NK_1 receptor activation in gerbils. Eur J Pharmacol 254:245–249

Buell G, Schultz MF, Arkinstall SJ (1992) Molecular characterisation, expression and localization of human neurokinin-3 receptor. FEBS Letts 299:90–95

Cascieri MA, Huang R-RC, Fong TM, Cheung AH, Sadowski S, Ber E, Strader CD (1992a) Determination of the amino acid residues in substance P conferring selectivity and specificity for the rat neurokinin receptors. Mol Pharmacol 41:1096–1099

Cascieri MA, Ber E, Fong TM, Sadowski S, Bansal A, Swain C, Seward E, Frances B, Burns D, Strader CD (1992b) Characterization of the binding of a potent, selective, radioiodinated antagonist to the human neurokinin-1 receptor. Mol Pharmacol 42:458–463

Cellier E, Barbot L, Regoli D, Couture R (1997) Cardiovascular and behavioural effects of intracerebroventricularly administered tachykinin NK_3 receptor antagonists in the conscious rat. Br J Pharmacol 122:643–654

Cheeta S, Tucci S, Sandhu J, Williams AR, Rupniak NM, File SE (2001) Anxiolytic actions of the substance P (NK_1) receptor antagonist L-760735 and the 5-HT_{1A} agonist 8-OH-DPAT in the social interaction test in gerbils. Brain Res 915:170–175

Chen B, Dowlatshahi D, MacQueen GM, Wang JF, Young LT (2001) Increased hippocampal BDNF immunoreactivity in subjects treated with antidepressant medication. Biol Psychiat 50:260–265

Coiro V, Capretti L, Volpi R, Davoli C, Marcato A, Cavazzini U, Caffarri G, Rossi G, Chiodera P (1992) Stimulation of ACTH/cortisol by intravenously infused substance P in normal men: inhibition by sodium valproate. Neuroendocrinology 56:459–463

Conley R, Cumberbatch MJ, Mason GS, Williamson DJ, Boyce S, Carlson EJ, Harrison T, Locker K, Swain CJ, Webb JK, Wheeldon A, Maubach KA, O'Donnell R, Rigby M, Hewson L, Smith D, Rupniak NMJ (2002) Substance P (NK$_1$) receptor antagonists enhance dorsal raphe neuronal activity. J Neurosci 22:7730–7736

Czéh B, Michaelis T, Watanabe T, Frahm J, de Biurrun G, van Kampen M, Bartolomucci A, Fuchs E (2001) Stress-induced changes in cerebral metabolites, hippocampal volume and cell proliferation are prevented by antidepressant treatment with tianeptine. Proc Natl Acad Sci USA 98:12796–12801

Dietl MM, Palacios JM (1991) Phylogeny of tachykinin receptor localization in the vertebrate central nervous system: apparent absence of neurokinin-2 and neurokinin-3 binding sites in the human brain. Brain Res 539:211–222

Drevets WC (1998) Functional neuroimaging studies of depression: the anatomy of melancholia. Annu Rev Med 49:341–361

Ebner K, Rupniak N, Saria A, Singewald N (2004) Substance P in the medial amygdala: emotional stress-sensitive release and modulation of anxiety-related behaviour in rats. Proc Natl Acad Sci USA. in press

File SE (1997) Anxiolytic action of a neurokinin-1 receptor antagonist in the social interaction test. Pharmacol Biochem Behav 58:747–752

File SE (2000) NKP608, an NK$_1$ receptor antagonist, has an anxiolytic action in the social interaction test. Psychopharmacology 152:105–109

Froger N, Gardier AM, Moratalla R, Alberti I, Lena I, de Felipe C, Rupniak NMJ, Hunt SP, Jacquot C, Hamon M, Lanfumey L (2001) 5-HT$_{1A}$ autoreceptor adaptive changes in substance P (NK$_1$) receptor knock-out mice mimic antidepressant-induced desensitization. J Neurosci 21:8188–8197

Gentsch C, Cutler M, Vassout A, Veenstra S, Brugger F (2002) Anxiolytic effect of NKP608, a NK$_1$-receptor antagonist, in the social investigation test in gerbils. Behav Brain Res 133:363–368

Graham EA, Turpin MP, Stubbs CM (1993) Characterisation of the tachykinin-induced hindlimb thumping response in gerbils. Neuropeptides 4:228

Gur RC, Erwin RJ, Gur RE, Zwil AS, Heimberg C, Kraemer HC (1992) Facial emotion discrimination: II. Behavioral findings in depression. Psychiat Res 42:241–251

Hagan RM, Beresford IJ, Stables J, Dupere J, Stubbs CM, Elliott PJ, Sheldrick RL, Chollet A, Kawashima E, McElroy AB (1993) Characterisation, CNS distribution and function of NK$_2$ receptors studied using potent NK$_2$ receptor antagonists. Regul Pept 46:9–19

Hahn MK, Bannon MJ (1999) Stress-induced c-fos expression in the rat locus coeruleus is dependent on neurokinin 1 receptor activation. Neuroscience 94:1183–1188

Heikkila L, Rimon R, Terenius L (1990) Dynorphin A and substance P in the cerebrospinal fluid of schizophrenic patients. Psychiat Res 34:229–236

Hargreaves R (2002) Imaging substance P receptors (NK$_1$) in the living human brain using positron emission tomography. J Clin Psychiatry 63 (Suppl 11): 18–24

Hökfelt T, Johansson O, Holets V, Meister B, Melander T (1987) Distribution of neuropeptides with special reference to their coexistence with classical transmitters. In HY Meltzer (ed) Psychopharmacology: The Third Generation of Progress. Raven, New York, pp 401–416

Itoi K, Tschope C, Jost N, Culman J, Lebrun C, Stauss B, Unger T (1992) Identification of the central tachykinin receptor subclass involved in substance P-induced cardiovascular and behavioral responses in conscious rats. Eur J Pharmacol 219:435–444

Jacobs BL, van Praag H, Gage FH (2000) Adult brain neurogenesis and psychiatry: a novel theory of depression. Mol Psychiat 5:262–269

Kamali F (2001) Osanetant Sanofi-Synthelabo. Curr Opin Investig Drugs 2:950–956

Koutcherov Y, Ashwell KW, Paxinos G (2000) The distribution of the neurokinin B receptor in the human and rat hypothalamus. NeuroReport 11:3127–3131

Kramer MS (2000) Update on Substance P (NK_1 receptor) antagonists in clinical trials for depression. Neuropeptides 34:256

Kramer MS, Cutler N, Feighner J, Shrivastava R, Carman J, Sramek JJ, Reines SA, Liu G, Snavely D, Wyatt-Knowles E, Hale JJ, Mills SG, MacCoss M, Swain CJ, Harrison T, Hill RG, Hefti FF, Scolnick EM, Cascieri MA, Chicchi GG, Sadowski S, Williams AR, Hewson L, Smith D, Carlson EJ, Hargreaves RJ, Rupniak NMJ (1998) Distinct mechanism for antidepressant activity by blockade of central substance P receptors. Science 281:1640–1645

Kramer MS, Winokur A, Kelsey J, Preskorn SH, Rothschild AJ, Snavely D, Ghosh K, Ball WA, Reines SA and the SPA Depression Study Group (2004) Demonstration of the efficacy and safety of a novel substance P (NK_1) receptor antagonist (SPA) in major depression. Neuropsychopharmacology 29:385–392

Krase W, Koch M, Schnizler H-U (1994) Substance P is involved in the sensitization of the acoustic startle response by foot shock in rats. Behav Brain Res 63:81–88

Li YH, Ku YH (2002) Involvement of rat lateral septum-acetylcholine pressor system in central amygdaloid nucleus-emotional pressor circuit. Neurosci Lett 323:60–64

Lieb K, Ahlvers K, Dancker K, Strohbusch S, Reincke M, Feige B, Berger M, Riemann D, Voderholzer U (2002) Effects of the neuropeptide substance P on sleep, mood, and neuroendocrine measures in healthy young men. Neuropsychopharmacology 27:1041–1049

Madsen TM, Treschow A, Bengzon J, Bolwig TG, Lindvall O, Tingstrom A (2000) Increased neurogenesis in a model of electroconvulsive therapy. Biol Psychiat 47:1043–1049

Malberg JE, Eisch AJ, Nestler EJ, Duman RS (2000) Chronic antidepressant treatment increases neurogenesis in adult rat hippocampus. J Neurosci 20:9104–9110

Mantyh PW, Hunt SP, Maggio JE (1984) Substance P receptors: localization by light microscopic autoradiography in rat brain using [^{3}H]SP as the radioligand. Brain Res 307:147–165

Maubach KA, Martin K, Chicchi G, Harrison T, Wheeldon A, Swain CJ, Cumberbatch MJ, Rupniak NMJ, Seabrook GR (2002) Chronic imipramine or substance P (NK_1) receptor antagonist treatment increases burst firing of monoamine neurons in the locus coeruleus. Neuroscience 109:609–617

Millan MJ, Lejeune F, de Nanteuil G, Gobert AJ (2001) Selective blockade of neurokinin (NK_1) receptors facilitates the activity of adrenergic pathways projecting to frontal cortex and dorsal hippocampus in rats. J Neurochem 76:1949–1954

Morcuende S, Gadd CA, Peters M, Moss A, Harris EA, Sheasby A, De Felipe C, Mantyh PW, Rupniak NMJ, Giese KP, Hunt SP (2003) Neurokinin-1 receptor modulates neurogenesis in adult hippocampus without influencing learning and memory. Eur J Neurosci 18:1828–1836

Navari RM, Reinhardt RR, Gralla RJ, Kris MG, Hesketh PJ, Khojasteh A, Kindler H, Grote TH, Pendergrass K, Grunberg SM, Carides AD, Gertz BJ, Antiemetic Trials Group (1999) Reduction of cisplatin-induced emesis by a selective neurokinin-1 receptor antagonist, L-754,030. New Engl J Med 340:190–195

Polidori C, Saija A, Perfumi M, Costa G, de Caro G, Massi M (1989) Vasopressin release induced by intracranial injection of tachykinins is due to activation of central neurokinin-3 receptors. Neurosci Lett 103:320–325

Preston Z, Lee K, Widdowson L, Richardson PJ, Pinnock RD (2000) Tachykinins increase [3H]acetylcholine release in mouse striatum through multiple receptor subtypes. Neuroscience 95:367–376

Quartara L, Maggi CA (1998) The tachykinin NK_1 receptor. Part II: Distribution and pathophysiological roles. Neuropeptides 32:1–49

Ribeiro SJ, Teixeira RM, Calixto JB, De Lima TC (1999) Tachykinin NK_3 receptor involvement in anxiety. Neuropeptides 33:181–188

Rimon R, le Greves P, Nyberg F, Heikkila L, Samela L, Terenius L (1984) Elevation of substance P-like peptides in the CSF of psychiatric patients. Biol Psychiat 19:509–516

Rosen A, Brodin K, Eneroth P, Brodin E (1992) Short-term restraint stress and s.c. saline injection alter the tissue levels of substance P and cholecystokinin in the peri-aqueductal grey and limbic regions of rat brain. Acta Physiol Scand 146:341–348

Rupniak NMJ, Carlson EC, Harrison T, Oates B, Seward E, Owen S, de Felipe C, Hunt S, Wheeldon A (2000) Pharmacological blockade or genetic deletion of substance P (NK_1) receptors attenuates neonatal vocalisation in guinea pigs and mice. Neuropharmacology 39:1413–1421

Rupniak NMJ, Carlson EJ, Shepheard S, Bentley G, Williams AR, Hill A, Swain C, Mills SG, Di Salvo J, Kilburn R, Cascieri MA, Kurtz MM, Tsao K-L, Gould SL, Chicchi GG (2003a) Comparison of the functional blockade of rat substance P (NK_1) receptors by GR205171, RP67580, SR140333 and NKP-608. Neuropharmacology 45:231–241

Rupniak NMJ, Kramer MS (1999) Discovery of the antidepressant and anti-emetic efficacy of substance P receptor (NK_1) antagonists. Trends Pharmacol Sci 20:485–490

Rupniak NMJ, Webb JK, Fisher A, Smith D, Boyce S (2003b) The substance P (NK_1) receptor antagonist L-760735 inhibits fear conditioning in gerbils. Neuropharmacology 44:516–523

Saffroy M, Beaujouan JC, Torrens Y, Besseyre J, Bergstrom L, Glowinski J (1987) Localization of tachykinin binding sites (NK_1, NK_2, NK_3 ligands) in the rat brain. Peptides 9:227–241

Santarelli L, Gobbi G, Debs PC, Sibille ET, Blier P, Hen R, Heath MJ (2001) Genetic and pharmacological disruption of neurokinin 1 receptor function decreases anxiety-related behaviors and increases serotonergic function. Proc Natl Acad Sci USA 98:1912–1917

Schedlowski M, Fluge T, Richter S, Tewes U, Schmidt RE, Wagner TO (1995) Beta-endorphin, but not substance-P, is increased by acute stress in humans. Psychoneuroendocrinology 20:103–110

Sergeyev V, Hlt T, Hurd Y (1999) Serotonin and substance P co-exist in dorsal raphe neurons of the human brain. NeuroReport 10:3967–3970

Shah PJ, Ebmeier KP, Glabus MF, Goodwin GM (1998) Cortical grey matter reductions associated with treatment-resistant chronic unipolar depression. Controlled magnetic resonance imaging study. Br J Psychiat 172:27–52

Shaikh MB, Steinberg A, Siegel A (1993) Evidence that substance P is utilized in the medial amygdaloid facilitation of defensive rage behavior in the cat. Brain Res 625:283–294

Sheline YI, Barch DM, Donnelly JM, Ollinger JM, Snyder AZ, Mintun MA (2001) Increased amygdala response to masked emotional faces in depressed subjects resolves with antidepressant treatment: an fMRI study. Biol Psychiat 50:651–658

Sheline YI, Wang PW, Gado MH, Csernansky JG, Vannier MW (1996) Hippocampal atrophy in recurrent major depression. Proc Natl Acad Sci USA 93:3908–2913

Siegel RA, Duker EM, Pahnke U, Wuttke W (1987) Stress-induced changes in cholecystokinin and substance P concentrations in discrete regions of the rat hypothalamus. Neuroendocrinology 46:75–81

Smith DW, Hewson L, Fuller P, Williams AR, Wheeldon A, Rupniak NM (1999) The substance P antagonist L-760,735 inhibits stress-induced NK_1 receptor internalisation in the basolateral amygdala. Brain Res 848:90–95

Steinberg R, Alonso R, Griebel G, Bert L, Jung M, Oury-Donat F, Poncelet M, Gueudet C, Desvignes C, Le Fur G, Soubrie P (2001) Selective blockade of neurokinin-2 receptors produces antidepressant-like effects associated with reduced corticotropin-releasing factor function. J Pharmacol Exp Ther 299:449–458

Steinberg R, Marco N, Voutsinos B, Bensaid M, Rodier D, Souilhac J, Alonso R, Oury-Donat F, Le Fur G, Soubrie P (1998) Expression and presence of septal neurokinin-2 receptors controlling hippocampal acetylcholine release during sensory stimulation in rat. Eur J Neurosci 10:2337–2345

Stoessl AJ, Dourish CT, Iversen SD (1988) Pharmacological characterization of the behavioural syndrome induced by the NK-3 tachykinin agonist senktide in rodents: evidence for mediation by endogenous 5-HT. Brain Res 517:111–116

Stratton SC, Beresford IJM, Hagan RM (1993) Anxiolytic activity of tachykinin NK_2 receptor antagonists in the mouse light-dark box. Eur J Pharmacol 250: R11–R12

Stratton SC, Beresford IJM, Hagan RM (1994) GR159897, a potent non-peptide tachykinin NK_2 receptor antagonist, releases suppressed behaviours in a novel aversive environment. Br J Pharmacol 112:49P

Stratton SC, Beresford IJM, Hagan RM (1996) Autoradiographic localization of tachykinin NK-2 receptors in adult rat brain using (^{3}H)SR48968. Br J Pharmacol 117:295P

Teixeira RM, Santos AR, Ribeiro SJ, Calixto JB, Rae GA, De Lima TC (1996) Effects of central administration of tachykinin receptor agonists and antagonists on plus-maze behaviour in mice. Eur J Pharmacol 311:7–14

Tousignant C, Chan C-C, Guevremont D, Brideau C, Hale JJ, MacCoss M, Rodger IW (1993) NK_2 receptors mediate plasma extravasation in guinea-pig lower airways. Br J Pharmacol 108:383–386

Unger T, Carolus S, Demmert G, Ganten D, Lang RE, Maser-Gluth C, Steinberg H, Veelken R (1988) Substance P induces a cardiovascular defense reaction in the rat: pharmacological characterization. Circ Res 63:812–820

Van Amerigen MV, Mancini C, Farvolden P, Oakman J (2002) Drugs in development for social anxiety disorder: more to social anxiety than meets the SSRI. Expert Opin Investig Drugs 9:2215–2231

Van der Hart M, Czéh B, de Biurrun G, Michaelis T, Watanabe T, Natt O, Frahm J, Fuchs E (2002) Substance P (NK_1) receptor antagonist and clomipramine prevent stress-induced alterations in cerebral metabolites, cytogenesis in the dentate gyrus and hippocampal volume. Mol Psychiat 7:933–941

Van Kampen M, Kramer M, Hiemke C, Flugge G, Fuchs E (2002) The chronic psychosocial stress paradigm in male tree shrews: evaluation of a novel animal model for depressive disorders. Stress 5:37–46

Varty GB, Cohen-Williams ME, Morgan CA, Pylak U, Duffy RA, Lachowicz JE, Carey GJ, Coffin VL (2002a) The gerbil elevated plus-maze II. Anxiolytic-like effects of selective neurokinin NK_1 receptor antagonists. Neuropsychopharmacology 27:371–379

Varty GB, Morgan CA, Cohen-Williams ME, Coffin VL, Carey GJ (2002b) The gerbil elevated plus-maze I: behavioral characterization and pharmacological validation. Neuropsychopharmacology 27:357–370

Vaupel R, Jarry H, Schlomer HT, Wuttke W (1988) Differential response of substance P-containing subtypes of adrenomedullary cells to different stressors. Endocrinology 123:2140–2145

Von Euler US, Gaddum JH (1931) An unidentified depressor substance in ceratin tissue extracts. J Physiol 72:74–87

Wahlestedt C (1998) Reward for persistence in substance P research. Science 281:1624–1625

Walsh DM, Stratton SC, Harvey FJ, Beresford IJM, Hagan RM (1995) The anxiolytic-like activity of GR159897, a non-peptide NK_2 receptor antagonist, in rodent and primate models of anxiety. Psychopharmacology 121:186–191

Weiss DW, Hirt R, Tarcic N, Berzon Y, Ben-Zur H, Breznitz S, Glaser B, Grover NB, Baras M, O'Dorisio TM (1996) Studies in psychoneuroimmunology: psychological, immunological, and neuroendocrinological parameters in Israeli civilians during and after a period of Scud missile attacks. Behav Med 22:5–14

The Role of Tachykinins and the Tachykinin NK₁ Receptor in Nausea and Emesis

P. L. R. Andrews[1] · J. A. Rudd[2]

[1] Department of Basic Medical Sciences, St George's Hospital Medical School,
Cranmer Terrace, London, SW17 0RE, UK
e-mail: pandrews@sghms.ac.uk
[2] Department of Pharmacology, Faculty of Medicine,
The Chinese University of Hong Kong, Shatin N.T., Hong Kong
e-mail: jar@cuhk.edu.hk

1	Introduction	361
2	The Biological and Clinical Problem of Emesis.	362
2.1	An Introduction to the Physiology of Nausea and Vomiting	363
2.1.1	Motor Components	363
2.1.2	Coordination of the Motor Components.	365
2.1.3	Triggering of the Emetic Reflex.	366
2.1.4	Nausea	369
3	Pre-clinical Evidence for the Involvement of Tachykinins in Emesis	370
3.1	Presence of Tachykinins in the Emetic Pathway.	370
3.1.1	Area Postrema	371
3.1.2	Brain Stem Nuclei	372
3.1.3	Vagal Afferents	373
3.1.4	Gastrointestinal Tract	373
3.2	Presence of Tachykinin Receptors in the Emetic Pathway	374
3.3	Induction of Emesis by Tachykinin Receptor Agonists	375
3.4	Release of Tachykinins by Emetic Stimuli	377
4	Implication of Tachykinins in the Emetic Pathway.	378
4.1	Pre-clinical Models for the Study of Emesis	378
4.2	Species Differences in the Pharmacology of NK₁ Receptors and the Potential Complicating Role of Receptor Conformers.	381
4.3	Studies of the Anti-emetic Effects of NK₁ Receptor Antagonists in Different Animal Models.	383
4.3.1	Studies in the Ferret.	383
4.3.2	Studies in the Dog	392
4.3.3	Studies in the Cat	394
4.3.4	Studies in *Suncus murinus* (the House Musk Shrew)	395
4.3.5	Studies in the Piglet	400
4.3.6	Studies in the Pigeon	401
4.3.7	Studies in the Rat	402
4.3.8	Summary of Pre-clinical Studies	403

5	Clinical Evidence for the Involvement of Tachykinins in Nausea and Vomiting.	405
5.1	Chemotherapy-Induced Nausea and Vomiting	406
5.1.1	CP122721 (Pfizer)	406
5.1.2	CJ11974 (Pfizer)	407
5.1.3	GR205171 (Glaxo-Wellcome)	408
5.1.4	L754030 and L758298 (Merck)	408
5.2	Post-operative Nausea and Vomiting.	413
5.3	Motion-Induced Emesis.	414
5.4	Summary of the Clinical Studies of the Anti-emetic Effects of NK_1 Receptor Antagonists	415
6	Site(s) and Mechanism(s) of the Anti-emetic Action of NK_1 Receptor Antagonists.	416
6.1	Central vs. Peripheral Site of Action	416
6.1.1	Spectrum of Anti-emetic Effects	417
6.1.2	Central Penetration	417
6.1.3	Central Administration	419
6.1.4	Precise Location	419
6.1.5	Mechanistic Considerations.	424
7	Closing Comments	426
References		427

Abstract Nausea and vomiting are both components of the body's defensive system to protect against the effects of accidentally ingested toxins. Whilst these responses have survival value in the wild, they can also be induced by diseases and disease treatments with one of the most unpleasant examples being the treatment of cancer using cytotoxic drugs and radiation. Understanding the mechanisms by which this occurs has been a major impetus to the identification of novel anti-emetic agents. The recent licensing of an NK_1 receptor antagonist for the treatment of chemotherapy-induced emesis provides the first example of a drug acting to block the effects of substance P. Whilst the blockade of emesis by selective NK_1 receptor antagonists provides the most powerful evidence implicating substance P in emesis there is a considerable body of supporting evidence including: presence of substance P (usually by immunohistochemistry) in relevant sites (e.g. vagal afferents, nucleus tractus solitarius, gastrointestinal mucosa); presence of NK_1 receptors in relevant sites (e.g. nucleus tractus solitarius); induction of emesis by administration of NK_1 receptor agonists. Pre-clinical studies in a variety of species revealed the broad-spectrum anti-emetic effects of NK_1 receptor antagonists against stimuli including the anti-cancer agent cisplatin (acute and delayed phases), radiation, opioids, copper sulphate, apomorphine, motion and electrical stimulation of abdominal vagal afferents. Species differences in response to NK_1 receptor antagonists and species-specific isoforms of the receptor are discussed and the potential implications for transfer of data from these animal models to humans reviewed. The spectrum of anti-emetic effects against stimuli acting via both peripheral and centrally acting emetic stimuli, a requirement for brain penetration and blockade of emesis by microinjection of antagonists into the brain stem all support a central site of action with the nucleus tractus solitarius and the vicinity of the Bötzinger complex being the favoured locations although definitive studies are awaited. Evidence for a contribution from a peripheral site in the delayed phase of cytotoxic drug-induced emesis is reviewed. The unique pre-clinical profile and especially the observation that NK_1 receptor antagonists could block both the acute and delayed phase of cisplatin-induced emesis prompted clinical trials of a number of agents [CJ11974 (ezlopitant), GR205171 (vofopitant), MK869/L754030 (aprepitant)] in patients undergoing chemotherapy. These studies and others in motion and post-operative nausea and vomiting are reviewed in detail. In general the trials in chemotherapy show NK_1 receptor antagonists have demonstrable efficacy against acute (first 24 h after therapy) emesis when given alone and enhance the efficacy of 5-hydroxytryptamine$_3$ receptor antagonists and dexamethasone when given in combination. Of particular clinical significance is the efficacy of NK_1 receptor antagonists given in combination with dexamethasone to reduce emesis in the delayed phase (days 2–5) as this phase of emesis is poorly controlled using current treatments. Efficacy against nausea has been reported but to date the effects appear less clear than against emesis and further studies are required. The availability of NK_1 receptor antagonists in the clinic will provide a useful tool to further investigate the involvement of NK_1 receptors in emesis and to explore the roles of central and peripheral substance P in health and disease.

Keywords Substance P · Tachykinin NK_1 receptor · Nausea · Vomiting · Emesis · Acute chemotherapy-induced emesis · Delayed chemotherapy-induced emesis · Post-operative nausea and vomiting (PONV) · Motion sickness · Ferret · Dog · Cat · *Suncus murinus* (house musk shrew) · Piglet · Pigeon · Rat · Aprepitant · Vofopitant · Ezlopitant · MK-869 · CP122721 · CP99994 · CJ11974 · GR205171 · L758298 · L754030 · Dexamethasone · Area postrema · Nucleus tractus solitarius · Vagus · Brain stem · 5-Hydroxytryptamine$_3$ receptor antagonists · Ondansetron · Granisetron

1
Introduction

Three publications in 1991 set the stage for the identification of the potential anti-emetic effects of NK_1 receptor antagonists: the cloning of the gene for the human NK_1 receptor (Gerard et al. 1991); the publication of the pharmacological properties of two potent, selective, non-peptide NK_1 receptor antagonists by scientists at Pfizer (CP96345, Snider et al. 1991) and Rhone-Poulenc Rorer (RP67580, Garret et al. 1991). Publications disclosing a number of other selective antagonists followed rapidly [e.g. CP99994, McLean et al. 1993; PD154075, Boyle et al. 1994; LY303870, Gitter et al. 1995; GR203040, Ward et al. 1995; GR205171, Gardner et al. 1996; CP-122721, McLean et al. 1996; MK869 (L754030), Hale et al. 1998; HSP117, Saito et al. 1998] providing a wide range of tools with which to investigate the therapeutic potential of this class of agent.

The discovery of the anti-emetic effects of the NK_1 receptor antagonists was one of the first clinical targets identified for this class of agent with the first pre-clinical studies being published in 1994 (see Sect. 5). This chapter will focus on the pre-clinical evidence for the involvement of substance P and NK_1 receptors in emesis and review the ever growing number of clinical studies demonstrating anti-emetic activity in humans and which led to the approval (April 2003) by the Food and Drug Administration (FDA) of a member of this class of agent (aprepitant, Emend™) for the treatment of acute and delayed chemotherapy-induced nausea and vomiting. To provide a framework for these discussions, the first part of the chapter will review a number of aspects relating to the current treatment of emesis and the physiology and pharmacology of the emetic reflex.

2
The Biological and Clinical Problem of Emesis

Although nausea (unpleasant sensation associated with the urge to vomit) and vomiting (forceful oral expulsion of gastric contents) may be most frequently encountered in a clinical context (Sleisenger 1993; Koch 1995; Quigley et al. 2001) as a symptom of disease (e.g. raised intracranial pressure, migraine, labyrinthitis, renal failure), as a symptom of viral (e.g. Norwalk virus) or bacterial infection (e.g. *Staphylococcal enterotoxin* B, *Salmonella typhimurium.*), as a 'side effect' of drug therapy [e.g. anti-tumour chemotherapy, morphine, oral L-DOPA, oral contraceptives, some antibiotics (e.g. erythromycin)], following accidental ingestion of plant (e.g. nicotine, emetine) or animal (e.g. batrachotoxin, ciguatoxins) toxins, or as sequelae to treatment [e.g. radiotherapy, post-operative nausea and vomiting (PONV)] it is important to remember that they are components of a biologically important, hierarchically organized system which, it is presumed, evolved to protect the body against the accidental ingestion of toxins.

When encountered in the natural environment nausea and vomiting are viewed as being 'appropriate' and relevant to survival but when induced by disease and as a side effect of therapy they are viewed as being 'inappropriate', particularly as the vomiting does not help rid the body of the inducing stimulus because it is in the circulation and not the gut lumen as would be the case initially with an ingested toxin. In addition, the nausea induced which would normally help the animal avoid the stimulus in the future, contributes to a learned aversion which manifests itself as anticipatory nausea and vomiting in the clinical setting, in which it is disadvantageous to avoid future courses of therapy, particularly in the case of treatment (radiotherapy, chemotherapy) for cancer. An understanding of the basic physiology and pharmacology of nausea and vomiting is essential to identification of novel anti-emetic strategies and this aspect is outlined below as a background for the detailed analysis of the role of substance P and NK_1 receptors.

In a provocative review Lembeck (1985) proposed a widespread neural, endocrine and immunological 'network of defence' involving tachykinins including substance P. Of the many defensive reflexes discussed emesis is notable by its absence.

2.1
An Introduction to the Physiology of Nausea and Vomiting

2.1.1
Motor Components

Irrespective of how the vomiting reflex is triggered in mammals, the motor components are the same and in this respect the reflex may be viewed as an example of a motor programme.

Pre-ejection or Prodromal Phase. Before expulsion of upper gut contents a considerable number of motor changes mediated by the autonomic nervous system have taken place, particularly in the gastrointestinal tract (Fig. 1) (for references see Andrews 1999). These include: (a) relaxation of the proximal stomach to confine the food together with any toxin to the stomach from where it can be ejected. This is mediated predominantly by vagal efferent activation of myenteric non-adrenergic, non-cholinergic inhibitory nerves, probably using nitric ox-

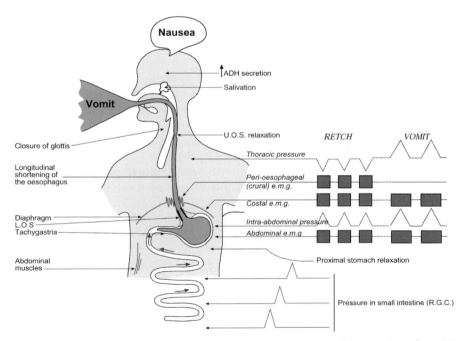

Fig. 1 A summary of some of the visceral and somatic motor components of the vomiting reflex. *UOS*, upper oesophageal sphincter; *LOS*, lower oesophageal sphincter; *RGC*, retrograde giant contraction

ide and vasoactive intestinal peptide (VIP) as transmitters. The vagal efferents supplying the gut arise from the dorsal motor vagal nucleus (DMVN) in the brain stem; a giant retrograde contraction originating in the small intestine is proposed to return any previously emptied and contaminated gastric contents to the stomach. It has also been proposed that it may propel alkaline pancreatic and intestinal fluid into the stomach to buffer gastric contents before ejection. This contraction is vagally mediated, mainly involving activation of atropine-sensitive cholinergic nerves. When there is expulsion these gastrointestinal changes always precede it, but they may also occur in the absence of vomiting. It is important to appreciate that blockade of these gastrointestinal events does not prevent retching and vomiting (Lang 1990).

Cutaneous vasoconstriction and sweating also occur in the prodromal phase giving rise to the characteristic pallor and cold sweaty appearance of nauseated individuals. This is mediated by the sympathetic nerves as is the tachycardia which accompanies nausea. Salivation may also occur due to activation of the parasympathetic supply to the salivary glands. Salivary secretion is of particular relevance to the present chapter as substance P is a potent sialogogue. Increases in the plasma levels of adrenaline, adrenocorticotropic hormone (ACTH), growth hormone, prolactin and vasopressin (ADH) have been reported during the prodromal phase. Some are likely to be a reflection of general stress (e.g. adrenaline), whereas others (e.g. vasopressin) may be specific to nausea (see Sect. 2.1.4).

Ejection Phase: Retching and Vomiting. The forceful oral expulsion of gastric contents (vomiting) is brought about by the co-ordinated rhythmic contraction of the diaphragm (innervated by the phrenic nerve originating in the spinal cord), external intercostals (inspiratory) and abdominal muscles (rectus abdominus, external oblique) compressing the stomach and ejecting the contents through the mouth via the oesophagus (Fig. 1) (for references see Bianchi et al. 1992). The propulsive pressure exceeds 100 mmHg. There is also some evidence that during expulsion there is longitudinal shortening of the oesophagus and an oesophago-pharyngeal retrograde contraction, regulated by the vagal and glossopharyngeal efferents, which could contribute to the oral expulsion of gastric contents (Lang 1990). Retching, which usually precedes vomiting and in contrast with vomiting occurs as multiple events, differs in that there is no oral ejection of gastric contents. This is thought to be because during retching the entire diaphragm contracts rhythmically under the influence of the phrenic nerve, whereas during vomiting the bulk of the diaphragm contracts but the peri-oesophageal region (crural fibres) is inactive to permit ejection of gastric contents. The lower oesophageal sphincter must also relax to facilitate expulsion of gastric contents, and studies in the ferret have examined the effect of NK_1 receptor antagonists on its control (Smid et al. 1998).

The function of retching is not known, although it is suggested that it helps to overcome the anti-reflux barrier present in the region of the gastro-oesophageal junction. The factors which regulate the number of retches in a burst of retching or which determine whether a vomit follows retching are not known, although studies in the ferret showed an inverse relationship between gastric volume and the number of retches prior to a vomit (Andrews et al. 1990). Blood gas levels have also been implicated in this transition (Fukuda and Koga 1995).

2.1.2
Coordination of the Motor Components

The motor systems that comprise the vomiting reflex are almost all employed in non-emetic reflexes and hence it is the magnitude, combination, and sequence of activation of these outputs that makes the reflex unique. The brain stem nuclei involved in the control of the autonomic and somatic motor components of emesis have largely been identified: for example, the DMVN and nucleus ambiguus provide the vagal output to the gut and heart; the pre-sympathetic neurons in the ventrolateral medulla regulate the drive to the pre-ganglionic sympathetic neurons in the spinal cord which supply the cutaneous vasculature; the dorsal respiratory group of neurons located in the ventral nucleus tractus solitarius (NTS) and the ventral respiratory group of neurons (e.g. nucleus retroambigualis, Bötzinger complex) provides the drive to the diaphragm (via the phrenic nerve nucleus in the spinal cord) and intercostal muscles involved in the somatic motor components of emesis. However, it is still not known exactly how the various components are coordinated. In particular, there must be interaction between the visceral and somatic motor nuclei to ensure that the gastrointestinal and other pre-ejection changes precede contraction of the diaphragm and abdominal muscles. Whether there is a single nucleus that corresponds to the old concept of a 'vomiting centre' is still a matter of discussion, but the parvicellular reticular formation, the Bötzinger complex (part of the ventral respiratory group of neurons) including adjacent regions and the NTS have all been proposed to fulfil this function to varying degrees (Miller and Grelot 1996; Yates and Miller 1998). The NTS is of particular interest because it is the major integrative nucleus for visceral information receiving, amongst other inputs, the abdominal vagal afferents capable of inducing emesis and inputs from the area postrema (AP) and vestibular nuclei. It also sends major projections to more rostral parts of the brain including the parabrachial nucleus, hypothalamus and amygdala. In the rat it is estimated that the NTS contains approximately 42,000 cells and has approximately 1,000,000 synapses (see Andresen and Kunze 1994). These central co-ordinating sites in the brain stem would appear to be an attractive site at which to target an anti-emetic but their involvement in life-critical functions such as regulation of respiration and blood pressure makes this problematic unless the emetic reflex has a unique neurotransmitter system or has a transmitter which makes a critical link in this reflex but not others.

2.1.3
Triggering of the Emetic Reflex

How the diverse range of stimuli that can evoke vomiting activate the central pathways outlined above is not known, but three major pathways have been identified and they provide a framework within which some emetic stimuli may be discussed (Figs. 2, 3). 'Higher' central nervous system inputs including those from the limbic system can also induce nausea and vomiting but these are poorly defined and hence are not discussed further but they may have an important role in anticipatory nausea and vomiting.

Gastrointestinal Afferents. Vomiting can be evoked by over-distension of the upper gut (e.g. overeating or motility disorder) and by the presence of chemical irritants (e.g. hypertonic solutions or copper sulphate) or toxins (e.g. plant alkaloids such as emetine, a constituent of the clinical emetic ipecacuanha, or bacterial toxins) in food. These stimuli activate mechanoreceptors or mucosal chemoreceptors with afferents principally in the vagus nerve, which projects to

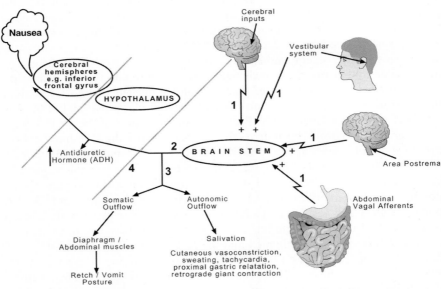

Fig. 2 A diagrammatic representation of the pathways for the induction of nausea and vomiting. *Numbers* refer to theoretical sites at which anti-emetics could act to have various effects: *1*, block at a peripheral or central site of a single pathway resulting in an anti-emetic (nausea and vomiting) effect only against stimuli acting via that pathway; *2*, block at a critical central point (e.g. NTS) prior to the site at which the emetic signal projects to rostral structures for the induction of nausea and to brainstem sites for activation of the visceral and somatic motor components of vomiting; *3*, block of the motor pathways resulting in selective blockade of the visceral and somatic components of emesis; *4*, block of the pathway leading from the brainstem to the more rostral parts of the brain for genesis of nausea. Block here would result in an effect on nausea but not vomiting

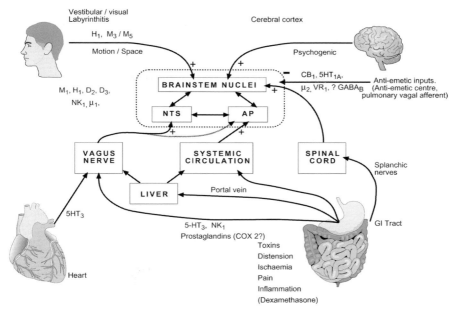

Fig. 3 Summary of the major transmitter systems, receptors and pathways involved in emesis. *AP*, area postrema; *CB₁*, cannabinoid 1 receptor; *COX 2*, cyclooxygenase 2; *D₂/D₃*, dopamine 2/3 receptors; *H₁*, histamine 1 receptor; *GABA_B*, gamma-aminobutyric acid B receptors; *M₁/M₃/M₅*, muscarinic 1/3/5 acetylcholine receptors; *NTS*, nucleus tractus solitarius; μ_1/μ_2, opioid μ_1/μ_2 receptors; *VR1*, vanilloid receptor 1. (Redrawn and modified from Sanger and Andrews 2001)

the medulla (NTS and AP), to trigger the emetic motor pathways. Biologically, this is probably the most important route for activation of emesis.

Studies in several species (dog, monkey, ferret and *Suncus murinus*) have implicated mucosal vagal afferents as the predominant pathway in the acute phase (first 24 h) of vomiting induced by systemic chemotherapy agents (e.g. cisplatin) used in the treatment of cancer and whole-body radiation (Andrews and Davis 1995). It is proposed that these stimuli induce the formation of free radicals in the enterochromaffin (EC) cells of the upper gastrointestinal tract, which triggers a calcium-dependent release of 5-hydroxytryptamine (5-HT) and probably other neuroactive substances (e.g. substance P, see Sect. 3.1.4), which activates 5-HT₃ receptors (a ligand-gated ion channel) located on the vagal afferents terminating in close proximity (Reynolds et al. 1995). It is this site at which 5-HT₃ receptor antagonists (granisetron, ondansetron, tropisetron, dolasetron, palonosetron) are thought to have their main anti-emetic effects against acute chemotherapy-induced nausea and vomiting and perhaps PONV. Vagal afferents from the heart can also evoke nausea and vomiting and are responsible for these symptoms during infarction.

Area Postrema. The AP is a circumventricular organ located at the caudal extremity of the fourth ventricle outside the blood–brain and cerebrospinal fluid–brain barriers and is therefore ideally suited to its proposed function as a chemoreceptor. Ablation studies implicated the AP in the mediation of emesis by a wide variety of systemic agents (e.g. cytotoxic drugs, the dopamine D_2 agonist apomorphine, the opioid receptor agonist morphine) and lead to it being called 'the chemoreceptor trigger zone for emesis'. Although there is no dispute that suitable activation can trigger emesis and perhaps influence the sensitivity of other emetic inputs, it is probably more accurate to regard the AP as having a wider chemoreceptive role with an involvement in conditioned taste aversion, salt balance and food intake. Because of the close proximity of the AP to the central emetic circuitry and sites of termination of abdominal vagal afferents (particularly the subnucleus gelatinosus of the NTS), collateral damage may be difficult to avoid and thus could confuse the interpretation of lesion studies. In addition, morphological studies have demonstrated that dendrites from NTS neurons project into the AP and could be stimulated by systemic agents accessing the AP. Also, there is some evidence from rats that capillaries from the AP may make vascular links with the dorsal region of the commissural subnucleus of the NTS (Roth and Yamamoto 1968; Gross et al. 1990) and that the NTS itself has an unexpectedly high capillary permeability. These studies taken together with the observation that substance P is able to cross the blood–brain barrier raise the possibility that substance P present in the circulation (Sect. 3.1.4) could act upon the AP or perhaps directly upon the NTS.

Teleologically, it may be argued that a toxin detector in the central nervous system would be of little use because after the toxin has been absorbed it cannot be expelled by vomiting. Hence, the emetic role of the AP is probably a reflection of its general chemoreceptive role and connectivity, a role supported by its presence in non-vomiting species (e.g. rat).

Vestibular System. Lesion studies of the vestibular system (otolith organs and semi-circular canal) demonstrated that it is critical for the induction of motion-induced emesis. These and other studies lead to the idea that over-stimulation was the major trigger, but this did not account for all the experimental observations. Current hypotheses are based on the idea of sensory conflict or rearrangement: some view this as a disparity between different sets of incoming information (e.g. from vision and the vestibular system) 'a sensory conflict', whereas others tend more toward the view that the conflict lies not between modalities but between what the body expects to happen in a given situation and what actually occurs: 'a sensory–motor' or 'efference–copy conflict' (Crampton 1990). Psychophysical studies support both hypotheses. The exact pathways by which the medullary emetic circuitry is activated are not known but both the lateral medullary reticular formation (lateral tegmental field) and the NTS receive vestibular inputs, and individual neurons in the NTS have been shown to receive vestibular and gastrointestinal inputs (Yates et al. 1998).

The main neurotransmitters involved in motion-induced emesis are acetylcholine and histamine as indicated by the efficacy of histamine H_1 (e.g. cinnarazine) and muscarinic, particularly M_3/M_5 (darifenacin but also the less selective agent scopolamine), receptor antagonists (Yates et al. 1998).

Sensitivity to motion sickness is a predictor of the emetic response to other emetic stimuli such as pregnancy sickness, PONV and anti-cancer chemotherapy-induced emesis. This suggests that the vestibular system may have a 'modulatory' influence on the brain stem emetic pathways setting their sensitivity and hence agents targeted at this pathway may have more widespread anti-emetic effects than expected. Based upon studies of different racial sensitivity to motion stimuli (orientals>westerners) there is some evidence for genetic determinants of sensitivity (see Lien et al. 2003).

2.1.4
Nausea

The term nausea is derived from the Greek word for 'ship' and well illustrates man's early experience of sea travel. Although we know that activation of the pathways described above, usually at a lower intensity than that required to induce vomiting, can induce nausea, the precise pathways in the central nervous system by which this sensation is generated are not known. If nausea is due to 'less intense' activation of the emetic pathways then one would predict that it should be easier to block than vomiting but clinically this is not the case. Magnetic source imaging has implicated the inferior frontal gyrus as a site involved in the processing of nauseogenic stimuli (Miller et al. 1996). Retching and vomiting are reflex motor responses to activation of pathways converging in the brain stem and can be elicited in decerebrate animals (i.e. lacking structures rostral to the colliculi) but nausea is a sensory experience and relies on the projection of information from the brain stem to 'higher' regions of the brain. The main reason for the relative paucity of knowledge of the mechanisms underlying nausea is that whereas in animals retching and vomiting are obviously the same as in man, in animals we can only study what we conclude are the 'behavioural equivalents' of nausea (e.g. conditioned taste aversion, pica and perhaps some behaviours seen prior to emesis) and not the actual sensation experience (pain studies). In the pre-ejection phase in man, increases in ADH may be many times maximal anti-diuretic levels and this has led to the suggestion that it may be a useful index of nausea, or may even be involved in its genesis. As nausea can be induced in patients with diabetes insipidus it is clear that ADH is not obligatory for its genesis (Nussey et al. 1988). The involvement of disturbed gastrointestinal motility (e.g. tachyantria) in the genesis of nausea from all causes is not resolved (Koch 1995), but in cases in which there is gastric stasis resulting in distension, it is possible that nausea (and vomiting) is caused by visceral afferent activation. Understanding nausea is still a major challenge but, fortunately, most drugs alleviating vomiting also mitigate nausea, although usually at a reduced efficacy. Because nausea and vomiting can be evoked by activation of the same inputs which converge in the brain stem and from which the motor (retching, vomiting and autonomic nervous system-mediated changes) and sensory (nausea) outputs diverge it is theoretically possible to have agents which could block both nausea and vomiting if they acted at the site of the emetic inputs (e.g. AP, vagal afferents, vestibular system) or at the central integrative site (e.g. NTS) before the pathways diverge. It is also of course possible to have agents which could affect either nausea or vomiting if they acted after the divergence point and the two pathways used different transmitter systems (Fig. 2).

Biologically, the function of nausea appears to be the induction of a learned aversion to the stimulus responsible (e.g. contaminated food) and whilst this is a useful survival strategy when the cause is an ingested toxin, if induced by a clinical intervention it may lead patients to decline further treatment. In the case of anti-cancer chemotherapy when multiple cycles of therapy are required, this manifests as 'anticipatory nausea and vomiting' and the probability of its occurrence is a direct function of the patient's emetic experience of each cycle of therapy (Morrow et al. 1998). This also emphasizes the importance of optimal anti-emetic therapy on each cycle of chemotherapy.

3
Pre-clinical Evidence for the Involvement of Tachykinins in Emesis

The major evidence implicating tachykinins in emesis comes from the pre-clinical studies of the anti-emetic effects of selective NK_1 receptor antagonists but there is also a considerable body of supporting evidence in the (older) literature. This is included below as it assists in understanding the site at which the antagonists act and also in the interpretation of the clinical studies. The focus of the sections below is on the brain stem as this is the site at which tachykinins are most implicated although an involvement of tachykinins and their receptors in the more rostral parts of the brain cannot be excluded, particularly in the genesis of nausea (see Rupniak and Kramer 1999 for a review of other central therapeutic targets for NK_1 receptor antagonists). As other chapters in this book deal with tachykinins in the gastrointestinal tract, only selected aspects directly related to emesis are discussed here.

3.1
Presence of Tachykinins in the Emetic Pathway

The major anatomical substrates of the emetic reflex are the AP, the NTS and other brain stem nuclei implicated in co-ordinating the reflex and vagal afferents, particularly those from the gastrointestinal tract. Each of these will be examined separately. It must be noted that whilst many of the early studies report the presence of substance P in extracts, chromatographically pure extracts were prepared only from 1953 onwards (Pernow 1953). Therefore, there is a possibility that some extracts also contain unknown amounts of other tachykinins. Similar arguments apply to some of the immunohistochemical studies using antibodies that may react with tachykinins (e.g. NKA) other than substance P.

3.1.1
Area Postrema

The brain was identified in early studies as a rich source of substance P (von Euler and Gaddum 1931) and studies of the dog, pig and cattle brain (Amin et al. 1954; Paasonen and Vogt 1956) reported some of the highest levels in the AP (100–480 units/g vs. 40–50 units/g in the floor of the fourth ventricle of the dog brain). Interestingly, in man Zettler and Schlosser (1955) found little substance P in the AP but high levels in the ala cinerea although a more recent study by Cooper et al. (1981) measured high levels of substance P (114±47 pmol/g wet tissue) in the human AP, although these levels were not higher than in other medullary structures. Substance P has also been found in relatively high levels (167±36 pmol/g wet tissue) in the AP of the rat, a species lacking a vomiting reflex (Douglas et al. 1982). This indicates that substance P in the AP is not exclusively involved in vomiting, although of course it could be involved in the genesis of nausea as, in response to emetic stimuli, rodents exhibit behaviours which are argued to be 'equivalents' of nausea (e.g. conditioned taste aversion, pica).

Borison and Wang pioneered the 'modern' understanding of the emetic pathways with a seminal paper in 1953 identifying the AP as the 'chemoreceptor trigger zone' for vomiting. The potential relevance of this finding was drawn attention to by Amin et al. in their 1954 paper, and in the discussion they comment that "the AP only contains active substances by virtue of its chemoreceptive properties... One of the functions of some parts of this tissue may be to act as a chemoreceptor for substances in the blood stream to convert messages received in this way into nervous impulses".

Immunohistochemical studies have been used extensively to map the distribution of substance P-containing neurons and nerve fibres in the dorsal brain stem. In the rat (Armstrong et al. 1981, 1982; Pickel and Armstrong 1984) substance P (like-immunoreactivity) was localized to varicose processes in the AP rather than in neuronal cell bodies, and electron microscopic studies demonstrated substance P-containing synapses terminating on the dendrites of intrinsic AP neurons. The apparent absence of substance P-containing cell bodies in the AP was confirmed in a study in the cat (Newton et al. 1985) that also found a network of substance P-containing fibres distributed to the ventral and lateral borders of the AP. In both the rat and cat the origin of the substance P containing axons was considered to be from outside the AP (e.g. NTS, vagal afferents) and this is discussed further below in the section on the brain stem nuclei.

The ferret has been used extensively as a model species to investigate the anti-emetic effects of NK_1 receptor antagonists and hence an extensive immunohistochemical study of the ferret dorsal brain stem is of particular relevance (Boissonade et al. 1996). A high density of immunoreactive fibres is present in the lateral but not medial borders of the AP and as in the rat and cat neuronal perikarya are not stained.

3.1.2
Brain Stem Nuclei

The main nucleus of interest is the NTS which is the major integrative nucleus in the dorsal brain stem for visceral afferent information and is critically involved in co-ordinating brain stem autonomic reflexes and respiration. The NTS has reciprocal links with other brain stem structures including the AP and parabrachial nucleus and is the major site for termination of vagal afferents. Hence, immunoreactive fibres seen in the NTS may represent projections from other sites rather than intrinsic neurons. Substance P-like immunoreactivity has been reported in both nerve fibres and neurons in the NTS in rat (Armstrong et al. 1981; Maley and Elde 1982; Kalia et al. 1984; Yamazoe et al. 1984; Helke et al. 1984) guinea pig, cat and man (Triepel et al. 1983; Maley 1985; Chigr et al. 1991; McRitchie and Tork 1994), and positive staining varicose fibres and terminals described in ferret, rhesus monkey and man (Maley and Elde 1982; Maley et al. 1987; Baude et al. 1989; Rikard Bell et al. 1990; Boissonade et al. 1996). Overall within the NTS the most prominent staining is observed in the medial (particularly the dorsal region along the border of the AP) and commissural subnuclei of the NTS. Dense staining is also reported in the subnucleus gelatinosus of the NTS, an area immediately subjacent to the AP. The presence of substance P-like immunoreactivity in the medial and subnucleus gelatinosus regions of the NTS is of particular interest as these regions are sites of termination of abdominal vagal afferents and are involved in regulation of gastrointestinal tract function. In view of this it is likely that some of the immunoreactive fibres in the NTS are the terminals of vagal afferents and this is supported by the study by Baude et al. (1989) in the cat showing a reduction in substance P like-immunoreactivity in the subnucleus gelatinosus (but not the medial subnucleus) by nodose ganglion removal (see below).

The DMVN is the major site of origin of vagal efferents supplying the viscera including the entire digestive tract (Travagli et al. 2001), and substance P-like immunoreactive fibres and/or neurons have been reported in several species including rat, cat, ferret and man (Del Fiacco et al. 1983; Boissonade et al. 1996; Baude et al. 1989; Ljungdahl et al. 1978; Maley and Elde 1981). The presence of substance P-like immunoreactivity in nerve fibres in the DMVN probably represents afferent projections to this structure involved in vago-vagal reflexes or descending modulation by other brain stem (e.g. NTS, parabrachial nucleus) or ventral forebrain nuclei (e.g. paraventricular nucleus, central nucleus of the amygdala, bed nucleus of the stria terminalis; Rogers and Herman 1992) whilst the presence of immunoreactive cell bodies indicates that substance P may be a transmitter in some pre-ganglionic vagal efferents but as yet there is little supportive functional evidence. A role for substance P as a transmitter to DMVN neurons projecting to the stomach is supported by studies in the rat in which substance P injected into either the dorsomedial NTS or the DMVN inhibited gastric motility, and in the case of the latter this was mediated by central NK_1 receptors (Spencer and Talman 1986; Kowicki and Hornby 2000).

Substance P-like immunoreactivity has also been reported in other nuclei implicated in emesis including the rostral nucleus ambiguus (a putative component of the 'central pattern generator' for emesis), parvocellular reticular nucleus and medial vestibular nucleus (Helke et al. 1984).

3.1.3
Vagal Afferents

Substance P was extracted from the vagus by von Euler (1936) and substance P-like immunoreactivity has been demonstrated in nerve fibres in the vagus and/or the nodose ganglion (the site of the cell bodies of vagal afferent fibres) in a variety of species including rat, guinea pig, cat and human (Lundberg et al. 1978, 1979; Lindh et al. 1983; Helke and Hill 1988; Baude et al. 1989). Additional studies in selected species demonstrated that substance P was synthesized in the nodose ganglion, axonally transported in the vagus and could be released from the ganglion by depolarizing stimuli (Gamse et al. 1979; Brimijon et al. 1980; MacLean and Lewis 1984a, 1984b; MacLean et al. 1990). Nausea and vomiting can only be induced by activation of vagal afferents from the heart and digestive tract (mainly stomach and upper small intestine). Whilst substance P immunoreactivity has been found in abdominal vagal branches in humans (Lundberg et al. 1979) and this is likely to be in the afferent fibres, the extent to which substance P is a central neurotransmitter in vagal afferents capable of inducing emesis is not known. The peripherally exported substance P plays an important role in 'axon' reflexes in the gut wall (e.g. increased blood flow, vascular permeability, motility; Holzer 1998) and substance P arising from axon collaterals of visceral afferents may be involved in the inflammatory-like responses evoked by some cytotoxic drugs (Alfieri and Gardner 1997).

3.1.4
Gastrointestinal Tract

Tachykinins, particularly substance P, are distributed extensively in the enteric nervous system of the gastrointestinal tract (see chapter by P. Holzer, this volume) and could be indirectly involved in the genesis of emesis if these pathways were involved in the genesis of abnormal motility patterns (e.g. delayed gastric emptying) associated with nausea and vomiting. Of particular relevance to emesis is the presence of substance P-like immunoreactivity in the EC cells of the intestinal mucosa. The mucosa of the dog gastrointestinal tract (particularly the duodenum) was recognized as a rich source of substance P in extraction studies carried out in the 1950s (Douglas et al. 1951; Pernow 1951). Immunohistochemical studies have shown the presence of substance P-like immunoreactivity in the intestinal mucosa of mouse, dog, marmoset and human (Heitz et al. 1976;

Alumets et al. 1977; Sundler et al. 1977; Keast et al. 1985). A population of EC cells has been shown to express substance P as well as 5-HT and has long been known to contain these mediators in high levels (Pearse and Polak 1975; Heitz et al. 1976; Alumets et al. 1977; Sundler et al. 1977). The latter is of particular relevance to emesis as the release of 5-HT from the EC cells has been shown to play an important part in the mechanisms by which cytotoxic anti-cancer therapies induce acute emesis although involvement of other locally released neuroactive agents including substance P have long been proposed to be involved (Andrews et al. 1988). The intestine is the major source of substance P in the circulation with the portal venous blood having a level four times that of the peripheral blood (Gamse et al. 1978). Thus, intestinal and hepatic afferents could be exposed to relatively high concentrations of substance P in the same way that they are exposed to 5-HT, providing a mechanism by which peripherally acting emetic agents could act.

3.2
Presence of Tachykinin Receptors in the Emetic Pathway

Evidence for the presence of tachykinin receptors in the emetic pathway comes from two main sources, ligand binding and electrophysiological studies (see also Sect. 4.2). Using either $[^{125}I]$-labelled or $[^{3}H]$-substance P (Bolton-Hunter substance P) high densities of binding sites have been identified in brain stem slices from rat and ferret (Helke et al. 1984; Maubach et al. 1995; Watson et al. 1995a; Ariumi et al. 2000). High levels of specific binding are found in the NTS and DMVN in both the ferret and rat. In the ferret, although binding is seen in both the dorsal and ventral regions of the NTS, one study using CP99994 to displace binding of $[^{125}I]$-labelled substance P (Watson et al. 1995a) reported binding in a higher density in the region corresponding to the subnucleus gelatinosus, the region where abdominal vagal afferents are known to terminate and a region immediately subjacent to the AP with which it connects. The AP in the ferret and rat has substantially lower levels of binding than the NTS (rat dorsal NTS 4.95 nCi/mg vs. AP 0.25 nCi/mg; ferret dorsal NTS 3.6 nCi/mg vs. AP 0.69 nCi/mg, Maubach et al. 1995; ferret NTS 0.125 OD units vs. AP 0.033 OD units, Watson et al. 1995a). Other regions with relatively high levels of binding in the ferret include the nucleus ambiguus, hypoglossal nucleus and the inferior olive nucleus. Studies in *Suncus* (K. Maubach, personal communication) used $[^{125}I]$-labelled substance P to demonstrate high levels of binding in the NTS and hypoglossal nucleus and substantially lower levels in the AP.

The most extensive electrophysiological studies have been undertaken in the rat brain stem and agonist studies have provided evidence that the NK_1 receptor is the receptor predominantly involved in the excitation of NTS and DMVN neurons (Maubach and Jones 1997). Interestingly both NK_2 and NK_3 receptors appear absent from the rat NTS (Maubach et al. 1994, 1995; Maubach and Jones 1997) but responses were evoked by NKA which were blocked by NK_1 receptor antagonists. The authors point out that there are other examples in the rat where excitatory effects of NKA are mediated via NK_1 receptors. The authors raise the interesting possibility that NKA might be the endogenous transmitter mediating tachykinin transmission in the NTS and as the selective NK_1 receptor antagonist CP99994 (at high concentrations) inhibited the responses to NKA in the NTS and DMVN they suggest that NKA rather than substance P may be important in the emetic pathways in the dorsal brain stem. Furthermore, they propose that CP99994 acts at the septide-sensitive receptor (see below). Inter-species differences (see Sect. 4.2) make the relevance of these results to the human difficult to assess and this is further compounded by the lack of an emetic reflex in the rat which may be contributed to by differences in brain stem neuroanatomy and pharmacology.

In the dog substance P was one of a large number of peptides to be shown to evoke excitatory responses when ionophoresed onto neurons in the AP (Carpenter et al. 1983a, 1983b). More recent studies in the dog demonstrated that the vanilloid receptor agonists capsaicin and resiniferatoxin (RTX), which are known to release substance P (and other substances) from unmyelinated afferents including those in the vagus, induced activation of neurons in the medial NTS and occasionally retching when applied to the fourth ventricle (Shiroshita et al. 1997). As the effect of selective NK_1 receptor antagonists was not investigated, this study only provides very indirect supportive evidence for the presence of NK_1 receptors in the dog NTS but other studies supporting this proposal are discussed below.

3.3
Induction of Emesis by Tachykinin Receptor Agonists

Relatively few studies have set out to investigate the ability of tachykinins to induce emesis and hence the majority of reports are en passant observations in publications with other major objectives. In addition, there are potential technical problems with ensuring that the tachykinin reaches the target site (putatively the brain stem) in sufficient concentration, and studies in man using substance P infusions show that responses (flushing, hypotension, increased bowel motility) exhibit rapid tachyphylaxis despite continued infusion (Powell et al. 1978). Whilst positive results are of assistance they must be treated with caution and negative results do not completely exclude an involvement of a tachykinin. The situation is analogous to that implicating $5-HT_3$ receptors in emesis where demonstrating the induction of emesis by 5-HT was very difficult to prove (Andrews 1994).

The earliest reports of induction of emesis induced by a tachykinin, of which we are aware, come from Erspamer and Glasser (1963) and from Bertaccini et al. (1965) using eledoisin and physalaemin. Eledoisin is extracted from the posterior salivary glands of the octopod *Eledone moschata* and physalaemin from the skin of a South American frog (*Physalaemus bilingonigerus* formerly *fuscumscultus*). Both are members of the phylogenetically conserved structurally related family of peptides found in mammals, non-mammalian vertebrates and invertebrates (Pernow 1983; Wieland and Bodanzky 1991; Severini et al. 2002). In the dog, subcutaneous (s.c.) injection of eledoisin (25–200 nmol/kg) induced profuse salivation and vomiting followed by defecation (Erspamer and Glasser 1963). Synthetic physalaemin (100–250 nmol/kg s.c.) induced vomiting of 'moderate severity' within a few minutes and a lower dose (25–50 nmol/kg) induced less severe emesis (Bertaccini et al. 1965). Emesis was also induced by intravenous (i.v.) administration which, in addition, induced salivation and defecation as did the s.c. route. Many peptides in amphibian skin are used to deter predators and it is tempting to suggest that the ability to induce nausea and vomiting is one of the functions of physalaemin. Subsequent studies in the dog by Carpenter et al. (1982, 1984, 1990) showed that substance P when given i.v. (0.03–0.20 mg/kg) could evoke emesis but this was also the case for a number of other substances including angiotensin, neurotensin, thyrotropin-releasing hormone, VIP, gastrin, ADH and enkephalin.

Although emesis could be induced by administration of vasopressin and angiotensin into the fourth ventricle of the dog, substance P (200–400 µg) did not induce an emetic response (Wu et al. 1985) and neither did it when injected into the ventricular system of the cat (30–50 units; 50–100 µg), although in one study some of the prodromal signs of emesis (licking, swallowing) were induced (Pernow and von Euler 1961; Yasnetsov et al. 1987). Interestingly, Yasnetsov et al. (1987) and Shashkov et al. (1988) whilst failing to demonstrate the induction of emesis in the cat by administration of substance P (50–100 µg) into the fourth ventricle found that it and some endorphins had an anti-emetic effect in preventing the 'vestibulo-vegetative syndrome' that accompanies motion exposure and the emetic response to beta-endorphin. Could this effect be due to desensitization of NK_1 receptors at part of the emetic pathway accessible from the fourth ventricle?

During the course of studies investigating the central control of gastric motility in the urethane-anaesthetized ferret Wood (1988) observed that topical application of high concentrations (0.1 mM) of substance P applied to the floor of the fourth ventricle in the region of the AP evoked retching after a latency of about 40 s. It was proposed that substance P induced emesis either by an action on the AP itself or by accessing the dorsal part of the NTS, particularly the subnucleus gelatinosus. Microinjection of substance P (10 µg in 2 µl) into the region of the NTS in conscious ferrets reliably evoked episodes of retching lasting <5 min after a latency of <1 min (Gardner et al. 1996).

Indirect, but supportive, evidence comes from studies of the emetic effects of capsaicin and RTX in *Suncus murinus* (Andrews et al. 2000) and the dog (Shiroshita et al. 1997, but see also Yamakuni et al. 2002). In the decerebrate dog capsaicin and RTX applied to the floor of the fourth ventricle induced a transient fictive retching response that was assumed to be due to the release of substance P (or other neroactive agents) from vagal C-fibres terminating on neurons in the medial NTS. It is also possible that these two vanilloid receptor agonists act directly on TRPV receptor channels located on NTS neurons. In *Suncus*, RTX administered s.c. (1–1000 µg/kg) induced a dose-related emetic response and c-*fos* immunoreactivity in the NTS and the AP. The emetic response was blocked by the NK_1 receptor antagonist CP99994 (10–20 mg/kg s.c.) or by neonatal treatment with capsaicin but was unaffected by chronic abdominal vagotomy. Both RTX (3–30 nmol) and capsaicin (10–100 nmol) also cause emesis following intracerebroventricular (i.c.v.) administration (Rudd and Wai 2001) as does substance P at 200 nmol (J.A. Rudd and M.K. Wai, unpublished results). In addition, RTX evoked emesis in the decerebrate working-heart brain stem preparation in *Suncus murinus* (Smith et al. 2002). Whilst not conclusive, these observations are consistent with the hypothesis that the release of substance P from sites in the dorsal brain stem can induce emesis. Complementary studies (N. Matsuki and P.L.R. Andrews, unpublished results) demonstrated release of immunoreactive substance P by RTX from *Suncus murinus* dorsal vagal complex in vitro. It must be noted that s.c. administration of RTX does not evoke emesis in the ferret (Andrews and Bhandari 1993) or the dog (Yamakuni et al. 2002).

Further studies of the induction of emesis using a range of NK_1 receptor agonists are needed, but will have to take account of the possibility that substance P may be enzymatically degraded (e.g. by neutral endopeptidases) before it reaches the active site in sufficient concentration to induce emesis and also that failure to induce emesis may be due to rapid desensitization of the NK_1 receptor.

3.4
Release of Tachykinins by Emetic Stimuli

This aspect has not really been addressed to date and probably reflects the technical problems involved in measuring the release of substance P or other tachykinins in the dorsal brain stem of animals during emesis.

Substance P is found in substantial quantities in the mucosa of the small intestine (see above) and if released could activate appropriately located NK_1 receptors. If mucosal substance P is involved in emesis it is most likely to be involved in the response to peripherally acting stimuli where activation of abdominal vagal afferents occurs (e.g. cytotoxic drugs, radiation, copper sulphate, hypertonic agents). Substance P released from the gut could enter the circulation to act on NK_1 receptors in the AP which may be located on dendrites of NTS neurons or access receptors in the NTS via permeable capillaries (Roth and Yamamoto 1968; Gross et al. 1990). This possibility needs to be investigated directly but appears unlikely as passage through the liver will metabolize substance P. A more likely possibility is that substance P is released from the basal surface of the EC cell to act upon receptors located on vagal afferents terminating in close proximity in a manner analogous to the interaction of 5-HT with the $5\text{-}HT_3$ receptor (Andrews and Davis 1995). Substance P has been shown to activate abdominal vagal afferents in the ferret and the response can be reduced in a dose-related manner by systemic CP99994 (Minami et al. 2001). Although this supports the hypothesis that peripheral substance P could activate vagal afferents and hence induce emesis, additional studies are needed before firmer conclusions can be drawn about this possibility.

A clinical report of plasma substance P levels in two patients with small cell lung cancer being treated with cisplatin provides some insight (Matsumoto et al. 1999). In one patient substance P levels increased 24 h after cisplatin and coincided with the onset of delayed emesis but although the substance P level declined emesis still continued. In the other patient substance P levels again increased at 24 h and coincided with the onset of emesis, however whilst substance P levels remained elevated this did not correlate with emesis. This study provides some very preliminary human evidence that plasma substance P levels rise following cisplatin, at a time coincident with the onset of delayed emesis but suggests that systemic elevation of substance P may not be necessary for the maintenance of the emetic response. Additional studies of this type would be of considerable interest.

4
Implication of Tachykinins in the Emetic Pathway

4.1
Pre-clinical Models for the Study of Emesis

One of the major impetuses behind the efforts to develop novel broad spectrum anti-emetic drugs was to improve the control of chemotherapy-induced acute and delayed emesis in man and for the treatment of PONV. In particular, there was a need to find a treatment for the emesis occurring during the delayed phase of chemotherapy-induced nausea and vomiting that was considered to be partially resistant to the 5-HT$_3$ receptor antagonists (Rizk and Hesketh 1999). Unfortunately, animal models of chemotherapy-induced acute and delayed emesis have been slow to develop and little is known about mechanisms involved (Naylor and Rudd 1996). Similarly, the mechanisms of PONV are poorly defined and currently there is no ideal animal model (Andrews 1992, 1999, 2003).

In the absence of suitable animal models, a major approach enabling the development of NK$_1$ receptor antagonists as anti-emetics has relied on screening a number of drug candidates against a range of emetogens or treatments that have disparate mechanisms of action. Obviously, the choice of the emetic stimulus depends on the preference of the investigator. Apomorphine is an example of one of the primary emetic stimuli that has been used in the screening process. It is an agonist that rapidly induces emesis in some animals and man via dopamine D$_2$ receptors located in the AP (Harding et al. 1985, 1987; Knox et al. 1993; Lindstrom and Brizzee 1962). This indirectly suggests that dopamine is a transmitter in the emetic reflex and that screening for novel drugs to prevent apomorphine-induced emesis in animals may eventually provide a clinically active anti-emetic compound (Niemegeers 1971, 1982). Other primary screens that have been used include models of loperamide-, copper sulphate-, or provocative motion-induced emesis; indeed the list is rather extensive (see below).

The AP represents the beginning of an afferent input into the emetic reflex (Andrews et al. 1998; Borison and Wang 1953), so it is no surprise that dopamine receptor antagonists do not have a broad utility to antagonize emesis. However, it should be appreciated that the use of apomorphine as part of a strategy to discover novel anti-emetic drugs is rational, since activation of dopamine receptors in the AP will cause a cascade of events that eventually culminates in emesis. Therefore, if a new drug treatment prevents apomorphine-induced emesis, and it is not a dopamine receptor antagonist, it creates immediate scientific interest, because it probably antagonizes emesis at a deeper point along the afferent pathway from the AP, or even possibly at an output from the central pattern generator for emesis. This leads to the possibility that the new drug could also reduce emesis induced by other drugs acting at the level of the AP (e.g. loperamide). Logically, the next step in the process of drug discovery may then be to examine if the new treatment could prevent the emetic response induced by other stimuli that act on other afferent systems. Using this approach, a vast amount of data can be collected relative to the anti-emetic spectrum of activity of a drug.

The cytotoxic anti-cancer agent cisplatin has been widely used as a primary screen to establish the anti-emetic potential of the NK_1 receptor antagonists. The early models used a high dose of cisplatin that induced a reliable emetic response in all animals following similar latencies observed in the clinic. The models were originally validated based on the anti-emetic action of metoclopramide (Florczyk et al. 1982; Gylys et al. 1979; Matsuki et al. 1988) and were subsequently used to discover the clinical utility of the $5-HT_3$ receptor antagonists (King 1990; Lucot 1989; Miner et al. 1987; Torii et al. 1991b). Indeed, the $5-HT_3$ receptor antagonists appeared highly effective against acute chemotherapy-induced emesis (Naylor and Rudd 1996).

There is little doubt that the $5-HT_3$ receptor antagonists revolutionized the treatment of chemotherapy-induced emesis in man (Fauser et al. 1999). However, the detailed and careful clinical evaluation of $5-HT_3$ receptor antagonists subsequently provided greater insight into the problem of chemotherapy-induced nausea and emesis that was not immediately realized at the time of the pre-clinical studies. Thus, whilst $5-HT_3$ receptor antagonists are highly effective in preventing emesis on the first day (i.e. acute emesis) of chemotherapy treatment in man (Fauser et al. 1999), they are less effective or even ineffective in reducing emesis on subsequent days (i.e. the 'delayed' phase) (Rizk and Hesketh 1999; Tavorath and Hesketh 1996). This realization prompted a re-evaluation of the original pre-clinical models. However, the initial cisplatin-induced emetic response appears to be mediated primarily by 5-HT acting upon $5-HT_3$ receptors (Barnes et al. 1988) but this is not the case with delayed emesis (Cubeddu 1996; Janes et al. 1998; Wilder-Smith et al. 1993). It must be noted that although $5HT_3$ receptors have a minimal involvement in the delayed phase of chemotherapy-induced emesis this does not exclude a role for 5-HT acting on one or more of the many subtypes of the 5-HT receptor.

There have been a number of advances in the understanding of the animal models of cisplatin-induced emesis that led to NK_1 receptor antagonists being tested under different conditions. Most of the initial studies with the NK_1 receptor antagonists have been performed in the ferret, using cisplatin at 10 mg/kg, during 4-h observation periods (see below). This model has little to do with mechanisms operating during delayed emesis, since the 5-HT_3 receptor antagonists and dexamethasone do not have additive anti-emetic actions to convincingly reduce retching and vomiting (Rudd et al. 1996a). There are other experiments where NK_1 receptor antagonists have been evaluated against cisplatin at approximate 10 mg/kg doses (i.e. approximately equivalent to 200 mg/m^2) but using extended 8.5–24-h observation times (see below). Again, ondansetron and dexamethasone do not interact to reduce retching and vomiting in the model (Rudd and Naylor 1997), limiting its usefulness to predict if a drug will have activity to prevent delayed emesis in man. The longer observation periods are useful, however, to collect valuable data on the duration of action of the anti-emetic compounds.

A simple step in the development of a ferret model of cisplatin-induced acute and delayed emesis was to reduce the dose of cisplatin to 5 mg/kg (Rudd et al. 1994; Rudd and Naylor 1994). This enabled the animals to tolerate cisplatin for 72 h and the profile of emesis is biphasic (Naylor and Rudd 1996). In the model, both 5-HT_3 receptor antagonists and dexamethasone have interactions that partially replicate the clinical situation (Rudd et al. 1996a). Acute (or early) and delayed cisplatin-induced emesis models have also been developed in the piglet (Grelot et al. 1996; Milano et al. 1995), pigeon (Tanihata et al. 2000, 2003) and dog (Fukui 1999; Yamakuni 2000). Each model has its advantages and disadvantages, but these will not be discussed in detail here. Nevertheless, the models of acute and delayed emesis provided important data to support the hypothesis that the broad inhibitory NK_1 receptor antagonists could have a utility to prevent both acute and delayed emesis in man (Tattersall et al. 2000).

It should be noted that the anti-emetic potential of NK_1 receptor antagonists has been evaluated in several laboratories. Each laboratory has tended to use different experimental protocols. It is difficult, therefore, to draw simple direct comparisons of anti-emetic activity between laboratories. In addition, some studies have admirably attempted to determine 50% inhibitory dose (ID_{50}) values of drugs to reduce emesis, but others have concentrated only on the anti-emetic potential of a blocker at one dose level. Therefore, in an attempt to enable a comparison of the data from all of the studies, we have decided that it may be more appropriate to compare the relative potency of the drugs to reduce emesis by at least 80% (ID_{80}): this criterion covers most of the drugs in the studies performed to date. Moreover, there is another important reason for comparing the anti-emetic potency of NK_1 receptor antagonists at the minimum ID_{80} dose. This is because if emesis is not usually prevented by more than 60–75%, the majority of NK_1 receptor antagonists fail to modify the latency to the first episode of retching and/or vomiting. It is not known why the NK_1 receptor antagonists fail to modify the latency of emesis at sub-ID_{80} doses. However, the failure may tell us something about the mechanism of the NK_1 receptors to inhibit emesis, relative to the position of NK_1 receptors in the emetic reflex. Alternatively, there may be other reasons for this unexplained phenomenon.

4.2
Species Differences in the Pharmacology of NK_1 Receptors and the Potential Complicating Role of Receptor Conformers

Substance P has long been suspected to be involved in the emetic reflex (see Sect. 3). However, it was difficult to identify its action precisely until the synthesis of selective tachykinin receptor antagonists. Although there are three major subtypes of tachykinin receptor (NK_1, NK_2 and NK_3), all of which are G protein-coupled (Maggi 1995; Regoli et al. 1994), only the role of the NK_1 receptor has been studied in detail in emesis control. This is surprising given that all three receptors may be in brain stem areas involved in the emetic reflex (see Sects. 2.1.2, 2.1.6) (Mazzone and Geraghty 2000a, 2000b). Moreover, it is not known if it is solely substance P that activates the NK_1 receptor in the emetic reflex, or if other endogenous agonists (e.g. neurokinin A or neurokinin B) are activating NK_1 receptors in addition to potentially activating their own preferential receptors. Such considerations and perhaps the possibility that substance P is released with other co-transmitters may explain some of the anomalies in the pre-clinical and clinical data (specifically with respect to the action of the antagonists on the latency to the first emetic episode). Indeed, it is not even established if an NK_1 receptor antagonist can block completely substance P-induced emesis or antagonize the potential emesis induced by septide (see below), neurokinin A (the preferred endogenous agonist for NK_2 receptors) or neurokinin B (the preferred endogenous agonist for NK_3 receptors). It should also be considered that if NK_1 receptors are blocked in the brain, substance P may have minor actions on NK_2 receptors to affect behaviour and cardiovascular reflexes (Picard et al. 1994); this phenomenon has not been explored with regard to the emetic reflex.

The purpose of this chapter is to focus on the characteristics of the NK_1 receptor system that may be involved in the emetic reflex. Some early studies to classify the pharmacology of NK_1 receptors identified major species differences that are now well known (Maggi 1995; Saria 1999). The ferret NK_1 receptor can be considered as 'human-like' (see Table 1) and enables a confident extrapolation of the findings of anti-emetic potency of NK_1 receptor antagonists in this species to man (Rupniak et al. 1997). The differences in affinity of NK_1 receptor antagonists between species seem to correlate well with the required doses to inhibit emesis. Out of the species studied that have the emetic reflex, *Suncus murinus* NK_1 receptors appear unique and antagonists are required at high doses to block emesis and they may loose specificity with regard to actions on ion channels, necessitating a careful analysis of the effects of pairs of enantiomers to establish the role of NK_1 receptors (Rudd et al. 1999b; Tattersall et al. 1995). Aside from the binding characteristics of the NK_1 receptor antagonists, one of the

Table 1 Species differences in the pharmacology of tachykinin NK_1 receptors

Compound	NK_1 receptor affinity (nM)				
	Human	Ferret	Rat	Cat	*Suncus murinus*
CP99994	0.3[a]	1.7[a]	111[a]	0.5[b]	12.0[c]
CJ11974	0.2[d]	0.6[d]			
RP67580	56[a]	111[a]	3[a]		>1,000[e]
GR205171	0.03[f]	0.16[f]	0.32[f]		
GR203040	~0.04[g]	0.08[g]	2.5[g]		
L741671	0.03[a]	0.7[a]	64[a]		
L743310	0.06[a]	0.1[a]	17[a]		
L758298	2.8[h]	1.1[h]	33[h]		
L754030	0.1[h]	0.7[h]	4[h]		
PD154075	0.84[i]	3.4[i]	302[i]		
R116301	0.45[j]	8.3[j]	98[j]		

[a] K_i values against [^{125}I]-labelled Tyr^8-substance P (Tattersall et al. 1996).
[b] IC_{50} values against [^{3}H]-substance P (Lucot et al. 1997).
[c] IC_{50} value (Tattersall et al. 1995).
[d] K_i values against [^{3}H]-substance P (Tsuchiya et al. 2002).
[e] K_i value (Tattersall et al. 1995).
[f] K_i values against [^{3}H]-substance P (Gardner et al. 1996).
[g] K_i values against [^{3}H]-substance P (Beattie et al. 1995).
[h] IC_{50} value against [^{125}I]-labelled substance P (Tattersall et al. 2000).
[i] IC_{50} value against [^{125}I]Bolton-Hunter Substance P (Singh et al. 1997).
[j] K_i values against [^{3}H]-substance P (Megens et al. 2002).

most important considerations necessary for broad inhibitory activity is that the compound must be able to penetrate the brain (Rupniak et al. 1997).

Example tests to estimate penetration include an ability to inhibit the foot tapping induced by the i.c.v. infusion of a selective NK_1 receptor agonist (e.g. [Sar^9, Met(O$_2$)^{11}]–substance P or GR73632) in gerbils (Rupniak et al. 1997; Tsuchiya et al. 2002) or the ability to reduce increases in locomotor activity induced by the NK_1 selective agonist, [Sar^9,Met(O$_2$)^{11}]–substance P in guinea pigs (McLean et al. 1996). However, ex vivo radioligand binding studies (Beattie et al. 1995) or direct measurement of the compound concentration in brain tissue (Singh et al. 1997; Tsuchiya et al. 2002) have also been performed. Such tests provide power to predict the potential of the compounds to antagonize emesis in animals and man (Rupniak et al. 1997).

It is clear that most antagonists of the NK_1 receptor have been classified with a focus on blocking the effects of substance P as a ligand/agonist in competition and functional studies. We have to consider if this approach is appropriate for anti-emetic development. Certainly, the concept that a pA_2 value for an antagonist acting at a receptor is constant, regardless of the identity of the agonist for the receptor, has been a cornerstone for receptor classification (Arunlakshana and Schild 1959). However, it may be significant that antagonists can display different pA_2 values at the NK_1 receptor depending on the nature of the agonist (Jenkinson et al. 1999). Are we then missing valuable data if it is not substance P acting in the emetic reflex? To explain the uniqueness of the NK_1 receptor system, one must consider that it appears to have two binding sites, or conformers, for agonists, and this is discussed elsewhere in this volume.

In summary, there are species differences in the pharmacology of tachykinin NK_1 receptors that are clearly defined, and reliable predictive screens have been utilized to optimize anti-emetic compound selection. However, it is possible that several endogenous agonists (e.g. substance P, neurokinin A and neurokinin B) could interact with the NK_1 receptors in the emetic reflex and this may also influence the characteristics of antagonism provided by the NK_1 receptor antagonists. Therefore, the situation may be complicated, particularly if the NK_1 receptors are located at different sites in the emetic reflex (e.g. NTS and pattern generators), are being activated by different substances, and must be simultaneously blocked to prevent emesis. Certainly, the nature of the endogenous agonist may potentially affect the dose and/or nature of competition provided by the anti-emetic NK_1 receptor antagonists. Little information is available that is relevant to the potential interaction of the newer NK_1 receptor antagonists for the septide-preferring conformer and this may be important, particularly if neurokinin A or neurokinin B is more relevant in the emetic reflex, or if substance P, neurokinin A and neurokinin B (or other substances) are co-released. This should be borne in mind when reading the sections on the pre-clinical and clinical data.

4.3
Studies of the Anti-emetic Effects of NK_1 Receptor Antagonists in Different Animal Models

4.3.1
Studies in the Ferret

The ferret has become one of the primary species with which to investigate the action of drugs on the emetic reflex. This is partly because its use contributed to successfully predicting the anti-emetic potential of the 5-HT$_3$ receptor antagonists to prevent chemotherapy- and radiotherapy-induced emesis. However, perhaps the most important reason for selecting this species is that human and ferret NK_1 receptors appear to have a similar pharmacological profile with respect to the affinity of NK_1 receptor antagonists (Table 1). Several classes of NK_1 receptor antagonists have been investigated for a potential to antagonize emesis in the ferret. The anti-emetic potential of the blockers has been investigated against drugs inducing emesis via an action at the level of the AP including the dopamine receptor agonist apomorphine (Andrews et al. 1990; Knox et al. 1993), and the opioid receptor agonist morphine (Thompson et al. 1992; Wynn et al. 1993) and loperamide (Bhandari et al. 1992). The anti-emetic potential of the NK_1 receptor antagonists has also been investigated for a potential to reduce the emesis induced by intragastric copper sulphate, which is considered to act via the vagus and splanchnic nerves as a function of its irritant action in the gastrointestinal tract (Andrews et al. 1990). It is important to consider the anti-emetic potential of the NK_1 receptor antagonists to inhibit emesis in each of these central and peripheral models, because the emesis induced by these challenges is resistant to 5-HT_3 receptor antagonists.

CP99994 can be considered to be a reference anti-emetic NK_1 receptor antagonist as it has been the most widely studied. One of the first investigations used racemic (±)CP99994 at a dose of 3 mg/kg intraperitoneally (i.p.) (Bountra et al. 1993). It produced impressive 83% and 85% reductions of the number of retches induced by morphine (0.5 mg/kg s.c.) and copper sulphate (40 mg/kg intragastrically), respectively. This was a remarkable antagonism of emesis, considering that (±)CP99994 was only administered at the same time as the emetogens, and that the onset of morphine- and copper sulphate-induced emesis is relatively short (usually 5–10 min). Retrospectively, the anti-emetic potency of (±)CP99994 may have been underestimated in these experiments since it may not have distributed fully from the injection site. Indeed, the active enantiomer, CP99994 at 2×1 mg/kg, or at 1×3 mg/kg, as a 30-min pre-treatment abolishes apomorphine (0.2–0.25 mg/kg s.c.)-induced emesis whilst the less active enantiomer, CP100263, had no effect on retching and vomiting (Tattersall et al. 1994; Watson et al. 1995a).

The anti-emetic potential of CP99994 (0.1–2 mg/kg) has also been investigated for a potential to reduce loperamide (0.25 mg/kg s.c.)-induced emesis and the emesis induced by a lower dose of copper sulphate (12.5 mg/kg intragastrically). As a 30-min s.c. pretreatment, CP99994 dose-dependently antagonizes loperamide-induced emesis by approximately 95% at 1 mg/kg (Watson et al. 1995a), which is similar to its potency to reduce apomorphine-induced emesis (see above). Again the effect of CP99994 was stereoselective, with CP100263 being inactive (Watson et al. 1995a). In addition, against a lower dose of copper sulphate (12.5 mg/kg), CP99994 produced a clear dose-related antagonism of emesis and completely prevented retching and vomiting at 2×1 mg/kg s.c. (Watson et al. 1995a). This apparent greater efficacy may be the result of testing the antagonist against a dose of copper sulphate (i.e. 12.5 mg/kg) that induces approximately 60–80% fewer retches and vomits than the higher dose (i.e. 40 mg/kg).

Similar studies using identical dosing schedules (Watson et al. 1995a) have confirmed the stereoselective anti-emetic action of CP99994 against loperamide (0.5 mg/kg s.c.)-induced emesis (Zaman et al. 2000). However, the studies encompassed a more extensive investigation of the effects of CP99994 on additional loperamide-induced behaviours. Whilst CP99994 prevented completely loperamide-induced emesis, it also reduced loperamide-induced lip-licking, mouth scratching, wet dog shakes and gags, but surprisingly failed to reduce loperamide-induced c-*fos* expression in the brain stem (mainly seen in the dorsomedial NTS). Such a failure to prevent c-*fos* expression induced by loperamide is a curiosity, considering the hypothesis surrounding the proposed mechanism of action of the NK_1 receptor antagonists to prevent emesis (discussed in Sect. 6) (Zaman et al. 2000).

CP122721 is a 2-series phenylpiperidine NK_1 receptor antagonist that is structurally similar to CP99994, but it is metabolically more stable with potent non-competitive actions at NK_1 receptors (McLean et al. 1996). It has been administered as a single s.c. pretreatment 30 min prior to loperamide (0.25 mg/kg s.c.) and copper sulphate (12.5 mg/kg intragastrically) (Gonsalves et al. 1996). Using this approach, CP122721 produced a predicted dose-related inhibition of emesis and its less active enantiomer, CP132687, had no discernible action. The dose of 0.1 mg/kg reduced retching induced by loperamide and copper sulphate by approximately 93% and 100%, respectively (Gonsalves et al. 1996). Based on ID_{50} values, CP122721 is approximately three times more potent than CP99994 to antagonize emesis, which is consistent with its three times higher affinity for human NK_1 receptors (Gonsalves et al. 1996; Watson et al. 1995a). However, what is also interesting from the above studies is that if CP99994 and CP122721 fail to prevent emesis completely (e.g. at ID_{50} doses), they do not significantly affect the latency to the first retching and vomiting episode (note, not all studies have reported the effect of CP99994 on drug-induced latencies to the first episode). It is also interesting that if CP122721 prevents emesis by only approximately 50%, it has no action to modify the gagging, lip-licking and backing that is induced by loperamide (Gonsalves et al. 1996) and abolished by CP99994 at doses completely preventing emesis (Zaman et al. 2000). Interestingly, CP99994 (1 mg/kg s.c., 15-min pretreatment) can also block emesis and the licking and swallowing movements induced by electrical stimulation (25 V, 0.5 ms, 40 Hz, for 30 s) of the vagus nerve in an anaesthetized ferret preparation (Watson et al. 1995a). Whilst the action of CP122721 to modify the loperamide-induced behaviours at ID_{80} doses was not reported, it has been suggested that such behaviours may be analogous to self-reported nausea in humans (Gonsalves et al. 1996).

The ability of an NK_1 receptor antagonist to have the same potency to reduce emesis induced by agents acting either centrally or peripherally suggests that the site of action is at a pivotal point in the emetic reflex such as the NTS, or closely associated structures, where NK_1 receptors are known to exist (Rupniak and Kramer 1999). However, only a few experiments have been conducted to test this hypothesis directly. One such experiment has involved injecting CP99994 and HSP117 directly into the AP as a 5-min pre-treatment prior to challenge with emetic drugs (Saito et al. 1999). It is technically very difficult to restrict the injection to the AP in this species because it is such a small brain area and because its caudal portion lies over the central canal. However, CP99994 at a dose of 7.5 μg prevented completely morphine (0.5 mg/kg s.c.)-induced emesis and, in another experiment, reduced copper sulphate (25 mg/kg intragastrically)-induced retching by approximately 62% (Saito et al. 1999). Conversely, HSP117 appeared almost equipotent against both stimuli, reducing retching by approximately 62% at 1 μg (the highest dose tested) (Saito et al. 1999). The separation of potency was expected, given that HSP117 has approximately 50 times higher affinity than CP99994 for human NK_1 receptors (Saito et al. 1999).

A halothane-anaesthetized ferret model has been used to examine the potential of several other NK_1 receptor antagonists to inhibit emesis induced by an intra-duodenal administration of hypertonic saline (1.5 M). Sendide (584 nmol), spantide (300 nmol) and CP96345 (309 nmol) were highly effective in blocking emesis (Davison et al. 1995). However, it was also noted that all compounds had additional activity to antagonize the saline-induced salivation and licking movements. CP95345 (4.12 μmol) but not its less active enantiomer, CP96344 (4.12 μmol), also blocked emesis when given via the i.v. route. Interestingly, however, CP96344 prevented emesis and salivation when administered i.c.v., but it was suggested that this action resulted from its additional calcium channel blocking activity (Davison et al. 1995).

If the site of CP99994 and HSP117 inhibition of emesis is the AP and/or the NTS, it would be expected that both compounds would be less potent if administered further away from the target area. Thus, an intraventricular administration should result in a lower concentration of the antagonists reaching the AP and/or NTS. Indeed, this may be the case, with CP99994 being required at 100 μg i.c.v. to antagonize morphine- and copper sulphate-induced emesis by at least 75%. HSP117 appeared slightly more active, since the dose of 100 μg i.c.v. produced an approximate 95% inhibition of the emesis induced by both challenges) (Ariumi et al. 2000).

Two of the most potent NK_1 receptor antagonists to inhibit morphine- and copper sulphate-induced emesis in the ferret following a peripheral administration are GR203040 and GR205171 (Gardner et al. 1995, 1996). These tetrazole-based NK_1 receptor antagonists both produce similar 98%–100% reductions of morphine (0.5 mg/kg s.c.)-induced retching when administered at does as low as 0.1 mg/kg s.c. (30-min pretreatment (Gardner et al. 1995). GR203040 at 0.1 mg/kg s.c. also reduces copper sulphate (40 mg/kg, intragastrically)-induced emesis by approximately 86% (Gardner et al. 1995) whilst GR205171 completely inhibits the response at the same dose level (Gardner et al. 1996). This is particularly impressive, considering that the tests were carried out over 2-h observation periods and the compounds were tested against the higher dose of copper sulphate (40 mg/kg). It is unfortunate, however, that the studies neglected to report the effect of the compounds on the latency of drug-induced emesis and that only one dose of each antagonist was investigated.

R116301 has been tested for a potential to antagonize loperamide (0.31 mg/kg s.c.)- and apomorphine (0.31 mg/kg s.c.)-induced emesis in ferrets (Megens et al. 2002). The ID_{50} value to inhibit loperamide emesis following a 1-h s.c. pretreatment was 3.2 mg/kg. It was estimated to be approximately 500 times less potent than GR203040 in this test and five times less potent than CP99994 (Megens et al. 2002). However, there is less of a difference in potency between R116301 and GR203040 following oral administration, with R116301 being only 12.5 times less potent. R116301 was at least four times more potent than CP99994 following oral administration, suggesting marked differences in the rates of metabolism of the compounds. L760735 was also compared in this series of experiments for a potential to inhibit loperamide-induced emesis following s.c. administration and this yielded an ID_{50} value of 0.31 mg/kg. L754030 yielded an ID_{50} value of 3.1 mg/kg following oral administration. The ID_{50} value for R116301 to inhibit apomorphine-induced emesis was 1.2 mg/kg per os (p.o.). Thus, this drug was approximately 35 times less potent than domperidone (Megens et al. 2002). Unfortunately, the doses of antagonists inhibiting emesis by 80%–100% were not discussed, nor were their effects on the latency to first retch or vomit.

The above studies clearly demonstrate that NK_1 receptor antagonists can prevent emesis induced by centrally acting and peripherally acting stimuli that are resistant to $5\text{-}HT_3$ receptor antagonists. Thus, NK_1 receptor antagonists may be useful, or have an advantage, in situations where $5\text{-}HT_3$ receptor antagonists lose efficacy (e.g. chemotherapy-induced delayed emesis). This was one of the initial aims (if not the major aim) of searching for broad inhibitory anti-emetic drugs. The models where the $5\text{-}HT_3$ receptor antagonists have efficacy include those involving chemotherapeutic drugs (Andrews et al. 1998), radiation treatment (Andrews and Hawthorn 1987), ipecacuanha (Andrews et al. 1990; Hasegawa et al. 2002) and type IV cyclic nucleotide phosphodiesterase (PDE IV) inhibitors (Robichaud et al. 1999). A discussion of the activity of the NK_1 receptor antagonists in these models is provided below.

The standard model of cisplatin-induced emesis in the ferret involves administering cisplatin at 10 mg/kg (i.v. or i.p.) and observing the animals for just a few hours (2–4 h) (see above). This model has proved reasonably predictive for the control of emesis over the first 24 h (acute emesis) following cisplatin treatment in man (Naylor and Rudd 1996). A model of cyclophosphamide-induce emesis, using doses in the region of 200 mg/kg and short observation times (up to 7 h) has also proved similarly predictive (Hawthorn et al. 1988). Obviously, the duration of the observation time of an experiment may influence the calculated $ID_{50/80}$ value for a particular antagonist when its elimination may potentially occur before the end of the emetic stimulus. Thus, in short 2–4 h experiments, CP99994 at 1–3 mg/kg reduces cisplatin-induced retching by approximately 80%–100% (Rupniak et al. 1997; Tattersall et al. 1993; Watson et al. 1995a) and CP122721 completely prevents retching at 0.3 mg/kg, these effects being stereoselective (Gonsalves et al. 1996; Watson et al. 1995a). The anti-emetic potency of the NK_1 receptor antagonists is practically identical to inhibit ipecacuanha-induced emesis (Gonsalves et al. 1996; Watson et al. 1995a). In other experiments, i.v. administered L742694 reduced cisplatin-induced emesis by approximately 80% at 1 mg/kg i.v. and almost abolished emesis at 3 mg/kg, while RPR100893 attenuated emesis by approximately 80% at >10 mg/kg (Rupniak et al. 1997).

The dose of a particular NK_1 receptor antagonist to reduce/prevent emesis induced by treatments involving the $5\text{-}HT_3$ receptor appears to be consistent in the short tests, regardless of the identity of the emetic stimulus. Indeed, the dose required appears to be closely related to their NK_1 receptor affinity and their ability to penetrate into the brain (Rupniak et al. 1997). Thus, consistent with the other tests described, GR203040 and GR205171 are the most potent antiemetic agents to antagonize cyclophosphamide, ipecacuanha and radiation-induced emesis, with ID_{80} doses being in the region of 0.03–0.1 mg/kg (Gardner et al. 1995, 1996).

There has also been a study of the anti-emetic activity of L741671 to prevent cisplatin-induced emesis in 4-h studies following peripheral and central administration (Tattersall et al. 1996). This compound is almost equipotent with CP99994 at ferret NK_1 receptors (see Table 1) and has almost an identical anti-emetic profile, preventing emesis at 3 mg/kg i.v. Conversely, a quaternarized compound, L743310, which is slightly more potent than both CP99994 and L743310 in in vitro tests, was without effect following peripheral administration (Tattersall et al. 1996); it also failed to affect the emesis induced by PDE IV inhibitors (RS14203, R-rolipram and CT2450), whereas CP99994 was highly effective (Table 2). Similarly, SR140333 and LY303870 that are reasonably potent NK_1 receptor antagonists, with reduced access into the central nervous system, fail to reduce cisplatin-induced emesis following i.v. administration (Rupniak et al. 1997). These data suggest that centrally located NK_1 receptors are important in the first 4 h of the emetic response induced by cisplatin (Tattersall et al. 1996).

This was confirmed in experiments using L743310 (30 μg) and L741671 (30 μg) injected into the vicinity of the NTS. Both compounds produced an almost equal approximate 67%–77% reduction of retching over a 2-h period (Tattersall et al. 1996). Interestingly, however, CP99994 (30 μg) was slightly more active than the other two compounds when administered into the NTS, since an approximate 90% reduction of retching was observed (Tattersall et al. 1996). This was unexpected considering that L743310 and L741671 have approximately 17 and three times higher affinity for human NK_1 receptors. However, CP99994 and its less active enantiomer (which was inactive to modify emesis) also appeared to cause motor impairment and it is not known if this has contributed to the apparent higher anti-emetic potency of CP99994 (Tattersall et al. 1996). In other experiments lasting 4 h, an intraventricular administration of CP99994 at 10 μg has been reported to reduce cisplatin-induced emesis by approximately 36%, which is comparable to the antagonism provided by GR82334 (an NK_1 receptor antagonist with approximately 60 times lower affinity for human NK_1 receptors than CP99994 (R.M. Hagen, personal communication) at 10 μg i.c.v. (Gardner et al. 1994). The apparent lower potency of CP99994 between experiments probably reflects a distribution away from the injection site over the longer time period.

Clearly, it is difficult to know what advantage the NK_1 receptor antagonists will have over the 5-HT_3 receptor antagonists from experiments that have been restricted to 4 h. However, some experiments have also evaluated NK_1 receptor antagonists over longer observation periods for a potential to reduce cisplatin-induced emesis. This rationale provides more information relevant to the duration of action of the compounds, and enables an assessment of the compound anti-emetic potential in situations where cisplatin-induced emesis is not totally

Table 2 Reported peripheral doses (approximates in mg/kg to reduce retching and/or vomiting by at least 80%; shown in parentheses) of tachykinin NK_1 receptor antagonists to inhibit emesis

Stimulus	Ferret	Dog	Cat	*Suncus murinus*
Central				
Apomorphine	R116301 [1.2][m]; CP99994 [~2–3][u,y]	CP99994 [0.04*][y]; R116301 [1.8][m]		
Morphine	GR203040 [0.1][i]; GR205171 [0.1][h]; (±)CP99994 [3][c];	GR205171 [0.03–0.05][f]		
Loperamide	CP122721 [0.1][j]; CP99994 [1][y,z]; R116301 [1.2][m]			
Nicotine				CP99994 [10][p,w]; CP122721 [10][p]; RP67580 [~30][p, w]
Xylazine			R116301 [0.6][m]	
Motion			CP99994 [0.3][l]	CP99994 [1–10][p]; GR203040 [~3][i]; GR205171 [3][h]
Peripheral				
CuSO$_4$	(±)CP99994 [0.3–3][c,z]; CP122721 [0.1][j]; GR203040 [0.1][g]; GR205171 [0.1][g]	CP99994 [1][a]		CP122721 [10][p]
Electrical stimulation of vagus	CP99994 [1][y]	GR205171 [0.05–0.7][e]		
Mixed/unknown				
Cisplatin (2–5 h)	GR203040 [0.1][q]; L754030 [0.3][s]; L758298 [0.3][s]; CP122721 [0.3][j]; L742694 [1][t]; CJ11974 [1][x]; CP99994 [1–3][z,q,u]; L741671 [3][t]; PD154075 [10][r]; RPR100893 [10][q];	CP122721 [0.1][j]		GR203040 [30][i]; GR205171 [30][h]
Cisplatin (8.5 h)	GR205171 [0.03][h]; GR203040 [0.1][i]; (±)CP99994 [3][c]			
Cyclophosphamide	GR203040 [0.1][i]; GR205171 [0.1][h]			
Radiation	GR205171 [0.03][h]; GR203040 [0.1][i]			
Ipecacuanha	GR205171 [0.1][h]; CP122721 [0.1][j]; GR203040 [0.1][i]; CP99994 [~0.3][y]; R116301 [1.2][m]	GR203040 [0.3][i]; GR205171 [0.1][h]	R116301 [1.2][13]	
PDE IV inhibitors (RS14203, R-rolipram, CT-2450)	CP99994 [3][n]			
TP agonist U46619				CP99994 [10][k]
Ethanol				CP99994 [10][d]
Naloxone				CP99994 [1–3][o]
Resiniferatoxin				CP99994 [20][b]
Halothane/N_2O				GR205171 [1][w]

An asterisk indicates followed by infusion.
References: [a] Andrews et al. 2001; [b] Andrews et al. 2000; [c] Bountra et al. 1993; [d] Chen et al. 1997; [e] Fukuda et al. 1998; [f] Fukuda et al. 1999; [g] Gardner and Perren 1998; [h] Gardner et al. 1996; [i] Gardner et al. 1995; [j] Gonsalves et al. 1996; [k] Kan et al. 2003; [l] Lucot et al. 1997; [m] Megens et al. 2002; [n] Robichaud et al. 1999; [o] Rudd et al. 1999a; [p] Rudd et al. 1999b; [q] Rupniak et al. 1997; [r] Singh et al. 1997; [s] Tattersall et al. 2000; [t] Tattersall et al. 1996; [u] Tattersall et al. 1993; [v] Tattersall et al. 1994; [w] Tattersall et al. 1995; [x] Tsuchiya et al. 2002; [y] Watson et al. 1995a; [z] Zaman et al. 2000.

controlled by a 5-HT$_3$ receptor antagonists or glucocorticoids (Rudd and Naylor 1997). Thus, longer observation periods of up to 8.5 h have been used. (±)CP99994 at 3 mg/kg s.c. reduces the cisplatin (200 mg/m^2)-induced retching response by approximately 90% during an 8.5-h observation period when administered at the first retch (Bountra et al. 1993). Certainly, CP99994 is much less active during experiments lasting 40 h, reducing retching and vomiting by only 54% at 10 mg/kg (administration being delayed until the first retch induced by cisplatin 10 mg/kg) (Rudd et al. 1996b). The reduced efficacy is probably a function of metabolism and this may have also limited the anti-emetic action of the peptide NK$_1$ receptor antagonist, sendide, at 3 mg/kg since it could only produce an approximate 44% reduction of cisplatin (10 mg/kg)-induced emesis (with no action on the latency to the first retch or vomit) in a 6 h experiment (Minami et al. 1998).

In the 8.5-h cisplatin (200 mg/m^2)-induced emesis model, GR203040 (0.1 mg/kg) reduced the response by approximately 80% when given as a 30-min s.c. pretreatment and by approximately 91% if administered at 1 mg/kg s.c. at 1 h post the onset of emesis (Gardner et al. 1995). GR205171 (0.1 mg/kg) also decreased retching by approximately 89% when given as a 30-min s.c. pretreatment (Gardner et al. 1996). It is apparent that there is some residual emesis following treatment with both antagonists, but this may be a function of the doses used, since GR205171 at the slightly higher dose of 0.3 mg/kg reduced retching during a 24-h period by approximately 96%, when administered concurrently with cisplatin (Gardner et al. 1996). Unfortunately, no mention was made on the effects of any of these drug treatments (excepting sendide, which failed to delay the onset of emesis) on the latency of cisplatin-induced emesis. Also, the use of cisplatin at 10 mg/kg and an extension of the observation periods of up to 40 h in this species still fails to replicate the clinical situation of cisplatin-induced acute and delayed emesis (Rudd et al. 1996b).

The first compound to be tested in the cisplatin (5 mg/kg i.p.)-induced acute and delayed emesis model was CP99994 (10 mg/kg i.p., every 8 h). It was effective in reducing the retching and vomiting occurring on day 1 by approximately 34%, but seemed more effective during the delayed phase (on days 2 and 3), providing an approximate 88% reduction of retching and vomiting (Rudd et al. 1996b). Interestingly, CP99994 also delayed the onset of cisplatin-induced emesis, but its action was only apparent for about 3–4 h following each injection (Rudd et al. 1996b). Again, the short duration of activity probably reflects a poor pharmacokinetic profile (see above), since when administered more frequently (every 4 h) as an intervention treatment it could abolish the delayed response (Fig. 4; Rudd et al. 1996b). In a similar experiment, CP99994, but not CP100263, dosed 30 min before and at 8, 24, 32, 48 and 56 h, was also effective to antagonize cisplatin (5 mg/kg i.p.)-induced acute and delayed emesis (Watson et al. 1995b). Interestingly, PD154075, a selective NK$_1$ receptor antagonist from Parke Davis, is approximately 10 times less potent than CP99994 in the 4 h cisplatin (10 mg/kg)-induced emesis model, but produces an approximate 96% reduction of the emesis occurring on day 1 when used at 10 mg/kg, administered

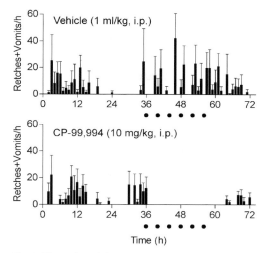

Fig. 4 The action of the NK_1 receptor antagonist CP99994 (10 mg/kg i.p.) to prevent cisplatin (5 mg/kg i.p.)-induced delayed emesis in the ferret. *Black dots* indicate the times at which either vehicle (*upper panel*) or CP99, 994 (*lower panel*) was administered. (Redrawn from Rudd et al. 1996b)

at 8-h intervals. The emesis occurring during the delayed phase was also reduced by approximately 98% (Singh et al. 1997). Again, the difference may be a reflection of the different pharmacokinetic profiles of the compounds in the model.

CJ11974, a novel quinuclidine derivative, has approximately three times higher affinity for ferret NK_1 receptors than CP99994, and a 1 mg/kg dose given s.c. 30 min prior to cisplatin 10 mg/kg is effective to reduce retching by approximately 93% during a 4-h observation period (i.e. it is about as potent as CP99994); it is also orally active (Tsuchiya et al. 2002). Interestingly, however, this dose did not affect significantly the latency of cisplatin (10 mg/kg) to induce emesis, with only one of five animals being totally protected from retching and/or vomiting (Tsuchiya et al. 2002). However, in the cisplatin (5 mg/kg i.p.) acute and delayed 72-h model, CJ11974 administered during the delayed phase at 1.2, 0.7, 0.7 and 0.7 mg/kg at 34, 37, 40 and 43 h, respectively, reduced emesis by approximately 98% (assessed over the drug administration periods) (Tsuchiya et al. 2002).

The only other compounds that have been investigated in both the cisplatin 10 mg/kg 4-h and cisplatin 5 mg/kg 72-h models are L754030 (also known as MK-869) and its water-soluble prodrug, L758298 (Tattersall et al. 2000). L754030 and L758298 have approximately 2.5 and 1.6 times higher affinity, respectively, than CP99994 for the ferret NK_1 receptor but at 0.3 mg/kg i.v. were active to cause >84% reductions of retching; L754030 was also orally active (1 mg/kg produced an approximate 87% reduction of retching activity and 3 mg/kg abolished emesis in two out of four ferrets). L754030 (0.1 mg/kg i.v.) had additive anti-emetic actions with ondansetron (0.1 mg/kg i.v.) and with an extremely high dose of dexamethasone (20 mg/kg i.v.) in the 4-h model (Tattersall et al. 2000). L745030 appears to be the most effective NK_1 receptor antagonist studied in the ferret cisplatin (5 mg/kg)-induced acute and delayed model of emesis, with the oral administration of 1 mg/kg/day antagonizing retching and vomiting in the acute and delayed phases by approximately 89%–99%; it is also highly active as an intervention treatment (Tattersall et al. 2000). Remarkably, L754030 also practically abolishes both phases of emesis following a single oral 10 mg/kg dose given 2 h prior to the injection of cisplatin. In this model of emesis, L754030 also delays the onset of the emetic response (Tattersall et al. 2000).

4.3.2
Studies in the Dog

The dog is considered to be a useful species to use in anti-emetic research based on its sensitivity to apomorphine and chemotherapeutic drugs (King 1990). It is also one of the species used to define much of the early knowledge of the physiology and pharmacology of the emetic reflex (Borison and Wang 1953). Further, it has a much larger brain than the ferret that more easily permits electrophysiological studies and a greater level of confidence regarding interpreting data from brain stem microinjection studies. The initial studies used CP99994 to investigate the potential role of NK_1 receptors in the mechanism of action of apomorphine (10 µg/kg i.v.) and copper sulphate (6 mg/kg intragastrically) to induce emesis (Watson et al. 1995a). In these studies, CP99994 at 40 µg/kg was administered i.v. as a loading dose, followed by a 5 µg/kg/min infusion. This dosing regime antagonized both apomorphine and copper sulphate-induced emesis by approximately 83%–95% (the effect on latency of the responding animals was not reported). However, a 10-min pretreatment with CP122721 at 0.1 mg/kg non-significantly reduced a higher dose of copper sulphate (20 mg/kg, intragastrically)-induced emesis by approximately 69%, without affecting the latency to onset of emesis (Yamakuni et al. 2000).

R116301 has also been investigated for a potential to reduce apomorphine (0.31 mg/kg s.c.)-induced emesis in the dog and the 50% effective dose (ED_{50}) for complete inhibition of emesis was 1.8 mg/kg, following a 2-h oral administration (Megens et al. 2002). Unfortunately, it is difficult to compare the relative potency of CP99994 and R116301 to prevent apomorphine-induced emesis since different administration protocols have been used. However, the ID_{50} doses for R116301 to antagonize apomorphine-induced emesis in the dog and ferret appear remarkably similar. In the ferret, the possible sites of action of the NK_1 receptor antagonists to inhibit emesis include the NTS (Gardner et al. 1994; Tattersall et al. 1996) and AP (Ariumi et al. 2000), which in the ferret are difficult to confidently dissociate by microinjection studies, since these brain areas are in close proximity and have intimate connections. However, an important experiment in the dog has demonstrated that surgical removal of the AP does not interfere with the ability of CP99994 (1 mg/kg s.c.) to prevent copper sulphate

(6 mg/kg intragastrically)-induced emesis (Andrews et al. 2001). These experiments point to important NK_1 receptors being possibly located in the NTS, or other associated structures.

More insight into the role of central NK_1 receptors in the emetic reflex of this species comes from studies using decerebrate dogs and GR205171. In these animals, fictive retching can be induced by electrical stimulation (10 Hz, 0.5 ms duration, at 25–30 V) of the peripheral vagus nerve and also by electrical stimulation (10 Hz, 0.2 ms, ±200–400 µA, biphasic pulses) of the medial NTS that is considered to be below the synapse where the vagal afferents terminate (Fukuda et al. 1998). Using this approach, it was evident that GR205171 (0.05–0.7 mg/kg) administered i.v. could rapidly prevent within 5 min the retching induced by stimulation of both pathways (Fukuda et al. 1998). The results suggest that the anti-emetic site of action of GR205171 is below the level of the medial NTS, rather than at a site that blocks information from the vagal afferents that synapse at the level of the medial NTS (Fukuda et al. 1998). Obviously, this would be consistent with a mechanism still operating in AP-ablated animals (see above).

More detailed investigations have examined the possibility that the NK_1 receptors critically involved in the broad inhibitory action of the NK_1 receptor antagonists are actually in the central pattern generator for emesis (Fukuda et al. 1999a, 1999b). The pattern generator for emesis is proposed to be located in the medullary reticular formation, dorsomedial to the retrofacial nucleus which receives some neuronal projections from the medial NTS. Apomorphine (0.2 mg/kg intramuscularly) and electrical stimulation (3.3–10 Hz, 0.5 ms duration, 25 or 30 V) of the vagus have been used to drive the pattern generator and some neuronal activity of this brain area, and fictive retching could be abolished by GR205171 (25–50 µg/kg, i.v.). However, some of the activity of the pattern generator persisted and may be mediated via another transmitter (Fukuda et al. 1999a). Clearly, the central pattern generator could represent a logical site at which NK_1 receptor antagonists act to block emesis (Fig. 5; Sect. 6).

Detailed intracerebral injection studies with GR205171 have since been performed using the vagally driven fictive retching model. Suppression of fictive retching could be achieved by injection of very low doses of GR205171 (0.1 ng) into the vicinity of a small medullary area that is dorsally adjacent to the semicompact part of the nucleus ambiguus (Fukuda et al. 1999b). Neurons in this area are about 4 mm below the floor of the fourth ventricle and are thought to receive inputs from the medial NTS that in turn may drive the central pattern generator for emesis. It was proposed that this interfacing brain area should be designated as the 'prodromal-sign centre', since GR205171 also reduced activation of the salivary centre. Thus, the prodromal-sign centre is thought to act as a relay from neurons exiting the medial NTS to both the central pattern generator for emesis and the salivary centre (Fukuda et al. 1999b).

Intravenous doses of GR205171 appear to act rapidly to prevent emesis induced by electrically stimulating the vagus nerve (see above), and there is also evidence that GR205171 and GR203040 prevent emesis induced by ipecacuanha (0.5 ml/kg; total alkaloid 0.7 mg/kg) that is probably mediated via the vagus nerves. Indeed, both drugs are effective to antagonize emesis following i.v. administration at 0.1 and 0.3 mg/kg (15 min pretreatment, 90 min observation time) (Gardner et al. 1995, 1996) and GR205171 also blocks emesis following a 0.2 mg/kg oral dose (Gardner et al. 1996). However, at sub-ID_{100} doses, both compounds are inactive in modifying the latency to the first episode of retching or vomiting (Gardner et al. 1995, 1996).

CP122721 has also been investigated in the dog for a potential to reduce emesis induced by chemotherapeutic drugs (Gonsalves et al. 1996). A dose of 0.1 mg/kg i.v. reduced cisplatin (3.2 mg/kg i.v.)-induced emesis during a 5-h period by approximately 87% and ondansetron and tropisetron produced approximate 91%–93% reductions (Gonsalves et al. 1996). However, CP122721 did not delay significantly the onset of cisplatin-induced emesis, whereas a number of dogs were completely protected from emesis when treated with the 5-HT_3 receptor antagonists (Gonsalves et al. 1996). Conversely, against methotrexate (2.5 mg/kg i.v.)-induced emesis the situation is different with CP122721 (0.1 mg/kg i.v. administered at 24, 36, 48 and 60 h) delaying the onset of emesis and both CP122721 and tropisetron antagonizing emesis by approximately 74%–78%, whereas ondansetron is essentially inactive (Yamakuni et al. 2000). The differential action of all three compounds in the cisplatin- and methotrexate-induced emesis models suggest fundamental differences in the mechanism of these chemotherapeutic drugs to cause emesis. The additional protection afforded by tropisetron relative to ondansetron was considered to be a function of its additional capacity to block 5-HT_4 receptors (Yamakuni et al. 2000).

4.3.3
Studies in the Cat

The cat is considered to be the most appropriate species to use for studies on motion sickness that involves the vestibular system and activation of central pathways in the emetic circuit (King 1990; Yates et al. 1998). It is also another species with historical importance in terms of providing important information relative to the organization of the emetic reflex (Borison and Wang 1953). A 30-min pretreatment with CP99994 at 0.3 mg/kg s.c. was highly effective in abolishing motion-induced emesis (Lucot et al. 1997). Conversely, CP100265, was inactive in this test consistent with its 900 times lower affinity for cat NK_1 receptors. Whilst the data indicate that NK_1 receptor antagonists may have a utility to antagonize motion-induced emesis in man, the studies found that sub-ID_{100} doses (30–170 µg/kg) were inactive to delay the first episode of emesis and all doses (including the ID_{100} dose) failed to modify the motion-induced epiphenomena that are presumed to reflect altered output from the autonomic nervous system (Lucot et al. 1997). It was hypothesized that the NK_1 receptor antagonists may act at a site very late in the emetic signalling pathway, possibly below the level of the medial NTS, or even outside of the NTS, in other emetic nuclei (Lucot et al. 1997).

There has also been a study of the anti-emetic potential of R116301 to inhibit emesis induced by xylazine (1.25 mg/kg s.c.) and ipecacuanha (1 ml/kg, intragastrically, total alkaloid content=1.4 mg/kg) (Megens et al. 2002). Xylazine is a potent α_2-adrenoceptor agonist that induces emesis centrally via the AP and the mechanism of ipecacuanha to induce emesis probably involves the vagus nerves (Colby et al. 1981; Hikasa et al. 1992). A 1-h oral pre-treatment with R116301 was effective in antagonizing emesis induced by both challenges, with ID_{50} values of 0.6 and 1.2 mg/kg to abolish xylazine- and ipecacuanha-induced vomiting, respectively (Megens et al. 2002). In these experiments R116301 was without effect on xylazine-induced sedation that is usually prevented by α_2-adrenoceptor antagonists, such as idazoxan. However, R116301 did prevent the ipecacuanha-induced meowing, which the investigators accepted as a potential of the compound to reduce nausea in this species (Megens et al. 2002). Unfortunately, the effect of R116301 on the latency of xylazine- or ipecacuanha-induced emesis was not reported.

4.3.4
Studies in *Suncus murinus* (the House Musk Shrew)

Suncus murinus has become established as a useful animal species to study the mechanisms involved in emesis control (Ueno et al. 1987). However, there are some differences in the sensitivity of the emetic reflex to several drug treatments. These include a reduced emetic sensitivity to apomorphine (Ueno et al. 1987), morphine, loperamide (Selve et al. 1994) and ipecacuanha (J.A. Rudd, unpublished results): the four challenges widely used to determine the anti-emetic potency of the NK_1 receptor antagonists in other species. The reason for the lack of sensitivity is unknown, but it may limit the capacity of this species in anti-emetic research. However, dopamine receptor antagonists have efficacy to reduce emesis in *Suncus murinus*, and morphine, loperamide and fentanyl have other actions on the emetic reflex that are prevented by naloxone (Rudd et al. 1999a; Selve et al. 1994), which also suggests the implication of functional dopamine and opioid receptors. Functional $5-HT_3$ receptors also clearly exist in the emetic reflex of *Suncus murinus* as $5-HT_3$ receptor agonists induce emesis (Torii et al. 1991a) and $5-HT_3$ receptor antagonists have an anti-emetic action against chemotherapeutic drugs (Torii et al. 1991b).

A wide variety of emetic stimuli have been used in this species to investigate the anti-emetic potential of NK_1 receptor antagonists. These include nicotine, which is presumed to act centrally at the AP (Beleslin and Krstic 1987), thus representing the substitute stimulus for apomorphine and morphine/loperamide in this species, while copper sulphate and cisplatin are presumed to act peripherally via the vagus nerves (albeit via different mechanisms).

Peripheral administration of CP99994 is approximately ten times less potent in *Suncus murinus* to antagonize nicotine (4–5 mg/kg s.c.)-induced emesis (Rudd et al. 1999b; Tattersall et al. 1995) than it is to antagonize apomorphine (0.25 mg/kg s.c.)-induced emesis in the ferret (Tattersall et al. 1994). The anti-emetic action was stereoselective and the difference in potency is presumed to relate to a lower affinity of CP99994 for *Suncus murinus* (IC_{50}=12 nM) versus ferret (IC_{50}=1.7–1.9 nM) NK_1 receptors (Table 1). However, whilst apomorphine-induced emesis in the ferret and dog can be prevented completely by CP99994 (Tattersall et al. 1994; Watson et al. 1995a), the emesis induced by nicotine in *Suncus murinus* is only maximally reduced by approximately 84%–89% (maximum reduction seen at 10 mg/kg). Further, against emesis induced by copper sulphate at a relatively high dose (120 mg/kg intragastrically), CP99994 produces only an approximate 53% reduction at 10 mg/kg (Rudd et al. 1999b). Whilst CP122721 is slightly more potent (maximum reduction observed at 10 mg/kg is approximately 80%) than CP99994 to antagonize copper sulphate-induced emesis, it is about equipotent to reduce nicotine-induced emesis (Rudd et al. 1999b). It is important to note, however, that none of the animals were protected completely from retching and vomiting induced by copper sulphate.

Using a wider range of compounds, the rank order of potency to reduce nicotine-induced emesis (based on ID_{50} doses) in this species was CP122721$\geq$CP99994>RP67580>FK888 (Rudd et al. 1999b). Indeed, RP67580 produces a significant reduction of nicotine-induced emesis only at very high doses (i.e. 30 mg/kg) that may relate to its low affinity for the *Suncus murinus* NK_1 receptor (>1 µM), and FK886 is essentially inactive. A similar rank order of inhibitory potency is seen for the compounds to antagonize copper sulphate-induced emesis (Rudd et al. 1999b). It is interesting to note that, of the above compounds studied, only CP122721 delayed significantly nicotine-induced emesis, but only at the high dose of 10 mg/kg which produces an approximate 80% reduction of emetic episodes, while all other compounds were inactive. Curiously, none of the compounds delayed significantly copper sulphate-induced emesis (J.A. Rudd, unpublished results).

It is possible that RP67580 and FK886 are less potent to antagonize emesis because of limited access into the brain, or low affinity for the *Suncus murinus* NK_1 receptor. To address this, the compounds (doses ranging from 1 to 30 µg) have been evaluated against nicotine (5 mg/kg)-induced emesis following intracerebral administration into the brain stem at the level of the dorsal vagal complex (Rudd et al. 1999b). These studies also investigated the anti-emetic potential of GR82334, which has been demonstrated to be less potent than CP99994 to antagonize cisplatin-induced emesis in the ferret following i.c.v. administration (Gardner et al. 1994). However, in *Suncus murinus*, GR82334 was slightly more potent than CP99994 (see above), with the complete rank order of inhibitory potency (based on ID_{50} values) being GR82334>CP122721$\geq$CP99994>FK888; RP67580 was inactive. The data support the hypothesis that the *Suncus murinus* tachykinin NK_1 receptor is unique (see Table 1) (Rudd et al. 1999b; Tattersall et al. 1995).

CP122721 has also been evaluated to inhibit nicotine (5 mg/kg s.c.)-induced emesis following i.c.v. administration (Fig. 5). Clearly, the latency of nicotine-induced emesis in these studies was not affected at the doses that failed to completely prevent emesis. For comparative purposes, the NK_2 receptor antagonists,

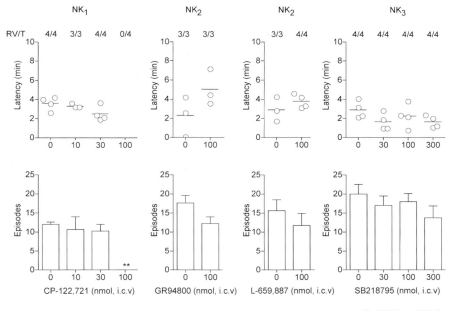

Fig. 5 Effect of centrally administered tachykinin receptor antagonists on peripherally administered nicotine-(5 mg/kg s.c.)-induced emesis in *Suncus murinus* (J.A. Rudd and M.K. Wai, unpublished results). RV/T indicates number of animals retching and/or vomiting out of the number of animals tested. **$P<0.01$ (ANOVA followed by Dunnett's multiple comparison test)

GR94800 and L659887, and the NK_3 receptor antagonist, SB218795, were also studied but they failed to affect the retching and vomiting response (Fig. 5). Whilst this may preclude a role for centrally located NK_2 and NK_3 receptors in the mechanism of nicotine-induced emesis, their involvement in the emesis induced by other treatments is unknown. Similarly, a role of peripheral NK_2 and NK_3 receptors in the emetic reflex remains to be investigated.

The NK_1 receptor antagonists have been less intensively studied in this species for the potential to antagonize cisplatin- and motion-induced emesis. The observation period that has been used to study cisplatin (at the very high dose of 80 mg/kg i.p.)-induced emesis was relatively short (3 h). Both GR203040 and GR205171 reduced the emetic response (Gardner et al. 1995, 1996). However, whilst GR203040 reduced motion (4-cm horizontal displacement for 5 min at 1 Hz)-induced emesis by approximately 78% at 3 mg/kg s.c., it antagonized cisplatin-induced emesis by only approximately 30%; a higher dose of 30 mg/kg reduces cisplatin-induced emesis by approximately 83% (Gardner et al. 1995). Conversely, GR205171 is required at a dose of 10 mg/kg to produce a similar 75% reduction of cisplatin-induced emesis but can reduce motion-induced emesis by approximately 71% at 1 mg/kg (Gardner et al. 1996). In comparison, CP99994 can completely prevent motion-induced emesis at 1 mg/kg in a 5-min test (Rudd et al. 1999b), but the level of antagonism is variable, with an approximate 74% reduction observed at 5 mg/kg i.p. in a 10 min test although, interestingly, the latency to onset of emesis is increased (Javid and Naylor 2002). No studies have been performed against cisplatin-induced emesis. These findings suggest that GR203040 and GR205171 are weak anti-emetics against cisplatin- and motion-induced emesis and again provide additional support for the hypothesis that the *Suncus murinus* NK$_1$ receptor is unique. However, it is interesting that GR205171 was capable of delaying the latency of motion to induce emesis, whilst it had no effect on the latency of cisplatin-induced emesis (Gardner et al. 1996); GR 203040 had no effect on the latency of motion-induced emesis but no report was made of its potential to modify the latency of cisplatin-induced emesis (Gardner et al. 1995). It is possible, therefore, that substance P (or another tachykinin) has a differential involvement in motion- and cisplatin-induced emesis in this species.

There are studies in other species (e.g. piglet) indicating that prostaglandins may be involved in the mechanism of cisplatin to induce emesis (Girod et al. 2002). It would be interesting to know, therefore, if NK$_1$ receptor antagonists have activity to prevent emesis induced by prostanoid receptor agonists. One of the most potent prostanoids to induce emesis in *Suncus murinus* is the prostanoid TP receptor agonist U46619 (0.3 mg/kg i.p.) (Kan et al. 2003a). The emetic mechanism involves the vagus nerves and is not prevented by adrenoceptor, 5-HT$_3$, dopamine, histamine and muscarinic receptor antagonists (Kan et al. 2003b). U46619-induced emesis is therefore resistant to conventional anti-emetics but CP99994 is active to antagonize the emetic response at 1 and 10 mg/kg by approximately 73% and 96%, respectively. However, in animals that were not completely prevented from retching and vomiting, the latency to the onset of emesis was not different from that of the control animals (Kan et al. 2003b).

The action of CP99994 has also been evaluated against emesis induced by a wide variety of other stimuli in this species, where the mechanism of emesis is not clearly defined. This includes the emesis induced by ethanol (40% v/v, i.p.) that probably involves metabolism to acetaldehyde to subsequently activate the emetic reflex (Chen et al. 1997). The precise sites of emetic action of ethanol are unknown but bilateral abdominal vagotomy does not prevent the response and emesis is not seen following intracerebral administration. Further, ethanol-induced emesis is unaffected by 5-HT$_3$ receptor antagonists, even though the mechanism may involve the generation of free radicals (Chen et al. 1997). However, CP99994 was effective to abolish emesis in four out of five animals at 10 mg/kg (Chen et al. 1997).

Another emetic stimulus that induces emesis via an unknown mechanism is high-dose naloxone (Rudd et al. 1999a). It should be emphasized that the emesis induced by naloxone (10–60 mg/kg s.c.) is not likely to be mediated via opioid receptors in this species since naltrexone does not modify the emetic response. However, it is known that the mechanism of naloxone to induce emesis does not involve dopamine, 5-HT$_3$, muscarinic or histamine H$_1$ receptors, but CP99994 at doses as low as 1 mg/kg antagonizes the emetic response (Rudd et al. 1999a). In these experiments, the emesis induced by naloxone was antagonised but not completely prevented in all animals even when the dose of CP99994 was increased to 30 mg/kg and there was no apparent trend to delay the latency to onset of emesis in those animals that were not completely protected from retching or vomiting (J.A. Rudd, unpublished results).

A treatment inducing emesis that may involve a peripheral and central mechanism is the ultra potent capsaicin analogue, resiniferatoxin (10–100 μg/kg s.c.). Thus, the emetic action is reduced by abdominal vagotomy, but not abolished, and it is known that intracerebroventricluar (i.c.v.) administration of resiniferatoxin also induces emesis (Andrews et al. 2000; Rudd and Wai 2001). The mechanism has been proposed to involve release of substance P or closely related tachykinins in the brain stem. Consistent with such a mechanism is the action of CP99994 to prevent emesis completely at 20 mg/kg s.c. (Andrews et al. 2000).

As NK_1 receptor antagonists clearly have broad anti-emetic effects in several species, it has been envisaged that they may be useful in circumstances where the mechanism of emesis is probably multifactorial. One such indication is to prevent the emesis occurring following surgery with anaesthesia (Andrews 1992, 1999, 2003). It is relatively straightforward to study the mechanisms of anaesthetic-induced emesis in *Suncus murinus*. Indeed, *Suncus murinus* is the only species where emesis following anaesthesia with halothane (1%) in a nitrous oxide (80%)/oxygen (20%) carrier occurs reliably. The model involves anaesthetizing the animals for 10 min and observing them for emesis during a 30-min recovery period (Gardner and Perren 1998). GR205171 at 1 and 3 mg/kg s.c. reduced significantly the number of emetic episodes by approximately 80%–87% and it significantly delayed the latency to the first emetic episode; these doses are similar to those required to prevent motion-induced emesis in this species (Gardner and Perren 1998). Unfortunately, there was no mention of the effect of the drug treatments on the duration of anaesthesia in the model, so the effect on latency must be viewed cautiously. Interestingly, ondansetron (3 mg/kg) also significantly reduced the number of emetic episodes by about 52%, but did not delay the latency to onset of emesis (Gardner and Perren 1998).

Emesis following surgery with anaesthesia in man may also involve mechanical stimulation of the pharynx during intubation (Andrews 1992). It is possible that the anaesthetic contributes to emesis if the anaesthetic distends the gastrointestinal tract. There have been many reports of the anti-emetic activity of several NK_1 receptor antagonists against stimuli that are presumed to have a peripheral mechanism involving the vagus nerves (see above). However, one of the most important investigations that may be relevant to PONV (and emesis involving gastric dysfunction) was a study investigating the mechanism of action of retching induced via mechanical stimulation of the upper gastrointestinal tract (Andrews et al. 1996). In these experiments, *Suncus murinus* was anaesthetized with pentobarbitone (30 mg/kg i.p.) and mechanical stimulation was effected by gentle probing of the pharynx and upper oesophagus. The retching response was antagonized by the 5-HT_{1A} receptor agonist, 8-OH-DPAT (0.1 mg/kg) but not by granisetron, (1–5 mg/kg), morphine (2 mg/kg s.c.) or CP99994

(20 mg/kg) (Andrews et al. 1996). These experiments clearly demonstrate an NK_1 receptor-independent form of emesis involving a pathway from the gastrointestinal tract. It is tempting to speculate that this pathway may also be involved in the mechanism of the residual emesis seen following copper sulphate and cisplatin in this species and it would be interesting to repeat the gastrointestinal distension experiments in other species. Nevertheless, the above experiments clearly demonstrate that NK_1 receptors are not the final receptor system to be activated to induce emesis in this species.

4.3.5
Studies in the Piglet

The piglet has been used to study the anti-emetic potential of GR205171 against cisplatin (5.5 mg/kg i.v.)- induced retching and vomiting during a 60-h observation period (Grelot et al. 1998). GR205171 was administered as an i.v. infusion 15 min prior to cisplatin and produced a delay in the onset of emesis (of approximately 28 h) that was significant at 0.3 and 1 mg/kg (Grelot et al. 1998). Interestingly, the authors classify the emesis occurring during the first 16 h as 'acute emesis' and that occurring during the 16–60-h period as delayed emesis (Milano et al. 1995). Clearly, GR205171 is highly effective in preventing acute emesis using this classification scheme (Grelot et al. 1998). It should be noted, however, that the classification scheme to define 'acute' and 'delayed' emesis in these experiments is somewhat limited: it has been based solely on the temporal appearance of emetic episodes. A more robust classification scheme would have used a pharmacological approach to define the response with 5-HT_3 receptor antagonists and glucocorticoids. Indeed, such experiments have been performed, but the drugs have not been administered past 36 h (Grelot et al. 1996). Nevertheless, GR205171 appears more potent than a 5-HT_3 receptor antagonist and dexamethasone in the model.

It is interesting that lower doses of GR205171 (0.01–0.1 mg/kg) that failed to reduce the initial emesis occurring during the first 16-h period by more than 95%, also failed to affect significantly the latency to onset of emesis (Grelot et al. 1998). In addition, the single administration of GR205171 at the start of the experiment was potent to antagonize delayed emesis, with 0.1 mg/kg reducing emesis by approximately 77%, while the higher doses of 0.1 and 1 mg/kg reduced emesis by approximately 72% and 86%, respectively (Grelot et al. 1998). This level of antagonism is particularly surprising considering that the plasma half-life of GR205171 is only 3.4±0.8 h in the piglet (Grelot et al. 1998). Nevertheless, in other experiments GR205171 (1 mg/kg) completely prevented the delayed emesis when given as a repeated 6-h injection starting at around 16 h post the start of the experiment (Grelot et al. 1998). Certainly, GR205171 seems more potent than 5-HT_3 receptor antagonists and dexamethasone in this model (Grelot et al. 1996).

It is worthy of comment that GR205171 was also active to antagonize chewing-like activity and the production of dense saliva that is usually associated with cisplatin-induced emesis (Grelot et al. 1998). The chewing-like activity and the production of dense saliva was interpreted as being representative of nausea, since it also occurs in piglets presenting with post-operative emesis and can be caused by intrathecally administered morphine (Grelot et al. 1998). However, these behaviours were usually only seen a few minutes or seconds prior to emetic episodes. The significance of the behaviours is therefore open to speculation and it is not even known if they involve the forebrain, which is thought to be integral to the mechanisms controlling nausea.

4.3.6
Studies in the Pigeon

The pigeon is a species that has been used recently in an attempt to develop a novel model of cisplatin (4 mg/kg i.v.)-induced acute and delayed emesis using 48–72 h observation periods (Tanihata et al. 2000). However, some $5-HT_3$ receptor antagonists induce emesis in this species (Navarra et al. 1992), and because of this the investigators have been careful to define the initial intense emetic action as an 'early' response, rather than an 'acute' response (Tanihata et al. 2000). The early response represents emetic episodes occurring during the first 8-h period, with subsequent episodes of emesis being defined as part of the 'delayed' response. However, the classification criteria are also based on the temporal appearance of the episodes and on the pharmacological sensitivity to parachlorophenylalanine, reserpine and dexamethasone. Thus, the early emesis in the pigeon is mediated partly via reserpine-sensitive monoaminergic mechanisms that include the 5-HT system, and the delayed response is associated with monoaminergic mechanisms but 5-HT systems are less important (Tanihata et al. 2000). In addition, both the early and the delayed response are partially mediated via the vagal nerves, with dexamethasone being active to reduce the emesis occurring during both the early and delayed phases (Tanihata et al. 2000).

The only NK_1 receptor antagonist evaluated for anti-emetic potential against cisplatin in the pigeon is GR205171. It was administered 1 h before cisplatin and reduced significantly the total emetic behaviours during the 0–8-h period by approximately 80% and 82% at 3 and 10 mg/kg, respectively; the emesis occurring in the remaining 8–24-h period was reduced by approximately 68% and 69%, respectively (Tanihata et al. 2003). Clearly, GR205171 has an anti-emetic potential in these studies but does not affect the latency to the first episode. A similar pattern of anti-emetic activity was seen following the i.c.v. administration of the compound, with a 30 μg/kg dose reducing the total emetic behaviours by approximately 59% and 93% during the 0–8-h and 8–24-h periods, respectively. Similarly, however, GR205171 did not affect the latency to the first episode. GR205171 was also administered a second time at 24 h during both experiments but only the intracerebral route was effective to reduce the total number of emetic behaviours (an approximate 95% reduction was observed at 30 μg) in the subsequent 24–48-h period.

In other studies, the investigators examined the anti-emetic activity of GR205171 against the emesis induced by an intracerebral administration of cisplatin (10 µg/kg). GR205171 was administered intramuscularly at 3 mg/kg or i.c.v. at 30 µg/kg and produced approximate 47% and 70% reductions in the total emetic behaviours occurring over a 24-h observation period. Clearly, GR205171 appears 100 times more potent to antagonize emesis when administered centrally. These findings demonstrate the relative importance of the central versus peripheral NK_1 receptors in the emetic reflex of this species. However, and importantly, it must be emphasized that none of the doses in any of the studies affected the latency of cisplatin to induce emesis (Tanihata et al. 2003).

4.3.7
Studies in the Rat

Common laboratory animals that do not possess the emetic reflex are also considered useful to anti-emetic research (Morita et al. 1988). In particular, the rat has been used in acute studies of drug- and motion-induced kaolin ingestion in an attempt to mimic mechanisms activated during vomiting (Takeda et al. 1993, 1995). Many toxins and emetic agents (including cisplatin) induce kaolin ingestion and the model also successfully identifies agents that have anti-emetic activity (Takeda et al. 1995). The kaolin ingestion model may offer a simplistic approach to studying the mechanisms that may be relevant to the emetic action of cisplatin.

Recently, it was shown that the rat exhibits acute (0–24 h) and delayed (48–72 h) pica activity following the i.p. administration of cisplatin (10 mg/kg; Saeki et al. 2001). In the model, daily injections of ondansetron (2 mg/kg i.p.), and ondansetron (2 mg/kg i.p.) plus dexamethasone (1 mg/kg i.p.), reduced the acute but not the delayed phase of kaolin consumption. CP99994 administered 1 h prior to cisplatin (10 mg/kg i.p.) did reduce kaolin consumption by approximately 60%–70% on days 1 and 2, but only at the extremely high dose of 60 mg/kg i.p. The antagonism of kaolin consumption at the higher doses may be non-specific as regards the NK_1 receptor. However, it also reduced acute kaolin consumption by approximately 40% when administered at 6 µg i.c.v.; only a dose of 600 µg was capable of reducing the kaolin consumption seen in both phases (Saeki et al. 2001). In the same model, HSP117 exhibited actions to reduce cisplatin-induced kaolin consumption, with effects being seen at doses as low as 3 mg/kg i.p. (an approximate 56% reduction of the acute kaolin consumption was observed). HSP117 was paradoxically less active following i.c.v. administration: only 400 µg produced a statistically significant effect (an approximate 88% reduction of cisplatin-induced acute kaolin consumption was observed) (Saeki et al. 2001).

GR205171 has been investigated for a potential to antagonize apomorphine (0.25 mg/kg s.c.) and amphetamine (0.5 mg/kg s.c.)-induced conditioned taste aversions (McAllister and Pratt 1998). GR205171 (0.2–1 mg/kg s.c.) dose-dependently antagonized the aversions induced by both stimuli, but so did ondansetron (0.01–0.1 mg/kg s.c.) (McAllister and Pratt 1998). The significance of the findings is difficult to explain since ondansetron does not prevent apomorphine-induced emesis in species with the vomiting reflex (Rudd et al. 1996a). Further, in the conditioned aversion paradigm, 5-HT$_3$ receptor antagonists fail to have actions to prevent cisplatin-induced conditioned taste aversions (Mele et al. 1992; Rudd et al. 1998), whereas they have well known actions to antagonize cisplatin-induced emesis in animals with the vomiting reflex (Andrews et al. 1988).

The mechanism of cisplatin to induce pica in rats is not fully understood but may involve activation of afferent pathways leading to the brain stem. However, cisplatin also has a 'delayed' toxic action to cause nephrotoxicity that may contribute indirectly to altered behaviour, and it is interesting that the structural damage to the renal tubules is antagonized by GR205171 (Alfieri and Cubeddu 2000). Unfortunately, the relevance to the anti-emetic potential of NK$_1$ receptor antagonists is unknown.

4.3.8
Summary of Pre-clinical Studies

It is clear that NK$_1$ receptor antagonists have an impressive spectrum of activity to antagonize emesis in at least five animal species. This includes an ability to antagonize emesis induced by centrally or peripherally acting stimuli, or by treatments that may have mixed actions to stimulate the emetic reflex (see Table 2). They also have a potential to reduce emetogen-induced conditioned taste aversions and pica in rodents. Other drug classes capable of exerting a broad inhibitory control of emesis include μ-opioid receptor (e.g. fentanyl and morphine) agonists (Rudd et al. 1999a), 5-HT$_{1A}$ receptor agonists (e.g. 8-OH-DPAT) (Kakimoto 1997; Yates et al. 1998) and vanilloid receptor agonists (e.g. resiniferatoxin) (Andrews et al. 2000). However, μ-opioid receptor agonists may be associated with respiratory depression (Martin 1983), 5-HT$_{1A}$ receptor agonists are not potent anti-emetics in all species and some have unwanted actions (Lucot 1994; Rudd et al. 1992) and some vanilloid receptor agonists produce hypothermia and other actions that are undesirable (Szallasi and Blumberg 1999). It appears that NK$_1$ receptor antagonists are distinguished as the only class of broad anti-emetics to date that achieve a useful control of emesis in the absence of major side effects. In addition, other broad-spectrum anti-emetics are receptor agonists.

It is apparent that several NK$_1$ receptor antagonists (e.g. CP122721, GR203040 and GR205171) can abolish emesis against centrally or peripherally acting stimuli, and this probably relates to the high affinity of these compounds for NK$_1$ receptors. The abolition of emesis is seen against practically all stimuli in the ferret, dog and cat, and in some experiments against motion-induced emesis in *Suncus murinus*. Indeed, in short experiments (where pharmacokinetic considerations are less of a concern), the dose of an antagonist in a particular species that is required to inhibit drug-induced emesis, or radiation-induced emesis, by at least 80%, is generally constant and not affected by the nature of the stimulus (Table 2). However, in *Suncus murinus*, whilst emesis due to many challenges is antagonized, there is a component of the emetic response that is resistant to, or not blocked at all by NK_1 receptor antagonists (i.e. gastrointestinal distension-induced emesis). The failure to completely abolish emesis in *Suncus murinus* probably reflects a different role, or position, of the NK_1 receptor in its emetic reflex. It is also clear in this species that slightly lower doses of NK_1 receptor antagonists are required to antagonize motion-induced emesis relative to drug-induced emesis. However, it is not known if this is an artefact because motion produces fewer emetic episodes, and consequently it may be easier to suppress the response.

It is also clear that against the majority of challenges, if the dose of the NK_1 receptor antagonist does not reduce emesis by approximately 80%, or more, it will not affect the latency to the first episode of emesis; unfortunately, not all studies reported the effect of the NK_1 receptor antagonists on the latency to the first emetic episode and this limits the power of the statement. The exceptions to this observation are the activity of the NK_1 receptor antagonists in ferret, dog and piglet models of acute and delayed emesis (it is not known why this is not extended to the pigeon model of cisplatin-induced early and delayed emesis) and the activity of GR205171 and CP99994 to antagonize motion-induced emesis in *Suncus murinus*.

There are some observations on the ability of NK_1 receptor antagonists to modify other behaviours associated with drug- and electrical stimulation-induced emesis (including fictive emesis) such as salivation, chewing, lip-licking and meowing, that have been taken as evidence that NK_1 receptor antagonists could have a potential to reduce nausea. However, many of these behaviours are seen in decerebrate or anaesthetized animals. Under such circumstances, the behaviours are clearly occurring independently of the forebrain, rather than as a manifestation of behaviours triggered by the forebrain as a consequence of nausea. Interestingly, however, NK_1 receptor antagonists do not block the motion-induced associated behavioural changes in the cat. Therefore, there may be differences between the motion experiments and the other situations where behavioural correlates of emesis have been recorded. Thus, it is tempting to speculate that motion-induced emesis involves several forebrain sensory systems and there may be a direct connection to areas involved in emotional control including nausea. Thus, the associated behavioural changes seen may occur independently of the brain stem although, undoubtedly, a concurrent activation of the brain stem pathways may be expected to return information rostrally.

What is also clear from the animal experiments is that the NK_1 receptors involved in the control of emesis are located centrally in the brain. This comes mainly from data showing a failure of potent non-brain penetrant NK_1 receptor antagonists to reduce cisplatin-induced emesis (Rupniak et al. 1997). However, this conclusion may be premature since data on the effect of these compounds to inhibit gastrointestinal irritation-induced emesis, where a potential release of substance P from EC cells may stimulate vagal afferents, is conspicuously absent. Thus, in some circumstances, peripheral NK_1 receptors may be involved (see Sect. 6). Other evidence for a role of central NK_1 receptors in the emetic reflex comes from studies administering agonists or antagonists into the vicinity of the NTS and AP, or by administration into the ventricular system of the brain. These studies clearly show that NK_1 receptor agonists can induce emesis and that NK_1 receptor antagonists reduce the emesis induced by cisplatin (ferret), intragastrically administered hypertonic saline (ferret) and nicotine (*Suncus murinus*). In the dog, centrally located NK_1 receptors below the level of the medial NTS, in the central prodromal-sign centre, appear to be critical for fictive emesis induced by electrical stimulation of the vagus. Some of the above points are considered in more detail in the closing section of the chapter in relation to the results from the clinical studies.

5
Clinical Evidence for the Involvement of Tachykinins in Nausea and Vomiting

The impressive and in many respects unique anti-emetic profile of selective NK_1 receptor antagonists reviewed above, together with the more diverse body of evidence implicating substance P (or a closely related tachykinin) in the emetic pathway, prompted clinical trials of a number of these agents to explore their potential to fulfil an unmet clinical need particularly in the delayed phase of chemotherapy-induced nausea and vomiting. The studies reviewed below cover relatively few compounds in specific sets of circumstances and hence may not be fully representative of this class of agent. In addition, ethical and clinical considerations have rightly determined the study design so that many of the studies in a chemotherapy setting have examined the agents in combination with current therapy rather than in isolation or as part of a placebo-controlled study as was possible when $5-HT_3$ receptor antagonists were at a similar stage of development. The studies have also been performed as part of the information required for licensing of a new drug and this is also likely to have influenced the study design. The information reviewed below has, where possible, used full papers rather than abstracts and has also used data from the FDA website following an Advisory Committee hearing. In view of these limitations this section should be regarded as indicating the clinical potential of the current NK_1 receptor antagonists rather than a definitive assessment which will have to await more extensive studies. The criteria by which each compound was selected for clinical trials is beyond the scope of this review but all compounds selected are highly efficacious in the ferret when given as single agents by either p.o., s.c. or i.v. routes and have a high affinity at the human and ferret NK_1 receptor (Tables 1, 2).

5.1
Chemotherapy-Induced Nausea and Vomiting

The identification in pre-clinical studies that NK_1 receptor antagonists could block both the acute and delayed phases of the emesis induced by the cytotoxic anti-cancer drug cisplatin, and hence could be differentiated from 5-HT_3 receptor antagonists, identified them as a potential advance in anti-emetic treatment. Therefore chemotherapy-induced nausea and vomiting has been the subject of the most extensive clinical investigations. Published studies are available for five NK_1 receptor antagonists (CP122721, CJ11974 [ezlopitant], GR205171 [vofopitant], L754030/MK869 [aprepitant], and L758298, the pro-drug for L754030), and studies with each will be reviewed separately. It must be noted that study design (e.g. double-blind, placebo), doses and combinations of cytotoxic drugs used, combination with and doses of concomitant medications (e.g. 5-HT_3 receptor antagonists and dexamethasone), methods of symptom assessment and clinical endpoints (e.g. criteria for complete control) vary between investigations.

5.1.1
CP122721 (Pfizer)

This agent was used in the first full paper investigating the anti-emetic effects of an NK_1 receptor antagonist in patients undergoing highly emetogenic cisplatin (≥ 80 mg/m^2) chemotherapy, a treatment regime inducing acute vomiting in 98% of patients not treated with an anti-emetic and delayed emesis in 89% of patients (Kris et al. 1985, 1989, 1997). This small study (17 patients in total) is of particular interest as CP122721 was given alone (50, 100 or 200 mg p.o.) as a single prophylactic dose and in combination with standard therapy of a 5-HT_3 receptor antagonist (unspecified, dose not stated) and dexamethasone (dose not stated). Patients were observed directly over the first 24 h to assess acute episodes of emesis but subsequent episodes were assessed by follow-up questioning. When given alone as a single prophylactic dose CP122721 (200 mg p.o.) blocked acute emesis in only 15% of patients but 71% of patients had two or fewer episodes representing a significant reduction from a historical control group. In patients given CP122721 prophylactically in combination with a 5-HT_3 receptor antagonist and dexamethasone no patient had acute emesis (compared with 78% in a subgroup in the absence of CP122721). Whilst these results show some efficacy of CP122721 in the control of acute emesis particularly when given in combination with other agents, the most striking results were in the delayed phase. In nine patients who had previously received cisplatin and had been given a 5-HT_3 receptor antagonist and dexamethasone, only 11% were free of emesis and this rose to 78% by the addition of a single prophylactic oral dose of CP122721. In a separate group of seven patients (four of whom had not previously received cisplatin) CP122721 blocked acute emesis in six (86%) patients. Taking all patients in the study together the complete control of delayed emesis rose from 17% to 83% by the addition of a single dose of CP122721.

Although the number of patients was small this study nevertheless provides evidence for the involvement of NK_1 receptors in chemotherapy-induced emesis in man and provides the first indication that this class of agent could be of particular benefit in the treatment of delayed emesis. Two issues of particular interest and which are recurring themes in many subsequent studies are: (a) How effective are NK_1 receptor antagonists when given alone in comparison to combination with current 'gold-standard' treatments? (b) What is the ideal dosing regime particularly for delayed emesis? In the above study a single oral dose given prior to cisplatin was efficacious in both the acute (24 -h) and delayed (2–5/7-day) phases.

5.1.2
CJ11974 (Pfizer)

A double-blind, randomized study was undertaken using CJ11974 in patients with cancer who were treated with cisplatin (>100 mg/m^2) on their first cycle of chemotherapy (Hesketh et al. 1999). Patients (n=61) were randomized to a group given a 5-HT$_3$ receptor antagonist (granisetron 10 µg/kg, i.v.) and dexamethasone (20 mg i.v.) immediately prior to cisplatin and a group given the same treatment but in combination with twice daily oral doses of CJ11974 (100 mg p.o.) for 5 days. Emetic episodes were counted by either clinical staff or patients and nausea quantified by patient self-assessment using a numerical scale. In the acute phase (first 24 h) 66.7% of the patients receiving a 5-HT$_3$ receptor antagonist + dexamethasone + oral placebo had no emesis. The addition of CJ11974 increased this incidence to 85.7% (P=0.09), and there was a significant increase in the percentage of patients reporting no nausea from 50% to 79% (P=0.024). CJ11974 significantly increased the percentage of patients free of emesis in the delayed phase (days 2–5) from 36.6% (placebo) to 67.8% (P=0.042). Although the percentage of patients free of nausea was higher in the CJ11974 group in the delayed phase (42.9% vs. 30%) this was not statistically significant. Over the entire 5 days following cisplatin CJ11974 increased the number of patients free of emesis from 30% to 64.3% (P=0.009).

This study clearly demonstrates the potential therapeutic use of an NK_1 receptor antagonist in the delayed phase of chemotherapy-induced emesis. However, it must be noted that this was a placebo-controlled study, and as the current treatment for delayed emesis is usually metoclopramide and dexamethasone, further studies are required. This study also showed a trend to improved control of acute emesis and also some indication that NK_1 receptor antagonists may have some beneficial effects against nausea.

5.1.3
GR205171 (Glaxo-Wellcome)

A small study (16 patients in total) investigated a single injection of GR205171 at one of two doses (5 and 25 mg) either alone and in combination with a single dose of ondansetron (8 mg i.v.) in patients receiving cisplatin-containing therapy (Fumoleau et al. 1998). GR205171 when given as a single agent to four patients who received a high dose of cisplatin (100 mg/m^2) showed no efficacy against acute nausea or emesis. As a result the additional studies were all performed in combination with ondansetron. In the 25-mg GR205171 group combined with ondansetron complete control of acute emesis was achieved in all five patients studied and in four out of five patients there was no nausea or rescue medication over the same time period. In three out of five patients complete control of emesis and in two out of five patients complete control of nausea was maintained over 72 h. The small numbers make interpretation difficult and without data on the number of emetic episodes it is difficult to assess whether GR205171 given alone is devoid of any efficacy. Although the patients in this study received a higher dose of cisplatin than in some other studies, this does not provide a satisfactory explanation. The combination study is consistent with other studies in appearing to show a benefit by the addition of an NK_1 receptor antagonist to a 5-HT_3 receptor antagonist regime.

5.1.4
L754030 and L758298 (Merck)

Published Studies. L758298 is the water soluble phosphoryl pro-drug of L754030 also designated as MK869 (aprepitant; Emend). Studies with both agents will be described and represent the largest series of published studies with NK_1 receptor antagonists. Navari et al. (1999) studied 159 patients who were receiving cisplatin for the first time at a dose of >70 mg/m^2. On the day of cisplatin administration all patients received the 5-HT_3 receptor antagonist granisetron (10 µg/kg i.v.) and dexamethasone (20 mg p.o.) and were then randomly assigned to one of three treatments: 400 mg L754030 p.o. just before cisplatin administration and 300 mg on days 2–5; 400 mg L754030 p.o. just before cisplatin treatment and p.o. placebo on days 2–5; placebo just before cisplatin and placebo on days 2–5 (i.e. this group only received a 5-HT_3 receptor antagonist on day 1). In the acute phase 67% of patients receiving granisetron and dexamethasone alone had no emesis, and this increased to 93% when L754030 was added to the regime ($P<0.001$) although there was no difference in the median nausea score between the groups. In the delayed phase only 33% of patients receiving granisetron and dexamethasone alone prior to cisplatin administration were free of emesis whereas in the group treated with L754030 for 5 days the figure was 87% and interestingly in the group given L754030 on day 1 only and placebo on subsequent days the figure was 78% ($P=0.001$). The median nausea rating was also significantly improved in the delayed phase in the L754030 groups. Although not part of the study the authors note that the values of complete protection obtained in the delayed phase of this study were up to 30 percentage points better than the most successful regimes with a 5-HT$_3$ receptor antagonist/metoclopramide + dexamethasone. This study demonstrates that emetic control in the acute phase can be improved by addition of an NK$_1$ receptor antagonist to a 'standard' regime and that control of delayed emesis and nausea is also markedly improved. However in the case of delayed emesis it is not clear from this study whether administration of an NK$_1$ receptor antagonist beyond day 1 confers any additional benefit (this has been studied in later trials) although the authors noted that in the group receiving L754030 for 5 days consistently more patients had no emesis, no rescue medication, lower nausea scores and higher global satisfaction scores than the group who received only L754030 on day 1. Examination of the daily median nausea scores (Fig. 2; Navari et al. 1999) suggests that if there is an effect of daily treatment it is at day 3 or beyond as the nausea ratings are virtually identical for days 1 and 2.

Campos et al. (2001) compared oral L754030/MK869 in combination with dexamethasone and a 5-HT$_3$ receptor antagonist for the treatment of acute and delayed emesis in 351 cisplatin ($\geq$70 mg/m^2)-naive patients who were randomized to four groups: granisetron (10 μg/kg i.v.) prior to cisplatin administration followed by placebo on days 2–5; granisetron (10 μg/kg i.v.) and MK869 (400 mg p.o.) pre-cisplatin and 300 mg p.o. MK869 on days 2–5; MK869 (400 mg p.o.) the evening before and pre-cisplatin followed by MK869 (300 mg p.o.) on days 2–5; MK869 (400 mg p.o.) followed by MK869 (300 mg p.o.) on days 2–5. All patients also received dexamethasone (20 mg p.o.) prior to cisplatin but did not receive dexamethasone on subsequent days. In the acute phase the best results were obtained for patients who received MK869, granisetron and dexamethasone with 80% of patients protected from emesis. In the other groups comparable levels of protection of 57%, 46% and 43% were obtained. This result essentially confirms those of previous studies that the optimal treatment for acute emesis is a combination of MK869, dexamethasone and a 5-HT$_3$ receptor antagonist. In the delayed phase of emesis all three groups in which MK869 was given on days 2–5 were superior (complete protection 63%, 57 %, 51%) to the placebo group (29%) and nausea scores were also lower in the MK869 group. This study supports the previous study which demonstrated efficacy of MK869 given alone in the treatment of delayed nausea and vomiting.

A randomized double-blind, randomized, active-agent (ondansetron)-controlled study of 53 cisplatin (50–100 mg/m^2)-naive patients was carried out using L758298, the pro-drug for L754030/MK869 (Cocquyt et al. 2001). Patients received either L758298 (60 or 100 mg i.v.) or ondansetron (32 mg i.v.) prior to cisplatin treatment with nausea and vomiting being assessed over 7 days. Note that in this study patients did not receive dexamethasone as part of the study regime. In the entire acute phase 37% of patients in the L758298 group were free of emesis whereas in the ondansetron group this percentage was 52%, this difference being not significant. Examination of the first 8 h following cisplatin revealed a significant difference (P=0.001) between the two treatments, with only

37% of patients being emesis-free in the L758298 group compared with 83% in the ondansetron group. Administration of L758298 on day 1 resulted in 72% of patients being free of emesis in the delayed phase compared to ondansetron where the percentage was 30% ($P=0.005$). There was also a significant difference in the percentage of patients free of emesis and who did not use rescue medication. In the acute phase nausea scores were not significantly different but as with emesis the scores were lower in the ondansetron group over the first 8 h but after that time the scores in the L758298 group were higher. The nausea scores in the delayed phase were not significantly different between the two groups although on day 2 only the scores for the L758298 group were significantly lower. This study demonstrates equivalence between an NK_1 receptor antagonist and ondansetron in the acute phase and superiority in the delayed phase.

A further study of MK869 was carried out in 177 patients receiving cisplatin (≥ 70 mg/m^2) to compare the effect of the addition of dexamethasone to L758298 (the pro-drug of MK869) with its addition to ondansetron (Van Belle et al. 2002). One group was given L758298 (100 mg i.v.) and dexamethasone (20 mg i.v.) prior to cisplatin and MK869 (300 mg p.o.) on days 2–5; a second group had the same regime on the day of cisplatin administration but was given oral placebo on days 2–5; the third group was treated with ondansetron (32 mg i.v.) and dexamethasone (20 mg i.v.) on the day of cisplatin administration and oral placebo on days 2–5. Emesis and nausea (visual analogue scale, abbreviated VAS) were assessed. In the acute phase, although the L758298 + dexamethasone combination had demonstrable efficacy, the ondansetron + dexamethasone combination was clearly superior when assessed using the proportion of patients with no emesis and who did not take rescue medication (combined L758298 + dexamethasone 40% vs. ondansetron + dexamethasone 83%, $P<0.001$) and this was also the case if the use of rescue medication was excluded. Nausea scores in the acute phase were also significantly lower in the ondansetron + dexamethasone group (median phase VAS score 1 vs. 11, $P<0.05$). In the delayed phase the proportion of patients with no emesis and no use of rescue medication was higher in the two groups given L758298 and dexamethasone on day 1 (59%, 46%) than in the ondansetron + dexamethasone group (38%) but only the group which received MK869 on days 2–5 was significantly different from the ondansetron group ($P<0.05$). If the use of rescue medication is excluded then both L758298 groups were significantly different from the ondansetron group (65% and 61% vs. 41%, $P<0.05$). No significant differences were found between the groups in the nausea score or incidence in the delayed phase (no or minimal nausea 38%, 46%, 50%). However, the authors argued that because the response to anti-emetic treatment in the delayed phase is to some extent dependent upon the side effects in the acute phase, the patients in the L758298 group would have been expected to have had worse nausea in the delayed phase than those in the ondansetron group which had substantially less nausea in the first 24 h. This was not the case and indicates that L758298 did have some effect against nausea in the delayed phase. This study essentially confirms the results of earlier studies in demonstrating a significant improvement in the treatment of delayed emesis in patients who received an NK_1 receptor antagonist as part of their anti-emetic therapy. A post hoc analysis examined the relationship between emesis in the acute phase and the delayed phase. This revealed that the percentage of patients who were without emesis in the delayed phase despite having had emesis in the acute phase was higher in the L758298 groups (22% and 26%) than in the ondansetron group (0%). In addition, examination of patients who had no acute emesis but then experienced emesis in the delayed phase revealed that these patients were least commonly encountered in either the group which received MK869 for 5 days (four patients) or L758298 for 1 day (seven patients) in comparison with the ondansetron group (25 patients).

These observations support the use of an NK_1 receptor antagonist on day 1 for the treatment of delayed emesis and argue for continued use to further increase efficacy. The lower efficacy of the NK_1 receptor antagonist + dexamethasone combination compared to the $5\text{-}HT_3$ receptor antagonist + dexamethasone combination in the acute phase is perhaps not surprising in the light of the earlier studies using a triple combination (Navari et al. 1999; Campos et al. 2001) but may be of interest from a mechanistic aspect as some of the anti-emetic effects of dexamethasone have been ascribed to an allosteric modulatory effect of ligand-gated ion channel receptors of which the $5\text{-}HT_3$ receptor is an example, as is the nicotinic acetylcholine receptor. The NK_1 receptor is a G-protein receptor and hence is unlikely to have an allosteric modulatory site although steroids may interact with tachykinin neurotransmission in other ways (see Sect. 6).

Hesketh et al. (2003b) have recently reviewed and reanalysed some of the data from the Phase II clinical trials of aprepitant (Cocquyt et al. 2001; Campos et al. 2001) to provide further insights into the clinical potential of this agent and the pathophysiology of chemotherapy-induced emesis. Analysis of the impact of aprepitant upon the time course of the emetic responses to cisplatin was undertaken using Kaplan–Meier curves and reveals a number of interesting effects of this NK_1 receptor antagonist in the acute phase. A comparison of a single dose of L758298 (the pro-drug for aprepitant) with ondansetron (32 mg i.v.) showed that in the ondansetron group there was a progressive decline in the number of patients completely protected from 82.6% at 8 h, 69.6% at 16 h and 52.2% at 24 h. In contrast with L758298, the corresponding percentages were 36.7%, 36.7% and 36.7%. Ondansetron is clearly superior in the early part of the acute phase but shows a progressive decline in efficacy in the latter half of the acute phase whereas L758298 maintained its efficacy and is relatively more effective between 8 and 24 h. A similar result was found in a study comparing oral aprepitant + dexamethasone (completely protected during 0–8 h 45.2%, 0–16 h 44%, 0–24 h 42.9%) with granisetron (10 μg/kg i.v.) + dexamethasone (0–8 h 92.2%, 0–16 h 81.1%, 0–24 h 56.7%). Addition of aprepitant to the granisetron + dexamethasone regime prevented this decline, with the 0–24-h figure being 79.8% for the triple therapy. This study also showed that the rapid drop (0–8 h) in the number of patients completely protected in the aprepitant group was not affected by giving an additional dose of aprepitant on the evening prior to cisplatin.

Studies on Aprepitant (Emend, L754030/MK869) Presented at the FDA Gastrointestinal Drugs Advisory Committee Meeting. Much of the material presented at the Advisory Committee Meeting was derived from the trials described above or from data presented in abstract form only, but additional material or analysis was presented and selected material will be included here to provide as complete a picture as possible of the only NK_1 receptor antagonist licensed to date[1]. The proposed indication is "Emend, in combination with other anti-emetic agents, is indicated for the prevention of acute and delayed nausea and vomiting associated with initial and repeat courses of highly emetogenic cancer chemotherapy, including high-dose cisplatin." A Phase III dose-finding study was undertaken using a capsule formulation of aprepitant which was found to have better than expected oral bioavailability. Aprepitant was studied in combination with ondansetron and dexamethasone on day 1 and dexamethsaone alone on subsequent days. A dose of 125 mg on day 1 followed by 80 mg on days 2 and 3 was selected based upon efficacy in terms of complete response in the acute (83%) and delayed (73%) phases and Kaplan–Meier analysis of the time to first emesis or rescue. This dose was confirmed to be effective in two multinational trials [2] of patients on high-dose cisplatin with complete response rates of 89% and 83% in the acute phase and 75% and 68% in the delayed phase of the first cycle of cisplatin. The additional benefit provided by combination of aprepitant with a $5-HT_3$ receptor antagonist and dexamethasone was maintained over the six chemotherapy cycles studied.

Although it is clear that aprepitant improves the efficacy of a 'conventional' anti-emetic regime in terms of control of emesis the data from the measurement of nausea using a VAS are less convincing as was the case in the published studies. The two Phase III studies showed that, although there was a clear trend in favour of the aprepitant regime, this was only significant for 'no nausea' (48% vs. 42%, $P<0.05$) and 'no significant nausea' (72% vs. 65%, $P<0.05$) when the results from the two studies were combined. However, aprepitant was statistically superior in both studies when the impact of nausea upon daily life was assessed. There was also less use of rescue medications in the aprepitant groups supporting the beneficial effects of the aprepitant regime.

Two particular aspects of the pharmacokinetic/pharmacodynamic studies undertaken during the development of aprepitant are of relevance to this review. Firstly, the oral dosing regime of 125 mg on day 1 followed by 80 mg on subsequent days was shown to provide peak plasma concentrations of aprepitant of ~1500 ng/ml and trough levels which did not fall below ~500 ng/ml. This value is of particular relevance when seen in the context of a human positron emission tomography (PET) study of NK_1 receptor occupancy in the corpus striatum, an area of the forebrain rich in NK_1 receptors (Hargreaves et al. 2002). Using a pos-

---

[1] Emend was approved by the European Medicines Evaluation Agency in July 2003 for the treatment of emesis induced by cisplatin chemotherapy.

[2] The results from the phase III studies used to support the FDA filing have now been published by Poli-Bigelli et al. (2003), Hesketh et al. (2003a) and de Wit et al. (2003, 2004).

itron emitting ligand which binds reversibly, quantitatively and with high affinity to NK_1 receptors in the human brain it was possible to demonstrate a sigmoid-shaped relationship between aprepitant plasma concentration and striatal NK_1 receptor occupancy, with plasma concentrations of aprepitant of ~10 ng/ml and ~100 ng/ml producing NK_1 receptor occupancies of ~50% and ~90%, respectively, in the striatum. Whilst the striatum is not implicated as a site at which these agents have their anti-emetic effects (see below) there is no reason to believe that the binding at this site is not representative of the brain stem which is proposed to be the major site of action, and as imaging techniques improve in resolution it should become possible to investigate this directly. Combining the results of this PET study with the trough plasma levels of aprepitant in patients on the 125/80-mg dosing regime and the efficacy of aprepitant found in the dose-ranging study indicated that brain occupancy levels of >95% are required for the maximum anti-emetic efficacy of aprepitant. This assumption is true only if the endogenous agonist for the NK_1 receptors at the site of action, presumed to be the brain stem (see Sect. 6), is the same as in the striatum. If this is not the case then the antagonist dose required to achieve an anti-emetic action may be different.

Secondly, aprepitant is metabolized by the cytochrome P450 system CYP3A4 but aprepitant is also a moderate (comparable to grapefruit juice, verapamil, diltiazem) net inhibitor of CYP3A4 in the initial 5 days of administration and this was found to have a significant effect on dexamethasone which is also a substrate for CYP3A4. The plasma area under the curve levels of dexamethasone were approximately doubled by co-administration with aprepitant (125/80 mg). As a result of this observation the dose of dexamethasone was reduced by approximately half (from 20 mg on day 1 and 8 mg on days 2–5 to 12 mg on day 1 and 4 mg on days 2–5) in the subsequent Phase III studies to match the dexamethasone exposure to that usually found in the absence of aprepitant. As both ondansetron and granisetron are metabolized in part by CYP3A4, a pharmacokinetic study was undertaken but this did not show a clinically significant effect and hence there is arguably no need for dose adjustment of these $5\text{-}HT_3$ receptor antagonists.

5.2
Post-operative Nausea and Vomiting

PONV is a major clinical problem with a multifactorial underlying mechanism (Andrews 1999, 2003) which can be treated to a limited extent using $5\text{-}HT_3$ receptor antagonists (Tramer et al. 1997). Studies of PONV have been undertaken in patients undergoing gynaecological surgery. One study compared the $5\text{-}HT_3$ receptor antagonist ondansetron given alone with CP122721 (200 mg p.o.) given alone and both drugs in combination. Whilst there were no differences in the nausea score (measured by VAS) between the three groups there was a significantly lower incidence of emetic episodes in the groups given CP122721 either alone or in combination with ondansetron and, although the combination was not superior to CP122721 alone in terms of emetic episodes, the combination was superior in increasing the time free of any symptoms of PONV (Gesztesi et al. 1998). A placebo-controlled study by Diemunsch et al. (1999) giving patients who experienced nausea and/or vomiting during recovery from surgery a single i.v. injection of GR205171 (25 mg) showed that the percentage of patients with complete control of emesis and nausea with no rescue medication or premature withdrawal was greater in the GR205171 group than in the placebo group. The number of emetic episodes was significantly lower in the GR205171 group 2 h after treatment and nausea was less severe (not significant) at all time points up to 24 h. The proportion of patients requiring rescue medication also differed in the two groups: at 6 h 44% of the GR205171 group vs. 67% of the placebo group; at 24 h 61% of the GR205171 group vs. 83% of the placebo group. In this study patients received at least one dose of opioid medication within 6 h of surgical recovery. Opioids are a risk factor for PONV and emesis associated with opioids is not blocked by 5-HT$_3$ receptor antagonists (Apfel et al. 2002).

Gesztesi et al. (2000) described a study of CP122721 in patients undergoing abdominal hysterectomy (nitrous oxide anaesthesia) comparing two oral doses of CP122721 (100 mg, 200 mg) with ondansetron (4 mg i.v.) and then the higher dose of CP122721 in combination with ondansetron. This study showed that 200 mg of CP122721 (p.o.) was more effective than 100 mg in reducing episodes of emesis. The dose of 200 mg CP122721 reduced the incidence of emetic episodes from 50% in the placebo group to 10% in the first 8 h after surgery and also reduced the percentage of patients requiring rescue medication (48% vs. 25%). Combination of CP122721 with ondansetron reduced the incidence of emesis in the 2 h following surgery from 17% to 2% and increased the median time patients remained emesis-free. The nausea scores were not significantly altered although there was a trend towards lower scores in the CP122721 group. There was no evidence for analgesic or opioid sparing in the CP122721 group when compared to ondansetron.

The results from these preliminary studies show that NK$_1$ receptor antagonists have demonstrable efficacy against emesis associated with recovery from surgery and whilst the effects on nausea were not significant the trend seen warrants further investigation.

5.3
Motion-Induced Emesis

Pre-clinical studies in the cat and *Suncus* demonstrated the potential of NK$_1$ receptor antagonists for the treatment of emesis induced by motion exposure. Using a well-established cross-coupled provocative motion stimulus with self-reported nausea on a scale of one to four as an endpoint, L758298 (60 mg i.v., the pro-drug for L754030/MK869) was not different from placebo and was less effective than scopolamine (Reid et al. 1998). The anti-nausea potential of GR205171 was tested in another motion study (Reid et al. 2000) that examined the compound both alone and in combination with a 5-HT$_3$ receptor antagonist (ondansetron). The positive control drug hyoscine bromide (0.6 mg p.o.) was superior to placebo but neither GR205171 alone (25 mg i.v.) or in combination with ondansetron (8 mg i.v.) was different from control. These two studies provide clear evidence that NK_1 receptor antagonists lack efficacy against motion-induced nausea; however, they do not address the issue of whether vomiting is affected.

5.4
Summary of the Clinical Studies of the Anti-emetic Effects of NK_1 Receptor Antagonists

Studies exploring the full clinical utility of NK_1 receptor antagonists are in their infancy but as the first member of this class has now been approved by the FDA (April 2003) and European Medicines Evaluation Agency (July 2003) this situation will change rapidly. The focus has been upon the emesis associated with anti-cancer chemotherapy, especially the delayed phase. These agents do provide an additional benefit (approximately 20% more patients are completely protected) when used in conjunction with the currently used anti-emetic regimes. There are encouraging data that chemotherapy-induced nausea is also improved but this requires considerable further investigation and, if one takes the current data at face value, in common with other anti-emetic regimes, the efficacy against nausea lags about 10% behind that for emesis.

Few studies have investigated the efficacy of NK_1 receptor antagonists given in isolation (there may be practical considerations which make this difficult) but without further studies of this aspect it will be impossible to establish the relative involvement of substance P and NK_1 receptors in the emetic pathway in man and hence to make comparisons with the animal models which may allow refinement for future research into anti-emetics. For example, the considerably greater efficacy of $5-HT_3$ receptor antagonists given alone when compared to the NK_1 receptor antagonists in the acute phase of chemotherapy-induced emesis implies that in man a $5-HT_3$ receptor-dependent pathway is active with a lesser involvement of NK_1 receptors whereas in the delayed phase this situation is reversed. In the ferret the acute phase of cisplatin-induced emesis can be blocked by either a $5-HT_3$ or an NK_1 receptor antagonist.

An interesting property of $5-HT_3$ receptors is their ability to stop ongoing emesis within a few minutes of i.v. injection in humans. Similar studies with NK_1 receptor antagonists comparing the acute and delayed phases would be of considerable interest and would shed light on the relative contributions of direct (e.g. blockade of neurotransmission) and indirect (e.g. tissue inflammation) effects to their anti-emetic action (see Sect. 6). The 'carry over' effect seen in some studies and the 'slow onset' (improving efficacy over 8–24 h and in the delayed phase) could more readily be explained by such indirect peripheral effects which are discussed in detail in Sect. 6.1.

The small studies in PONV are encouraging that these agents may be of benefit but more extensive studies are needed to establish their role in treatment of

PONV and to understand the underlying mechanisms of PONV (Andrews 2003). Again the results against nausea were equivocal.

The involvement of NK_1 receptors in motion sickness in humans is difficult to judge from studies that looked only at nausea but not vomiting. From the more extensive studies of chemotherapy-induced emesis (and to some extent the pre-clinical studies) one would have predicted that there would be effects against vomiting but that effects against nausea may be fewer or absent. The results from the motion-induced nausea studies may be important in understanding the site of action of the NK_1 receptor antagonists as, if nausea is unaffected, they may indicate that the site of action is 'distal' to the divergence of the motor (retching and vomiting) and sensation (nausea) parts of the central pathway. Alternatively, nausea may represent a special case where the pathway for the genesis of nausea does not pass via the brain stem as is the case for the other major inputs of the AP and vagal afferents. It is to be hoped that further studies of motion sickness will be undertaken to answer this question with wider implications.

With the exception of the motion sickness study published, studies on the anti-emetic effects of NK_1 receptor antagonists in human volunteers are notably lacking. Emesis induced by systemic apomorphine and oral ipecac is well established and the latter was used in the development of the $5\text{-}HT_3$ receptor antagonist ondansetron (Minton et al. 1993; Foster et al. 1994) as it may provide a model for a component of the acute emetic response to cytotoxic drugs. Such studies could provide important insights into the site and mechanism of action (see Sect. 6) of this class of agent in humans and enable a reduction in the use of animals or refinement of existing models.

6
Site(s) and Mechanism(s) of the Anti-emetic Action of NK_1 Receptor Antagonists

Two major issues need to be considered here: the site(s) at which NK_1 receptor antagonists act and the mechanism(s) underlying their actions. The majority of the conclusions drawn are based upon pre-clinical studies. Discussion of these issues will be considered under two headings: central versus peripheral, and then the precise location with relevant mechanistic comments included in both sections.

6.1
Central vs. Peripheral Site of Action

Several key experiments and pieces of supportive evidence provide a convincing case for a central site of action in the brain and, whilst this is certainly the major site, there are a number of observations which suggest that a peripheral contribution should not be ignored.

6.1.1
Spectrum of Anti-emetic Effects

The broad spectrum anti-emetic effects demonstrated against stimuli acting through central (e.g. AP, vestibular system) and peripheral (e.g. vagal afferents) pathways in the pre-clinical studies is very hard to explain without invoking a central site of action at some critical central 'choke' point where either the input pathways converge, the motor outputs originate or where the two connect (Figs. 2, 3).

6.1.2
Central Penetration

Experimental studies using potent, selective NK_1 receptor antagonists which are peripherally restricted provides some of the most convincing evidence that the anti-emetic effects are mediated at a site within the blood–brain barrier. All studies of this type have been performed in the ferret (Gardener et al. 1994; Watson et al. 1995a; Tattersall et al. 1996), but are supported by studies in the rat measuring brain penetration of compounds (Tattersall et al. 1996) and studies in the gerbil using foot-tapping induced by a selective NK_1 agonist as an assay of antagonist brain penetration (Rupniak et al. 1997). Watson et al. (1995b) reported that a series of non-brain penetrant NK_1 receptor antagonists were not effective against intragastric ipecacuanha-induced emesis, and both Gardner et al. (1994) and Tattersall et al. (1996) reported that the non-penetrant compounds GR82334 and L743310 showed no anti-emetic activity against acute cisplatin-induced emesis when given systemically. However, as is the case with $5-HT_3$ receptor antagonists, this only excludes an involvement of the specific receptor in the response to that specific emetic stimulus. Caution should be exercised in generalizing the results. Anti-emetic effects could be demonstrated against cisplatin when GR82334 and L743310 were given either intra-cisternally or directly into the dorsal brain stem. These studies provide strong evidence for a central site of action but several issues need to be raised.

Area Postrema. The AP is a brain stem region where the blood–brain barrier is relatively permeable and where agents unable to access the remainder of the brain would be expected to reach. NK_1 receptor antagonist binding has been reported in the AP (Ariumi et al. 2000) and even peripherally restricted agents would be expected to penetrate here so their lack of efficacy would suggest that the AP NK_1 receptors do not play a significant role in cisplatin-induced emesis in the ferret.

Peripheral Anti-inflammatory Action. The peptide NK_1 receptor antagonist sendide (Tyr-D-Phe-Phe-D-His-Leu-Met-NH$_2$) which is presumed to be poorly brain penetrant, reduced but did not block cisplatin-induced emesis when given systemically in the ferret (Minami et al. 1998). Whilst it is possible that sendide may have gained access to the dorsal brain stem, a peripheral action appears a more likely explanation and is supported to some extent by afferent recordings showing effects of sendide on substance P-induced vagal afferent activation. Whilst the apparent anomaly with the sendide study is not yet resolved, it and other studies (e.g. Tattersall et al. 1996; Minami et al. 2001) prompt a consideration of the peripheral actions of the centrally penetrant agents with anti-emetic effects and whether such actions could either contribute to the anti-emetic effects or have additional beneficial effects in patients treated with cytotoxic drugs.

Watson et al. (1995a, 1995b) pointed out that although NK_1 receptor antagonists such as CP99994 most likely had their site of action in the brain stem, a peripheral site on vagal afferents responding to mucosally released substance P could be involved, and Minami et al. (2001) reiterated the argument that a peripheral site of action would be obscured by the central action. There are two sources of substance P in the mucosa: axon collaterals of sensory afferents and EC cells (see Sects. 3.1.3, 3.1.4), and there is circumstantial evidence to suggest that they should not be discounted as relevant to emesis. For example, Tattersall et al. (1996) have shown that both brain-penetrant (L741671) and non-brain-penetrant (L743310) compounds reduced plasma extravasation induced by resiniferatoxin in the guinea pig oesophagus. Plasma extravasation is one component of the inflammatory response which leads to localized oedema. This vascular effect results from an increase in the permeability of the microvasculature, an effect that can also be induced by substance P released from sensory afferent terminals ('axon reflex'). A reduction in inflammation and oedema are well known effects of the corticosteroid dexamethasone, and hence NK_1 receptor antagonists may be beneficial in situations in which dexamethasone is effective (e.g. delayed emesis). Recording afferent activity from abdominal vagal afferents in the ferret, Minami et al. (2001) provided functional evidence that injection of 5-HT evoked an afferent discharge that was significantly reduced by CP99994 as was the response to substance P itself. They proposed that as both 5-HT and substance P are known to be involved in epithelial inflammatory responses, and gut inflammation has been implicated in the delayed emetic response to cytotoxic drugs, an action of NK_1 receptor antagonists modulating the responses mediated by 5-HT and substance P could indirectly contribute to its efficacy in the delayed phase of cisplatin-induced emesis. This possibility requires direct experimental investigation. Blockade of these peripheral effects may also be important in reducing the possibility of up-regulation of NK_1 (or other) receptors in the NTS in a way analogous to that seen in the spinal cord in chronic pain studies.

Peripheral Motor Action. An additional peripheral action which could contribute to an anti-emetic effect is on gastrointestinal motility. NK_1 receptors are located in the enteric nervous system (Southwell et al. 1998) and are expressed by $19\pm7\%$ of vagal efferents (pre-ganglionic) supplying the stomach and $46\pm7\%$ supplying the duodenum (Blondeau et al. 2002). Substance P-induced relaxation of the lower oesophageal sphincter (an event occurring during vomiting) can be blocked by CP99994 (Smid et al. 1998). It is not possible to predict what effect blockade of these receptors will have on the motility patterns associated with nausea (e.g. antral tachygastria, delayed gastric emptying) or vomiting (lower oesophageal sphincter relaxation, retrograde giant contraction in the small intestine), but this merits investigation. Early studies implicated substance P in the genesis of intestinal retroperistalsis, but this has not been studied recently (Niel 1991). Disordered motility has been implicated in the mechanism of delayed emesis induced by cytotoxic drugs based upon the efficacy of the prokinetic agent metoclopramide (Gralla 1981).

6.1.3
Central Administration

Intracerebroventricular administration of the peptide NK_1 receptor antagonist sendide (ED_{100} 584 nM) reduced the retching and vomiting response to intraduodenal hypertonic saline in halothane-anaesthetized ferrets and also additional physical signs such as licking and salivation as well as reducing the accompanying rise in blood pressure (Davison et al. 1995). Similar effects were obtained using the non-peptide antagonist CP96345 (309 nM) in the ferret, and in the dog i.c.v. administration of CP99994 blocked the emetic response to abdominal radiation (Otterson et al. 1997). In the latter study the retrograde giant contractions which are one of the vagally mediated gastrointestinal motor events which normally occur prior to retching and vomiting were also blocked whereas the giant migrating contractions associated with diarrhoea were not. The emetic response to morphine and copper sulphate in the ferret was also blocked by i.c.v. administration of the NK_1 receptor antagonists HSP117 and CP99994 (Saito et al. 1998).

Several studies in which either brain-penetrant (e.g. L741671, CP99994, GR205171, HSP117) or non-brain-penetrant (e.g. GR82334, L743310) NK_1 receptor antagonists were injected into the brain stem have demonstrated anti-emetic effects against stimuli including cisplatin, electrical stimulation of abdominal vagal afferents, morphine and copper sulphate (Gardner et al. 1994; Tattersall et al. 1996; Fukuda et al. 1998; Ariumi et al. 2000). These studies provide evidence for a central site of action of NK_1 receptor antagonists in the brain stem. Attempts to identify the precise location are reviewed next.

6.1.4
Precise Location

The precise site of the anti-emetic action of NK_1 receptor antagonists within the brain stem has not been identified but it is possible to identify the most likely site(s) based upon a number of neurophysiological studies in the dog and also by re-considering some of the neuroanatomical, physiological and pharmacological studies described in other sections. In attempting to identify the site, it must be borne in mind that the majority of pre-clinical studies have used retching and vomiting, the somatomotor expressions of activation of the emetic reflex, as the index of antagonist activity. Some clinical studies have investigated the effects against nausea and a few pre-clinical studies have reported effects of NK_1 receptor antagonists on the behaviours which accompany emesis but which may arguably be surrogates of nausea, although this is highly controversial. Finally, very few studies have investigated whether the visceral (e.g. salivation, proximal gastric relaxation, retrograde giant contraction of the small intestine) and other autonomic nervous system prodromata (e.g. tachycardia) of emesis are affected. A knowledge of the effect of NK_1 receptor antagonists on each of these parameters would allow better mapping of the site(s) of action.

Prior to discussing the possible sites of action it is necessary to briefly outline the functional neuroanatomy of the brain stem emetic circuitry (see Fig. 2; Grelot and Miller 1997; Grelot and Bianchi 1997). The reflex acts of retching and vomiting including the accompanying visceral events are coordinated entirely within the brain stem, but the sensation of nausea requires more rostral brain structures. In the brain stem two main sets of events need to be coordinated independently and with each other: the respiratory and somatomotor events comprising retching and vomiting and the visceral events mediated by the autonomic nervous system.

The NTS is a complex structure which is the major integrative site for visceral afferent information and receives inputs from other key sites involved in emesis including the AP, vestibular system (middle and lateral regions), hypothalamus and higher brain regions. The ventral part of the NTS contains a group of neurons, often called the dorsal respiratory group, that is one component of the respiratory central pattern generator (CPG). The NTS also contains a CPG for swallowing (a frequent prodrome to emesis), and a pattern generator for vomiting in the medial subnucleus, and sends outputs to the major groups of autonomic motor neurons involved in the regulation of the gastrointestinal (e.g. vagus) and cardiovascular (e.g. sympathetic nerves) systems as well as projections to the reticular area dorsomedial to the retrofacial nucleus in the region of the ventrolateral brain stem (Bötzinger complex) implicated as the CPG for vomiting. A second CPG for respiration known as the ventral respiratory group is located in the ventro-lateral medulla and involves the retrofacial nucleus, nucleus retroambigualis and nucleus ambiguus; the latter is also implicated as a pattern generator for swallowing. The dorsal and ventral respiratory group pattern generators communicate with each other. These CPGs for normal respiration are also involved in non-respiratory acts such as coughing, sneezing and vomiting by 'reshaping' the outputs to the respiratory motor neurons, and similar arguments could apply to the autonomic nervous system pattern generators that may be envisaged by considering the autonomic responses to food as opposed to an ingested toxin.

Substance P injected into various regions of the brain stem has been shown to have marked effects on the gastrointestinal (inhibition of gastric motility, Krowicki and Hornby 2000), cardiovascular (bradycardia and hypotension,

Paton 1998) and respiratory (increased ventilation, Ptak et al. 2002) systems. In addition selective lesions of NK_1 receptor expression of neurons in the pre-Bötzinger complex of adult rats produced marked changes in their normal breathing pattern (Gray et al. 2001). In view of this, one may have expected that administration of an NK_1 receptor antagonist would have serious consequences, but the absence of obvious effects other than anti-emesis suggests either a very specific site and mode of action or that there are multiple neurotransmitters and pathways mediating these crucial functions so that they continue when one of the transmitters (e.g. substance P) is blocked. An additional explanation is suggested by studies of NK_1 receptor knockout animals in which, although resting respiratory patterns did not differ between knockout and wild-type animals, the knockout animals had a blunted response to short-lasting hypoxia. This may indicate that effects of blockade are only revealed when the NK_1 receptor-containing pathway is challenged such as it occurs in emesis and perhaps cough (Bolser et al. 1997). There are two main 'sites' at which the antagonists could act but it must be emphasized that studies of this problem are still in progress.

The Dorsal Brain Stem. The NTS is the point of convergence between the main emetic inputs and whilst it is theoretically possible that each input utilizes substance P as its transmitter, this appears unlikely. In addition, in the dog whilst systemic GR205171 blocked the 'fictive' retching response to electrical stimulation of abdominal vagal afferents, it did not block the activation of neurons in the medial NTS by the vagus (Fukuda et al. 1998). Yet, retching induced by stimulation of the medial NTS was blocked by systemic administration of GR205171. The authors concluded that GR205171 did not block transmission between the vagal afferents and the medial NTS but that the link between the NTS and the central pattern generator was a more likely site. However, this study does not exclude the possibility of a blockade within the complex neural circuitry of the NTS after the integration of the incoming emetic signals, and prior to the genesis of the output to the appropriate motor nuclei/pattern generators.

The absence of a block of transmission between vagal afferents and their primary NTS projection is consistent with the finding by Watson et al. (1995a) that although systemic CP99994 blocked the emetic response to abdominal vagal afferent stimulation, it did not block the accompanying rise in blood pressure. A c-*fos* study in the ferret (Zaman et al. 2000) examined the pattern of activation of the NTS in response to the centrally acting opiate receptor agonist loperamide. Whilst loperamide induced an increase in c-*fos* expression in the NTS, this was not affected by treatment with CP99994, although this agent abolished the emetic response. This study again supports the proposal that NK_1 receptor antagonists act at a site distal to the primary input (the AP in this case) but within the NTS, and this is supported by recordings from unidentified neurons in the NTS of the ferret responding to substance P which had their evoked, but not their spontaneous, discharge blocked by the NK_1 receptor antagonist HSP117 (Saito et al. 1998). The swallowing seen prior to the onset of emesis induced by vagal afferent stimulation was blocked by CP99994, and as the NTS is one site of a swallowing pattern generator this arguably provides further support for a site of action there.

A site within the NTS is further supported by the limited evidence that some of the visceral autonomically regulated components of the emetic reflex are blocked in addition to the more overt retching and vomiting. For example Otterson et al. (1997) demonstrated a block of the vagally mediated retrograde giant contraction associated with emesis and Furukawa et al. (1998) reported that systemic GR205171 blocked fictive retching and antral contractions induced by vagal afferent activation and reduced the increase in salivation and associated efferent nerve activity. Gastric corpus relaxation associated with emesis was unaffected by GR205171. An alternative explanation is that the visceral events which accompany emesis are coordinated at sites outside the NTS in an area of the ventral brain stem termed the 'prodromal sign centre' (see Fukuda et al. 1999b).

A site of action within the NTS is a likely site especially bearing in mind the distribution of substance P and NK_1 receptors in the brain stem. However, a site here poses a number of problems. Firstly, emesis is not the only reflex in which the NTS participates so how is it possible that only the emetic reflex is blocked? Watson et al. (1995a) showed that doses of CP99994, which blocked the retching response to abdominal vagal afferent stimulation in the anaesthetized ferret, left the vagally mediated cardio-pulmonary reflex evoked by jugular injection of a $5-HT_3$ receptor agonist unaffected. Gag, another reflex mediated via the NTS was also unaffected by CP99994. These observations would imply that the emetic reflex is uniquely chemically coded within the NTS with NK_1 receptors having a substantially more important role in this reflex than in other NTS-mediated responses. Alternatively, if there is a very discrete group of neurons in the NTS serving to integrate and transmit the emetic signal to other brain stem regions and dependent upon substance P, then other non-emetically related pathways would be unaffected.

Secondly, if the emetic signal is blocked within the NTS, then a blockade of both nausea and vomiting would be expected as it is proposed that whilst retching and vomiting are co-ordinated entirely within the brain stem, genesis of nausea requires rostral projection of information from the NTS. Although the vestibular system must eventually send projections to the brain stem for the induction of emesis it is unclear whether induction of nausea by motion requires such a routing. The animal data on whether NK_1 receptor antagonists have effects against 'nausea' are inconsistent, which may be related to the use of different 'motor behaviours' (e.g. licking, chewing, salivation) as surrogate markers of nausea (see Lucot et al. 1997; Gonsalves et al. 1996; Grelot et al. 1998; Zaman et al. 2000). The clinical studies indicate that NK_1 receptor antagonists, whilst showing some efficacy against nausea, are clearly less effective than against emesis. These observations would argue that if the NK_1 receptor antagonists act within the NTS then its greatest influence is at a site distal to the divergence of the nausea and vomiting outputs. It must be emphasized that any anti-nausea effect of NK_1 receptor antagonists could also be due to an action at more rostral sites in the brain as NK_1 receptors are distributed extensively and have been shown to have effects on seizures (Zachrisson et al. 1998) and depression (Rupniak and Kramer 1999).

The Ventral Brain Stem. The main studies supporting a site here are from Fukuda and colleagues (1999b) who used microinjection of GR205171 to investigate three possible sites within the brain stem for their ability to block the 'fictive' emetic response to electrical stimulation of vagal afferents in the dog. They investigated an area ventrolateral to the solitary complex, an area dorsal to the retrofacial nucleus and an area adjacent to the semicompact part of the nucleus ambiguus. Whilst some blockade was demonstrated at the two former sites the most effective site for producing a blockade in terms of dose and latency was in the semicompact part of the nucleus ambiguus. This region is part of the Bötzinger complex and the authors argue that the groups of neurons blocked by the NK_1 receptor antagonists in this region of the brain stem which they estimate to be 2 mm^3 is also the location of the 'prodromal centre'. In support of this they reported that the salivary response was also blocked by injection of GR205171 here. Such a discrete site of action is conceptually very attractive and could explain the lack of general effects on respiration and other systems utilized in a unique way during emesis. A single site of action here would make effects on nausea difficult to explain.

The studies by Fukuda et al. (1998, 1999a, 1999b) are the most extensive detailed studies of the site of action but they have focused on vagal activation of emesis and need to be extended using other stimuli. In addition, of necessity they have looked at 'fictive' emesis in decerebrate animals with blood gases adjusted to lower the threshold for emesis. Blockade under such conditions may represent a more challenging target than in conscious animals. It is perhaps conceivable that whilst experimental studies are searching for a single site of action there are, in reality, several sites which contribute to the overall effect, with some sites being more sensitive to blockade than others depending upon the input activated and the intensity of activation.

If a single or predominant site is identified in pre-clinical studies, this would have major implications for anti-emetic research with early questions being 'do the same neurons in humans subserve the same function, and more significantly, do they utilize the same neurotransmitters?' The answers to these questions could have substantial implications for understanding the pathophysiology of emesis; for example, could a defect in these neurons explain the puzzling cyclical vomiting syndrome and account for individual differences in emetic sensitivity? Additionally a crucial central role for substance P would raise the potential of finding other (more effective?) ways of modulating substance P transmission. A number of selective receptor agonists (e.g. opioid, $5-HT_{1A}$, $GABA_B$) have been shown to have broad anti-emetic effects which could be explained by suppression of substance P transmission.

6.1.5
Mechanistic Considerations

Whilst the site of the anti-emetic action of NK_1 receptor antagonists is most consistent with them acting at a brain stem site to antagonize the action of substance P at an NK_1 receptor, there are several issues relating to their mechanism of action which need consideration as they may explain some of the discrepancies between pre-clinical and clinical studies and allow their clinical use to be optimized.

Transmission and Co-transmission. The identification of the potent anti-emetic effects of NK_1 receptor antagonists provided a crucial piece of evidence implicating either substance P or a closely related tachykinin in the emetic reflex. Neurophysiological studies by Fukuda et al. (1998) revealed that neurons in the medial NTS activated by vagal afferent stimulation exhibited a 'wind up' of their discharge, which was unaffected by GR205171 even though the fictive emetic response was abolished. This observation shows that NK_1 receptors are not involved in the link between these neurons and vagal afferents but does not exclude an involvement of substance P (or related tachykinin, e.g. NKA) acting on another tachykinin receptor or another transmitter entirely. If the neurons 'wound up' by the vagal input sent projections rostrally as well as to the CPGs for emesis, this could account for the lower efficacy of NK_1 receptor antagonists against nausea in situations where vagal afferents are activated (e.g. chemotherapy-induced emesis). Glutamate and N-methyl-D-aspartate (NMDA)-type glutamate receptors have been implicated in the 'wind up' effect in the spinal cord, and Furukawa et al. (2001) have shown that the non-NMDA receptor antagonist NBQX applied to the IVth ventricle abolished vagally induced retching in the dog, implicating glutamate as the neurotransmitter. This is supported by other studies implicating glutamate as a transmitter in the dorsal brain stem (Hornby 2001). Glutamate is a common excitatory neurotansmitter providing 'fast' synaptic input but usually co-transmits with another substance (e.g. substance P, calcitonin gene-related peptide) that provides 'slow' neuromodulation.

Recordings made from CPG neurons in the Bötzinger complex showed an immediate onset transient increase in discharge at the onset of vagal afferent stimulation (Fukuda et al. 1999a, 1999b), followed by a slower sustained rise which when a threshold level was achieved changed into an oscillatory pattern coincident with episodes of retching reflected in a phasic pattern of activity in the phrenic and abdominal motorneurons. The transient rapid onset discharge was unaffected by GR205171, whereas the slower rise and retching were abolished. This provides indirect evidence that this pathway contains 'fast' and 'slow' elements and that it is the latter which is crucial for induction or generation of the emetic pattern. 'Wind up' of activity could be induced in CPG neurons by bursts of vagal afferent activity which, in contrast with the same phenomenon in NTS neurons, was blocked by GR205171. Fukuda et al. (1998) proposed that the slow rise in discharge, which can be 'wound up', is essential for generation of the emetic pattern and this is mediated by substance P.

The observation that the immediate onset response of CPG neurons to vagal stimulation was not blocked by GR205171 perhaps provides a basis for the differences between efficacy against emesis and nausea as it shows that transmission of an emetic signal beyond the NTS may not be blocked by GR205171. As small molecule neuotransmitters can be released by a few action potentials, whereas neuropeptides require trains of impulses for their release, this could account for the apparent contradiction that whilst nausea can be evoked by less intense stimulation of a pathway than vomiting, it is more difficult to treat.

Other Tachykinins, Tachykinin Receptors and Receptor Isoforms. The inhibition of the emetic response to a variety of stimuli by NK_1 receptor antagonists from several structural series provides strong supportive evidence that substance P acting at NK_1 receptors is a critical neurotransmitter in the emetic reflex with a central site being implicated (see above). Unfortunately, this hypothesis is rather difficult to prove, as it is possible that other endogenous tachykinins could activate the emetic reflex, or that substance P could be acting at other receptors. It is also possible that different tachykinins activate the NK_1 receptors at different points in the emetic reflex but that a block of all of the NK_1 receptors is important for an anti-emetic action.

The complexity of the issue has been revealed by an electrophysiological study of rat brain stem neurons. It seems that a septide-sensitive NK_1 receptor site is involved in the excitation of both NTS and DMNV neurons (substance P being much less potent than other agonists in the NTS) and that a 'classical' NK_1 receptor may play more of a role in the DMVN. A third unknown site may be responsible for a depolarizing response to substance P in the NTS that is not likely to involve tachykinin receptors (Maubach and Jones 1997). In this study neurokinin A was interpreted to cause excitatory responses by acting at the septide-sensitive NK_1 site, and it was suggested that the anti-emetic action of NK_1 receptor antagonists may be due to blockade of this action (Maubach and Jones 1997). However, these studies need to be repeated in a species that has the vomiting reflex.

The issue of the identity of the agonist activating NK_1 receptors during emesis has already been raised in Sect. 4.2. However, what the above studies also show is that substance P may have other actions in the NTS that are not blocked by NK_1 receptor antagonists. It was pointed out that substance P may interact with nicotinic and bombesin receptors, but it is not certain if this occurs in the emetic reflex (Maubach and Jones 1997). Clearly, further studies are required to address these issues. It would be desirable to investigate the potential of a number of tachykinin receptor agonists to induce emesis, with additional studies to assay endogenous tachykinin levels in the brain stem during retching and vomiting episodes. Molecular studies of brain stem nuclei involved in emesis are needed to investigate the tachykinin genes, in particular the preprotachykinin A gene and the factors regulating post-translational processing of α-, β-, γ- and δ-preprotachykinin.

Substance P and Steroid Interactions. In the treatment of chemotherapy-induced emesis, particularly in the delayed phase, the glucocorticoid dexamethasone is used as part of the anti-emetic regime. Whilst it is clear that this steroid has independent anti-emetic effects (see Andrews and Davis 1993; Sanger 1993 for discussion of mechanisms) there is some evidence that there may be clinically relevant interactions between substance P and corticosteroids in addition to the effect of the NK_1 receptor antagonist aprepitant on dexamethasone metabolism. Using the synthesis of substance P by the rat nodose ganglion as a model it was shown that chronic (14 days) ACTH or corticosterone treatment (MacLean 1987; MacLean et al. 1987) reduced the anterograde transport of substance P by 30%–40%. Comparable effects are also seen with 7 days' corticosterone treatment. Using human IM9 lymphoblasts cells to study NK_1 receptor mRNA expression Gerard et al. (1991) found that after incubation with dexamethasone (100 μm) for 48 h there was an approximately sevenfold reduction in NK_1 receptor binding sites. In rat pancreatic acinar AR42J cells an almost complete suppression of NK_1 receptor expression was reported with incubation for 4 h with dexamethasone (Ihara and Nakanishi 1990).

Taking these observations together it is possible that dexamethasone (which may have its concentration increased by aprepitant, see Sect. 5.1.4.2) can enhance the efficacy of other anti-emetics by reducing NK_1 receptor expression and substance P synthesis in vagal afferents (and perhaps other neurons). Such effects are most likely to be relevant in the delayed phase of chemotherapy-induced emesis.

7
Closing Comments

Since its discovery almost three-quarters of a century ago, substance P has proven an elusive therapeutic target but the identification of the anti-emetic effects of selective NK_1 receptor antagonists has provided the first example of a clinically useful drug acting to antagonize an effect of substance P. As with the $5\text{-}HT_3$ receptor antagonists, pre-clinical studies, particularly in the ferret, played a key role in elucidating the clinical potential, revealing underlying mechanisms and differentiating the spectrum of the anti-emetic effects of NK_1 receptor antagonists from other agents. The latter studies reveal a unique broad-spectrum anti-emetic effect for an antagonist acting at a single receptor; other agents with equally broad spectra are all agonists. Identification of the site(s) of action has provided novel insights into the neuropharmacology of emesis and has also demonstrated that brain stem nuclei are tractable targets for drug therapy despite their involvement in critical functions such as cardiovascular and respiratory control. Despite the relative wealth of pre-clinical evidence supporting the involvement of substance P and NK_1 receptors in emesis there are several unresolved issues which need investigation including: (a) Is substance P the only endogenous tachykinin involved in emesis and what is the role of other tachykinin receptors? (b) What is the explanation for the apparent lack of effect of NK_1 receptor antagonists on the latency to emesis? (c) Do peripheral NK_1 receptors contribute to the anti-emetic effect to some stimuli (e.g. cytotoxic drugs)? (d) What are the pharmacological interactions between dexamethasone, 5-HT_3 and NK_1 receptors in vivo? (e) What is the co-transmitter of substance P?

Whilst the current clinical indication for NK_1 receptor antagonists is as part of the treatment regime for acute and delayed emesis induced by highly emetogenic chemotherapy, the availability of such agents will provide a powerful tool to further explore not only the neuropharmacology of emesis in humans but other possible pathophysiological roles. Whilst the pre-clinical studies have shown a broad spectrum anti-emetic effect this important aspect has been little explored in the clinic. The clinical studies provide a unique opportunity to investigate the involvement of NK_1 receptors in the pathway leading to the genesis of nausea. An important question is whether NK_1 receptor antagonists have a differential effect on nausea induced by motion (see Sect. 5.3) as opposed to cisplatin (see Sect. 5.1). Clinical studies clearly demonstrate that this class of agent will have an important role in the clinic in acute and delayed phases of chemotherapy-induced emesis. However, in some aspects the pre-clinical studies have not been fully translated to the clinic, and studies investigating the reasons (e.g. other tachykinin receptors, co-transmitters, species-specific isoforms, pathways and transmitters in primates vs. non-primates) are required to both refine models and to further improve clinical efficacy.

Acknowledgements. We would like to acknowledge a number of colleagues for stimulating discussions on emesis and NK_1 receptors over the years: Dr. Keith T Bunce, Dr. Chris Davis, Dr. Richard Hargreaves, Dr. Kevin Horgan, Dr. Chris Jordan, Prof. Robert J. Naylor, Dr. Gareth Sanger, Dr. David Tattersall, and Dr. John Watson.

References

Alfieri AB, Cubbedu LX (2000) Role of NK_1 receptors on cisplatin-induced nephrotoxicity in the rat. Naunyn-Schmiedeberg's Arch Pharmacol 361:334–338

Alfieri AB, Gardner CJ (1995) The NK_1 antagonist GR203040 inhibits cyclophosphamide-induced damage in the rat and ferret bladder. Gen Pharmacol 29:245–250

Alumets J, Hakanson R, Ingemansson S et al. (1977) Substance P and 5-HT in granules isolated from an intestinal argentaffin carcinoid. Histochemistry 52:217–222

Amin AH, Crawford TBB, Gaddum JH (1954) The distribution of substance P and 5-hydroxytryptamine in the central nervous system of the dog. J Physiol (Lond) 126:596–618

Andresen MC, Kunze DL (1994) Nucleus tractus solitarius—gateway to neural circulatory control. Ann Rev Physiol 56:93–116

Andrews PLR (1992) Physiology of nausea and vomiting. Br J Anaesth 69:2S–19S

Andrews PLR (1994) 5-HT_3 receptor antagonists and antiemesis. In: King FD, Jones BJ, Sanger GJ (eds) 5-Hydroxytryptamine-3 Receptor Antagonists. CRC Press, Boca Raton, USA, pp 255–317

Andrews PLR (1999) Postoperative nausea and vomiting. In: Herbert MK, Holzer P, Roewer N (eds) Problems of the Gastrointestinal Tract in Anesthesia, the Perioperative Period, and Intensive Care. Springer, Berlin, pp 267–288

Andrews PLR (2003) Approaching and understanding of the mechanism of post-operative nausea and vomiting. In: Strunin L, Rowbotham D, Miles A (eds) The Effective Prevention and Managemnt of Post-operative Nausea and Vomiting. Aesculapius Medical Press, London, UK, pp 3–28

Andrews PLR, Bhandari P (1993) Resinferatoxin, an ultrapotent capsaicin analogue, has anti-emetic properties in the ferret. Neuropharmacology 32:799–806

Andrews PLR, Davis CJ (1995) The physiology of emesis induced by anti-cancer therapy. In: Reynolds DJM, Andrews PLR, Davis CJ (eds) Serotonin and the Scientific Basis of Anti-emetic Therapy. Oxford Clinical Communications, Oxford, UK, pp 25–49

Andrews PLR, Davis CJ, Bingham S et al. (1990) The abdominal visceral innervation and the emetic reflex: pathways, pharmacology, and plasticity. Can J Physiol Pharmacol 68:325–345

Andrews PLR, Hawthorn J (1987) Evidence for an extra-abdominal site of action for the 5-HT$_3$ receptor antagonist BRL24924 in the inhibition of radiation-evoked emesis in the ferret. Neuropharmacology 26:1367–1370

Andrews PLR, Hawthorn J (1988) The neurophysiology of vomiting. Clin Gastroenterol 2:141–168

Andrews PLR, Kovacs M, Watson JW (2001) The anti-emetic action of the neurokinin-1 receptor antagonist CP-99,994 does not require the presence of the area postrema in the dog. Neurosci Lett 314:102–104

Andrews PLR, Naylor RJ, Joss RA (1998) Neuropharmacology of emesis and its relevance to anti-emetic therapy. Consensus and controversies. Support Care Cancer 6:197–203

Andrews PLR, Okada F, Woods AJ et al. (2000) The emetic and anti-emetic effects of the capsaicin analogue resiniferatoxin in *Suncus murinus*, the house musk shrew. Br J Pharmacol 130:1247–1254

Andrews PLR, Rapeport WG and Sanger GJ (1988) Neuropharmacology of emesis induced by anti-cancer therapy. Trends Pharmacol Sci 9:334–341

Andrews PLR, Torii Y, Saito H et al. (1996) The pharmacology of the emetic response to upper gastrointestinal tract stimulation in *Suncus murinus*. Eur J Pharmacol 307:305–313

Andrews PLR, Davis CJ (1993) The mechanism of emesis induced by anti-cancer therapies. In: Andrews PLR, Sanger GJ (eds) Emesis in Anti-cancer Therapy: Mechanisms and Treatment. Chapman and Hall Medical, London, UK, pp 113–161

Apfel CC, Katz MH, Kranke P et al. (2002) Volatile anaesthetics may be the main cause of early but not delayed postoperative vomiting: a random controlled trial of factorial design. Br J Anaesth 88:1–10

Ariumi H, Saito R, Nago S et al. (2000) The role of tachykinin NK-1 receptors in the area postrema of ferrets in emesis. Neurosci Lett 286:123–126

Armstrong DM, Pickel VM, Joh TH et al. (1981) Immunocytochemical localization of catecholamine-synthesizing enzymes and neuropeptides in area postrema and medial nucleus tractus solitarius of rat brain. J Comp Neurol 196:505–517

Armstrong DM, Pickel VM, Reis DJ (1982) Electron microscopic immunocytochemical localization of substance P in the area postrema of rat. Brain Res 243:141–146

Arunlakshana O, Schild HO (1959) Some quantitative uses of drug antagonists. Br J Pharmacol 14:48–58

Barnes JM, Barnes NM, Costall B et al. (1988) Reserpine, para-chlorophenylalanine and fenfluramine antagonise cisplatin-induced emesis in the ferret. Neuropharmacology 27:783–790

Bartho L, Holzer P (1985) Search for a physiological role of substance P in gastrointestinal motility. Neuroscience 16:1–32

Baude A, Lanoir J, Vernier P et al. (1989) Substance P-immuno-reactivity in the dorsal medial region of the medulla in the cat: effects of nodosectomy. J Chem Neuroanat 2:67–81

Beattie DT, Beresford IJ, Connor HE et al. (1995) The pharmacology of GR203040, a novel, potent and selective non-peptide tachykinin NK_1 receptor antagonist. Br J Pharmacol 116:3149–3157

Beleslin DB, Krstic SK (1987) Further studies on nicotine-induced emesis: nicotinic mediation in area postrema. Physiol Behav 39:681–686

Bertaccini G, Cei JM, Erspamer V (1965) Occurrence of physalaemin in extracts of the skin of Physalaemus fuscumaculatus and its pharmacological actions of extravascular smooth muscle. Br J Pharmacol 25:363–379

Bhandari P, Bingham S, Andrews PL (1992) The neuropharmacology of loperamide-induced emesis in the ferret: the role of the area postrema, vagus, opiate and $5\text{-}HT_3$ receptors. Neuropharmacology 31:735–742

Bianchi AL, Grelot L, Miller AD et al. (1992) Mechanisms and Control of Emesis. Colloque INSERM: John Libbey Eurotext, France

Blackshaw LA, Dent J (1997) Lower oesophageal sphincter responses to noxious oesophageal chemical stimuli in the ferret: involvement of tachykinin receptors. J Auton Nerv Syst 66:189–200

Blondeau C, Clerc N, Baude A (2002) Neurokinin-1 and neurokinin-3 receptors are expressed in vagal efferent neurons that innervate different parts of the gastro-intestinal tract. Neuroscience 110:339–349

Boissonade FM, Davison JS, Egizii R et al. (1996) The dorsal vagal complex of the ferret: anatomical and immunohistochemical studies. Neurogastroenterol Motil 8:255–272

Bolser DC, DeGennaro FC, O'Reilly SO et al. (1997) Central antitussive activity of the NK_1 and NK_2 tachykinin receptor antagonist, CP-99,994 and SR48968, in the guinea-pig and cat. Br J Pharmacol 121:165–170

Borison HL, Wang SC (1953) Physiology and pharmacology of vomiting. Pharmacol Rev 5:193–230

Bountra C, Bunce K, Dale T et al. (1993) Anti-emetic profile of a non-peptide neurokinin NK_1 receptor antagonist, CP-99,994, in ferrets. Eur J Pharmacol 249:R3–R4

Boyle S, Guard S, Higginbottom M, Horwell DC et al. (1994) Rational design of high affinity tachykinin NK_1 receptor antagonists. Bioorg Med Chem 2:357–370

Brimijoin S, Lundberg JM, Brodin E et al. (1980) Axonal transport of substance P in the vagus and sciatic nerves of the guinea pig. Brain Res 191:443–457

Campos D, Rodrigues-Pereira J, Reinhardt RR et al. (2001). Prevention of cisplatin-induced emesis by the oral neurokinin-1 antagonist, MK-869, in combination with granisteron and dexamethasone or with dexamethasone alone. J Clin Oncol 19:1759–1767

Carpenter DO, Briggs DB, Strominger N (1983a) Responses of neurons of canine area postrema to neurotransmitters and peptides. Cell Mol Neurobiol 3:113–126

Carpenter DO, Briggs DB, Strominger N (1983b) Responses of neurons of the canine area postrema to neurotransmitters and peptides. Final Report No. USAFSAM-tr-83-37 (October 1981–September 1982), USAF School of Aerospace Medicine, Texas and SE Center for Electrical Engineering Education, Florida, USA

Carpenter DO, Briggs DB, Strominger N (1984) Behavioural and electrophysiological studies of peptide-induced emesis in dogs. Fed Proc 43:2952–2954

Carpenter DO (1990) Neural mechanism of emesis. Can J Physiol Pharmacol 68:230–236

Carraway R, Leeman SE (1979) The amino acid sequence of bovine hypothalamic substance P. Identity to substance P from colliculi and small intestine. J Biol Chem 254:2944–2945

Chang MM, Leeman SE (1970) Isolation of a sialogogic peptide from bovine hypothalamic tissue and its characterization as substance P. J Biol Chem 245:4784–4790

Chang MM, Leeman SE, Niall HD (1971) Amino acid sequence of substance P. Nature New Biol 232:86–87

Chen Y, Saito H, Matsuki N et al. (1997) Ethanol-induced emesis in the house musk shrew, Suncus murinus. Life Sci 60:253–261

Chigr F, Najimi M, Leduque P et al. (1991) Anatomical distribution of substance P-like immunoreactive neurons in the human brainstem during the first postnatal year. Brain Res Bull 26:515–523

Cocquyt V, Van Belle S, Reinhardt RR et al. (2001) Comparison of L-758,298, a prodrug for the selective neurokinin-1 antagonist, L-754–030, with ondansetron for the prevention of cisplatin-induced emesis. Eur J Cancer 37:835–842

Colby ED, McCarthy LE, Borison HL (1981) Emetic action of xylazine on the chemoreceptor trigger zone for vomiting in cats. J Vet Pharmacol Ther 4:93–96

Cooper PE, Fernstrom MH, Rorstad OP et al. (1981) The regional distribution of somatostatin, substance P and neurotensin in human brain. Brain Res 218:219–232

Crampton GH (1990) Neurophysiology of motion sickness. In: Crampton GH (ed) Motion and Space Sickness. CRC Press, Boca Raton, FL, USA, pp 29–44

Cubeddu LX (1996) Serotonin mechanisms in chemotherapy-induced emesis in cancer patients. Oncology 53:18–25

Davison JS, Oland L, Boissonade F (1995) The effects of centrally injected NK_1 receptor antagonists on emesis in the ferret. Gastroenterology 108:A589

Del Fiacco M, Dessi ML, Levanti MC (1983) Immunohistochemical localization of substance P in the human central nervous system: the brainstem and hippocampal formation. In: Skrabanek P, Powell D (eds) Substance P. Proceedings of the International Symposium—Dublin 1983. Boole Press, Dublin, Ireland, pp 261–262

Diemunsch P, Schoeffler P, Bryssine B et al. (1999) Antiemetic activity of the NK_1 receptor antagonist GR205171 in the treatment of established postoperative nausea and vomiting after major gynaecological surgery. Br J Anaesth 82:274–276

Douglas FL, Palkovits M, Brownstein MJ (1982) Regional distribution of substance P-like immunoreactivity in the lower brainstem of the rat. Brain Res 245:376–378

Douglas WW, Feldberg W, Paton WDM et al. (1951) Distribution of histamine and substance P in the wall of the dog's digestive tract. J Physiol (Lond) 115:163–176

Erspamer V, Glasser A (1963) The action of eledoisin on systemic arterial blood pressure of some experimental animals. Br J Pharmacol 20:516–527

Fauser AA, Fellhauer M, Hoffmann M et al. (1999) Guidelines for anti-emetic therapy: acute emesis. Eur J Cancer 35:361–370

Florczyk AP, Schurig JE, Bradner WT (1982) Cisplatin-induced emesis in the ferret: a new animal model. Cancer Treat Rep 66:187–189

Forster ER, Palmer JL, Bedding AW et al. (1994) Syrup of ipecacuanha-induced nausea and emesis is mediated by $5-HT_3$ receptors in man. J Physiol (Lond) 477:72P

Fukuda H, Koga T (1995) Activation of peripheral and/or central chemoreceptors changes retching activities of Bötzinger complex neurons and induces expulsion in decerebrate dogs. Neurosci Res 23:171–183

Fukuda H, Koga T, Furukawa N et al. (1998) The tachykinin NK_1 receptor antagonist GR205171 prevents vagal stimulation-induced retching but not neuronal transmission from emetic vagal afferents to solitary nucleus neurons in dogs. Brain Res 802:221–231

Fukuda H, Koga T, Furukawa N et al. (1999a) The tachykinin NK_1 receptor antagonist GR205171 abolishes the retching activity of neurons comprising the central pattern generator for vomiting in dogs. Neurosci Res 33:25–32

Fukuda H, Nakamura E, Koga T et al. (1999b) The site of the anti-emetic action of tachykinin NK_1 receptor antagonists may exist in the medullary area adjacent to the semicompact part of the nucleus ambiguus. Brain Res 818:439–449

Fukui H, Yamamoto M (1999) Methotrexate produces delayed emesis in dogs: a potential model of delayed emesis induced by chemotherapy. Eur J Pharmacol 372:261–267

Fumoleau P, Graham E, Giovanni M et al. (1998) Control of acute cisplatin-induced emesis and nausea with the NK_1 receptor antagonist GR205171 in combination with ondansetron. Proc Am Soc Clinical Oncol 17:225
Furukawa N, Fukuda H, Hatano M et al. (1998) A neurokinin-1 receptor antagonist reduced hypersalivation and gastric contractility related to emesis in dogs. Am J Physiol 275:G1193–G1201
Furukawa N, Hatano M, Fukuda H (2001) Glutamatergic vagal afferents may mediate both retching and gastric adaptive relaxation in dogs. Auton Neurosci Basic Clin 93:21–30
Gamse R, Lembeck F, Cuello AC (1979) Substance P in the vagus nerve. Immunochemical and immunohistochemical evidence for axoplasmic transport. Naunyn-Schmiedeberg's Arch Pharmacol 306:37–44
Gamse R, Mroz E, Leeman SE et al. (1978) The intestine as source of immunoreactive sustance P in plasma of the cat. Naunyn-Schmiedeberg's Arch Pharmacol 305:17–21
Gardner C, Perren M (1998) Inhibition of anaesthetic-induced emesis by a NK_1 or 5-HT_3 receptor antagonist in the house musk shrew, *Suncus murinus*. Neuropharmacology 37:1643–1644
Gardner CJ, Armour DR, Beattie DT et al. (1996) GR205171: a novel antagonist with high affinity for the tachykinin NK_1 receptor, and potent broad-spectrum anti-emetic activity. Regul Pept 65:45–53
Gardner CJ, Bountra C, Nbunce KT et al. (1994) Anti-emetic activity of neurokinin NK_1 receptor antagonists is mediated centrally in the ferret. Br J Pharmacol 112:516P
Gardner CJ, Twissell DJ, Dale TJ et al. (1995) The broad-spectrum anti-emetic activity of the novel non-peptide tachykinin NK_1 receptor antagonist GR203040. Br J Pharmacol 116:3158–3163
Garret C, Carruette A, Fardin V et al. (1991) Pharmacological properties of a potent and selective nonpeptide substance P antagonist. Proc Natl Acad Sci USA 88:10208–10212
Gerard NP, Garraway LA, Eddy RL et al. (1991) Human substance P receptor (NK-1): organization of the gene, chromosome localization, and functional expression of cDNA clones. Biochemistry 30:10640–10646
Gesztesi ZS, Song D, White PF (1998) Comparison of a new NK-1 antagonist (CP122,721) to ondansetron in the prevention of postoperative nausea and vomiting. Anesth Analg 86:S32
Gesztesi ZS, Scudieri PE, White PF et al. (2000) Substance P (neurokinin-1) antagonists prevents postoperative vomiting after abdominal hysterectomy procedures. Anesthesiology 93:931–937
Girod V, Dapzol J, Bouvier M et al. (2002) The COX inhibitors indomethacin and meloxicam exhibit anti-emetic activity against cisplatin-induced emesis in piglets. Neuropharmacology 42:428–436
Gitter BD, Bruns RF, Howbert JJ et al. (1995) Pharmacological characterization of LY303870: a novel, potent and selective nonpeptide substance P (neurokinin-1) receptor antagonist. J Pharmacol Exp Ther 275:737–744
Gonsalves S, Watson J, Ashton C (1996) Broad spectrum antiemetic effects of CP-122,721, a tachykinin NK_1 receptor antagonist, in ferrets. Eur J Pharmacol 305:181–185
Gralla RJ, Itri LM, Pisko SE et al. (1981) Antiemetic efficacy of high-dose metoclopramide: randomized trials with placebo and prochlorperazine in patients with chemotherapy-induced nausea and vomiting. New Engl J Med 305:905–909
Gray PA, Janczewski WA, Mellen N et al. (2001) Normal breathing requires pre-Bötzinger complex neurokinin-1 receptor expressing neurons. Nat Neurosci 4:927–930
Grelot L, Bianchi AL (1997) Multifunctional medullary respiratory neurons. In: Miller AD, Bianchi AL, Bishop BP (eds) Neural Control of Respiratory Muscles. CRC Press, Boca Raton, USA, pp 297–304

Grelot L, Dapzol J, Esteve E et al. (1998) Potent inhibition of both the acute and delayed emetic responses to cisplatin in piglets treated with GR205171, a novel highly selective tachykinin NK_1 receptor antagonist. Br J Pharmacol 124:1643–1650

Grelot L, Le Stunff H, Milano S et al. (1996) Repeated administration of the 5-HT_3 receptor antagonist granisetron reduces the incidence of delayed cisplatin-induced emesis in the piglet. J Pharmacol Exp Ther 279:255–261

Grelot L, Miller AD (1997) Neural control of respiratory muscle activation during vomiting. In: Miller AD, Bianchi AL, Bishop BP (eds). Neural Control of the Respiratory Muscles. CRC Press, Boca Raton, USA, pp 239–248

Gross PM, Wainman DS, Shaver SW et al. (1990) Metabolic activation of efferent pathways from the rat area postrema. Am J Physiol 258:R788–R797

Gylys JA, Doran KM, Buyniski JP (1979) Antagonism of cisplatin induced emesis in the dog. Res Commun Chem Pathol Pharmacol 23:61–68

Hakanson R, Beding B, Ljungqvist A et al. (1988) Blockade of sensory nerve mediated contraction in rabbit iris sphincter mediated by a series of novel tachykinin antagonists. Regul Pept 20:99–105

Hale JJ, Mills SG, MacCoss M et al. (1998) Structural optimization affording 2-(R)-(1-(R)-3,5-Bis (trifluoromethyl)pheylethoxy)-3-(S)-(4-fluoro)phenyl-4-(3-oxo-1,2,4-triazol-5-yl)methylmorpholone, a potent, orally active, long-acting morpholine acetyl human NK-1 receptor antagonist. J Med Chem 41:4607–4614

Harding RK, Hugenholtz H, Keaney M et al. (1985) Discrete lesions of the area postrema abolish radiation-induced emesis in the dog. Neurosci Lett 53:95–100

Harding RK, Hugenholtz H, Kucharczyk J et al. (1987) Central mechanisms for apomorphine-induced emesis in the dog. Eur J Pharmacol 144:61–65

Hargreaves R (2002) Imaging substance P (NK_1) receptors in the living human brain using positron emission tomography. J Clin Psychiatry 63(Suppl 11):18–24

Harmar AJ (1984) Three tachykinins in mammalian brain. Trends Neurosci 7:57–60

Hasegawa M, Sasaki T, Sadakane K et al. (2002) Studies for the emetic mechanisms of ipecac syrup (TJN-119) and its active components in ferrets: involvement of 5-hydroxytryptamine receptors. Jpn J Pharmacol 89:113–119

Hawthorn J, Ostler KJ, Andrews PLR (1988) The role of the abdominal visceral innervation and 5-hydroxytryptamine M-receptors in vomiting induced by the cytotoxic drugs cyclophosphamide and cis-platin in the ferret. Q J Exp Physiol 73:7–21

Heitz P, Polak JM, Timson CM et al. (1976) Enterochromaffin cells as the endocrine source of gastrointestinal substance P. Histochemistry 49:343–347

Helke CJ, Hill KM (1988) Immunohistochemical study of neuropeptides in vagal and glossopharyngeal afferent neurons in the rat. Neuroscience 26:539–551

Helke CJ, Shults CW, Chase TN et al. (1984) Autoradiographic localization of substance P receptors in rat medulla: effect of vagotomy and nodose ganglionectomy. Neuroscience 12:215–223

Hesketh PJ, Gralla RJ, Webb RT et al. (1999) Randomized phase II study of the neurokinin 1 receptor antagonist CJ-11974 in the control of cisplatin-induced emesis. J Clin Oncol 17:338–343

Hesketh PJ, Grunberg SM, Gralla RJ et al. (2003a) The oral neurokinin-1 antagonist aprepitant for the prevention of chemotherapy -induced nausea and vomiting: A multinational, randomised , double blind, placebo-controlled trial in patients receiving high-dose cisplatin-the aprepitant protocol 052 study group. J Clin Oncol 21:4112–4119

Hesketh PJ, Van Belle S, Aapro M et al. (2003b) Differential involvement of neurotransmitters through the time course of cisplatin-induced emesis as revealed by therapy with specific receptor antagonists. Eur J Cancer 39:1074–1080

Hikasa Y, Akiba T, Iino Y et al. (1992) Central alpha-adrenoceptor subtypes involved in the emetic pathway in cats. Eur J Pharmacol 229:241–251

Holzer P (1998) Implications of tachykinins and calcitonin gene- related peptide in inflammatory bowel disease. Digestion 59:269-283

Hornby PJ (2001) Receptors and transmission in the brain-gut axis: potential for novel therapies. II. Excitatory amino acid receptors in the brain gut axis. Am J Physiol 2001:G1055-G1060

Ihara H, Nakanishi S (1990) Selective inhibition of expression of the substance P receptor mRNA in pancreatic acinar AR42J cells by glucocorticoids. J Biol Chem 265:22441-22445

Janes RJ, Muhonen T, Karjalainen UP et al. (1998) Urinary 5-hydroxyindoleacetic acid (5-HIAA) excretion during multiple-day high-dose chemotherapy. Eur J Cancer 34:196-198

Javid FA, Naylor RJ (2002) The effect of serotonin and serotonin receptor antagonists on motion sickness in *Suncus murinus*. Pharmacol Biochem Behav 73:979-989

Jenkinson KM, Southwell BR, Furness JB (1999) Two affinities for a single antagonist at the neuronal NK_1 tachykinin receptor: evidence from quantitation of receptor endocytosis. Br J Pharmacol 126:131-136

Kakimoto S, Saito H, Matsuki N (1997) Antiemetic effects of morphine on motion- and drug-induced emesis in *Suncus murinus*. Biol Pharm Bull 20:739-742

Kalia M, Fuxe K, Hökfelt T et al. (1984) Distribution of neuropeptide immunoreactive nerve terminals within the subnuclei of the nucleus of the tractus solitarius of the rat. J Comp Neurol 222:409-444

Kan KKW, Jones RL, Ngan MP et al. (2003a) Action of prostanoids on the emetic reflex of *Suncus murinus* (the house musk shrew). Eur J Pharmacol 477:247-251

Kan KKW, Jones RL, Ngan MP et al. (2003b) Emetic action of the TP prostanoid receptor agonist, U46619, in *Suncus murinus* (the house musk shrew). Eur J Pharmacol 482:297-304

Keast JR, Furness JB, Costa M (1985) Distribution of certain peptide-containing nerve fibres and endocrine cells in the gastrointestinal mucosa in five mammalian species. J Comp Neurol 236:403-422

King GL (1990) Animal models in the study of vomiting. Can J Physiol Pharmacol 68:260-268

Knox AP, Strominger NL, Battles AH et al. (1993) Behavioral studies of emetic sensitivity in the ferret. Brain Res Bull 31:477-484

Koch KL (1995) Approach to the patient with nausea and vomiting. In: Yamada T (ed) Textbook of Gastroenterology vol 1. Lippincott, Philadelphia, USA, pp 731-749

Kowicki ZK, Hornby PJ (2000) Substance P in the dorsal motor nucleus of the vagus evokes gastric motor inhibition via neurokinin-1 receptor in rat. J Pharmacol Exp Ther 293:214-221

Kris MG, Gralla RJ, Clark RA (1985) Incidence, course, and severity of delayed nausea and vomiting following the administration of high-dose cisplatin. J Clin Oncol 3:1379-1384

Kris MG, Gralla RJ, Tyson LB et al. (1989) Controlling delayed vomiting: double-blind randomized trial comparing placebo, dexamethasone alone, and metoclopramide plus dexamethsaone in patienst receiving cisplatin. J Clin Oncol 7:108-114

Kris MG, Pisters KMW, Hinkley L (1994) Delayed emesis following anti-cancer chemotherapy. Support Care Cancer 2:297-300

Kris MG, Radford J, Pizzo BA et al. (1997) Control of emesis following cisplatin by CP-122,721, a selective NK_1 receptor antagonist. J Natl Cancer Inst 89:817-818.

Lang IM (1990) Digestive tract motor correlates of vomiting and nausea. Can J Physiol Pharmacol 68:242-253

Lembeck, F (1986) A network of defence. In: Henry JL, Couture R, Cuello AC, Pelletier G, Quirion R, Regoli D (eds) Substance P and Neurokinins—Montreal '86. Springer, New York, Berlin, Heidelberg, pp 380-386

Lien HC, Sun WM, Chen YH et al. (2003) Effects of ginger on motion sickness and gastric slow- wave dysrhythmias induced by circular vection. Am J Physiol 284:G481–G489

Lindh B, Dalsgaard CJ, Elfvin LG et al. (1983) Evidence of substance P immunoreactive neurons in dorsal root ganglia and vagal ganglia projecting to the guinea-pig pylorus. Brain Res 269:365–369

Lindstrom PA, Brizzee KR (1962) Relief of intractable vomiting from surgical lesion in the area postrema. J Neurosurg 19:228–236

Ljungdahl A, Hökfelt T, Nilsson G (1978) Distribution of substance P-like immunoreactivity in the central nervous system of the rat. I. Cell bodies and nerve terminals. Neuroscience 3:861–943

Lucot JB (1989) Blockade of 5-hydroxytryptamine$_3$ receptors prevents cisplatin-induced but not motion- or xylazine-induced emesis in the cat. Pharmacol Biochem Behav 32:207–210

Lucot JB (1994) Antiemetic effects of flesinoxan in cats: Comparisons with 8-hydroxy-2-(di-n-propylamino)tetralin. Eur J Pharmacol 253:53–60

Lucot JB, Obach RS, McLean S et al. (1997) The effect of CP-99,994 on the responses to provocative motion in the cat. Br J Pharmacol 120:116–120

Lundberg JM, Hökfelt T, Kewenter J et al. (1979) Substance P-, VIP- and enkephalin-like immunoreactivity in the human vagus nerve. Gastroenterology 77:468–471

Lundberg JM, Hökfelt T, Nilsson G et al. (1978) Peptide neurons in the vagus, splanchnic and sciatic nerves. Acta Physiol Scand 104:499–501

MacLean DB (1987) Adrenocorticotropin-adrenal regulation of transported substance P in the vagus nerve of the rat. Endocrinology 121:1540–1547

MacLean DB, Lewis SF (1984) Axoplasmic transport of somatostatin and substance P in the vagus nerve of the cat, guinea-pig and rat. Brain Res 307:135–145

MacLean DB, Wheeler F, Hayes K (1990) Basal and stimulated release of substance P from dissociated cultures of vagal sensory neurons. Brain Res 519:308–314

Maggi CA (1995) The mammalian tachykinin receptors. Gen Pharmacol 26:911–944

Maley BE (1985) The ultrastructural localization of enkephalin and substance P immunoreactivities in the nucleus tractus solitarii of the cat. J Comp Neurol 233:490–496

Maley BE, Elde R (1981) Localisation of substance P-like immunoreactivity in cell bodies of the feline dorsal vagal nucleus. Neurosci Lett 27:187–191

Maley BE, Elde R (1982) Immunohistochemical localization of putative neurotransmitters within the feline nucleus tractus solitarii. Neuroscience 7:2469–2490

Maley BE, Newton BW, Howes KA et al. (1987) Immunohistochemical localization of substance P and enkephalin in the nucleus tractus solitarii of the rhesus monkey, *Macaca mulatta*. J Comp Neurol 260:483–490

Martin WR (1983) Pharmacology of opioids. Pharmacol Rev 35:283–323

Matsuki N, Ueno S, Kaji T et al. (1988) Emesis induced by cancer chemotherapeutic agents in the *Suncus murinus*: a new experimental model. Jpn J Pharmacol 48:303–306

Matsumoto S, Kawasaki Y, Mikami M et al. (1999) Relationship between cancer chemotherapeutic drug-induced delayed emesis and plasma levels of substance P in two patients with small cell lung cancer. Jpn J Cancer Chemother 26:535–538

Maubach K, Hagan RM, Jones RSG (1994) Responses of neurones in the nucleus tractus solitarius to NK$_1$ receptor agonists. Can J Physiol Pharmacol 72(Suppl 1):460

Maubach KA, Jones RS (1997) Electrophysiological characterisation of tachykinin receptors in the rat nucleus of the solitary tract and dorsal motor nucleus of the vagus *in vitro*. Br J Pharmacol 122:1151–1159

Maubach KA, Jones RSG, Stratton SC et al. (1995) Autoradiographic distribution of substance P binding sites in the brainstem of the rat and the ferret. Br J Pharmacol 116:249P

Mayer DJ (2000) Acupuncture: an evidence-based review of the clinical literature. Annu Rev Med 51:49–63

Mazzone SB, Geraghty DP (2000a) Characterization and regulation of tachykinin receptors in the nucleus tractus solitarius. Clin Exp Pharmacol Physiol 27:939–942

Mazzone SB, Geraghty DP (2000b) Respiratory actions of tachykinins in the nucleus of the solitary tract: characterization of receptors using selective agonists and antagonists. Br J Pharmacol 129:1121–1131

McAllister KHM, Pratt JA (1998) GR205171 blocks apomorphine and amphetamine-induced conditioned taste aversions. Eur J Pharmacol 353:141–148

McLean S, Ganong A, Seymour PA et al. (1993) Pharmacology of CP-99,994; a nonpeptide antagonist of the tachykinin neurokinin-1 receptor. J Pharmacol Exp Ther 267:472–479

McLean S, Ganong A, Seymour PA et al. (1996) Characterization of CP-122,721; a nonpeptide antagonist of the neurokinin NK_1 receptor. J Pharmacol Exp Ther 277:900–908

McRitchie DA, Tork I (1994) Distribution of substance P-like immunoreactive neurones and terminals throughout the nucleus of the solitary tract in the human brain stem. J Comp Neurol 343:83–101

Megens AAHP, Ashton D, Vermeire JCA et al. (2002) Pharmacological profile of (2r-trans)-4-[1-[3,5-bis(trifluromethyl)benzoyl]-2-(phenylmethyl)-4-piperidinyl]-n-(2,6-dimethylphenyl)-1-acetamide (s)-hydroxybutanedioate (R116301), an orally and centrally active neurokinin-1 receptor antagonist. J Pharmacol Exp Ther 302:696–709

Mele PC, McDonough JR, McLean DB et al. (1992) Cisplatin-induced conditioned taste aversion: attenuation by dexamethasone but not zacopride or GR38032F. Eur J Pharmacol 218:229–236

Milano S, Blower P, Romain D et al. (1995) The piglet as a suitable animal model for studying the delayed phase of cisplatin-induced emesis. J Pharmacol Exp Ther 274:951–961

Miller AD, Grelot L (1996) The neural basis of nausea and vomiting. In Yates BJ, Miller AD (eds) Vestibular Autonomic Regulation. CRC Press, Boca Raton, USA, pp 85–94

Miller AD, Rowley HA, Roberts TPL et al. (1996) Human cortical activity during vestibular- and drug-induced nausea. Ann New York Acad Sci 781:670–672

Minami M, Endo T, Kikuchi K et al. (1998) Antiemetic effects of sendide, a peptide tachykinin NK_1 receptor antagonist, in the ferret. Eur J Pharmacol 363:49–55

Minami M, Endo T, Yokota H et al. (2001) Effects of CP-99,994, a tachykinin receptor antagonist, on abdominal afferent vagal activity in ferrets: evidence for involvement of NK_1 and $5-HT_3$ receptors. Eur J Pharmacol 428:215–220

Miner WD, Sanger GJ, Turner DH (1987) Evidence that 5-hydroxytryptamine$_3$ receptors mediate cytotoxic drug and radiation-evoked emesis. Br J Cancer 56:159–162

Minton N, Swift R, Lawlor C (1993) Ipecacuanha-induced emesis: a human model for testing anti-emetic drug activity. Clin Pharmacol Ther 54:53–57

Miolan JP, Niel JP (1988) Non-cholinergic ascending excitatory response in the cat small intestine: possible involvement of substance P. Neuropeptides 12:243–248

Morita M, Takeda N, Kubo T et al. (1988) Pica as an index of motion sickness in rats. ORL J Otorhinolaryngol Relat Spec 50:188–192

Morrow GR, Rosco JA, Hynes He et al. (1998) Progress in reducing anticipatory nausea and vomiting: a study of community practice. Supp Care Cancer 6:46–50

Navari RM, Reinhardt RR, Gralla RJ et al. (1999) Reduction of cisplatin-induced emesis by a selective neurokinin-1 receptor antagonist. N Engl J Med 340:190–195

Navarra P, Martire M, del Carmine R et al. (1992) A dual effect of some $5-HT_3$ receptor antagonists on cisplatin-induced emesis in the pigeon. Toxicol Lett 64–65(Spec No):745–749

Naylor RJ, Rudd JA (1996) Mechanisms of chemotherapy/radiotherapy-induced emesis in animal models. Oncology 53:8–17

Newton BW, Maley B, Traurig H (1985) The distribution of substance P, enkephalin, and serotonin immunoreactivities in the area postrema of the rat and cat. J Comp Neurol 234:87–104

Niel JP (1991) Rôle de la substance P dans le contrôle nerveux de la motricité digestive. Association des physiologistes, Nancy (Septembre 1991):A65–A76

Niemegeers CJ (1971) The apomorphine antagonism test in dogs. Experimental evidence and critical considerations on specific methodological criteria. Pharmacology 6:353–364

Niemegeers CJ (1982) Antiemetic specificity of dopamine antagonists. Psychopharmacology (Berl) 78:210–213

Nussey SS, Hawthorn J, Page SR et al. (1988) Responses of plasma oxytocin and arginine vasopressin to nausea induced by apomorphine and ipecacuanha. Clin Endocrinol (Oxf) 28:297–304

Otterson MF, Leming SC, Moulder JE (1997) Central NK_1 receptors mediate radiation induced emesis. Gastroenterology 112:A801

Paasonen MK, Vogt M (1956) The effect of drugs on the amounts of substance P and 5-hydroxytryptamine in mammalian brain. J Physiol (Lond) 131:617–626

Paton JF (1998) Importance of neurokinin-1 receptors in the nucleus tractus solitarii of mice for the integration of cardiac vagal inputs. Eur J Neurosci 10:2261–2275

Pearse AGE, Polak JM (1975) Immunochemical localization of substance P in mammalian intestine. Histochemistry 41:373–375

Pernow B (1951) Substamce P distribution in the digestive tract. Acta Physiol Scand 24:97–102

Pernow B (1983) Substance P. Pharmacol Rev 35:85–141

Pernow B (1953) Studies on substance P. Purification, occurrence and biological actions. Acta Physiol Scand 29:1–90

Pernow B, von Euler US (1961) Effect of intraventricular administration of substance P in the unanaesthetized cat. In: Stern P (ed) Proceedings of the Symposium on Substance P, Sarajevo. Scientific Society of Bosnia and Herzegovina, Yugoslavia, p 82

Picard P, Regoli D, Couture R (1994) Cardiovascular and behavioural effects of centrally administered tachykinins in the rat: characterization of receptors with selective antagonists. Br J Pharmacol 112:240–249

Pickel VM, Armstrong D (1984) Ultrastructural localization of monoamines and peptides in rat area postrema. Fed Proc 43:2949–2951

Poli-Bigelli S, Rodrigues-Pereira J, Carides, AD et al. (2003) Addition of the neurokinin 1 receptor antagonist aprepitant to standard antiemetic therapy improves control of chemotherapy-induced nausea and vomiting. Results from a randomised double-blind, placebo-controlled trial in latin America. Cancer 97:3090–3098

Powell D, Cannon D, Skrabanek P et al. (1978) The pathophysiology of substance P in man. In: Bloom SR, Grossman MI (eds) Gut Hormones. Churchhill Livingstone, Edinburgh, pp 524–529

Ptak K, Burnet H, Blanchi B et al. (2002) The murine neurokinin NK_1 receptor gene contributes to the adult hypoxic facilitation of ventilation. Eur J Neurosci 16:2245–2252

Quigley EMM, Hasler WL, Parkman HP (2001) AGA technical review on nausea and vomiting. Gastroenterology 120:263–286

Quirion R (1985) Multiple tachykinin receptors. Trends Neurosci 8:183–185

Regoli D, Drapeau G, Dion S et al. (1988) New selective agonists for neurokinin receptors: pharmacological tools for receptor characterization. Trends Pharmacol Sci 9:290–295

Regoli D, Rouissi N, D'Orleans-Juste P (1994) Pharmacological characterization of receptor types. In: Buck SH (eds) The Tachykinin Receptors. Humana Press, Totowa, New Jersey, USA, pp 367–393

Reid K, Palmer JL, Wright RJ et al. (2000) Comparison of the neurokinin-1 antagonist GR205171, alone and in combination with the $5-HT_3$ antagonist ondansetron, hyo-

scine and placebo in prevention of motion-induced nausea in man. Br J Clin Pharmacol 50:61–64
Reid K, Sciberras DG, Gertz BJ et al. (1998) Comparison of a neurokinin-1 antagonist, L-758,298, and scopolamine with placebo in the prevention of motion-induced nausea in man. Br J Clin Pharmacol 45:282P
Reynolds DJM (1995) Where do 5-HT$_3$ receptor antagonists act as anti-emetics? In: Reynolds DJM, Andrews PLR, Davis CJ (eds) Serotonin and the Scientific Basis of Anti-emetic Therapy. Oxford Clinical Communications, Oxford, UK, pp 111–126
Rikard Bell GC, Tork I, Sullivan C et al. (1990) Distribution of substance P-like immunoreactive fibres and terminals in the medulla oblongata of the human infant. Neuroscience 34:133–148
Rizk AN, Hesketh PJ (1999) Antiemetics for cancer chemotherapy-induced nausea and vomiting. A review of agents in development. Drugs Res Dev 2:229–235
Robichaud A, Tattersall FD, Choudhury I et al. (1999) Emesis induced by inhibitors of type IV cyclic nucleotide phosphodiesterase (PDE IV) in the ferret. Neuropharmacology 38:289–297
Rogers RC, Herman GE (1992) Central regulation of brainstem gastric vago-vagal control circuits. In: Ritter S, Ritter RC, Barnes, CD (eds) Neuroanatomy and Physiology of Abdominal Vagal Afferents. CRC Press, Boca Raton, USA, pp 99–134
Roth GI, Yamamoto WS (1968) The microcirculation of the area postrema in the rat. J Comp Neurol 133:329–340
Rudd JA, Bunce KT, Naylor RJ (1992) Effect of 8-OH-DPAT on drug-induced emesis in the ferret. Br J Pharmacol 106:101P
Rudd JA, Bunce KT, Naylor RJ (1996a) The interaction of dexamethasone with ondansetron on drug-induced emesis in the ferret. Neuropharmacology 35:91–97
Rudd JA, Cheng CH, Naylor RJ et al. (1999a) Modulation of emesis by fentanyl and opioid receptor antagonists in *Suncus murinus* (house musk shrew). Eur J Pharmacol 374:77–84
Rudd JA, Jordan CC, Naylor RJ (1994) Profiles of emetic action of cisplatin in the ferret: a potential model of acute and delayed emesis. Eur J Pharmacol 262:R1–R2
Rudd JA, Jordan CC, Naylor RJ (1996b) The action of the NK$_1$ tachykinin receptor antagonist, CP-99,994, in antagonizing the acute and delayed emesis induced by cisplatin in the ferret. Br J Pharmacol 119:931–936
Rudd JA, Naylor RJ (1994) Effects of 5-HT$_3$ receptor antagonists on models of acute and delayed emesis induced by cisplatin in the ferret. Neuropharmacology 33:1607–1608
Rudd JA, Naylor RJ (1997) The actions of ondansetron and dexamethasone to antagonise cisplatin-induced emesis in the ferret. Eur J Pharmacol 322:79–82
Rudd JA, Naylor RJ (1996) An interaction of ondansetron and dexamethasone antagonizing cisplatin-induced acute and delayed emesis in the ferret. Br J Pharmacol 118:209–214
Rudd JA, Ngan MP, Wai MK (1998) 5-HT$_3$ receptors are not involved in conditioned taste aversions induced by 5-hydroxytryptamine, ipecacuanha or cisplatin. Eur J Pharmacol 352:143–149
Rudd JA, Ngan MP, Wai MK (1999b) Inhibition of emesis by tachykinin NK$_1$ receptor antagonists in *Suncus murinus* (house musk shrew). Eur J Pharmacol 366:243–252
Rudd JA, Wai MK (2001) Genital grooming and emesis induced by vanilloids in *Suncus murinus*, the house musk shrew. Eur J Pharmacol 422:185–195
Rupniak NM, Kramer MS (1999) Discovery of the anti-depressant and anti-emetic efficacy of substance P receptor (NK$_1$) antagonists. Trends Pharmacol Sci 20:485–490
Rupniak NM, Tattersall FD, Williams AR et al. (1997) *In vitro* and *in vivo* predictors of the anti-emetic activity of tachykinin NK$_1$ receptor antagonists. Eur J Pharmacol 326:201–209

Saeki M, Sakai M, Saito R et al. (2001) Effects of HSP-117, a novel tachykinin NK_1-receptor antagonist, on cisplatin-induced pica as a new evaluation of delayed emesis in rats. Jpn J Pharmacol 86:359–362

Saito R, Ariumi H, Kubota H et al. (1999) The role of tachykinin NK-1 receptors in emetic action in the area postrema of ferrets. Nippon Yakurigaku Zasshi 114:209P–214P

Saito R, Suehiro Y, Ariumi H et al. (1998) Anti-emetic effects of a novel NK-1 receptor antagonist HSP-117 in ferrets. Neurosci Lett 254:169–172

Sakaruda T, Manome Y, Tan-no K et al. (1992) A selective and extremely potent anatagonist of the neurokinin-1 receptor. Brain Res 593:319–322

Sanger GJ (1993) The pharmacology of anti-emetic agents. In: Andrews PLR, Sanger GJ (eds) Emesis in Anti-cancer Therapy: Mechanisms and Treatment. Chapman and Hall, London, UK, pp 179–210

Saria A (1999) The tachykinin NK_1 receptor in the brain: pharmacology and putative functions. Eur J Pharmacol 375:51–60

Selve N, Friderichs E, Reimann W et al. (1994) Absence of emetic effects of morphine and loperamide in *Suncus murinus*. Eur J Pharmacol 256:287–293

Severini C, Improta G, Falconieri-Erspamer G et al. (2002) The tachykinin peptide family. Pharmacol Rev 54:285–322

Shashkov VS, Iasnetsov VV, Drozd IV et al. (1988) Neuropharmacology of the autonomic vestibular syndrome (Russian). Farmakol Toksikol 51:30–36

Shiroshita Y, Koga T, Fukuda H (1997) Capsaicin in the 4th ventricle abolishes retching and transmission of emetic vagal afferents to solitary nucleus neurons. Eur J Pharmacol 339:183–192

Singh L, Field MJ, Hughes J et al. (1997) The tachykinin NK_1 receptor antagonist PD154075 blocks cisplatin-induced delayed emesis in the ferret. Eur J Pharmacol 321:209–216

Sleisenger MH (ed.) (1993) The Handbook of Nausea and Vomiting. Caduceus Medical Publishers, Pathenon Publ Group, New York, USA

Smid SD, Lynn PA, Templeman R et al. (1998) Activation of non-adrenergic non-cholinergic inhibitory pathways by endogenous and exogenous tachykinins in the ferret lower oesophageal sphincter. Neurogastroenterol Motil 10:149–156

Smith JE, Paton JFR, Andrews PLR (2002) An arterially perfused decerebrate preparation of *Suncus murinus* (house musk shrew) for the study of emesis and swallowing. Exp Physiol 87:563–574

Snider RM, Constantine JW, Lowe JA et al. (1991) A potent nonpeptide antagonist of the substance P (NK_1) receptor. Science 251:435–437

Sothwell BR, Seynold VS, Woodman HL et al. (1998) Quantitation of neurokinin 1 receptor internalisation and recycling in guinea-pig myenteric neurons. Neuroscience 87:925–931

Spencer SE, Talman WT (1986) Central modulation of gastric pressure by substance P: a comparison with glutamate and acetylcholine. Brain Res 385:371–374

Sundler F, Håkanson R, Larsson LI et al. (1977) Substance P in the gut: an immunohistochemical study of its distribution and development. In: von Euler US, Pernow B (eds) Substance P, Raven Press, New York, USA, pp 59–65

Szallasi A, Blumberg PM (1999) Vanilloid (capsaicin) receptors and mechanisms. Pharmacol Rev 51:159–212

Takeda N, Hasegawa S, Morita M et al. (1993) Pica in rats is analogous to emesis: an animal model in emesis research. Pharmacol Biochem Behav 45:817–821

Takeda N, Hasegawa S, Morita M et al. (1995) Neuropharmacological mechanisms of emesis. I. Effects of antiemetic drugs on motion- and apomorphine-induced pica in rats. Methods Find Exp Clin Pharmacol 17:589–590

Tanihata S, Igarashi H, Suzuki M et al. (2000) Cisplatin-induced early and delayed emesis in the pigeon. Br J Pharmacol 130:132–138

Tanihata S, Oda S, Kakuta S et al. (2003) Antiemetic effect of a tachykinin NK_1 receptor antagonist GR205171 on cisplatin-induced early and delayed emesis in the pigeon. Eur J Pharmacol 461:197–206

Tattersall FD, Rycroft W, Cumberbatch M et al. (2000) The novel NK_1 receptor antagonist MK-0869 (L-754,030) and its water soluble phosphoryl prodrug, L-758,298, inhibit acute and delayed cisplatin-induced emesis in ferrets. Neuropharmacology 39:652–663

Tattersall FD, Rycroft W, Francis B et al. (1996) Tachykinin NK_1 receptor antagonists act centrally to inhibit emesis induced by the chemotherapeutic agent cisplatin in ferrets. Neuropharmacology 35:1121–1129

Tattersall FD, Rycroft W, Hargreaves RJ et al. (1993) The tachykinin NK_1 receptor antagonist CP-99,994 attenuates cisplatin induced emesis in the ferret. Eur J Pharmacol 250:R5–R6

Tattersall FD, Rycroft W, Hill RG et al. (1994) Enantioselective inhibition of apomorphine-induced emesis in the ferret by the neurokinin1 receptor antagonist CP-99,994. Neuropharmacology 33:259–260

Tattersall FD, Rycroft W, Marmont N et al. (1995) Enantiospecific inhibition of emesis induced by nicotine in the house musk shrew (*Suncus murinus*) by the neurokinin 1 (NK_1) receptor antagonist CP-99,994. Neuropharmacology 34:1697–1699

Tavorath R, Hesketh PJ (1996) Drug treatment of chemotherapy-induced delayed emesis. Drugs 52:639–648

Thompson PI, Bingham S, Andrews PL et al. (1992) Morphine 6-glucuronide: a metabolite of morphine with greater emetic potency than morphine in the ferret. Br J Pharmacol 106:3–8

Torii Y, Saito H, Matsuki N (1991a) 5-hydroxytryptamine is emetogenic in the house musk shrew, *Suncus murinus*. Naunyn-Schmiedeberg's Arch Pharmacol 344:564–567

Torii Y, Saito H, Matsuki N (1991b) Selective blockade of cytotoxic drug-induced emesis by $5-HT_3$ receptor antagonists in *Suncus murinus*. Jpn J Pharmacol 55:107–113

Tramer MR, Moore RA, Reylods DJM (1997) A quantitiative systematic review of ondansetron in treatment of established postoperative nausea and vomiting. Br Med J 314:1088–1092

Travagli RA, Rogers RC (2001) Receptors and transmission in the brain-gut axis: potential for novel therapies. V. Fast and slow extrinsic modulation of dorsal vagal complex circuits. Am J Physiol 281:G595–G601

Triepl J, Weindl A, Reinecke M et al. (1983) The distribution of substance P in the spinal cord and brain stem of cat, guinea-pig and rat. A comparative immunohistochemical investigation. In: Skrabanek P, Powell D (eds) Substance P. Proceedings of the International Symposium—Dublin 1983. Boole Press, Dublin, Ireland, pp 267–268

Tsuchiya M, Fujiwara Y, Kanai Y et al. (2002) Anti-emetic activity of the novel nonpeptide tachykinin NK_1 receptor antagonist ezlopitant (CJ-11,974) against acute and delayed cisplatin-induced emesis in the ferret. Pharmacology 66:144–152

Ueno S, Matsuki N, Saito H (1987) *Suncus murinus*: A new experimental model in emesis research. Life Sci 41:513–518

Van Belle S, Lichinitser MR, Navari RM et al. (2002). Prevention of cisplatin-induced acute and delayed emesis by the selective neurokinin-1 antagonists, L-758,298 and MK-869. Cancer 94:3032–3041

von Euler US (1936) Untersuchungen über Substanz P, die atropinfeste, darmerregende und gefässerweiternde Substanz aus Darm und Gehirn. Naunyn-Schmiedeberg's Arch Exp Pathol Pharmakol 181:181–197

von Euler US, Gaddum JH (1931) An unidentified depressor substance in certain tissue extracts. J Physiol (Lond) 72:74–87

von Euler US, Pernow B (eds) (1976) Substance P. Nobel Symposium 37, Raven Press, New York, USA

Ward P, Armour DR, Bays DE et al. (1995) Discovery of an orally bioavailable NK_1 receptor antagonist, (2S,3S)-(2-methoxy-5-tetrazol-1-ylbenzyl)(2-phenylpiperidin-3-yl)-amine (GR203040), with potent antiemetic activity. J Med Chem 38:4985–4992

Watson JW, Gonsalves SF, Fossa AA et al. (1995a) The anti-emetic effects of CP-99,994 in the ferret and the dog: Role of the NK_1 receptor. Br J Pharmacol 115:84–94

Watson JW, Gonsalves SF, Fossa AA et al. (1995b) The tachykinins and emesis: Towards complete control? In: Reynolds DJM, Andrews PLR, Davis CJ (eds) Serotonin and the Scientific Basis of Anti-emetic Therapy. Oxford Clinical Communications, Oxford, UK, pp 233–238

Wieland T, Bodanszky M (1991) Biologically active fragments of proteins. Chapter 8 in The World of Peptides, Springer, Berlin

Wilder-Smith OH, Borgeat A, Chappuis P et al. (1993) Urinary serotonin metabolite excretion during cisplatin chemotherapy. Cancer 72:2239–2241

de Wit R, Herrstedt J, Rapoport B et al. (2004) The oral NK(1) antagonist, aprepitant, given with standard antiemetics provides protection against nausea and vomiting over multiple cycles of cisplatin-based chemotherapy: a combined analysis of two randomised, placebo-controlled phase III clinical trials. Eur J Cancer 40:403–410

de Wit R, Herrstedt J, Rapoport B et al. (2003) Addition of the oral NK1 antagonist aprepitant to standard antiemetics provides protection against nausea and vomiting during multiple cycles of cisplatin-based chemotherapy. J Clin Oncol 21:4105–4111

Wood KL (1988) Aspects of the central control of gastric motility in the ferret and the rat. PhD Thesis, University of London, UK

Wu M, Harding K, Hugenholtz H et al. (1985) Emetic effects of centrally-administered angiotensin II, arginin vasopressin and neurotensin in the dog. Peptides 6:173–175

Wynn RL, Essien E, Thut PD (1993) The effects of different antiemetic agents on morphine-induced emesis in ferrets. Eur J Pharmacol 241:47–54

Yamakuni H, Sawai H, Maeda Y et al. (2000) Probable involvement of the 5-hydroxytryptamine-4 receptor in methotrexate-induced delayed emesis in dogs. J Pharmacol Exp Ther 292:1002–1007

Yamanuki H, Sawai-Nakayama H, Imazumi K et al. (2002) Resiniferatoxin antagonises cisplatin-induced emesis in dogs and ferrets. Eur J Pharmacol 442:273–278

Yamazoe M, Shiosaka S, Shibasaki T et al. (1984) Distribution of six neuropeptides in the nucleus tractus solitarii of the rat: an immuno-histochemical analysis. Neuroscience 13:1243–1266

Yasnetsov VV, Drozd YV, Shashkov VS (1987) Emetic and anti-emetic properties of some regulatory peptides. Byulleten Eksperimental'noi Biologii I Meditsiny 103:586–588

Yates BJ, Miller AD, Lucot JB (1998) Physiological basis and pharmacology of motion sickness: an update. Brain Res Bull 47:395–406

Zachrisson O, Lindefors N, Brene S (1998) A tachykinin NK_1 receptor antagonist, CP-122,721-1, attenuates kainic acid-induced seizure activity. Mol Brain Res 60:291–295

Zaman S, Woods AJ, Watson JW et al. (2000) The effect of the NK_1 receptor antagonist CP-99,994 on emesis and c-fos protein induction by loperamide in the ferret. Neuropharmacology 39:316–323

Zettler G, Schlosser L (1955) Über die Verteilung von Substanz P und Cholinacetylase im Gehirn. Naunyn-Schmiedeberg's Arch Exp Pathol Pharmakol 224:159–175

Substance P (NK$_1$) Receptor Antagonists—Analgesics or Not?

S. Boyce · R. G. Hill

Merck Sharp and Dohme Research Laboratories, Neuroscience Research Centre,
Terlings Park, Harlow, CM20 2QR, UK
e-mail: susan_boyce@merck.com
e-mail: raymond_hill@merck.com

1	Substance P and Pain Transmission	441
2	NK$_1$ Receptor Antagonists in Preclinical Assays of Pain	442
2.1	In Vivo Studies	442
2.2	Studies in NK$_1$ Receptor Knockout Mice	446
3	Clinical Trials with NK$_1$ Receptor Antagonists	447
4	Discrepant Results of Preclinical and Clinical Trials	448
4.1	Have NK1 Receptor Antagonists Been Tested in the Right Clinical Trials?	448
4.2	Other Reasons for Lack of Clinical Efficacy	450
5	Substance P Antagonists as Adjuncts to Existing Analgesics	452
	References	453

Abstract Over the last two decades much research has focused on the role of substance P in pain and on the development of substance P antagonists as novel analgesics. Despite the identification of high affinity and selective substance P (NK$_1$) receptor antagonists and a plethora of preclinical data supporting an analgesic profile of these agents, the outcome from clinical trials has been extremely disappointing with no clear analgesic efficacy being observed in a variety of pain states. This has led the pain community to seriously question the predictability and utility of preclinical pain assays, especially for novel targets. This chapter will review the animal studies and clinical trials with NK$_1$ receptor antagonists and suggests possible reasons for the apparent mismatch between preclinical and clinical studies in pain.

Keywords Pain · Hyperalgesia · Behaviour · Stress

1
Substance P and Pain Transmission

Substance P has long been considered as a prime candidate as a pain transmitter. Evidence supporting this role comes from anatomical and immunochemical studies showing that substance P is expressed in small unmyelinated sensory fibres (Nagy et al. 1981) which transmit noxious information to the spinal cord. Using antibody microprobes, Duggan and colleagues (1987) showed that substance P is released into the dorsal horn of the spinal cord following intense noxious stimulation. In addition, substance P when applied onto the dorsal horn neurons, produces prolonged excitation which resembles the activation observed following noxious stimulation (Henry 1976) and given intrathecally produces behavioural hyperalgesia (Cridland and Henry 1986). More recently, it has been shown that following peripheral noxious stimulation NK_1 receptors become internalized on dorsal horn neurons and this effect can be blocked by NK_1 receptor antagonists (Mantyh et al. 1995).

In addition to its effects on spinal nociceptive processing, substance P has also been implicated in the pain associated with migraine. C-fibre sensory afferents that innervate meningeal tissues contain substance P and other neuropeptides (e.g., calcitonin gene related peptide, CGRP) and it has been suggested that release of these neuropeptides causes neurogenic inflammation which could lead to activation of nociceptive afferents projecting to the brain stem and, consequently, pain (Shepheard et al. 1995). Based on such evidence, the expectation was that centrally acting NK_1 receptor antagonists would be antinociceptive in animals, analgesic in man and constitute a novel class of analgesic drug.

2
NK_1 Receptor Antagonists in Preclinical Assays of Pain

2.1
In Vivo Studies

One of the initial problems encountered for NK_1 receptor antagonists was the marked species difference in NK_1 receptor pharmacology. For example, compounds which have been optimized for NK_1 receptors expressed in humans typically have low affinity for rats or mice, species used for antinociceptive studies (see Table 1). Consequently, early studies with the NK_1 receptor antagonists which were performed in rats and mice, required high doses to observe antinociceptive effects and so interpretation of the data was confounded by potential off-target activity such as ion channel effects (Rupniak et al. 1993). There is now considerable evidence generated from well controlled studies using enantiomeric pairs, one having high affinity for the NK_1 receptor and the other low affinity, to control for non-specific effects and in appropriate species (gerbils, guinea pigs), to demonstrate unequivocally that NK_1 receptor antagonists possess antinociceptive effects in animals.

Conventional conscious animal nociception tests which are well known to be sensitive to analgesic effects of opioids like morphine, such as hot plate, tail or paw flick and paw pressure tests, are ineffective at identifying antinociceptive effects of NK_1 receptor antagonists even when administered at high doses (Rupniak et al. 1993). Similarly, NK_1 receptor antagonists have little effect on baseline spinal nociceptive reflexes in anaesthetized animals elicited by electri-

Table 1 Species variants in NK_1 receptor pharmacology and central nervous system penetration of NK_1 receptor antagonists

Compound	IC$_{50}$ for inhibition of [^{125}I]-SP binding (nM)					CNS penetration
	Human	Gerbil	Guinea pig	Rabbit	Rat	Gerbil ID50 mg/kg i.v.
CP96345[a]	28.8	32.3	31.6	41.6	5,888	–
Aprepitant	0.1	0.3	0.31	–	–	0.3
L733060	0.87	0.36	0.3	–	550	0.2
L733061	350	370	240	–	>1,000	>1,000
L760735	0.3	0.5	0.34	–	10	0.1
GR205171	0.08	0.06	0.09	–	1.4	0.02
LY303870	0.15	–	0.22	–	8.7	~10
RPR100893	30	–	–	–	1417	>10
RP67580[b]	–	–	–	–	10	–
TAK637[c]	–	–	–	–	87	–
CI1021[d]	0.84	2.5	6.2	–	302	~10 s.c. versus SarMet SP[e]

[^{125}I]-SP binding assays were performed as described by Cascieri et al. (1992). Cloned human NK_1 and rat NK_1 receptors were stably expressed in CHO cells. For other species, membrane homogenates were prepared from cerebral cortex. Central nervous system penetration was determined by the inhibition of foot tapping induced by intracerebroventricular injection of the NK1 receptor agonist GR73632 in gerbils as described by Rupniak and Williams et al. (1994).
Data taken from [a] Beresford et al. (1991), [b] Garret et al. (1992), [c] Natsugari et al. (1999) and [d] Singh et al. (1997).
s.c., Subcutaneous; i.v., intravenous.

cal stimulation (Laird et al. 1993). The first clear antinociceptive effect of NK_1 receptor antagonists came from in vivo electrophysiological studies on anaesthetized/spinalized or decerebrate/spinalized animals. For example, CP-96,345 and LY303870, but not their less active enantiomers, CP-96,344 and LY396155, have been shown to be potent inhibitors of the excitation of dorsal horn neurons elicited by prolonged noxious mechanical or thermal peripheral stimulation or by iontophoretic application of substance P in cats (Radhakrishnan and Henry 1991; Radhakrishnan et al. 1998) showing that these effects are due to a specific blockade of NK_1 receptors. In addition, aprepitant (MK0869) or CP-99,994, but not its less active enantiomer CP-100,263, inhibited the facilitation or wind up of a spinal flexion reflex produced by C-fibre conditioning stimulation in decerebrate/spinalized rabbits (Boyce et al. 1993; see Boyce and Hill 2000) and RP67580 which has high affinity for rat NK_1 receptors (Table 1), but not its less active enantiomer RP68651, inhibited facilitation of the hind limb flexor reflex in anaesthetized/spinalized rats (Laird et al. 1993).

In conscious animals, the first demonstration of a clear enantioselective analgesic effect came from studies using L-733,060, a highly selective and brain penetrant NK_1 receptor antagonist with long duration of action. This compound, but not its less active enantiomer L-733,061, was able to inhibit the late phase nociceptive responses to intraplantar injection of formalin in gerbils (Rupniak et al. 1996). In the same study, the poorly brain penetrant compound L-743,310, a potent inhibitor of peripherally mediated NK_1 receptor agonist induced chromodacryorrhoea, failed to inhibit the late phase response indicating that the antinociceptive effect of L-733,060 was via blockade of central NK_1 receptors. Consistent with a central antinociceptive action, intrathecal injection of CP-96,345, but not its less active enantiomer CP-96,344, attenuated the late phase response in rats (Yamamoto and Yaksh 1991). LY303870 also blocked the late phase of the formalin test in rats (Iyengar et al. 1997), although it is not clear whether this effect is mediated centrally or peripherally (Rupniak et al. 1997; Iyengar et al. 1997; see Table 1). Oral administration of CP-99,994, SDZNKT343 (Novartis) or LY303870 has also been shown to attenuate mechanical hyperalgesia induced by carrageenan in guinea pigs (Patel et al. 1996; Urban et al. 1999). The effect appears to be mediated via blockade of spinal NK_1 receptors, as intrathecal but not intraplantar injection of SDZNKT343 reduced the hyperalgesia. We have found that L-733,060, but not its inactive isomer L-733,061, reversed carrageenan-induced mechanical hyperalgesia in guinea pigs (Boyce and Hill 2000).

In addition to their effects in assays of inflammatory hyperalgesia, NK_1 receptor antagonists are effective in a number of neuropathic pain assays. Urban and colleagues (1999) using partial sciatic nerve ligation in guinea pigs showed that SDZNKT343 and LY303870 reduced established mechanical hyperalgesia following either oral or intrathecal administration. In contrast, RPR100893 was active only following intrathecal administration (Urban al. 1999) probably due to poor brain penetration (Rupniak et al. 1997; Table 1). Likewise, CI1-021 was effective in reversing mechanical hypersensitivity (reduction in weight bearing) in guinea pigs following sciatic nerve constriction injury (CCI; Gonzalez et al. 2000) and reduced mechanical hypersensitivity in rats following streptozotocin treatment (Field et al. 1998). A central site of action again appears to be important for the antinociceptive effects of NK_1 receptor antagonists in diabetic rats as intrathecal administration of RP67580, but not its less active enantiomer, reduced mechanical hyperalgesia (Coudore-Civiale et al. 2000). Administration of a non-brain penetrant antagonist, PD156982, had no effect at doses shown to block peripheral NK_1 receptors (Field et al. 1998). Studies performed in our laboratories have shown that GR205171 which has nanomolar affinity at rat NK_1 receptors (see Table 1), reversed both mechanical hypersensitivity and the increase in receptive field size of dorsal horn neurons in rats following loose ligation of the sciatic nerve; these effects were not observed with its inactive enantiomer L-796,325 (Cumberbatch et al. 1998). In contrast, using the spinal nerve ligation (Chung) model, we were unable to demonstrate an anti-algesic effect of GR205171 at doses that were effective in rats with nerve constriction (M. Sablad, A. Hama and M. Urban, personal communication; Fig. 1). This disparity in the effectiveness of GR205171 in the two neuropathic pain models may relate to differences in the underlying pathophysiology. As well as the nerve damage associated with constriction injury, a marked neurogenic inflammation also develops (Daemen et al. 1998) which could contribute to the development of hyperalgesia and allodynia. The analgesic effects of NK_1 receptor antagonists in CCI and par-

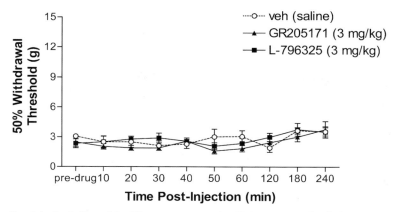

Fig. 1 Lack of effect of the NK_1 receptor antagonist GR205171 on tactile allodynia in rats following spinal nerve ligation. Tactile withdrawal thresholds were determined using von Frey filaments 7 days after spinal nerve ligation. GR205171 (3 mg/kg i.v.), its less active enantiomer L-796,325 (3 mg/kg i.v.) or vehicle was administered and thresholds again determined. The dose of GR205171 had previously been shown to completely reverse mechanical hyperalgesia in rats following chronic constriction injury (Cumberbatch et al. 1998). Data are expressed as 50% withdrawal threshold (g)

tial nerve ligation models may therefore relate to anti-inflammatory actions and not specifically to effects on neuropathic pain.

Although only a limited number of studies have been performed to date, there is evidence to suggest that NK_1 receptor antagonists may be effective against visceral pain. Thus, CP-99,994 inhibited the nociceptive reflex response (depressor effect) to jejunal distension in rats (McLean et al. 1998) and TAK637, but not its less active enantiomer, and CP-99,994 reduced the number of abdominal contractions induced by colorectal distension in rabbits following sensitization to acetic acid (Okano et al. 2002). In common with the above mentioned studies on neuropathic and inflammatory pain, the analgesic effects appear to be centrally mediated as intrathecal administration of TAK637 and CP-99,994 inhibited abdominal contractions. In contrast with these findings, Julia et al. (1994) demonstrated that the NK_2 receptor antagonist SR48968, but not the NK_1 receptor antagonists CP-96,345 or RP67580, inhibited abdominal contractions to rectal distension in rats. However, the NK_1 receptor antagonists did inhibit distension-induced inhibition of colonic motility in this investigation.

There is also preclinical evidence suggesting that NK_1 receptor antagonists are effective in inflammation associated with arthritic changes. In a collaborative study with S. Cruwys and B. Kidd (Inflammation Research Group, London Hospital Medical College), we found that, in a similar manner to indomethacin, daily administration of the NK_1 receptor antagonist L-760,735 (3 mg/kg subcutaneously) for 21 days reduced paw oedema and the associated thermal and mechanical hyperalgesia in Freund's adjuvant arthritic guinea pigs (see Boyce and Hill 2000). Consistent with these findings, Binder et al. (1999) showed that repeated administration of GR205171, but not its less active enantiomer, also reduced arthritic joint damage (joint swelling, synovitis and bone demineralization) caused by complete Freund's adjuvant in rats. These findings suggest that NK_1 receptor antagonists may possess anti-inflammatory as well as antinociceptive activity. Other data suggest that NK_1 receptor antagonists are extremely potent inhibitors of neurogenic inflammation. The NK_1 receptor antagonists RP67580, CP-99,994, LY30380 and aprepitant are highly potent at blocking neurogenic plasma extravasation in the dura following trigeminal ganglion stimulation in rats or guinea pigs (Shepheard et al. 1993, 1995; Phebus et al. 1997; Boyce and Hill 2000). In addition to its actions on neurogenic extravasation, CP-99,994 has also been shown to reduce c-*fos* mRNA expression in the trigeminal nucleus caudalis in rats after trigeminal ganglion stimulation (Shepheard et al. 1995). Based on these findings, it was hypothesized that brain penetrant NK_1 receptor antagonists may have anti-migraine effects peripherally through blockade of dural extravasation and centrally by inhibition of nociceptive pathways.

2.2
Studies in NK_1 Receptor Knockout Mice

As well as using NK_1 receptor antagonists to determine the pharmacological significance of NK_1 receptors in pain processing, another approach taken was to use mice in which the receptor has been deleted. Like the studies with the non-peptide antagonists, $NK_1^{-/-}$ mice exhibit little or no changes in acute nociception tests such as the hot plate, thermal paw withdrawal or responses to von Frey filaments (De Felipe et al. 1998; Mansikka et al. 1999). In addition, acute nociceptive responses to intraplantar injection of chemical stimuli such as formalin or capsaicin, as well as the resultant mechanical/heat hypersensitivity are attenuated in $NK_1^{-/-}$ mice (De Felipe et al. 1998; Laird et al. 2001; Mansikka et al. 1999). However, these mice do develop hyperalgesia after induction of hind paw inflammation with complete Freund's adjuvant (De Felipe et al. 1998) which contrasts with the findings observed with NK_1 receptor antagonists (see above). Other studies investigating the role of NK_1 receptors in neuropathic pain have also revealed differences between the profile of non-peptide antagonists and $NK_1^{-/-}$ mice. Again, Martinez-Caro and Laird (2000) failed to demonstrate any difference between wild-type (WT) and $NK_1^{-/-}$ mice following partial sciatic nerve ligation, whereas NK_1 receptor antagonists were effective in the same model in guinea pigs. Using the L5 spinal ligation (modified Chung) model, Mansikka et al. (2000) found that $NK_1^{-/-}$ mice did not develop mechanical hypersensitivity, and hyperalgesic responses to thermal stimuli (radiant heat or cooling) were unaltered in $NK_1^{-/-}$ compared to WT mice. Data from $NK_1^{-/-}$ mice also support a role of NK_1 receptors in visceral pain, particularly associated with neurogenic inflammation. Thus, instillation of capsaicin, which evokes neurogenic inflammation, into the colon of $NK_1^{-/-}$ mice produced fewer abdominal contractions than were observed in WT mice and they failed to develop referred hyperalgesia (Laird et al. 2000). Similarly, behavioural responses to cyclophosphamide and the acute nociceptive (pressor) reflex response or primary hyperalgesia following intracolonic acetic acid were impaired in $NK_1^{-/-}$ mice. In contrast, nociceptive responses to intracolonic mustard oil which was found to evoke direct tissue damage, were unchanged in $NK_1^{-/-}$ mice (Laird et al. 2000). Electrophysiological studies have also shown that the characteristic amplification ('wind up') of spinal nociceptive reflexes to repetitive high frequency electrical stimulation is absent in $NK_1^{-/-}$ mice (De Felipe et al. 1998), which is in agreement with the findings from studies with NK_1 receptor antagonist drugs.

The evidence obtained from studies with $NK_1^{-/-}$ mice supporting a role of substance P and NK_1 receptors in pain is less compelling than has been reported for non-peptide NK_1 receptor antagonists, and it is interesting to speculate whether the search for such compounds as analgesics would have been so widely followed by the pharmaceutical industry if the knockout data had been available before the discovery of non-peptide antagonists. It is not clear why there should be differences in the antinociceptive profile of NK_1 receptor antagonists and $NK_1^{-/-}$ mice. Non-specific actions of NK_1 receptor antagonists contributing to the antinociceptive effects can be ruled out as the studies outlined above are well-controlled with many demonstrating marked enantioselective inhibition. More likely, these differences may relate to compensatory changes in the knockout mice as a result of the life-long absence of NK_1 receptors.

In summary, there is a wealth of preclinical data that are supportive of an analgesic potential of NK_1 antagonists, particularly in pain conditions associated with inflammation or nerve injury, and there is growing evidence for a potential utility in treating some visceral pain conditions. Their analgesic profile is likely to be more similar to that of the non-steroidal anti-inflammatory drugs (NSAIDs) than that of opioid analgesics.

3
Clinical Trials with NK₁ Receptor Antagonists

Despite the convincing evidence from animal studies for antinociceptive effects of NK_1 receptor antagonists, only one clinical study, with CP-99,994 in postoperative dental pain, has demonstrated analgesic activity in man (Dionne et al. 1999). In this study, CP-99,994 was administered as an intravenous infusion over 5 h, starting 30 min prior to surgery (total dose 0.75 mg/kg). CP-99,994 had comparable clinical efficacy to ibuprofen (Dionne et al. 1998). In contrast with these findings, the long acting orally active NK_1 receptor antagonist aprepitant [300 mg per os (p.o.) given 2 h prior to surgery, a dose established to be antiemetic in man; Navari et al. 1999] was ineffective in postoperative dental pain (Reinhardt et al. 1998; Fig. 1). Similarly, CP-122,721 (200 mg p.o; also an antiemetic dose in man; Gesztesi et al. 1998) was reported to be without effect in postoperative dental pain (Gesztesi et al. 1998). NK_1 receptor antagonists have also been evaluated in patients with neuropathic pain, and aprepitant (300 mg p.o. for 2 weeks) was ineffective in patients with established post-herpetic neuralgia (PHN; duration of 6 months to 6 years; Block et al. 1998). Lanepitant (LY303870; 50 mg, 100 mg or 200 mg p.o. twice a day for 8 weeks) had no significant effect on pain intensity (daytime or night time) when compared to placebo in patients with painful diabetic neuropathy (Goldstein et al. 1999). Lanetipant (10 mg, 30 mg, 100 mg, or 300 mg p.o. for 3 weeks) was also without effect in patients with moderate to severe osteoarthritis (Goldstein et al. 1998). Finally, clinical trials with NK_1 receptor antagonists for acute migraine and migraine prophylaxis have also been disappointing. L-758,298, an intravenous prodrug of aprepitant [20 mg, 40 mg or 60 mg intravenously (i.v.)] failed to abort migraine pain as measured either by the time to meaningful relief or the number of patients reporting pain relief within 4 h (Norman et al. 1998). Similarly, GR205171 (25 mg i.v.; Connor et al. 1998) and lanepitant (30 mg, 80 mg or 240 mg p.o.; Goldstein et al. 1997) were ineffective as abortive treatments for migraine headache. Furthermore, prophylactic administration of lanepitant (200 mg p.o. per day) for 1 month had no effect on migraine frequency and severity compared to placebo (Goldstein et al. 1999).

The lack of clinical efficacy of aprepitant and GR205171 in pain or migraine trials is not due to insufficient dose or lack of brain penetration. At the dose used in the analgesia trials, aprepitant was found to produce in excess of 90% NK_1 receptor occupancy using positron emission tomography (PET; Hargreaves 2002), has been shown to produce antidepressant effects in patients with moderate to severe depression (Kramer et al. 1998) and is anti-emetic in cancer patients following chemotherapy (Navari et al. 1999). Similarly, the dose of GR205171 used in the migraine trial was based on NK_1 receptor adequate occupancy calculated from PET studies (Connor et al. 1998). The negative findings with lanepitant may be inconclusive, however, as it has not been published that functional blockade of central NK_1 receptors was achieved at the doses used in the clinical trials. In view of these largely negative findings it has been concluded that NK_1 receptor antagonists are not effective as analgesic agents in man.

4
Discrepant Results of Preclinical and Clinical Trials

4.1
Have NK1 Receptor Antagonists Been Tested in the Right Clinical Trials?

Data obtained from preclinical studies with the antagonists or from $NK_1^{-/-}$ mice indicate that NK_1 receptor antagonists are likely to be more effective in inflammatory and nerve injury pain assays, yet, the majority of clinical trials have been acute pain studies (dental pain or migraine). The dental pain model is therefore unlikely to be ideal for testing the analgesic effects of NK_1 receptor antagonists, especially as these agents are not effective against acute noxious stimuli in rodents (Urban and Fox 2000). A recent study which has utilized NK_1 receptor internalization to determine the extent of substance P release in the spinal cord following dental tooth extraction in rodents would support such a notion (Sabino et al. 2002). Although it was possible to demonstrate NK_1 receptor internalization in neurons in the trigeminal nucleus caudalis and cervical spinal cord (100% of neurons within lamina I and 65% in lamina III-V) within 5 min of incisor extraction, the effect was relatively short-lived with only few neurons remaining internalized at 60 min. This suggests that substance P may play only a minor role in mediating pain associated with tooth extraction and this would be limited to the initial phase of pain which is not typically assessed in the clinical studies. A more appropriate model to evaluate NK_1 receptor antagonists may be dental pain associated with pulpal inflammation where elevated levels of substance P are found in the dental pulp (Awawdeh et al. 2002).

The hypothesis that NK_1 receptor antagonists would be effective anti-migraine agents was based on their ability to inhibit dural plasma extravasation (Shepheard et al. 1993), an effect shared with the modern antimigraine drugs, such as 5-HT(1B/D) agonists ('triptans') (Williamson et al. 1997). Calcitonin gene related peptide (CGRP) has also been implicated in the pathogenesis of migraine headache (Edvinsson 2001; Brain et al. 2002). For example, i.v. administration of CGRP induces a migraine-like headache in migraineurs (Lassen et al. 2002), levels of CGRP are elevated in the external jugular vein during a migraine headache, and following treatment with 'triptans' the plasma levels of CGRP return to control levels with successful amelioration of the headache (Gallai et al. 1995). With the recent disclosure by Boehringer Ingelheim that the selective CGRP receptor antagonist BIBN4096BS is effective in relieving headache in migraineurs (Boehringer Ingelheim Annual Press Conference, April 2002), CGRP appears to be the major peptide mediating the pain associated with migraine, and CGRP, not substance P, receptor antagonists are likely to be effective antimigraine agents.

To date, lanepitant and aprepitant are the only NK_1 receptor antagonists which have been evaluated in chronic pain conditions. Since the data with lanepitant are inconclusive because of the reasons outlined above (Sect. 3), we feel that NK_1 receptor antagonists may not have been adequately evaluated in chronic conditions such as osteoarthritis and diabetic neuropathy. The negative data obtained with aprepitant in PHN is not due to insufficient dose as the same dose was shown to block central NK_1 receptors in man (Hargreaves et al. 2002). Destruction of peptidergic C-fibre function has been reported in patients with long-established disease, and this loss in substance P function could therefore account, at least in part, for the failure of aprepitant to produce analgesia in patients with PHN.

The recent findings that NK_1 receptor antagonists appear to reduce arthritic joint damage and resultant hypersensitivity in animals with adjuvant arthritis suggest that these agents may be worth evaluating in rheumatoid arthritis. This concept is supported by the high expression of NK_1 receptor mRNA in synovia taken from patients with rheumatoid arthritis and that NK_1 mRNA positively correlated with serum C-reactive protein levels and radiological grade of joint destruction (Sakai et al. 1998), suggesting that NK_1 receptor gene expression may reflect the disease progression in rheumatoid arthritis. NK_1 receptor antagonists may also be effective in alleviating other bone-generated pain including bone cancer pain and fractures. A preclinical study using a murine model of cancer pain has shown substantial internalization of NK_1 receptors on NK_1 receptor-expressing neurons in lamina I of the spinal cord following non-noxious palpation of the tumorous bone, which was positively correlated with the extent of bone destruction (Schwei et al. 1999). In addition, substance P, acting via the NK_1 receptor, stimulates osteoclast (OCL) formation and activates OCL bone resorption (Goto et al. 2001). Since bisphosphonates and osteoprotogerin reduce tumour-induced bone pain by blocking bone resorption (Honore and Mantyh 2000), NK_1 receptor antagonists could be effective in reducing bone cancer pain by blocking substance P-mediated bone resorption, as well as the spinal effects of substance P. Other chronic pain conditions in which NK_1 receptor antagonists may be worth testing include fibromyalgia, a syndrome in which elevated levels of substance P are found in the cerebrospinal fluid (Russell et al. 1994). Recently, Littman et al. (1999) reported that the NK_1 receptor antagonist CJ-11974 (50 mg p.o. twice daily for 4 weeks) was able to reduce dysaesthesias in patients with fibromyalgia, and in a subset of patients there was some improvement in pain severity, morning stiffness and sleep disturbances. These findings warrant further investigation.

Of the clinical trials published to date, all have focused on somatic pain and none have reported on the effects of NK_1 receptor antagonists in visceral pain. The distribution of substance P certainly favours a major role in visceral rather than somatic pain as a greater number of visceral primary afferents (>80%) express substance P compared with only 25% in cutaneous afferents (Laird et al. 2001), and laminae I and X of the spinal dorsal horn, which receive afferents from the viscera, show the highest density of NK_1 receptors in the spinal cord (Li et al. 1998). The preclinical findings also support a potential utility of NK_1 receptor antagonists in visceral pain conditions, although their ability to alleviate abdominal pain may be dependent on a significant neurogenic component to the pain (see Sect. 2.2). Conditions such as irritable bowel syndrome (IBS) in which stress is strongly associated with the occurrence of diarrhoea, constipation and abdominal pain may be an appropriate visceral pain condition with which to investigate the effects of NK_1 receptor antagonists, particularly in view of the reported antidepressant effects of these agents (Kramer et al. 1998). Interestingly, symptoms of IBS are closely associated with the severity of depression (Drossman 1999) and antidepressants are effective in relieving these symptoms (Clouse 1994). Phase II trials are reported to be underway with TAK637 for IBS as well as depression but no data is yet available. The outcome of these trials is awaited with interest.

4.2
Other Reasons for Lack of Clinical Efficacy

The negative data obtained in the clinical pain trials have made scientists and clinicians extremely cautious of the preclinical assays in which NK_1 receptor antagonists were active and raised concerns as to their ability to predict clinical analgesic efficacy, particularly for novel targets like substance P. It is clear that most assays have been extremely good at predicting analgesic activity of other agents such as COX-2 inhibitors, gabapentin and antagonists of N-methyl-D-aspartate (NMDA)-type glutamate receptors. The animal data strongly argue for an analgesic effect of NK_1 receptor antagonists against inflammatory pain, and the outcome of trials in rheumatoid arthritis, IBS or other visceral pain conditions will be important to determine whether these preclinical assays are truly predictive, or not, of analgesic efficacy in man. With regard to neuropathic pain conditions, a number of NK_1 receptor antagonists exhibited antinociceptive activity in sciatic nerve injury models (CCI and partial ligation), yet, recent data have shown this not to be the case in the spinal nerve ligation (Chung) model. It is interesting to speculate, therefore, whether the Chung model may be more predictive of analgesic efficacy in man for this class of agents.

The anomaly between preclinical and clinical studies might also be explained by species differences in the physiology of substance P or distribution of NK_1 receptors or by differences between clinical pain and the responses to noxious stimuli measured in small animals. Whereas in rats substance P is co-expressed in 5-HT dorsal raphe descending fibres (Neckers et al. 1979), in man it is restricted to dorsal raphe ascending fibres (Sergeyev et al. 1999) suggesting that it may play a greater role in supraspinal functions in man. Most preclinical pain assays, however, rely on spinal nociceptive reflexes to assess pain which might account for the greater analgesic effect observed with NK_1 receptor antagonists in animals. Also, substance P may be a key neurotransmitter in pain processing in lower species but play a less important role as one moves up the evolutionary ladder. Even across different mouse strains there are marked genetic differences in the sensitivity to a range of analgesics (Wilson et al. 2003). Alternatively, if one regards the noxious stimuli used in animal studies as stressful stimuli (see Fig. 2), one can link the work on nociception with studies supporting an antide-

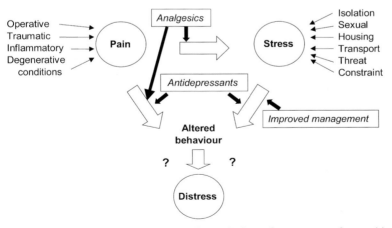

Fig. 2 Behaviour is an imprecise measure of pain. (Redrawn from a concept first used by Headley and Livingston 1989)

pressant action of NK_1 receptor antagonists (see chapter by N.M.J. Rupniak, this volume). These data suggest that NK_1 receptor antagonists may reliably inhibit a variety of stressors but not be sufficient to produce clinical analgesia.

5
Substance P Antagonists as Adjuncts to Existing Analgesics

It must be remembered that substance P is only one of many neurotransmitters expressed in primary sensory afferent neurons and that only a small proportion of these fibres (<25% of cutaneous fibres) contain substance P. Blocking the actions of substance P or NK_1 receptors alone might not be sufficient to produce clinical analgesia. It is interesting to note that animals injected intrathecally with saporin toxin conjugated to substance P to kill NK_1 receptor-expressing cells within the dorsal horn, display a more pronounced antinociception than was achieved by blocking NK_1 receptors (Mantyh et al. 1997), presumably by blocking all inputs to these neurons and not just by inhibiting the actions of substance P. Other studies have shown that the most important action of substance P might be to modulate the action of other transmitters in the spinal cord, particularly glutamate (Juranek and Lembeck 1997). The NMDA receptor antagonist ketamine is analgesic in man but its use is limited by unwanted adverse effects. The ability of substance P to modulate the function of glutamate raises the possibility that NK_1 receptor antagonists may be effective as adjuncts to potentiate the analgesic effects of NMDA receptor antagonists or other existing analgesics. In support of this notion, a combination of RP67580 (but not its inactive enantiomer) with the NMDA/glycine receptor antagonist (+)-HA966 enhanced antinociception in the rat formalin paw test (Seguin and Millan 1994). Similarly, Field and colleagues (2002) demonstrated a marked synergy of the antinociceptive effects of gabapentin with the NK_1 receptor antagonists CI-1021 or CP-99,994 in neuropathic (chronic constriction injury or streptozotocin) rats.

In conclusion, although substance P acting at NK_1 receptors appears to play an important role in pain transmission in animals, it is clear that NK_1 receptor antagonists are not likely to be useable as simple analgesic drugs in the way that, for example, opioids and NSAIDs are in the clinic. However, it is worth noting that a number of existing antidepressant drugs are already known to be useful in treating patients with a variety of painful conditions either as monotherapy or in combination. This is particularly the case in conditions such as IBS or fibromyalgia where the pain experience has a pronounced affective component. It is therefore possible that in a subset of pain patients the antidepressant actions of the NK_1 receptor antagonists (see chapter by N.M.J. Rupniak, this volume) will be beneficial as part of a pain control regimen.

Acknowledgements. We thank our colleagues M. Sablad, A. Hama and M. Urban for the preclinical data on GR205171 in the rat model (Chung) of neuropathic pain.

References

Ardid D, Guilbaud G (1992) Antinociceptive effects of acute and 'chronic' injections of tricyclic antidepressant drugs in a new model of mononeuropathy in rats. Pain 49:279–287

Awawdeh L, Lundy FT, Shaw C, Lamey PJ, Linden GJ, Kennedy JG (2002) Quantitative analysis of substance P, neurokinin A and calcitonin gene-related peptide in pulp tissue from painful and healthy human teeth. Int Endod J 35:30–36

Beresford IJ, Birch PJ, Hagan RM, Ireland SJ (1991) Investigation into species variants in tachykinin NK_1 receptors by use of the non-peptide antagonist, CP-96,345. Br J Pharmacol 104:292–293

Binder W, Scott C, Walker JS (1999) Involvement of substance P in the anti-inflammatory effects of the peripherally selective kappa-opioid asimadoline and the NK_1 antagonist GR 205171. Eur J Neurosci 11:2065–2072

Block GA, Rue D, Panebianco D et al. (1998) The substance P receptor antagonist L-754,030 (MK-0869) is ineffective in he treatment of postherpetic neuralgia. Neurology 4:A225

Boyce S, Hill RG (2000) Discrepant results from preclinical and clinical studies on the potential of substance P-receptor antagonists compounds as analgesics. In Devor M et al (eds) Proceedings of the 9th World Congress on Pain. IASP press, Seattle, pp 313–324

Boyce S, Laird JMA, Tattersall FD et al (1993) Antinociceptive effects of NK_1 receptor antagonists: comparison of behavioural and electrophysiological tests. 7th World Congress on Pain, abstract 641

Brain SD, Poyner DR, Hill RG (2002) CGRP receptors: a headache to study, but will antagonists prove therapeutic in migraine? Trends Pharmacol Sci 23:51–53

Cascieri MA, Ber E, Fong TM, Sadowski S, Bansal A, Swain C, Seward E, Frances B, Burns D, Strader CD (1992) Characterization of the binding of a potent, selective, radioiodinated antagonist to the human neurokinin-1 receptor. Mol Pharmacol 42:458–463

Clouse RE (1994) Antidepressants for functional gastrointestinal syndromes. Dig Dis Sci 39:2352–2363

Connor HE, Bertin L, Gillies et al. (1998) Clinical evaluation of a novel, potent, CNS penetrating NK_1 receptor antagonist in the acute treatment of migraine. Cephalalgia 18:392

Coudore-Civiale M, Courteix C, Boucher M, Fialip J, Eschalier A (2000) Evidence for an involvement of tachykinins in allodynia in streptozocin-induced diabetic rats. Eur J Pharmacol 401:47–53

Cridland RA, Henry JL (1986) Comparison of the effects of substance P, neurokinin A, physalaemin and eledoisin in facilitating a nociceptive reflex in the rat. Brain Res 381:93–99

Cumberbatch MJ, Carlson E, Wyatt A, Boyce S, Hill RG, Rupniak NM (1998) Reversal of behavioural and electrophysiological correlates of experimental peripheral neuropathy by the NK_1 receptor antagonist GR205171 in rats. Neuropharmacol 37:1535–1543.

Daemen MA, Kurvers HA, Kitslaar PJ, Slaaf DW, Bullens PH, Van den Wildenberg FA (1998) Neurogenic inflammation in an animal model of neuropathic pain. Neurol Res 20:41–45

De Felipe C, Herrero JF, O'Brien JA, Palmer JA, Doyle CA, Smith AJ, Laird JM, Belmonte C, Cervero F, Hunt SP (1998) Altered nociception, analgesia and aggression in mice lacking the receptor for substance P. Nature 392:394–397

Dionne RA (1999) Clinical analgesic trials of NK_1 antagonists. Curr Opin CPNS Invest Drugs 1:82–85

Dionne RA, Max MB, Gordon SM, Parada S, Sang C, Gracely RH, Sethna NF, MacLean DB (1998) The substance P receptor antagonist CP-99,994 reduces acute postoperative pain. Clin Pharmacol Ther 64:562–568

Drossman DA (1999) Do psychosocial factors define symptom severity and patient status in irritable bowel syndrome? Am J Med 107:41S–50S

Duggan AW, Morton CR, Zhao ZQ, Hendry IA (1987) Noxious heating of the skin releases immunoreactive substance P in the substantia gelatinosa of the cat: a study with antibody microprobes. Brain Res 403:345–349

Edvinsson L (2001) Calcitonin gene-related peptide (CGRP) and the pathophysiology of headache: therapeutic implications. CNS Drugs 15:745–753

Field MJ, McCleary S, Boden P, Suman-Chauhan N, Hughes J, Singh L (1998) Involvement of the central tachykinin NK_1 receptor during maintenance of mechanical hypersensitivity induced by diabetes in the rat. J Pharmacol Exp Ther 285:1226–1232

Field MJ, Gonzalez MI, Tallarida RJ, Singh L (2002) Gabapentin and the neurokinin$_1$ receptor antagonist CI-1021 act synergistically in two rat models of neuropathic pain. J Pharmacol Exp Ther 303:730–735

Garret C, Carruette A, Fardin V, Moussaoui S, Peyronel JF, Blanchard JC, Laduron PM (1992) RP 67580, a potent and selective substance P non-peptide antagonist. C R Acad Sci III 314:199–204

Gallai V, Sarchielli P, Floridi A, Franceschini M, Codini M, Glioti G, Trequattrini A, Palumbo R (1995) Vasoactive peptide levels in the plasma of young migraine patients with and without aura assessed both interictally and ictally. Cephalalgia 15:384–390

Gesztesi ZS, Song D, White PF (1998) Comparison of a new NK-1 antagonist (CP-122,721) to ondansetron in the prevention of postoperative nausea and vomiting. Anesth Analg 86:S32

Goldstein DJ, Offen WW, Klein EG (1999) Lanepitant, a NK_1 antagonist, in migraine prophylaxis. Clin Pharmacol Therap 65:Abstract

Goldstein DJ, Wang O (1999) Lanepitant, a NK_1 antagonist, in painful diabetic neuropathy. Clin Pharmacol Therap 65:Abstract

Goldstein DJ, Wang O, Saper JR, Stoltz R, Silberstein SD, Mathew NT (1997) Ineffectiveness of neurokinin-1 antagonist in acute migraine: a crossover study. Cephalalgia 17:785–790

Goldstein DJ, Wang O, Todd TE (1998) Lanepitant in osteoarthritis pain. Clin Pharmacol Ther 63:168

Gonzalez MI, Field MJ, Hughes J, Singh L (2000) Evaluation of selective NK_1 receptor antagonist CI-1021 in animal models of inflammatory and neuropathic pain. J Pharmacol Exp Ther 294:444–450

Goto T, Tanaka T (2002) TAchykinins and tachykinin receptors in bone. Microscopy Res and Technique 58:91–97

Hargreaves R (2002) Imaging substance P receptors (NK_1) in the living human brain using positron emission tomography. J Clin Psychiatry 63 Suppl:18–24

Headley PM, Livingston A (1989) Pain and stress in animals: problems of assessment and treatment. Front Pain 1:1–4

Henry JL (1976) Effects of substance P on functionally identified units in cat spinal cord. Brain Res 114:439–451

Hill RG, Rupniak NMJ (1999) Tachykinin receptors and the potential of tachykinin antagonists as clinically effective analgesics and anti-inflammatory agents. In Brain SB and Moore PK (eds) Pain and neurogenic inflammation. Birkhäuser, Basel, pp 313–333

Honore P, Mantyh P (2000) Bone cancer pain: from mechanism to model to therapy. Pain Med 1:303–309

Iyengar S, Hipskind PA, Gehlert DR, Schober D, Lobb KL, Nixon JA, Helton DR, Kallman MJ, Boucher S, Couture R, Li DL, Simmons RM (1997) LY303870, a centrally active neurokinin-1 antagonist with a long duration of action. J Pharmacol Exp Ther 280:774–785

Juranek I, Lembeck F (1997) Afferent C-fibres release substance P and glutamate. Can J Physiol Pharmacol 75:661–664

Kramer MS, Cutler N, Feighner J, Shrivastava R, Carman J, Sramek JJ, Reines SA, Liu G, Snavely D, Wyatt-Knowles E, Hale JJ, Mills SG, MacCoss M, Swain CJ, Harrison T, Hill RG, Hefti F, Scolnick EM, Cascieri MA, Chicchi GG, Sadowski S, Williams AR, Hewson L, Smith D, Rupniak NM (1998) Distinct mechanism for antidepressant activity by blockade of central substance P receptors. Science 281:1640–1645

Laird JM, Hargreaves RJ, Hill RG (1993) Effect of RP 67580, a non-peptide neurokinin$_1$ receptor antagonist, on facilitation of a nociceptive spinal flexion reflex in the rat. Br J Pharmacol 109:713–718

Laird JM, Olivar T, Roza C, De Felipe C, Hunt SP, Cervero F (2000) Deficits in visceral pain and hyperalgesia of mice with a disruption of the tachykinin NK_1 receptor gene. Neuroscience 98:345–352

Laird JM, Roza C, De Felipe C, Hunt SP, Cervero F (2001) Role of central and peripheral tachykinin NK_1 receptors in capsaicin-induced pain and hyperalgesia in mice. Pain 90:97–103

Lassen LH, Haderslev PA, Jacobsen VB, Iversen HK, Sperling B, Olesen J (2002) CGRP may play a causative role in migraine. Cephalalgia 22:54–61

Li JL, Ding YQ, Xiong KH, Li JS, Shigemoto R, Mizuno N (1998) Substance P receptor (NK_1)-immunoreactive neurons projecting to the periaqueductal gray: distribution in the spinal trigeminal nucleus and the spinal cord of the rat. Neurosci Res 30:219–225

Littman B, Newton FA, Russell IJ (1999) Substance P antagonism in fibromyalgia: a trial with CJ-11,974. Proceedings of the 9th World Congress on Pain. IASP Press, Seattle, p67

Mansikka H, Sheth RN, DeVries C, Lee H, Winchurch R, Raja SN (2000) Nerve injury-induced mechanical but not thermal hyperalgesia is attenuated in neurokinin-1 receptor knockout mice. Exp Neurol 162:343–349

Mansikka H, Shiotani M, Winchurch R, Raja SN (1999) Neurokinin-1 receptors are involved in behavioral responses to high-intensity heat stimuli and capsaicin-induced hyperalgesia in mice. Anesthesiology 90:1643–1649

Mantyh PW, DeMaster E, Malhotra A, Ghilardi JR, Rogers SD, Mantyh CR, Liu H, Basbaum AI, Vigna SR, Maggio JE, Simone DA (1995) Receptor endocytosis and dendrite reshaping in spinal neurons after somatosensory stimulation. Science 268:1629–1632

Mantyh PW, Hunt SP (1985) The autoradiographic localization of substance P receptors in the rat and bovine spinal cord and the rat and cat spinal trigeminal nucleus pars caudalis and the effects of neonatal capsaicin. Brain Res 332:315–324

Mantyh PW, Rogers SD, Honore P, Allen BJ, Ghilardi JR, Li J, Daughters RS, Lappi DA, Wiley RG, Simone DA (1997) Inhibition of hyperalgesia by ablation of lamina I spinal neurons expressing the substance P receptor. Science 278:275–279

Martinez-Caro L, Laird JM (2000) Allodynia and hyperalgesia evoked by sciatic mononeuropathy in NK_1 receptor knockout mice. Neuroreport 11:1213–1217

Max MB, Schafer SC, Culnane M, Dubner R, Gracely RH (1988) Association of pain relief with drug side-effects in postherpetic neuralgia: a single-dose study clonidine, codeine, ibuprofen and placebo. Clin Pharmacol Ther 43:363–371

McLean S, Ganong AH, Seeger TF, Bryce DK, Pratt KG, Reynolds LS, Siok CJ, Lowe JA 3rd, Heym J (1991) Activity and distribution of binding sites in brain of a nonpeptide substance P (NK_1) receptor antagonist. Science 251:437–439

Molander C, Ygge J, Dalsgaard CJ (1987) Substance P-, somatostatin- and calcitonin gene-related peptide-like immunoreactivity and fluoride resistant acid phosphatase-activity in relation to retrogradely labeled cutaneous, muscular and visceral primary sensory neurons in the rat. Neurosci Lett 74:37–42

Nagy JI, Hunt SP, Iversen LL, Emson PC (1981) Biochemical and anatomical observations on the degeneration of peptide-containing primary afferent neurons after neonatal capsaicin. Neuroscience 6:1923–1934

Natsugari H, Ikeura Y, Kiyota Y, Ishichi Y, Ishimaru T, Saga O, Shirafuji H, Tanaka T, Kamo I, Doi T et al (1995) Novel, potent, and orally active substance P antagonists: synthesis and antagonist activity of N-benzylcarboxamide derivatives of pyrido[3,4-b]pyridine. J Med Chem 38:3106–3120

Navari RM, Reinhardt RR, Gralla RJ, Kris MG, Hesketh PJ, Khojasteh A, Kindler H, Grote TH, Pendergrass K, Grunberg SM, Carides AD, Gertz BJ (1999) Reduction of cisplatin-induced emesis by a selective neurokinin-1-receptor antagonist. L-754,030 Antiemetic Trials Group. N Engl J Med 340:190–195

Neckers LM, Schwartz JP, Wyatt RJ, Speciale SG (1979) Substance P afferents from the habenula innervate the dorsal raphe nucleus. Exp Brain Res 37:619–23

Norman B, Panebianco D, Block GA (1998) A controlled, in clinic study to explore the preliminary safety and efficacy of intravenous L-758,298 (a prodrug of the NK1 receptor antagonist L-754,030) in the acute treatment of migraine. Cephalalgia 18:4507–442

Okano S, Ikeura Y, Inatomi N (2002) Effects of tachykinin NK1 receptor antagonists on the viscerosensory response caused by colorectal distention in rabbits. J Pharmacol Exp Ther 300:925–931

Okayama Y, Ono Y, Nakazawa T, Church MK, Mori M (1998) Human skin mast cells produce TNF-alpha by substance P. Int Arch Allergy Immunol 117 Suppl 1:48–51

Patel S, Gentry CT, Campbell EA (1996) A model for in vivo evaluation of tachykinin NK_1 receptor antagonists using carrageenan-induced hyperalgesia in the guinea pig paw. Br J Pharmacol 117:248P

Phebus LA, Johnson KW, Stengel PW, Lobb KL, Nixon JA, Hipskind PA (1997) The nonpeptide NK-1 receptor antagonist LY303870 inhibits neurogenic dural inflammation in guinea pigs. Life Sci 60:1553–1561

Radhakrishnan V, Henry JL (1991) Novel substance P antagonist, CP-96,345, blocks responses of cat spinal dorsal horn neurons to noxious cutaneous stimulation and to substance P. Neurosci Lett 132:39–43

Radhakrishnan V, Iyengar S, Henry JL (1998) The nonpeptide NK-1 receptor antagonists LY303870 and LY306740 block the responses of spinal dorsal horn neurons to substance P and to peripheral noxious stimuli. Neuroscience 83:1251–1260

Reinhardt RR, Laub JB, Fricke JR et al. (1998) Comparison of the neurokinin-1 antagonist, L-754,030, to placebo, acetaminophen and ibuprofen in the dental pain model. Clin Pharmacol Ther 63:168

Rupniak NM, Boyce S, Williams AR, Cook G, Longmore J, Seabrook GR, Caeser M, Iversen SD, Hill RG (1993) Antinociceptive activity of NK_1 receptor antagonists: nonspecific effects of racemic RP67580. Br J Pharmacol 110:1607–1613

Rupniak NM, Carlson E, Boyce S, Webb JK, Hill RG (1996) Enantioselective inhibition of the formalin paw late phase by the NK_1 receptor antagonist L-733,060 in gerbils. Pain 67:189–195

Rupniak NM, Tattersall FD, Williams AR, Rycroft W, Carlson EJ, Cascieri MA, Sadowski S, Ber E, Hale JJ, Mills SG, MacCoss M, Seward E, Huscroft I, Owen S, Swain CJ, Hill RG, Hargreaves RJ (1997) In vitro and in vivo predictors of the anti-emetic activity of tachykinin NK_1 receptor antagonists. Eur J Pharmacol 326:201–209

Russell IJ, Orr MD, Littman B, Vipraio GA, Alboukrek D, Michalek JE, Lopez Y, MacKillip F (1994) Elevated cerebrospinal fluid levels of substance P in patients with the fibromyalgia syndrome. Arthritis Rheum 37:1593–1601

Sabino MA, Honore P, Rogers SD, Mach DB, Luger NM, Mantyh PW (2002) Tooth extraction-induced internalization of the substance P receptor in trigeminal nucleus and spinal cord neurons: imaging the neurochemistry of dental pain. Pain 95:175–186

Sakai K, Matsuno H, Tsuji H, Tohyama M (1998) Substance P receptor (NK$_1$) gene expression in synovial tissue in rheumatoid arthritis and osteoarthritis. Scand J Rheumatol 27:135–141

Schwei MJ, Honore P, Rodgers SD, Salak-Johnson JL, Finke MP, Ramnaraine ML, Clohisy DR, Mantyh PW (1999) Neurochemical and cellular reorganization of the spinal cord in a murine model of bone cancer pain. J Neurosci 19:10886–10897

Seguin L, Millan MJ (1994) The glycine B receptor partial agonist, (+)-HA966, enhances induction of antinociception by RP 67580 and CP-99,994. Eur J Pharmacol 253:R1–R3

Sergeyev V, Hökfelt T, Hurd Y (1999) Serotonin and substance P co-exist in dorsal raphe neurons of the human brain. Neuroreport 10:3967–3970

Shepheard SL, Williamson DJ, Hill RG, Hargreaves RJ (1993) The non-peptide neurokinin$_1$ receptor antagonist, RP 67580, blocks neurogenic plasma extravasation in the dura mater of rats. Br J Pharmacol 108:11–12

Shepheard SL, Williamson DJ, Williams J, Hill RG, Hargreaves RJ (1995) Comparison of the effects of sumatriptan and the NK$_1$ antagonist CP-99,994 on plasma extravasation in Dura mater and c-fos mRNA expression in trigeminal nucleus caudalis of rats. Neuropharmacology 34:255–261

Urban L, Gentry C, Patel S et al. (1999) Selective NK$_1$ receptor antagonists block neuropathic and inflammatory pain in the guinea pig. Proceedings of the 9th World Congress on Pain, IASP Press, Seattle, abstract 125, p 40

Urban LA, Fox AJ (2000) NK$_1$ receptor antagonists—are they really without effect in the pain clinic? Trends Pharmacol Sci 21:462–464

Williamson DJ, Shepheard SL, Hill RG, Hargreaves RJ (1997) The novel anti-migraine agent rizatriptan inhibits neurogenic dural vasodilation and extravasation. Eur J Pharmacol 328:61–64

Wilson SG, Bryant CD, Lariviere, Olsen MS, Giles BE, Chesler EJ, Mogil JS (2003) The heritability of antinociception II: pharmacogenetic mediation of three over-the-counter analgesics in mice. J Pharmacol Exp Ther 305:755–764

Yamamoto T, Yaksh TL (1991) Stereospecific effects of a nonpeptidic NK$_1$ selective antagonist, CP-96,345: antinociception in the absence of motor dysfunction. Life Sci 49:1955–1963

Yasuda T, Iwamoto T, Ohara M, Sato S, Kohri H, Noguchi K, Senba E (1999) The novel analgesic compound OT-7100 (5-n-butyl-7-(3,4,5-trimethoxybenzoylamino)pyrazolo[1,5-a]pyrimid ine) attenuates mechanical nociceptive responses in animal models of acute and peripheral neuropathic hyperalgesia. Jpn J Pharmacol 79:65–73

Role of Tachykinins in Neurogenic Inflammation of the Skin and Other External Surfaces

A. Rawlingson · S. K. Costa · S. D. Brain

Centre for Cardiovascular Biology and Medicine, King's College London, Guy's Campus, London, SE1 1UL, UK
e-mail: sue.brain@kcl.ac.uk

1	Introduction	460
2	Localization of Tachykinins, Their Receptors and Metabolizing Enzymes in Skin	461
2.1	Neuron-Derived Tachykinins	461
2.2	Concept of Non-neuronal Sources	462
2.3	Receptors	462
2.4	Release of Tachykinins in Skin	463
2.5	Tachykinin Degradation in Skin	463
3	The Classical Concept of Neurogenic Inflammation in Skin	463
4	The Relevance of the Acute Neurogenic Inflammatory Response in Skin	464
4.1	Effects on Mast Cells	464
4.2	Cutaneous Blood Flow and Neurogenic Vasodilation	465
4.3	Effects on Cutaneous Microvascular Permeability and Oedema Formation	466
5	Effect of Tachykinins on Leukocyte–Endothelial Cell Interactions	467
5.1	Role of Mast Cells in Polymorphonuclear Leukocyte Accumulation	467
5.2	Direct Neutrophil Activation by Tachykinins	468
5.3	Tachykinins and Eosinophils	469
5.4	Tachykinins and Endothelial Cell Adhesion Molecules	469
5.5	NK_1 Receptor-Mediated Neutrophil Recruitment in Animal Models	469
6	Tachykinins and their Contribution to Ongoing Immune Processes in Skin	470
6.1	Lymphocytes	470
6.2	Monocytes	471
6.3	Keratinocytes	471
6.4	Fibroblasts	472
7	Tachykinins in Non-mammalian Skin and Naturally Occurring Toxins and Venoms	472
7.1	Arachnids	472
7.2	Reptiles and Amphibians	473
7.3	Microbial Toxins	474
8	Evidence for Tachykinin Involvement in Human Skin Disease	474
8.1	Psoriasis	475
8.2	Dermatitis	475

8.3	Pruritus	476
8.4	Urticaria	477
8.5	Other Skin Diseases	477
9	**Wound Healing**	478
10	**Conclusions**	479
References		479

Abstract There is evidence for the presence of tachykinins at neuronal and non-neuronal sites in skin. The major mammalian peripheral tachykinins, substance P and neurokinin A, are best known for their localization in the cutaneous sensory nerves. Their release from these nerves to mediate neurogenic inflammation is well documented and the contribution of tachykinins to the acute inflammatory components of oedema formation, increased blood flow and skin diseases is further discussed, including the major involvement of the tachykinin NK_1 receptor and the range of toxins and venoms that contain or can release tachykinin-like peptides. Although substance P may have a transient role in mediating increased blood flow in the skin, the more prominent effect in acute situations is that of mediating inflammatory oedema formation, although the relevance of this to pathophysiological situations is poorly understood. Evidence for the involvement of the tachykinins in more chronic inflammatory events, including inflammatory cell accumulation, is detailed and emphasis put on the range of mechanisms by which substance P acts to influence polymorphonuclear leukocyte accumulation and activity in the skin. In addition, the concept that tachykinins from non-neuronal sources may be involved in influencing the inflammatory response, especially that involving immune cells, is raised. Possible roles for the tachykinins in human skin diseases and the evidence for an involvement in specific conditions are also given.

Keywords Tachykinins · Substance P · Skin · Neurogenic inflammation

1
Introduction

The skin has an essential role in providing environmental information to the rest of the body that relates to chemical, thermal and mechanical stimuli. It is designed to do this with an elaborate sensory nerve network. These nerves act in concert with a range of chemical mediators and the cutaneous microvasculature to mediate pain and inflammation, as well as influencing normal physiological function. The precise role of the tachykinins in skin and their importance in inflammatory skin disease remains unclear. However, the recent development of selective, non-peptide tachykinin antagonists and tachykinin receptor knockout mice has furthered understanding of the tachykinins in cutaneous physiology and pathophysiology. It is suggested that the tachykinins act as important signalling molecules that mediate neuroendocrine–immune system interactions in the skin and may therefore represent a novel target for the treatment of human skin disease.

This review provides an overview of tachykinin biology in normal skin, the role of the tachykinins in the acute inflammatory response with particular emphasis on neurogenic inflammation, tachykinin involvement in chronic inflammatory skin diseases and where appropriate, interaction of the tachykinins with leukocytes and other chemical mediator systems. It builds upon recent reviews by Ansel et al. (1996), Brain (1997), Maggi (1997), Holzer (1998) and Scholzen et al. (1998).

2
Localization of Tachykinins, Their Receptors and Metabolizing Enzymes in Skin

2.1
Neuron-Derived Tachykinins

Substance P and neurokinin A (NKA) are most usually found in peripheral sensory C-fibres (Hökfelt et al. 1975; Brain 1997). They are synthesized in the dorsal root ganglion and are transported to their peripheral sites of release. They are commonly co-localized with the vasodilator peptide calcitonin gene-related peptide (CGRP). Treatment of rats with capsaicin, which first activates and then desensitizes sensory neurons, causes a 70% depletion of substance P indicating that the major source of substance P in skin is primary afferent sensory C-fibres (Holzer et al. 1991). There is a dense innervation of skin tissues, especially dermal blood vessels. It has been recently demonstrated that whilst substance P-containing and autonomic fibres were found in the lower dermis, only substance P-containing fibres are found in the upper dermis of the rat and monkey lower lip (Ruocco et al. 2002). Indeed, a three-layered sensory nerve plexus has been observed in the deep, superficial and subepidermal layers of normal guinea pig (Ishihara et al. 2002). The deep layers have been suggested to contain mast cells whilst the greatest density of sensory nerves is found in the superficial dermis (Kowalski et al. 1990). Despite this, there is evidence of a direct contact between sensory nerve endings and mast cells in skin (Alving et al. 1991). The subepidermal layer includes single nerve fibres, containing substance P, NKA and CGRP, that branch out and apparently terminate as free nerve endings of approximately 0.5 μm diameter (Dalsgaard et al. 1985; Ishihara et al. 2002). Immunohistochemically, substance P and CGRP appear as 'globules', as previously described in the peribulbar skin of the human scalp (Hordinsky et al. 1999). These free nerve endings are in direct contact with basal keratinocytes (Hartschuh et al. 1983; Reilly et al. 1997; Schulze et al. 1997). The epithelial layer also contains Merkel cells, possibly derived from basal keratinocytes, which contain neuropeptides including substance P (Misery 1997).

2.2
Concept of Non-neuronal Sources

There has been increasing reference to non-neuronal sources of tachykinins in recent years, although it is well established that capsaicin treatment depletes a large proportion of substance P in skin. It is possible that these non-neuronal sources become of functional importance in pathological conditions. It is suggested that substance P is contained within specific granules of human skin mast cells and that immunoreactivity is increased in patients with atopic dermatitis (Toyoda et al. 2000). Alternatively, the substance P identified may have actually resided in closely associated C-fibre nerves. By comparison, there is good evidence of substance P-like immunoreactivity in endothelial cells of the rat hind limb (Ralevic et al. 1990). Furthermore, substance P expression has been shown to occur in human eosinophils (Metwali et al. 1994), T lymphocytes (Lai et al. 1998) and rodent and human monocytes (Bost et al. 1992; Ho et al. 1997). Of potential importance is the demonstration that human immunodeficiency virus (HIV) enhances substance P expression in human monocytes and lymphocytes (Ho et al. 2002) and that the substance P produced amplifies HIV-1 replication in human monocytes (Lai et al. 2001).

2.3
Receptors

There are three established tachykinin receptors; the NK_1, NK_2 and NK_3 receptors (Regoli et al. 1997). The NK_1 receptor is by far the most studied in skin. It is widely distributed and there is good evidence for its presence on post-capillary venule endothelial cells, the primary site of inflammatory oedema formation and polymorphonuclear leukocyte accumulation (Brain 1997). A high affinity NK_1 receptor has been characterized in the abdominal skin of healthy human volunteers (Bianchi et al. 1999). NK_1 receptors have also been shown to be present on skin cells that include keratinocytes, mast cells, fibroblasts (Parenti et al. 1996), Merkel cells, Langerhans' cells (Staniek et al. 1997) and circulating immune cells (for reviews see Ansel et al. 1996 and Scholzen et al. 1998). This receptor is also considered to be present on unmyelinated cutaneous axons and to be more densely expressed during inflammation (Carlton and Coggeshall 2002).

Less is known about the existence of NK_2 and NK_3 receptor in skin cells. There have been suggestions in the literature that these receptors can contribute to neurogenic oedema formation, but this has not been substantiated. Although it is generally considered that keratinocytes express the NK_1 receptor (Staniek et al. 1998; Burbach et al. 2001), the NK_2 receptor has been shown to be present on the murine keratinocyte (Song et al. 2000). Cultured human dermal microvascular endothelial cells (HDMEC) express mRNA for all of the tachykinin receptors (Quinlan et al. 1998), but functional studies have revealed a role for the NK_1 receptor only.

2.4
Release of Tachykinins in Skin

The capsaicin-sensitive vanilloid receptor VR1 may be involved in the activation of sensory nerves in response to a variety of agents such as low pH, some eicosanoids and anandamide (for a review see Szallasi 2002). In addition, cutaneous afferent sensory neurons express specific receptors for neuropeptides, prostaglandins, histamine, kinins, and neurotrophins (see Lundberg 1996 and Brain 2000). Furthermore, Steinhoff et al. (2000) have demonstrated that proteinase-activated receptor 2 is expressed on cutaneous sensory nerves, where it is activated by trypsin to release vasoactive neuropeptides. Thus, there is a range of chemical mediators that, if present in skin, will act either alone or in a synergistic manner to stimulate the release of substance P from sensory nerves in normal and inflamed skin.

The ability of environmental stimulants to activate sensory nerves is particularly relevant to discussion of the skin. The skin is the first line of defence against ultraviolet radiation and there is evidence to suggest that substance P contributes to both the acute cutaneous inflammation and the chronic immunosuppression that is caused by ultraviolet exposure (Scholzen et al. 1999). In addition the skin is a natural target for a range of injurious agents including poisons (e.g. spider and snake venoms) and noxious temperature changes (e.g. thermal burn injuries). These aspects of cutaneous pathophysiology are further discussed in Sects. 7 and 8, respectively.

2.5
Tachykinin Degradation in Skin

The major enzyme responsible for degradation of substance P and NKA is neutral endopeptidase (NEP), a membrane-bound zinc metalloprotease present on vascular endothelial cells and keratinocytes (see Olerud et al. 1999). It metabolizes a range of peptides by hydrolysing amino acid bonds at the N terminus, and the activity of substance P is prolonged in guinea pig skin treated with NEP inhibitors (Iwamoto et al. 1989). The functional importance of NEP in skin inflammation and wound healing is only now being understood with the development of NEP knockout mice (see Sects. 8, 9).

3
The Classical Concept of Neurogenic Inflammation in Skin

The acute vasodilation observed in skin as erythema, local reddening or flare and plasma extravasation that results from the release of neuropeptides from stimulated sensory nerves is commonly defined as 'neurogenic inflammation'. This theory was developed by Bayliss (1901) who demonstrated that antidromic stimulation of sensory nerves resulted in an increased blood flow in skin. Bruce (1913) then demonstrated that mustard oil stimulates an acute inflammatory response that is not observed when the sensory nerve supply to the skin is ligated. Thomas Lewis (1927) extended these studies by establishing the concept of the 'triple response' to skin injury. The mechanisms involved in the triple response are still debated today. The response consists of a wheal, local reddening and a flare that can spread up to several centimetres from the site of a pin-point injury in humans (e.g. insect bites or injection of mediators that include histamine and substance P). The flare component is also present in pig skin (Jancso et al. 1993) but less prominent in the skin of other species (Lynn and Shakhanbeh 1988).

Jancso extended the original observation of Bruce (1913) by demonstrating that capsaicin acts selectively to stimulate and then deplete sensory nerves (Jancso et al. 1967, 1977). Sensory nerves play a major role in mediating the flare component of the triple response since this component is lost following topical capsaicin pretreatment (Foreman and Jordan 1983). It would appear that substance P stimulates the triple response via its ability to activate mast cells, when injected intradermally (see Sect. 4). The flare component of the triple response is abolished by histamine H_1 receptor antagonists, although the wheal component is less substantially inhibited (Hägermark et al. 1978; Foreman and Jordan 1983). It is considered that H_1 receptors on cutaneous sensory nerves are activated by mast cell-derived histamine. These sensory nerves, as first suggested by Lewis, then send signals in an antidromic manner via the connected collateral fibre network to release the vasodilator that mediates the flare. The mediator of the flare is not universally agreed upon, although substance P remains a candidate (Lembeck and Holzer 1979), along with CGRP (Brain et al. 1986) as discussed in detail in Sect. 4.

4
The Relevance of the Acute Neurogenic Inflammatory Response in Skin

4.1
Effects on Mast Cells

Substance P has been shown to stimulate the degranulation of mast cells from several species and tissues including human skin in vitro (Fewtrell et al. 1982; Benyon et al. 1987). The intradermal injection of substance P into human skin induces a wheal and flare response, as discussed above, where the flare is abolished by H_1 receptor antagonists (Hägermark et al. 1978; Foreman and Jordan 1983). Despite this, it is suggested that the acute neurogenic inflammatory response consisting of vasodilation and plasma exudation takes place in the superficial epidermis and does not involve mast cells (Kowalski et al. 1990). Substance P can activate mast cells independently of the tachykinin NK_1 receptor via direct interaction of the basic arginine residue of the peptide with membrane components of the mast cell. It would appear that relatively large amounts of substance P are required for this response compared with vasoactive responses. Indeed, release of endogenous substance P by capsaicin does not induce a subsequent release of histamine in skin slices (Tausk and Undem 1995) or in microdialysis studies (Huttunen et al. 1996; Petersen et al. 1997). However, it has been suggested that the close anatomical contact of substance P-containing nerves with mast cells could lead to the mast cell being exposed to high concentrations of substance P under certain circumstances. This may be of relevance to the time-dependent ability of substance P to influence neutrophil accumulation via secondary mediators (Okabe et al. 2001; see Sect. 5). Interestingly, there is little evidence to show that NKA, which does not contain a basic arginine residue, can activate mast cells and its intradermal injection in human skin is not associated with formation of a flare (Devillier et al. 1986; Fuller et al. 1987).

4.2
Cutaneous Blood Flow and Neurogenic Vasodilation

Substance P was originally considered to be the primary mediator of neurogenic vasodilation (Lembeck and Holzer 1979). Indeed, the intravenous administration of substance P is associated with hypotension and substance P has been shown to act via nitric oxide (NO)-dependent (Whittle et al. 1989) and -independent (Santicioli et al. 1993) mechanisms. However, the importance of substance P as the mediator of neurogenic vasodilation has been the subject of much debate since the discovery of the neuropeptide vasodilator, CGRP which is often co-localized with substance P in sensory nerves (Brain and Williams 1985). This is mainly due to studies in rats where the vasodilator activity of endogenous tachykinins and CGRP have been compared. Typically, cutaneous neurogenic vasodilation induced by electrical stimulation can be inhibited by the CGRP antagonist, $CGRP_{8-37}$, but not by NK_1 receptor antagonists (Escott et al. 1995). The release of CGRP, but not substance P, in hairy rat skin in response to low pH has further substantiated the functional importance of CGRP as a vasodilator when compared with substance P (Averbeck and Reeh 2001). However, it has been suggested that NK_1 receptor antagonists can modulate neurogenic vasodilator responses in porcine nasal mucosa (Rinder and Lundberg 1996).

Häbler et al. (1999) have suggested that whilst endogenous tachykinins have a minor role in mediating neurogenic vasodilation in hairy skin, they act synergistically with CGRP in hairless skin (e.g. plantar glabrous skin). This interesting study revealed that a CGRP antagonist could reduce the amplitude and duration of vasodilation induced by electrical stimulation of the saphenous nerve whilst an NK_1 receptor antagonist delayed the onset of the response, but did not alter its amplitude in hairless skin. The authors wondered whether a subset of $A\delta$-fibres that contain CGRP, but not substance P, play a more important role in hairy skin. Analysis of the contribution of endogenous tachykinins to neurogenic vasodilation in skin has now been extended to include studies in genetically modified mice that lack the tachykinin NK_1 receptor. Surprisingly, CGRP blockade had no effect on capsaicin-induced blood flow in the wild-type mice, although it attenuated this parameter in both NK_1 receptor knockout mice and wild-type mice treated with an NK_1 receptor antagonist (Grant et al. 2002). The results of these studies led the authors to speculate on the presence of presynaptic NK_1 receptors on sensory nerves that act to modulate neuropeptide release. This hypothesis may be related to the demonstration of NK_1 receptors on the sensory neuron which are up-regulated in rat skin during inflammation (Carlton and Coggeshall 2002).

Schmelz and co-workers have established a microdialysis technique to enable the study of sensory neuropeptides in human skin. They have demonstrated that both substance P and CGRP induced dose-dependent vasodilation in perfused skin (Weidner et al. 2000). The response to substance P was rapid in onset, but then declined as infusion continued. This desensitization is similar to that which had been observed in the human forearm upon brachial artery infusion (McEwan et al. 1988). Interestingly, the microdialysis technique was employed to demonstrate that release of CGRP, but not substance P, is observed in the flare component of the triple response (Schmelz et al. 1997). Thus, it would appear from present evidence that CGRP, which is a more potent and prolonged vasodilator (Brain et al. 1986), is the more likely to be the primary mediator of flare. Despite this, there is clear evidence that substance P can increase blood flow, albeit in a transient manner, in skin. However, it would appear that substance P works in a facilitatory manner with CGRP.

The functional significance of the flare component is unknown. It enables adjacent skin to respond to injury by increasing blood flow and enhancing nociceptive sensitivity (for a review see Brain 1997). The widespread use of anti-histamines to treat acute skin conditions is probably related to the inhibition of histamine-mediated flare in addition to other histamine-mediated effects (e.g. itch). However, capsaicin acts directly on skin, presumably via vanilloid receptors to induce a histamine-independent flare (Barnes et al. 1986) and, thus, it may be hypothesized that endogenous vanilloids may activate sensory nerves and axon reflex flares directly. Morhenn (2000) has attempted to correlate the vasodilation observed in response to stroking of the skin with substance P release. He suggests that the pleasurable sensation of massaging and grooming is associated with the ability of substance P to inhibit a peptidase that degrades the endogenous morphine-like opiate enkephalin (Xu et al. 1995).

4.3
Effects on Cutaneous Microvascular Permeability and Oedema Formation

The most obvious response to substance P and NKA when injected intradermally into a range of species is that of inflammatory swelling or oedema formation. The mechanisms underlying this response have been extensively investigated in human and rodent skin. Neurogenic oedema formation is considered to be mediated primarily via stimulation of NK_1 receptors. The NK_1 receptors are situated on endothelial cells and act to make the post-capillary venule layer leaky and thus permeable to plasma proteins. A range of non-peptide NK_1 receptor antagonists have been used in rat models, including CP96345 (Lembeck et al. 1992; Xu et al. 1992), RP67580 (Garret et al. 1991) and SR140333 (Emonds-Alt et al. 1993) to show that neurogenic oedema formation is mediated by the NK_1 receptor with little involvement of the NK_2 receptor (Xu et al. 1992). More recently studies in the wild-type and NK_1 knockout mouse have confirmed the importance of the NK_1 receptor in mediating the neurogenic oedema formation (Bozic et al. 1996; Cao et al. 1998, 1999). This contrasts with earlier studies that demonstrated a role of all three tachykinin receptors in mediating cutaneous oedema formation (Inoue et al. 1996). Current evidence also indicates that NK_1 receptor-mediated inflammatory oedema formation is involved in the tissue response to traumatic injury (e.g. thermal burn injuries; Saria 1984; Rawlingson et al. 2001) and experimental inflammation (e.g. carragheenan-induced; Birch et al. 1992).

The functional significance of endogenous substance P-mediated oedema formation in human skin is difficult to assess. Topical application of capsaicin in humans is not associated with oedema formation, despite substance P release (Schmelz et al. 1997), although capsaicin may desensitise sensory nerves before sufficient concentrations of substance P are released to induce oedema formation. Certainly, the perfusion of human skin with substance P (10^{-8}–10^{-5} M) leads to oedema formation, with no involvement of histamine until the higher concentrations are reached indicating that functional NK_1 receptors exist in human skin (Weidner et al. 2000).

5
Effect of Tachykinins on Leukocyte–Endothelial Cell Interactions

It is not clear whether substance P has a direct effect on polymorphonuclear leukocytes and endothelial cells or whether it acts via release of secondary mediators. Furthermore, data regarding the concentrations of substance P required to trigger these effects vary widely and endogenous substance P may not be released in high enough quantities to mimic some of the effects of exogenous substance P which is often administered at supraphysiological concentrations. It is therefore possible that multiple mechanisms are involved in mediating substance P-induced neutrophil and eosinophil accumulation in vivo.

5.1
Role of Mast Cells in Polymorphonuclear Leukocyte Accumulation

There is much experimental evidence to show that exogenous substance P can activate mast cells to release multiple factors that are chemotactic for neutrophils. However, as with oedema formation, the ability of the endogenous tachykinins to act via these mechanisms in the cutaneous microvasculature is disputed. In particular, Saban and co-workers have provided evidence that substance P can trigger a cascade of events leading to leukotriene and cytokine production in mouse skin. However, the model used in this study is associated with abnormal pathology and an increased abundance of dendritic cells (Saban et al. 1997). Thus, data provided by studies of this kind should not be over-interpreted.

The intradermal injection of relatively high concentrations of substance P causes mast cell degranulation and usually neutrophil accumulation in animal

(Iwamoto et al. 1993; Walsh et al. 1995) and human (Smith et al. 1993) skin. The involvement of mast cells is confirmed by the absence of the responses in mast cell deficient mice (Matsuda et al. 1989; Yano et al. 1989). Mast cells release a range of agents that contribute to neutrophil accumulation in vivo and these are thought to include the 5-lipoxygenase metabolite, leukotriene B_4 (LTB_4). Interestingly, LTB_4 receptor blockade or inhibition of LTB_4 synthesis attenuates substance P-induced neutrophil accumulation (Iwamoto et al. 1993; Walsh et al. 1995). Recently, Okabe et al. (2001) investigated the ability of substance P, at micromolar concentrations, to stimulate LTB_4 release from human skin slices and found that functional heterogeneity exists amongst human skin mast cells. The results varied widely, with large amounts of LTB_4 released within minutes from some, but not other, skin samples. Furthermore, the release was not blocked by NK_1 receptor antagonists, suggesting a receptor-independent mechanism.

The cytokines, tumour necrosis factor-α (TNFα) and interleukin (IL)-8 are also released by human skin slices following exposure to substance P, but their production is dependent upon de novo protein synthesis, thus incubation periods of several hours are required before seeing this effect (Ansel et al. 1993; Okayama et al. 1998; Okabe et al. 2000). Cytokine release is associated with mast cell degranulation under such circumstances and TNFα is able to promote inflammatory cell recruitment via its ability to induce endothelial cell activation. By comparison, IL-8, like LTB_4 is directly chemotactic for neutrophils (Brandolini et al. 1997).

5.2
Direct Neutrophil Activation by Tachykinins

Substance P has been shown to enhance human neutrophil–endothelial cell interactions in vitro in a manner that is independent of cellular adhesion molecule expression (Zimmerman et al. 1992). Substance P has also been suggested to function as a priming agent for human neutrophils and to increase the production of LTB_4 and TNFα by these cells (Wozniak et al. 1989; Lloyds and Hallett 1993; Saban et al. 1997). In addition, nanomolar concentrations of substance P have been shown to increase the production of reactive oxygen species in human neutrophils via an up-regulation of constitutive NO synthase (cNOS) and through calmodulin-dependent NADPH oxidase generation of superoxide and hydrogen peroxide (Sterner-Kock et al. 1999). Furthermore, tyrosine phosphorylation and raised intracellular calcium appear to be involved in this process (Lloyds and Hallett 1993), and substance P acting via the NK_1 receptor has recently been shown to augment neutrophil intracellular calcium concentrations following exposure to IL-8 (Dianzani et al. 2001). Substance P acting via the NK_1 receptor has also been shown to delay neutrophil apoptosis (Bockmann et al. 2001) and this may be another mechanism by which the peptide can increase neutrophil activity in inflamed skin.

5.3
Tachykinins and Eosinophils

There is less information available relating to the ability of substance P to influence eosinophil accumulation. However, it has been shown that substance P-induced neutrophil accumulation, but not eosinophil accumulation, is blocked by inhibition of the activity of LTB_4 in mouse and guinea pig skin (Iwamoto et al. 1993; Walsh et al. 1995). This may be attributed to the fact that substance P can stimulate human eosinophil migration via a phospholipase C (PLC)-dependent mechanism (El Shazly et al. 1997). The ability of substance P to influence equine eosinophil-endothelial cell interactions in vitro has recently been investigated, but substance P was found to have weak activity when compared with cytokines (Bailey and Cunningham 2001).

5.4
Tachykinins and Endothelial Cell Adhesion Molecules

It has been shown that substance P has the ability to stimulate expression of the endothelial adhesion molecules ICAM-1 and VCAM-1 in cultured HDMEC (Quinlan et al. 1998; Quinlan et al. 1999) and this may, in part, explain why intradermal injection of substance P stimulates the recruitment of several classes of leukocyte in human skin (Smith et al. 1993). Interestingly, the induction of VCAM-1 was seen to be ten-fold more sensitive to substance P (10 nM) than ICAM-1 in this study and occurred in a more rapid manner. Moreover, HDMEC are reported to express the NK_1 receptor (Quinlan et al. 1998) and NK_1 receptor-mediated up-regulation of adhesion molecules in these cells appears to be specific to substance P since NKA and neurokinin B (NKB) were without effect (Quinlan et al. 1999). The results have been extended by studies in healthy human volunteers, where the topical application of capsaicin was seen to increase dermal ICAM-1 immunoreactivity (see Lindsey et al. 2000). It is suggested that the preferential up-regulation of VCAM-1 is relevant to the selective accumulation of eosinophils that has been observed in human skin in response to substance P (Smith et al. 1993; Quinlan et al. 1999). Quinlan and co-workers have examined the signal transduction mechanisms involved in this process and suggest that PLC activation and increased intracellular calcium levels are associated with NFκB- and NF-AT-mediated up-regulation of VCAM-1 and ICAM-1 respectively (Quinlan et al. 1999).

5.5
NK_1 Receptor-Mediated Neutrophil Recruitment in Animal Models

An involvement of substance P in the accumulation of granulocytes, principally neutrophils, has been discussed for some time. However, intradermal injection of substance P, at vasoactive doses is not associated with neutrophil accumulation in normal rat (Pinter et al. 1999) and mouse (Cao et al. 2000) skin. On the other hand, there is good evidence of an endogenous tachykinin component in the neutrophil recruitment observed in inflamed tissues (e.g. immune complex-mediated inflammation; Bozic et al. 1996; Ahluwalia et al. 1998). Moreover, a neurogenic component has been observed in carrageenan-induced inflammation in the rat as mentioned above (Birch et al. 1992). This is in keeping with the observation that carrageenan-induced neutrophil accumulation is suppressed in both wild-type mice treated with an NK_1 receptor antagonist and NK_1 knockout mice (Cao et al. 2000). This response is also inhibited by bradykinin receptor antagonists, indicating a close relationship between dermal NK_1 receptor activation and kinin generation in vivo (Cao et al. 2000). Interestingly, McLean and coworkers have demonstrated that the kinin B_1 receptor is involved in IL-1β-induced neutrophil rolling and recruitment in the mouse mesentery and that an NK_1 receptor antagonist inhibited this response (McLean et al. 2000).

6
Tachykinins and their Contribution to Ongoing Immune Processes in Skin

There is increasing evidence to suggest that substance P can affect the development of chronic inflammatory responses through its ability to stimulate cytokine release. Much of this evidence comes from cell culture studies performed in vitro and one must therefore consider whether the concentrations of substance P used are physiologically or, at least, pathophysiologically relevant. However, these findings are significant in that they demonstrate the ability of substance P to influence the ongoing inflammatory process and this may well have implications for the therapy of some forms of dermatological disease.

6.1
Lymphocytes

Substance P has a direct effect on T-cell activation (Scicchitano et al. 1988). At concentrations of 10^{-13}–10^{-11} M it can enhance mitogen- and antigen-induced proliferation of T lymphocytes and IL-2 production by these cells in culture (Payan et al. 1983; Calvo et al. 1992; Nio et al. 1993; Rameshwar et al. 1993). Substance P has also been shown to up-regulate functional macrophage inflammatory protein-1α expression in human T-lymphocytes via an NK_1 receptor-dependent mechanism (Guo et al. 2002).

Herzberg et al. (1995) have observed the migration of CD4+ lymphocytes into normal rat skin after stimulation of the sensory sciatic nerve. Furthermore, substance P derived from non-neuronal sources such as T lymphocytes can also contribute to the amplification of inflammation in disease (Lambrecht 2001). Thus, for example, HIV-1 enhances substance P expression in human monocytes and lymphocytes (Ho et al. 2002) and the substance P produced can then amplify viral replication in human monocytes (Lai et al. 2001).

There is good evidence for the presence of NK_1 receptors, but not other tachykinin receptors, on human and mouse lymphocytes (Stanisz et al. 1987;

Cook et al. 1994; Lai et al. 1998). Results from studies carried out in capsaicin-treated rats indicate that endogenous substance P can contribute to lymphocyte responses via the tachykinin NK_1 receptor (Santoni et al. 1999). In keeping with this suggestion, the tachykinin NK_1 receptor is up-regulated in lymphocytes, especially CD4+ lymphocytes, in response to the respiratory syncytial virus (Tripp et al. 2002).

6.2
Monocytes

Substance P has long been known to stimulate cytokine release from monocytic cells (Kimball et al. 1988; Lotz et al. 1988; Laurenzi et al. 1990). However, it has recently been suggested that these results could be attributed to endotoxin contamination (Maggi 1997), although some more recent studies have extended these positive findings (Guo et al. 2002).

6.3
Keratinocytes

Substance P has been shown to influence the growth and activity of keratinocytes (Tanaka et al. 1988; Rabier et al. 1993; Paus et al. 1995). Human keratinocytes express both NK_1 and NK_2 receptors (Burbach et al. 2001), whilst murine keratinocytes may only express NK_2 receptors (Song et al. 2000). It is becoming clear that the tachykinins can stimulate the up-regulation and release of a range of dermatologically important mediators from keratinocytes. Nerve growth factor (NGF) is important for the maintenance and growth of nerves, including the sensory nerves that contain and release tachykinins. Interestingly, substance P and NKA can induce NGF expression and release in cultured human and murine keratinocytes, and topically applied capsaicin has been shown to induce NGF production in murine skin (Burbach et al. 2001). Furthermore, NK_1 receptor blockade is reported to prevent substance P-mediated NGF expression in rat hind paw skin, but not that induced by carrageenan or allergens (Amann et al. 2000). However, the relevance of these findings to wound healing and inflammatory skin disease remains unknown.

The tachykinins can also stimulate the up-regulation and release of pro-inflammatory cytokines from keratinocytes. For example, substance P can activate keratinocyte cell lines to up-regulate IL-1α gene expression via stimulation of the NK_2 receptor (Song et al. 2000). In addition, substance P can act via keratinocyte NK_1 receptors to enhance the production of interferon-induced protein 10, a member of the CXC chemokine family, involved in the recruitment of T-helper lymphocytes into injured skin (Kanda and Watanabe 2002).

6.4
Fibroblasts

It is well established that substance P can induce DNA synthesis in fibroblasts (Nilsson et al. 1985). Indeed, it is effective in stimulating the proliferation of cultured skin fibroblasts and their ability to release prostaglandins (Kahler et al. 1993). It has been suggested that these effects are important for the growth of new blood vessels (Ziche et al. 1990) and more recently it has been shown that the NK_1 receptor can stimulate fibroblast migration (Parenti et al. 1996).

7
Tachykinins in Non-mammalian Skin and Naturally Occurring Toxins and Venoms

Given the confusion with venom, poison and toxin terminology, it is important to clarify that technically all venoms are poisons, but not all poisons are venoms. In most instances, toxins or poisons are a single substance or closely related group of substances, whereas venoms tend to be a mixture of substances. Venoms most commonly are released from the fangs of biting animals such as spiders and snakes, while microbes produce a large proportion of naturally occurring toxins. Many of the substances that constitute venoms and toxins have tachykinin-like properties or have the ability to activate sensory neurons, thereby causing endogenous tachykinin release and neurogenic inflammation. In addition, the study of tachykinin-like peptides in non-mammalian skin has contributed much to this field of research and, in particular, the evolutionary significance of the tachykinins.

7.1
Arachnids

Spiders and scorpions are the main representatives of the *Arachnida* class of animals. Spider venoms, like snake or scorpion venoms, contain a complex cocktail of biologically active components, including proteins, peptides, neurotoxins, nucleic acids, inorganic acids and monoamines (Jackson and Parks 1989). The precise nature of these components varies amongst and within the same species. An aggressive solitary spider named *Phoneutria nigriventer*, found in South America, contains a potent venom that may cause direct activation of sensory neurons. Bites by this species produce a range of symptoms, including excruciating pain and oedema formation, both at the site of exposure and in remote organs (Bucaretchi et al. 2000). Intradermal injection of *Phoneutria nigriventer* venom (PNV) induces a marked plasma extravasation in the dorsal skin of both rats and rabbits (Antunes et al. 1992). This response is partially inhibited by the NK_1 receptor antagonist, SR140333 (Palframan et al. 1996), and capsaicin pretreatment has been shown to attenuate PNV-induced oedema formation in rat dorsal and paw skin (Costa et al. 1997; Costa et al. 2001). It has therefore been suggested that PNV directly activates sensory nerve endings to cause the release of endogenous tachykinins and in particular, substance P. Whether this response can occur via the direct interaction of PNV with the VR1 receptor remains to be determined. However, recent studies indicate that the 5-HT$_4$ receptor may be involved in this process, since blockade of this receptor inhibits PNV-induced plasma extravasation in rodent skin and reduces PNV-induced depolarization of the rat vagus nerve (Costa et al. 2003). Moreover, it is interesting to speculate that PNV-mediated neurogenic inflammation is involved in pulmonary oedema formation, a serious and sometimes fatal complication of *Phoneutria nigriventer* bite.

In contrast to the neurostimulatory actions of PNV, charybdotoxin, isolated from the venom of the Israeli scorpion *Leiurus quinquestriatus* is a potassium channel blocker (Miller et al. 1985). Charybdotoxin antagonizes the inhibitory effect of morphine on substance P release evoked by antidromic stimulation of the sciatic nerve and may therefore serve as a useful experimental tool for the study of neurogenic inflammation (Yonehara and Takiuchi 1997).

7.2
Reptiles and Amphibians

Tachykinin-like substances are found in the skin of lizards (Lembeck et al. 1985) and have a wide anatomical distribution in crocodiles, alligators and turtles (Buchan et al. 1983; Reiner et al. 1984; Karila et al. 1995). Amphibian skin, such as that of frogs and toads possesses a large number of biologically active substances, including peptides belonging to the tachykinin or physalaemin-like peptide family, e.g. physalaemin, kassinin, hylambatin, PG-SPI, PGII, PG-KI, PG-KII and PG-KIII (Nakajima et al. 1980; Simmaco et al. 1990; Severini et al. 2002). Physalaemin, like substance P, is an undecapeptide (Erspamer et al. 1964; Chang et al. 1971), while kassinin and hylambatin, tachykinin-like peptides isolated from the skin of the African frog, *Kassina senegalensis* and *Hylambates maculates*, respectively, are dodecapeptides (Nakajima et al. 1980). In mammalian skin, physalaemin acts in a similar manner to substance P, causing plasma protein extravasation (Gamse et al. 1987). However, in contrast with substance P, physalaemin does not cause histamine release from human skin mast cells (Lowman et al. 1988).

In frog skin (*Rana esculenta*), physalaemin, substance P, and to a lesser extent, NKB stimulate ion transport (Lippe et al. 1994). In this tissue, physalaemin-induced effects were inhibited by an NK$_1$ receptor antagonist, but not by an NK$_2$ receptor antagonist (Lobasso et al. 1998), indicating that an NK$_1$-like receptor mediates ion transport in frog skin. The C-terminal sequence, Phe-X-Gly-Leu-Met-NH$_2$ and at least one proline residue in the N-terminal sequence of these peptides are likely to be the minimal structural requirements for stimulating ion transport under such circumstances (Lippe et al. 1998). The rank order of tachykinin potency in the frog skin is as follows PG-KI > uperolein > hylambatin > kassinin > phyllomedusin >[Sar9-Met(O2)11]-substance P > ranatachykinin A > physalaemin > ranakinin > substance P and eledoisin > NKA.

The tetradecapeptide bombesin was first isolated from the European frog skin (*Bombina bombina*) over 30 years ago (Anastasi et al. 1971). Since then, several bombesin forms and bombesin-related peptides have been isolated and/or cloned from amphibian skin (Nagalla et al. 1994; Lai et al. 2002). Smooth muscle contraction (Tomomasa et al. 1989), gastrohormonal secretion (Ghatei et al. 1982) and altered neurotransmission (Ladenheim and Ritter 1993) are the main pharmacological effects evoked by bombesin, however, the physiological role of the peptide has yet to be determined since bombesin receptor antagonists are, at present, lacking (Cowan et al. 1985). Interestingly and in contrast with mammals, it seems that all tachykinin-like peptides found in the skin of amphibians are of non-neuronal origin (Severini et al. 2002).

7.3
Microbial Toxins

Enterotoxins (A-E and G-I) belong to a family of exoproteins secreted by different strains of *Staphylococcus aureus* (Bergdoll 1989). Besides their role in acute food poisoning (Iandolo 1989; Muller-Alouf et al. 2001), these toxins have also been implicated in inflammatory skin diseases such as eczema (Abeck and Mempel 1998) and psoriasis (Bour et al. 1995). Although the exact relationship between staphylococcal endotoxins and human skin disease is unclear, there is evidence to suggest that these toxins can cause sensory neuron activation and tachykinin release under some circumstances. Gottfried et al. (1989) have shown that staphylococcal endotoxin B (SEB)-induced skin reactions in primates can be attributed to mast cell activation caused by substance P released from sensory neurons. Furthermore, intraplantar injection of SEB in the mouse hind paw causes a plasma extravasation that is partially inhibited by capsaicin pretreatment (Desouza et al. 1996) or SR140333 (Linardi et al. 2000). Similarly, *Clostridium perfringens* β-toxin causes skin inflammation when injected intradermally in to rodent skin and this response may also be attributable to neurogenic mechanisms (Nagahama et al. 2003).

8
Evidence for Tachykinin Involvement in Human Skin Disease

Given that cutaneous tissues are densely innervated by substance P-containing sensory nerve fibres, one might expect tachykininergic activity to play some role in the development of human skin disease and, in particular, conditions which have a strong inflammatory component. The use of animal models has provided much evidence to support this hypothesis, however, a precise role for the tachykinins in mediating the pain and inflammation caused by skin disease remains unknown. Nonetheless, there is significant evidence to implicate substance P in the pathogenesis of psoriasis, dermatitis and pruritus, and tachykininergic activity has been associated with alopecia areata and urticaria, as well as bacterial and traumatic skin disease.

8.1
Psoriasis

Psoriasis is a non-infectious, chronic inflammatory skin disease that is characterized by erythematous plaques and the formation of silvery scales. There is a significant familial component to this condition and it is precipitated by a variety of factors, notably skin trauma, ultraviolet radiation and adverse drug reactions. In addition, stress and psychological trauma have been implicated in the development of this disease (Seville 1989). This has prompted investigation of neuroendocrine–immune system interactions in psoriasis and several authors report an increase in the number and staining density of substance P-containing neurons in lesions collected from psoriatic patients. Typically, this is seen in the epidermis of symptomatic psoriatic skin (Naukkarinen et al. 1989; Chan et al. 1997) and to a lesser extent in the papillary dermis (Naukkarinen et al. 1993; Chan et al. 1997). An increase in neuron–mast cell contacts has also been observed within psoriatic plaques (Naukkarinen et al. 1993) and this may have important implications for disease pathogenesis since substance P has the ability to cause mast cell degranulation leading to the release of histamine and other mediators (see Sect. 4). This in turn, may contribute to the maintenance of inflammation in psoriasis. Nonetheless, it should be noted that some investigators report comparable levels of substance P immunoreactivity in normal, non-lesional and lesional psoriatic skin (Anand et al. 1991; Pincelli et al. 1992) indicating that idiopathic or possibly time-dependent factors influence neurogenic involvement in the condition. With respect to therapeutics, it has been known for some time that topical capsaicin treatment alleviates the scaling, erythema (Bernstein et al. 1986) and pruritus (Ellis et al. 1993) associated with psoriasis and this further supports a role for substance P in its pathogenesis.

8.2
Dermatitis

Dermatitis or eczema is a term used to describe a range of inflammatory skin conditions with diverse aetiology, but similar symptoms. Patients with dermatitis commonly exhibit erythema, spongiosis (the accumulation of oedematous fluid between keratinocytes), pruritus, exudation, scaling and thickening of the dermis and epidermis. Atopic dermatitis is the most common form of the disease and occurs as a result of IgE overproduction and basophil hypersensitivity. Clinical evidence of tachykinin involvement in atopic dermatitis comes largely from immunohistochemical studies that show an increase in substance P expression throughout affected skin. However, in contrast to psoriasis, much of this increased expression is localized to the dermis rather than the epidermis (Pincelli et al. 1990; Sugiura et al. 1997). Furthermore, Toyoda et al. (2002a) report that the plasma concentration of substance P in individuals with atopic dermatitis correlates well with the severity of the disease state and suggest that this may prove to be a useful marker in clinical practice. It is also of interest to note that substance P-induced wheal and flare responses are suppressed in the skin of asymptomatic atopic patients when compared to healthy individuals (Giannetti and Girolomoni 1989) and one might conclude that this is indicative of a tachyphylaxis caused by abnormally high concentrations of the peptide in atopic skin.

Allergic contact dermatitis is another common form of dermatitis. It is brought about by a delayed, type IV hypersensitivity reaction following exposure of sensitized individuals to a specific antigen or hapten. Allergic contact dermatitis is relatively easy to mimic in laboratory animals and a number of studies implicate neuropeptides in its pathogenesis. Topical application of substance P in combination with oxazolone has been shown to exacerbate oedema formation (Gutwald et al. 1991), leukocyte recruitment and adhesion molecule expression in oxazolone-sensitized mice and localized substance P immunoreactivity is reported to increase with the development of dermatitis in these animals (Goebeler et al. 1994). Moreover, Scholzen et al. (2001) have demonstrated that allergen-, but not irritant-induced dermatitis is augmented in NEP-deficient mice in a manner that can be reversed by NK_1 receptor blockade or capsaicin pretreatment indicating a specific role for substance P in mediating allergic contact dermatitis. With respect to clinical studies, capsaicin pretreatment has been shown to ablate flare and itch responses caused by intradermal injection of rat allergen extract into the forearms of sensitized individuals (Lundblad et al. 1987) and the NK_1 receptor antagonist spantide is reported to inhibit nickel-induced dermatitis in patients with an allergy to nickel (Wallengren and Moller 1988). The latter study is of particular significance since nickel is a common cause of allergic contact dermatitis in the population at large due to its presence in everyday objects such as jewellery and silver coins.

8.3
Pruritus

As mentioned, itch or pruritus is a prominent feature of both psoriasis and dermatitis. In addition, pruritus accompanies several other forms of inflammatory skin disease (e.g. urticaria), as well as being a symptom of certain, more generalized systemic conditions (e.g. renal disease, hyperthyroidism). The pathophysiology of pruritus is surprisingly complex and appears to involve a subset of primary afferent C-fibres that are functionally distinct from those that transmit nociceptive responses to the central nervous system (Greaves and Wall 1996). Nonetheless, both forms of neurotransmission are understood to terminate in the thalamus and cortex where signalling is susceptible to modulation by a variety of factors (e.g. opiate peptides; Twycross et al. 2003). In peripheral tissues, histamine has long been regarded as the primary mediator of itch, however, current evidence suggests that other mediators and, in particular, the tachykinins may also play some role. Intradermal injection of NK_1 receptor agonists, but not NK_2 or NK_3 receptor agonists is reported to cause pruritus in the mouse (Andoh et al. 1998). Interestingly, this response occurs in a mast cell-independent manner (Andoh et al. 1998) and is not affected by histamine receptor antagonists (Andoh et al. 2001). Moreover, substance P-induced scratching behaviour in the mouse is dependent upon the production of LTB_4, and substance P has been shown to increase LTB_4 synthesis in both murine skin and cultured keratinocytes (Andoh et al. 2001). The production of lipoxygenase products may therefore be an essential element of substance P signalling under such circumstances. In humans, intradermal injection of substance P causes pain (Jensen et al. 1991) and enhances sodium lauryl sulphate-induced itch (Thomsen et al. 2002), however, the perfusion of human skin with substance P at vasoactive doses fails to cause pruritus (Weidner et al. 2000). Regardless, double-blind, placebo-controlled trials indicate that topically applied capsaicin-containing creams are of benefit in psoriasis- (Bernstein et al. 1986; Ellis et al. 1993), haemodialysis- (Breneman et al. 1992; Tarng et al. 1996; Cho et al. 1997) and water-induced pruritus (Lotti et al. 1994) and such preparations are routinely used in some areas of clinical practice. Similarly, capsaicin is reported to reduce itching and enhance wound healing in patients with prurigo nodularis (Stander et al. 2001), an intensely pruritic condition that is associated with elevated substance P production (Vaalasti et al. 1989; Abadia et al. 1992).

8.4
Urticaria

Urticaria can be described as focal dermal oedema caused by an increase in microvascular permeability and surrounded by a zone of erythema. It is a recurrent-transient condition, lasting less than 24 h and can be brought about by a variety of stimuli, most notably, allergens, adverse drug reactions, heat, cold and autoimmune factors (e.g. IgG overproduction). Although histamine release is thought to be important in the development of urticaria, H_1 receptor antagonists are of limited value in treating the condition clinically indicating that other mediators are involved. Sensory neuropeptides, and in particular the tachykinins, are probable candidates since capsaicin pretreatment attenuates heat and cold urticaria in susceptible individuals (Jancso et al. 1985), and Wallengren (1991) reports that NK_1 receptor blockade can prevent the development of allergen-induced urticaria. In addition, substance P, but not CGRP production is increased in cold urticaria (Wallengren et al. 1987) and patients with chronic idiopathic urticaria exhibit enhanced sensitivity to intradermally injected substance P (Borici-Mazi et al. 1999).

8.5
Other Skin Diseases

In addition to the conditions discussed above, tachykininergic activity has been associated with the development of acne vulgaris, alopecia areata and heat- and ultraviolet B (UVB)-induced skin inflammation. Acne vulgaris is a chronic inflammatory disease of the sebaceous follicles that commonly affects the face and upper back of adolescents and young adults. The pathophysiology of acne vulgaris is complex and multifactorial, but sebum overproduction by the sebaceous glands is a key stage in its development. Toyoda and Morohashi (2001) report that substance P increases the sebum production of cultured sebaceous glands and that substance P, but not CGRP, vasoactive intestinal polypeptide or neuropeptide Y, increases the growth and differentiation of sebaceous glands in vitro. Moreover, an increased density of substance P-containing nerve fibres is observed in the skin of patients with acne vulgaris, especially in areas of close proximity to the sebaceous glands (Toyoda et al. 2002b). Similarly, an increase in substance P immunoreactivity has been documented in the scalp of patients with alopecia areata, a stress-related, autoimmune condition that results in a sudden, irregular loss of hair (Toyoda et al. 2001). In this instance, the authors suggested that substance P might act as an initiator of localized inflammation during the onset of alopecia areata.

Animal studies have provided much evidence to support a role for the tachykinins in mediating the profound inflammatory response that occurs following thermal burn injuries. Capsaicin pretreatment (Saria 1984; Yonehara et al. 1987) and NK_1 receptor antagonists (Siney and Brain 1996; Yonehara and Yoshimura 2000) are widely reported to suppress post-burn oedema formation in the rat and a comparable response is seen in NK_1 receptor knockout mice (Rawlingson et al. 2001). It is also of interest to note that both human (Dunnick et al. 1996) and guinea pig (Kishimoto 1984) skin exhibits a temporary loss of substance P-containing nerve fibres following deep partial thickness burn injury and this may be explained by a rapid depletion of localized tachykinin stores under such circumstances. In addition, substance P has been shown to promote wound contraction in the chronic phase of burn injury (Khalil and Helme 1996).

Since UVB (290–320 nm) is known to be responsible for the erythema and inflammation associated with acute sunburn, several investigators have examined the effects of UVB on tachykininergic activity within the skin. Eschenfelder et al. (1995) report that exposure of rat skin to UVB causes oedema formation and erythema that can be inhibited by NK_1 receptor antagonists indicating a specific role for substance P and/or NKA in the development of sunburn. In this study, UVB was also shown to cause a prolonged and direct activation of sensory C-fibres in rat skin and an increase in the staining density of substance P-containing nerve fibres. Comparable results were obtained by Legat et al. (2002) who demonstrated that repetitive, low dose exposure to UVB caused an increase in dermal substance P content and enhanced responsiveness to mustard oil in the rat. With respect to human studies, a single exposure of UVB has been shown to increase substance P-specific binding in the epidermis of healthy volunteers, indicating an up-regulation of the NK_1 receptor under such circumstances (Viac et al. 1997).

9
Wound Healing

The role of the tachykinins in wound healing is only now being understood with the development of tachykinin receptor and NEP knockout mice. Multiple studies support a role for substance P in promoting wound healing (Khalil and Helme 1996; Scholzen et al. 1998) and NEP knockout mice exhibit enhanced healing in response to skin trauma indicating that tachykinin metabolism may be detrimental to this process (Scholzer et al. 2001). Nonetheless, recent studies performed in our own laboratory have demonstrated that the closure of a simple experimental laceration occurred in a comparable manner in NK_1 receptor knockout and wild-type mice (Cao et al. 2001).

10
Conclusions

In this review, we have summarized the current literature relating to the involvement of the tachykinins in the acute, dermal inflammatory response, as well as more chronic inflammatory skin conditions. Evidence has been presented to show that substance P can increase skin blood flow, but appears to act in a transient manner when compared with its co-localized neuropeptide, CGRP. The tachykinins have potent oedema-inducing properties and animal studies indicate that this is an important component of the dermatological response to injury. However, the functional significance of tachykinin release in normal human skin is questioned, as is the role of substance P-mediated mast cell degranulation since recent evidence indicates that endogenous tachykinins do not reach a sufficiently high concentration to cause mast cell degranulation or have vasoactive effects. Similarly, substance P-induced leukocyte recruitment in human skin must be queried if this too is proven to be a mast cell-dependent effect. Nonetheless, preliminary studies demonstrate that substance P can initiate neutrophil priming independently of the mast cell and bring about leukocyte recruitment via its effects on endothelial cell adhesion molecule expression. In addition, the production of non-neuronal substance P and NK_1 receptor up-regulation may be highly relevant to the pathophysiology of dermatological disease, especially in conditions where the expression of substance P and/or the density of substance P-containing nerve fibres is also increased. However, the present lack of information relating to the effect of NK_1 receptor antagonists on human skin disease makes the importance of these observations difficult to assess, and the publication of such investigations is eagerly awaited.

References

Abadia MF, Burrows NP, Jones RR, Terenghi G, Polak JM (1992) Increased sensory neuropeptides in nodular prurigo: a quantitative immunohistochemical analysis. Br J Dermatol 127:344–351

Abeck D, Mempel M (1998) Staphylococcus aureus colonization in atopic dermatitis and its therapeutic implications. Br J Dermatol 139(Suppl 53):13–16

Ahluwalia A, Giuliani S, Scotland R, Maggi CA (1998) Ovalbumin-induced neurogenic inflammation in the bladder of sensitized rats. Br J Pharmacol 124:190–196

Alving K, Sundstrom C, Matran R, Panula P, Hökfelt T, Lundberg JM (1991) Association between histamine-containing mast cells and sensory nerves in the skin and airways of control and capsaicin-treated pigs. Cell Tissue Res 264:529–538

Amann R, Egger T, Schuligoi R (2000) The tachykinin NK_1 receptor antagonist SR140333 prevents the increase of nerve growth factor in rat paw skin induced by substance P or neurogenic inflammation. Neuroscience 100:611–615

Anand P, Springall DR, Blank MA, Sellu D, Polak JM, Bloom SR (1991) Neuropeptides in skin disease: increased VIP in eczema and psoriasis but not axillary hyperhidrosis. Br J Dermatol 124:547–549

Anastasi A, Erspamer V, Bucci M (1971) Isolation and structure of bombesin and alytesin, 2 analogous active peptides from the skin of the European amphibians Bombina and Alytes. Experientia 27:166–167

Andoh T, Katsube N, Maruyama M, Kuraishi Y (2001) Involvement of leukotriene B_4 in substance P-induced itch-associated response in mice. J Invest Dermatol 117:1621–1626

Andoh T, Nagasawa T, Satoh M, Kuraishi Y (1998) Substance P induction of itch-associated response mediated by cutaneous NK_1 tachykinin receptors in mice. J Pharmacol Exp Ther 286:1140–1145

Ansel JC, Brown JR, Payan DG, Brown MA (1993) Substance P selectively activates TNF-alpha gene expression in murine mast cells. J Immunol 150:4478–4485

Ansel JC, Kaynard AH, Armstrong CA, Olerud J, Bunnett N, Payan D (1996) Skin-nervous system interactions. J Invest Dermatol 106:198–204

Antunes E, Marangoni RA, Brain SD, De Nucci G (1992) Phoneutria nigriventer (armed spider) venom induces increased vascular permeability in rat and rabbit skin *in vivo*. Toxicon 30:1011–1016

Averbeck B, Reeh PW (2001) Interactions of inflammatory mediators stimulating release of calcitonin gene-related peptide, substance P and prostaglandin E_2 from isolated rat skin. Neuropharmacology 40:416–423

Bailey SR, Cunningham FM (2001) Inflammatory mediators induce endothelium-dependent adherence of equine eosinophils to cultured endothelial cells. J Vet Pharmacol Ther 24:209–214

Bayliss W.M (1901) On the origin from the spinal cord of the vasodilator fibres of the hind limb, and on the nature of these fibres. J Physiol 26:173–209

Barnes PJ, Brown MJ, Dollery CT, Fuller RW, Heavey DJ, Ind PW (1986) Histamine is released from skin by substance P but does not act as the final vasodilator in the axon reflex. Br J Pharmacol 88:741–745

Benyon RC, Lowman MA, Church MK (1987) Human skin mast cells: their dispersion, purification, and secretory characterization. J Immunol 138:861–867

Bergdoll MS (1989) Regulation and control of toxic shock syndrome toxin 1: overview. Rev Infect Dis 11(Suppl 1):S142–S144

Bernstein JE, Parish LC, Rapaport M, Rosenbaum MM, Roenigk HH, Jr. (1986) Effects of topically applied capsaicin on moderate and severe psoriasis vulgaris. J Am Acad Dermatol 15:504–507

Bianchi B, Matucci R, Danesi A, Rossi R, Ipponi P, Giannotti B, Johansson O, Cappugi P (1999) Characterization of [3^H]substance P binding sites in human skin. J Eur Acad Dermatol Venereol 12:6–10

Birch PJ, Harrison SM, Hayes AG, Rogers H, Tyers MB (1992) The non-peptide NK_1 receptor antagonist, (+/-)-CP-96,345, produces antinociceptive and anti-oedema effects in the rat. Br J Pharmacol 105:508–510

Bockmann S, Seep J, Jonas L (2001) Delay of neutrophil apoptosis by the neuropeptide substance P: involvement of caspase cascade. Peptides 22:661–670

Borici-Mazi R, Kouridakis S, Kontou-Fili K (1999) Cutaneous responses to substance P and calcitonin gene-related peptide in chronic urticaria: the effect of cetirizine and dimethindene. Allergy 54:46–56

Bost KL, Breeding SA, Pascual DW (1992) Modulation of the mRNAs encoding substance P and its receptor in rat macrophages by LPS. Reg Immunol 4:105–112

Bour H, Demidem A, Garrigue JL, Krasteva M, Schmitt D, Claudy A, Nicolas JF (1995) *In vitro* T cell response to staphylococcal enterotoxin B superantigen in chronic plaque type psoriasis. Acta Derm Venereol 75:218–221

Bozic CR, Lu B, Hopken UE, Gerard C, Gerard NP (1996) Neurogenic amplification of immune complex inflammation. Science 273:1722–1725

Brain SD (1997) Sensory neuropeptides: their role in inflammation and wound healing. Immunopharmacology 37:133–152

Brain SD (2000) New feelings about the role of sensory nerves in inflammation. Nat Med 6:134–135

Brain SD, Tippins JR, Morris HR, MacIntyre I, Williams TJ (1986) Potent vasodilator activity of calcitonin gene-related peptide in human skin. J Invest Dermatol 87:533–536

Brain SD, Williams TJ (1985) Inflammatory oedema induced by synergism between calcitonin gene-related peptide (CGRP) and mediators of increased vascular permeability. Br J Pharmacol 86:855–860

Brandolini L, Sergi R, Caselli G, Boraschi D, Locati M, Sozzani S, Bertini R (1997) Interleukin-1 beta primes interleukin-8-stimulated chemotaxis and elastase release in human neutrophils via its type I receptor. Eur Cytokine Netw 8:173–178

Breneman DL, Cardone JS, Blumsack RF, Lather RM, Searle EA, Pollack VE (1992) Topical capsaicin for treatment of hemodialysis-related pruritus. J Am Acad Dermatol 26:91–94

Bruce AN (1913) Vasodilator axon reflexes. Q J Physiol 6:339–354.

Bucaretchi F, Deus Reinaldo CR, Hyslop S, Madureira PR, De Capitani EM, Vieira RJ (2000) A clinico-epidemiological study of bites by spiders of the genus Phoneutria. Rev Inst Med Trop Sao Paulo 42:17–21

Buchan AM, Lance V, Polak JM (1983) Regulatory peptides in the gastrointestinal tract of Alligator mississipiensis. An immunocytochemical study. Cell Tissue Res 231:439–449

Burbach GJ, Kim KH, Zivony AS, Kim A, Aranda J, Wright S, Naik SM, Caughman SW, Ansel JC, Armstrong CA (2001) The neurosensory tachykinins substance P and neurokinin A directly induce keratinocyte nerve growth factor. J Invest Dermatol 117:1075–1082

Calvo CF, Chavanel G, Senik A (1992) Substance P enhances IL-2 expression in activated human T cells. J Immunol 148:3498–3504

Cao T, Gerard NP, Brain SD (1999) Use of NK_1 knockout mice to analyze substance P-induced edema formation. Am J Physiol 277:R476–R481

Cao T, Grant AD, Gerard NP, Brain SD (2001) Lack of a significant effect of deletion of the tachykinin neurokinin-1 receptor on wound healing in mouse skin. Neuroscience 108:695–700

Cao T, Pinter E, Al Rashed S, Gerard N, Hoult JR, Brain SD (2000) Neurokinin-1 receptor agonists are involved in mediating neutrophil accumulation in the inflamed, but not normal, cutaneous microvasculature: an *in vivo* study using neurokinin-1 receptor knockout mice. J Immunol 164:5424–5429

Cao YQ, Mantyh PW, Carlson EJ, Gillespie AM, Epstein CJ, Basbaum AI (1998) Primary afferent tachykinins are required to experience moderate to intense pain. Nature 392:390–394

Carlton SM, Coggeshall RE (2002) Inflammation-induced up-regulation of neurokinin 1 receptors in rat glabrous skin. Neurosci Lett 326:29–32

Chan J, Smoller BR, Raychaudhuri SP, Jiang WY, Farber EM (1997) Intraepidermal nerve fiber expression of calcitonin gene-related peptide, vasoactive intestinal peptide and substance P in psoriasis. Arch Dermatol Res 289:611–616

Chang MM, Leeman SE, Niall HD (1971) Amino-acid sequence of substance P. Nat New Biol 232:86–87

Cho YL, Liu HN, Huang TP, Tarng DC (1997) Uremic pruritus: roles of parathyroid hormone and substance P. J Am Acad Dermatol 36:538–543

Cook GA, Elliott D, Metwali A, Blum AM, Sandor M, Lynch R, Weinstock JV (1994) Molecular evidence that granuloma T lymphocytes in murine schistosomiasis mansoni express an authentic substance P (NK_1) receptor. J Immunol 152:1830–1835

Costa SKP, De Nucci G, Antunes E, Brain SD (1997) Phoneutria nigriventer spider venom induces oedema in rat skin by activation of capsaicin sensitive sensory nerves. Eur J Pharmacol 339:223–226

Costa SKP, Esquisatto LC, Camargo E, Gambero A, Brain SD, De Nucci G, Antunes E (2001) Comparative effect of Phoneutria nigriventer spider venom and capsaicin on the rat paw oedema. Life Sci 69:1573–1585

Costa SKP, Brain SD, Docherty, RJ (2002) Novel mechanisms involving $5-HT_4$ receptors for the activation of sensory nerves. Neuropeptides 36(Suppl):460

Costa SKP, Brain SD, Antunes E, De Nucci G, Docherty RJ (2003) *Phoneutria nigriventer* spider venom activates $5-HT_4$ receptors in rat isolated vagus nerve. Br J Pharmacol 139:59–64

Cowan A, Khunawat P, Zhu XZ, Gmerek DE (1985) Effects of bombesin on behavior. Life Sci 37:135–145

Dalsgaard CJ, Haegerstrand A, Theodorsson-Norheim E, Brodin E, Hökfelt T (1985) Neurokinin A-like immunoreactivity in rat primary sensory neurons; coexistence with substance P. Histochemistry 83:37–39

Desouza IA, Bergdoll MS, Ribeiro-DaSilva G (1996) Pharmacological characterization of mouse paw edema induced by Staphyloccocal enterotoxin B. J Nat Toxins 5:61–75

Devillier P, Regoli D, Asseraf A, Descours B, Marsac J, Renoux M (1986) Histamine release and local responses of rat and human skin to substance P and other mammalian tachykinins. Pharmacology 32:340–347

Dianzani C, Lombardi G, Collino M, Ferrara C, Cassone MC, Fantozzi R (2001) Priming effects of substance P on calcium changes evoked by interleukin-8 in human neutrophils. J Leukoc Biol 69:1013–1018

Dunnick CA, Gibran NS, Heimbach DM (1996) Substance P has a role in neurogenic mediation of human burn wound healing. J Burn Care Rehabil 17:390–396

El Shazly AE, Masuyama K, Ishikawa T (1997) Mechanisms involved in activation of human eosinophil exocytosis by substance P: an in vitro model of sensory neuroimmunomodulation. Immunol Invest 26:615–629

Ellis CN, Berberian B, Sulica VI, Dodd WA, Jarratt MT, Katz HI, Prawer S, Krueger G, Rex IH, Jr., Wolf JE (1993) A double-blind evaluation of topical capsaicin in pruritic psoriasis. J Am Acad Dermatol 29:438–442

Emonds-Alt X, Doutremepuich JD, Heaulme M, Neliat G, Santucci V, Steinberg R, Vilain P, Bichon D, Ducoux JP, Proietto V, . (1993) In vitro and in vivo biological activities of SR140333, a novel potent non-peptide tachykinin NK_1 receptor antagonist. Eur J Pharmacol 250:403–413

Eschenfelder CC, Benrath J, Zimmermann M, Gillardon F (1995) Involvement of substance P in ultraviolet irradiation-induced inflammation in rat skin. Eur J Neurosci 7:1520–1526

Escott KJ, Beattie DT, Connor HE, Brain SD (1995) Trigeminal ganglion stimulation increases facial skin blood flow in the rat: a major role for calcitonin gene-related peptide. Brain Res 669:93–99

Fewtrell CM, Foreman JC, Jordan CC, Oehme P, Renner H, Stewart JM (1982) The effects of substance P on histamine and 5-hydroxytryptamine release in the rat. J Physiol 330:393–411

Foreman J, Jordan C (1983) Histamine release and vascular changes induced by neuropeptides. Agents Actions 13:105–116

Fuller RW, Conradson TB, Dixon CM, Crossman DC, Barnes PJ (1987) Sensory neuropeptide effects in human skin. Br J Pharmacol 92:781–788

Gamse R, Posch M, Saria A, Jancso G (1987) Several mediators appear to interact in neurogenic inflammation. Acta Physiol Hung 69:343–354

Garret C, Carruette A, Fardin V, Moussaoui S, Peyronel JF, Blanchard JC, Laduron PM (1991) Pharmacological properties of a potent and selective nonpeptide substance P antagonist. Proc Natl Acad Sci USA 88:10208–10212

Ghatei MA, Jung RT, Stevenson JC, Hillyard CJ, Adrian TE, Lee YC, Christofides ND, Sarson DL, Mashiter K, MacIntyre I, Bloom SR (1982) Bombesin: action on gut hormones and calcium in man. J Clin Endocrinol Metab 54:980–985

Giannetti A, Girolomoni G (1989) Skin reactivity to neuropeptides in atopic dermatitis. Br J Dermatol. 121:681–688

Goebeler M, Henseleit U, Roth J, Sorg C (1994) Substance P and calcitonin gene-related peptide modulate leukocyte infiltration to mouse skin during allergic contact dermatitis. Arch Dermatol Res 286:341–346

Gottfried A, Scheuber PH, Bernhard R, Sailer-Kramer B, Hartmann A (1989) Role of substance P in immediate-type skin reactions induced by Staphyloccocal enterotoxin B unsensitized monkeys. J Allergy Clin 84:880–885.

Grant AD, Gerard NP, Brain SD (2002) Evidence of a role for NK_1 and CGRP receptors in mediating neurogenic vasodilatation in the mouse ear. Br J Pharmacol 135:356–362

Greaves MW, Wall PD (1996) Pathophysiology of itching. Lancet 348:938–940

Guo CJ, Lai JP, Luo HM, Douglas SD, Ho WZ (2002) Substance P up-regulates macrophage inflammatory protein-1beta expression in human T lymphocytes. J Neuroimmunol 131:160–167

Gutwald J, Goebeler M, Sorg C (1991) Neuropeptides enhance irritant and allergic contact dermatitis. J Invest Dermatol 96:695–698

Häbler HJ, Timmermann L, Stegmann JU, Jänig W (1999) Involvement of neurokinins in antidromic vasodilatation in hairy and hairless skin of the rat hindlimb. Neuroscience 89:1259–1268

Hägermark O, Hökfelt T, Pernow B (1978) Flare and itch induced by substance P in human skin. J Invest Dermatol 71:233–235

Hartschuh W, Weihe E, Reinecke M (1983) Peptidergic (neurotensin, VIP, substance P) nerve fibres in the skin. Immunohistochemical evidence of an involvement of neuropeptides in nociception, pruritus and inflammation. Br J Dermatol 109(Suppl 25):14–17

Herzberg U, Murtaugh MP, Mullet MA, Beitz AJ (1995) Electrical stimulation of the sciatic nerve alters neuropeptide content and lymphocyte migration in the subcutaneous tissue of the rat hind paw. Neuroreport 6:1773–1777

Ho WZ, Lai JP, Li Y, Douglas SD (2002) HIV enhances substance P expression in human immune cells. FASEB J 16:616–618

Ho WZ, Lai JP, Zhu XH, Uvaydova M, Douglas SD (1997) Human monocytes and macrophages express substance P and neurokinin-1 receptor. J Immunol 159:5654–5660

Hökfelt T, Kellerth JO, Nilsson G, Pernow B (1975) Substance P: localization in the central nervous system and in some primary sensory neurons. Science 190:889–890

Holzer P (1998) Neurogenic vasodilatation and plasma leakage in the skin. Gen Pharmacol 30:5–11

Holzer P, Lippe II, Raybould HE, Pabst MA, Livingston EH, Amann R, Peskar BM, Peskar BA, Tache Y, Guth PH (1991) Role of peptidergic sensory neurons in gastric mucosal blood flow and protection. Ann NY Acad Sci 632:272–282

Hordinsky M, Ericson M, Snow D, Boeck C, Lee WS (1999) Peribulbar innervation and substance P expression following nonpermanent injury to the human scalp hair follicle. J Invest Dermatol Symp Proc 4:316–319

Huttunen M, Harvima IT, Ackermann L, Harvima RJ, Naukkarinen A, Horsmanheimo M (1996) Neuropeptide- and capsaicin-induced histamine release in skin monitored with the microdialysis technique. Acta Derm Venereol 76:205–209

Iandolo JJ (1989) Genetic analysis of extracellular toxins of Staphylococcus aureus. Annu Rev Microbiol 43:375–402

Inoue H, Nagata N, Koshihara Y (1996) Involvement of tachykinin receptors in oedema formation and plasma extravasation induced by substance P, neurokinin A, and neurokinin B in mouse ear. Inflamm Res 45:316–323

Ishihara M, Endo R, Rivera MR, Mihara M (2002) Bird's eye view observations of the subepidermal nerve network of normal guinea pig skin. Arch Dermatol Res 294:281–285

Iwamoto I, Tomoe S, Yoshida S (1993) Role of leukotriene B4 in substance P-induced granulocyte infiltration in mouse skin. Regul Pept 46:225–227

Iwamoto I, Ueki IF, Borson DB, Nadel JA (1989) Neutral endopeptidase modulates tachykinin-induced increase in vascular permeability in guinea pig skin. Int Arch Allergy Appl Immunol 88:288–293

Jackson H, Parks TN (1989) Spider toxins: recent applications in neurobiology. Annu Rev Neurosci 12:405–414

Jancso N, Jancso-Gabor A, Szolcsanyi J (1967) Direct evidence for neurogenic inflammation and its prevention by denervation and by pretreatment with capsaicin. Br J Pharmacol 31:138–151

Jancso G, Kiraly E, Jancso-Gabor A (1977) Pharmacologically induced selective degeneration of chemosensitive primary sensory neurones. Nature 270:741–743

Jancso G, Obal F, Toth-Kasa I, Katona M, Husz S (1985) The modulation of cutaneous inflammatory reactions by peptide-containing sensory nerves. Int J Tissue React 7:449–457

Jancso G, Pierau FK, Sann H (1993) Mustard oil-induced cutaneous inflammation in the pig. Agents Actions 39:31–34

Jensen K, Tuxen C, Pedersen-Bjergaard U, Jansen I (1991) Pain, tenderness, wheal and flare induced by substance-P, bradykinin and 5-hydroxytryptamine in humans. Cephalalgia 11:175–182

Kahler CM, Herold M, Wiedermann CJ (1993) Substance P: a competence factor for human fibroblast proliferation that induces the release of growth-regulatory arachidonic acid metabolites. J Cell Physiol 156:579–587

Kanda N, Watanabe S (2002) Substance P Enhances the Production of Interferon-induced Protein of 10 kDa by Human Keratinocytes in Synergy with Interferon-gamma. J Invest Dermatol 119:1290–1297

Karila P, Axelsson M, Franklin CE, Fritsche R, Gibbins IL, Grigg GC, Nilsson S, Holmgren S (1995) Neuropeptide immunoreactivity and co-existence in cardiovascular nerves and autonomic ganglia of the estuarine crocodile, Crocodylus porosus, and cardiovascular effects of neuropeptides. Regul Pept 58:25–39

Khalil Z, Helme R (1996) Sensory peptides as neuromodulators of wound healing in aged rats. JGerontol A Biol Sci Med Sci 51:B354–B361

Kimball ES, Persico FJ, Vaught JL (1988) Neurokinin-induced generation of interleukin-1 in a macrophage cell line. Ann NY Acad Sci 540:688–690

Kishimoto S (1984) The regeneration of substance P-containing nerve fibers in the process of burn wound healing in the guinea pig skin. J Invest Dermatol 83:219–223

Kowalski ML, Sliwinska-Kowalska M, Kaliner MA (1990) Neurogenic inflammation, vascular permeability, and mast cells. II. Additional evidence indicating that mast cells are not involved in neurogenic inflammation. J Immunol 145:1214–1221

Ladenheim EE, Ritter RC (1993) Caudal hindbrain participation in the suppression of feeding by central and peripheral bombesin. Am J Physiol 264:R1229–R1234

Lai JP, Douglas SD, Ho WZ (1998) Human lymphocytes express substance P and its receptor. J Neuroimmunol 86:80–86

Lai JP, Ho WZ, Zhan GX, Yi Y, Collman RG, Douglas SD (2001) Substance P antagonist (CP-96,345) inhibits HIV-1 replication in human mononuclear phagocytes. Proc Natl Acad Sci USA 98:3970–3975

Lai R, Liu H, Lee WH, Zhang Y (2002) A novel proline rich bombesin-related peptide (PR-bombesin) from toad Bombina maxima. Peptides 23:437–442

Lambrecht BN (2001) Immunologists getting nervous: neuropeptides, dendritic cells and T cell activation. Respir Res 2:133–138

Laurenzi MA, Persson MA, Dalsgaard CJ, Haegerstrand A (1990) The neuropeptide substance P stimulates production of interleukin 1 in human blood monocytes: activated cells are preferentially influenced by the neuropeptide. Scand J Immunol 31:529–533

Legat FJ, Griesbacher T, Schicho R, Althuber P, Schuligoi R, Kerl H, Wolf P (2002) Repeated subinflammatory ultraviolet B irradiation increases substance P and calcitonin gene-related peptide content and augments mustard oil-induced neurogenic inflammation in the skin of rats. Neurosci Lett 329:309–313

Lembeck F, Bernatzky G, Gamse R, Saria A (1985) Characterization of substance P-like immunoreactivity in submammalian species by high performance liquid chromatography. Peptides 6(Suppl 3):231–236

Lembeck F, Donnerer J, Tsuchiya M, Nagahisa A (1992) The non-peptide tachykinin antagonist, CP-96,345, is a potent inhibitor of neurogenic inflammation. Br J Pharmacol 105:527–530

Lembeck F, Holzer P (1979) Substance P as neurogenic mediator of antidromic vasodilation and neurogenic plasma extravasation. Naunyn Schmiedeberg's Arch Pharmacol 310:175–183

Lewis T (1927) The blood vessels of the human skin and their response. Shaw, London, UK

Linardi A, Costa SK, da Silva GR, Antunes E (2000) Involvement of kinins, mast cells and sensory neurons in the plasma exudation and paw oedema induced by staphylococcal enterotoxin B in the mouse. Eur J Pharmacol 399:235–242

Lindsey KQ, Caughman SW, Olerud JE, Bunnett NW, Armstrong CA, Ansel JC (2000) Neural regulation of endothelial cell-mediated inflammation. J Invest Dermatol Symp Proc 5:74–78

Lippe C, Bellantuono V, Castronuovo G, Ardizzone C, Cassano G (1994) Action of capsaicin and related peptides on the ionic transport across the skin of Rana esculenta. Arch Int Physiol Biochim Biophys 102:51–54

Lippe C, Lobasso S, Cassano G, Bellantuono V, Ardizzone C (1998) Actions of tachykinins on the ion transport across the frog skin. Peptides 19:1435–1438

Lloyds D, Hallett MB (1993) Activation and priming of the human neutrophil oxidase response by substance P: distinct signal transduction pathways. Biochim Biophys Acta 1175:207–213

Lobasso S, Lippe C, Bellantuono V, Ardizzone C (1998) Action of physalaemin on the ionic transport across the frog skin. Arch Physiol Biochem 105:329–336

Lotti T, Teofoli P, Tsampau D (1994) Treatment of aquagenic pruritus with topical capsaicin cream. J Am Acad Dermatol 30:232–235

Lotz M, Vaughan JH, Carson DA (1988) Effect of neuropeptides on production of inflammatory cytokines by human monocytes. Science 241:1218–1221

Lowman MA, Benyon RC, Church MK (1988) Characterization of neuropeptide-induced histamine release from human dispersed skin mast cells. Br J Pharmacol 95:121–130

Lundberg JM (1996) Pharmacology of cotransmission in the autonomic nervous system: integrative aspects on amines, neuropeptides, adenosine triphosphate, amino acids and nitric oxide. Pharmacol Rev 48:113–178

Lundblad L, Lundberg JM, Anggard A, Zetterstrom O (1987) Capsaicin-sensitive nerves and the cutaneous allergy reaction in man. Possible involvement of sensory neuropeptides in the flare reaction. Allergy 42:20–25

Lynn B, Shakhanbeh J (1988) Substance P content of the skin, neurogenic inflammation and numbers of C-fibres following capsaicin application to a cutaneous nerve in the rabbit. Neuroscience 24:769–775

Maggi CA (1997) The effects of tachykinins on inflammatory and immune cells. Regul Pept 70:75–90

Matsuda H, Kawakita K, Kiso Y, Nakano T, Kitamura Y (1989) Substance P induces granulocyte infiltration through degranulation of mast cells. J Immunol 142:927–931

McEwan JR, Benjamin N, Larkin S, Fuller RW, Dollery CT, MacIntyre I (1988) Vasodilatation by calcitonin gene-related peptide and by substance P: a comparison of their effects on resistance and capacitance vessels of human forearms. Circulation 77:1072–1080

McLean PG, Ahluwalia A, Perretti M (2000) Association between kinin B(1) receptor expression and leukocyte trafficking across mouse mesenteric postcapillary venules. J Exp Med 192:367–380

Metwali A, Blum AM, Ferraris L, Klein JS, Fiocchi C, Weinstock JV (1994) Eosinophils within the healthy or inflamed human intestine produce substance P and vasoactive intestinal peptide. J Neuroimmunol 52:69–78

Miller C, Moczydlowski E, Latorre R, Phillips M (1985) Charybdotoxin, a protein inhibitor of single Ca^{2+}-activated K^+ channels from mammalian skeletal muscle. Nature 313:316–318

Misery L (1997) Skin, immunity and the nervous system. Br J Dermatol 137:843–850

Morhenn VB (2000) Firm stroking of human skin leads to vasodilatation possibly due to the release of substance P. J Dermatol Sci 22:138–144

Muller-Alouf H, Carnoy C, Simonet M, Alouf JE (2001) Superantigen bacterial toxins: state of the art. Toxicon 39:1691–1701

Nagahama M, Morimitsu S, Kihara A, Akita M, Setsu K, Sakurai J (2003) Involvement of tachykinin receptors in Clostridium perfringens beta-toxin-induced plasma extravasation. Br J Pharmacol 138:23–30

Nagalla SR, Barry BJ, Spindel ER (1994) Cloning of complementary DNAs encoding the amphibian bombesin-like peptides Phe8 and Leu8 phyllolitorin from Phyllomedusa sauvagei: potential role of U to C RNA editing in generating neuropeptide diversity. Mol Endocrinol 8:943–951

Nakajima T, Yasuhara T, Erspamer V, Erspamer GF, Negri L, Endean R (1980) Physalaemin- and bombesin-like peptides in the skin of the Australian leptodactylid frog Uperoleia rugosa. Chem Pharm Bull (Tokyo) 28:689–695

Naukkarinen A, Harvima I, Paukkonen K, Aalto ML, Horsmanheimo M (1993) Immunohistochemical analysis of sensory nerves and neuropeptides, and their contacts with mast cells in developing and mature psoriatic lesions. Arch Dermatol Res 285:341–346

Naukkarinen A, Nickoloff BJ, Farber EM (1989) Quantification of cutaneous sensory nerves and their substance P content in psoriasis. J Invest Dermatol 92:126–129

Nilsson J, von Euler AM, Dalsgaard CJ (1985) Stimulation of connective tissue cell growth by substance P and substance K. Nature 315:61–63

Nio DA, Moylan RN, Roche JK (1993) Modulation of T lymphocyte function by neuropeptides. Evidence for their role as local immunoregulatory elements. J Immunol 150:5281–5288

Okabe T, Hide M, Koro O, Nimi N, Yamamoto S (2001) The release of leukotriene B4 from human skin in response to substance P: evidence for the functional heterogeneity of human skin mast cells among individuals. Clin Exp Immunol 124:150–156

Okabe T, Hide M, Koro O, Yamamoto S (2000) Substance P induces tumor necrosis factor-alpha release from human skin via mitogen-activated protein kinase. Eur J Pharmacol 398:309–315

Okayama Y, Ono Y, Nakazawa T, Church MK, Mori M (1998) Human skin mast cells produce TNF-alpha by substance P. Int Arch Allergy Immunol 117(Suppl 1):48–51

Olerud JE, Usui ML, Seckin D, Chiu DS, Haycox CL, Song IS, Ansel JC, Bunnett NW (1999) Neutral endopeptidase expression and distribution in human skin and wounds. J Invest Dermatol 112:873–881

Palframan RT, Costa SK, Wilsoncroft P, Antunes E, De Nucci G, Brain SD (1996) The effect of a tachykinin NK_1 receptor antagonist, SR140333, on oedema formation induced in rat skin by venom from the Phoneutria nigriventer spider. Br J Pharmacol 118:295-298

Parenti A, Amerini S, Ledda F, Maggi CA, Ziche M (1996) The tachykinin NK_1 receptor mediates the migration-promoting effect of substance P on human skin fibroblasts in culture. Naunyn Schmiedebergs Arch Pharmacol 353:475-481

Paus R, Heinzelmann T, Robicsek S, Czarnetzki BM, Maurer M (1995) Substance P stimulates murine epidermal keratinocyte proliferation and dermal mast cell degranulation in situ. Arch Dermatol Res 287:500-502

Payan DG, Brewster DR, Goetzl EJ (1983) Specific stimulation of human T lymphocytes by substance P. J Immunol 131:1613-1615

Petersen LJ, Winge K, Brodin E, Skov PS (1997) No release of histamine and substance P in capsaicin-induced neurogenic inflammation in intact human skin in vivo: a microdialysis study. Clin Exp Allergy 27:957-965

Pincelli C, Fantini F, Massimi P, Girolomoni G, Seidenari S, Giannetti A (1990) Neuropeptides in skin from patients with atopic dermatitis: an immunohistochemical study. Br J Dermatol 122:745-750

Pincelli C, Fantini F, Romualdi P, Sevignani C, Lesa G, Benassi L, Giannetti A (1992) Substance P is diminished and vasoactive intestinal peptide is augmented in psoriatic lesions and these peptides exert disparate effects on the proliferation of cultured human keratinocytes. J Invest Dermatol 98:421-427

Pinter E, Brown B, Hoult JR, Brain SD (1999) Lack of evidence for tachykinin NK_1 receptor-mediated neutrophil accumulation in the rat cutaneous microvasculature by thermal injury. Eur J Pharmacol 369:91-98

Quinlan KL, Song IS, Bunnett NW, Letran E, Steinhoff M, Harten B, Olerud JE, Armstrong CA, Wright CS, Ansel JC (1998) Neuropeptide regulation of human dermal microvascular endothelial cell ICAM-1 expression and function. Am J Physiol 275:C1580-C1590

Quinlan KL, Song IS, Naik SM, Letran EL, Olerud JE, Bunnett NW, Armstrong CA, Caughman SW, Ansel JC (1999) VCAM-1 expression on human dermal microvascular endothelial cells is directly and specifically up-regulated by substance P. J Immunol 162:1656-1661

Rabier MJ, Farber EM, Wilkinson DI (1993) Neuropeptides modulate leukotriene B4 mitogenicity toward cultured human keratinocytes. J Invest Dermatol 100:132-136

Ralevic V, Milner P, Hudlicka O, Kristek F, Burnstock G (1990) Substance P is released from the endothelium of normal and capsaicin-treated rat hind-limb vasculature, *in vivo*, by increased flow. Circ Res 66:1178-1183

Rameshwar P, Gascon P, Ganea D (1993) Stimulation of IL-2 production in murine lymphocytes by substance P and related tachykinins. J Immunol 151:2484-2496

Rawlingson A, Gerard NP, Brain SD (2001) Interactive contribution of NK_1 and kinin receptors to the acute inflammatory oedema observed in response to noxious heat stimulation: studies in NK_1 receptor knockout mice. Br J Pharmacol 134:1805-1813

Regoli D, Nguyen K, Calo G (1997) Neurokinin receptors. Comparison of data from classical pharmacology, binding, and molecular biology. Ann NY Acad Sci 812:144-146

Reilly DM, Ferdinando D, Johnston C, Shaw C, Buchanan KD, Green MR (1997) The epidermal nerve fibre network: characterization of nerve fibres in human skin by confocal microscopy and assessment of racial variations. Br J Dermatol 137:163-170

Reiner A, Krause JE, Keyser KT, Eldred WD, McKelvy JF (1984) The distribution of substance P in turtle nervous system: a radioimmunoassay and immunohistochemical study. J Comp Neurol 226:50-75

Rinder J, Lundberg JM (1996) Effects of hCGRP 8-37 and the NK_1-receptor antagonist SR 140.333 on capsaicin-evoked vasodilation in the pig nasal mucosa in vivo. Acta Physiol Scand 156:115-122

Ruocco I, Cuello AC, Parent A, Ribeiro-da-Silva A (2002) Skin blood vessels are simultaneously innervated by sensory, sympathetic, and parasympathetic fibers. J Comp Neurol 448:323–336

Saban MR, Saban R, Bjorling D, Haak-Frendscho M (1997) Involvement of leukotrienes, TNF-alpha, and the LFA-1/ICAM-1 interaction in substance P-induced granulocyte infiltration. J Leukoc Biol 61:445–451

Santicioli P, Giuliani S, Maggi CA (1993) Failure of L-nitroarginine, a nitric oxide synthase inhibitor, to affect hypotension and plasma protein extravasation produced by tachykinin NK_1 receptor activation in rats. J Auton Pharmacol 13:193–199

Santoni G, Perfumi MC, Spreghini E, Romagnoli S, Piccoli M (1999) Neurokinin type-1 receptor antagonist inhibits enhancement of T cell functions by substance P in normal and neuromanipulated capsaicin-treated rats. J Neuroimmunol 93:15–25

Saria A (1984) Substance P in sensory nerve fibres contributes to the development of oedema in the rat hind paw after thermal injury. Br J Pharmacol 82:217–222

Schmelz M, Luz O, Averbeck B, Bickel A (1997) Plasma extravasation and neuropeptide release in human skin as measured by intradermal microdialysis. Neurosci Lett 230:117–120

Scholzen T, Armstrong CA, Bunnett NW, Luger TA, Olerud JE, Ansel JC (1998) Neuropeptides in the skin: interactions between the neuroendocrine and the skin immune systems. Exp Dermatol 7:81–96

Scholzen TE, Brzoska T, Kalden DH, O'Reilly F, Armstrong CA, Luger TA, Ansel JC (1999) Effect of ultraviolet light on the release of neuropeptides and neuroendocrine hormones in the skin: mediators of photodermatitis and cutaneous inflammation. J Invest Dermatol Symp Proc 4:55–60

Scholzen TE, Steinhoff M, Bonaccorsi P, Klein R, Amadesi S, Geppetti P, Lu B, Gerard NP, Olerud JE, Luger TA, Bunnett NW, Grady EF, Armstrong CA, Ansel JC (2001) Neutral endopeptidase terminates substance P-induced inflammation in allergic contact dermatitis. J Immunol 166:1285–1291

Schulze E, Witt M, Fink T, Hofer A, Funk RH (1997) Immunohistochemical detection of human skin nerve fibers. Acta Histochem 99:301–309

Scicchitano R, Biennenstock J, Stanisz AM (1988) *In vivo* immunomodulation by the neuropeptide substance P. Immunology 63:733–735

Severini C, Improta G, Falconieri-Erspamer G, Salvadori S, Erspamer V (2002) The tachykinin peptide family. Pharmacol Rev 54:285–322

Seville RH (1989) Stress and psoriasis: the importance of insight and empathy in prognosis. J Am Acad Dermatol 20:97–100

Simmaco M, Severini C, De Biase D, Barra D, Bossa F, Roberts JD, Melchiorri P, Erspamer V (1990) Six novel tachykinin- and bombesin-related peptides from the skin of the Australian frog Pseudophryne guntheri. Peptides 11:299–304

Siney L, Brain SD (1996) Involvement of sensory neuropeptides in the development of plasma extravasation in rat dorsal skin following thermal injury. Br J Pharmacol 117:1065–1070

Smith CH, Barker JN, Morris RW, Macdonald DM, Lee TH (1993) Neuropeptides induce rapid expression of endothelial cell adhesion molecules and elicit granulocytic infiltration in human skin. J Immunol 151:3274–3282

Song IS, Bunnett NW, Olerud JE, Harten B, Steinhoff M, Brown JR, Sung KJ, Armstrong CA, Ansel JC (2000) Substance P induction of murine keratinocyte PAM 212 interleukin 1 production is mediated by the neurokinin 2 receptor (NK_2R). Exp Dermatol 9:42–52

Stander S, Luger T, Metze D (2001) Treatment of prurigo nodularis with topical capsaicin. J Am Acad Dermatol 44:471–478

Staniek V, Liebich C, Vocks E, Odia SG, Doutremepuich JD, Ring J, Claudy A, Schmitt D, Misery L (1998) Modulation of cutaneous SP receptors in atopic dermatitis after UVA irradiation. Acta Derm Venereol 78:92–94

Staniek V, Misery L, Peguet-Navarro J, Abello J, Doutremepuich JD, Claudy A, Schmitt D (1997) Binding and *in vitro* modulation of human epidermal Langerhans cell functions by substance P. Arch Dermatol Res 289:285–291

Stanisz AM, Scicchitano R, Dazin P, Bienenstock J, Payan DG (1987) Distribution of substance P receptors on murine spleen and Peyer's patch T and B cells. J Immunol 139:749–754

Steinhoff M, Vergnolle N, Young SH, Tognetto M, Amadesi S, Ennes HS, Trevisani M, Hollenberg MD, Wallace JL, Caughey GH, Mitchell SE, Williams LM, Geppetti P, Mayer EA, Bunnett NW (2000) Agonists of proteinase-activated receptor 2 induce inflammation by a neurogenic mechanism. Nat Med 6:151–158

Sterner-Kock A, Braun RK, van d, V, Schrenzel MD, McDonald RJ, Kabbur MB, Vulliet PR, Hyde DM (1999) Substance P primes the formation of hydrogen peroxide and nitric oxide in human neutrophils. J Leukoc Biol 65:834–840

Sugiura H, Omoto M, Hirota Y, Danno K, Uehara M (1997) Density and fine structure of peripheral nerves in various skin lesions of atopic dermatitis. Arch Dermatol Res 289:125–131

Szallasi A (2002) Vanilloid (capsaicin) receptors in health and disease. Am J Clin Pathol 118:110–121

Tanaka T, Danno K, Ikai K, Imamura S (1988) Effects of substance P and substance K on the growth of cultured keratinocytes. J Invest Dermatol 90:399–401

Tarng DC, Cho YL, Liu HN, Huang TP (1996) Hemodialysis-related pruritus: a double-blind, placebo-controlled, crossover study of capsaicin 0.025% cream. Nephron 72:617–622

Tausk F, Undem B (1995) Exogenous but not endogenous substance P releases histamine from isolated human skin fragments. Neuropeptides 29:351–355

Thomsen JS, Sonne M, Benfeldt E, Jensen SB, Serup J, Menne T (2002) Experimental itch in sodium lauryl sulphate-inflamed and normal skin in humans: a randomized, double-blind, placebo-controlled study of histamine and other inducers of itch. Br J Dermatol 146:792–800

Tomomasa T, Yagi H, Kimura S, Snape WJ, Hyman PE (1989) Developmental changes in agonist-mediated gastric smooth muscle contraction in the rabbit. Pediatr.Res 26:458–461

Toyoda M, Makino T, Kagoura M, Morohashi M (2000) Immunolocalization of substance P in human skin mast cells. Arch Dermatol Res 292:418–421

Toyoda M, Makino T, Kagoura M, Morohashi M (2001) Expression of neuropeptide-degrading enzymes in alopecia areata: an immunohistochemical study. Br J Dermatol 144:46–54

Toyoda M, Morohashi M (2001) Pathogenesis of acne. Med Electron Microsc 34:29–40

Toyoda M, Nakamura M, Makino T, Hino T, Kagoura M, Morohashi M (2002a) Nerve growth factor and substance P are useful plasma markers of disease activity in atopic dermatitis. Br J Dermatol 147:71–79

Toyoda M, Nakamura M, Makino T, Kagoura M, Morohashi M (2002b) Sebaceous glands in acne patients express high levels of neutral endopeptidase. Exp Dermatol 11:241–247

Tripp RA, Barskey A, Goss L, Anderson LJ (2002) Substance P receptor expression on lymphocytes is associated with the immune response to respiratory syncytial virus infection. J Neuroimmunol 129:141–153

Twycross R, Greaves MW, Handwerker H, Jones EA, Libretto SE, Szepietowski JC, Zylicz Z (2003) Itch: scratching more than the surface. QJM 96:7–26

Vaalasti A, Suomalainen H, Rechardt L (1989) Calcitonin gene-related peptide immunoreactivity in prurigo nodularis: a comparative study with neurodermatitis circumscripta. Br J Dermatol 120:619–623

Viac J, Goujon C, Misery L, Staniek V, Faure M, Schmitt D, Claudy A (1997) Effect of UVB 311 nm irradiation on normal human skin. Photodermatol Photoimmunol Photomed 13:103–108

Wallengren J (1991) Substance P antagonist inhibits immediate and delayed type cutaneous hypersensitivity reactions. Br J Dermatol 124:324–328

Wallengren J, Moller H (1988) Some neuropeptides as modulators of experimental contact allergy. Contact Dermatitis 19:351–354

Wallengren J, Moller H, Ekman R (1987) Occurrence of substance P, vasoactive intestinal peptide, and calcitonin gene-related peptide in dermographism and cold urticaria. Arch Dermatol Res 279:512–515

Walsh DT, Weg VB, Williams TJ, Nourshargh S (1995) Substance P-induced inflammatory responses in guinea-pig skin: the effect of specific NK_1 receptor antagonists and the role of endogenous mediators. Br J Pharmacol 114:1343–1350

Weidner C, Klede M, Rukwied R, Lischetzki G, Neisius U, Skov PS, Petersen LJ, Schmelz M (2000) Acute effects of substance P and calcitonin gene-related peptide in human skin—a microdialysis study. J Invest Dermatol 115:1015–1020

Whittle BJ, Lopez-Belmonte J, Rees DD (1989) Modulation of the vasodepressor actions of acetylcholine, bradykinin, substance P and endothelin in the rat by a specific inhibitor of nitric oxide formation. Br J Pharmacol 98:646–652

Wozniak A, McLennan G, Betts WH, Murphy GA, Scicchitano R (1989) Activation of human neutrophils by substance P: effect on FMLP-stimulated oxidative and arachidonic acid metabolism and on antibody-dependent cell-mediated cytotoxicity. Immunology 68:359–364

Xu XJ, Dalsgaard CJ, Maggi CA, Wiesenfeld-Hallin Z (1992) NK_1, but not NK_2, tachykinin receptors mediate plasma extravasation induced by antidromic C-fiber stimulation in rat hindpaw: demonstrated with the NK_1 antagonist CP-96,345 and the NK_2 antagonist Men 10207. Neurosci Lett 139:249–252

Xu Y, Wellner D, Scheinberg DA (1995) Substance P and bradykinin are natural inhibitors of CD13/aminopeptidase N. Biochem Biophys Res Commun 208:664–674

Yano H, Wershil BK, Arizono N, Galli SJ (1989) Substance P-induced augmentation of cutaneous vascular permeability and granulocyte infiltration in mice is mast cell dependent. J Clin Invest 84:1276–1286

Yonehara N, Shibutani T, Inoki R (1987) Contribution of substance P to heat-induced edema in rat paw. J Pharmacol Exp Ther 242:1071–1076

Yonehara N, Takiuchi S (1997) Involvement of calcium-activated potassium channels in the inhibitory prejunctional effect of morphine on peripheral sensory nerves. Regul Pept 68:147–153

Yonehara N, Yoshimura M (2000) Interaction between nitric oxide and substance P on heat-induced inflammation in rat paw. Neurosci Res 36:35–43

Ziche M, Morbidelli L, Pacini M, Geppetti P, Alessandri G, Maggi CA (1990) Substance P stimulates neovascularization *in vivo* and proliferation of cultured endothelial cells. Microvasc Res 40:264–278

Zimmerman BJ, Anderson DC, Granger DN (1992) Neuropeptides promote neutrophil adherence to endothelial cell monolayers. Am J Physiol 263:G678–G682

Role of Tachykinins in Obstructive Airway Disease

G. F. Joos

Department of Respiratory Diseases, Ghent University Hospital, De Pintelaan 185, 9000 Ghent, Belgium
e-mail: guy.joos@ugent.be

1	Introduction	492
2	Tachykinins in Human Airways	493
3	Airway Tachykinin Receptors	495
4	Bronchoconstrictor Effects of Tachykinins	496
5	Airway Plasma Extravasation	498
6	Tachykinins and Allergen-Induced Airway Responses	499
7	Involvement of Tachykinins in Animal Models of Asthma and COPD	500
8	Plasticity of Tachykinergic Airway Innervation	500
9	Tachykinin Receptor Antagonists as a Potential New Treatment for Obstructive Airway Diseases	501
10	Conclusions and Future Perspectives	503
	References	504

Abstract Asthma and chronic obstructive pulmonary disease (COPD) are characterized by airway obstruction, hyperresponsiveness and inflammation. The tachykinins substance P and neurokinin A are present in the human airways where they are found in sensory nerves and immune cells. Tachykinins are released in the airways after inhalation of ozone, cigarette smoke or allergen. They interact in the airways with tachykinin NK_1, NK_2 and NK_3 receptors to cause bronchoconstriction, plasma protein extravasation, and mucus secretion and to attract and activate immune cells. In preclinical studies they have been implicated in the pathophysiology of asthma and COPD, including allergen-and cigarette smoke-induced airway inflammation and bronchial hyperresponsiveness. So, tachykinin receptor antagonists have potential in the treatment of asthma and COPD. With the arrival of more potent dual and triple tachykinin receptor antagonists the tools to obtain a more definite judgment about the role of tachykinins in airway diseases such as asthma and COPD are now available.

Keywords Tachykinins · Asthma · COPD · Airway inflammation · Tachykinin receptors

1
Introduction

Asthma and chronic obstructive pulmonary disease (COPD) are two major respiratory diseases characterized by airways obstruction, airway inflammation and bronchial hyperresponsiveness (Bousquet 2000; Pauwels et al. 2001). The Global Initiative for Asthma (GINA) uses the following operational description of asthma: " Asthma is a chronic inflammatory disorder of the airways in which many cells and cellular elements play a role. The chronic inflammation causes an associated increase in airway hyperresponsiveness that leads to recurrent episodes of wheezing, breathlessness, chest tightness, and coughing, particularly at night or in the early morning. These episodes are usually associated with widespread but variable airflow obstruction that is often reversible either spontaneously or with treatment" (Fig. 1). According to the Global Initiative for Chronic Obstructive Lung Disease (GOLD), COPD is " a disease state characterized by airflow limitation that is not fully reversible. The airflow limitation is usually both progressive and associated with an abnormal response of the lungs to noxious particles or gases". Figure 2 summarizes the current thinking on the pathogenesis of COPD.

At the present time bronchodilators and inhaled corticosteroids form the cornerstone of the pharmacological treatment for these two diseases. The tachykinins substance P and neurokinin A have various effects that could contribute to the pathophysiological changes observed in asthma and COPD. Hence, interfering with the action of tachykinins on the various targets in the airways is one of the approaches that is considered for the development of new treatments for asthma and COPD (Joos and Pauwels 2000).

Substance P and neurokinin A are members of the tachykinin peptide family which consists of mammalian and non-mammalian tachykinins (Severini et al. 2002). These neuropeptides are synthesized by sensory neurons and subsequently stored in the terminal parts of the axon collaterals. The tachykinins are

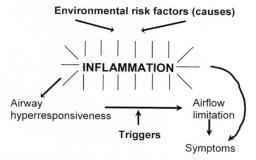

Fig. 1 Current concept of the pathogenesis of asthma. Airway inflammation, hyperresponsiveness and obstruction are essential components (see International Guidelines on the Diagnosis and Treatment of Asthma—www.ginasthma.com)

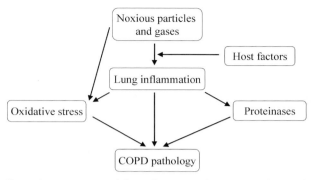

Fig. 2 Current concept of the pathogenesis of COPD (according to the GOLD guidelines—www.gold-copd.com)

potent vasodilators and contractors of smooth muscle. In studies on rodent airways substance P and neurokinin A have been implicated as the neurotransmitters mediating the excitatory part of the non-adrenergic non-cholinergic (NANC) nervous system. Non-cholinergic excitatory nerves can be activated by mechanical and chemical stimuli, generating antidromic impulses and a local axon reflex which leads to non-cholinergic bronchoconstriction and neurogenic inflammation (Barnes 1986).

Substance P and neurokinin A have various effects that could contribute to the changes observed in asthma and COPD. These include smooth muscle contraction, submucosal gland secretion, vasodilatation, increase in vascular permeability, stimulation of cholinergic nerves, stimulation of mast cells, stimulation of B and T lymphocytes, stimulation of macrophages, chemo-attraction of eosinophils and neutrophils and the vascular adhesion of neutrophils (Maggi 1997; Joos et al. 2000). Substance P and neurokinin A interact with the different targets in the airways by stimulation of tachykinin NK_1, NK_2 and NK_3 receptors (Maggi 2000). The physiological activity of both exogenously administered and endogenously released neuropeptides is modulated through enzymatic cleavage and inactivation by peptidases (Di Maria et al. 1998). In this chapter the evidence in favour of a role of tachykinins and their receptors in the pathogenesis of asthma and COPD is reviewed, and the potential of tachykinin receptor antagonists as a potential treatment for these diseases is discussed.

2
Tachykinins in Human Airways

SP and neurokinin A are contained in a distinct subpopulation of primary afferent nerves that are characterized by sensitivity to capsaicin. Substance P and neurokinin A are present in nerve profiles, found beneath and within the epithelium, around blood vessels and submucosal glands and within the bronchial smooth muscle layer (Lundberg et al. 1984; Luts et al.1993). Using confocal laser

Table 1 Concentrations of substance P in human airways (modified from Joos et al. 2000)

	Healthy controls	Asthma	COPD	References
Trachea (pmol/g)	13.0 (5.0)	3.0 (3.0)		Lilly et al. 1995
Induced sputum (fmol/ml)	1.1 (0.4)	17.7 (2.4)	25.6 (5.5)	Tomaki et al. 1995
BAL (fmol/ml)	30.0 (7.0); further increase after ozone exposition	185.0 (22.0); further increase after allergen instillation		Nieber et al. 1992; Hazbun et al. 1993

BAL, Bronchoalveolar lavage.

scanning microscopy and immunohistochemistry for the protein gene product 9.5 (PGP-IR), Lamb and Sparrow (2002) demonstrated an arrangement of epithelial nerves in human bronchi, which was very similar to that of porcine nerves containing sensory neuropeptides. An apical layer of varicose processes terminated in enlarged varicosities, and these fibres encircled goblet cells. These processes arose from fibres that had crossed the epithelium from a basal plexus that was supplied by nerve bundles in the lamina propria. As in other studies substance P-immunoreactive nerves were faintly stained in human bronchi.

Tachykinins have been measured in bronchoalveolar lavage (BAL), induced sputum and plasma (Table 1). Various studies have shown that tachykinins can be released into the airways after exposure to allergen, ozone or hypertonic saline. Nieber et al. (1992) found a significantly larger amount of substance P in BAL fluid of atopic compared to non-allergic subjects. After intrasegmental provocation with allergen, a significant increase in BAL substance P levels was observed. Heaney et al. (1998) also recovered neurokinin A from BAL fluid of normal and asthmatic patients. Neurokinin A was increased in the asthmatic patients 4 h after inhalation challenge with house dust mite. Exposure of healthy individuals to ozone increased the concentration of substance P in BAL fluid (Hazbun et al. 1993) and decreased the immunoreactivity of substance P assessed on bronchial biopsies (Krishna et al. 1997). Nasal application of hypertonic saline also induced release of substance P (Baraniuk et al. 1999).

Although increased amounts of tachykinins have been recovered from airways, there has been a debate about a possible up-regulation of substance P-containing nerves in patients with asthma or COPD. Ollerenshaw et al. (1991) reported that in tissue obtained at autopsy, after lobectomy and at bronchoscopy, both the number and the length of substance P-immunoreactive nerve fibres was increased in airways of subjects with asthma, when compared to airways from subjects without asthma. However, Howarth et al. (1995) could not identify any substance P-containing nerves in endobronchial biopsies from patients with mild asthma. In a study on 49 asthmatic patients, including 16 patients with severe asthma, Chanez et al. (1998) did not find evidence for an up-regulation of substance P-containing nerves in asthma. Nerves were present in most of the endobronchial biopsies, as demonstrated by the neural marker PGP 9.5, and were found within and below the epithelium and adjacent to smooth muscle, glands and blood vessels. However, nerves positive for substance P and calcitonin gene-related peptide (CGRP) were rarely found in the biopsy specimens. Moreover, no increase in peptide immunoreactivity was observed in patients with asthma. On the other hand, substance P-like immunoreactivity was decreased in tracheal tissue of asthmatic subjects studied at autopsy. This may reflect augmented release of substance P, followed by degradation (Lilly et al. 1995). Lucchini et al. (1997) studied lung tissue obtained at thoracotomy from patients with chronic bronchitis. The lung tissue in chronic bronchitis did not exhibit a larger number of nerves and did not show a change in the amount of substance P- or CGRP-containing nerves. In contrast, the density of vasoactive intestinal peptide (VIP)-containing nerves was significantly higher in the glands of patients with chronic bronchitis compared with those of control subjects.

In recent years, it has become clear from both animal and human studies that immune cells may form an additional source of tachykinins (Maggi 1997; Joos and Pauwels 2000). Evidence for the production of substance P by eosinophils (mouse, man), monocytes and macrophages (rat, man), lymphocytes (man) and dendritic cells (mice) has been reported. Inflammatory stimuli such as lipopolysaccharide (LPS) can up-regulate the concentration of tachykinins in these cells (Germonpre et al.1999; Lambrecht et al. 1999). These findings can help to explain the paradox between the relatively few SP- and neurokinin A-containing nerves observed in human airways and the increased recovery of SP and neurokinin A from sputum and BAL (see above). Thus, increased amounts of immune cells attracted to the inflamed airways might be responsible for an increased content of substance P and neurokinin A in patients with asthma or COPD. Moreover, the release of tachykinins from these inflammatory cells might further stimulate and activate these cells in an autocrine or paracrine fashion. Indeed, these inflammatory cells produce and secrete substance P and possess tachykinin NK_1 receptors on their membrane (Germonpre et al. 1999).

Epithelial or endothelial cells might be an additional non-neuronal source of substance P (Maggi 1997). Chu et al. (2000) reported staining for substance P in the epithelium on human biopsy specimens. In comparison to control subjects, patients with asthma showed a significantly higher expression of substance P in the epithelium but not in the submucosa. It is of interest to note that in their study the epithelial expression of substance P was correlated with the epithelial mucus content.

3
Airway Tachykinin Receptors

Most of the biological actions of tachykinins are mediated via activation of one of three tachykinin receptors, denoted NK_1, NK_2 and NK_3. The NK_1, NK_2 and NK_3 tachykinin receptor have the highest affinity for substance P, neurokinin A and neurokinin B, respectively. There is, however, a wide cross-talk between natural tachykinin peptides and their receptors: for instance it is presently accepted that neurokinin A is a high-affinity and effective endogenous ligand for NK_1 receptors at many synapses and/or neuroeffector junctions (Maggi 2000).

The tachykinins SP and neurokinin A interact with the targets on the airways by specific tachykinin receptors. The tachykinin NK_1 and NK_2 receptor have been characterized in human airways, both pharmacologically and by cloning. In in vitro studies on isolated human bronchi and bronchioli, tachykinin NK_2 receptors were found to be present on smooth muscle of both large and small airways and to mediate part of the bronchoconstrictor effect of the tachykinins (Advenier et al. 1999). Tachykinin NK_1 receptors are localized on smooth muscle of small airways and are responsible for a transient, low-intensity contraction (Naline et al. 1996). Most of the proinflammatory effects of substance P are mediated by the tachykinin NK_1 receptor (reviewed in Advenier et al. 1997). The tachykinin NK_3 receptor has been demonstrated in guinea pig airways, where it mediates citric acid-induced cough and changes in airway responsiveness induced by substance P and citric acid (Daoui et al. 1998, 2000). Moreover, the tachykinin NK_3 receptor mediates the neurokinin-induced facilitation of synaptic neurotransmission in parasympathetic ganglia (Canning et al. 2002).

Various studies have now explored possible changes in the expression of airway tachykinin receptors as consequence of inflammatory changes in the airways. Smoking increases the tachykinin NK_1 receptor mRNA (Adcock et al. 1993). In asthmatics the expression of the tachykinin NK_2 receptor mRNA is enhanced (Bai et al. 1995). Glucocorticoids can reduce the level of tachykinin receptor mRNA (Katsunuma et al. 1998). Using antibodies to the tachykinin NK_1 and NK_2 receptors, Mapp et al. (2000) found expression of both tachykinin receptors in bronchial glands, bronchial vessels, and bronchial smooth muscle. Receptors were occasionally found in nerves (NK_1) and in inflammatory cells (NK_2) such as T lymphocytes, macrophages, and mast cells. The distribution of both tachykinin NK_1 and NK_2 receptors was similar in the tissues examined from non-smokers, asymptomatic smokers, symptomatic smokers with normal lung function and symptomatic smokers with chronic airflow limitation. In a study on endobronchial biopsies by Chu et al. (2000) immunoreactivity for the tachykinin NK_1 receptor was found in the epithelium and submucosa. The NK_1 receptor expression was mainly seen on cell surfaces of the upper half of the epithelial layer. Goblet cells appeared to be the cells with the strongest staining. In the submucosa the tachykinin NK_1 receptor was localized primarily on the endothelial cells of the blood vessels, the surfaces of inflammatory cells, and some smooth muscle cells. In this study the expression of substance P as well as the tachykinin NK_1 receptor was significantly higher in the epithelium of asthmatic subjects.

4
Bronchoconstrictor Effects of Tachykinins

Tachykinins are potent contractors of airways. The in vitro contractile effect of substance P and neurokinin A has been studied extensively. Substance P contracts human bronchi and bronchioli (Finney et al.1985; Advenier et al. 1987), but is less potent than histamine or acetylcholine (Martling et al. 1987). Neurokinin A is a more potent constrictor of human bronchi than substance P and was reported to be, on a molar base, two to three orders of magnitude more potent than histamine or acetylcholine.

It has long been thought that only tachykinin NK_2 receptors are involved in contraction of isolated human airways (Advenier et al. 1992). However, in small-diameter bronchi (~1 mm in diameter) tachykinins also cause contraction via NK_1 tachykinin receptor stimulation. The NK_1 receptor-mediated contraction of small bronchi appears to be mediated by prostanoids (Naline et al. 1996). In a study on medium-sized human isolated bronchi (2–5 mm in diameter) the specific NK_1 receptor agonist [Sar^9,$Met(O_2)^{11}$]SP was found to induce contraction in about 60% of the preparations, an effect that was not mediated by prostanoids but resulted from a direct activation of smooth muscle receptors and release of inositol phosphate (Amadesi et al. 2001). So, part of the airway contraction induced by tachykinins in man is mediated by tachykinin NK_1 receptors.

Several groups have studied the in vivo bronchoconstrictor effect of substance P and neurokinin A, administered via inhalation or intravenous infusion. Neurokinin A was found to be a more potent bronchoconstrictor than substance P. Patients with asthma are hyper-responsive to substance P and neurokinin A (reviewed in Joos et al. 1994). The bronchoconstrictor effect of inhaled substance P and neurokinin A in asthmatics can be prevented by pretreatment with sodium cromoglycate and nedocromil sodium (Crimi et al.1988; Joos et al. 1989). This has led us to postulate that substance P and neurokinin A are indirect broncoconstrictors in man. This indirect bronchoconstrictor effect could arise from an effect on inflammatory cells (e.g. mast cells) and/or nerves (Van Schoor et al. 2000). In experimental animals, tachykinins are able to cause acetylcholine release from postganglionic cholinergic airway nerve endings (Szarek et al. 1993; Tanaka and Grunstein 1984; Tournoy et al. 2003) and to activate mast cells (Fewtrell et al. 1982; Joos and Pauwels 1993; Joos et al. 1997; van der Kleij al. 2003). The exact mechanism of tachykinin-induced bronchoconstriction in man is, however, not yet known. In some patients cholinergic mechanisms do play a role (Crimi et al. 1990; Joos et al. 1988). Histamine, however, does not seem to be involved as pretreatment with different H_1 receptor antagonists (astemizole and terfenadine) did not inhibit neurokinin A-induced bronchoconstriction (Crimi et al. 1990, 1993). It was recently demonstrated that leukotrienes mediate part of the bronchoconstrictor effect of neurokinin A: the cysLT1 receptor antagonists zafirlukast and montelukast partially inhibited bronchoconstriction induced by inhaled neurokinin A in patients with asthma (Joos et al. 2001; Crimi et al. 2003). So it would seem that neurokinin A causes bronchoconstriction in patients with asthma by indirect mechanisms, involving mast cells, leukotrienes and cholinergic nerves (Joos et al. 2003).

5
Airway Plasma Extravasation

Ablation of sensory nerves by capsaicin pretreatment abolishes airway plasma extravasation induced by a variety of mediators and stimuli, including histamine, serotonin and cigarette smoke (Lundberg and Saria 1983; Saria et al. 1983). Tachykinin NK_1 receptors that mediate neurogenic plasma extravasation have been visualized on postcapillary endothelial cells (Bowden et al. 1994). Once plasma extravasation occurs, leukocytes initiate a process that results in slowing down their velocity, rolling on and adhering to the venular endothelium. The involvement of tachykinin NK_1 receptors in the neurogenic plasma extravasation in the central airways of rats and guinea pigs has been demonstrated by the use of selective tachykinin receptor agonists and antagonists. The tachykinin NK_1 receptors are also implicated in the extravasation caused by hypertonic saline, bradykinin, antigen challenge or acetylcholine. Tachykinin NK_2 receptors mediate part of the neurogenic plasma extravasation in the secondary bronchi and intraparenchymal airways of the guinea pig (Bertrand et al.1993b; Tousignant et al. 1993).

Bowden et al. (1994) have demonstrated that in the rat trachea tachykinin NK_1 receptors are present on endothelial cells of postcapillary venules. These receptors are internalized in endosomes upon binding with substance P. They also observed an increase in the number of tachykinin NK_1 receptor immunoreactive endosomes when the vagus nerve is electrically stimulated, indicating that substance P released by activation of the sensory nerves has a direct effect on the tachykinin NK_1 receptors on postcapillary venular endothelium. In addition to the direct effects of tachykinins on the venular endothelium, indirect mechanisms involving mast cell activation and serotonin (5-HT) release participate in the tachykinin-induced plasma exudation in the respiratory tract of some animal species. In rabbit airways, involvement of 5-HT receptors and arachidonic acid derivates in the tachykinin-induced increase in vascular permeability has been suggested (Delaunois et al. 1993a, 1993b). Neurogenic plasma protein extravasation in rat airways involves the activation of tachykinin NK_1 receptors. In F344 but not in BDE rats, an additional indirect mechanism, involving mast cell activation, 5-HT release and stimulation of $5-HT_2$ receptors, participates in this process (Germonpre et al. 1995, 1997).

Microvascular leakage can now be measured in human airways. A particular interesting approach is the 'dual induction' model: first substance P is inhaled and then sputum is induced by inhalation of hypertonic saline. In this way, Van Rensen et al. (2002) demonstrated that inhalation of substance P induced significant increases in the levels of alpha2-macroglobulin, ceruloplasmin and albumin in induced sputum.

6
Tachykinins and Allergen-Induced Airway Responses

Tachykinins have been found to be involved in antigen-induced bronchoconstriction, airway inflammation and enhanced bronchial responsiveness in various animal models. A combination of a tachykinin NK_1 receptor antagonist, CP96345 (Snider et al. 1991), and a tachykinin NK_2 receptor antagonist, SR48968 (Emonds-Alt et al. 1992), inhibited bronchoconstriction produced by ovalbumin challenge in sensitized guinea pigs (Bertrand et al. 1993c). The NK_1 receptor antagonist CP96345 was also able to limit antigen-induced plasma extravasation (Bertrand et al. 1993a). By using NK_1/NK_2 receptor blockade the involvement of endogenous tachykinins in antigen-induced bronchial hyperresponsiveness was demonstrated (Kudlacz et al. 1996).

The relative contribution of the NK_1 and NK_2 receptor in antigen-induced airway changes has been studied in guinea pigs and rats. In conscious, unrestrained guinea pigs the NK_1 receptor is involved in both the development of antigen-induced airway hyperresponsiveness to histamine and the antigen-induced infiltration of eosinophils, neutrophils and lymphocytes (Schuiling et al. 1999b). On the other hand, the NK_2 receptor is involved in the development of the antigen-induced late reaction (Schuiling et al. 1999a). The effect of allergen challenge has also been studied in the Brown Norway (BN) rat model (Maghni et al. 2000). Substance P was found to increase 2.4-fold in BAL after challenge with ovalbumin. The tachykinin NK_1 receptor antagonist CP99994 and the tachykinin NK_2 receptor antagonist SR48968 were not able to reduce the early airway response to ovalbumin, but both antagonists reduced the ovalbumin-induced late airway responses. An interesting finding in this study was that the NK_2 tachykinin receptor antagonist decreased the number of eosinophils in BAL fluid and decreased the expression of both Th1 (interferon-γ) and Th2 [interleukin (IL)-4 and IL-5] cytokines in BAL cells.

Thus, from animal studies it would appear that both the tachykinin NK_1 and the NK_2 receptor are involved in antigen-induced airway effects. In addition, it is possible that tachykinin NK_3 receptors are implicated in this process: administration of the tachykinin NK_3 receptor antagonist SR142801 via aerosol caused a significant reduction in neutrophil and eosinophil influx in the airways of ovalbumin sensitized and challenged mice (Nénan et al. 2001).

At present it is not known whether tachykinins and their receptors play a role in allergen-induced changes in human airways. An answer to this question could be provided by studying the effect of a dual NK_1/NK_2 or triple $NK_1/NK_2/NK_3$ tachykinin receptor antagonist on allergen challenge in patients with asthma. In this clinical model inhalation of an allergen induces an early and in about half of the patients a late asthmatic reaction that is accompanied by an increase in sputum eosinophilia and an increase in bronchial responsiveness to histamine or methacholine (Hansel et al. 2002).

7
Involvement of Tachykinins in Animal Models of Asthma and COPD

In animal models tachykinins and their receptors have been involved in airway responses to non-specific stimuli. Both tachykinin NK_1 and NK_2 receptors have been involved in airway contraction induced by cold-air (Yoshihara et al. 1996; Yang et al. 1997), hyperventilation and cigarette smoke (Wu and Lee 1999), in plasma extravasation induced by hypertonic saline (Piedimonte et al. 1993; Pedersen et al. 1998), and in airway hyperresponsiveness induced by viruses (Piedimonte et al. 1999), IL-5 and nerve growth factor (Kraneveld et al. 1997; De Vries et al. 1999). In addition, the tachykinin NK_3 receptor was found to be involved in citric acid-induced cough and enhanced bronchial responsiveness (Daoui et al. 1998).

Neurogenic mucus secretion results from the contribution of different components and shows marked variation among species. Adrenergic and cholinergic agonists may stimulate mucus secretion in the ferret and human airways. Tachykinins cause also marked mucus secretion in the ferret, an effect exclusively mediated by NK_1 receptors. Endogenous tachykinins mediate mucus secretion induced by electrical stimuli and part of the response produced by exposure to antigen via NK_1 receptor activation (Khan et al. 2001).

8
Plasticity of Tachykinergic Airway Innervation

Allergic airway inflammation may increase the amount of substance P and neurokinin A in the airways, by increasing the number of nerves, and/or by increasing the non-neuronal sources of these tachykinins. In a guinea pig model a three- to fourfold increase of substance P, neurokinin A and calcitonin gene-related peptide measured in lung tissue was seen 24 h following antigen challenge. Moreover, increased levels of preprotachykinin (PPT)-I mRNA were found in the nodose ganglia of these animals (Fischer et al. 1996). In the guinea pig, both allergic airway inflammation and viral respiratory infection lead to a qualitative change in the vagal afferent innnervation such that both small diameter nociceptive-like neurons and large diameter non-nociceptive neurons express tachykinins (Carr et al. 2002).

One of the factors known to increase the expression of substance P is nerve growth factor (NGF) which belongs to the family of neurotrophins. Neurotrophins control the survival, differentiation and maintenance of neurons in the peripheral and central nervous system. Under physiological conditions, neurotrophins are produced by nerve-associated cells like glia cells or Schwann cells and by nerve cells themselves, while during inflammation neurotrophins can also be produced by fibroblasts, mast cells, macrophages, and T and B cells (for review see Bonini et al. 1999; Braun et al. 2000).

Elevated levels of NGF have been reported in asthma (Bonini et al. 1996). In a study reported by Höglund et al. (2002), an intense NGF-like immunoreactivity was observed in bronchial epithelium, smooth muscle cells and infiltrating inflammatory cells in the submucosa, and to a lesser extent in the connective tissue. In patients with asthma a higher number of NGF-immunoreactive infiltrating cells in the bronchial mucosa was observed compared to control subjects. So, NGF is produced by both structural cells and infiltrating inflammatory cells in the human bronchus in vivo. Moreover, allergens are able to enhance the expression of NGF in the airways: in a study performed by Virchow et al. (1998) a significant increase in the neurotrophins NGF, brain-derived neurotrophic factor, and neurotrophin-3 was observed in BAL fluid obtained 18 h after segmental allergen provocation in patients with asthma.

The role of NGF and its receptors has been studied in transgenic mice, overexpressing NGF, and in mice that were either knockout for the pan-neurotrophin receptor p75, or were treated with antibodies to NGF. Mice with NGF over-expression from a lung-specific promoter were shown to have an increased number of tachykinin-containing sensory nerve fibres in the airways (Hoyle et al. 1998; Graham et al. 2001). Compared to wild-type mice, mice over-expressing NGF exhibited a twofold increase in the number of lung lavage neutrophil levels in response to ozone inhalation (Graham et al. 2001). When NGF is injected in the tracheal wall of guinea pigs, the substance P expression in airway neurons is enhanced. Moreover, a phenotypic switch in the nature of vagal sensory neurons producing this neuropeptide was observed: large, capsaicin-insensitive nodose neurons become substance P positive (Hunter et al. 2000).

One of the factors that might up-regulate NGF expression in human airways is inflammation. Indeed, incubation of isolated human airway smooth muscle cells with the pro-inflammatory cytokine interleukin-1 beta increased NGF production in a dose-dependent manner (Freund et al. 2002). Another factor that might be of importance is a respiratory infection. In a study on rat airways, infection with respiratory syncytial virus (RSV) doubled the expression of NGF and its receptors, the high affinity tyrosine kinase receptor trkA, and the low-affinity receptor p75 (Hu et al. 2002).

9
Tachykinin Receptor Antagonists as a Potential New Treatment for Obstructive Airway Diseases

Studies on rodents have demonstrated that a number of strategies are possible to interfere with the action of sensory neuropeptides in the airways: (1) depletion of neuropeptides within nerves (e.g. by the neurotoxin capsaicin); (2) inhibition of the release of sensory neuropeptides [e.g. by β_2-adrenoceptor agonists, theophylline, cromoglycate or phophodiesterase (PDE4) inhibitors]; and (3) inhibition of tachykinin receptors by receptor antagonists (Lundberg 1996). In animal models tachykinins and their NK_1 and NK_2 receptors have been involved in airway responses to non-specific stimuli (Joos et al. 2001). The tachykinin NK_1 receptor also promotes ozone-induced lung inflammation (Graham et al. 2001). Viral infection causes an up-regulation of the tachykinin NK_1 receptor, leading to enhanced neurogenic inflammation (Piedimonte 2001). In recent years, the tachykinin NK_3 receptor has been implicated in citric acid-induced cough and enhanced bronchial responsiveness (see above).

At the present time clinical trials with at least seven tachykinin receptor antagonists have been performed and reported. Reports studying the effects of FK224, a cyclic peptide antagonist for NK_1 and NK_2 receptors, CP99994, a non-peptide NK_1 receptor antagonist, FK888, a peptidic NK_1 receptor antagonist, and SR48968 (saredutant), a non-peptide NK_2 receptor antagonist have been reported. In our centre we recently completed three additional studies, one with the bicyclic peptidic NK_2 receptor antagonist MEN11420 (nepadutant), one with the dual NK_1/NK_2 tachykinin receptor antagonist DNK333A, and one with the triple $NK_1/NK_2/NK_3$ receptor antagonist CS003.

FK244, 4 mg given by a metered dose inhaler, was shown to inhibit bradykinin-induced bronchoconstriction and cough in nine asthmatics (Ichinose et al. 1992). These findings suggested that bradykinin causes bronchoconstriction by release of tachykinins from sensory nerves. However, in a similar study, using a lower dose, FK224 had no effect on bradykinin-induced bronchoconstriction (Schmidt et al. 1996). Moreover, we have studied the effect of inhaled FK224 on neurokinin A-induced bronchoconstriction in ten patients with mild asthma(Joos et al. 1996). Inhaled FK224 had no effect on baseline lung function and offered no protection against neurokinin A-induced bronchoconstriction.

Fahy et al. (1995) studied the effect of the non-peptide NK_1 receptor antagonist CP99994 on hypertonic saline-induced bronchoconstriction in 14 mild asthmatics. Compared to placebo, CP99994 had no effect on the bronchoconstriction and cough induced by hypertonic saline. However it remains to be proven whether this compound at the given dose, is able to block the airway responses induced by inhaled SP or neurokinin A.

The NK_1 receptor antagonist FK888 has been studied in exercise-induced asthma (Ichinose et al. 1996). Inhalation of FK888 in nine asthmatic subjects had no effect on baseline lung function. While pretreatment with FK888 did not attenuate the maximal decrease in lung function, it did reduce the time of recovery from exercise-induced bronchoconstriction.

The NK_2 receptor antagonist SR48968 (saredutant; 100 mg per os) caused a significant inhibition of neurokinin A-induced bronchoconstriction, with a mean three- to fivefold shift of the dose–response curve for neurokinin A to the right (Van Schoor et al. 1998). Administration of saredutant during 9 days (100 mg per os, each day) did, however, not affect baseline airway calibre or bronchial responsiveness to adenosine (Kraan et al. 2001). In a recently completed study we found that the NK_2 tachykinin receptor antagonist MEN11420 (nepadutant; 2 and 8 mg intravenously) shifted the dose–response curve for neurokinin A to the right, in a similar way as observed in the study with SR48968 (Joos et al. 2001). These studies are the first to demonstrate activity of a tachykinin NK_2 receptor antagonist in the airways of man. It is however clear that although these agents are potent antagonists in vitro (Advenier et al. 1992; Catalioto et al. 1998), their potential to inhibit the bronchoconstrictor effect of neurokinin A in patients with asthma is rather limited. This can be explained by an additional involvement of tachykinin NK_1 receptors in the bronchoconstrictor effect of neurokinin (Naline et al. 1996; Amadesi et al. 2001). This idea is supported by our recent findings with a dual NK_1/NK_2 receptor antagonist. In a study on 19 patients with asthma, DNK333A, a newly developed NK_1/NK_2 receptor tachykinin receptor antagonist, was able to shift the dose–response curve to inhaled neurokinin A to a large extent (4.8 doubling doses) (Joos et al. 2004). So, by inhibiting both the tachykinin NK_1 and the NK_2 receptor a better protection against the bronchoconstrictor effect of neurokinin A was observed.

10
Conclusions and Future Perspectives

Tachykinins are produced in human airways and mimic various features of asthma and COPD. So, tachykinin receptor antagonists have potential in the treatment of asthma and COPD. However, the development of tachykinin receptor antagonists for airway diseases has been rather slow and up to now somehow disappointing. This is in contrast with the extensive and overwhelming preclinical data suggesting a role for tachykinins in asthma and possibly COPD. There are several explanations for this apparent paradox (Joos and Pauwels 2001):

1. The lack of efficacy can be easily explained by the low potency or defective pharmacokinetics of the compounds tested so far. Potent tachykinin receptor antagonists have only recently become available and some of them have not been considered for application in airway diseases, but for depression or emesis (Diemunsch and Grelot 2000).
2. Blocking either the NK_1 or the NK_2 receptor is probably an insufficient approach, as most of the effects of tachykinins in the airways are mediated by more than one tachykinin receptor.
3. In the application of a new tachykinin receptor antagonist to airway diseases, it is crucial that one first demonstrates that the antagonist is indeed able to block airway effects of an agonist (e.g. substance P or neurokinin A). This allows the determination of the in vivo activity of the antagonist under consideration and the dose and dosing frequency for further clinical study.

So, in conclusion, the arrival of more potent, and low molecular weight single and dual or even triple tachykinin receptor antagonists (Rumsey et al. 2001; Anthes et al. 2002) should allow us to obtain a final judgment on the role of tachykinins in airway diseases such as asthma and COPD (Schelfhout et al. 2003).

References

Adcock IM, Peters M, Gelder C, Shirasaki H, Brown CR, Barnes PJ (1993) Increased tachykinin receptor gene expression in asthmatic lung and its modulation by steroids. J Mol Endocrinol 11:1–7

Advenier C, Joos G, Molimard M, Lagente V, Pauwels R (1999) Role of tachykinins as contractile agonists of human airways in asthma. Clin Exp Allergy 29:579–584

Advenier C, Lagente V, Boichot E (1997) The role of tachykinin receptor antagonists in the prevention of bronchial hyperresponsiveness, airway inflammation and cough. Eur Respir J 10:1892–906

Advenier C, Naline E, Drapeau G, Regoli D (1987) Relative potencies of neurokinins in guinea-pig trachea and human bronchus. Eur J Pharmacol 139:133–137

Advenier C, Naline E, Toty L, Bakdach H, Edmonds-Alt X, Vilain P, Breliere J-C, Le Fur G (1992) Effects on the isolated human bronchus of SR 48968, a potent and selective nonpeptide antagonist of the neurokinin A (NK_2) receptors. Am Rev Respir Dis 146:1177–1181

Amadesi S, Moreau J, Tognetto M, Springer J, Trevisani M, Naline E, Advenier C, Fischer A, Vinci D, Mapp CE, Miotto D, Geppetti P (2001) NK_1 receptor stimulation causes contraction and inositol phosphate increase in medium-size human isolated bronchi. Am J Respir Crit Care Med 163:1206–1211

Anthes JC, Chapman RW, Richard C, Eckel S, Corboz M, Hey JA, Fernandez X, Greenfeder S, McLeod R, Sehring S, Rizzo C, Crawley Y, Shih NY, Piwinski J, Reichard G, Ting P, Carruthers N, Cuss FM, Billah M, Kreutner W, Egan RW (2002) SCH 206272: a potent, orally active tachykinin NK_1, NK_2 and NK_3 receptor antagonist. Eur J Pharmacol 450:191–202

Bai TR, Zhou D, Weir T, Walker B, Hegele R, Hayashi S, McKay K, Bondy GP, Fong T (1995) Substance P (NK_1)- and neurokinin A (NK_2)-receptor gene expression in inflammatory airway disease. Am J Physiol 269:L309–L317

Baraniuk JN, Ali M, Yuta A, Fang SY, Naranch K (1999) Hypertonic saline nasal provocation stimulates nociceptive nerves, substance P release, and glandular mucous exocytosis in normal humans. Am J Respir Crit.Care Med 160:655–662

Barnes PJ (1986) Asthma as an axon reflex. Lancet i:242–244

Bertrand C, Geppetti P, Baker J, Yamawaki I, Nadel JA (1993a) Role of neurogenic inflammation in antigen-induced vascular extravasation in guinea pig trachea. J Immunol 150:1479–1485

Bertrand C, Geppetti P, Baker J, Yamawaki I, Nadel JA (1993b) Tachykinins, via NK_1 receptor activation, play a relevant role in plasma protein extravasation evoked by allergen challenge in the airways of sensitized guinea-pigs. Regul Pept 46:214–216

Bertrand C, Geppetti P, Graf PD, Foresi A, Nadel JA (1993c) Involvement of neurogenic inflammation in antigen-induced bronchoconstriction in guinea pigs. Am J Physiol 265:L507–L511

Bonini S, Lambiase A, Angelucci F, Magrini L, Manni L, Aloe L (1996) Circulating nerve growth factor levels are increased in humans with allergic diseases and asthma. Proc Natl Acad Sci USA 93:10955–10960

Bonini S, Lambiase A, Levi-Schaffer F, Aloe L (1999) Nerve growth factor: an important molecule in allergic inflammation and tissue remodelling. Int Arch Allergy Immunol 118:159–162

Bousquet J (2000) Global initiative for asthma (GINA) and its objectives. Clin Exp Allergy 30(Suppl 1):2–5

Bowden JJ, Garland AM, Baluk P, Lefevre P, Grady EF, Vigna SR, Bunnett NW, McDonald DM (1994) Direct observation of substance P-induced internalization of neurokinin 1 (NK_1) receptors at sites of inflammation. Proc Natl Acad Sci USA 91:8964–8968

Braun A, Lommatzsch M, Renz H (2000) The role of neurotrophins in allergic bronchial asthma. Clin Exp.Allergy 30:178–186

Canning BJ, Reynolds SM, Anukwu LU, Kajekar R, Myers AC (2002) Endogenous neurokinins facilitate synaptic transmission in guinea pig airway parasympathetic ganglia. Am J Physiol 283:R320–R330

Carr MJ, Hunter DD, Jacoby DB, Undem BJ (2002) Expression of tachykinins in nonnociceptive vagal afferent neurons during respiratory viral infection in guinea pigs. Am J Respir Crit Care Med 165:1071–1075

Catalioto R-M, Criscuoli M, Cucchi P, Giachetti A, Giannotti D, Giuliani S, Lecci A, Lippi A, Patacchini R, Quartara L, Renzetti AR, Tramontana M, Arcamone F, Maggi CA (1998) MEN 11420 (Nepadutant), a novel glycosylated bicyclic peptide tachykinin NK_2 receptor antagonist. Br J Pharmacol 123:81–91

Chanez P, Springall D, Vignola AM, Moradoghi-Hattvani A, Polak JM, Godard P, Bousquet J (1998) Bronchial mucosal immunoreactivity of sensory neuropeptides in severe airway diseases. Am J Respir Crit Care Med 158:985–990

Chu HW, Kraft M, Krause JE, Rex MD, Martin RJ (2000) Substance P and its receptor neurokinin 1 expression in asthmatic airways. J Allergy Clin Immunol 106:713–722

Crimi N, Oliveri R, Polosa R, Palermo F, Mistretta A (1993) The effect of oral terfenadine on neurokinin-A induced bronchoconstriction. J Allergy Clin Immunol 91:1096–1098

Crimi N, Pagano C, Palermo F, Mastruzzo C, Prosperini G, Pistorio MP, Vancheri C (2003) Inhibitory effect of a leukotriene receptor antagonist (montelukast) on neurokinin A-induced bronchoconstriction. J Allergy Clin.Immunol 111:833–839

Crimi N, Palermo F, Oliveri R, Palermo B, Vancheri C, Polosa R, Mistretta A (1988) Effect of nedocromil on bronchospasm induced by inhalation of substance P in asthmatic subjects. Clin Allergy 18:375–382

Crimi N, Palermo F, Oliveri R, Palermo B, Vancheri C, Polosa R, Mistretta A (1990) Influence of antihistamine (astemizole) and anticholinergic drugs (ipratropium bromide) on bronchoconstriction induced by substance P. Ann Allergy 65:115–120

Daoui S, Cognon C, Naline E, Edmonds-Alt X (1998) Involvement of tachykinin NK_3 receptors in citric acid-induced cough and bronchial responses in guinea pigs. Am J Respir Crit Care Med 158:42–48

Daoui S, Naline E, Lagente V, Emonds-Alt X, Advenier C (2000) Neurokinin B- and specific tachykinin NK_3 receptor agonists-induced airway hyperresponsiveness in the guinea-pig. Br J Pharmacol 130:49–56

De Vries A, Dessing MC, Engels F, Henricks PAJ, Nijkamp FP (1999) Nerve growth factor induces a neurokinin-1 receptor-mediated airway hyperresponsiveness in guinea pigs. Am J Respir Crit Care Med 159:1541–1544

Delaunois A, Gustin P, Ansay M (1993a) Effects of capsaicin on the endothelial permeability in isolated and perfused rabbit lungs. Fund Clin Pharmacol 7:81–91

Delaunois A, Gustin P, Ansay M (1993b) Role of neuropeptides in acetylcholine-induced edema in isolated and perfused rabbit lungs. J Pharmacol Exp Ther 266:483–491

Di Maria GU, Bellofiore S, Geppetti P (1998) Regulation of airway neurogenic inflammation by neutral endopeptidase. Eur Respir J 12:1454–1462

Diemunsch P, Grelot L (2000) Potential of substance P antagonists as antiemetics. Drugs 60:533–546

Emonds-Alt X, Vilain P, Goulaouic P, Proietto V, Van Broeck D, Advenier C, Naline E, Neliat G, Le Fur G, Breliere JC (1992) A potent and selective non-peptide antagonist of the neurokinin A (NK_2) receptor. Life Sci 50:L101–L106

Fahy JV, Wong HH, Geppetti P, Reis JM, Harris SC, Maclean DB, Nadel JA, Boushey HA (1995) Effect of an NK_1 receptor antagonist (CP-99,994) on hypertonic saline-induced bronchoconstriction and cough in male asthmatic subjects. Am J Respir Crit Care Med 152:879–884

Fewtrell CMS, Foreman JC, Jordan CC, Oehma P, Renner H, Stewart JM (1982) The effects of substance P on histamine and 5-hydroxytryptamine release in the rat. J Physiol 330:393–411

Finney MJB, Karlsson JA, Persson CGA (1985) Effects of bronchoconstrictors and bronchodilators on a novel human small airway preparation. Br J Pharmacol 85:29–36

Fischer A, McGregor GP, Saria A, B Philippin, Kummer W (1996) Induction of tachykinin gene and peptide expression in guinea pig nodose primary afferent neurons by allergic airway inflammation. J Clin Invest 98:2284–2291

Freund V, Pons F, Joly V, Mathieu E, Martinet N, Frossard N (2002) Upregulation of nerve growth factor expression by human airway smooth muscle cells in inflammatory conditions. Eur Respir J 20:458–463

Germonpre PR, Bullock GR, Lambrecht BN, Van De Velde V, Luyten WH, Joos GF, Pauwels RA (1999) Presence of substance P and neurokinin 1 receptors in human sputum macrophages and U-937 cells. Eur Respir J 14:776–782

Germonpre PR, Joos GF, Everaert E, Kips JC, Pauwels RA (1995) Characterization of neurogenic inflammation in the airways of two highly inbred rat strains. Am.J.Respir.-Crit.Care Med 152:1796–1804

Germonpre PR, Joos GF, Mekeirele K, Pauwels RA (1997) Role of the 5-HT receptor in neurogenic inflammation in Fisher 344 rat airways. Eur.J.Pharmacol 324:249–255

Graham RM, Friedman M, Hoyle GW (2001) Sensory nerves promote ozone-induced lung inflammation in mice. Am J Respir Crit Care Med 164:307–313

Hansel TT, Erin EM, Barnes PJ (2002) The allergen challenge. Clin.Exp.Allergy 32:162–167

Hazbun ME, Hamilton R, Holian A, Eschenbacher WL (1993) Ozone-induced increases in substance P and 8-epi-prostaglandin F2a in the airways of human subjects. Am J Respir Cell Mol Biol 9:568–572

Heaney LG, Cross LJM, McGarvey LPA, Buchanan KD, Ennis M, Shaw C (1998) Neurokinin A is the predominant tachykinin in human bronchoalveolar lavage fliud in normal and asthmatic subjects. Thorax 53:357–362

Höglund C, de Blay F, Oster J-P, Duvernelle C, Kassel O, Pauli G, Frossard N (2002) Nerve growth factor levels and localisation in human asthmatic bronchi. Eur Respir J 20:1110–1116

Howarth PH, Springall DR, Redington AE, Djukanovic R, Holgate S, Polak J (1995) Neuropeptide-containing nerves in endobronchial biopsies from asthmatic and nonasthmatic subjects. Am J Respir Cell Mol Biol 3:288–296

Hoyle GW, Graham RG, Finkelstein JB, Thi Nguyen K-P, Gozal D, Friedman M (1998) Hyperinnervation of the airways in transgenic mice overexpressing nerve growth factor. Am J Respir Cell Mol Biol 18:149–157

Hu C, Wedde-Beer K, Auais A, Rodriguez MM, Piedimonte G (2002) Nerve growth factor and nerve growth factor receptors in respiratory syncytial virus-infected lungs. Am J Physiol 283:L494–L502

Hunter DD, Myers AC, Undem BJ (2000) Nerve Growth Factor-Induced Phenotypic Switch in Guinea Pig Airway Sensory Neurons. Am J Respir Crit.Care Med 161:1985–1990

Ichinose M, Miura M, Yamauchi H, Kageyama N, Tomaki M, Oyake T, Ohuchi Y, Hida W, Miki H, Tamura G, Shirato K (1996) A neurokinin 1-receptor antagonist improves exercise-induced airway narrowing in asthmatic patients. Am J Respir Crit Care Med 153:936–941

Ichinose M, Nakajima N, Takahashi T, Yamauchi H, Inoue H, Takishima T (1992) Protection against bradykinin-induced bronchoconstriction in asthmatic patients by neurokinin receptor antagonist. Lancet 340:1248–1251

Joos G, Schelfhout V, Van De Velde V, Pauwels R (2001) The effect of the tachykinin NK_2 receptor antagonist MEN11420 (nepadutant) on neurokinin A-induced bronchoconstriction in patients with asthma. Am J Respir Crit Care Med 163:A628

Joos G, Van De Velde V, Schelfhout V, Pauwels R (2001) The leukotriene receptor antagonist zafirlukast inhibits neurokinin-A induced bronchoconstriction in patients with asthma. Am J Respir Crit Care Med 163:A418

Joos GF and Pauwels RA (2001) Tachykinin receptor antagonists: potential in airways disease. Curr Opin Pharmacol 1:235–241

Joos GF, Germonpre PR, Pauwels RA (2000) Role of tachykinins in asthma. Allergy 55:321–337

Joos GF, O'Connor B, Anderson SD, Chung F, Cockcroft DW, Dahlen B, DiMaria G, Foresi A, Hargreave FE, Holgate ST, Inman M, Lötvall, J., Magnussen, H., Polosa, R., Postma, D. S., and Riedler J (2003) Indirect airway challenges. Eur Respir J 21:1050–1068

Joos GF, Vincken W, Louis RE, Schelfhout VJ, Shaw MJ, Wang J, Della Ciopa G, Pauwels RA (2004) The dual NK_1/NK_2 tachykinin receptor antagonist DNK333 inhibits neurokinin A (NKA)-induced bronchoconstriction in patients with asthma. Eur Respir J 23:76–81

Joos G, Pauwels R, Van Der Straeten M (1989) The effect of nedocromil sodium on the bronchoconstrictor effect of neurokinin A in subjects with asthma. J Allergy Clin Immunol 83:663–668

Joos GF, De Swert KO, Pauwels RA (2001) Airway inflammation and tachykinins: prospects for the development of tachykinin receptor antagonists. Eur J Pharmacol 429:239–250

Joos GF, Germonpre PR, Kips JC, Peleman RA, Pauwels RA (1994) Sensory neuropeptides and the human lower airways: present state and future directions. Eur Respir J 7:1161–1171

Joos GF, Germonpre PR, Pauwels RA (2000) Role of tachykinins in asthma. Allergy 55:321–337

Joos GF, Lefebvre RA, Bullock G, Pauwels RA (1997) Role of 5-hydroxytryptamine and mast cells in the tachykinin-induced contraction of rat trachea in vitro. Eur J Pharmacol 338:259–268

Joos GF, Pauwels RA (1993) The in vivo effect of tachykinins on airway mast cells in the rat. Am Rev Respir Dis 148:922–926

Joos GF, Pauwels RA (2000) Pro-inflammatory effects of substance P: new perspectives for the treatment of airway diseases? Trends Pharmacol Sci 21:131–133

Joos GF, Pauwels RA, Van Der Straeten ME (1988) The effect of oxitropiumbromide on neurokinin A-induced bronchoconstriction in asthmatics. Pulmon Pharmacology 1:41–45

Joos GF, Van Schoor J, Kips JC, Pauwels RA (1996) The effect of inhaled FK244, a tachykinin NK-1 and NK-2 receptor antagonist, on neurokinin A-induced bronchoconstriction in asthmatics. Am J Respir Crit Care Med 153:1781–1784

Katsunuma T, Mak JCW, Barnes PJ (1998) Glucocorticoids reduce tachykinin NK_2 receptor expression in bovine tracheal smooth muscle. Eur J Pharmacol 344:99–106

Khan S, Liu Y-C, Khawaja AM, Manzini S, Rogers DF (2001) Effect of the long-acting tachykinin NK_1 receptor antagonist MEN 11467 on tracheal mucus secretion in allergic ferrets. Br J Pharmacol 132:189–196

Kraan J, Vink-Klooster H, Postma DS (2001) The NK-2 receptor antagonist SR 48968C does not improve adenosine hyperresponsiveness and airway obstruction in allergic asthma. Clin Exp Allergy 31:274–278

Kraneveld AD, Nijkamp FP, Van Oosterhout AJM (1997) Role for neurokinin-2 receptor in interleukin-5-induced airway hyperresponsiveness but not eosinophilia in guinea pigs. Am J Respir Crit Care Med 156:367–374

Krishna MT, Springall DR, Meng Q-H, Whiters N, Macleod D, Biscione G, Frew A, Polak J, Holgate S (1997) Effects of ozone on epithelium and sensory nerves in the bronchial mucosa of healthy humans. Am J Respir Crit Care Med 156:943–950

Kudlacz EM, Knippenberg RW, Logan DE, Burkholder TP (1996) Effect of MDL 105,212, a nonpeptide NK-1/NK-2 receptor antagonist in an allergic guinea pig model. J Pharmacol Exp Ther 279:732–739

Lamb JP, Sparrow MP (2002) Three-dimensional mapping of sensory innervation with substance p in porcine bronchial mucosa: comparison with human airways. Am.J Respir Crit Care Med 166:1269–1281

Lambrecht BN, Germonpre PR, Everaert EG, Carro-Muino I, De Veerman M, De Felipe C, Hunt SP, Thielemans K, Joos GF, Pauwels RA (1999) Endogenously produced substance P contributes to lymphocyte proliferation induced by dendritic cells and direct TCR ligation. Eur J Immunol 29:3815–3825

Lilly CM, Bai TR, Shore SA, Hall AE, Drazen JM (1995) Neuropeptide content of lungs from asthmatic and nonasthmatic patients. Am J Respir Crit Care Med 151:548–553

Lucchini RE, Facchini F, Turato G, Saetta M, Caramori G, Ciaccia A, Maestrelli P, Springall DR, Polak JM, Fabbri L, Mapp CE (1997) Increased VIP-positive nerve fibers in the mucous glands of subjects with chronic bronchitis. Am J Respir Crit Care Med 156:1963–1968

Lundberg JM (1996) Pharmacology of cotransmission in the autonomic nervous system: integrative aspects on amines, neuropeptides, adenosine triphosphate, amino acids and nitric oxide. Pharmacol Rev 48:113–178

Lundberg JM, Hökfelt T, Martling C-R, Saria A, Cuello C (1984) Substance P-immunoreactive sensory nerves in the lower respiratory tract of various mammals including man. Cell Tissue Res 235:251–261

Lundberg JM, Saria A (1983) Capsaicin-induced desensitization of airway mucosa to cigarette smoke, mechanical and chemical irritants. Nature 302:251–253

Luts A, Uddman R, Alm P, Basterra J, Sundler F (1993) Peptide-containing nerve fibers in human airways: distribution and coexistence pattern. Int Arch Allergy Immunol 101:52–60

Maggi CA (1997) The effects of tachykinins on inflammatory and immune cells. Regul Pept 70:75–90

Maggi CA (2000) The troubled story of tachykinins and neurokinins. Trends Pharmacol Sci 21:173–175

Maghni K, Taha R, Afif W, Hamid Q, Martin JG (2000) Dichotomy between neurokinin receptor actions in modulating allergic airway responses in an animal model of helper T Cell type 2 cytokine-associated inflammation. Am J Respir Crit Care Med 162:1068–1074

Mapp CE, Miotto D, Braccioni F, Saetta M, Turato G, Maestrelli P, Krause JE, Karpitskiy V, Boyd N, Geppetti P, Fabbri LM (2000) The distribution of neurokinin-1 and neurokinin-2 receptors in human central airways. Am J Respir Crit Care Med 161:207–215

Martling C-R, Theodorsson-Norheim E, Lundberg JM (1987) Occurrence and effects of multiple tachykinins; substance P, neurokinin A and neuropeptide K in human lower airways. Life Sci 40:1633–1643

Naline E, Molimard M, Regoli D, Emonds-Alt X, Bellamy JF, Advenier C (1996) Evidence for functional tachykinin NK_1 receptors on human isolated small bronchi. Am J Physiol 271:L763–L767

Nénan S, Germain N, Lagente V, Edmonds-Alt X, Advenier C, Boichot E (2001) Inhibition of inflammatory cell recruitment by the tachykinin NK_3-receptor antagonist, SR 14280, in a murine model of asthma. Eur J Pharmacol 421:210–205

Nieber K, Baumgarten CR, Rathsack R, Furkert J, Oehme P, Kunkel G (1992) Substance P and ß-endorphine-like immunoreactivity in lavage fluids of subjects with and without allergic asthma. J Allergy Clin Immunol 90:646–652

Ollerenshaw SL, Jarvis D, Sullivan CE, Woolcock AJ (1991) Substance P immunoreactive nerves in airways from asthmatics and nonasthmatics. Eur Respir J 4:673–682

Pauwels RA, Buist AS, Calverley PM, Jenkins CR, Hurd SS (2001) Global strategy for the diagnosis, management, and prevention of chronic obstructive pulmonary disease. NHLBI/WHO Global Initiative for Chronic Obstructive Lung Disease (GOLD) Workshop summary. Am J Respir Crit Care Med 163:1256–1276

Pedersen KE, Meeker SN, Riccio MM, Undem BJ (1998) Selective stimulation of jugular ganglion afferent neurons in guinea pig airways by hypertonic saline. J Appl Physiol 84:499–506

Piedimonte G (2001) Neural mechanisms of respiratory syncytial virus-induced inflammation and prevention of respiratory syncytial virus sequelae. Am J Respir Crit Care Med 163:S18–S21

Piedimonte G, Hoffman JI, Husseini WK, Snider RM, Desai MC, Nadel JA (1993) NK_1 receptors mediate neurogenic inflammatory increase in blood flow in rat airways. J Appl Physiol 74:2462–2468

Piedimonte G, Rodriguez MM, King KA, McLean S, Jiang X (1999) Respiratory syncytial virus upregulates expression of the substance P receptor in rat lungs. Am J Physiol 277:L831–L840

Rumsey WL, Aharony D, Bialecki RA, Abbott BM, Barthlow HG, Caccese R, Ghanekar S, Lengel D, McCarthy M, Wenrich B, Undem B, Ohnmacht C, Shenvi A, Albert JS, Brown F, Bernstein PR, Russell K (2001) Pharmacological characterization of zd6021: a novel, orally active antagonist of the tachykinin receptors. J Pharmacol Exp Ther 298:307–315

Saria A, Lundberg JM, Skofitsch G, Lembeck F (1983) Vascular protein leakage in various tissues induced by substance P, capsaicin, bradykinin, serotonin, histamine and by antigen challenge. Naunyn-Schmiedeberg's Arch Pharmacol 324:212–218

Schmidt D, Jorres RA, Rabe KF, Magnussen H (1996) Reproducibility of airway response to inhaled bradykinin and effect of the neurokinin receptor antagonist FK-224 in asthmatic subjects. Eur J Clin Pharmacol 50:269–273

Schelfhout, Louis R, Lenz W, Pauwels R, Joos G (2003) The triple neurokinin receptor antagonist CS-003 inhibits neurokinin A (NKA)-induced bronchoconstriction in patients with asthma. Eur Respir J 22(Suppl 45): 415S

Schuiling M, Zuidhof A, Meurs H, Zaagsma J (1999a) Role of tachykinin NK_2-receptor activation in the allergen-induced late asthmatic reaction, airway hyperreactivity and airway inflammation cell influx in conscious, unrestrained guinea-pigs. Br J Pharmacol 127:1030–1038

Schuiling M, Zuidhof A, Zaagsma J, Meurs H (1999b) Involvement of tachykinin NK_1 receptor in the development of allergen-induced airway hyperreactivity and airway inflammation in conscious, unrestrained guinea-pigs. Am J Respir Crit Care Med 159:423–430

Severini C, Improta G, Falconieri-Erspamer G, Salvadori S, Erspamer V (2002) The tachykinin peptide family. Pharmacol Rev 54:285–322

Snider RM, Constantine JW, Lowe III JA, Longo KP, Lebel WS, Woody HA, Drzda SE, Desai MC, Vinick FJ, Spence RW, Hess HJ (1991) A potent nonpeptide antagonist of the substance P (NK_1) receptor. Science 251:435–437

Szarek JL, Zhang JZ, Gruetter CA (1993) 5-HT2 receptors augment cholinergic nerve-mediated contraction of rat bronchi. Eur J Pharmacol 231:339–346

Tanaka DT, Grunstein MM (1984) Mechanisms of substance P-induced contraction of rabbit airway smooth muscle. J Appl Physiol 57:1551–1557

Tomaki M, Ichinose M, Miura M, Hirayama Y, Yamauchi H, Nakajima N, Shirato K (1995) Elevated substance P content in induced sputum from patients with asthma and patients with chronic bronchitis. Am J Respir Crit Care Med 151:613–617

Tournoy KG, De Swert KO, Leclere PG, Lefebvre RA, Pauwels RA, Joos GF (2003) Modulatory role of tachykinin NK_1 receptor in cholinergic contraction of mouse trachea. Eur Respir J 21:3–10

Tousignant C, Chan CC, Guevremont D, Brideau C, Hale JJ, Maccoss M, Rodger IW (1993) NK_2 receptors mediate plasma extravasation in guinea-pig lower airways. Br J Pharmacol 108:383–386

van der Kleij HP, Ma D, Redegeld FA, Kraneveld AD, Nijkamp FP, Bienenstock J (2003) Functional expression of neurokinin 1 receptors on mast cells induced by IL-4 and stem cell factor. J Immunol 171:2074–2079

Van Rensen EL, Hiemstra PS, Rabe KF, Sterk PJ (2002) Assessment of microvascular leakage via sputum induction: the role of substance P and neurokinin A in patients with asthma. Am J Respir Crit Care Med 165:1275–1279

Van Schoor J, Joos G, Chasson B, Brouard R, Pauwels R (1998) The effect of the NK_2 receptor antagonist SR48968 (saredutant) on neurokinin A-induced bronchoconstrction in asthmatics. Eur Respir J 12:17–23

Van Schoor J, Joos GF, Pauwels RA (2000) Indirect bronchial hyperresponsiveness in asthma: mechanisms, pharmacology and implications for clinical research. Eur Respir J 16:514–533

Virchow JC, Julius P, Lommatzsch M, Luttmann W, Renz H, Braun A (1998) Neurotrophins are increased in bronchoalveolar lavage fluid after segmental allergen provocation. Am J Respir Crit Care Med 158:2002–2005

Wu ZX, Lee LY (1999) Airway hyperresponsiveness induced by chronic exposure to cigarette smoke in guinea pigs: role of tachykinins. J Appl.Physiol 87:1621–1628

Yang XX, Powell WS, Hojo M, Martin JG (1997) Hyperpnea-induced bronchoconstriction is dependent on tachykinin-induced cysteinyl leukotriene synthesis. J Appl Physiol 82:538–544

Yoshihara S, Geppetti P, Hara M, Linden A, Ricciardolo FL, Chan B, Nadel JA (1996) Cold air-induced bronchoconstriction is mediated by tachykinin and kinin release in guinea pigs. Eur J Pharmacol 296:291–296

Role of Tachykinins in the Gastrointestinal Tract

P. Holzer

Department of Experimental and Clinical Pharmacology, Medical University of Graz, Universitätsplatz 4, 8010 Graz, Austria
e-mail: peter.holzer@meduni-graz.at

1	Introduction	512
2	Expression of Tachykinin Genes in the Gut	513
3	Cellular Localization of Tachykinins in the Gut	514
3.1	Neuronal Sources of Tachykinins in the GI Tract	514
3.2	Non-neuronal Sources of Tachykinins in the GI Tract	517
4	Cellular Expression of Tachykinin Receptors in the Gut	518
5	Functional Implications of Tachykinins in the Gut	519
5.1	Transmitter Function	519
5.2	Motor Regulation	521
5.3	Secretion of Hormones, Enzymes, Electrolytes and Fluid	523
5.4	Blood Flow, Vascular Permeability and Immune Cell Function	525
6	Tachykinins and Tachykinin Receptors in Gastrointestinal Disorders	526
6.1	Pathological Changes in Tachykinin Expression	526
6.2	Pathological Changes in Tachykinin Receptor Expression	528
6.3	Motor Disturbances	529
6.4	Hypersecretion and Inflammation	532
6.5	Hyperalgesia and Pain	536
7	Summary	538
	References	539

Abstract The preprotachykinin-A gene-derived peptides substance (SP) and neurokinin A (NKA) are expressed in distinct neural pathways of the mammalian gut. When released from intrinsic enteric or extrinsic primary afferent neurons, they have the potential to influence most digestive effector systems by interaction with tachykinin NK_1, NK_2 and NK_3 receptors. Within the enteric nervous system, SP and NKA mediate slow synaptic transmission and modulate neuronal excitability via stimulation of NK_3 and NK_1 receptors. As NK_1 and NK_2 receptors are expressed on muscle and epithelial cells, tachykinins can also directly stimulate gastrointestinal motor activity and facilitate the secretion of fluid and electrolytes. In addition, SP and NKA utilized by extrinsic afferent neurons participate in inflammatory and nociceptive processes. Various gastrointestinal disorders are associated with distinct changes in the expression of tachykinins and their receptors, and there is increasing experimental evidence that tachykinins participate in the dysmotility, hypersecretion, vascular and immunological disturbances associated with infection and inflammation. While tachykinin receptor antagonists are little active in the normal gut, they are able to correct disturbed motility and secretion and to inhibit inflammation and hyperalgesia. Extrapolation of these experimental findings to disorders of the human digestive system identifies tachykinin receptors as novel targets for gastroenterological therapy.

Keywords Tachykinins · Substance P · Neurokinin A · Tachykinin NK_1, NK_2 and NK_3 receptors · Enteric nervous system · Sensory neurons · Gastrointestinal motility · Gastrointestinal secretion · Inflammatory bowel disease · Abdominal hyperalgesia · Irritable bowel syndrome

1
Introduction

In 1931 von Euler and Gaddum reported that the gut contained a factor, later to be named substance P (SP), which caused contraction of the intestine and was different from any of the endogenous compounds known at that time (von Euler and Gaddum 1931). Pernow (1951) was the first to investigate the occurrence of SP in the gastrointestinal (GI) tract of mammals by bioassay in a systematic manner. It was, however, only after the identification of SP as an undekapeptide and the development of immunochemical methods for its measurement in the tissue that the distribution of this peptide and its functional implications in the gut became more fully understood. Additional tachykinins, particularly neurokinin A (NKA), were found to occur in the gut, and all three major tachykinin receptors, termed NK_1, NK_2 and NK_3 receptors, are now known to be expressed in the alimentary canal in a cell-specific manner (Holzer and Holzer-Petsche 1997a, 1997b). As the pathophysiological implications of tachykinins in the GI tract are uncovered it becomes evident that tachykinin receptor antagonists hold particular potential in the treatment of various GI disorders including inflammatory bowel disease (IBD), irritable bowel syndrome (IBS), vomiting and abdominal pain.

The objective of this chapter is to review the presence of tachykinins and their receptors in the digestive tract, to summarize their diverse biological actions in this organ system, to detail their physiological and pathophysiological implications and to assess their potential as pharmacological targets in the treatment of GI disease. Within the gut, tachykinins are primarily messenger molecules of enteric neurons that are intrinsic to the gut and that control various aspects of digestive activity. Regulation of motor activity by tachykinins is an area that has been studied most intensively ever since SP was discovered to occur in the gut, but it is now obvious that tachykinins also influence secretory activity, vascular and immune function as well as GI sensitivity and pain. As the pertinent literature has been repeatedly reviewed (Pernow 1983; Barthó and Holzer

1985; Otsuka and Yoshioka 1993; Holzer-Petsche 1995; Shuttleworth and Keef 1995; Holzer and Holzer-Petsche 1997a, 1997b, 2001; Maggi 2000), the current chapter will build on these foundations and update the current knowledge of tachykinin actions and implications in the gut.

2
Expression of Tachykinin Genes in the Gut

Besides SP, the mammalian gut contains NKA and various N-terminally extended forms of NKA, which all are derived from alternative splicing of the primary transcripts of the preprotachykinin A (*PPTA*, *TAC1*) gene. In the rat enteric nervous system (ENS) it is the γ-*PPTA* transcript that accounts for as much as 80%–90% of the tachykinin-encoding mRNA, while β-PPTA comprises about 10%–20% and α-PPTA less than 1% of the total SP/NKA-encoding mRNA (Sternini et al. 1989). There is, in addition, a fourth form of mRNA, δ-PPTA, which occurs in the rat small and large intestine but is less abundant than γ-PPTA and β-PPTA (Khan and Collins 1994). Because they are both derived from PPTA, SP and NKA are colocalized in and coreleased from the same synaptic vesicles of the guinea pig, porcine and canine ENS (Deacon et al. 1987; Christofi et al. 1990; McDonald et al. 1990; Schmidt et al. 1991). Neuropeptide K (NPK) which is produced by translation of β-PPTA mRNA is thought to be a precursor which during packaging into storage vesicles for axonal transport is converted to NKA (Deacon et al. 1987; Hellström et al. 1991). This inference is consistent with the finding that, unlike SP and NKA, NPK is apparently not released from enteric neurons (Christofi et al. 1990; Hellström et al. 1991), although NPK and NPK-(1–24) are liberated from carcinoid tumour cells (Ahlman et al. 1988; Conlon et al. 1988). The expression of neuropeptide γ is species-specific in as much as this product of γ-PPTA is present in the rabbit, but not guinea pig, intestine (Kage et al. 1988).

It is widely held that neurokinin B (NKB) is practically absent from the human, porcine, guinea pig and rat gut (Deacon et al. 1987; Too et al. 1989; Christofi et al. 1990; McDonald et al. 1990; Takeda et al. 1990; Schmidt et al. 1991; Shuttleworth et al. 1991; Moussaoui et al. 1992). This contention is in keeping with the failure to detect mRNA of the *PPTB* (*TAC3*) gene, wherefrom NKB is exclusively derived, in the ENS of the rat intestine (Sternini 1991). However, there are reports that the *PPTB* gene is expressed in the bovine gut (Kotani et al. 1986) and that the human (Kishimoto et al. 1991) and rat (Tateishi et al. 1990) intestine contain minute amounts of NKB which have been localized to a few myenteric and submucosal neurons that also contain SP (Yunker et al. 1999). It would hence appear that, in the gut, the *PPTB* gene is only weakly expressed in a species- and condition-dependent manner.

Hemokinin-1 is a newly identified mammalian tachykinin that is encoded by the *PPTC* (*TAC4*) gene. Originally discovered in haematopoietic cells (Zhang et al. 2000), hemokinin-1 is also expressed in the mouse and human GI tract (Kurtz et al. 2002). Although its cellular sources in the gut remain to be defined, hemokinin-1 is a high affinity ligand for NK_1 receptors (Bellucci et al. 2002; Camarda et al. 2002; Kurtz et al. 2002) and thus may function as a messenger of those cells that release this tachykinin. Other tachykinin-like products of the *TAC4* gene, i.e. endokinin A, B, C and D, are apparently absent from the human and murine stomach (Page et al. 2003).

3
Cellular Localization of Tachykinins in the Gut

3.1
Neuronal Sources of Tachykinins in the GI Tract

Most of the SP and NKA present in the gut is derived from two groups of neurons: intrinsic enteric neurons and extrinsic primary afferent nerve fibres (Holzer and Holzer-Petsche 1997a, 1997b). The quantitatively most important source of tachykinins is the ENS which has its cell bodies in the myenteric and submucosal (submucous) plexuses and supplies all GI effector systems (Furness et al. 1991; Costa et al. 1996; Holzer and Holzer-Petsche 1997a). However, the extent of the tachykininergic innervation and the characteristics of the tachykininergic neurons show some variation between different regions and layers of the GI tract as well as between different species.

A large number of studies (Holzer et al. 1981, 1982; Brodin et al. 1983; McGregor et al. 1984; Keast et al. 1985; Ferri et al. 1987; Tateishi et al. 1990; Timmermans et al. 1990; Kishimoto et al. 1991; McGregor and Conlon 1991; Schmidt et al. 1991; Burns and Cummings 1993; Wattchow et al. 1997; Clerc et al. 1998; Neunlist et al. 1999b; Hens et al. 2000, 2001, 2002; Lomax and Furness 2000; Schemann et al. 2001) has shown that in the human, rabbit, rat and guinea pig gut the tissue concentrations of SP and NKA are low in the oesophagus, intermediate in the stomach and high in the intestine, the highest levels usually being measured in the small intestine. In contrast, the distribution of SP and NKA in the porcine and feline gut shows little variation and is fairly even between the oesophagogastric region and the rectum. Species differences also exist with regard to the concentrations of SP in the different layers of the GI wall. The levels of SP and NKA in the external muscle layer of the guinea pig, rat, rabbit, ferret and feline gut are considerably higher than in the mucosa/submucosa layer, whereas the mucosa/submucosa layer of the human and equine intestine contains similar or even higher concentrations of SP than the external muscle layer (Holzer et al. 1981, 1982; Brodin et al. 1983; Llewellyn-Smith et al. 1984; Deacon et al. 1987; Ferri et al. 1987; Greenwood et al. 1990; Burns and Cummings 1993).

The projections of enteric tachykinin-expressing neurons have been identified with neuroanatomical lesion and tracing techniques, and their chemical coding (that is, the characteristic combination of coexisting transmitters, neuropeptides and other neuronal markers) has been revealed by immunocytochemistry. These studies have identified several classes of enteric tachykininer-

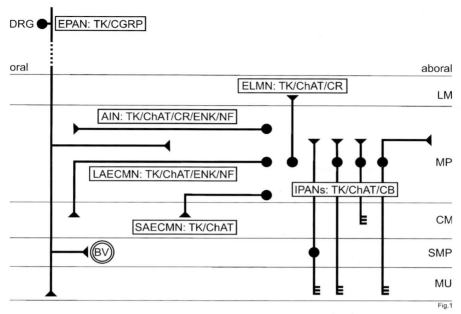

Fig. 1 Schematic summary of the important classes of tachykinin (*TK*)-immunoreactive neurons, their chemical coding and projections within the guinea pig gut. *AIN*, Ascending interneuron; *BV*, blood vessel; *CB*, calbindin; *CGRP*, calcitonin gene-related peptide; *ChAT*, choline acetyltransferase; *CM*, circular muscle; *CR*, calretinin; *DRG*, dorsal root ganglion; *ELMN*, excitatory longitudinal muscle motor neuron; *ENK*, enkephalin; *EPAN*, extrinsic primary afferent neuron; *IPANs*, intrinsic primary afferent neurons; *LAECMN*, long ascending excitatory circular muscle motor neuron; *LM*, longitudinal muscle; *MP*, myenteric plexus; *MU*, mucosa; *NF*, neurofilament protein; *SAECMN*, short ascending excitatory circular muscle motor neuron; *SMP*, submucosal plexus

gic neurons which differ with regard to morphology, chemical coding, projection and function (Costa et al. 1996; Furness 2000; Brookes 2001; Schemann et al. 2001; Furness and Sanger 2002). In the guinea pig intestine, SP is typically present in intrinsic primary afferent neurons (IPANs) of the myenteric and submucosal plexus and, within the myenteric plexus, in ascending interneurons (Brookes et al. 1991) as well as in excitatory motor neurons to the circular and longitudinal muscle (Fig. 1). Tachykinin-positive nerve fibres thus connect the ganglia within the myenteric plexus (Brookes et al. 1991; Wattchow et al. 1997), link the two plexuses with each other (Messenger 1993), innervate the longitudinal muscle (Shuttleworth et al. 1991; Brookes et al. 1992; Messenger 1993), issue ascending projections of short and intermediate length to the circular muscle (Costa et al. 1996), come in close contact with interstitial cells of Cajal (ICCs) in the deep muscular plexus (Lavin et al. 1998; Vannucchi et al. 1999; Wang et al. 1999) and supply the muscularis mucosae (Steele and Costa 1990; Messenger 1993) and mucosa (Holzer and Holzer-Petsche 1997b; Furness et al. 1998; Neunlist et al. 1999a). SP-containing submucosal neurons also project to submucosal arterioles of the guinea pig colon, whereas these projections are largely absent from the small intestine (Galligan et al. 1988; Vanner and Surprenant 1991; Jiang and Surprenant 1992).

Characteristic of most enteric neurons containing SP in the guinea pig GI tract is that they coexpress choline acetyltransferase, which means that tachykinins are cotransmitters of cholinergic neurons (Furness 2000; Brookes 2001; Schemann et al. 2001). These tachykininergic neurons can be further subgrouped by their content of calbindin/calretinin, neurofilament protein triplet and dynorphin/enkephalin-like immunoreactivity (Fig. 1), whereas vasoactive intestinal polypeptide (VIP) and nitric oxide synthase do not coexist with tachykinins in the same enteric neurons of the guinea pig, murine and canine gut (Brookes et al. 1991, 1992; Schemann et al. 1995; Sang and Young 1996; Costa et al. 1996). Coexpression of enkephalin-like immunoreactivity and SP has also been found in enteric neurons of the human GI tract (Wattchow et al. 1988). Tachykininergic enteric neurons in the murine small and large intestine coexpress calretinin, γ-aminobutyric acid or 5-hydroxytryptamine (5-HT) to various degrees and show some distinct differences from the chemical coding of SP neurons in the guinea pig gut (Sang and Young 1996). By a synopsis of their chemical codes, morphological characteristics and projections it has been possible to deduce the functional identity of tachykininergic enteric neurons as sensory neurons, interneurons and motor neurons (Costa et al. 1996; Furness 2000; Brookes 2001; Schemann et al. 2001).

Extrinsic afferent neurons are a further source of SP and NKA in the gut (Fig. 1). Most of these neurons originate from dorsal root ganglia and reach the gut via sympathetic (splanchnic, colonic and hypogastric) and sacral parasympathetic (pelvic) nerves while passing through prevertebral ganglia and forming collateral synapses with sympathetic ganglion cells (Green and Dockray 1988; Holzer and Holzer-Petsche 1997a). In contrast, vagal afferents emanating from the nodose ganglia make a relatively small contribution to the SP content of the GI tract (Green and Dockray 1988; Bäck et al. 1994; Holzer and Holzer-Petsche 1997a; Suzuki et al. 1997). The spinal afferents project primarily to submucosal arteries and arterioles where they form a para- and perivascular network of axons, although some axons also supply mucosa, enteric nerve plexuses and muscle layers (Furness et al. 1982; Green and Dockray 1988; Holzer and Holzer-Petsche 1997a). Tachykininergic extrinsic afferent neurons differ from intrinsic enteric neurons expressing SP in their chemical coding and sensitivity to capsaicin. Characteristically, many spinal afferents containing SP coexpress calcitonin gene-related peptide (CGRP), a combination of peptides that is not found in enteric neurons of the rodent and canine gut (Gibbins et al. 1987; Green and Dockray 1988; Sternini 1992; Perry and Lawson 1998).

Apart from the alimentary canal, SP and NKA have been localized to the salivary glands, where they stimulate salivary secretion (Holzer and Holzer-Petsche 1997a, 1997b) as well as to the hepatobiliary system and pancreas where they are present in both intrinsic neurons and extrinsic primary afferent nerve fibres (Holzer and Holzer-Petsche 1997a, 1997b). The SP-expressing somata in these organs are considered to be extensions of the enteric nerve plexuses in the gut (Goehler et al. 1988; Mawe and Gershon 1989; Kirchgessner and Gershon 1990; Huchtebrock et al. 1991; Sand et al. 1993; Ekblad et al. 1994; De Giorgio et al. 1995).

Their localization to distinct neurons make SP and NKA neuropeptides of the GI tract. A messenger role is consistent with their vesicular localization in neurons and their release upon stimulation. Ultrastructurally, SP has been found to occur in the varicose (vesiculate) as well as intervaricose (nonvesiculate) portions of nerve processes (Llewellyn-Smith et al. 1984, 1988, 1989) where it is present both in large granular vesicles (Probert et al. 1983; Llewellyn-Smith et al. 1984, 1988, 1989) and in the cytoplasm outside the synaptic vesicles. Although SP-positive varicosities make extensive contacts with other nerve processes or somata in the human and guinea pig small intestine, only a very small proportion of the contacts show synaptic specializations (Llewellyn-Smith et al. 1984, 1989; Portbury et al. 2001). Furthermore, SP-containing varicose axons are closely apposed to surface mucous cells, mucosal myofibroblasts and muscularis mucosae cells in the human gastric antrum (Ericson et al. 2002).

As is expected for transmitters with a vesicular localization, depolarizing stimuli cause release of SP and NKA from intrinsic enteric and extrinsic primary afferent neurons as has been shown by in vitro and in vivo studies (Holzer and Holzer-Petsche 1997a, 1997b). The calcium dependency of the release process (Holzer 1984; Maggi et al. 1994a; Lippi et al. 1998) points to an exocytotic mechanism, which is an important criterion for establishing tachykinins as GI neurotransmitters. Apart from direct immunochemical evidence for a stimulus-evoked release of SP and NKA, there is a wealth of indirect evidence that tachykinins are released in response to a variety of stimuli as revealed by pharmacological blockade of the actions of endogenous tachykinins. Tachykinin release from extrinsic afferents can specifically be probed by capsaicin (Holzer 1991; Holzer and Holzer-Petsche 1997a) because functional receptors for this drug (vanilloid receptors of type 1) are expressed by spinal (Caterina et al. 1997) and vagal (Helliwell et al. 1998) afferent neurons but have not yet been proved to occur on enteric neurons.

3.2
Non-neuronal Sources of Tachykinins in the GI Tract

The GI mucosa of humans and some other mammals contains SP-positive endocrine cells which belong to the 5-HT-containing enterochromaffin cells (Heitz et al. 1976; Sundler et al. 1977; Sjölund et al. 1983; Spångéus et al. 2001). In addition, some epithelial cells which in the human colon express SP may represent a separate population of endocrine cells that are distinct from enterochromaffin cells (Sokolski and Lechago 1984). SP-containing endocrine cells have also been localized to the mouse, rabbit, dog, cow, goat and sheep gut but are absent from the rat, guinea pig, feline and porcine intestine (Aiken et al. 1994 Calingasan et al. 1984; Grönstad et al. 1985; Keast et al. 1987; Wathuta 1986; Schmidt et al. 1991; Aiken et al. 1994; Mimoda et al. 1998).

Blood-derived or resident immune cells are a further source of SP in the lamina propria of the GI mucosa. Specifically, SP has been found in macrophages of the rat small intestine (Castagliuolo et al. 1997), in eosinophils of the inflamed human large intestine (Metwali et al. 1994), in eosinophils of the intestine of *Schistosoma*-infected mice (Weinstock and Blum 1990) and in T lymphocytes of the mouse intestine (Qian et al. 2001).

4
Cellular Expression of Tachykinin Receptors in the Gut

Tachykinin NK_1, NK_2 and NK_3 receptors are expressed by enteric neurons and effector cells (Fig. 2) in a cell- and region-specific manner (Holzer and Holzer-Petsche 2001; Furness and Sanger 2002; Tonini et al. 2002). In the guinea pig, rat and mouse gut, NK_1 receptors are present on IPANs, excitatory and inhibitory motor neurons, secretomotor neurons, ICCs, longitudinal and circular muscle cells, endothelial cells, epithelial cells as well as some granulocytes and lympho-

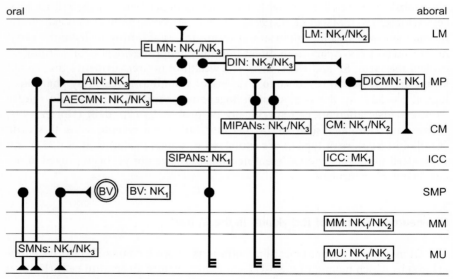

Fig. 2 Schematic summary of the cellular expression of tachykinin NK_1, NK_2 and NK_3 receptors in the guinea pig gut, based on immunocytochemical and pharmacological evidence. *Filled circles* depict neuronal somata, *filled triangles* nerve endings. *AECMN*, Ascending excitatory circular muscle motor neuron; *AIN*, ascending interneuron; *BV*, blood vessel; *CM*, circular muscle; *DICMN*, descending inhibitory circular muscle motor neuron; *DIN*, descending interneuron; *ELMN*, excitatory longitudinal muscle motor neuron; *ICC*, interstitial cells of Cajal; *LM*, longitudinal muscle; *MIPANs*, myenteric intrinsic primary afferent neurons; *MM*, muscularis mucosae; *MP*, myenteric plexus; *MU*, mucosa; *SIPANs*, submucosal intrinsic primary afferent neurons; *SMN*, secretomotor neuron; *SMP*, submucosal plexus

cytes (Sternini et al. 1995; Grady et al. 1996; Legat et al. 1996; Southwell et al. 1998b; Portbury et al. 1996b, 2001; Lavin et al. 1998; Lomax et al. 1998; Mann et al. 1999a, 1999b; Bian et al. 2000; Vannucchi and Faussone-Pellegrini 2000; Höckerfelt et al. 2001; Qian et al. 2001; Southwell and Furness 2001; Bayguinov et al. 2003; Yip et al. 2003). A similar distribution of NK_1 receptors applies to the human GI tract in which this type of tachykinin receptor has been found in the muscularis propria, myenteric plexus, muscularis mucosae and media of submucosal blood vessels as well as on endothelial cells of capillaries and venules, epithelial cells and some mononuclear and lymphoid cells (Goode et al. 1998, 2000; Smith et al. 2000; Renzi et al. 2000; Rettenbacher and Reubi 2001). The cell line Caco-2 derived from the human colonic epithelium has also been shown to express functional NK_1 receptors (Böckmann 2002).

NK_2 receptors are typically expressed by cells of the longitudinal and circular muscle layer and the muscularis mucosae (Grady et al. 1996; Portbury et al. 1996a; Renzi et al. 2000; Vannucchi et al. 2000). In the rat and guinea pig small intestine NK_2 receptors are in addition present on epithelial cells and on enteric nerve endings (Fig. 2) in the nerve plexuses and muscle layers (Grady et al. 1996; Portbury et al. 1996a; Vannucchi and Faussone-Pellegrini 2000; Vannucchi et al. 2000). The NK_2 receptor-bearing fibres are processes of descending interneurons as well as motor neurons to the circular and longitudinal muscle, the latter location being not depicted in Fig. 2 because the identity of the motor neurons (excitatory or inhibitory) remains to be determined.

NK_3 receptors are largely confined to enteric neurons (Fig. 2) that in the rat, mouse and guinea pig small bowel comprise IPANs, ascending and descending interneurons, excitatory and inhibitory motor neurons as well as secretomotor neurons (Grady et al. 1996; Mann et al. 1997, 1999a; Neunlist et al. 1999a; Jenkinson et al. 1999a, 2000; Vannucchi and Faussone-Pellegrini 2000; Wang et al. 2002). While myenteric IPANs in the guinea pig small intestine extensively express NK_3 receptors, this type of tachykinin receptor seems to be absent from submucosal IPANs (Jenkinson et al. 1999a). In addition, NK_3 receptors have been localized to muscle cells of the rat oesophagus and the mouse oesophagus, stomach and intestine (Mann et al. 1997; Vannucchi and Faussone-Pellegrini 2000; Wang et al. 2002) and to freshly isolated or cultured ICCs from the murine stomach and small intestine (Epperson et al. 2000).

5
Functional Implications of Tachykinins in the Gut

5.1
Transmitter Function

SP and NKA are cotransmitters of enteric neurons as well as of extrinsic afferent nerve fibres (Holzer and Holzer-Petsche 1997a, 1997b). Transmission through endogenously released tachykinins can be visualized via internalization (endocytosis) of activated tachykinin receptors as seen by confocal microscopy

(Southwell et al. 1996, 1998a, 1998b; Lavin et al. 1998; Jenkinson et al. 1999b, 2000; Mann et al. 1999b; Southwell and Furness 2001). Electrophysiological investigations have proved that, within the enteric nerve plexuses, tachykinins participate in slow excitatory synaptic transmission, a process that is accomplished via slow postsynaptic excitatory potentials and mediated both by NK_1 and NK_3 receptors in a pathway-specific manner (Holzer and Holzer-Petsche 1997b; Alex et al. 2001, 2002; Manning and Mawe 2001; Furness and Sanger 2002; Thornton and Bornstein 2002).

This type of communication is particularly relevant to the transmission between IPANs which form a self-reinforcing network of neurons, and from IPANs to interneurons and motor neurons (Furness et al. 1998; Alex et al. 2001, 2002; Furness and Sanger 2002). In agreement with the predominance of NK_3 receptors on myenteric IPANs of the guinea pig small bowel, with some contribution from NK_1 receptors (Lomax et al. 1998; Jenkinson et al. 1999a), slow IPAN to IPAN signalling and reinforcement of IPAN excitation is brought about predominantly by NK_3 receptors (Alex et al. 2001, 2002; Furness and Sanger 2002). Functional evidence suggests that NK_3 receptors also participate in the transmission from IPANs to ascending and descending interneurons as well as from ascending interneurons to excitatory circular muscle motor neurons (Johnson et al. 1996, 1998). While NK_1 receptors do not seem to make a contribution to transmission between IPANs and descending interneurons, they play an important role in the communication between myenteric IPANs with long anal projections and inhibitory motor neurons (Fig. 2) which bear NK_1 receptors (Portbury et al. 1996b; Johnson et al. 1998; Lomax et al. 1998; Alex et al. 2002; Furness and Sanger 2002; Thornton and Bornstein 2002). This junction is thus far the only example where slow transmission is mediated exclusively by tachykinins acting via NK_1 receptors, without any contribution made by acetylcholine (Alex et al. 2002; Furness and Sanger 2002). Since NK_1 receptors are not confined to synaptic specializations of tachykinin-immunoreactive enteric neurons (where their density is actually reduced) it appears that tachykininergic transmission in the ENS is determined by diffusion of the released tachykinins to nearby tachykinin receptors, given that cells expressing tachykinin receptors are often closely associated with tachykinin-containing nerve fibres (Sternini et al. 1995; Grady et al. 1996; Portbury et al. 2001).

Tachykinins acting via NK_1 and NK_2 receptors participate in excitatory neuromuscular transmission, although in this instance they are subordinate to the principal transmitter acetylcholine (Holzer and Holzer-Petsche 1997). The relative contributions of NK_1 and NK_2 receptors to muscular excitation vary along the GI tract and depend on the modality of stimulation (Maggi 2000; Lecci et al. 2002). In the guinea pig colon, the temporal and pharmacological characteristics of NK_1 and NK_2 receptor-mediated muscle contraction differ most likely because NK_1 receptors are prominently located on ICCs while NK_2 receptors are primarily found on smooth muscle cells (Lecci et al. 2002).

Since NK_1, NK_2 and NK_3 receptors are present on SP-expressing neurons (Sternini et al. 1995; Grady et al. 1996; Portbury et al. 1996b; Lomax et al. 1998;

Jenkinson et al. 1999a; Mann et al. 1999a; Maggi 2000; Vannucchi and Faussone-Pellegrini 2000; Vannucchi et al. 2000; Yip et al. 2003), it appears likely that presynaptic tachykinin autoreceptors exert a feedback control of the transmission process (Patacchini et al. 2000b). In addition, activation of prejunctional NK_1 receptors inhibits acetylcholine release from myenteric motor neurons to the longitudinal muscle of the guinea pig ileum (Kilbinger et al. 1986).

5.2
Motor Regulation

Tachykinins can both stimulate and inhibit motility in the mammalian gut, the net response depending on the type and site of tachykinin receptors that are activated (Fig. 2). Tachykinin-evoked facilitation of GI motor activity or muscle contraction is mediated by activation of NK_1 receptors on IPANs, ICCs and muscle cells, NK_2 receptors on muscle cells, and NK_3 receptors on IPANs, cholinergic interneurons and cholinergic motor neurons. The activation of NK_1 receptors on ICCs (Sternini et al. 1995; Grady et al. 1996; Portbury et al. 1996b; Lavin et al. 1998; Vannucchi et al. 1999) enforces motility by prolonging the duration of the slow waves generated by these cells (Huizinga et al. 1997). In addition, NK_1 receptors contribute to the nerve-dependent and nerve-independent effect of tachykinins to contract smooth muscle in the guinea pig intestine (Legat et al. 1996; Holzer and Holzer-Petsche 1997a; Goldhill et al. 1999; Lecci et al. 1999; Bian et al. 2000; Southwell and Furness 2001; Venkova et al. 2002), a response in which muscular NK_2 receptors also participate (Maggi et al. 1990).

The tachykinin-induced muscle contraction in the human oesophagus as well as in the human and rat intestine is mediated prominently by NK_2 receptors on myocytes (Giuliani et al. 1991; Huber et al. 1993; Croci et al. 1998; Warner et al. 1999; Cao et al. 2000; Maggi 2000; Mulé et al. 2000; Patacchini et al. 2000a, 2001; Krysiak and Preiksaitis 2001). In the circular muscle of the human sigmoid it appears as if tachykinins acting via NK_2 receptors are the main excitatory neurotransmitters (Cao et al. 2000), while in the rat colon both NK_1 and NK_2 receptors mediate the prominent non-cholinergic excitation of the circular muscle (Serio et al. 1998) and NK_2 receptor activation can, in addition, stimulate cholinergic motor neurons (Carini et al. 2001). In vivo, NK_2 receptors make a major contribution to the effect of SP and NKA to stimulate intestinal motility in the rat and to replace the regular pattern of interdigestive motor activity by a pattern of irregular activity reminiscent of postprandial motility, although some participation of NK_1 and NK_3 receptors is also discernible (Lördal et al. 1998; Schmidt et al. 2002). The motor effects of tachykinins on the human gut are very similar (Lördal et al. 1997), with NK_2 receptors also playing a predominant role (Lördal et al. 2001). In contrast, the contractile responses to NK_3 receptor stimulation are predominantly mediated by cholinergic neurons (Holzer & Holzer-Petsche 1997a).

The presence of NK_1 and NK_3 receptors on inhibitory motor pathways within the ENS of the guinea pig, rat, mouse and rabbit intestine enables tachykinins to cause inhibition of GI motor activity via release of nitric oxide and adenosine triphosphate (Jin et al. 1993; Maggi et al. 1993; Holzer 1997; Holzer and Holzer-Petsche 1997a; Lecci et al. 1999; Bian et al. 2000; Mulé et al. 2000; Saban et al. 2000; Alex et al. 2002; Thornton and Bornstein 2002; Onori et al. 2003). Accordingly, peristalsis in the guinea pig isolated small intestine is first stimulated, and then inhibited, by SP (Holzer et al. 1995; Holzer 1997). While NK_2 and NK_3 receptors are responsible for the stimulant response, it is NK_1 receptors that mediate the anti-peristaltic effect of SP through activation of inhibitory motor pathways releasing nitric oxide. A similar nitric oxide-dependent mechanism has been described in the rabbit isolated distal colon, where NK_1 receptor agonists slow and NK_1 receptor antagonists enhance the velocity of propulsion (Onori et al. 2003).

There is considerable in vitro evidence that tachykinins participate in neuroneuronal and neuromuscular transmission processes relevant to GI motor control (Johnson et al. 1996, 1998; Alex et al. 2001, 2002; Furness and Sanger 2002). Analysis of ascending excitatory pathways within the guinea pig intestinal ENS has revealed that NK_1, NK_2 and NK_3 receptor-mediated transmission processes participate in the circular muscle contraction orally to a distension stimulus (Holzer et al. 1993; Maggi et al. 1994b; Johnson et al. 1996, 1998; Furness et al. 2002). While NK_3 receptors contribute to neuroneuronal communication (Johnson et al. 1996, 1998), NK_2 receptors and to some extent NK_1 receptors mediate neuromuscular transmission. When released from excitatory motor neurons, SP and NKA synergize with acetylcholine in the neural excitation of smooth muscle in the guinea pig intestine (Grider 1989; Holzer et al. 1993; Maggi et al. 1994b; Holzer and Holzer-Petsche 1997a; Lecci et al. 1998; Maggi 2000). Acetylcholine and tachykinins acting via NK_2 and NK_1 receptors also mediate the contraction of the guinea pig pylorus in response to duodenal distension (Yuan et al. 2001), and a similar cotransmission process is involved in the descending longitudinal muscle contraction that in the rat colon is elicited by muscle stretch or mucosal stimulation (Grider 2003). The role of tachykinins in the rabbit colon is different in as much as stimulation of NK_2 and NK_3 receptors on descending pathways activates a nitrergic unit and thereby slows peristalsis, whereas stimulation of NK_3 receptors on ascending pathways facilitates peristalsis (Onori et al. 2000, 2001).

Despite the high pharmacological potency of tachykinins in modifying GI motility it is important to realize that under physiological conditions tachykinin receptor antagonists have little effect on GI motor performance. Thus, canine gastric tone and compliance during fasting is not altered by combined blockade of NK_1, NK_2 and NK_3 receptors (Crema et al. 2002). NK_1 receptor antagonists cause a minor stimulation of peristaltic motility in the guinea pig isolated small intestine (Holzer et al. 1995) and enhance the velocity of propulsion in the rabbit isolated distal colon (Onori et al. 2003), which is consistent with their ability to cause mild diarrhoea in humans (Campos et al. 2001; Goldstein et al. 2001). NK_2 receptor antagonists, to the contrary, do not alter small intestinal motility in humans (Lördal et al. 2001) but lead to a minute inhibition of peristalsis in the guinea pig small bowel (Holzer et al. 1995). Interdigestive motility in the canine small and large intestine remains unaltered after treatment with an NK_1, NK_2 or NK_3 receptor antagonist (De Ponti et al. 2001; Giuliani et al. 2001). Only when the overwhelming cholinergic component in the neural activation of smooth muscle has been compromised, does blockade of NK_1 or NK_2 receptors impair peristalsis in the guinea pig and porcine small intestine, whereas blockade of NK_3 receptors has comparatively little effect (Holzer and Maggi 1994; Holzer et al. 1998; Tonini et al. 2002; Schmidt and Holst 2002). Tachykinins thus function as a backup system in the neural activation of GI muscle during peristalsis, a role that in the porcine ileum is brought about by NK_1 receptors (Schmidt and Holst 2002), in the guinea pig small intestine by both NK_1 and NK_2 receptors (Holzer et al. 1998) and in the rabbit colon as well as in the human oesophagus and intestine primarily by NK_2 receptors (Onori et al. 2000; Al-Saffar and Hellström 2001; Krysiak and Preiksaitis 2001; Mitolo-Chieppa et al. 2001).

When, however, all three tachykinin receptors are blocked simultaneously, peristalsis in the guinea pig distal colon is significantly depressed even without concomitant blockade of cholinergic transmission (Tonini et al. 2001). Similarly, NK_1, NK_2 and NK_3 receptors participate in the effect of tachykinins to delay gastric emptying and accelerate intestinal transit in the rat (Chang et al. 1999) and in the contractile response of the guinea pig oesophagus, bile duct and small intestine to stimulation of extrinsic neurons by capsaicin (Barthó et al. 1999; Patacchini et al. 1999). The formation of the migrating motor complex in the mouse isolated colon is inhibited by a combination of NK_1 and NK_2 receptor antagonists (Brierley et al. 2001), whereas interdigestive motility in the canine bowel is not affected by any tachykinin receptor antagonist (De Ponti et al. 2001; Giuliani et al. 2001). However, endogenous tachykinins acting via NK_2 receptors and, to some extent, NK_1 and NK_3 receptors contribute to the prokinetic action of a 5-HT_4 receptor agonist in the canine colon, but not small bowel (De Ponti et al. 2001). The non-cholinergic component in the contractile response of the guinea pig proximal colon to 5-HT_3 and 5-HT_4 stimulation is also inhibited by an NK_1 receptor antagonist (Briejer and Schuurkes 1996).

Taken together, the findings related to GI motor control by tachykinins raise two important issues. Firstly, tachykinins appear to regulate motility primarily in the colon, a hypothesis that has not yet been systematically tested. Secondly, multi- or pan-tachykinin receptor antagonists are predicted to be more efficacious than mono-receptor antagonists in modifying GI motor activity and, eventually, GI motor disorders.

5.3
Secretion of Hormones, Enzymes, Electrolytes and Fluid

SP and NKA modify endocrine and exocrine secretory processes in the GI tract including the stomach and pancreas (Holzer and Holzer-Petsche 1997b). Thus, the gastric secretion of acid, bicarbonate and pepsinogen in certain mammalian species is altered by tachykinins, although the physiological relevance of these actions awaits to be elucidated. In the pig, SP and NKA stimulate the gastric secretion of acid and pepsinogen (Schmidt et al. 1999a) and, via activation of NK_1 receptors, the intestinal secretion of VIP, glucagon-like peptide-1 and somatostatin (Schmidt et al. 1999b). Exocrine and endocrine secretion in the porcine pancreas is also stimulated following NK_1 receptor activation by exogenous and endogenous tachykinins (Schmidt et al. 2000), whereas in the rat pancreas cholecystokinin- and secretin-evoked secretion of amylase and fluid is inhibited by SP through an action on both NK_1 and NK_2 receptors (Kirkwood et al. 1999). Selective activation of NK_3 receptors on acinar cells of the rat pancreas, though, leads to a modest stimulation of amylase secretion (Linari et al. 2002).

Chloride, bicarbonate and fluid output in the mammalian small and large intestine is stimulated through activation of tachykinin receptors on enteric secretomotor pathways (NK_1 and NK_3 receptors) and epithelial cells (NK_1 and NK_2 receptors) (Holzer and Holzer-Petsche 1997b; Schemann et al. 2001). Although SP and NKA can act directly on NK_1 receptor-expressing enterocytes in the guinea pig colon (Cooke et al. 1997a; Hosoda et al. 2002) and NK_2 receptor-bearing epithelial cells in the rat small intestine (Hällgren et al. 1997), tachykinins modify exocrine secretion in the intestine primarily by a neural site of action (Holzer and Holzer-Petsche 2001). Thus, tachykinin-induced secretory processes in the guinea pig colon are mediated predominantly by NK_1 and NK_3 receptors on cholinergic and non-cholinergic neurons (releasing nitric oxide and VIP) of the submucosal plexus, whereas NK_2 receptor stimulation is without effect (Cooke et al. 1997b; Moore et al. 1997; MacNaughton et al. 1997; Frieling et al. 1999; Goldhill et al. 1999). Epithelial ion transport in the human colon is enhanced by both NK_1 and NK_2 receptor activation and subsequent stimulation of enteric neurons (Riegler et al. 1999a; Moriarty et al. 2001a; Tough et al. 2003). Ion transport in the rat colon can be enhanced by activation of all three tachykinin receptors, the NK_2 receptor-mediated effect depending on prostaglandin formation (Moriarty et al. 2001b; Patacchini et al. 2001).

The neurogenic actions of SP and NKA to elicit electrolyte and fluid secretion are consistent with the concept that tachykinins are transmitters of enteric secretory reflex pathways. Some of these reflexes are initiated by activation of enterochromaffin cells which function as sensory transducers of luminal stimuli and activate IPANs via 5-HT acting on $5-HT_{1P}$ receptors (Cooke et al. 1997b, 1997c). The release of 5-HT from enterochromaffin cells is under the inhibitory control of tachykinins working through NK_1 and, indirectly, NK_3 receptors (Ginap and Kilbinger 1997). Activation of IPANs via 5-HT, prostaglandins (Cooke et al. 1997b) or distension (Weber et al. 2001) can induce secretory activity via two different mechanisms. On the one hand, IPANs activate secretory reflex pathways involving interneurons and both cholinergic and non-cholinergic secretomotor neurons (Cooke et al. 1997b, 1997c; Weber et al. 2001). Tachykinins acting via NK_1 and NK_3 receptors participate in the transmission to secretomotor neurons which cause ion secretion through release of acetylcholine or nitric oxide and VIP (Cooke et al. 1997b; MacNaughton et al. 1997; Weber et al. 2001). On the other hand, SP (and other transmitters such as acetylcholine) can be released from axon collaterals of IPANs close to the epithelial effector cells and elicit chloride secretion via an axon reflex type of mechanism (Cooke et al. 1997a). Finally, tachykinins released from extrinsic afferent neurons in response to capsaicin, *Clostridium difficile* toxin A (CDTA) or distension can also stimulate enteric secretomotor neurons through activation of NK_1 and NK_3 receptors (Mantyh et al. 1996a; MacNaughton et al. 1997; Moriarty et al. 2001a, 2001b; Weber et al. 2001).

5.4
Blood Flow, Vascular Permeability and Immune Cell Function

SP and NKA are vasoactive peptides and may induce vasodilatation or vasoconstriction in the digestive tract, the type of action depending on the vascular bed and species under study (Holzer and Holzer-Petsche 1997b). The tachykinin-evoked vasodilatation in the intestine of cat, dog and guinea pig is mediated by NK_1 receptors (Hellström et al. 1991; Ito et al. 1993; Prokopiw and McDonald 1994; Vanner 1994). NK_1 receptors are also present in the media of submucosal blood vessels in the human colon (Goode et al. 2000; Renzi et al. 2000), and both SP and NKA are able to enhance blood flow in the proximal small intestine of humans, probably through activation of NK_1 receptors (Schmidt et al. 2003). In the rat intestine it has proved difficult to demonstrate a vasodilator action of tachykinins (Prokopiw and McDonald 1994; Holzer and Holzer-Petsche 1997b), although there is evidence that SP can dilate the superior mesenteric artery of the rat and that tachykinins acting via NK_1 receptors contribute to the mesenteric hyperaemia evoked by duodenal acidification (Lagaud et al. 1999; Leung et al. 2003). Conversely, blood flow in the rat gastric mucosa is diminished by tachykinins via constriction of collecting venules (Katori et al. 1993; Heinemann et al. 1996; Stroff et al. 1996). Further analysis has revealed that NKA reduces gastric blood flow by activation of NK_1 receptors and inhibits the vasodilator response to ethanol/acid challenge of the gastric mucosa through an NK_2 receptor-mediated action (Heinemann et al. 1996, 1997). SP likewise inhibits the hyperaemic reaction to hypertonic saline/acid challenge of the rat gastric mucosa by a mechanism that involves release of proteases from mast cells (Grønbech and Lacy 1994; Rydning et al. 1999). The physiological relevance of these findings and the functional implication of the NK_3 receptor-positive vasodilator/secretomotor neurons in the submucosal plexus of the guinea pig small intestine (Jenkinson et al. 1999a) await to be explored.

Another effect of SP is to increase venular permeability in the gut and thereby to enhance the extravasation of plasma proteins and cause tissue oedema (Holzer and Holzer-Petsche 1997b). Such an effect has been observed in the GI tract and pancreas of the rat and mouse, where it is probably mediated by NK_1 receptors on the venular endothelium (Kraneveld et al. 1995; Lördal et al. 1996; Figini et al. 1997; Sturiale et al. 1999; Maa et al. 2000a). The activity of tachykinins to enhance venular permeability is normally kept low by the cell surface enzyme neutral endopeptidase (EC 3.4.24.11), given that inhibition or genetic deletion of this enzyme exacerbates tachykinin-induced plasma protein exudation (Sturiale et al. 1999; Grady et al. 2000; Maa et al. 2000a, 2000b).

Tachykinins have the capacity to influence the function of various immune cells in the gut (Holzer and Holzer-Petsche 1997b; Maggi 1997a). The tachykinin-elicited increase in venular permeability facilitates not only the extravasation of proteins and fluid but also that of leukocytes whose activity, like that of mast cells, may be under the direct influence of tachykinins. NK_1 and NK_2 receptors have been localized to monocytes/macrophages, granulocytes, lymphoid cells and eosinophils (Bost et al. 1992; Forsgren et al. 2000; Goode et al. 2000; Renzi et al. 2000; Höckerfelt et al. 2001), and tachykinin receptor stimulation can lead to recruitment and activation of granulocytes as well as mast cells in the GI tract (McCafferty et al. 1994; Pothoulakis et al. 1994; Maggi 1997a).

6
Tachykinins and Tachykinin Receptors in Gastrointestinal Disorders

6.1
Pathological Changes in Tachykinin Expression

An increasing number of studies shows that GI infection, inflammation and mucosal injury are associated with time-related changes in the expression of tachykinins and tachykinin receptors in the gut. It has thus been postulated that, under conditions of GI disease where the tachykinin system is upregulated, tachykinin receptor antagonists may provide significant therapeutic benefit (Holzer 1998; Evangelista 2001). However, the alterations in SP and NKA expression in GI inflammation are contradictory (Table 1), given that the intestinal tachykinin levels in patients with IBD have been reported to be either decreased, increased or unchanged (Koch et al. 1987; Bernstein et al. 1993; Kimura et al. 1994; Keranen et al. 1995; Lee et al. 2002). Importantly, whole-mount immunohistochemistry of the myenteric plexus has shown that the chemical coding of enteric neurons is markedly altered in ulcerative colitis in as much as the proportion of neurons containing SP plus choline acetyltransferase is significantly increased whereas that of SP-negative cholinergic neurons is decreased (Neunlist et al. 2003). Several animal studies have attempted to characterize infection- and inflammation-induced changes in the GI tachykinin system (Table 1) and thus to establish experimental models with which to study the pathophysiological mechanisms behind the observed neuropeptide perturbations. Although some of the experimentally evoked alterations mirror those seen in IBD (Table 1), in several cases it is not known whether the changes are primary or secondary to the insult and whether they reflect changes in the transcriptional, translational or metabolic fate of tachykinins, changes in nerve activity or changes in peptide release (Holzer and Holzer-Petsche 1997b; Holzer 1998). The results are conclusive only when time-related alterations in the tachykinin levels have been related to time-related alterations in gene transcription and peptide release.

Table 1 Alterations of tachykinin (SP, NKA) expression in experimental models of gastrointestinal disorders and in gastrointestinal disease

Species and region	Insult or disease	Change	References
Human stomach	Gastro-oesophageal reflux	Decrease	Wattchow et al. 1992
	Nonulcer dyspepsia	Increase	Kaneko et al. 1993
Mouse stomach	*Helicobacter pylori*	Increase	Bercik et al. 2002
Human pancreas	Pancreatitis	Increase	Di Sebastiano et al. 2000
Human ileum	Pouchitis	Increase	Keranen et al. 1996
Rat small bowel	Irradiation	Decrease	Esposito et al. 1996
		Increase	Höckerfelt et al. 2000, 2001
	E. coli endotoxin	Decrease	Hellström et al. 1997
	C. difficile toxin A	Increase	Castagliuolo et al. 1997
	N. brasiliensis	Decrease	Faussone-Pellegrini et al. 2002
		Increase	Masson et al. 1996
	Trichinella spiralis	Increase	Swain et al. 1992; De Giorgio et al. 2001
Mouse small bowel	Diabetes	Decrease	Spångéus et al. 2001
	Trichinella spiralis	Increase	Agro and Stanisz 1993
	Schistosoma mansoni	Decrease	Varilek et al. 1991
Guinea-pig small bowel	*Trichinella spiralis*	Decrease	Palmer and Koch 1995
	TNBS ileitis	Decrease	Miller et al. 1993
Ferret small bowel	*Trichinella spiralis*	Decrease	Palmer and Greenwood 1993
Human colon	Crohn disease	Decrease	Kimura et al. 1994
		No change	Lee et al. 2002
	Ulcerative colitis	Increase	Koch et al. 1987; Bernstein et al. 1993; Keranen et al. 1995; Neunlist et al. 2003
		No change	Lee et al. 2002
		Decrease	Kimura et al. 1994
	Irradiation	Increase	Forsgren et al. 2000>
	Chronic obstipation	Decrease	Hutson et al. 1996; Tzavella et al. 1996; Treepongkaruna et al. 2001
Rat colon	TNBS colitis	Decrease	Renzi et al. 1992
	Zymosan colitis	Decrease	Traub et al. 1999
	Dextran sulphate colitis	Increase	Kishimoto et al. 1994; Björck et al. 1997
Mouse colon	*Trypanosoma cruzi*	Decrease	Maifrino et al. 1999
Rabbit colon	Immune complex colitis	Decrease	Eysselein et al. 1991; Reinshagen et al. 1995
	TNBS colitis	Decrease	Depoortere et al. 2001
Pig colon	*Schistosoma japonicum*	Increase	Balemba et al. 2001

C. difficile, *Clostridium difficile*; *E. coli*, *Escherichia coli*; *N. brasiliensis*, *Nippostrongylus brasiliensis*; TNBS, trinitrobenzene sulphonic acid.

There is some evidence that the loss of SP from the rat and rabbit colon affected by inflammation due to trinitrobenzene sulphonic acid (TNBS) is the sequel of enhanced peptide release during the initial phase of the inflammatory reaction (Renzi et al. 1992; Depoortere et al. 2001). A similar conclusion applies to the depletion of SP from the rabbit colon in immune complex-induced inflammation (Eysselein et al. 1991), given that the expression of β-PPTA mRNA remains unaltered (Reinshagen et al. 1995). Inflammation-induced release of SP is probably reflected by the elevated SP plasma concentrations that accompany the increase of SP synthesis in rat myenteric neurons following dextran sulphate-induced colitis (Kishimoto et al. 1994). Peptide release from extrinsic spinal afferents likewise seems to account for the loss of SP from the large intestine of rats with zymosan-induced colitis (Traub et al. 1999).

Conversely, infection with *Salmonella dublin* leads to up-regulation of β- and γ-PPTA mRNA in macrophages of rat GI lymphoid organs (Bost et al. 1992; Bost 1995), and macrophages in the lamina propria of the rat ileum treated with CDTA release greater amounts of SP than macrophages from normal ileum (Castagliuolo et al. 1997). The levels of SP in enteric neurons and enterochromaffin cells in the colon of pigs are elevated in inflammation due to infection with *Schistosoma japonicum* (Balemba et al. 2001). Infection of mice with *Helicobacter pylori* leads to an increase in the density of SP- and CGRP-immunoreactive nerves fibres in the stomach and spinal cord, a change that probably reflects alterations in the extrinsic innervation and does not reverse upon eradication of the bacterium (Bercik et al. 2002). Irradiation-induced inflammation and injury is associated with an increase in the expression of SP in the rat duodenum (Höckerfelt et al. 2000; Höckerfelt et al. 2001) and in submucosal neurons and mucosal polymorphonuclear leukocytes of the human colon (Forsgren et al. 2000). The nerve fibres that exhibit an increased content of SP in the chronically inflamed human pancreas seem to be of extrinsic afferent origin (Di Sebastiano et al. 2000). Animal models of type 1 and 2 diabetes, i.e. non-obese diabetic mice and obese diabetic mice, display a reduced number of SP-positive enterochromaffin cells in the proximal but not distal gut (Spångéus et al. 2001), whereas streptozotocin-induced diabetes does not change the SP levels in the rat stomach (Miyamoto et al. 2001).

6.2
Pathological Changes in Tachykinin Receptor Expression

Perturbations of tachykinin receptor expression in GI disease are of particular relevance to disease mechanisms and may provide important therapeutic clues. In this context it is especially worth noting that GI inflammation seems to involve an exaggerated and imbalanced function of the tachykinin system in the gut (Holzer 1998), given that IBD is associated with an up-regulation and ectopic expression of SP binding sites on small blood vessels, lymphoid aggregates and enteric neurons of the small and large bowel (Mantyh et al. 1988, 1995). While the ectopic expression of SP binding sites in ulcerative colitis is confined to active, pathologically positive specimens of the colon, up-regulation of SP binding sites in Crohn disease is seen in both pathologically positive and negative samples of the small and large intestine (Mantyh et al. 1995). These autoradiographic observations have been corroborated by immunohistochemistry showing that IBD causes up-regulation of NK_1 receptor expression in the inflamed and non-inflamed regions of the human ileum and colon (Goode et al. 2000; Renzi et al. 2000). The rise in NK_1 receptor density has been localized to lymphoid, epithelial and endothelial cells and, in patients with Crohn disease, to the myenteric plexus, whereas in ulcerative colitis the number of epithelial NK_1 receptors may be decreased (Goode et al. 2000; Renzi et al. 2000). In addition, the number of NK_2 receptors on eosinophils of the lamina propria is enhanced both in ulcerative colitis and Crohn disease (Renzi et al. 2000). The functional implication of the up-regulated NK_1 receptors in the disease process remains to be elucidated as does the source of tachykinins that target these receptors.

Like IBD, pseudomembranous colitis due to *Clostridium difficile* infection is associated with an up-regulation and ectopic expression of SP binding sites on small blood vessels and lymphoid aggregates (Mantyh et al. 1996a), and experimental enteritis evoked by CDTA in the rat causes rapid expression of NK_1 receptors on epithelial cells (Pothoulakis et al. 1998). Similarly, intestinal inflammation in *Cryptosporidium parvum*-infected mice leads to up-regulation of NK_1 receptors (Sonea et al. 2002). Mononuclear cells in the human colonic mucosa represent another cell system that bears NK_1 receptors (Goode et al. 1998), and *Salmonella dublin* infection enhances the expression of NK_1 receptor mRNA on macrophages of lymphoid organs in the rat gut (Bost et al. 1992). Other experimental models of GI inflammation, though, have failed to reproduce the up-regulation and ectopic expression of NK_1 receptors that occurs in IBD. Thus, irradiation-induced inflammation and injury in the rat duodenum is not followed by any consistent change in the expression of NK_1 receptors (Höckerfelt et al. 2001), and inflammation induced by *Nippostrongylus brasiliensis* leads to a reduction in the number of NK_1 receptors on myenteric neurons and of NK_2 receptors on circular muscle cells in the rat jejunum (Faussone-Pellegrini et al. 2002). Trinitrobenzene sulphonic acid-induced colitis in the rat and rabbit also decreases NK_1 and NK_2 receptor mRNA expression in vasculature, muscle and nerve (Evangelista et al. 1996; Depoortere et al. 2000), a change that is thought to reflect a consequence, not cause, of the inflammatory reaction.

6.3
Motor Disturbances

The concept that tachykinins contribute to pathological disturbances of GI motility (and other digestive functions) is based on three lines of evidence. Firstly, some pathologies with a derangement of GI motility are associated with alterations in the expression of tachykinins (Table 1) and tachykinin receptors. Secondly, the ability of tachykinins to influence GI motor activity is changed in a number of GI motility disorders. Thirdly, tachykinin receptor antagonists are able to normalize motility in some experimental models of GI motor disturbances. Chronic idiopathic constipation, ulcerative colitis and Crohn disease have been found to blunt the ability of the small and large intestine to contract in response to tachykinin receptor activation (Tomita et al. 2000; Al-Saffar and

Hellström 2001; Menzies et al. 2001; Mitolo-Chieppa et al. 2001). Specifically, the efficacy of NK_2 receptor agonists to stimulate the colonic circular muscle is decreased in specimens taken from patients with Crohn disease and ulcerative colitis (Al-Saffar and Hellström 2001; Menzies et al. 2001). The observations made in chronic idiopathic constipation are contradictory, given that the efficacy of NK_2 receptor agonists to contract the colonic circular muscle has been reported to be either increased (Menzies et al. 2001) or unchanged (Mitolo-Chieppa et al. 2001) and their potency to be unaltered (Menzies et al. 2001) or decreased (Mitolo-Chieppa et al. 2001), respectively. While Menzies et al. (2001) hold that NK_1 and NK_3 receptor agonists are unable to activate the circular muscle of the normal colon, Mitolo-Chieppa et al. (2001) report that NK_1 and NK_3 receptor agonists do contract the normal colon and that this activity is depressed in chronic idiopathic constipation.

Alterations in the motor effects of tachykinins have also been observed in experimental models of GI inflammation. While TNBS-evoked colitis in the rat and rabbit blunts the ability of the muscle to contract in response to tachykinin receptor activation (Di Sebastiano et al. 1999; Depoortere et al. 2000; Lecci et al. 2001), ricin-evoked ileitis in the rabbit amplifies neurogenic contractions that are mediated by tachykinins (Goldhill et al. 1997). Irradiation-evoked inflammation likewise enhances the sensitivity of the rat jenunum to contract in response to SP (Esposito et al. 1996).

A role of tachykinins in experimentally induced derangements of GI motor activity is conclusively deduced from the ability of tachykinin receptor antagonists to improve disturbed motility (Table 2). These studies have revealed that NK_1 and NK_2 receptors play differential roles in the GI motor pathologies in as much as NK_1 receptor antagonists are useful in alleviating pathological inhibition of GI motor activity whereas NK_2 receptor antagonism is beneficial in conditions of pathological hypermotility. Specifically, NK_1 receptors are implicated in the hypomotility and muscular hyporesponsiveness associated with anaphylaxis, inflammation and pain (Fargeas et al. 1993; Julia et al. 1994; Espat et al. 1995; Blackshaw and Dent 1997; Di Sebastiano et al. 1999) but also contribute to stress-induced defecation (Ikeda et al. 1995; Castagliuolo et al. 1996; Okano et al. 2001). While peritonitis-evoked gastroparesis is attenuated by NK_1 receptor antagonists (Holzer-Petsche and Rordorf-Nikolic 1995; Julia and Bueno 1997), acute postoperative ileus in the rat is ameliorated by both NK_1 and NK_2 receptor antagonists and the postoperative recovery of myoelectric activity in the jejunum is hastened by an NK_2 receptor antagonist (Espat et al. 1995; Toulouse et al. 2001).

NK_2 receptor antagonists appear to be particularly useful in suppressing the hypermotility associated with colonic inflammation because they are spasmolytic without causing constipation, at least in the rat colon (Croci et al. 1997). Thus, the ability of luminal acetic acid to trigger high-amplitude colonic motility in the rat colon is reduced by the NK_2 receptor antagonist nepadutant (Carini et al. 2001; Lecci et al. 2001). Importantly, the giant colonic contractions associated with castor oil-induced inflammation and diarrhoea (Table 2) are prevent-

Table 2 Tachykinin receptor implications in pathological disturbances of gastrointestinal motility

Stimulus or insult	Motor dysfunction	Tachykinin receptor implication	References
Luminal acidification	Relaxation of ferret lower oesophageal sphincter	NK_1 receptors	Blackshaw and Dent 1997
Intraperitoneal irritation	Inhibition of rat gastric motility and emptying (peritoneogastric reflex)	NK_1 receptors	Holzer-Petsche and Rordorf-Nikolic 1995; Julia and Bueno 1997
Abdominal surgery	Inhibition of rat gastrointestinal transit (intestinointestinal reflex)	NK_1 and NK_2 receptors	Espat et al. 1995; Toulouse et al. 2001
Ovalbumin anaphylaxis	Disruption of migrating motor complex in rat small intestine	NK_1 receptors	Fargeas et al. 1993
TNBS inflammation	Hyporesponsiveness of rat colon muscle to excitatory stimuli	NK_1 receptors	Di Sebastiano et al. 1999
Castor oil-induced inflammatory diarrhoea	Giant contractions in rat colon	NK_2 and, partly, NK_1 receptors	Croci et al. 1997
Acetic acid irritation	High-amplitude motility in rat colon	NK_2 receptors	Carini et al. 2001; Lecci et al. 2001
Rectal distension	Inhibition of rat colonic motility (rectocolonic reflex)	NK_1 receptors	Julia et al. 1994
Restraint stress	Increased defecation in the rat and Mongolian gerbil	NK_1 receptors	Ikeda et al. 1995; Castagliuolo et al. 1996; Okano et al. 2001

TNBS, Trinitrobenzene sulphonic acid.

ed by an NK_2 receptor antagonist and reduced by an NK_1 receptor antagonist (Croci et al. 1997). This finding is consistent with the activity of SP to stimulate migrating giant contractions of the canine colon via an NK_1 receptor-mediated mechanism, an activity that is enhanced after induction of colonic inflammation by ethanol/acetic acid (Tsukamoto et al. 1997). The usefulness of NK_3 receptor antagonists in pathological disturbances of GI motility remains to be explored.

Although the source of the tachykinins and the location of the receptors whose blockade by the antagonists is beneficial are often not known, ENS and GI muscle are very probably the primary target of intervention. For instance, the site of action whereby NK_1 receptor antagonists prevent stress-induced defecation in the Mongolian gerbil seems to be in the ENS (Okano et al. 2001) and, possibly, on colonic mast cells (Bradesi et al. 2002). In contrast, the anti-hypermotility effect of NK_2 receptor antagonists probably takes place at the level of GI smooth muscle (Croci et al. 1997; Lecci et al. 2001). A further site of action concerns tachykininergic extrinsic afferent neurons which can contribute to GI motor dysfunction in two distinct ways. On the one hand, these afferents are likely to participate in autonomic intestinointestinal reflexes in which SP and NKA, released from the central endings of afferent neurons in the spinal cord or brainstem, mediate transmission to the efferent reflex arc. Such a central role is reflected by the contribution which tachykinins make to the peritoneogastric reflex, the rectocolonic reflex and postoperative ileus (Table 2). Tachykinins may also participate in short-loop sympathetic reflexes relayed in the prevertebral ganglia, because the sympathetic neurons in these ganglia receive not only preganglionic but also primary afferent input (Otsuka and Yoshioka 1993). On the other hand, tachykinins released from the peripheral terminals of afferent neurons in the gut can interfere with GI motility (Holzer and Barthó 1996), an instance that may be reflected by the ability of NK_1 receptor antagonists to ameliorate motor dysfunctions caused by oesophageal acidification, anaphylaxis and local inflammation (Table 2).

6.4
Hypersecretion and Inflammation

Pathological changes in GI ion and fluid secretion are frequent manifestations of infection, inflammation and mucosal injury, and there is increasing evidence that SP and NKA participate in a variety of hypersecretory and inflammatory reactions of the gut (Holzer 1998; Evangelista 2001). Although the SP release in response to electrical stimulation or capsaicin exposure is enhanced in the colonic mucosa taken from patients with Crohn disease, but not ulcerative colitis, the secretory response to SP is blunted in tissues from both Crohn disease and ulcerative colitis patients (Moriarty et al. 2001a). Conversely, NK_3 receptor-mediated ion transport in the rat colon is enhanced during stress (Moriarty et al. 2001b).

An implication of tachykinins in GI mucosal pathologies is inferred from the ability of tachykinin receptor antagonists to exert beneficial effects in various models of experimental diarrhoea, GI inflammation and injury (Table 3). Thus, NK_1 receptor antagonists attenuate the granulocyte infiltration and inflammation induced by TNBS or dextran sulphate in the rat colon (McCafferty et al. 1994; Mazelin et al. 1998; Wallace et al. 1998; Di Sebastiano et al. 1999; Stucchi et al. 2000) but not necessarily the TNBS-evoked damage in the rat and guinea pig colon (McCafferty et al. 1994; Wallace et al. 1998). The hypersecretion and inflammation associated with dinitrobenzene sulphonic acid treatment, delayed-type hypersensitivity and infection with *Trichinella spiralis* or *Cryptosporidium parvum* in the mouse intestine (Kataeva et al. 1994; Kraneveld et al. 1995; Sturiale et al. 1999; Sonea et al. 2002) and the allergic hypersecretion in the guinea pig jejunum (Gay et al. 1999) are likewise diminished by NK_1 receptor antagonists. Tachykinins play a particular role in the inflammation (granulocyte, mast cell and macrophage activation) and hypersecretion evoked by CDTA in the rat ileum, which involves activation of capsaicin-sensitive afferent neurons, release of SP and activation of NK_1 receptors on enteric neurons (Castagliuolo et al. 1994; Pothoulakis et al. 1994; Mantyh et al. 1996b). This concept has been corroborated by genetic deletion of the NK_1 receptor, which protects from the secretory and inflammatory responses to CDTA in the mouse ileum (Castagliuolo et al. 1998). Similarly, the increase in vascular permeability,

Table 3 Tachykinin implications in intestinal hypersecretion and inflammation

Stimulus or insult	Dysfunction	Tachykinin receptor implication	References
Caerulein	Acute pancreatitis in the mouse	NK_1 receptors	Bhatia et al. 1998; Grady et al. 2000
Choline-deficient diet	Necrotizing pancreatitis in the mouse	NK_1 receptors	Maa et al. 2000b
Ischaemia and reperfusion	Inflammation and haemorrhage in rat duodenum	NK_1 receptors	Souza et al. 2002
Trichinella spiralis	Inflammation in mouse small intestine	NK_1 receptors	Kataeva et al. 1994
Milk protein allergy	Hypermastocytosis and hypersecretion in rat jejunum	NK_1 receptors	Gay et al. 1999
Cholera toxin	Hypersecretion in rat jejunum	NK_1 and NK_2 receptors	Turvill et al. 2000
C. difficile toxin A	Inflammation and hypersecretion in rat and mouse small intestine	NK_1 receptors	Pothoulakis et al. 1994; Mantyh et al. 1996b; Castagliuolo et al. 1998
DNBS	Mast cell degranulation and plasma leakage in mouse small intestine[a]	NK_1 receptors	Kraneveld et al. 1995
TNBS	Granulocyte infiltration and damage in guinea-pig ileum	NK_2 and NK_3 receptors	Mazelin et al. 1998
	Granulocyte infiltration and damage in rat colon	NK_1 and NK_2 receptors	McCafferty et al. 1994; Mazelin et al. 1998; Wallace et al. 1998; Di Sebastiano et al. 1999
	Damage in mouse colon	NK_1 receptors	Castagliuolo et al. 2002
DNBS	Colitis in the mouse	NK_1 receptors	Sturiale et al. 1999
Dextrane sulphate	Colitis in the rat	NK_1 receptors	Stucchi et al. 2000
Acetic acid	Rectocolitis in the guinea-pig	NK_1 and NK_2 receptors	Cutrufo et al. 1999, 2000
Anti-IgE	Mast cell degranulation in human colon mucosa	NK_1 receptors	Moriarty et al. 2001a
Cryptosporidium parvum	Colitis in the mouse	NK_1 receptors	Sonea et al. 2002
Interleukin-1β	Hypersecretion in rat colon	NK_1 and NK_2 receptors	Eutamene et al. 1995
Castor oil	Diarrhoea in the rat	NK_2 and, partly, NK_1 receptors	Croci et al. 1997
C. difficile toxins A plus B	Diarrhoea in the mouse	NK_2 receptors	Lecci et al. 2001
E. coli toxin Sta	Diarrhoea in the mouse	NK_2 receptors	Lecci et al. 2001
Rectal distension	Hypersecretion in rat colon	NK_1, NK_2 and NK_3 receptors	Eutamene et al. 1997

Anti-IgE, Anti-immunoglobulin E; C. difficile, Clostridium difficile; DNBS, dinitrobenzene sulphonic acid; E. coli, Escherichia coli; TNBS, trinitrobenzene sulphonic acid.
[a] Delayed-type hypersensitivity to DNBS after exposure to dinitrofluorobenzene.

granulocyte recruitment, release of tumour necrosis factor-α and vascular damage induced by ischaemia/reperfusion in the rat duodenum depends on activation of capsaicin-sensitive afferent neurons, release of SP and activation of NK_1 receptors on the venular endothelium (Souza et al. 2002).

GI hypersecretion and inflammation are not only diminished by NK_1 but also by NK_2 receptor antagonists (Table 3). Thus, the diarrhoea evoked by *Escherichia coli* toxin Sta and *Clostridium difficile* toxin A plus B in the mouse is attenuated by an NK_2 receptor antagonist (Lecci et al. 2001). The hypersecretion evoked by *Cholera* toxin in the rat jejunum is reduced by NK_1 and NK_2 receptor antagonists (Turvill et al. 2000) whereas the hypersecretion in the ileum remains unaltered by an NK_1 receptor antagonist (Pothoulakis et al. 1994). Likewise, the diarrhoea induced by castor oil and rectal distension involves activation of multiple tachykinin receptors as does the TNBS-induced rectocolitis in the rat which is diminished by NK_1 and NK_2 receptor antagonists (Table 3). In contrast, the TNBS-evoked ileitis in the guinea pig is attenuated by NK_2 and NK_3, but not NK_1, receptor antagonists (Mazelin et al. 1998). It appears as if tachykinins participate primarily in the initial stage of inflammation, given that NK_1 receptor antagonists inhibit granulocyte infiltration in the rat colon only during the first 12 h after TNBS administration (McCafferty et al. 1994; Wallace et al. 1998), possibly because of NK_1 receptor down-regulation (Evangelista et al. 1996). A similar situation applies to the acetic acid-induced rectocolitis in the guinea pig, which during its early phase is ameliorated by NK_1 and NK_2 receptor antagonists (Cutrufo et al. 1999, 2000). Genetic deletion of NK_1 receptors attenuates the early damage caused by TNBS in the mouse colon, whereas at later stages colonic damage due to TNBS and dextran sulphate is enhanced in NK_1 receptor knockout mice (Castagliuolo et al. 2002). This observation may reflect a facilitatory role of tachykinins in the recovery from colonic injury, given that SP can stimulate intestinal fibroblast proliferation via an NK_1 receptor-mediated mechanism (Castagliuolo et al. 2002).

Tachykinins exert a proinflammatory action not only in the alimentary canal but also in the pancreas, given that experimental pancreatitis evoked by caerulein (Bhatia et al. 1998; Grady et al. 2000) or a choline-deficient diet (Maa et al. 2000b) is ameliorated in NK_1 receptor knockout mice. Inflammation is facilitated through activation of endothelial NK_1 receptors, which causes a rise of venular permeability and subsequent oedema formation. This effect of tachykinins is enhanced in the inflamed tissue because inflammation leads to a down-regulation of neutral endopeptidase which in the normal tissue maintains low levels of SP in the extracellular fluid and thus limits its proinflammatory effects (Sturiale et al. 1999; Grady et al. 2000; Maa et al. 2000a; Kirkwood et al. 2001). Genetic deletion or pharmacological inhibition of neutral endopeptidase enhances basal vascular permeability and exaggerates experimental pancreatitis as well as dinitrobenzene sulphonic acid-evoked colitis and CDTA-induced ileitis in the mouse, these changes being reversed by an NK_1 receptor antagonist (Sturiale et al. 1999; Grady et al. 2000; Maa et al. 2000b; Kirkwood et al. 2001).

Tachykinins are messengers at the interface between the nervous and immune system and it seems as if mast cells, lymphocytes, granulocytes and macrophages in the gut are under the influence of tachykinins (Maggi 1997a). SP-positive nerve fibres lie in close proximity to mucosal mast cells (Stead et al. 1987) from which histamine and other factors are released by the peptide (Shanahan et al. 1985; Lowman et al. 1988), while mast cell stimulation evokes release of SP from the rat and human colonic mucosa (Moriarty et al. 2001a, 2001b). There is increasing awareness that this bidirectional interplay between mast cells and tachykininergic neurons is of relevance to GI disease. Thus, in female rats subjected to stress, but not in controls, SP-induced release of histamine from colonic mast cells is inhibited by an NK_1 receptor antagonist in a female sex steroid-dependent manner (Bradesi et al. 2002). The hypermastocytosis and hypersecretion evoked by food allergy in the guinea pig jejunum (Gay et al. 1999), the secretory response to mast cell stimulation in the human colon (Moriarty et al. 2001a), the plasma protein leakage associated with delayed-type hypersensitivity in the mouse small intestine (Kraneveld et al. 1995) and the ileitis induced by CDTA in the rat and mouse (Pothoulakis et al. 1994; Wershil et al. 1998) involve both tachykinins acting via NK_1 receptors and mast cell-derived factors. Similarly, the NK_1 receptor-mediated secretion in the human and rabbit colonic mucosa involves mast cells, histamine and prostaglandins (Riegler et al. 1999a, 1999b). In addition, NK_1 receptor activation aggravates experimental injury of the rat gastric mucosa through mast cell degranulation (Karmeli et al. 1991; Grønbech and Lacy 1994; Rydning et al. 1999), while NK_2 receptor activation enhances gastric mucosal resistance to injury (Stroff et al. 1996; Improta et al. 1997).

Other tachykinin-responsive immune cells in the gut include lymphocytes (Stanisz et al. 1987; Qian et al. 2001), which is fitting with the ability of an NK_1 receptor antagonist to attenuate lymphocyte proliferation in the small intestine of *Trichinella spiralis*-infected mice (Kataeva et al. 1994). A regulatory influence of tachykinins on granulocytes is suggested by the finding that the granulocyte infiltration caused by CDTA, TNBS and ischaemia/reperfusion in the rat intestine is attenuated by an NK_1 receptor antagonist (McCafferty et al. 1994; Pothoulakis et al. 1994; Souza et al. 2002). The interrelationship between the tachykinin and immune system is bidirectional, since the hypersecretory reaction of the rat colon to interleukin-1β depends on tachykinins (Eutamene et al. 1995). It is also necessary to consider that immune cells are not only targets, at which SP acts to modify immune responses, but under pathological conditions can themselves be induced to synthesize and release tachykinins. This is true for rat peritoneal macrophages exposed to bacterial endotoxin (Bost et al. 1992), macrophages in the mucosa of the rat ileum exposed to CDTA (Castagliuolo et al. 1997), eosinophils from intestinal granulomas of *Schistosoma*-infected mice (Weinstock and Blum 1990) and eosinophils from the mucosa of the inflamed human colon (Metwali et al. 1994).

6.5
Hyperalgesia and Pain

Since the majority of spinal afferents supplying the rodent GI tract expresses SP (Perry and Lawson 1998), tachykinin receptor antagonists are explored for their therapeutic potential in abdominal pain and functional bowel disorders such as IBS (Bueno et al. 2000; Camilleri 2001). A double-blind pilot study has in fact shown that the NK_1 receptor antagonist CJ11974 reduces IBS symptoms and attenuates the emotional response to rectosigmoid distension (Lee et al. 2000). Tachykinins could contribute to abdominal hyperalgesia and pain both indirectly and directly, by enhancing the gain of nociceptors through their proinflammatory, motor and secretory actions in the gut and by being transmitters of nociceptive pathways (Fig. 3). Experimental studies in mice have shown that genetic deletion of NK_1 receptors prevents intracolonic acetic acid and capsaicin from inducing primary mechanical hyperalgesia in the colon and referred mechanical hyperalgesia in the abdominal skin (Laird et al. 2000). In addition, NK_1 receptor-deficient mice fail to respond to intracolonic acetic acid and capsaicin with cardiovascular responses indicative of pain, whereas the reaction to distension is normal (Laird et al. 2000). Thus, NK_1 receptor knockout mice lack pseudoaf-

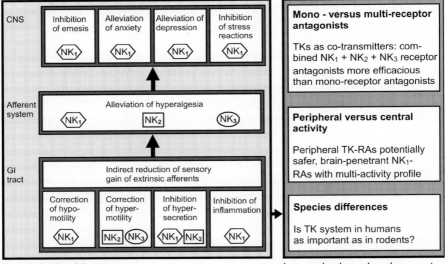

Fig. 3 Summary of the possible utility of tachykinin receptor antagonists (*TK-RAs*) in the treatment of irritable bowel syndrome (*IBS*). The *left panel* highlights the functional implications of the tachykinin (*TK*) system in the rodent gastrointestinal (*GI*) tract and summarizes preclinical evidence for a role of tachykinin NK_1, NK_2 and NK_3 receptors in IBS manifestations at the level of gut, primary afferent neurons and central nervous system (*CNS*). The *right panel* presents some considerations that are relevant to the extrapolation of preclinical data to the development of TK-RAs as IBS therapeutics

fective pain responses to intraluminal stimuli that induce neurogenic inflammation in the colon and do not develop hyperalgesia in response to inflammation.

These findings are complemented by studies with tachykinin receptor antagonists, which show that behavioural pain responses to colorectal (Julia et al. 1994, 1999; Toulouse et al. 2000) and gastric (Toulouse et al. 2001) distension in the rat are attenuated by systemic administration of an NK_2 and NK_3 receptor antagonist but are left unaltered by an NK_1 receptor antagonist. Importantly, the inflammation- and stress-induced hypersensitivity to rectal distension is largely prevented by an NK_2 receptor antagonist (Toulouse et al. 2000). The cardiovascular reaction to jejunal distension, which is exaggerated in rats infected with *Nippostrongylus brasiliensis*, is likewise normalized by an NK_2 receptor antagonist (McLean et al. 1997). However, the reaction to jejunal distension in the absence of intestinal infection is attenuated by both NK_1 and NK_2 receptor antagonists (McLean et al. 1998), and the pseudoaffective responses to peritoneal irritation are reduced either by an NK_1 (Holzer-Petsche and Rordorf-Nikolic 1995) or an NK_2 receptor antagonist (Julia and Bueno 1997). In an attempt to elucidate the site of the antihyperalgesic action of tachykinin receptor antagonists, Kamp et al. (2001) administered the test drugs intrathecally to rats made hyperalgesic by intracolonic instillation of zymosan. The visceromotor response to noxious colorectal distension was not affected by NK_1 and NK_2 receptor antagonists but attenuated by an NK_3 receptor antagonist which was antinociceptive even in rats without colonic inflammation. However, combined blockade of intraspinal NK_1 plus NK_2 receptors was as effective as an NK_3 receptor antagonist in reducing colonic hyperalgesia (Kamp et al. 2001).

The concept that the antinociceptive efficacy of drugs targeting two or all three tachykinin receptors (multi/pan-tachykinin receptor antagonists) is superior to that of selective mono-receptor antagonists (Fig. 3) is supported by other findings. Thus, the afferent signalling of a noxious acid stimulus from the rat stomach to the brainstem is inhibited only by a combination of an NK_1, an NK_2 and an ionotropic *N*-methyl-D-aspartate-type glutamate receptor antagonist (MK801) but not when the drugs are given singly by a systemic route (Jocic et al. 2001). It needs to be considered, though, that chronic pain states are associated with changes in tachykinin receptor expression as is the case in the hyperalgesia associated with experimental pancreatitis in which the density of SP immunoreactivity in the dorsal horn of the rat spinal cord is increased (Vera-Portocarrero et al. 2003). Furthermore, zymosan-induced colitis in the rat leads to de novo expression of NK_1 receptors on dorsal column neurons in the spinal cord (Palecek et al. 2003). In contrast, infection of rats with *Trichinella spiralis* has been found to down-regulate NK_1 receptors in the dorsal horn of the thoracic spinal cord and to attenuate the bradycardic response to gastric distension (De Giorgio et al. 2001).

An additional antinociceptive mechanism is envisaged from the possibility that tachykinin receptor antagonists can act directly on nociceptive afferents, although the issue of tachykinin receptor expression by these nerve fibres (Maggi 1997b; McCarson 1999; Bueno et al. 2000; Blondeau et al. 2002) in health and disease awaits to be clarified. Thus, the NK_2 receptor antagonist nepadutant has been observed to inhibit the enhanced firing which lumbosacral afferent neurons exhibit after distension of the experimentally inflamed rat colon (Laird et al. 2001). Since the activity in pelvic and somatic afferent neurons is not affected, nepadutant has been proposed to be antihyperalgesic by a peripheral action on hypersensitive afferents supplying the colon (Laird et al. 2001). In addition, peripheral nepadutant suppresses the effect of acute TNBS irritation of the colon, but not non-noxious colorectal distension, to stimulate neurons in the spinal cord as visualized by c-*fos* expression (Birder et al. 2003). The effect of intraperitoneally injected tachykinin receptor agonists to elicit pseudoaffective reactions (Kishimoto et al. 1994; Julia and Bueno 1997) as well as the ability of intravenous SP to stimulate mesenteric afferent nerve fibres (Jiang et al. 2002) and to increase vagal afferent nerve activity via an NK_1 receptor-mediated mechanism (Minami et al. 2001) also point to a peripheral site of action.

The tachykinin receptors relevant to abdominal hyperalgesia vary between species. For instance, the behavioural pain responses to colorectal distension in the rat are preferentially inhibited by systemically administered NK_2 and NK_3 receptor antagonists, whereas NK_1 receptor antagonists are ineffective (Julia et al. 1994, 1999; Toulouse et al. 2000, 2001; Laird et al. 2001). In contrast, systemically administered NK_1 receptor antagonists are active in the rabbit, given that TAK637 antagonizes the mechanical hyperalgesia, which in the rabbit colon is seen after acetic acid-induced inflammation (Okano et al. 2002). The antihyperalgesic effect of the compound is due to a site of action in the spinal cord (Okano et al. 2002). In extrapolating these observations to the development of effective drugs for IBS therapy it needs to be clarified whether the tachykinin system in humans is as important in the manifestation of visceral hyperalgesia as in rats, mice, guinea pigs and rabbits (Fig. 3). Inherent in this quest is the issue which types of tachykinin receptors are most relevant in the initiation and/or maintenance of IBS symptoms in humans.

7
Summary

The tachykinins SP and NKA participate in the physiological regulation of various digestive functions, an implication that is portrayed by the cell-specific expression of these peptides and their receptors in the gut (Figs. 1, 2). Gut disorders of various aetiology, particularly those due to infection and inflammation, are related to changes in the expression of tachykinins and their receptors, and there is mounting evidence that tachykinins are causally involved in inflammation-induced perturbations of digestive function. It is hypothesized, therefore, that the contribution of the tachykinin system to normal GI physiology is out of balance in the diseased gut (Holzer 1998, 2000). Remodelling of the ENS away from cholinergic towards tachykininergic innervation and regulation takes place in ulcerative colitis as well as in experimental infection and inflammation of the intestine (Masson et al. 1996; Neunlist et al. 2003). In terms of motor implications, there is evidence that tachykinins play a more prominent transmitter role in the colon than in more proximal parts of the GI tract (Serio et al. 1998; Cao et al. 2000). These experimental findings identify tachykinin receptors as novel targets for gastroenterological therapy. Since tachykinins are cotransmitters of enteric and primary afferent neurons, and multiple tachykinin receptors mediate their messenger role, it appears likely that antagonists targeting two or three tachykinin receptor types are more efficacious than selective mono-receptor antagonists.

Acknowledgements. I should like to thank Professor John B. Furness for helpful discussions and Evelin Painsipp for her expert help in preparing the figures. Work in the author's laboratory is currently supported by the Austrian Research Funds (FWF grant 14295), the Austrian National Bank (grant 9858) and the Zukunftsfonds Steiermark (grant 262).

References

Agro A, Stanisz AM (1993) Inhibition of murine intestinal inflammation by anti-substance P antibody. Reg Immunol 5:120–126

Ahlman H, Ahlund L, Nilsson O, Skolnik G, Theodorsson E, Dahlström A (1988) Carcinoid tumour cells in long-term culture: release of serotonin but not of tachykinins on stimulation with adrenoceptor agonists. Int J Cancer 42:506–510

Aiken KD, Kisslinger JA, Roth KA (1994) Immunohistochemical studies indicate multiple enteroendocrine cell differentiation pathways in the mouse proximal small intestine. Dev Dyn 201:63–70

Alex G, Kunze WA, Furness JB, Clerc N (2001) Comparison of the effects of neurokinin-3 receptor blockade on two forms of slow synaptic transmission in myenteric AH neurons. Neuroscience 104:263–269

Alex G, Clerc N, Kunze WA, Furness JB (2002) Responses of myenteric S neurones to low frequency stimulation of their synaptic inputs. Neuroscience 110:361–373

Al-Saffar A, Hellström PM (2001) Contractile responses to natural tachykinins and selective tachykinin analogs in normal and inflamed ileal and colonic muscle. Scand J Gastroenterol 36:485–493

Bäck N, Ahonen M, Häppölä O, Kivilaakso E, Kiviluoto T (1994) Effect of vagotomy on expression of neuropeptides and histamine in rat oxyntic mucosa. Digt Dis Sci 39:353–361

Balemba OB, Semuguruka WD, Hay-Schmidt A, Johansen MV, Dantzer V (2001) Vasoactive intestinal peptide and substance P-like immunoreactivities in the enteric nervous system of the pig correlate with the severity of pathological changes induced by Schistosoma japonicum. Int J Parasitol 31:1503–1514

Barthó L, Holzer P (1985) Search for a physiological role of substance P in gastrointestinal motility. Neuroscience 16:1–32

Barthó L, Lénárd L, Patacchini R, Halmai V, Wilhelm M, Holzer P, Maggi CA (1999) Tachykinin receptors are involved in the 'local efferent' motor response to capsaicin in the guinea-pig small intestine and oesophagus. Neuroscience 90:221–228

Bayguinov O, Hagen B, Sanders KM (2003) Substance P modulates localized calcium transients and membrane current responses in murine colonic myocytes. Br J Pharmacol 138:1233–1243

Bellucci F, Carini F, Catalani C, Cucchi P, Lecci A, Meini S, Patacchini R, Quartara L, Ricci R, Tramontana M, Giuliani S, Maggi CA (2002) Pharmacological profile of the novel mammalian tachykinin, hemokinin 1. Br J Pharmacol 135:266–274

Bercik P, De Giorgio R, Blennerhassett P, Verdu EF, Barbara G, Collins SM (2002) Immune-mediated neural dysfunction in a murine model of chronic Helicobacter pylori infection. Gastroenterology 123:1205–1215

Bernstein CN, Robert ME, Eysselein VE (1993) Rectal substance P concentrations are increased in ulcerative colitis but not in Crohn's disease. Am J Gastroenterol 88:908–913

Bhatia M, Saluja AK, Hofbauer B, Frossard JL, Lee HS, Castagliuolo I, Wang CC, Gerard N, Pothoulakis C, Steer ML (1998) Role of substance P and the neurokinin 1 receptor in acute pancreatitis and pancreatitis-associated lung injury. Proc Natl Acad Sci USA 95:4760–4765

Bian XC, Bertrand PP, Furness JB, Bornstein JC (2000) Evidence for functional NK_1 tachykinin receptors on motor neurones supplying the circular muscle of guinea-pig small and large intestine. Neurogastroenterol Motil 12:307–315

Birder LA, Kiss S, de Groat WC, Lecci A, Maggi CA (2003) Effect of nepadutant, a neurokinin 2 tachykinin receptor antagonist, on immediate-early gene expression after trinitrobenzenesulfonic acid-induced colitis in the rat. J Pharmacol Exp Ther 304:272–276

Björck S, Jennische E, Dahlström A, Ahlman H (1997) Influence of topical rectal application of drugs on dextran sulfate-induced colitis in rats. Digest Dis Sci 42:824–832

Blackshaw LA, Dent J (1997) Lower oesophageal sphincter responses to noxious oesophageal chemical stimuli in the ferret: involvement of tachykinin receptors. J Auton Nerv Syst 66:189–200.

Blondeau C, Clerc N, Baude A (2002) Neurokinin-1 and neurokinin-3 receptors are expressed in vagal efferent neurons that innervate different parts of the gastro-intestinal tract. Neuroscience 110:339–349

Böckmann S (2002) Substance P (NK_1) receptor expression by human colonic epithelial cell line Caco-2. Peptides 23:1783–1791

Bost KL (1995) Inducible preprotachykinin mRNA expression in mucosal lymphoid organs following oral immunization with Salmonella. J Neuroimmunol 62:59–67

Bost KL, Breeding SA, Pascual DW (1992) Modulation of the mRNAs encoding substance P and its receptor in rat macrophages by LPS. Reg Immunol 4:105–112

Bradesi S, Eutamene H, Fioramonti J, Bueno L (2002) Acute restraint stress activates functional NK1 receptor in the colon of female rats: involvement of steroids. Gut 50:349–354

Briejer MR, Schuurkes JA (1996) $5-HT_3$ and $5-HT_4$ receptors and cholinergic and tachykininergic neurotransmission in the guinea-pig proximal colon. Eur J Pharmacol 308:173–180

Brierley SM, Nichols K, Grasby DJ, Waterman SA (2001) Neural mechanisms underlying migrating motor complex formation in mouse isolated colon. Br J Pharmacol 132:507–517

Brodin E, Sjölund K, Håkanson R, Sundler F (1983) Substance P-containing nerve fibers are numerous in human but not in feline intestinal mucosa. Gastroenterology 85:557–564

Brookes SJH (2001) Classes of enteric nerve cells in the guinea-pig small intestine. Anat Rec 262:58–70

Brookes SJ, Steele PA, Costa M (1991) Calretinin immunoreactivity in cholinergic motor neurones, interneurones and vasomotor neurones in the guinea-pig small intestine. Cell Tissue Res 263:471–481

Brookes SJ, Song ZM, Steele PA, Costa M (1992) Identification of motor neurons to the longitudinal muscle of the guinea pig ileum. Gastroenterology 103:961–973

Bueno L, Fioramonti J, Garcia-Villar R (2000) Pathobiology of visceral pain: molecular mechanisms and therapeutic implications. III. Visceral afferent pathways: a source of new therapeutic targets for abdominal pain. Am J Physiol 278:G670–G676

Burns GA, Cummings JF (1993) Neuropeptide distributions in the colon, cecum, and jejunum of the horse. Anat Rec 236:341–350

Calingasan NY, Kitamura N, Yamada J, Oomori Y, Yamashita T (1984) Immunocytochemical study of the gastroenteropancreatic endocrine cells of the sheep. Acta Anat Basel 118:171–180

Camarda V, Rizzi A, Calo G, Guerrini R, Salvadori S, Regoli D (2002) Pharmacological profile of hemokinin 1: a novel member of the tachykinin family. Life Sci 71:363–370

Camilleri M (2001) Management of the irritable bowel syndrome. Gastroenterology 120:652–668

Campos D, Pereira JR, Reinhardt RR, Carracedo C, Poli S, Vogel C, Martinez-Cedillo J, Erazo A, Wittreich J, Eriksson LO, Carides AD, Gertz BJ (2001) Prevention of cisplatin-induced emesis by the oral neurokinin-1 antagonist, MK-869, in combination with granisetron and dexamethasone or with dexamethasone alone. J Clin Oncol 19:1759–1767

Cao W, Pricolo VE, Zhang L, Behar J, Biancani P, Kirber MT (2000) G_q-linked NK_2 receptors mediate neurally induced contraction of human sigmoid circular smooth muscle. Gastroenterology 119:51–61

Carini F, Lecci A, Tramontana M, Giuliani S, Maggi CA (2001) Tachykinin NK_2 receptors and enhancement of cholinergic transmission in the inflamed rat colon: an in vivo motility study. Br J Pharmacol 133:1107–1113

Castagliuolo I, LaMont JT, Letourneau R, Kelly C, O'Keane JC, Jaffer A, Theoharides TC, Pothoulakis C (1994) Neuronal involvement in the intestinal effects of Clostridium difficile toxin A and Vibrio cholerae enterotoxin in rat ileum. Gastroenterology 107:657–665

Castagliuolo I, LaMont JT, Qiu BS, Fleming SM, Bhaskar KR, Nikulasson ST, Kornetsky C, Pothoulakis C (1996) Acute stress causes mucin release from rat colon: role of corticotropin releasing factor and mast cells. Am J Physiol 271:G884–G892

Castagliuolo I, Keates AC, Qiu BS, Kelly CP, Nikulasson S, Leeman SE, Pothoulakis C (1997) Increased substance P responses in dorsal root ganglia and intestinal macrophages during Clostridium difficile toxin A enteritis in rats. Proc Natl Acad Sci USA 94:4788–4793

Castagliuolo I, Riegler M, Pasha A, Nikulasson S, Lu B, Gerard C, Gerard NP, Pothoulakis C (1998) Neurokinin-1 (NK_1) receptor is required in Clostridium difficile-induced enteritis. J Clin Invest 101:1547–1550

Castagliuolo I, Morteau O, Keates AC, Valenick L, Wang CC, Zacks J, Lu B, Gerard NP, Pothoulakis C (2002) Protective effects of neurokinin-1 receptor during colitis in mice: role of the epidermal growth factor receptor. Br J Pharmacol 136:271–279

Caterina MJ, Schumacher MA, Tominaga M, Rosen TA, Levine JD, Julius D (1997) The capsaicin receptor: a heat-activated ion channel in the pain pathway. Nature 389:816–824

Chang FY, Lee SD, Yeh GH, Wang PS (1999) Rat gastrointestinal motor responses mediated via activation of neurokinin receptors. J Gastroenterol Hepatol 14:39–45

Christofi FL, McDonald TJ, Cook MA (1990) Adenosine receptors are coupled negatively to release of tachykinin(s) from enteric nerve endings. J Pharmacol Exp Ther 253:290–295

Clerc N, Furness JB, Li ZS, Bornstein JC, Kunze WA (1998) Morphological and immunohistochemical identification of neurons and their targets in the guinea-pig duodenum. Neuroscience 86:679–94

Conlon JM, Deacon CF, Grimelius L, Cedermark B, Murphy RF, Thim L, Creutzfeldt W (1988) Neuropeptide $K_{(1-24)}$-peptide: storage and release by carcinoid tumors. Peptides 9:859–866

Cooke HJ, Sidhu M, Fox P, Wang YZ, Zimmermann EM (1997a) Substance P as a mediator of colonic secretory reflexes. Am J Physiol 272:G238–G245

Cooke HJ, Sidhu M, Wang YZ (1997b) 5-HT activates neural reflexes regulating secretion in the guinea-pig colon. Neurogastroenterol Motil 9:181–186

Cooke HJ, Sidhu M, Wang YZ (1997c) Activation of 5-HT$_{1P}$ receptors on submucosal afferents subsequently triggers VIP neurons and chloride secretion in the guinea-pig colon. J Auton Nerv Syst 66:105–110

Costa M, Brookes SJ, Steele PA, Gibbins I, Burcher E, Kandiah CJ (1996) Neurochemical classification of myenteric neurons in the guinea pig ileum. Neuroscience 75:949–967

Crema F, Moro E, Nardelli G, de Ponti F, Frigo G, Crema A (2002). Role of tachykininergic and cholinergic pathways in modulating canine gastric tone and compliance in vivo. Pharmacol Res 45:341–347

Croci T, Landi M, Emonds-Alt X, Le Fur G, Maffrand J-P, Manara L (1997) Role of tachykinins in castor oil diarrhoea in rats. Br J Pharmacol 121:375–380

Croci T, Aureggi G, Manara L, Emonds-Alt X, Le Fur G, Maffrand JP, Mukenge S, Ferla G (1998) In vitro characterization of tachykinin NK$_2$ receptors modulating motor responses of human colonic muscle strips. Br J Pharmacol 124:1321–1327

Cutrufo C, Evangelista S, Cirillo R, Ciucci A, Conte B, Lopez G, Manzini S, Maggi CA (1999) Effect of MEN 11467, a new tachykinin NK$_1$ receptor antagonist, in acute rectocolitis induced by acetic acid in guinea-pigs. Eur J Pharmacol 374:277–283

Cutrufo C, Evangelista S, Cirillo R, Ciucci A, Conte B, Lopez G, Manzini S, Maggi CA (2000) Protective effect of the tachykinin NK$_2$ receptor antagonist nepadutant in acute rectocolitis induced by diluted acetic acid in guinea-pigs. Neuropeptides 34:355–359

Deacon CF, Agoston DV, Nau R, Conlon JM (1987) Conversion of neuropeptide K to neurokinin A and vesicular colocalization of neurokinin A and substance P in neurons of the guinea pig small intestine. J Neurochem 48:141–146

De Giorgio R, Zittel TT, Parodi JE, Becker JM, Brunicardi FC, Go VL, Brecha NC, Sternini C (1995) Peptide immunoreactivities in the ganglionated plexuses and nerve fibers innervating the human gallbladder. J Auton Nerv Syst 51:37–47

De Giorgio R, Barbara G, Blennerhassett P, Wang L, Stanghellini V, Corinaldesi R, Collins SM, Tougas G (2001) Intestinal inflammation and activation of sensory nerve pathways: a functional and morphological study in the nematode infected rat. Gut 49:822–827

De Ponti F, Crema F, Moro E, Nardelli G, Croci T, Frigo GM (2001) Intestinal motor stimulation by the 5-HT4 receptor agonist ML10302: differential involvement of tachykininergic pathways in the canine small bowel and colon. Neurogastroenterol Motil 13:543–553

Depoortere I, Thijs T, van Assche G, Keith JC, Peeters TL (2000) Dose-dependent effects of recombinant human interleukin-11 on contractile properties in rabbit 2,4,6-trinitrobenzene sulfonic acid colitis. J Pharmacol Exp Ther 294:983–990

Depoortere I, Thijs T, Thielemans L, Keith JC, Van Assche G, Peeters TL (2001) Effect of recombinant human interleukin-11 on motilin and substance P release in normal and inflamed rabbits. Regul Pept 97:111–119

Di Sebastiano P, Grossi L, Di Mola FF, Angelucci D, Friess H, Marzio L, Innocenti P, Büchler MW (1999) SR140333, a substance P receptor antagonist, influences morphological and motor changes in rat experimental colitis. Dig Dis Sci 44:439–444

Di Sebastiano P, di Mola FF, Di Febbo C, Baccante G, Porreca E, Innocenti P, Friess H, Büchler MW (2000) Expression of interleukin 8 (IL-8) and substance P in human chronic pancreatitis. Gut 47:423–428

Ekblad E, Alm P, Sundler F (1994) Distribution, origin and projections of nitric oxide synthase-containing neurons in gut and pancreas. Neuroscience 63:233–248

Epperson A, Hatton WJ, Callaghan B, Doherty P, Walker RL, Sanders KM, Ward SM, Horowitz B (2000) Molecular markers expressed in cultured and freshly isolated interstitial cells of Cajal. Am J Physiol 279:C529–C539

Ericson AC, Kechagias S, Öqvist G, Sjöstrand SE (2002) Morphological examination of the termination pattern of substance P-immunoreactive nerve fibers in human antral mucosa. Regul Pept 107:79–86

Espat NJ, Cheng G, Kelley MC, Vogel SB, Sninsky CA, Hocking MP (1995) Vasoactive intestinal peptide and substance P receptor antagonists improve postoperative ileus. J Surg Res 58:719–723

Esposito V, Linard C, Maubert C, Aigueperse J, Gourmelon P (1996) Modulation of gut substance P after whole-body irradiation: a new pathological feature. Digest Dis Sci 41:2070–2077

Euler US von, Gaddum JH (1931) An unidentified depressor substance in certain tissue extracts. J Physiol (Lond) 72:74–87

Eutamene H, Theodorou V, Fioramonti J, Bueno L (1995) Implication of NK_1 and NK_2 receptors in rat colonic hypersecretion induced by interleukin-1β: role of nitric oxide. Gastroenterology 109:483–489

Eutamene H, Theodorou V, Fioramonti J, Bueno L (1997) Rectal distention-induced colonic net water secretion in rats involves tachykinins, capsaicin sensory, and vagus nerves. Gastroenterology 112:1595–1602

Evangelista S (2001) Involvement of tachykinins in intestinal inflammation. Curr Pharm Des 7:19–30

Evangelista S, Maggi M, Renzetti AR (1996) Down-regulation of substance P receptors during colitis induced by trinitrobenzene sulfonic acid in rats. Neuropeptides 30:425–428

Eysselein VE, Reinshagen M, Cominelli F, Sternini C, Davis W, Patel A, Nast CC, Bernstein D, Anderson K, Khan H, Snape WJ (1991) Calcitonin gene-related peptide and substance P decrease in the rabbit colon during colitis. A time study. Gastroenterology 101:1211–1219

Fargeas MJ, Fioramonti J, Bueno L (1993) Involvement of capsaicin-sensitive afferent nerves in the intestinal motor alterations induced by intestinal anaphylaxis in rats. Int Arch Allergy Immunol 101:190–195

Faussone-Pellegrini MS, Gay J, Vannucchi MG, Corsani L, Fioramonti J (2002) Alterations of neurokinin receptors and interstitial cells of Cajal during and after jejunal inflammation induced by Nippostrongylus brasiliensis in the rat. Neurogastroenterol Motil 14:83–95

Ferri GL, Adrian TE, Ghatei MA, Soimero L, Rebecchi L, Biliotti G, Polak JM, Bloom SR (1987) Intramural distribution of regulatory peptides in the human stomach and duodenum. Hepatogastroenterology 34:81–85

Figini M, Emanueli C, Grady EF, Kirkwood K, Payan DG, Ansel J, Gerard C, Geppetti P, Bunnett NW (1997) Substance P and bradykinin stimulate plasma extravasation in the mouse gastrointestinal tract and pancreas. Am J Physiol 272:G785–G793

Forsgren S, Höckerfelt U, Norrgård Ö, Henriksson R, Franzén L (2000) Pronounced substance P innervation in irradiation-induced enteropathy—a study on human colon. Regul Pept 88:1–13

Frieling T, Dobreva G, Weber E, Becker K, Rupprecht C, Neunlist M, Schemann M (1999) Different tachykinin receptors mediate chloride secretion in the distal colon through activation of submucosal neurones. Naunyn Schmiedebergs Arch Pharmacol 359:71–79

Furness JB (2000) Types of neurons in the enteric nervous system. J Auton Nerv Syst 81:87–96

Furness JB, Sanger GJ (2002) Intrinsic nerve circuits of the gastrointestinal tract: identification of drug targets. Curr Opin Pharmacol 2:612–622

Furness JB, Papka RE, Della NG, Costa M, Eskay RL (1982) Substance P-like immunoreactivity in nerves associated with the vascular system of guinea-pigs. Neuroscience 7:447–459

Furness JB, Lloyd KC, Sternini C, Walsh JH (1991) Evidence that myenteric neurons of the gastric corpus project to both the mucosa and the external muscle: myectomy operations on the canine stomach. Cell Tissue Res 266:475–481

Furness JB, Kunze WA, Bertrand PP, Clerc N, Bornstein JC (1998) Intrinsic primary afferent neurons of the intestine. Prog Neurobiol 54:1–18

Furness JB, Kumano K, Larsson H, Murr E, Kunze WA, Vogalis F (2002) Sensitization of enteric reflexes in the rat colon in vitro. Auton Neurosci 97:19–25

Galligan JJ, Costa M, Furness JB (1988) Changes in surviving nerve fibers associated with submucosal arteries following extrinsic denervation of the small intestine. Cell Tissue Res 253:647–656

Gay J, Fioramonti J, Garcia-Villar R, Emonds-Alt X, Bueno L (1999) Involvement of tachykinin receptors in sensitisation to cow's milk proteins in guinea pigs. Gut 44:497–503

Gibbins IL, Furness JB, Costa M (1987) Pathway-specific patterns of the co-existence of substance P, calcitonin gene-related peptide, cholecystokinin and dynorphin in neurons of the dorsal root ganglia of the guinea-pig. Cell Tissue Res 248:417–437

Ginap T, Kilbinger H (1997) NK_1- and NK_3-receptor mediated inhibition of 5-hydroxytryptamine release from the vascularly perfused small intestine of the guinea-pig. Naunyn-Schmiedeberg's Arch Pharmacol 356:689–693

Giuliani S, Barbanti G, Turini D, Quartara L, Rovero P, Giachetti A, Maggi CA (1991) NK_2 tachykinin receptors and contraction of circular muscle of the human colon: characterization of the NK_2 receptor subtype. Eur J Pharmacol 203:365–370

Giuliani S, Guelfi M, Toulouse M, Bueno L, Lecci A, Tramontana M, Criscuoli M, Maggi CA (2001) Effect of a tachykinin NK_2 receptor antagonist, nepadutant, on cardiovascular and gastrointestinal function in rats and dogs. Eur J Pharmacol 415:61–71

Goehler LE, Sternini C, Brecha NC (1988) Calcitonin gene-related peptide immunoreactivity in the biliary pathway and liver of the guinea-pig: distribution and colocalization with substance P. Cell Tissue Res 253:145–150

Goldhill JM, Shea-Donohue T, Ali N, Pineiro-Carrero VM (1997) Tachykininergic neurotransmission is enhanced in small intestinal circular muscle in a rabbit model of inflammation. J Pharmacol Exp Ther 282:1373–1378

Goldhill J, Porquet MF, Selve N (1999) Antisecretory and relaxatory effects of tachykinin antagonists in the guinea-pig intestinal tract. J Pharm Pharmacol 51:1041–1048

Goldstein DJ, Wang O, Gitter BD, Iyengar S (2001) Dose–response study of the analgesic effect of lanepitant in patients with painful diabetic neuropathy. Clin Neuropharmacol 24:16–22

Goode T, O'Connell J, Sternini C, Anton P, Wong H, O'Sullivan GC, Collins JK, Shanahan F (1998) Substance P (neurokinin-1) receptor is a marker of human mucosal but not peripheral mononuclear cells: molecular quantitation and localization. J Immunol 161:2232–2240

Goode T, O'Connell J, Anton P, Wong H, Reeve J, O'Sullivan GC, Collins JK, Shanahan F (2000) Neurokinin-1 receptor expression in inflammatory bowel disease: molecular quantitation and localisation. Gut 47:387–396

Grady EF, Baluk P, Böhm S, Gamp PD, Wong H, Payan DG, Ansel J, Portbury AL, Furness JB, McDonald DM, Bunnett NW (1996). Characterization of antisera specific to NK_1, NK_2 and NK_3 neurokinin receptors and their utilization to localize receptors in the rat gastrointestinal tract. J Neurosci 16:6975–6986

Grady EF, Yoshimi SK, Maa J, Valeroso D, Vartanian RK, Rahim S, Kim EH, Gerard C, Gerard N, Bunnett NW, Kirkwood KS (2000) Substance P mediates inflammatory oedema in acute pancreatitis via activation of the neurokinin-1 receptor in rats and mice. Br J Pharmacol 130:505–512

Green T, Dockray GJ (1988) Characterization of the peptidergic afferent innervation of the stomach in the rat, mouse and guinea-pig. Neuroscience 25:181–193
Greenwood B, Doolittle T, See NA, Koch TR, Dodds WJ, Davison JS (1990) Effects of substance P and vasoactive intestinal polypeptide on contractile activity and epithelial transport in the ferret jejunum. Gastroenterology 98:1509–1517
Grider JR (1989) Tachykinins as transmitters of ascending contractile component of the peristaltic reflex. Am J Physiol 257:G709–G714
Grider JR (2003) Reciprocal activity of longitudinal and circular muscle during intestinal peristaltic reflex. Am J Physiol 284:G768–G775
Grønbech JE, Lacy ER (1994) Substance P attenuates gastric mucosal hyperemia after stimulation of sensory neurons in the rat stomach. Gastroenterology 106:440–449
Grönstad KO, Ahlman H, Nilsson O, Dahlström A, Florence L, Zinner MJ, Jaffe BM (1985) The influence of cholinergic and adrenergic blockade on the portal release of substance P in the cat. J Surg Res 39:316–321
Hällgren A, Flemström G, Hellström PM, Lördal M, Hellgren S, Nylander O (1997) Neurokinin A increases duodenal mucosal permeability, bicarbonate secretion, and fluid output in the rat. Am J Physiol 273:G1077–G1086
Heitz P, Polak JM, Kasper M, Timson CM, Pearse AG (1977) Immunoelectron cytochemical localization of motilin and substance P in rabbit bile duct enterochromaffin (EC) cells. Histochemistry 50:319–325
Heinemann A, Jocic M, Herzeg G, Holzer P (1996) Tachykinin inhibition of acid-induced gastric hyperaemia in the rat. Br J Pharmacol 119:1525–1532
Heinemann A, Sattler V, Jocic M, Holzer P (1997) Inhibition of acid-induced hyperaemia in the rat stomach by endogenous NK_2 receptor ligands. Neurosci Lett 237:133–135
Helliwell RJA, McLatchie LM, Clarke M, Winter J, Bevan S, McIntyre P (1998) Capsaicin sensitivity is associated with the expression of the vanilloid (capsaicin) receptor (VR1) mRNA in adult rat sensory ganglia. Neurosci Lett 250:177–180
Hellström PM, Söder O, Theodorsson E (1991) Occurrence, release, and effects of multiple tachykinins in cat colonic tissues and nerves. Gastroenterology 100:431–440
Hellström PM, Al Saffar A, Ljung T, Theodorsson E (1997) Endotoxin actions on myoelectric activity, transit, and neuropeptides in the gut: role of nitric oxide. Digest Dis Sci 42:1640–1651
Hens J, Schrodl F, Brehmer A, Adriaensen D, Neuhuber W, Scheuermann DW, Schemann M, Timmermans JP (2000) Mucosal projections of enteric neurons in the porcine small intestine. J Comp Neurol 421:429–436
Hens J, Vanderwinden JM, De Laet MH, Scheuermann DW, Timmermans JP (2001) Morphological and neurochemical identification of enteric neurones with mucosal projections in the human small intestine. J Neurochem 76:464–471
Hens J, Gajda M, Scheuermann DW, Adriaensen D, Timmermans JP (2002) The longitudinal smooth muscle layer of the pig small intestine is innervated by both myenteric and submucous neurons. Histochem Cell Biol 117:481–492
Höckerfelt U, Franzén L, Kjörell U, Forsgren S (2000) Parallel increase in substance P and VIP in rat duodenum in response to irradiation. Peptides 21:271–281
Höckerfelt U, Franzén L, Forsgren S (2001) Substance P (NK_1) receptor in relation to substance P innervation in rat duodenum after irradiation. Regul Pept 98:115–126
Holzer P (1984) Characterization of the stimulus-induced release of immunoreactive substance P from the myenteric plexus of the guinea-pig small intestine. Brain Res 297:127–136
Holzer P (1991) Capsaicin: cellular targets, mechanisms of action, and selectivity for thin sensory neurons. Pharmacol Rev 43:143–201
Holzer P (1997) Involvement of nitric oxide in the substance P-induced inhibition of intestinal peristalsis. NeuroReport 8:2857–2860
Holzer P (1998) Implications of tachykinins and calcitonin gene-related peptide in inflammatory bowel disease. Digestion 59:269–283

Holzer P (2002) Sensory neurone responses to mucosal noxae in the upper gut: relevance to mucosal integrity and gastrointestinal pain. Neurogastroenterol Motil 14:459–475

Holzer P, Barthó L (1996) Sensory neurons in the intestine. In: Geppetti P, Holzer P (eds) Neurogenic Inflammation. CRC Press, Boca Raton, Florida, pp 153–167

Holzer P, Holzer-Petsche U (1997a) Tachykinins in the gut. Part I. Expression, release and motor function. Pharmacol Ther 73:173–217

Holzer P, Holzer-Petsche U (1997b) Tachykinins in the gut. Part II. Roles in neural excitation, secretion and inflammation. Pharmacol Ther 73:219–263

Holzer P, Holzer-Petsche U (2001) Tachykinin receptors in the gut: physiological and pathological implications. Curr Opin Pharmacol 1:583–590

Holzer P, Maggi CA (1994) Synergistic role of muscarinic acetylcholine and tachykinin NK-2 receptors in intestinal peristalsis. Naunyn Schmiedeberg's Arch Pharmacol 349:194–201

Holzer P, Emson PC, Iversen LL, Sharman DF (1981) Regional differences in the response to substance P on the longitudinal muscle and the concentration of substance P in the digestive tract of the guinea-pig. Neuroscience 6:1433–1441

Holzer P, Bucsics A, Saria A, Lembeck F (1982) A study of the concentrations of substance P and neurotensin in the gastrointestinal tract of various mammals. Neuroscience 7:2919–2924

Holzer P, Schluet W, Maggi CA (1993) Ascending enteric reflex contraction: roles of acetylcholine and tachykinins in relation to distension and propagation of excitation. J Pharmacol Exp Ther 264:391–396

Holzer P, Schluet W, Maggi CA (1995) Substance P stimulates and inhibits intestinal peristalsis via distinct receptors. J Pharmacol Exp Ther 274:322–328

Holzer P, Lippe IT, Heinemann A, Barthó L (1998) Tachykinin NK_1 and NK_2 receptor-mediated control of peristaltic propulsion in the guinea-pig small intestine in vitro. Neuropharmacology 37:131–138

Holzer-Petsche U (1995) Tachykinin receptors in gastrointestinal motility. Regul Pept 57:19–42

Holzer-Petsche U, Rordorf-Nikolic T (1995) Central versus peripheral site of action of the tachykinin NK_1-antagonist RP 67580 in inhibiting chemonociception. Br J Pharmacol 115:486–490

Hosoda Y, Karaki S, Shimoda Y, Kuwahara A (2002) Substance P-evoked Cl⁻ secretion in guinea pig distal colonic epithelia: interaction with PGE_2. Am J Physiol 283:G347–G356

Huber O, Bertrand C, Bunnett NW, Pellegrini CA, Nadel JA, Debas HT, Geppetti P (1993) Tachykinins contract the circular muscle of the human esophageal body in vitro via NK_2 receptors. Gastroenterology 105:981–987

Huchtebrock HJ, Niebel W, Singer MV, Forssmann WG (1991) Intrinsic pancreatic nerves after mechanical denervation of the extrinsic pancreatic nerves in dogs. Pancreas 6:1–8

Huizinga JD, Thuneberg L, Vanderwinden JM, Rumessen JJ (1997) Interstitial cells of Cajal as targets for pharmacological intervention in gastrointestinal motor disorders. Trends Pharmacol Sci 18:393–403

Hutson JM, Chow CW, Borg J (1996) Intractable constipation with a decrease in substance P-immunoreactive fibres: is it a variant of intestinal neuronal dysplasia? J Pediatr Surg 31:580–583

Ikeda K, Miyata K, Orita A, Kubota H, Yamada T, Tomioka K (1995) RP67580, a neurokinin$_1$ receptor antagonist, decreased restraint stress-induced defecation in rat. Neurosci Lett 198:103–106

Improta G, Broccardo M, Tabacco A, Evangelista S (1997) Central and peripheral antiulcer and antisecretory effects of Ala^5-NKA-(4–10), a tachykinin NK_2 receptor agonist, in rats. Neuropeptides 31:399–402

Ito S, Ohta T, Honda H, Nakazato Y, Ohga A (1993) Gastric vasodilator and motor responses to splanchnic stimulation and tachykinins in the dog. Gen Pharmacol 24:291–298

Jenkinson KM, Morgan JM, Furness JB, Southwell BR (1999a) Neurons bearing NK_3 tachykinin receptors in the guinea-pig ileum revealed by specific binding of fluorescently labelled agonists. Histochem Cell Biol 112:233–246

Jenkinson KM, Southwell BR, Furness JB (1999b) Two affinities for a single antagonist at the neuronal NK_1 tachykinin receptor: evidence from quantitation of receptor endocytosis. Br J Pharmacol 126:131–136

Jenkinson KM, Mann PT, Southwell BR, Furness JB (2000) Independent endocytosis of the NK_1 and NK_3 tachykinin receptors in neurons of the rat myenteric plexus. Neuroscience 100:191–199

Jiang MM, Surprenant A (1992) Re-innervation of submucosal arterioles by myenteric neurones following extrinsic denervation. J Auton Nerv Syst 37:145–154

Jiang W, Kirkup AJ, Bunnett NW, Grundy D (2002) Role of substance P in mesenteric afferent excitation evoked by proteinase-activated receptor subtype 2 agonist. Gastroenterology 122:A-160

Jin JG, Misra S, Grider JR, Makhlouf GM (1993) Functional difference between SP and NKA: relaxation of gastric muscle by SP is mediated by VIP and NO. Am J Physiol 264:G678–G685

Jocic M, Schuligoi R, Schöninkle E, Pabst MA, Holzer P (2001) Cooperation of NMDA and tachykinin NK_1 and NK_2 receptors in the medullary transmission of vagal afferent input from the acid-threatened rat stomach. Pain 89:147–157

Johnson PJ, Bornstein JC, Yuan SY, Furness JB (1996) Analysis of contributions of acetylcholine and tachykinins to neuro-neuronal transmission in motility reflexes in the guinea-pig ileum. Br J Pharmacol 118:973–983

Johnson PJ, Bornstein JC, Burcher E (1998) Roles of neuronal NK_1 and NK_3 receptors in synaptic transmission during motility reflexes in the guinea-pig ileum. Br J Pharmacol 124:1375–1384

Julia V, Bueno L (1997) Tachykininergic mediation of viscerosensitive responses to acute inflammation in rats: role of CGRP. Am J Physiol 272:G141–G146

Julia V, Morteau O, Bueno L (1994) Involvement of neurokinin 1 and 2 receptors in viscerosensitive response to rectal distension in rats. Gastroenterology 107:94–102

Julia V, Su X, Bueno L, Gebhart GF (1999) Role of neurokinin 3 receptors on responses to colorectal distention in the rat: electrophysiological and behavioral studies. Gastroenterology 116:1124–1131

Kage R, McGregor GP, Thim L, Conlon JM (1988) Neuropeptide-γ: a peptide isolated from rabbit intestine that is derived from γ-preprotachykinin. J Neurochem 50:1412–1417

Kamp EH, Beck DR, Gebhart GF (2001) Combinations of neurokinin receptor antagonists reduce visceral hyperalgesia. J Pharmacol Exp Ther 299:105–113

Kaneko H, Mitsuma T, Uchida K, Furusawa A, Morise K (1993) Immunoreactive-somatostatin, substance P, and calcitonin gene-related peptide concentrations of the human gastric mucosa in patients with nonulcer dyspepsia and peptic ulcer disease. Am J Gastroenterol 88:898–904

Karmeli F, Eliakim R, Okon E, Rachmilewitz D (1991) Gastric mucosal damage by ethanol is mediated by substance P and prevented by ketotifen, a mast cell stabilizer. Gastroenterology 100:1206–1216

Kataeva G, Agro A, Stanisz AM (1994) Substance P-mediated intestinal inflammation: inhibitory effects of CP96,345 and SMS 201-995. Neuroimmunomodulation 1:350–356

Katori M, Ohno T, Nishiyama K (1993) Interaction of substance P and leukotriene C_4 in ethanol-induced mucosal injury of rat stomach. Regul Pept 46:241–243

Keast JR, Furness JB, Costa M (1985) Distribution of certain peptide-containing nerve fibres and endocrine cells in the gastrointestinal mucosa in five mammalian species. J Comp Neurol 236:403–422

Keast JR, Furness JB, Costa M (1987) Distribution of peptide-containing neurons and endocrine cells in the rabbit gastrointestinal tract, with particular reference to the mucosa. Cell Tissue Res 248:565–577

Keranen U, Kiviluoto T, Järvinen H, Bäck N, Kivilaakso E, Soinila S (1995) Changes in substance P-immunoreactive innervation of human colon associated with ulcerative colitis. Digest Dis Sci 40:2250–2258

Keranen U, Järvinen H, Kiviluoto T, Kivilaakso E, Soinila S (1996) Substance P- and vasoactive intestinal polypeptide-immunoreactive innervation in normal and inflamed pouches after restorative proctocolectomy for ulcerative colitis. Digest Dis Sci 41:1658–1664

Khan I, Collins SM (1994) Fourth isoform of preprotachykinin messenger RNA encoding for substance P in the rat intestine. Biochem Biophys Res Commun 202:796–802

Kilbinger H, Stauss P, Erlhof I, Holzer P (1986) Antagonist discrimination between subtypes of tachykinin receptors in the guinea-pig ileum. Naunyn-Schmiedeberg's Arch Pharmacol 334:181–187

Kimura M, Masuda T, Hiwatashi N, Toyota T, Nagura H (1994) Changes in neuropeptide-containing nerves in human colonic mucosa with inflammatory bowel disease. Pathol Int 44:624–634

Kirchgessner AL, Gershon MD (1990) Innervation of the pancreas by neurons in the gut. J Neurosci 10:1626–1642

Kirkwood KS, Kim EH, He XD, Calaustro EQ, Domush C, Yoshimi SK, Grady EF, Maa J, Bunnett NW, Debas HT (1999) Substance P inhibits pancreatic exocrine secretion via a neural mechanism. Am J Physiol 277:G314–G320

Kirkwood KS, Bunnett NW, Maa J, Castagliolo I, Liu B, Gerard N, Zacks J, Pothoulakis C, Grady EF (2001) Deletion of neutral endopeptidase exacerbates intestinal inflammation induced by Clostridium difficile toxin A. Am J Physiol 281:G544–G551

Kishimoto S, Tateishi K, Kobayashi H, Kobuke K, Hagio T, Matsuoka Y, Kajiyama G, Miyoshi A (1991) Distribution of neurokinin A-like and neurokinin B-like immunoreactivity in human peripheral tissues. Regul Pept 36:165–171

Kishimoto S, Kobayashi H, Machino H, Tari A, Kajiyama G, Miyoshi A (1994) High concentrations of substance P as a possible transmission of abdominal pain in rats with chemical induced ulcerative colitis. Biomed Res 15(Suppl 2):133–140

Koch TR, Carney JA, Go VL (1987) Distribution and quantitation of gut neuropeptides in normal intestine and inflammatory bowel diseases. Digest Dis Sci 32:369–376

Kotani H, Hoshimaru M, Nawa H, Nakanishi S (1986) Structure and gene organization of bovine neuromedin K precursor. Proc Natl Acad Sci USA 83:7074–7078

Kraneveld AD, Buckley TL, van Heuven-Nolsen D, van Schaik Y, Koster AS, Nijkamp FP (1995) Delayed-type hypersensitivity-induced increase in vascular permeability in the mouse small intestine: inhibition by depletion of sensory neuropeptides and NK_1 receptor blockade. Br J Pharmacol 114:1483–1489

Krysiak PS, Preiksaitis HG (2001) Tachykinins contribute to nerve-mediated contractions in the human esophagus. Gastroenterology 120:39–48

Kurtz M, Wang R, Clements M, Cascieri M, Austin C, Cunningham B, Chicchi G, Liu Q (2002) Identification, localization and receptor characterization of novel mammalian substance P-like peptides. Gene 296:205–212

Lagaud GJ, Skarsgard PL, Laher I, van Breemen C (1999) Heterogeneity of endothelium-dependent vasodilation in pressurized cerebral and small mesenteric resistance arteries of the rat. J Pharmacol Exp Ther 290:832–839

Laird JM, Olivar T, Roza C, De Felipe C, Hunt SP, Cervero F (2000) Deficits in visceral pain and hyperalgesia of mice with a disruption of the tachykinin NK_1 receptor gene. Neuroscience 98:345–352

Laird JM, Olivar T, Lopez-Garcia JA, Maggi CA, Cervero F (2001) Responses of rat spinal neurons to distension of inflamed colon: role of tachykinin NK_2 receptors. Neuropharmacology 40:696–701

Lavin ST, Southwell BR, Murphy R, Jenkinson KM, Furness JB (1998) Activation of neurokinin-1 receptors on interstitial cells of Cajal of the guinea-pig small intestine by substance P. Histochem Cell Biol 110:263–271

Lecci A, Giuliani S, Tramontana M, Giorgio RD, Maggi CA (1998) The role of tachykinin NK_1 and NK_2 receptors in atropine-resistant colonic propulsion in anaesthetized guinea-pigs. Br J Pharmacol 124:27–34

Lecci A, De Giorgio R, Barthó L, Sternini C, Tramontana M, Corinaldesi R, Giuliani S, Maggi CA (1999) Tachykinin NK_1 receptor-mediated inhibitory responses in the guinea-pig small intestine. Neuropeptides 33:91–97

Lecci A, Carini F, Tramontana M, D'Aranno V, Marinoni E, Crea A, Bueno L, Fioramonti J, Criscuoli M, Giuliani S, Maggi CA (2001) Nepadutant pharmacokinetics and dose-effect relationships as tachykinin NK_2 receptor antagonist are altered by intestinal inflammation in rodent models. J Pharmacol Exp Ther 299:247–254

Lecci A, Santicioli P, Maggi CA (2002) Pharmacology of transmission to gastrointestinal muscle. Curr Opin Pharmacol 2:630–641

Lee O-Y, Munakata J, Naliboff BD, Chang L, Mayer EA (2000) A double blind parallel group pilot study of the effects of CJ-11,974 and placebo on perceptual and emotional responses to rectosigmoid distension in IBS patients. Gastroenterology 118:A-846

Lee CM, Kumar RK, Lubowski DZ, Burcher E (2002) Neuropeptides and nerve growth in inflammatory bowel diseases: a quantitative immunohistochemical study. Dig Dis Sci 47:495–502

Legat FJ, Althuber P, Maier R, Griesbacher T, Lembeck F (1996) Evidence for the presence of NK_1 and NK_3 receptors on cholinergic neurones in the guinea-pig ileum. Neurosci Lett 207:125–128

Leung FW, Iwata F, Seno K, Leung JWC (2003) Acid-induced mesenteric hyperemia in rats. Role of CGRP, substance P, prostaglandin, adenosine, and histamine. Dig Dis Sci 48:523–532

Linari G, Broccardo M, Nucerito V, Improta G (2002) Selective tachykinin NK_3 receptor agonists stimulate in vitro exocrine pancreatic secretion in the guinea pig. Peptides 23:947–953

Lippi A, Santicioli P, Criscuoli M, Maggi CA (1998) Depolarization evoked co-release of tachykinins from enteric nerves in the guinea-pig proximal colon. Naunyn-Schmiedeberg's Arch Pharmacol 357:245–251

Llewellyn-Smith IJ, Furness JB, Murphy R, O'Brien PE, Costa M (1984) Substance P-containing nerves in the human small intestine. Distribution, ultrastructure, and characterization of the immunoreactive peptide. Gastroenterology 86:421–435

Llewellyn-Smith IJ, Furness JB, Gibbins IL, Costa M (1988) Quantitative ultrastructural analysis of enkephalin-, substance P-, and VIP-immunoreactive nerve fibers in the circular muscle of the guinea pig small intestine. J Comp Neurol 272:139–148

Llewellyn-Smith IJ, Furness JB, Costa M (1989) Ultrastructural analysis of substance P-immunoreactive nerve fibers in myenteric ganglia of guinea pig small intestine. J Neurosci 9:167–174

Lomax AE, Furness JB (2000) Neurochemical classification of enteric neurons in the guinea-pig distal colon. Cell Tissue Res 302:59–72

Lomax AE, Bertrand PP, Furness JB (1998) Identification of the populations of enteric neurons that have NK_1 tachykinin receptors in the guinea-pig small intestine. Cell Tissue Res 294:27–33

Lördal M, Hällgren A, Nylander O, Hellström PM (1996) Tachykinins increase vascular permeability in the gastrointestinal tract of the rat. Acta Physiol Scand 156:489–494

Lördal M, Theodorsson E, Hellström PM (1997) Tachykinins influence interdigestive rhythm and contractile strength of human small intestine. Dig Dis Sci 42:1940–1949

Lördal M, Branström R, Hellström PM (1998) Mediation of irregular spiking activity by multiple neurokinin receptors in the small intestine of the rat. Br J Pharmacol 123:63-70

Lördal M, Navalesi G, Theodorsson E, Maggi CA, Hellström PM (2001) A novel tachykinin NK_2 receptor antagonist prevents motility-stimulating effects of neurokinin A in small intestine. Br J Pharmacol 134:215-223

Lowman MA, Rees PH, Benyon RC, Church MK (1988) Human mast cell heterogeneity: histamine release from mast cells dispersed from skin, lung, adenoids, tonsils, and colon in response to IgE-dependent and nonimmunologic stimuli. J Allergy Clin Immunol 81:590-597

Maa J, Grady EF, Kim EH, Yoshimi SK, Hutter MM, Bunnett NW, Kirkwood KS (2000a) NK-1 receptor desensitization and neutral endopeptidase terminate SP-induced pancreatic plasma extravasation. Am J Physiol 279:G726-G732

Maa J, Grady EF, Yoshimi SK, Drasin TE, Kim EH, Hutter MM, Bunnett NW, Kirkwood KS (2000b) Substance P is a determinant of lethality in diet-induced hemorrhagic pancreatitis in mice. Surgery 128:232-239

MacNaughton W, Moore B, Vanner S (1997) Cellular pathways mediating tachykinin-evoked secretomotor responses in guinea pig ileum. Am J Physiol 273:G1127-G1134

Maggi CA (1997a) The effects of tachykinins on inflammatory and immune cells. Regul Pept 70:75-90

Maggi CA (1997b) Tachykinins as peripheral modulators of primary afferent nerves and visceral sensitivity. Pharmacol Res 36:153-169

Maggi CA (2000) Principles of tachykininergic co-transmission in the peripheral and enteric nervous system. Regul Pept 93:53-64

Maggi CA, Patacchini R, Giachetti A, Meli A (1990) Tachykinin receptors in the circular muscle of the guinea-pig ileum. Br J Pharmacol 101:996-1000

Maggi CA, Patacchini R, Meini S, Giuliani S (1993) Nitric oxide is the mediator of tachykinin NK_3 receptor-induced relaxation in the circular muscle of the guinea-pig ileum. Eur J Pharmacol 240:45-50

Maggi CA, Holzer P, Giuliani S (1994a) Effect of ω-conotoxin on cholinergic and tachykininergic excitatory neurotransmission to the circular muscle of the guinea-pig colon. Naunyn-Schmiedeberg's Arch Pharmacol 350:529-536

Maggi CA, Patacchini R, Barthó L, Holzer P, Santicioli P (1994b) Tachykinin NK_1 and NK_2 receptor antagonists and atropine-resistant ascending excitatory reflex to the circular muscle of the guinea-pig ileum. Br J Pharmacol 112:161-168

Maifrino LB, Liberti EA, de Souza RR (1999) Vasoactive-intestinal-peptide- and substance P-immunoreactive nerve fibres in the myenteric plexus of mouse colon during the chronic phase of Trypanosoma cruzi infection. Ann Trop Med Parasitol 93:49-56

Mann PT, Southwell BR, Ding YQ, Shigemoto R, Mizuno N, Furness JB (1997) Localisation of neurokinin 3 (NK_3) receptor immunoreactivity in the rat gastrointestinal tract. Cell Tissue Res 289:1-9

Mann PT, Furness JB, Southwell BR (1999a) Choline acetyltransferase immunoreactivity of putative intrinsic primary afferent neurons in the rat ileum. Cell Tissue Res 297:241-248

Mann PT, Southwell BR, Furness JB (1999b) Internalization of the neurokinin 1 receptor in rat myenteric neurons. Neuroscience 91:353-362

Manning BP, Mawe GM (2001) Tachykinins mediate slow excitatory postsynaptic transmission in guinea pig sphincter of Oddi ganglia. Am J Physiol 281:G357-G364

Mantyh CR, Gates TS, Zimmerman RP, Welton ML, Passaro EP Jr, Vigna SR, Maggio JE, Kruger L, Mantyh PW (1988) Receptor binding sites for substance P, but not substance K or neuromedin K, are expressed in high concentrations by arterioles, venules, and lymph nodules in surgical specimens obtained from patients with ulcerative colitis and Crohn disease. Proc Natl Acad Sci USA 85:3235-3239

Mantyh CR, Vigna SR, Bollinger RR, Mantyh PW, Maggio JE, Pappas TN (1995) Differential expression of substance P receptors in patients with Crohn's disease and ulcerative colitis. Gastroenterology 109:850–860

Mantyh CR, Maggio JE, Mantyh PW, Vigna SR, Pappas TN (1996a) Increased substance P receptor expression by blood vessels and lymphoid aggregates in Clostridium difficile-induced pseudomembranous colitis. Dig Dis Sci 41:614–620

Mantyh CR, Pappas TN, Lapp JA, Washington MK, Neville LM, Ghilardi JR, Rogers SD, Mantyh PW, Vigna SR (1996b) Substance P activation of enteric neurons in response to intraluminal Clostridium difficile toxin A in the rat ileum. Gastroenterology 111:1272–1280

Masson SD, McKay DM, Stead RH, Agro A, Stanisz A, Perdue MH (1996) Nippostrongylus brasiliensis infection evokes neuronal abnormalities and alterations in neurally regulated electrolyte transport in rat jejunum. Parasitology 113:173–182

Mawe GM, Gershon MD (1989) Structure, afferent innervation, and transmitter content of ganglia of the guinea pig gallbladder: relationship to the enteric nervous system. J Comp Neurol 283:374–390

Mazelin L, Theodorou V, More J, Emonds-Alt X, Fioramonti J, Bueno L (1998) Comparative effects of nonpeptide tachykinin receptor antagonists on experimental gut inflammation in rats and guinea-pigs. Life Sci 63:293–304

McCafferty DM, Sharkey KA, Wallace JL (1994) Beneficial effects of local or systemic lidocaine in experimental colitis. Am J Physiol 266:G560–G567

McCarson KE (1999) Central and peripheral expression of neurokinin-1 and neurokinin-3 receptor and substance P-encoding messenger RNAs: peripheral regulation during formalin-induced inflammation and lack of neurokinin receptor expression in primary afferent sensory neurons. Neuroscience 93:361–370

McDonald TJ, Ahmad S, Allescher HD, Kostka P, Daniel EE, Barnett W, Brodin E (1990) Canine myenteric, deep muscular, and submucosal plexus preparations of purified nerve varicosities: content and chromatographic forms of certain neuropeptides. Peptides 11:95–102

McGregor GP, Conlon JM (1991) Regulatory peptide and serotonin content and brush-border enzyme activity in the rat gastrointestinal tract following neonatal treatment with capsaicin; lack of effect on epithelial markers. Regul Pept 32:109–119

McGregor GP, Bishop AE, Blank MA, Christofides ND, Yiangou Y, Polak JM, Bloom SR (1984) Comparative distribution of vasoactive intestinal polypeptide (VIP), substance P and PHI in the enteric sphincters of the cat. Experientia 40:469–471

McLean PG, Picard C, Garcia-Villar R, More J, Fioramonti J, Bueno L (1997) Effects of nematode infection on sensitivity to intestinal distension: role of tachykinin NK_2 receptors. Eur J Pharmacol 337:279–282

McLean PG, Garcia-Villar R, Fioramonti J, Bueno L (1998) Effects of tachykinin receptor antagonists on the rat jejunal distension pain response. Eur J Pharmacol 345:247–252

Menzies JR, McKee R, Corbett AD (2001) Differential alterations in tachykinin NK_2 receptors in isolated colonic circular smooth muscle in inflammatory bowel disease and idiopathic chronic constipation. Regul Pept 99:151–156

Messenger JP (1993) Immunohistochemical analysis of neurons and their projections in the proximal colon of the guinea-pig. Arch Histol Cytol 56:459–474

Metwali A, Blum AM, Ferraris L, Klein JS, Fiocchi C, Weinstock JV (1994) Eosinophils within the healthy or inflamed human intestine produce substance P and vasoactive intestinal peptide. J Neuroimmunol 52:69–78

Miller MJ, Sadowska-Krowicka H, Jeng AY, Chotinaruemol S, Wong M, Clark DA, Ho W, Sharkey KA (1993) Substance P levels in experimental ileitis in guinea pigs: effects of misoprostol. Am J Physiol 265:G321–G330

Mimoda T, Kitamura N, Hondo E, Yamada J (1998) Immunohistochemical colocalization of serotonin, substance P and Met-enkephalin-Arg^6-Gly^7-Leu^8 in the endocrine cells of the ruminant duodenum. Anat Histol Embryol 27:65–69

Minami M, Endo T, Yokota H, Ogawa T, Nemoto M, Hamaue N, Hirafuji M, Yoshioka M, Nagahisa A, Andrews PL (2001) Effects of CP-99, 994, a tachykinin NK_1 receptor antagonist, on abdominal afferent vagal activity in ferrets: evidence for involvement of NK_1 and $5-HT_3$ receptors. Eur J Pharmacol 428:215–220

Mitolo-Chieppa D, Mansi G, Nacci C, De Salvia MA, Montagnani M, Potenza MA, Rinaldi R, Lerro G, Siro-Brigiani G, Mitolo CI, Rinaldi M, Altomare DF, Memeo V (2001) Idiopathic chronic constipation: tachykinins as cotransmitters in colonic contraction. Eur J Clin Invest 31:349–355

Miyamoto Y, Yoneda M, Morikawa A, Itoh H, Makino I (2001) Gastric neuropeptides and gastric motor abnormality in streptozotocin-induced diabetic rats: observation for four weeks after streptozotocin. Dig Dis Sci 46:1596–1603

Moore BA, Vanner S, Bunnett NW, Sharkey KA (1997) Characterization of neurokinin-1 receptors in the submucosal plexus of guinea pig ileum. Am J Physiol 273:G670–G678

Moriarty D, Goldhill J, Selve N, O'Donoghue DP, Baird AW (2001a) Human colonic antisecretory activity of the potent NK_1 antagonist, SR140333: assessment of potential anti-diarrhoeal activity in food allergy and inflammatory bowel disease. Br J Pharmacol 133:1346–1354

Moriarty D, Selve N, Baird AW, Goldhill J (2001b) Potent NK_1 antagonism by SR-140333 reduces rat colonic secretory response to immunocyte activation. Am J Physiol 280:C852–C858

Moussaoui SM, Le Prado N, Bonici B, Faucher DC, Cuine F, Laduron PM, Garret C (1992) Distribution of neurokinin B in rat spinal cord and peripheral tissues: comparison with neurokinin A and substance P and effects of neonatal capsaicin treatment. Neuroscience 48:969–978

Mulé F, D'Angelo S, Tabacchi G, Serio R (2000) Involvement of tachykinin NK_2 receptors in the modulation of spontaneous motility in rat proximal colon. Neurogastroenterol Motil 12:459–466

Neunlist M, Dobreva G, Schemann M (1999a) Characteristics of mucosally projecting myenteric neurones in the guinea-pig proximal colon. J Physiol (London) 517:533–546

Neunlist M, Reiche D, Michel K, Pfannkuche H, Hoppe S, Schemann M (1999b) The enteric nervous system: region and target specific projections and neurochemical codes. Eur J Morphol 37:233–240

Neunlist M, Aubert P, Toquet C, Oreshkova T, Barouk J, Lehur PA, Schemann M, Galmiche JP (2003) Changes in chemical coding of myenteric neurones in ulcerative colitis. Gut 52:84–90

Okano S, Nagaya H, Ikeura Y, Natsugari H, Inatomi N (2001) Effects of TAK-637, a novel neurokinin-1 receptor antagonist, on colonic function in vivo. J Pharmacol Exp Ther 298:559–564

Okano S, Ikeura Y, Inatomi N (2002) Effects of tachykinin NK_1 receptor antagonists on the viscerosensory response caused by colorectal distention in rabbits. J Pharmacol Exp Ther 300:925–931

Onori L, Aggio A, Taddei G, Tonini M (2000) Contribution of NK_2 tachykinin receptors to propulsion in the rabbit distal colon. Am J Physiol 278:G137–G147

Onori L, Aggio A, Taddei G, Ciccocioppo R, Severi C, Carnicelli V, Tonini M (2001) Contribution of NK_3 tachykinin receptors to propulsion in the rabbit isolated distal colon. Neurogastroenterol Motil 13:211–219

Onori L, Aggio A, Taddei G, Loreto MF, Ciccocioppo R, Vicini R, Tonini M (2003) Peristalsis regulation by tachykinin NK_1 receptors in the rabbit isolated distal colon. Am J Physiol 285:G325–G331

Otsuka M, Yoshioka K (1993) Neurotransmitter functions of mammalian tachykinins. Physiol Rev 73:229–308

Page NM, Bell NJ, Gardiner SM, Manyonda IT, Brayley KJ, Strange PG, Lowry PJ (2003) Characterization of the endokinins: human tachykinins with cardiovascular activity. Proc Natl Acad Sci USA 100:6245–6250

Palecek J, Paleckova V, Willis WD (2003) Postsynaptic dorsal column neurons express NK_1 receptors following colon inflammation. Neuroscience 116:565–572

Palmer JM, Greenwood B (1993) Regional content of enteric substance P and vasoactive intestinal peptide during intestinal inflammation in the parasitized ferret. Neuropeptides 25:95–103

Palmer JM, Koch TR (1995) Altered neuropeptide content and cholinergic enzymatic activity in the inflamed guinea pig jejunum during parasitism. Neuropeptides 28:287–297

Patacchini R, Barthó L, De Giorgio R, Lénárd L, Stanghellini V, Barbara G, Lecci A, Maggi CA (1999) Involvement of endogenous tachykinins and CGRP in the motor responses produced by capsaicin in the guinea-pig common bile duct. Naunyn-Schmiedeberg's Arch Pharmacol 360:344–353

Patacchini R, Giuliani S, Turini A, Navarra G, Maggi CA (2000a) Effect of nepadutant at tachykinin NK_2 receptors in human intestine and urinary bladder. Eur J Pharmacol 398:389–397

Patacchini R, Maggi CA, Holzer P (2000b) Tachykinin autoreceptors in the gut. Trends Pharmacol Sci 21:166

Patacchini R, Cox HM, Stahl S, Tough IR, Maggi CA (2001) Tachykinin NK_2 receptor mediates contraction and ion transport in rat colon by different mechanisms. Eur J Pharmacol 415:277–283

Pernow P (1951) Substance P distribution in the digestive tract. Acta Physiol Scand 24:97–102

Pernow B (1983) Substance P. Pharmacol Rev 35:85–141

Perry MJ, Lawson SN (1998) Differences in expression of oligosaccharides, neuropeptides, carbonic anhydrase and neurofilament in rat primary afferent neurons retrogradely labelled via skin, muscle or visceral nerves. Neuroscience 85:293–310

Portbury AL, Furness JB, Southwell BR, Wong H, Walsh JH, Bunnett NW (1996a) Distribution of neurokinin-2 receptors in the guinea-pig gastrointestinal tract. Cell Tissue Res 286:281–292

Portbury AL, Furness JB, Young HM, Southwell BR, Vigna SR (1996b) Localisation of NK_1 receptor immunoreactivity to neurons and interstitial cells of the guinea-pig gastrointestinal tract. J Comp Neurol 367:342–351

Portbury AL, Grkovic I, Young HM, Furness JB (2001) Relationship between postsynaptic NK_1 receptor distribution and nerve terminals innervating myenteric neurons in the guinea-pig ileum. Anat Rec 263:248–254

Pothoulakis C, Castagliuolo I, LaMont JT, Jaffer A, O'Keane JC, Snider RM, Leeman SE (1994) CP-96,345, a substance P antagonist, inhibits rat intestinal responses to Clostridium difficile toxin A but not cholera toxin. Proc Natl Acad Sci USA 91:947–951

Pothoulakis C, Castagliuolo I, Leeman SE, Wang CC, Li H, Hoffman BJ, Mezey E (1998) Substance P receptor expression in intestinal epithelium in Clostridium difficile toxin A enteritis in rats. Am J Physiol 275:G68–G75

Probert L, De Mey J, Polak JM (1983) Ultrastructural localization of four different neuropeptides within separate populations of p-type nerves in the guinea pig colon. Gastroenterology 85:1094–1104

Prokopiw I, McDonald TJ (1994) Effects of tachykinins on the splanchnic circulation: a comparison of the dog and rat. Peptides 15:1189–1194

Qian BF, Zhou GQ, Hammarström ML, Danielsson A (2001) Both substance P and its receptor are expressed in mouse intestinal T lymphocytes. Neuroendocrinology 73:358–368

Reinshagen M, Adler G, Eysselein VE (1995) Substance P gene expression in acute experimental colitis. Regul Pept 59:53–58

Renzi D, Tramontana M, Panerai C, Surrenti C, Evangelista S (1992) Decrease of calcitonin gene-related peptide, but not vasoactive intestinal polypeptide and substance P, in the TNB-induced experimental colitis in rats. Neuropeptides 22:56–57

Renzi D, Pellegrini B, Tonelli F, Surrenti C, Calabrò A (2000) Substance P (neurokinin-1) and neurokinin A (neurokinin-2) receptor gene and protein expression in the healthy and inflamed human intestine. Am J Pathol 157:1511–1522

Rettenbacher M, Reubi JC (2001) Localization and characterization of neuropeptide receptors in human colon. Naunyn-Schmiedeberg's Arch Pharmacol 364:291–304

Riegler M, Castagliuolo I, So PT, Lotz M, Wang C, Wlk M, Sogukoglu T, Cosentini E, Bischof G, Hamilton G, Teleky B, Wenzl E, Matthews JB, Pothoulakis C (1999a) Effects of substance P on human colonic mucosa in vitro. Am J Physiol 276:G1473–G1483

Riegler M, Castagliuolo I, Wlk M, Pothoulakis C (1999b) Substance P causes a chloride-dependent short-circuit current response in rabbit colonic mucosa in vitro. Scand J Gastroenterol 34:1203–1211

Rydning A, Lyng O, Aase S, Grønbech JE (1999) Substance P may attenuate gastric hyperemia by a mast cell-dependent mechanism in the damaged gastric mucosa. Am J Physiol 277:G1064–G1073

Saban R, Nguyen N-B, Saban MR, Gerard NP, Pasricha PJ (1999) Nerve-mediated motility of ileal segments isolated from NK_1 receptor knockout mice. Am J Physiol 277:G1173–G1179

Sand J, Tainio H, Nordback I (1993) Neuropeptides in pig sphincter of Oddi, bile duct, gallbladder, and duodenum. Dig Dis Sci 38:694–700

Sang Q, Young HM (1996) Chemical coding of neurons in the myenteric plexus and external muscle of the small and large intestine of the mouse. Cell Tissue Res 284:39–53

Schemann M, Schaaf C, Mader M (1995) Neurochemical coding of enteric neurons in the guinea pig stomach. J Comp Neurol 353:161–178

Schemann M, Reiche D, Michel K (2001) Enteric pathways in the stomach. Anat Rec 262:47–57

Schmidt PT, Holst JJ (2002) Tachykinin NK_1 receptors mediate atropine-resistant net aboral propulsive complexes in porcine ileum. Scand J Gastroenterol 37:531–535

Schmidt P, Poulsen SS, Rasmussen TN, Bersani M, Holst JJ (1991) Substance P and neurokinin A are codistributed and colocalized in the porcine gastrointestinal tract. Peptides 12:963–973

Schmidt PT, Rasmussen TN, Holst JJ (1999a) Tachykinins stimulate acid and pepsinogen secretion in the isolated porcine stomach. Acta Physiol Scand 166:335–340

Schmidt PT, Rickelt LF, Holst JJ (1999b) Tachykinins stimulate release of peptide hormones (glucagon-like peptide-1) and paracrine (somatostatin) and neurotransmitter (vasoactive intestinal polypeptide) from porcine ileum through NK-1 receptors. Dig Dis Sci 44:1273–1281

Schmidt PT, Tornoe K, Poulsen SS, Rasmussen TN, Holst JJ (2000) Tachykinins in the porcine pancreas: potent exocrine and endocrine effects via NK-1 receptors. Pancreas 20:241–247

Schmidt PT, Bozkurt A, Hellström PM (2002) Tachykinin-stimulated small bowel myoelectric pattern: sensitization by NO inhibition, reversal by neurokinin receptor blockade. Regul Pept 105:15–21

Schmidt PT, Lördal M, Gazelius B, Hellström PM (2003) Tachykinins potently stimulate human small bowel blood flow: a laser Doppler flowmetry study in humans. Gut 52:53–56

Serio R, Mulé F, Bonvissuto F, Postorino A (1998) Tachykinins mediate noncholinergic excitatory neural responses in the circular muscle of rat proximal colon. Can J Physiol Pharmacol 76:684–689

Shanahan F, Denburg JA, Fox J, Bienenstock J, Befus D (1985) Mast cell heterogeneity: effects of neuroenteric peptides on histamine release. J Immunol 135:1331–1337

Shuttleworth CW, Keef KD (1995) Roles of peptides in enteric neuromuscular transmission. Regul Pept 56:101–120
Shuttleworth CW, Murphy R, Furness JB, Pompolo S (1991) Comparison of the presence and actions of substance P and neurokinin A in guinea-pig taenia coli. Neuropeptides 19:23–34
Sjölund K, Sanden G, Håkanson R, Sundler F (1983) Endocrine cells in human intestine: an immunocytochemical study. Gastroenterology 85:1120–1130
Smith VC, Sagot MA, Wong H, Buchan AMJ (2000) Cellular expression of the neurokinin 1 receptor in the human antrum. J Auton Nerv Syst 79:165–172
Sokolski KN, Lechago J (1984) Human colonic substance P-producing cells are a separate population from the serotonin-producing enterochromaffin cells. J Histochem Cytochem 32:1066–1074
Sonea IM, Palmer MV, Akili D, Harp JA (2002) Treatment with neurokinin-1 receptor antagonist reduces severity of inflammatory bowel disease induced by Cryptosporidium parvum. Clin Diagn Lab Immunol 9:333–340
Southwell BR, Furness JB (2001) Immunohistochemical demonstration of the NK_1 tachykinin receptor on muscle and epithelia in guinea pig intestine. Gastroenterology 120:1140–1151
Southwell BR, Woodman HL, Murphy R, Royal SJ, Furness JB (1996) Characterisation of substance P-induced endocytosis of NK_1 receptors on enteric neurons. Histochem Cell Biol 106:563–571
Southwell BR, Seybold VS, Woodman HL, Jenkinson KM, Furness JB (1998a) Quantitation of neurokinin 1 receptor internalization and recycling in guinea-pig myenteric neurons. Neuroscience 87:925–931
Southwell BR, Woodman HL, Royal SJ, Furness JB (1998b) Movement of villi induces endocytosis of NK_1 receptors in myenteric neurons from guinea-pig ileum. Cell Tissue Res 292:37–45
Souza DG, Mendonca VA, de A Castro MS, Poole S, Teixeira MM (2002) Role of tachykinin NK receptors on the local and remote injuries following ischaemia and reperfusion of the superior mesenteric artery in the rat. Br J Pharmacol 135:303–312
Spångéus A, Forsgren S, El-Salhy M (2001) Effect of diabetic state on co-localization of substance P and serotonin in the gut in animal models. Histol Histopathol 16:393–398
Stanisz AM, Scicchitano R, Dazin P, Bienenstock J, Payan DG (1987) Distribution of substance P receptors on murine spleen and Peyer's patch T and B cells. J Immunol 139:749–754
Stead RH, Tomioka M, Quinonez G, Simon GT, Felten SY, Bienenstock J (1987) Intestinal mucosal mast cells in normal and nematode-infected rat intestines are in intimate contact with peptidergic nerves. Proc Natl Acad Sci USA 84:2975–2979
Steele PA, Costa M (1990) Opioid-like immunoreactive neurons in secretomotor pathways of the guinea-pig ileum. Neuroscience 38:771–786
Sternini C (1991) Tachykinin and calcitonin gene-related peptide immunoreactivities and mRNAs in the mammalian enteric nervous system and sensory ganglia. In: Costa M, Surrenti C, Gorini S, Maggi CA, Meli A (eds) Sensory Nerves and Neuropeptides in Gastroenterology. From Basic Science to Clinical Perspectives. Plenum Press, New York, pp 39–51
Sternini C (1992) Enteric and visceral afferent CGRP neurons. Targets of innervation and differential expression patterns. Ann New York Acad Sci 657:170–186.
Sternini C, Anderson K, Frantz G, Krause JE, Brecha N (1989) Expression of substance P/neurokinin A-encoding preprotachykinin messenger ribonucleic acids in the rat enteric nervous system. Gastroenterology 97:348–356
Sternini C, Su D, Gamp PD, Bunnett NW (1995) Cellular sites of expression of the neurokinin-1 receptor in the rat gastrointestinal tract. J Comp Neurol 358:531–540

Stroff T, Plate S, Ebrahim JS, Ehrlich KH, Respondek M, Peskar BM (1996) Tachykinin-induced increase in gastric mucosal resistance: role of primary afferent neurons, CGRP, and NO. Am J Physiol 271:G1017–G1027

Stucchi AF, Shofer S, Leeman S, Materne O, Beer E, McClung J, Shebani K, Moore F, O'Brien M, Becker JM (2000) NK-1 antagonist reduces colonic inflammation and oxidative stress in dextran sulfate-induced colitis in rats. Am J Physiol 279:G1298–G1306

Sturiale S, Barbara G, Qiu B, Figini M, Geppetti P, Gerard N, Gerard C, Grady EF, Bunnett NW, Collins SM (1999) Neutral endopeptidase (EC 3.4.24.11) terminates colitis by degrading substance P. Proc Natl Acad Sci USA 96:11653–11658

Sundler F, Alumets J, Håkanson R (1977) 5-Hydroxytryptamine-containing enterochromaffin cells: storage site of substance P. Acta Physiol Scand Suppl 452:121–123

Suzuki T, Kagoshima M, Shibata M, Inaba N, Onodera S, Yamaura T, Shimada H (1997) Effects of several denervation procedures on distribution of calcitonin gene-related peptide and substance P immunoreactive fibers in rat stomach. Dig Dis Sci 42:1242–1254

Swain MG, Agro A, Blennerhassett P, Stanisz,A, Collins SM (1992) Increased levels of substance P in the myenteric plexus of Trichinella-infected rats. Gastroenterology 102:1913–1919

Takeda Y, Takeda J, Smart BM, Krause JE (1990) Regional distribution of neuropeptide å and other tachykinin peptides derived from the substance P gene in the rat. Regul Pept 28:323–333

Tateishi K, Kishimoto S, Kobayashi H, Kobuke K, Matsuoka Y (1990) Distribution and localization of neurokinin A-like immunoreactivity and neurokinin B-like immunoreactivity in rat peripheral tissue. Regul Pept 30:193–200

Thornton PD, Bornstein JC (2002) Slow excitatory synaptic potentials evoked by distension in myenteric descending interneurones of guinea-pig ileum. J Physiol (London) 539:589–602

Timmermans JP, Scheuermann DW, Stach W, Adriaensen D, De Groodt-Lasseel MH (1990) Distinct distribution of CGRP-, enkephalin-, galanin-, neuromedin U-, neuropeptide Y-, somatostatin-, substance P-, VIP- and serotonin-containing neurons in the two submucosal ganglionic neural networks of the porcine small intestine. Cell Tissue Res 260:367–379

Tomita R, Tanjoh K, Fujisaki S, Fukuzawa M (2000) Peptidergic nerves in the colon of patients with ulcerative colitis. Hepatogastroenterology 47:400–404

Tonini M, Spelta V, De Ponti F, De Giorgio R, D'Agostino G, Stanghellini V, Corinaldesi R, Sternini C, Crema F (2001) Tachykinin-dependent and -independent components of peristalsis in the guinea pig isolated distal colon. Gastroenterology 120:938–945

Tonini M, De Ponti F, Frigo G, Crema F (2002) Pharmacology of the enteric nervous system. In: Brookes S, Costa M (eds) Innervation of the Gastrointestinal Tract. Taylor & Francis, London, pp 213–294

Too HP, Cordova JL, Maggio JE (1989) A novel radioimmunoassay for neuromedin K. I. Absence of neuromedin K-like immunoreactivity in guinea pig ileum and urinary bladder. II. Heterogeneity of tachykinins in guinea pig tissues. Regul Pept 26:93–105

Tough IR, Lewis CA, Fozard J, Cox HM (2003) Dual and selective antagonism of neurokinin NK_1 and NK_2 receptor-mediated responses in human colon mucosa. Naunyn-Schmiedeberg's Arch Pharmacol 367:104–108

Toulouse M, Coelho AM, Fioramonti J, Lecci A, Maggi CA, Bueno L (2000) Role of tachykinin NK_2 receptors in normal and altered rectal sensitivity in rats. Br J Pharmacol 129:193–199

Toulouse M, Fioramonti J, Maggi C, Bueno L (2001) Role of NK_2 receptors in gastric barosensitivity and in experimental ileus in rats. Neurogastroenterol Motil 13:45–53

Traub RJ, Hutchcroft K, Gebhart GF (1999) The peptide content of colonic afferents decreases following colonic inflammation. Peptides 20:267–273

Treepongkaruna S, Hutson JM, Hughes J, Cook D, Catto-Smith AG, Chow CW, Oliver MR (2001) Gastrointestinal transit and anorectal manometry in children with colonic substance P deficiency. J Gastroenterol Hepatol 16:624–630

Tsukamoto M, Sarna SK, Condon RE (1997) A novel motility effect of tachykinins in normal and inflamed colon. Am J Physiol 272:G1607–G1614

Turvill JL, Connor P, Farthing MJ (2000) Neurokinin 1 and 2 receptors mediate cholera toxin secretion in rat jejunum. Gastroenterology 119:1037–1044

Tzavella K, Riepl RL, Klauser AG, Voderholzer WA, Schindlbeck NE, Müller-Lissner SA (1996) Decreased substance P levels in rectal biopsies from patients with slow transit constipation. Eur J Gastroenterol Hepatol 8:1207–1211

Vanner S (1994) Corelease of neuropeptides from capsaicin-sensitive afferents dilates submucosal arterioles in guinea pig ileum. Am J Physiol 267:G650–G655

Vanner S, Surprenant A (1991) Cholinergic and noncholinergic submucosal neurons dilate arterioles in guinea pig colon. Am J Physiol 261:G136–G144

Vannucchi M-G, Faussone-Pellegrini M-S (2000) NK_1, NK_2 and NK_3 tachykinin receptor localization and tachykinin distribution in the ileum of the mouse. Anat Embryol 202:247–255

Vannucchi MG, Corsani L, Faussone-Pellegrini MS (1999) Substance P immunoreactive nerves and interstitial cells of Cajal in the rat and guinea-pig ileum. A histochemical and quantitative study. Neurosci Lett 268:49–52

Vannucchi M-G, Corsani L, Faussone-Pellegrini M-S (2000) Co-distribution of NK_2 tachykinin receptors and substance P in nerve endings of guinea-pig ileum. Neurosci Lett 287:71–75

Varilek GW, Weinstock JV, Williams TH, Jew J (1991) Alterations of the intestinal innervation in mice infected with Schistosoma mansoni. J Parasitol 77:472–478

Venkova K, Sutkowski-Markmann DM, Greenwood-Van Meerveld B (2002) Peripheral activity of a new NK_1 receptor antagonist TAK-637 in the gastrointestinal tract. J Pharmacol Exp Ther 300:1046–1052

Vera-Portocarrero LP, Lu Y, Westlund KN (2003) Nociception in persistent pancreatitis in rats: effects of morphine and neuropeptide alterations. Anesthesiology 98:474–484

Wallace JL, McCafferty DM, Sharkey KA (1998) Lack of beneficial effect of a tachykinin receptor antagonist in experimental colitis. Regul Pept 73:95–101

Wang H, Zhang YQ, Ding YQ, Zhang JS (2002) Localization of neurokinin B receptor in mouse gastrointestinal tract. World J Gastroenterol 8:172–175

Wang XY, Sanders KM, Ward SM (1999) Intimate relationship between interstitial cells of cajal and enteric nerves in the guinea-pig small intestine. Cell Tissue Res 295:247–256

Warner FJ, Comis A, Miller RC, Burcher E (1999) Characterization of the ^{125}I-neurokinin A binding site in the circular muscle of human colon. Br J Pharmacol 127:1105–1110

Wathuta EM (1986) The distribution of vasoactive intestinal polypeptide-like, substance P-like and bombesin-like immunoreactivity in the digestive system of the sheep. Q J Exp Physiol 71:615–631

Wattchow DA, Furness JB, Costa M (1988) Distribution and coexistence of peptides in nerve fibers of the external muscle of the human gastrointestinal tract. Gastroenterology 95:32–41

Wattchow DA, Jamieson GG, Maddern GJ, Furness JB, Costa M (1992) Distribution of peptide-containing nerve fibers in the gastric musculature of patients undergoing surgery for gastroesophageal reflux. Ann Surg 216:153–160

Wattchow DA, Porter AJ, Brookes SJ, Costa M (1997) The polarity of neurochemically defined myenteric neurons in the human colon. Gastroenterology 113:497–506

Weber E, Neunlist M, Schemann M, Frieling T (2001) Neural components of distension-evoked secretory responses in the guinea-pig distal colon. J Physiol (London) 536:741–751

Weinstock JV, Blum AM (1990) Release of substance P by granuloma eosinophils in response to secretagogues in murine schistosomiasis Mansoni. Cell Immunol 125:380–385

Wershil BK, Castagliuolo I, Pothoulakis C (1998) Direct evidence of mast cell involvement in Clostridium difficile toxin A-induced enteritis in mice. Gastroenterology 114:956–964

Yip L, Kwok YN, Buchan AM (2003) Cellular localization and distribution of neurokinin-1 receptors in the rat stomach. Auton Neurosci 104:95–108

Yuan SY, Costa M, Brookes SJ (2001) Neuronal control of the pyloric sphincter of the guinea-pig. Neurogastroenterol Motil 13:187–198

Yunker AMR, Krause JE, Roth KA (1999) Neurokinin B- and substance P-like immunoreactivity are co-localized in enteric nerves of rat ileum. Regul Pept 80:67–74

Zhang Y, Lu L, Furlonger C, Wu GE, Paige CJ (2000) Hemokinin is a hematopoietic-specific tachykinin that regulates B lymphopoiesis. Nat Immunol 1:392–739

Subject Index

abdominal
- hyperalgesia 536
- pain 512

acetylcholine 2, 17, 122, 126, 258, 351, 369, 520–522, 524
acne vulgaris 477
addiction 314
adenosine 522
adjuncts 452
adrenaline 91
adrenocorticotrophic hormone (ACTH) 7
Advisory Committee Meeting 412
airway
- dysfunction 277
- hyperresponsiveness 280, 492
- inflammation 499
- obstruction 492
- edema 183
- ganglia 277

allergen challenge 499
allodynia 202
alopecia areata 474
amino acid 16
ammonium chloride 165
amphetamine 314, 402
amygdala 301, 310, 347
amygdalectomy 348
anaesthesia 399
analgesic effect 452
anandamide 463
anaphylaxis 530
angiotensin 376
- receptor antagonist 254

antidepressant efficacy 314
antidromic stimulation 463
antinociceptive action 40
anxiety 103, 197, 205, 309, 344, 348
apomorphine 378, 379, 383, 386, 393, 395, 402, 416
aprepitant 103, 343–345, 362, 411–413
arachidonic acid 175
- cascade 42
- mobilization 31
area postrema (AP) 365, 368, 371, 417
arrestin 145
- β-arrestin 143, 145, 147, 150, 152, 157, 164
- – MAP kinase signaling 160
arthritis 445
ascending interneuron 520
ascorbic acid 7
asthma 103, 184, 257, 276, 289, 290, 481
- hyperreactivity 206
atopic dermatitis 462
atropine 2, 122
autonomic reflex 13, 15
axon reflex 418

baclofen 158
bacterial artificial chromosome 304
bafilomycin A1 165
basal ganglia 87
BDNF 313, 349
benzylamino piperidines 196
blood

– flow 525
– pressure 41–45, 47
^{125}I-Bolton-Hunter 70
– reagent 31, 42
– scyliorhinin II 37
bombesin 425, 474
Bötzinger complex 420, 423
bradycardia 38, 41, 44, 46
bradykinin 4, 124, 289
brain
– penetration 417
– stem 349, 365
– – nuclei 372
breathing pattern 277
bronchial
– hyperreactivity 206
– responsiveness 499
bronchoconstriction 39, 44, 47, 183, 184, 192, 206, 258, 276, 277, 288
– antigen-induced 499
– effects 496
bronchodilator 492
bronchospasm 42, 278
bufokinin 42, 43

C fiber 12
calcitonin gene-related peptide (CGRP) 9, 449, 461, 465, 515, 516
calcium
– intracellular ion concentration 49
– channel 189
– mobilization 42, 48
calretinin 515, 516
cAMP 175
– accumulation 31
cancer chemotherapy 195
capsaicin 10, 13, 188, 191, 193, 203, 274, 277, 278, 288, 298, 375, 376, 461, 516, 517, 523, 525, 532, 536
carboxyl tail 153
carcinoid tumor 30, 34, 44, 513
cardiovascular
– effect 40
– response 41, 44, 313
– studies 49
– system 41
castor oil 534
cat studies 394
CDTA 528, 529

central nervous system (CNS) 276, 342
central pattern generator (CPG) 420
cerebrospinal fluid 34, 44
c-fos 377
– expression 384, 421
CGP49823 200, 287
charybdotoxin 473
chemical coding 514–516, 526
chemotherapy 406, 407
Cholera toxin 534
choline acetyltransferase 515, 516, 526
cholinergic
– motor neuron 521
– neuron cotransmitter 516
chronic
– bronchitis 495
– idiopathic constipation 529
– inflammation 492
– obstructive pulmonary disease (COPD) 257, 491
cigarette smoke 500
circular muscle 515, 521, 522, 529, 530
– cells 518
– layer 519
cisplatin 197, 202, 378–380, 390, 391, 396, 401, 409
citric acid 279
CJ11974 391, 407, 536
clathrin 149
Clostridium difficile 159
– infection 529
– toxin A (CDTA) 525
cluster headache 193
CNS disorders 261
colchicine 68
cold air 500
colitis
– dextran sulphate-induced 528
– TNBS-evoked 530
– zymosan-induced 528, 537
colon 523, 524, 527–530, 532, 535
concanavalin A 165
conditioned place
– aversion 315
– preference (CPP) 314
conscious animal 443
constitutive NO synthase (cNOS) 468
copper sulphate 384, 385, 392, 396
corticosteroid 492
corticosterone 426

corticotropin-releasing factor (CRF) 7, 69
cotransmission 522
cough 280
CP100263 390
CP122721 190, 344, 385, 387, 394, 396, 406, 413, 414
CP96345 79, 175, 187, 189, 278, 386, 466
CP99994 189, 278, 279, 374, 377, 384–388, 390, 392, 396, 398, 404, 418, 419
– antiemetic activity 190
Crohn disease 528, 529, 532
Cryptosporidium parvum 529, 532
CS003 502
cultured human dermal microvascular endothelial cells (HDMEC) 462
cyclophosphamide 387
cysLT1 receptor antagonist 497
cytochrome P450 413

defecation 376
– stress-induced 530, 531
dental pain 447
depression 202, 205, 309
dermatitis 474
descending
– facilitation 308
– interneurons 519, 520
desensitization 143
dexamethasone 380, 392, 401, 402, 407–410, 418, 426
dextran sulphate 532
diabetes 528
diabetic neuropathy 448
diarrrhoea 522, 530, 532, 534
diazepam 350
3,4-dichlorophenyl 284, 285
diencephalon 77, 87
dinitrobenzene sulphonic acid 532, 534
distension 522, 524, 534, 536, 537
DNK333 284, 287
dog studies 367, 392
dopamine 66, 87, 91, 262, 383
dorsal root ganglia 516
downregulation 163, 166
drug reward 314
dual
– receptor antagonist 289

– receptor blockade 280
– antagonist 274
dynamin 150
dysmotility disorder 257
dyspnea 258

Escherichia coli toxin 534
eczema 474
E-enamide 285
electrolyte secretion 524
electrophysiological studies 443
eledoisin 6, 37, 123, 125, 375
Eledone moschata 6, 123
Emend 343, 362, 412
emesis 190, 195, 197, 205, 368, 376
– acute type 400, 409
– delayed type 400
– motion-induced 414
emetic pathway 374
endocrine cells 517
endocytic domain 151
endocytosis 143, 150, 158, 164, 519
– clathrin-mediated 149
– agonist-induced 151
endokinin 514
– A 135, 136
– B 135
– C 26, 29, 49, 135
– D 26, 29, 135
endopeptidase 276
endosomes 156, 16, 165
endothelial cell 518, 529
enkephalin 515
enteric
– nervous system (ENS) 511, 513–515, 521, 531, 538
– neuron 519, 524, 526, 528
enterochromaffine cell 517, 524, 528
enterotoxin 474
eosinophils 518, 526, 529, 535
epithelial
– cell 511, 518, 519, 524
– ion transport 524
epithelium 275
ERK1/2 161
ethanol 398
excitatory
– motor neuron 515
– neuromuscular transmission 520

experimental infection 538
extravasation 525
extrinsic
– afferent neurons 531
– primary afferent
– – nerve 514
– – neuron 511

feedback control 521
fentanyl 395, 403
ferret studies 367
fibromyalgia 450
FK224 502
FK886 396
FK888 183, 502
fluid secretion 524, 532
fluoxetine 349, 350
Aδ fibers 465
food intake 45, 46
foot tapping 310
formalin 158, 187
FR113680 183

gamma-aminobutyric acid (GABA) 92, 157
gastric
– emptying 523
– secretion 523
gastrin releasing peptide receptor 163
gastrointestinal (GI)
– afferents 366
– disorder 257
– tract 373, 512, 517, 519, 523, 526, 539
gastroparesis 530
glutamate 16, 19, 424
G-protein-coupled receptor (GPCR) 142
– lysosomal degradation 164
– recycling 163
– trafficking 143, 163
– B 136
GR100679 221, 351
GR112000 221
GR138676 248
GR159897 232, 351
GR203040 193, 386, 388, 390, 393, 397, 398
GR205171 195, 310, 388, 390, 393, 397, 398, 400, 401, 404, 408, 414, 421

GR71251 181
GR82334 183, 388, 417
GR83074 221
GR94800 221, 397
granisetron 408, 409
granulocytes 518, 526, 535

habenula 301
halothane 399, 419
heart
– disease 259
– rate 41, 42, 45
Helicobacter pylori 528
hemokinin 1 (HK-1) 26, 28, 37, 48, 49, 133, 136, 513
hepatobiliary system 516
hippocampus 348
histamine 2, 183, 369, 535
homologous recombination 304
Hox gene 29
HSP117 386
5-HT 349, 352, 374, 375
– neurons 91
– 5-HT$_3$ receptor 415
human brain 83
human immunodeficiency virus (HIV) 462
hyoscine bromide 415
hyperalgesia 159, 187, 191, 201, 202, 308, 512, 536–538
– inflammatory form 444
hypercortisolemia 347
hypermastocytosis 535
hypermotility 530
hypersecretion 512, 532, 534, 535
hypersensitivity 202, 537
– delayed type 532, 535
– type IV 476
hypertension 259
hypertonic
– saline 500
– sucrose 157
hyperventilation 500
hypomotility 530
hypotension 42, 44, 48, 49, 135, 192
hypothalamus 352
hypothermia 346
hypoxia 421
hysterectomy 414

ICC   518, 519
idazoxan   395
ileitis
- CDTA-induced   534
- ricin-evoked   530
- TNBS-evoked   534
imipramine   350
immune
- cell   518, 526, 535
- system   535
immunohistochemistry   65–67, 101
in situ hybridization   66, 68, 101
infection   512, 526, 528, 537, 538
inflammation   192, 205, 512, 526, 528, 530, 532, 534, 537, 538
- castor oil-induced   530
- neurogenic inflammation   11
inflammatory bowel disease (IBD)   206, 257, 512, 526, 528, 529
inflammatory pain   158, 260
inhibitory motor neuron   520
inositol monophosphate   42, 43, 48, 192
interdigestive
- motility   523
- motor activity   521
internalization   519
interneuron   516, 520, 524
interstitial cells of Cajal   515
intestinal pain   257
intestine   514, 519, 523–525, 528, 529, 535
intestinointestinal reflex   531
intrinsic
- enteric neurons   514
- primary afferent neuron (IPAN)   515, 518
ion secretion   532
ipecacuanha   387, 395, 417
irritable bowel syndrome (IBS)   190, 450, 512, 538

kaolin   402
Kaplan–Meier curve   411
kassinin   125
keratinocytes   461
L659837   222
L659874   221
L659877   222, 397
L668169   181

L709210   196
L732138   198
L733060   196, 310
L741671   388
L742694   196
L743310   388, 417
L743986   256
L754030   197, 391, 408, 409
L758298   391, 408–410
L759274   344, 345
lamina I   308
lanepitant   198
leukocytes   526
leukotriene B_4 468
L-glutamate   8
light–dark shuttle box   310
linker   284
lipophilicity   183
locus coeruleus   301
longitudinal muscle   515, 518, 519
long-term potentiation   310
loperamide   383–386, 395
losartan   251
lower brain stem   88
lung   275
- cancer   378
LY303870 (lanepitant)   198, 388
lymphocytes   518, 535
lymphoid cell   526
lysine residues   163
lysosomes   163, 164

macrophages   526, 528, 529, 535
major depressive disorder   197, 343, 344
MDL103220   284
MDL103392   284
MDL105172A   286
MDL105212   256, 284, 286
MDL29913   226
melanin-concentrating hormone (MCH)   103
MEN10207   220
MEN10376   220
MEN10456   220
MEN10573   226
MEN10627   223, 226
MEN10930   184
MEN11420   224, 502

MEN11467 184
mesencephalon 77
methacholine 276, 386
methioninamide 6
methotrexate 394
methyl amide 284
metoclopramide 379, 407
micturition reflex 42, 203
migraine 192, 195, 200, 448, 449
migrating motor complex 523
mitogen-activated protein kinase (MAP kinase) 160
– activation 162, 498
mitogenic signaling 161
– arrestin-dependent 161
MK869 309, 343, 408–410
monensin 157, 165
monkey
– brain 89
– studies 367
monoamine reuptake inhibitor 350
mono-receptor antagonist 537, 539
Montreal Nomenclature 127, 128, 136
morphine 383, 385, 392, 395, 403
– withdrawal 314, 315
Morris water maze 317
motility 521, 522, 529, 532
motor
– component coordination 365
– disturbance 529
– neurons 516, 518–520
– regulation 521
mucosa 515, 516, 518, 525, 529, 535
mucosal injury 526, 532
mucus secretion 277
muscle
– cell 521
– contraction 520, 521
muscularis mucosae 515, 518, 519
mustard oil 463
myenteric
– ganglia 17
– neuron 529
– plexus 515, 518, 519, 526, 529

naloxone 343, 395, 398
naphtalene 285
nausea 190, 195, 343, 362, 369, 407
– pathways for the induction 366

– physiology 363
nepadutant 224, 530, 538
nerve
– growth factor (NGF) 471, 500
– terminal, SP-positive 97
neuroblastoma 36
neuroendocrine reflex 13
neurogenesis 310, 348
neurogenic
– inflammation 11, 537
– mucus secretion 500
– plasma extravasation 446, 498
neurokinin
– A (NKA) 15, 18, 65, 124, 149, 175, 248, 258, 342, 381, 383, 492, 511, 512
– – cell bodies 71
– B (NKB) 15, 36, 47, 65, 124, 175, 247, 248, 342, 381, 383, 513
– – mRNA 88
neuromedin
– K 36
– L 33
neuromuscular transmission 522
neuron
– sensory neuron 34
– SP-positive multipolar neuron 85
Neuronorm 227
neuropeptide
– γ 25, 35, 39, 45, 65, 149, 513
– K (NPK) 25, 34, 65, 70, 149, 513
– Y 102
neurotransmission 415
neurotrophin 500
neutral endopeptidase 526, 534
nicotine 396, 397, 425
nifedipine 191
Nippostrongylus brasiliensis 529, 537
nitric oxide (NO) 465, 522, 524
NK_1 receptor 175, 297, 361, 495, 497, 511
– antagonist 342
– – ^{11}C-labeled 100
– – peptidase-resistant 31, 39
– – therapeutic perspectives 204
– distribution 96
– gene 298
– intraspecies heterogeneity 177
– knockout mice 534
– locus 303
– mRNA

– – telencephalon 96
– nonpeptide antagonist 187, 361
– protein
– – amino acid sequence 176
– selective nonpeptide antagonist 248
– species difference in pharmacology
– variants 131
NK$_2$ receptor 132, 495, 497, 511, 523
– antagonist 220, 342, 351
– selective nonpeptide antagonist 248
– subtypes 132
– variants 131
NK$_3$ receptor 248, 495, 511, 523
– antagonist 246, 256, 342, 352
– – clinical applications 256
– distribution 96
– nonselective antagonist 250
– selective agonist 247, 249
– variants 133
NK$_3$-positive cell 97
NK5807 234
NKP608 201
N-methyl-D-aspartate (NMDA) 16, 157, 424
nociception 205, 260, 305
– visceral nociception 308
nodose ganglia 516
nonpeptide
– angiotensin 1 receptor antagonist 251
– antagonist 220, 286
noradrenaline 17
norepinephrine 350
NPK 44–46
nucleus basalis magnocellularis 310
null alleles 297

obstructive airway disease 257
oedema 525, 534
oesophagus 514, 519, 523
okadaic acid 165
olfactory bulb 301
ondansetron 380, 392, 394, 402, 403, 408, 410, 414
opiate addiction 315
μ-opioid receptor 315
opsin kinase 145
osanetant 252, 352
osteoarthritis 448
ovalbumin 279

oxazolidine 285, 288
oxazolone 476
oxime 285
oxytocin 14

pain 192, 261, 536
– neuropathic assays 444
– transmission 12, 123
pancreas 516, 524, 525, 534
pancreatitis 534, 537
pan-tachykinin 280, 281
– receptor antagonist 289, 537
PAR2 161, 164
parabrachial nucleus 301
parachlorophenylalanine 401
paroxetine 346
PD147714 222
PD154075 202
PD154740 250
PD157672 250
PD161182 254
pentobarbitone 399
peptide-based antagonist 220
peptides 3, 16
periaqueductal grey 301
peristalsis 522, 523
peritoneogastric reflex 532
PGE$_2$ 193
phenylarsine oxide 157
pheochromocytoma 36
Phoneutria nigriventer venom (PNV) 472
phosphatidyl inositol 175
– turnover 35
phosphodiesterase 387
phosphoinositol turnover 31
phosphorylation 146, 151
– of the carboxyl tail 148
physalaemin 6, 123, 125, 375, 473
Physalaemus
– *biligoniderus* 123
– *fuscomaculata* 6
pigeon studies 401
piglet studies 400
piperidine 283
place preference 314
plasma protein extravasation 184, 192, 206
polyploidization 30

PONV 378, 413
post-herpetic neuralgia 447
postoperative ileus 530, 532
postprandial motility 521
prazosin 259
pre-cisplatin 409
pre-eclampsia 47, 259
preprotachykinin
– A (PPTA) 26, 124, 513
– – mRNA 84
– – α 28, 513
– – β 26, 513, 528
– – δ 28, 513
– – γ 28, 513, 528
– – gene 28, 29, 35, 301
– B (PPTB) 124
– – mRNA 47
– – gene 25, 23, 28, 36
– gene 68, 70
– C (PPTC) 28, 133
– – gene 28, 29, 38, 134
– I 274
– II 274
– mRNA 87
pressor response 46, 47
prostaglandin 398, 524, 535
prostanoid 398
protease 525
protein kinase
– C (PKC) 145
– cyclic-AMP-dependent (PKA) 145
proteinase-activated receptor 463
proteinuria 259
pruritus 474
pseudoaffective response 537, 538
pseudomembranous colitis 529
pseudo-tetra-peptide antagonist 288
psoriasis 474
psychiatric disorder 342
pulpal inflammation 449

4-quinolinecarboxamide 254
quinuclidine 391

R116301 204, 386
R396 221
rab5a 157

radioimmunoassay (RIA) 64
radioimmunocytochemistry 67
ranakinin 42, 44
rat studies 402
receptor phosphorylation 143
rectocolitis, TNBS-induced 534
rectocolonic reflex 532
recycling 143, 153, 164
resensitization 149, 153, 164
reserpine 401
resiniferatoxin 197, 279, 375, 399, 403, 418
respiratory disease 290
retching 364, 369
reward 314
reward-related behaviour 315
rhodopsin 143, 175
– kinase 145
rhombencephalon 79
RP67580 175, 191, 396, 466
– calcium blocking activity 191
RPR100893 191, 387

S18523 184
salivary
– secretion 516
– gland 516
salivation 376
Salmonella dublin 528, 529
saporin 159
saredutant 227, 351
SB14240 233
SB218795 397
SB222200 255, 260
SB223412 (talnetant) 254, 258
SB235375 255
SB400238 256
scaffold protein 160
SCH206272 256, 285, 287
Schistosoma japonicum 528
schizophrenia 103, 254, 262
scopolamine 414
scyliorhinin
– I 32
– II 37, 48
– – [^{125}I]-Bolton-Hunter
SDZNKT343 201
sebaceous gland 477
sebum 477

second messenger kinase 147
secretomotor neuron 519, 519, 524
secretory diarrhoea 257
selective norepinephrine reuptake inhibitor (SNRI) 346
selective serotonin reuptake inhibitor (SSRI) 346
self-administration 314
sendide 186, 386, 418
senktide 26, 35, 37, 39, 43, 46, 470 126, 247, 252, 255, 258, 259, 262, 287, 352
sensitization
– central sensitization 308
– to morphine 315
sensory neuron 34, 516
septide 26, 31, 32, 39, 131, 177, 178
serine 146
– serine–threonine kinase 160
serotonin 4, 91
– release 498
sexual dysfunction 344
slow wave 521
smooth muscle of large and small airways 496
social interaction test 310
social phobia 352
sodium channel 189
spantide 179, 386, 476
spinal afferents 536
spinal-bulbo-spinal facilitatory loop 308
SR140333 175, 192, 279, 388, 466
SR142801 (osanetant) 252, 258, 259, 261, 352
SR144190 232, 351
SR48968 (saredutant) 227, 248, 255, 278, 279, 280, 289, 351, 502
Src kinase 160
SSR146977 254
SSR240600 193
stomach 514, 519, 528
Streptomyces violaceoniger 288
stress 313, 345, 348
– neurotransmitter 347
striatum 301, 413
submucosal
– arterioles 515
– plexus 515, 518, 524, 525
substance P (SP) 15, 64, 122, 125, 146, 174, 260, 342, 364, 370, 398, 405, 492, 511
– (1–7) 26
– – fragment 40
– (1–8) 41
– [^{125}I]-Bolton-Hunter 275
– [Sar^9,Met(O$_2$)^{11}] 31, 45
– [sar^9] sulfone 38, 39
– amino acid 15
– – sequence 7
– antagonists 18
– – early peptide-based 179
– as stress neurotransmitter 345
– autonomic reflexes 13
– cat brain 90
– cell bodies 71
– – distribution 71
– classical transmitters 81
– discovery 1
– distribution 83
– – in the brain 70
– – in the spinal cord 70
– E type 125
– ferret brain 91
– fibers 85
– glutamate 16
– guinea pig brain 90
– hamster brain 91
– immunohistochemistry 8, 17, 75, 77, 79
– immunoreactive fibers 75, 77, 79
– – fiber tracts 81
– immunoreactive neurons 89
– mouse brain 90
– mRNA 88
– nerve terminals 83
– neuroendocrine reflexes 13
– P type 125
– pathway 81, 82
– peptides 15
– rabbit brain 91
– radioimmunoassay 8
– salivary secretion 6
– saporin conjugate 159
– spinal cord 8
– striatonigral neurons 82
substantia nigra 11
Suncus murinus 37, 367, 376, 377, 380, 395, 404
superior colliculus 301
synaptic specialization 517, 520
synergism 280

TAC363   222
tachycardia   44, 49, 135, 420
tachygastria   419
tachykinergic innervation   514
tachykinin (TK)   9, 14, 15, 64, 174, 247, 274, 495
– 4 (TAC4)   133
– – gene   134
– agonist-induced trafficking   143
– amino acid sequence   123
– autoreceptors   521
– carboxyl terminus   148
– circuitries   103
– conformer   177
– endogenous tachykinins   255
– expression   526
– family   65
– in emesis   370
– in the lung   275
– neurons
– – immunohistochemistry   67
– NK$_1$ receptor antagonist   174
– peptides   153
– ppt-a-derived   36
– receptors (TACR)   102, 521
– – agonist-induced endocytosis   149
– – agonist-induced phosphorylation   146
– – agonist-induced trafficking   154
– – antagonist   502, 512
– – desensitization   143
– – downregulation   166
– – endocytic domains   151
– – G protein-coupled   381
– – mediated desensitization   147
– – nonselective nonpeptide antagonist   255
– – recycling   165
– – resensitization   165
– – trafficking in the nervous system   156
– – early classification   126
– – nomenclature   127
– – terminology   129
tachyphylaxis   375
TAK637   203, 538
talnetant   254
telencephalon   75, 85
threonine   146
– threonine–tyrosine kinase   160
tissue inflammation   415

TNBS   532
trace fear conditioning   317
transcription–polymerase chain reaction   351
transferrin receptor   157
transmission   519, 523, 524
– slow excitatory synaptic   520
Trichinella spiralis   532, 535, 537
trinitrobenzene sulphonic acid (TNBS)   527, 529
triple response   464
tropisetron   394
trypsin   2
tryptophan   202
tyrosine   152
– kinase   161
– hydroxylase   79

ubiquitin   163
UK224671   234
ulcerative colitis   526, 528, 529, 532, 538
ulcerogenic factor   43
ultraviolet B (UVB)   477
urticaria   474

vagal afferents   516
vagus   373
vanilloid receptor   463
varicosity   517
vasoconstriction   49, 525
vasodilatation   49, 525
vasopressin   376
– V$_2$ receptor   156
vasopressor response   46
velocity of propulsion   522
venular permeability   525, 534
verapamil   191
vesicular localization   517
vestibular system   368
vestibulo-vegetative syndrome   376
VIP   524
visceral pain   445
visceromotor response   537
volume transmission   102
vomiting   343, 362, 364, 368, 369
– pathways for the induction   366
– physiology   363

wet-dog shakes 39, 44, 45, 46
WIN51078 204

xylazine 395

YM38336 234

ZD6021 256, 287
ZD7944 234, 279
ZM253270 231

Printing: Saladruck, Berlin
Binding: Stein+Lehmann, Berlin